# Beginning Algebra

## Ninth Edition

## John Tobey

*North Shore Community College*
Danvers, Massachusetts

## Jeffrey Slater

*North Shore Community College*
Danvers, Massachusetts

## Jamie Blair

*Orange Coast College*
Costa Mesa, California

## Jennifer Crawford

*Normandale Community College*
Bloomington, Minnesota

**PEARSON**

Boston   Columbus   Indianapolis   New York   San Francisco
Amsterdam   Cape Town   Dubai   London   Madrid   Milan   Munich   Paris   Montréal   Toronto
Delhi   Mexico City   São Paulo   Sydney   Hong Kong   Seoul   Singapore   Taipei   Tokyo

Editorial Director, Mathematics: *Christine Hoag*
Editor in Chief: *Michael Hirsch*
Editorial Assistant: *Megan Tripp*
Development Editor: *Elaine Page*
Project Management Team Lead: *Christina Lepre*
Project Manager: *Ron Hampton*
Program Management Team Lead: *Karen Wernholm*
Program Manager: *Beth Kaufman*
Interior Design: *Tamara Newnam*
Interior Design Supervision/Cover Design: *Barbara T. Atkinson*
Design Manager: *Andrea Nix*
Manager, Course Production: *Ruth Berry*
Senior Multimedia Producer: *Vicki Dreyfus*
Media Producer: *Nicholas Sweeny*
Senior Marketing Manager: *Rachel Ross*
Marketing Assistant: *Alexandra Habashi*
Senior Author Support/Technology Specialist: *Joe Vetere*
Procurement Manager: *Carol Melville*
Rights Manager: *Gina Cheselka*
Production Management, Composition, and Art: *Integra*
Cover Images: *Illustration by Amy DeVoogd*

**Library of Congress Cataloging-in-Publication Data**

Tobey, John,
    Beginning algebra / John Tobey, Jr., North Shore Community College, Jeffrey Slater, North Shore Community College, Jamie Blair, Orange Coast College, Jennifer Crawford, Normandale Community College. — 9th edition.
        pages cm
    ISBN 0-13-418779-2
1. Mathematics — Textbooks.   I. Slater, Jeffrey,   II. Blair, Jamie.
III. Crawford, Jennifer   IV. Title.
    QA152.3.T63 2017
    512.9 — dc23                                    2015007589

Photo credits are located on page C-1 and represent an extension of this copyright.

2   16

www.pearsonhighered.com

0-13-418779-2 (Student Edition paperback)
978-0-13-418779-2 (Student Edition paperback)

This book is dedicated to the memory of
Lexie Tobey and John Tobey, Sr.
They have left a legacy of love, a memory of four
decades of faithful teaching, and a sense of helping
others that will influence generations to come.
For their grandchildren, they have left an inspiring
model of a loving family, true character,
and service to God and community.

# Contents

# Preface

## TO THE INSTRUCTOR

Developmental mathematics course structures, trends, and dynamics continue to evolve and change, as **course redesign trends** continue to evolve and change, including the introduction of **new pathways-type courses.** Developmental mathematics instructors are increasingly challenged with helping their students **navigate career-oriented math tracks (including non-STEM and STEM pathways),** plus helping students think about **selecting a major** and **work-force readiness**. To help instructors on this front, with this revision of *Beginning Algebra,* you'll find a **new emphasis on, and integration of, Career Explorations** throughout the text and MyMathLab course.

Additionally, the program retains its hallmark characteristics that have always made the text so easy to learn and teach from, including its building-block organization. Each section is written to stand on its own, and every homework set is completely self-testing. Exercises are paired and graded and are of varying levels and types to ensure that all skills and concepts are covered. As a result, the text offers students an effective and proven learning program suitable for a variety of course formats—including lecture-based classes; computer-lab based or hybrid classes; discussion-oriented, activity-driven classes; modular and/or self-paced programs; and distance-learning, online programs.

We have visited and listened to teachers across the country and have incorporated a number of suggestions into this edition to help you with the particular learning-delivery system at your school. The following pages describe the key changes in this ninth edition.

## WHAT'S NEW IN THE NINTH EDITION?

### New Career Explorations Interactions for Students

Each chapter begins with a **Career Opportunities** feature that enables students to personally investigate possible future career options while putting the math into context. Students are asked simple, interactive questions prompting them to consider employment opportunities that perhaps they had never thought possible.

Then, the students are directed to the corresponding **Career Exploration Problems** where they can actually solve problems that help them visualize what work would be like in that career field. This feature opens up possibilities for personal success in future employment.

The Career Exploration Problems are also assignable in MyMathLab, allowing this feature to be seamlessly integrated with the technology. The problems help to foster active learning and better understanding of the math concepts.

### New Guided Learning Videos

Faculty have asked for specific interactive videos that will clearly show each step of the **key concepts** of each chapter. With this revision, you'll find a new series of **Guided Learning Videos** that show in a powerful, interactive way **how to solve the most important types of problems contained in each chapter.** For student ease, icons throughout the eText indicate where the videos are available. The eText is clickable, opening the videos on the spot. Plus, a new *Video Workbook with the Math Coach* allows students to take notes and practice by studying and solving problems.

### Expanded Video Program

In addition to the new Guided Learning Videos with icons throughout the eText, objective-level video clips have also been added to the MyMathLab course with accompanying icons throughout the eText. These video additions expand upon an already complete video lecture series available in MyMathLab. Students and instructors will also find complete Section Lecture Videos, Math Coach Videos, and Chapter Test Prep Videos.

- **The Math Coach** has been expanded within the MyMathLab course, with even more stepped-out, guided Math Coach problems assignable in MyMathLab. Within the text, following each Chapter Test, the **Math Coach** provides students with a personal office-hour experience by walking them through some helpful hints to keep them from making common errors on test problems. For additional help, students can also watch the authors work through these problems on the accompanying Math Coach videos in the MyMathLab course. Instructors can also assign the Math Coach problems in MyMathLab and use the companion *Video Workbook with the Math Coach* for additional practice and to serve as the foundation for a course notebook.

- Fifteen percent of the exercises throughout the text have been refreshed.

- Real-world application problems have been updated throughout the text.

- **New Use Math to Save Money Animations** have been added to the MyMathLab course. The animations expand upon a favorite feature from the text, allowing students to put the math they just learned into context. These newly created animations are set to music and depict real-life scenarios and real-life people using math to cut costs and spend less. To ensure that students watch and understand the animations, there are accompanying Use Math to Save Money homework assignments available in MyMathLab, which are prebuilt for instructor convenience.

Additionally, we've created an even stronger connection between the approach that is used to teach the concepts in the text, and the media assets and assignable exercises within the accompanying MyMathLab course.

To make sure you and your students are getting the most out of the text *and* the MyMathLab course, see the following MyMathLab feature descriptions.

# Get the most out of
# MyMathLab®

MyMathLab is the world's leading online resource for teaching and learning mathematics. MyMathLab helps students and instructors improve results and provides engaging experiences and personalized learning for each student so learning can happen in any environment. Plus, MyMathLab offers flexible and time-saving course-management features to allow instructors to easily manage their classes while remaining in complete control, regardless of course format.

## Personalized Support for Students

- MyMathLab comes with many learning resources—eText, animations, videos, and more—all designed to support your students as they progress through their course.

- The Adaptive Study Plan acts as a personal tutor, updating in real time based on student performance to provide personalized recommendations on what to work on next. With the new Companion Study Plan assignments, instructors can now assign the Study Plan as a prerequisite to a test or quiz, helping to guide students through concepts they need to master.

- Personalized Homework allows instructors to create homework assignments tailored to each student's specific needs by focusing on just the topics students have not yet mastered.

Used by nearly 4 million students each year, the MyMathLab and MyStatLab family of products delivers consistent, measurable gains in student learning outcomes, retention, and subsequent course success.

# Resources for Success

## MyMathLab® Online Course

### *Beginning Algebra* by Tobey/Slater/Blair/Crawford

(access code required)

MyMathLab is available to accompany Pearson's market-leading text offerings. To give students a consistent tone, voice, and teaching method, each text's approach is tightly integrated throughout the accompanying MyMathLab course, making learning the material as seamless as possible.

### New Career Explorations Interactions

A new integration of Career Explorations has been added throughout the text and MyMathLab course in an interactive format that engages students and gets them thinking about future career possibilities. Each chapter starts with a **Career Opportunities** feature that puts the math into context and ends with multiple **Career Exploration Problems** that are also assignable in MyMathLab!

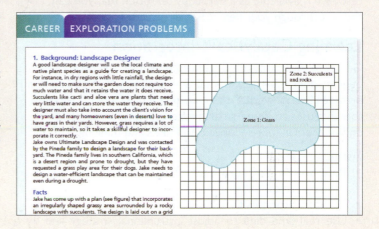

### New Guided Learning Videos, Objective-Level Video Clips, and Video Workbook

New Guided Learning Videos show in a powerful, interactive way how to solve the most important types of problems in each chapter. Icons throughout the eText indicate where videos are available. The eText is clickable, opening videos on the spot. Plus, a new *Video Workbook with the Math Coach* ties it all together and provides opportunity for extra practice.

### New Use Math to Save Money Animations

These newly created animations, which have been added to the MyMathLab course, are set to music and depict real-life scenarios in which people use math to cut costs and spend less. Accompanying Use Math to Save Money homework assignments are available in MyMathLab to help further students' understanding.

# Resources for Success

With MyMathLab, students and instructors get a robust course-delivery system, the full Tobey/Slater/Blair/Crawford eText, and many assignable exercises and media assets. Additionally, MyMathLab also houses these additional instructor and student resources, making the entire set of resources available in one easy-to-access online location.

## Instructor Resources

### Annotated Instructor's Edition

This version of the text includes answers to all exercises presented in the book, as well as helpful teaching tips. This resource is available as a hardcopy textbook that you can request through your Pearson sales representative.

### Learning Catalytics™ Integration

Generate class discussion, guide your lecture, and promote peer-to-peer learning with real-time analytics. MyMathLab now provides Learning Catalytics—an interactive student-response tool that uses students' smartphones, tablets, or laptops to engage them in more sophisticated tasks and thinking.

Instructors, can
- Pose a variety of open-ended questions that help students develop critical-thinking skills.
- Monitor responses to find out where students are struggling.
- Use real-time data to adjust instructional strategy and try other ways of engaging students during class.
- Manage student interactions by automatically grouping students for discussion, teamwork, and peer-to-peer learning.

### Instructor's Solutions Manual

The *Instructor's Solutions Manual* is available for download from the Pearson Instructor's Resource Center or within the MyMathLab course, and it includes detailed, step-by-step solutions to the even-numbered section exercises as well as solutions to every exercise (odd and even) in the Classroom Quiz, mid-chapter reviews, chapter reviews, chapter tests, cumulative tests, and practice final.

### Instructor's Resource Manual with Tests and Mini Lectures

Also available for download from the Pearson Instructor's Resource Center and within the MyMathLab course, the *Instructor's Resource Manual* includes a mini lecture for each text section, two short group activities per chapter, three forms of additional practice exercises, two pretests, six tests, and two final exams for every chapter, both free response and multiple choice, as well as two cumulative tests for every even numbered chapter. The *Instructor's Resource Manual* also contains the answers to all items.

### PowerPoint Lecture Slides

Available through www.pearsonhighered.com and in MyMathLab, these fully editable lecture slides include definitions, key concepts, and examples for use in a lecture setting.

## TestGen

TestGen® (www.pearsoned.com/testgen) enables instructors to build, edit, print, and administer tests using a computerized bank of questions developed to cover all the objectives of the text. TestGen is algorithmically based, allowing instructors to create multiple but equivalent versions of the same question or test with the click of a button. Instructors can also modify test bank questions or add new questions. The software and test bank are available for download from Pearson's Instructor Resource Center.

## Student Resources

### Student Solutions Manual

The *Student Solutions Manual* provides worked-out solutions to all odd-numbered section exercises, even and odd exercises in the Quick Quiz, mid-chapter reviews, chapter reviews, chapter tests, Math Coach, and cumulative reviews. Instructors have the option to make an electronic version available to students within the MyMathLab course, or students can purchase it separately in printed form.

### New Video Workbook with the Math Coach

The new *Video Workbook with the Math Coach* expands upon the popular *Math Coach* workbook format and is correlated with the new Guided Learning Videos to serve as a video note-taking and practice guide for students. It is available to students in electronic form within the MyMathLab course, and students can also purchase it separately in printed form.

### Student Success Module in MyMathLab

This new interactive module is available in the left-hand navigation of MyMathLab and includes videos, activities, and post-tests for these three student-success areas:

- **Math-Reading Connections**, including topics such as "Using Word Clues" and "Looking for Patterns."
- **Study Skills**, including topics such as "Time Management" and "Preparing for and Taking Exams."
- **College Success**, including topics such as "College Transition" and "Online Learning."

Instructors can assign these videos and/or activities as media assignments, along with prebuilt post-tests to make sure students learn and understand how to improve their skills in these areas. Instructors can integrate these assignments with their traditional MyMathLab homework assignments to incorporate student success topics into their course, as they deem appropriate.

# Diagnostic Pretest: Beginning Algebra

*Follow the directions for each question. Simplify each answer.*

## Chapter 0

1. Add. $3\frac{1}{4} + 2\frac{3}{5}$

2. Multiply. $\left(1\frac{1}{6}\right)\left(2\frac{2}{3}\right)$

3. Divide. $\frac{15}{4} \div \frac{3}{8}$

4. Multiply. $(1.63)(3.05)$

5. Divide. $120 \div 0.0006$

6. Find 7% of 64,000.

## Chapter 1

7. Add. $-3 + (-4) + (+12)$

8. Subtract. $-20 - (-23)$

9. Combine. $5x - 6xy - 12x - 8xy$

10. Evaluate $2x^2 - 3x - 4$ when $x = -3$.

11. Remove the grouping symbols. $2 - 3\{5 + 2[x - 4(3 - x)]\}$

12. Evaluate. $-3(2 - 6)^2 + (-12) \div (-4)$

## Chapter 2

*In questions 13–16, solve each equation for x.*

13. $40 + 2x = 60 - 3x$

14. $7(3x - 1) = 5 + 4(x - 3)$

15. $\frac{2}{3}x - \frac{3}{4} = \frac{1}{6}x + \frac{21}{4}$

16. $\frac{4}{5}(13x + 4) = 20$

17. Solve for $p$. $A = \frac{1}{2}(13p - 4f)$

18. Solve for $x$ and graph the result. $42 - 18x < 48x - 24$

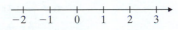

## Chapter 3

19. The length of a rectangle is 7 meters longer than twice the width. The perimeter is 46 meters. Find the dimensions.

20. One side of a triangle is triple the second side. The third side is 3 meters longer than double the second side. Find each side of the triangle if the perimeter of the triangle is 63 meters.

21. Hector has four test scores of 80, 90, 83, and 92. What does he need to score on the fifth test to have an average of 86 on the five tests?

22. Marcia invested $6000 in two accounts. One earned 5% interest, while the other earned 7% interest. After one year, she earned $394 in interest. How much did she invest in each account?

1. _____

2. _____

3. _____

4. _____

5. _____

6. _____

7. _____

8. _____

9. _____

10. _____

11. _____

12. _____

13. _____

14. _____

15. _____

16. _____

17. _____

18. _____

19. _____

20. _____

21. _____

22. _____

23. _____

24. _____

25. _____

26. _____

27. _____

28. _____

29. _____

30. _____

31. _____

32. _____

33. _____

34. _____

35. _____

36. _____

37. _____

38. _____

39. _____

40. _____

41. _____

42. _____

**23.** Melissa has three more dimes than nickels. She has twice as many quarters as nickels. The value of the coins is \$4.20. How many of each coin does she have?

**24.** The drama club put on a play for Thursday, Friday, and Saturday nights. The total attendance for the three nights was 6210. Thursday night had 300 fewer people than Friday night. Saturday night had 510 more people than Friday night. How many people came each night?

## Chapter 4

**25.** Multiply. $(-2xy^2)(-4x^3y^4)$

**26.** Divide. $\dfrac{36x^5y^6}{-18x^3y^{10}}$

**27.** Raise to the indicated power. $(-2x^3y^4)^5$

**28.** Evaluate. $(-3)^{-4}$

**29.** Multiply. $(3x^2 + 2x - 5)(4x - 1)$

**30.** Divide. $(x^3 + 6x^2 - x - 30) \div (x - 2)$

## Chapter 5

*Factor completely.*

**31.** $5x^2 - 5$

**32.** $x^2 - 12x + 32$

**33.** $8x^2 - 2x - 3$

**34.** $3ax - 8b - 6a + 4bx$

*Solve for x.*

**35.** $16x^2 - 24x + 9 = 0$

**36.** $\dfrac{x^2 + 8x}{5} = -3$

## Chapter 6

**37.** Simplify. $\dfrac{x^2 + 3x - 18}{2x - 6}$

**38.** Multiply. $\dfrac{6x^2 - 14x - 12}{6x + 4} \cdot \dfrac{x + 3}{2x^2 - 2x - 12}$

**39.** Divide and simplify. $\dfrac{x^2}{x^2 - 4} \div \dfrac{x^2 - 3x}{x^2 - 5x + 6}$

**40.** Add. $\dfrac{3}{x^2 - 7x + 12} + \dfrac{4}{x^2 - 9x + 20}$

**41.** Solve for $x$. $2 - \dfrac{5}{2x} = \dfrac{2x}{x + 1}$

**42.** Simplify. $\dfrac{3 + \dfrac{1}{x}}{\dfrac{9}{x} + \dfrac{3}{x^2}}$

# Chapter 7

**43.** Graph. $y = 2x - 4$

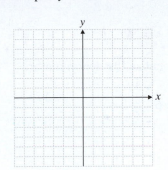

**44.** Graph. $3x + 4y = -12$

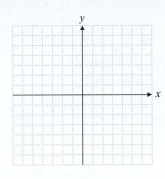

**45.** What is the slope of a line passing through $(6, -2)$ and $(-3, 4)$?

**46.** If $f(x) = 2x^2 - 3x + 1$, find $f(3)$.

**47.** Graph the region. $y \geq -\frac{1}{3}x + 2$

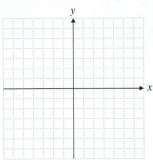

**48.** Find the equation of a line with a slope of $\frac{3}{5}$ that passes through the point $(-1, 3)$.

# Chapter 8

*Solve each system by the appropriate method.*

**49.** Substitution method

$x + y = 17$
$2x - y = -5$

**50.** Addition method

$-5x + 4y = 8$
$2x + 3y = 6$

**51.** Any method

$2(x - 2) = 3y$
$6x = -3(4 + y)$

**52.** Any method

$x + \frac{1}{3}y = \frac{10}{3}$

$\frac{3}{2}x + y = 8$

**53.** Is $(2, -3)$ a solution for the following system?

$3x + 5y = -9$
$2x - 3y = 13$

**54.** A man bought three pairs of gloves and four scarves for $53. A woman bought two pairs of the same-priced gloves and three of the same-priced scarves for $38. How much did each item cost?

43. _____

44. _____

45. _____

46. _____

47. _____

48. _____

49. _____

50. _____

51. _____

52. _____

53. _____

54. _____

XV

55.

56.

57.

58.

59.

60.

61.

62.

63.

64.

65.

66.

## Chapter 9

**55.** Evaluate. $\sqrt{121}$

**56.** Simplify. $\sqrt{125x^3y^5}$

**57.** Multiply and simplify.
$$\left(\sqrt{2} + \sqrt{6}\right)\left(2\sqrt{2} - 3\sqrt{6}\right)$$

**58.** Rationalize the denominator.
$$\frac{\sqrt{5} - \sqrt{3}}{\sqrt{6}}$$

**59.** In the right triangle with sides $a$, $b$, and $c$, find side $c$ if side $a = 4$ and side $b = 6$.

**60.** $y$ varies directly with $x$. When $y = 56$, then $x = 8$. Find $y$ when $x = 11$.

## Chapter 10

*In questions 61–64, solve for x.*

**61.** $14x^2 + 21x = 0$

**62.** $2x^2 + 1 = 19$

**63.** $2x^2 - 4x - 5 = 0$

**64.** $x^2 - x + 8 = 5 + 6x$

**65.** Graph the equation. $y = x^2 + 8x + 15$

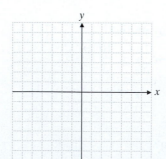

**66.** A rectangle is 4 inches longer in length than in width. The area of the rectangle is 96 square inches. Find the length and the width of the rectangle.

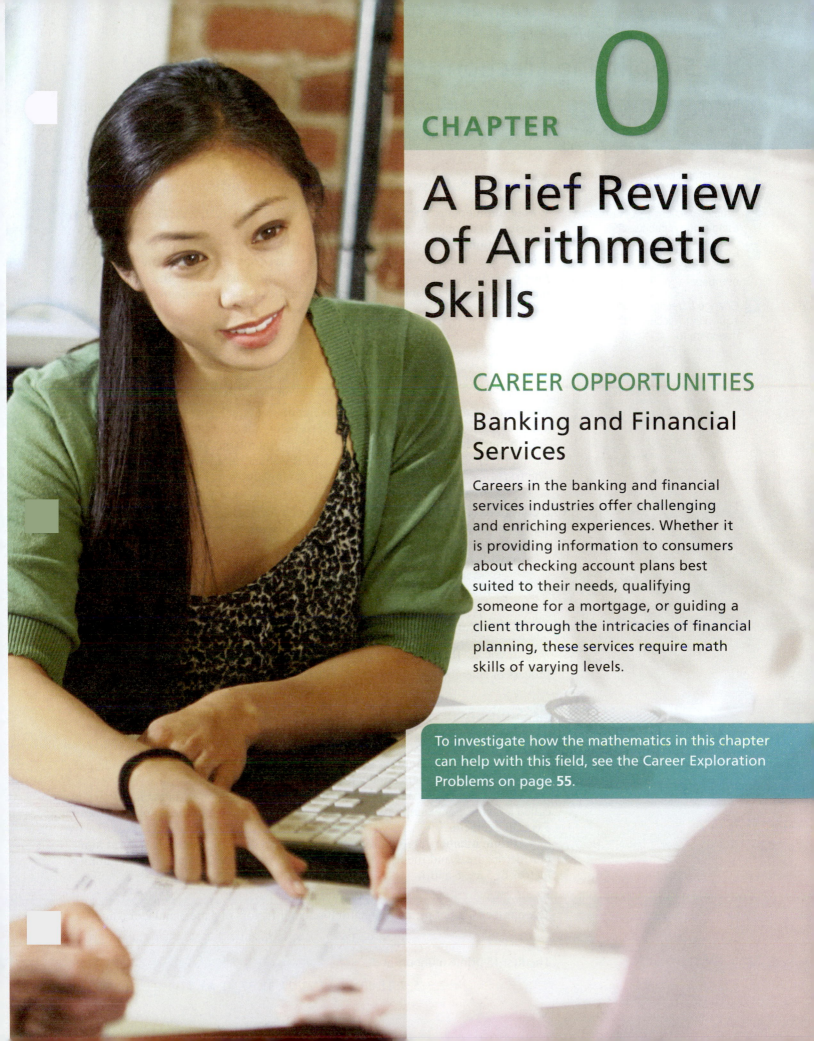

CHAPTER 0

# A Brief Review of Arithmetic Skills

## CAREER OPPORTUNITIES

### Banking and Financial Services

Careers in the banking and financial services industries offer challenging and enriching experiences. Whether it is providing information to consumers about checking account plans best suited to their needs, qualifying someone for a mortgage, or guiding a client through the intricacies of financial planning, these services require math skills of varying levels.

To investigate how the mathematics in this chapter can help with this field, see the Career Exploration Problems on page **55**.

# 0.1 Simplifying Fractions

**Student Learning Objectives**

After studying this section, you will be able to:

1. Understand basic mathematical definitions. ●

2. Simplify fractions to lowest terms using prime numbers. ●

3. Convert between improper fractions and mixed numbers. ●

4. Change a fraction to an equivalent fraction with a given denominator. ●

Chapter 0 is designed to give you a mental "warm-up." In this chapter you'll be able to step back a bit and tone up your math skills. This brief review of arithmetic will increase your math flexibility and give you a good running start into algebra.

## 1 Understanding Basic Mathematical Definitions ●

**Whole numbers** are the set of numbers 0, 1, 2, 3, 4, 5, 6, 7, . . . . They are used to describe whole objects, or entire quantities.

**Fractions** are a set of numbers that are used to describe parts of whole quantities. In the object shown in the figure there are four equal parts. The *three* of the *four* parts that are shaded are represented by the fraction $\frac{3}{4}$. In the fraction $\frac{3}{4}$ the number 3 is called the **numerator** and the number 4, the **denominator.**

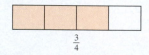
$$\frac{3}{4}$$

$\underline{3} \leftarrow$ *Numerator* is on the top
$4 \leftarrow$ *Denominator* is on the bottom

The *denominator* of a fraction shows the number of equal parts in the whole and the *numerator* shows the number of these parts being talked about or being used.

**Numerals** are symbols we use to name numbers. There are many different numerals that can be used to describe the same number. We know that $\frac{1}{2} = \frac{2}{4}$. The fractions $\frac{1}{2}$ and $\frac{2}{4}$ both describe the same number.

Usually, we find it more useful to use fractions that are simplified. A fraction is considered to be in **simplest form** or **reduced form** when the numerator (top) and the denominator (bottom) have no common divisor other than 1, and the denominator is greater than 1.

$\frac{1}{2}$ is in simplest form.

$\frac{2}{4}$ is *not* in simplest form, since the numerator and the denominator can both be divided by 2.

If you get the answer $\frac{2}{4}$ to a problem, you should state it in simplest form, $\frac{1}{2}$. The process of changing $\frac{2}{4}$ to $\frac{1}{2}$ is called **simplifying** or **reducing** the fraction.

## 2 Simplifying Fractions to Lowest Terms Using Prime Numbers ●

**Natural numbers** or **counting numbers** are the set of whole numbers excluding 0. Thus the natural numbers are the numbers 1, 2, 3, 4, 5, 6, . . . .

When two or more numbers are multiplied, each number that is multiplied is called a **factor.** For example, when we write $3 \times 7 \times 5$, each of the numbers 3, 7, and 5 is called a factor.

**Prime numbers** are natural numbers greater than 1 whose only natural number factors are 1 and themselves. The number 5 is prime. The only natural number factors of 5 are 5 and 1.

$$5 = 5 \times 1$$

The number 6 is not prime. The natural number factors of 6 are 3 and 2 or 6 and 1.

$$6 = 3 \times 2 \qquad 6 = 6 \times 1$$

The first 15 prime numbers are

2, 3, 5, 7, 11, 13, 17, 19, 23, 29, 31, 37, 41, 43, 47.

Any natural number greater than 1 either is prime or can be written as the product of prime numbers. For example, we can take each of the numbers 12, 30, 14, 19, and 29 and either indicate that they are prime or, if they are not prime, write them as the product of prime numbers. We write as follows:

$$12 = 2 \times 2 \times 3 \qquad 30 = 2 \times 3 \times 5 \qquad 14 = 2 \times 7$$

19 is a prime number.    29 is a prime number.

To reduce a fraction, we use prime numbers to factor the numerator and the denominator. Write each part of the fraction (numerator and denominator) as a product of prime numbers. Note any *factors* that appear in both the *numerator* (top) and *denominator* (bottom) of the fraction. If we divide numerator and denominator by these values we will obtain an equivalent fraction in *simplest form*. When the new fraction is simplified, it is said to be in **lowest terms.** Throughout this text, to *simplify* a fraction will always mean to write the fraction in lowest terms.

## Example 1 Simplify each fraction.

(a) $\dfrac{14}{21}$ 　　　(b) $\dfrac{15}{35}$ 　　　(c) $\dfrac{20}{70}$

### Solution

(a) $\dfrac{14}{21} = \dfrac{\cancel{7} \times 2}{\cancel{7} \times 3} = \dfrac{2}{3}$ 　　We factor 14 and factor 21. Then we divide numerator and denominator by 7.

(b) $\dfrac{15}{35} = \dfrac{\cancel{5} \times 3}{\cancel{5} \times 7} = \dfrac{3}{7}$ 　　We factor 15 and factor 35. Then we divide numerator and denominator by 5.

(c) $\dfrac{20}{70} = \dfrac{2 \times \cancel{2} \times \cancel{5}}{7 \times \cancel{2} \times \cancel{5}} = \dfrac{2}{7}$ 　　We factor 20 and factor 70. Then we divide numerator and denominator by both 2 and 5. 　　☐

 **Student Practice 1** Simplify each fraction.

(a) $\dfrac{10}{16}$ 　　　(b) $\dfrac{24}{36}$ 　　　(c) $\dfrac{36}{42}$

Sometimes when we simplify a fraction, all the prime factors in the top (numerator) are divided out. When this happens, we must remember that a 1 is left in the numerator.

## Example 2 Simplify each fraction.

(a) $\dfrac{7}{21}$ 　　　　　　　(b) $\dfrac{15}{105}$

### Solution

(a) $\dfrac{7}{21} = \dfrac{\cancel{7} \times 1}{\cancel{7} \times 3} = \dfrac{1}{3}$ 　　　　(b) $\dfrac{15}{105} = \dfrac{\cancel{5} \times \cancel{3} \times 1}{7 \times \cancel{5} \times \cancel{3}} = \dfrac{1}{7}$ 　　☐

**Student Practice 2** Simplify each fraction.

(a) $\dfrac{4}{12}$ 　　　(b) $\dfrac{25}{125}$ 　　　(c) $\dfrac{73}{146}$

If all the prime numbers in the bottom (denominator) are divided out, we do not need to leave a 1 in the denominator, since we do not need to express the answer as a fraction. The answer is then a whole number and is not usually expressed as a fraction.

**Example 3** Simplify each fraction.

(a) $\dfrac{35}{7}$

(b) $\dfrac{70}{10}$

**Solution**

(a) $\dfrac{35}{7} = \dfrac{5 \times \cancel{7}}{\cancel{7} \times 1} = 5$

(b) $\dfrac{70}{10} = \dfrac{7 \times \cancel{5} \times \cancel{2}}{\cancel{5} \times \cancel{2} \times 1} = 7$ ☐

**Student Practice 3** Simplify each fraction.

(a) $\dfrac{18}{6}$

(b) $\dfrac{146}{73}$

(c) $\dfrac{28}{7}$

Sometimes the fraction we use represents how many of a certain thing are successful. For example, if a baseball player was at bat 30 times and achieved 12 hits, we could say that he had a hit $\frac{12}{30}$ of the time. If we reduce the fraction, we could say he had a hit $\frac{2}{5}$ of the time.

**Example 4** Cindy got 48 out of 56 questions correct on a test. Write this as a fraction in simplest form.

**Solution** Express as a fraction in simplest form the number of correct responses out of the total number of questions on the test.

$$48 \text{ out of } 56 \rightarrow \dfrac{48}{56} = \dfrac{2 \times 3 \times \cancel{2} \times \cancel{2} \times \cancel{2}}{7 \times \cancel{2} \times \cancel{2} \times \cancel{2}} = \dfrac{6}{7}$$

Cindy answered the questions correctly $\frac{6}{7}$ of the time. ☐

**Student Practice 4** The major league pennant winner in 1917 won 56 games out of 154 games played. Express as a fraction in simplest form the number of games won in relation to the number of games played.

The number *one* can be expressed as $1, \frac{1}{1}, \frac{2}{2}, \frac{6}{6}, \frac{8}{8}$, and so on, since

$$1 = \dfrac{1}{1} = \dfrac{2}{2} = \dfrac{6}{6} = \dfrac{8}{8}.$$

We say that these numerals are *equivalent ways* of writing the number *one* because they all express the same quantity even though they appear to be different.

### Sidelight: The Multiplicative Identity

When we simplify fractions, we are actually using the fact that we can multiply any number by 1 without changing the value of that number. (Mathematicians call the number 1 the **multiplicative identity** because it leaves any number it multiplies with the same identical value as before.)

Let's look again at one of the previous examples.

$$\dfrac{14}{21} = \dfrac{7 \times 2}{7 \times 3} = \dfrac{7}{7} \times \dfrac{2}{3} = 1 \times \dfrac{2}{3} = \dfrac{2}{3}$$

So we see that

$$\dfrac{14}{21} = \dfrac{2}{3}$$

When we simplify fractions, we are using this property of multiplying by 1.

### 3 Converting Between Improper Fractions and Mixed Numbers ▶

If the numerator is less than the denominator, the fraction is a **proper fraction.** A proper fraction is used to describe a quantity smaller than a whole.

Fractions can also be used to describe quantities larger than a whole. The following figure shows two bars that are equal in size. Each bar is divided into 5 equal pieces. The first bar is shaded completely. The second bar has 2 of the 5 pieces shaded.

The shaded-in region can be represented by $\frac{7}{5}$ since 7 of the pieces (each of which is $\frac{1}{5}$ of a whole box) are shaded. The fraction $\frac{7}{5}$ is called an improper fraction. An **improper fraction** is one in which the numerator is larger than or equal to the denominator.

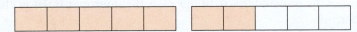

The shaded-in region can also be represented by 1 whole added to $\frac{2}{5}$ of a whole, or $1 + \frac{2}{5}$. This is written as $1\frac{2}{5}$. The fraction $1\frac{2}{5}$ is called a mixed number. A **mixed number** consists of a whole number added to a proper fraction (the numerator is smaller than the denominator). The addition is understood but not written. When we write $1\frac{2}{5}$, it represents $1 + \frac{2}{5}$. The numbers $1\frac{7}{8}$, $2\frac{3}{4}$, $8\frac{1}{3}$, and $126\frac{1}{10}$ are all mixed numbers. From the preceding figure it seems clear that $\frac{7}{5} = 1\frac{2}{5}$. This suggests that we can change from one form to the other without changing the value of the fraction.

From a picture it is easy to see how to *change improper fractions to mixed numbers*. For example, suppose we start with the fraction $\frac{11}{3}$ and represent it by the following figure (where 11 of the pieces, each of which is $\frac{1}{3}$ of a box, are shaded). We see that $\frac{11}{3} = 3\frac{2}{3}$, since 3 whole boxes and $\frac{2}{3}$ of a box are shaded.

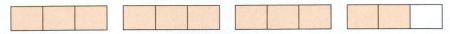

**Changing Improper Fractions to Mixed Numbers**    You can follow the same procedure without a picture. For example, to change $\frac{11}{3}$ to a mixed number, we can do the following:

$$\frac{11}{3} = \frac{3}{3} + \frac{3}{3} + \frac{3}{3} + \frac{2}{3} \qquad \text{Use the rule for adding fractions (which is discussed in detail in Section 0.2).}$$

$$= 1 + 1 + 1 + \frac{2}{3} \qquad \text{Write 1 in place of } \frac{3}{3}, \text{ since } \frac{3}{3} = 1.$$

$$= 3 + \frac{2}{3} \qquad \text{Write 3 in place of } 1 + 1 + 1.$$

$$= 3\frac{2}{3} \qquad \text{Use the notation for mixed numbers.}$$

Now that you know how to change improper fractions to mixed numbers and why the procedure works, here is a shorter method.

---

### TO CHANGE AN IMPROPER FRACTION TO A MIXED NUMBER

1. Divide the denominator into the numerator.
2. The quotient is the whole-number part of the mixed number.
3. The remainder from the division will be the numerator of the fraction. The denominator of the fraction remains unchanged.

---

We can write the fraction as a division statement and divide. The arrows show how to write the mixed number.

$$\frac{7}{5} \qquad \begin{array}{r} 1 \\ 5\overline{)7} \\ 5 \\ \hline 2 \end{array}$$

Whole-number part      Numerator of fraction

$\longrightarrow 1\frac{2}{5}$

Remainder

Thus, $\frac{7}{5} = 1\frac{2}{5}$.

$$\frac{11}{3} \qquad \begin{array}{r} 3 \\ 3{\overline{\smash{)}11}} \\ \underline{9} \\ 2 \end{array}$$

Whole-number part $\longrightarrow 3\frac{2}{3}\longleftarrow$ Numerator of fraction

Remainder

Thus, $\frac{11}{3} = 3\frac{2}{3}$.

Sometimes the remainder is 0. In this case, the improper fraction changes to a whole number.

**Example 5** Change to a mixed number or to a whole number.

(a) $\frac{7}{4}$

(b) $\frac{15}{3}$

**Solution**

(a) $\frac{7}{4} = 7 \div 4 \qquad \begin{array}{r} 1 \\ 4{\overline{\smash{)}7}} \\ \underline{4} \\ 3 \end{array}$ Remainder

(b) $\frac{15}{3} = 15 \div 3 \qquad \begin{array}{r} 5 \\ 3{\overline{\smash{)}15}} \\ \underline{15} \\ 0 \end{array}$ Remainder

Thus $\frac{7}{4} = 1\frac{3}{4}$.

Thus $\frac{15}{3} = 5$.

**Student Practice 5** Change to a mixed number or to a whole number.

(a) $\frac{12}{7}$

(b) $\frac{20}{5}$

**Changing Mixed Numbers to Improper Fractions** It is not difficult to see how to change mixed numbers to improper fractions. Suppose that you wanted to write $2\frac{2}{3}$ as an improper fraction.

$$2\frac{2}{3} = 2 + \frac{2}{3} \qquad \text{The meaning of mixed number notation}$$

$$= 1 + 1 + \frac{2}{3} \qquad \text{Since } 1 + 1 = 2$$

$$= \frac{3}{3} + \frac{3}{3} + \frac{2}{3} \qquad \text{Since } 1 = \frac{3}{3}$$

When we draw a picture of $\frac{3}{3} + \frac{3}{3} + \frac{2}{3}$, we have this figure:

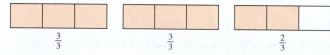

$$\frac{3}{3} \qquad \frac{3}{3} \qquad \frac{2}{3}$$

If we count the shaded parts, we see that

$$\frac{3}{3} + \frac{3}{3} + \frac{2}{3} = \frac{8}{3}. \quad \text{Thus} \quad 2\frac{2}{3} = \frac{8}{3}.$$

Now that you have seen how this change can be done, here is a shorter method.

**TO CHANGE A MIXED NUMBER TO AN IMPROPER FRACTION**

1. Multiply the whole number by the denominator.
2. Add this to the numerator. The result is the new numerator. The denominator does not change.

**Example 6** Change to an improper fraction.

(a) $3\frac{1}{7}$ 

(b) $5\frac{4}{5}$

**Solution**

(a) $3\frac{1}{7} = \frac{(3 \times 7) + 1}{7} = \frac{21 + 1}{7} = \frac{22}{7}$

(b) $5\frac{4}{5} = \frac{(5 \times 5) + 4}{5} = \frac{25 + 4}{5} = \frac{29}{5}$

**Student Practice 6** Change to an improper fraction.

(a) $3\frac{2}{5}$ 

(b) $1\frac{3}{7}$ 

(c) $2\frac{6}{11}$ 

(d) $4\frac{2}{3}$

## 4 Changing a Fraction to an Equivalent Fraction with a Given Denominator

A fraction can be changed to an equivalent fraction with a different denominator by multiplying both numerator and denominator by the same number.

$$\frac{5}{6} = \frac{5 \times 2}{6 \times 2} = \frac{10}{12} \quad \text{and} \quad \frac{3}{7} = \frac{3 \times 3}{7 \times 3} = \frac{9}{21} \quad \text{so}$$

$$\frac{5}{6} \text{ is equivalent to } \frac{10}{12} \quad \text{and} \quad \frac{3}{7} \text{ is equivalent to } \frac{9}{21}.$$

We often multiply in this way to obtain an equivalent fraction with a *particular denominator*.

**Example 7** Find the missing numerator.

(a) $\frac{3}{5} = \frac{?}{25}$ 

(b) $\frac{4}{7} = \frac{?}{14}$ 

(c) $\frac{2}{9} = \frac{?}{36}$

**Solution**

(a) $\frac{3}{5} = \frac{?}{25}$   Observe that we need to multiply the denominator by 5 to obtain 25. So we multiply the numerator 3 by 5 also.

$\frac{3 \times 5}{5 \times 5} = \frac{15}{25}$   The desired numerator is 15.

(b) $\frac{4}{7} = \frac{?}{14}$   Observe that $7 \times 2 = 14$. We need to multiply the numerator by 2 to get the new numerator.

$\frac{4 \times 2}{7 \times 2} = \frac{8}{14}$   The desired numerator is 8.

(c) $\frac{2}{9} = \frac{?}{36}$   Observe that $9 \times 4 = 36$. We need to multiply the numerator by 4 to get the new numerator.

$\frac{2 \times 4}{9 \times 4} = \frac{8}{36}$   The desired numerator is 8.

**Student Practice 7** Find the missing numerator.

(a) $\frac{3}{8} = \frac{?}{24}$ 

(b) $\frac{5}{6} = \frac{?}{30}$ 

(c) $\frac{2}{7} = \frac{?}{56}$

## 0.1 Exercises  MyMathLab®

**Verbal and Writing Skills, Exercises 1–4**

1. In the fraction $\frac{12}{13}$, what number is the numerator?

2. In the fraction $\frac{13}{17}$, what number is the denominator?

3. What is a factor? Give an example.

4. Give some examples of the number 1 written as a fraction.

5. Draw a diagram to illustrate $2\frac{2}{3}$.

6. Draw a diagram to illustrate $3\frac{3}{4}$.

*Simplify each fraction.*

7. $\frac{9}{15}$  8. $\frac{20}{24}$  9. $\frac{12}{36}$  10. $\frac{8}{48}$  11. $\frac{60}{12}$  12. $\frac{72}{18}$

13. $\frac{24}{36}$  14. $\frac{32}{64}$  15. $\frac{30}{85}$  16. $\frac{33}{55}$  17. $\frac{42}{54}$  18. $\frac{63}{81}$

*Change to a mixed number.*

19. $\frac{17}{6}$  20. $\frac{19}{5}$  21. $\frac{47}{5}$  22. $\frac{54}{7}$  23. $\frac{38}{7}$  24. $\frac{41}{6}$

25. $\frac{41}{2}$  26. $\frac{25}{3}$  27. $\frac{32}{5}$  28. $\frac{79}{7}$  29. $\frac{111}{9}$  30. $\frac{124}{8}$

*Change to an improper fraction or whole number.*

31. $3\frac{1}{5}$  32. $4\frac{2}{5}$  33. $6\frac{3}{5}$  34. $5\frac{1}{12}$  35. $1\frac{2}{9}$  36. $1\frac{5}{6}$

37. $8\frac{3}{7}$  38. $6\frac{2}{3}$  39. $24\frac{1}{4}$  40. $10\frac{1}{9}$  41. $\frac{72}{9}$  42. $\frac{78}{6}$

*Find the missing numerator.*

43. $\frac{3}{8} = \frac{?}{64}$  44. $\frac{5}{9} = \frac{?}{54}$  45. $\frac{3}{5} = \frac{?}{35}$

46. $\frac{5}{9} = \frac{?}{45}$  47. $\frac{4}{13} = \frac{?}{39}$  48. $\frac{13}{17} = \frac{?}{51}$

49. $\frac{3}{7} = \frac{?}{49}$  50. $\frac{10}{15} = \frac{?}{60}$  51. $\frac{3}{4} = \frac{?}{20}$

52. $\frac{7}{8} = \frac{?}{40}$  53. $\frac{35}{40} = \frac{?}{80}$  54. $\frac{45}{50} = \frac{?}{100}$

## Applications

*Solve.*

55. **Women's Professional Basketball** During the 2014 WNBA basketball season, Maya Moore of the Minnesota Lynx scored 799 points in 34 games. Express as a mixed number in simplified form how many points she averaged per game.

56. **Kentucky Derby Nominations** In 2014, 424 horses were nominated to compete in the Kentucky Derby. Only 20 horses were actually chosen to compete in the Derby. What simplified fraction shows what portions of the nominated horses actually competed?

**57. *Income Tax*** Last year, my parents had a combined income of $64,000. They paid $13,200 in federal income taxes. What simplified fraction shows how much my parents spent on their federal taxes?

**58. *Employment*** A large employment agency was able to find jobs within 6 months for 1400 people out of 2420 applicants who applied at one of its branches. What simplified fraction shows what portion of applicants gained employment?

***Trail Mix*** *The following chart gives recipes for two trail mix blends.*

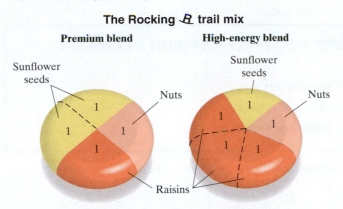

**The Rocking *R* trail mix**

**59.** What fractional part of the premium blend is nuts?

**60.** What fractional part of the high-energy blend is raisins?

**61.** What fractional part of the premium blend is not sunflower seeds?

**62.** What fractional part of the high-energy blend does not contain nuts?

***College Enrollment*** *The following chart provides statistics about the total enrollment of male and female students in U.S. colleges for specific years during the period from 1984 to 2014.*

**63.** What fractional part of the number of students enrolled in 2014 are female?

**64.** What fractional part of the number of students enrolled in 1994 are male?

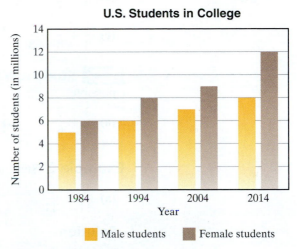

**U.S. Students in College**

Source: Digest of Education statistics, 2014

---

## Quick Quiz 0.1

**1.** Simplify. $\dfrac{84}{92}$

**2.** Write as an improper fraction. $6\dfrac{9}{11}$

**3.** Write as a mixed number. $\dfrac{103}{21}$

**4. Concept Check** Explain in your own words how to change a mixed number to an improper fraction.

## 0.2 Adding and Subtracting Fractions ▶

### 1 Adding or Subtracting Fractions with a Common Denominator ▶

If fractions have the same denominator, the numerators may be added or subtracted. The denominator remains the same.

> **TO ADD OR SUBTRACT TWO FRACTIONS WITH A COMMON DENOMINATOR**
>
> 1. Add or subtract the numerators.
> 2. Keep the same (common) denominator.
> 3. Simplify the answer whenever possible.

**Example 1** Add the fractions. Simplify your answer whenever possible.

(a) $\dfrac{5}{7} + \dfrac{1}{7}$    (b) $\dfrac{2}{3} + \dfrac{1}{3}$    (c) $\dfrac{1}{8} + \dfrac{3}{8} + \dfrac{2}{8}$    (d) $\dfrac{3}{5} + \dfrac{4}{5}$

**Solution**

(a) $\dfrac{5}{7} + \dfrac{1}{7} = \dfrac{5+1}{7} = \dfrac{6}{7}$      (b) $\dfrac{2}{3} + \dfrac{1}{3} = \dfrac{2+1}{3} = \dfrac{3}{3} = 1$

(c) $\dfrac{1}{8} + \dfrac{3}{8} + \dfrac{2}{8} = \dfrac{1+3+2}{8} = \dfrac{6}{8} = \dfrac{3}{4}$      (d) $\dfrac{3}{5} + \dfrac{4}{5} = \dfrac{3+4}{5} = \dfrac{7}{5}$ or $1\dfrac{2}{5}$ □

**▷ Student Practice 1** Add the fractions. Simplify your answer whenever possible.

(a) $\dfrac{3}{6} + \dfrac{2}{6}$    (b) $\dfrac{3}{11} + \dfrac{8}{11}$    (c) $\dfrac{1}{8} + \dfrac{2}{8} + \dfrac{1}{8}$    (d) $\dfrac{5}{9} + \dfrac{8}{9}$

**Example 2** Subtract the fractions. Simplify your answer whenever possible.

(a) $\dfrac{9}{11} - \dfrac{2}{11}$        (b) $\dfrac{5}{6} - \dfrac{1}{6}$

**Solution**

(a) $\dfrac{9}{11} - \dfrac{2}{11} = \dfrac{9-2}{11} = \dfrac{7}{11}$      (b) $\dfrac{5}{6} - \dfrac{1}{6} = \dfrac{5-1}{6} = \dfrac{4}{6} = \dfrac{2}{3}$ □

**▷ Student Practice 2** Subtract the fractions. Simplify your answer whenever possible.

(a) $\dfrac{11}{13} - \dfrac{6}{13}$        (b) $\dfrac{8}{9} - \dfrac{2}{9}$

Although adding and subtracting fractions with the same denominator is fairly simple, most problems involve fractions that do not have a common denominator. Fractions and mixed numbers such as halves, fourths, and eighths are often used. To add or subtract such fractions, we begin by finding a common denominator.

### 2 Using Prime Factors to Find the Least Common Denominator of Two or More Fractions ▶

Before you can add or subtract fractions, they must have the same denominator. To save work, we select the smallest possible common denominator. This is called the **least common denominator** or LCD (also known as the *lowest common denominator*).

The LCD of two or more fractions is the smallest whole number that is exactly divisible by each denominator of the fractions.

**Example 3** Find the LCD. $\frac{2}{3}$ and $\frac{1}{4}$

**Solution** The numbers are small enough to find the LCD by inspection. The LCD is 12, since 12 is exactly divisible by 4 and by 3. There is no smaller number that is exactly divisible by 4 and 3. □

 **Student Practice 3** Find the LCD. $\frac{1}{8}$ and $\frac{5}{7}$

In some cases, the LCD cannot easily be determined by inspection. If we write each denominator as the product of prime factors, we will be able to find the LCD. We will use ( · ) to indicate multiplication. For example, $30 = 2 \cdot 3 \cdot 5$. This means $30 = 2 \times 3 \times 5$.

**PROCEDURE TO FIND THE LCD USING PRIME FACTORS**

1. Write each denominator as the product of prime factors.
2. The LCD is a product containing each different factor.
3. If a factor occurs more than once in any one denominator, the LCD will contain that factor repeated the greatest number of times that it occurs in any one denominator.

**Example 4** Find the LCD of $\frac{5}{6}$ and $\frac{1}{15}$ using the prime factor method.

**Solution**

$$6 = 2 \cdot 3$$
$$15 = \ \ 3 \cdot 5$$
Write each denominator as the product of prime factors.

$$LCD = 2 \cdot 3 \cdot 5$$ The LCD is a product containing each different prime factor.
$$LCD = 2 \cdot 3 \cdot 5 = 30$$ The different factors are 2, 3, and 5, and each factor appears at most once in any one denominator. □

 **Student Practice 4** Find the LCD of $\frac{8}{35}$ and $\frac{6}{15}$ using the prime factor method.

Great care should be used to determine the LCD in the case of repeated factors.

**Example 5** Find the LCD of $\frac{4}{27}$ and $\frac{5}{18}$.

**Solution**

$$27 = 3 \cdot 3 \cdot 3$$
Write each denominator as the product of prime factors. We observe that the factor 3 occurs three times in the factorization of 27.

$$18 = \ \ 3 \cdot 3 \cdot 2$$
$$LCD = 3 \cdot 3 \cdot 3 \cdot 2$$
$$LCD = 3 \cdot 3 \cdot 3 \cdot 2 = 54$$

The LCD is a product containing each different factor. The factor 3 *occurred most* in the factorization of 27, where it occurred *three* times. Thus the LCD will be the product of *three* 3s and *one* 2. □

**Student Practice 5** Find the LCD of $\frac{5}{12}$ and $\frac{7}{30}$.

**Example 6** Find the LCD of $\frac{5}{12}, \frac{1}{15}$, and $\frac{7}{30}$.

**Solution**

$$12 = 2 \cdot 2 \cdot 3$$
$$15 = \quad\quad 3 \cdot 5$$
$$30 = \quad 2 \cdot 3 \cdot 5$$

Write each denominator as the product of prime factors. Notice that the only repeated factor is 2, which occurs twice in the factorization of 12.

$$LCD = 2 \cdot 2 \cdot 3 \cdot 5$$
$$LCD = 2 \cdot 2 \cdot 3 \cdot 5 = 60$$

The LCD is the product of each different factor, with the factor 2 appearing twice since it occurred twice in one denominator.

 **Student Practice 6** Find the LCD of $\frac{2}{27}, \frac{1}{18}$, and $\frac{5}{12}$.

### 3 Adding or Subtracting Fractions with Different Denominators

Before you can add or subtract them, fractions must have the same denominator. Using the LCD will make your work easier. First you must find the LCD. Then change each fraction to a fraction that has the LCD as the denominator. Sometimes one of the fractions will already have the LCD as the denominator. Once all the fractions have the same denominator, you can add or subtract. Be sure to simplify the fraction in your answer if this is possible.

> **TO ADD OR SUBTRACT FRACTIONS THAT DO NOT HAVE A COMMON DENOMINATOR**
>
> 1. Find the LCD of the fractions.
> 2. Change each fraction to an equivalent fraction with the LCD for a denominator.
> 3. Add or subtract the fractions.
> 4. Simplify the answer whenever possible.

Let us return to the two fractions of Example 3. We have previously found that the LCD is 12.

**Example 7** Bob picked $\frac{2}{3}$ of a bushel of apples on Monday and $\frac{1}{4}$ of a bushel of apples on Tuesday. How much did he pick in total?

**Solution** To solve this problem we need to add $\frac{2}{3}$ and $\frac{1}{4}$, but before we can do so, we must change $\frac{2}{3}$ and $\frac{1}{4}$ to fractions with the same denominator. We change each fraction to an equivalent fraction with a common denominator of 12, the LCD.

$$\frac{2}{3} = \frac{?}{12} \quad\quad \frac{2 \times 4}{3 \times 4} = \frac{8}{12} \quad so \quad \frac{2}{3} = \frac{8}{12}$$

$$\frac{1}{4} = \frac{?}{12} \quad\quad \frac{1 \times 3}{4 \times 3} = \frac{3}{12} \quad so \quad \frac{1}{4} = \frac{3}{12}$$

Then we rewrite the problem with common denominators and add.

$$\frac{2}{3} + \frac{1}{4} = \frac{8}{12} + \frac{3}{12} = \frac{8+3}{12} = \frac{11}{12}$$

In total Bob picked $\frac{11}{12}$ of a bushel of apples.

 **Student Practice 7** Carol planted corn in $\frac{5}{7}$ of the farm fields at the Old Robinson Farm. Connie planted soybeans in $\frac{1}{8}$ of the farm fields. What fractional part of the farm fields of the Old Robinson Farm was planted in corn or soybeans?

Sometimes one of the denominators is the LCD. In such cases the fraction that has the LCD for the denominator will not need to be changed. If every other denominator divides into the largest denominator, the largest denominator is the LCD.

**Example 8** Find the LCD and then add. $\dfrac{3}{5} + \dfrac{7}{20} + \dfrac{1}{2}$

**Solution** We can see by inspection that both 5 and 2 divide exactly into 20. Thus 20 is the LCD. Now add.

$$\frac{3}{5} + \frac{7}{20} + \frac{1}{2}$$

We change $\frac{3}{5}$ and $\frac{1}{2}$ to equivalent fractions with a common denominator of 20, the LCD.

$$\frac{3}{5} = \frac{?}{20} \qquad \frac{3 \times 4}{5 \times 4} = \frac{12}{20} \quad \text{so} \quad \frac{3}{5} = \frac{12}{20}$$

$$\frac{1}{2} = \frac{?}{20} \qquad \frac{1 \times 10}{2 \times 10} = \frac{10}{20} \quad \text{so} \quad \frac{1}{2} = \frac{10}{20}$$

Then we rewrite the problem with common denominators and add.

$$\frac{3}{5} + \frac{7}{20} + \frac{1}{2} = \frac{12}{20} + \frac{7}{20} + \frac{10}{20} = \frac{12 + 7 + 10}{20} = \frac{29}{20} \quad \text{or} \quad 1\frac{9}{20} \qquad \square$$

**Student Practice 8** Find the LCD and then add.

$$\frac{4}{5} + \frac{6}{25} + \frac{1}{50}$$

Now we turn to examples where the selection of the LCD is not so obvious. In Examples 9 through 11 we will use the prime factorization method to find the LCD.

**Example 9** Add. $\dfrac{7}{18} + \dfrac{5}{12}$

**Solution** First we find the LCD.

$$18 = 3 \cdot 3 \cdot 2$$
$$12 = \quad 3 \cdot 2 \cdot 2$$
$$\downarrow \downarrow \downarrow \downarrow$$
$$\text{LCD} = 3 \cdot 3 \cdot 2 \cdot 2 = 36$$

Now we change $\frac{7}{18}$ and $\frac{5}{12}$ to equivalent fractions that have the LCD.

$$\frac{7}{18} = \frac{?}{36} \qquad \frac{7 \times 2}{18 \times 2} = \frac{14}{36}$$

$$\frac{5}{12} = \frac{?}{36} \qquad \frac{5 \times 3}{12 \times 3} = \frac{15}{36}$$

Now we add the fractions.

$$\frac{7}{18} + \frac{5}{12} = \frac{14}{36} + \frac{15}{36} = \frac{29}{36} \qquad \text{This fraction cannot be simplified.} \qquad \square$$

**Student Practice 9** Add.

$$\frac{1}{49} + \frac{3}{14}$$

**Example 10** Subtract. $\dfrac{25}{48} - \dfrac{5}{36}$

**Solution**   First we find the LCD.

$$48 = 2 \cdot 2 \cdot 2 \cdot 2 \cdot 3$$
$$36 = \quad\;\; 2 \cdot 2 \cdot 3 \cdot 3$$
$$\text{LCD} = 2 \cdot 2 \cdot 2 \cdot 2 \cdot 3 \cdot 3 = 144$$

Now we change $\dfrac{25}{48}$ and $\dfrac{5}{36}$ to equivalent fractions that have the LCD.

$$\frac{25}{48} = \frac{?}{144} \qquad \frac{25 \times 3}{48 \times 3} = \frac{75}{144}$$

$$\frac{5}{36} = \frac{?}{144} \qquad \frac{5 \times 4}{36 \times 4} = \frac{20}{144}$$

Now we subtract the fractions.

$$\frac{25}{48} - \frac{5}{36} = \frac{75}{144} - \frac{20}{144} = \frac{55}{144} \quad \text{This fraction cannot be simplified.} \qquad \square$$

 **Student Practice 10**   Subtract.

$$\frac{1}{12} - \frac{1}{30}$$

**Example 11** Combine. $\dfrac{1}{5} + \dfrac{1}{6} - \dfrac{3}{10}$

**Solution**   First we find the LCD.

$$5 = 5$$
$$6 = \quad 2 \cdot 3$$
$$10 = 5 \cdot 2$$
$$\text{LCD} = 5 \cdot 2 \cdot 3 = 30$$

Now we change $\frac{1}{5}, \frac{1}{6}$, and $\frac{3}{10}$ to equivalent fractions that have the LCD for a denominator.

$$\frac{1}{5} = \frac{?}{30} \qquad \frac{1 \times 6}{5 \times 6} = \frac{6}{30}$$

$$\frac{1}{6} = \frac{?}{30} \qquad \frac{1 \times 5}{6 \times 5} = \frac{5}{30}$$

$$\frac{3}{10} = \frac{?}{30} \qquad \frac{3 \times 3}{10 \times 3} = \frac{9}{30}$$

Now we combine the three fractions.

$$\frac{1}{5} + \frac{1}{6} - \frac{3}{10} = \frac{6}{30} + \frac{5}{30} - \frac{9}{30} = \frac{2}{30} = \frac{1}{15}$$

Note the important step of simplifying the fraction to obtain the final answer.   $\square$

**Student Practice 11**   Combine.

$$\frac{2}{3} + \frac{3}{4} - \frac{3}{8}$$

## 4 Adding or Subtracting Mixed Numbers ▶

If your addition or subtraction problem has mixed numbers, change them to improper fractions first and then combine (add or subtract). As a convention in this book, if the original problem contains mixed numbers, express the result as a mixed number rather than as an improper fraction.

**Example 12** Combine. Simplify your answer whenever possible.

(a) $5\frac{1}{2} + 2\frac{1}{3}$      (b) $2\frac{1}{5} - 1\frac{3}{4}$      (c) $1\frac{5}{12} + \frac{7}{30}$

**Solution**

(a) First we change the mixed numbers to improper fractions.

$$5\frac{1}{2} = \frac{5 \times 2 + 1}{2} = \frac{11}{2} \qquad 2\frac{1}{3} = \frac{2 \times 3 + 1}{3} = \frac{7}{3}$$

Next we change each fraction to an equivalent form with the common denominator of 6.

$$\frac{11}{2} = \frac{?}{6} \qquad \frac{11 \times 3}{2 \times 3} = \frac{33}{6}$$

$$\frac{7}{3} = \frac{?}{6} \qquad \frac{7 \times 2}{3 \times 2} = \frac{14}{6}$$

Finally, we add the two fractions and change our answer to a mixed number.

$$\frac{33}{6} + \frac{14}{6} = \frac{47}{6} = 7\frac{5}{6}$$

Thus $5\frac{1}{2} + 2\frac{1}{3} = 7\frac{5}{6}$.

(b) First we change the mixed numbers to improper fractions.

$$2\frac{1}{5} = \frac{2 \times 5 + 1}{5} = \frac{11}{5} \qquad 1\frac{3}{4} = \frac{1 \times 4 + 3}{4} = \frac{7}{4}$$

Next we change each fraction to an equivalent form with the common denominator of 20.

$$\frac{11}{5} = \frac{?}{20} \qquad \frac{11 \times 4}{5 \times 4} = \frac{44}{20}$$

$$\frac{7}{4} = \frac{?}{20} \qquad \frac{7 \times 5}{4 \times 5} = \frac{35}{20}$$

Now we subtract the two fractions.

$$\frac{44}{20} - \frac{35}{20} = \frac{9}{20}$$

Thus $2\frac{1}{5} - 1\frac{3}{4} = \frac{9}{20}$.

*Note:* It is not necessary to use these exact steps to add and subtract mixed numbers. If you know another method and can use it to obtain the correct answers, it is all right to continue to use that method throughout this chapter.

(c) Now we add $1\frac{5}{12} + \frac{7}{30}$.

The LCD of 12 and 30 is 60. Why? Change the mixed number to an improper fraction. Then change each fraction to an equivalent form with a common denominator.

$$1\frac{5}{12} = \frac{17 \times 5}{12 \times 5} = \frac{85}{60} \qquad \frac{7 \times 2}{30 \times 2} = \frac{14}{60}$$

*Continued on next page*

Then add the fractions, simplify, and write the answer as a mixed number.

$$\frac{85}{60} + \frac{14}{60} = \frac{99}{60} = \frac{33}{20} = 1\frac{13}{20}$$

Thus $1\frac{5}{12} + \frac{7}{30} = 1\frac{13}{20}$.

**Student Practice 12** Combine. Simplify your answer whenever possible.

(a) $1\frac{2}{3} + 2\frac{4}{5}$　　　(b) $5\frac{1}{4} - 2\frac{2}{3}$

▲ **Example 13** Manuel is enclosing a triangle-shaped exercise yard for his new dog. He wants to determine how many feet of fencing he will need. The sides of the yard measure $20\frac{3}{4}$ feet, $15\frac{1}{2}$ feet, and $18\frac{1}{8}$ feet. What is the perimeter of (total distance around) the triangle?

**Solution** *Understand the problem.* Begin by drawing a picture.

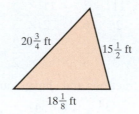

We want to add up the lengths of all three sides of the triangle. This distance around the triangle is called the **perimeter.**

$$20\frac{3}{4} + 15\frac{1}{2} + 18\frac{1}{8} = \frac{83}{4} + \frac{31}{2} + \frac{145}{8}$$

$$= \frac{166}{8} + \frac{124}{8} + \frac{145}{8} = \frac{435}{8} = 54\frac{3}{8}$$

He will need $54\frac{3}{8}$ feet of fencing.

▲ **Student Practice 13** Find the perimeter of a rectangle with sides of $4\frac{1}{5}$ cm and $6\frac{1}{2}$ cm. Begin by drawing a picture. Label the picture by including the measure of *each* side.

## 👣 STEPS TO SUCCESS Faithful Class Attendance Is Well Worth It.

**If you attend a traditional mathematics class that meets one or more times each week:**

*Get started in the right direction.* Make a personal commitment to attend class every day, beginning with the first day of class. Teachers and students all over the country have discovered that faithful class attendance and good grades go together.

*The vital content of class.* What goes on in class is designed to help you learn more quickly. Each day significant information is given that will truly help you to understand concepts. There is no substitute for this firsthand learning experience.

*Meet a friend.* You will soon discover that other students are also coming to class every single class period. It is easy to strike up a friendship with students like you who have this common commitment. They will usually be available to answer a question after class and give you an additional source of help when you encounter difficulty.

**Making it personal:** Write down what you think is the most compelling reason to come to every class meeting. Make that commitment and see how much it helps you. ▼

## 0.2 Exercises   MyMathLab®

### Verbal and Writing Skills, Exercises 1 and 2

1. Explain why the denominator 8 is the least common denominator of $\frac{3}{4}$ and $\frac{5}{8}$.

2. What must you do before you add or subtract fractions that do not have a common denominator?

*Find the LCD (least common denominator) of each set of fractions. Do not combine the fractions; only find the LCD.*

3. $\frac{4}{9}$ and $\frac{5}{12}$

4. $\frac{21}{30}$ and $\frac{17}{20}$

5. $\frac{7}{10}$ and $\frac{1}{4}$

6. $\frac{5}{18}$ and $\frac{1}{24}$

7. $\frac{5}{18}$ and $\frac{7}{54}$

8. $\frac{5}{16}$ and $\frac{7}{48}$

9. $\frac{1}{15}$ and $\frac{4}{21}$

10. $\frac{11}{12}$ and $\frac{7}{20}$

11. $\frac{17}{40}$ and $\frac{13}{60}$

12. $\frac{7}{30}$ and $\frac{8}{45}$

13. $\frac{2}{5}, \frac{3}{8}$, and $\frac{5}{12}$

14. $\frac{1}{7}, \frac{3}{14}$, and $\frac{9}{35}$

15. $\frac{5}{6}, \frac{9}{14}$, and $\frac{17}{26}$

16. $\frac{3}{8}, \frac{5}{12}$, and $\frac{11}{42}$

17. $\frac{1}{2}, \frac{1}{18}$, and $\frac{13}{30}$

18. $\frac{5}{8}, \frac{3}{14}$, and $\frac{11}{16}$

*Combine. Be sure to simplify your answer whenever possible.*

19. $\frac{3}{8} + \frac{2}{8}$

20. $\frac{3}{11} + \frac{5}{11}$

21. $\frac{5}{14} - \frac{1}{14}$

22. $\frac{11}{15} - \frac{2}{15}$

23. $\frac{5}{12} + \frac{5}{8}$

24. $\frac{3}{20} + \frac{13}{15}$

25. $\frac{5}{7} - \frac{2}{9}$

26. $\frac{4}{5} - \frac{3}{7}$

27. $\frac{1}{3} + \frac{2}{5}$

28. $\frac{3}{8} + \frac{1}{3}$

29. $\frac{5}{9} + \frac{5}{12}$

30. $\frac{2}{15} + \frac{7}{10}$

31. $\frac{11}{15} - \frac{31}{45}$

32. $\frac{21}{12} - \frac{23}{24}$

33. $\frac{16}{24} - \frac{1}{6}$

34. $\frac{13}{15} - \frac{1}{5}$

35. $\frac{3}{8} + \frac{4}{7}$

36. $\frac{7}{4} + \frac{5}{9}$

37. $\frac{2}{3} + \frac{7}{12} + \frac{1}{4}$

38. $\frac{4}{7} + \frac{7}{9} + \frac{1}{3}$

39. $\frac{5}{30} + \frac{3}{40} + \frac{1}{8}$

40. $\frac{1}{12} + \frac{3}{14} + \frac{4}{21}$

41. $\frac{1}{3} + \frac{1}{12} - \frac{1}{6}$

42. $\frac{1}{5} + \frac{2}{3} - \frac{11}{15}$

17

**43.** $\dfrac{5}{36} + \dfrac{7}{9} - \dfrac{5}{12}$

**44.** $\dfrac{5}{24} + \dfrac{3}{8} - \dfrac{1}{3}$

**45.** $4\dfrac{1}{3} + 3\dfrac{2}{5}$

**46.** $3\dfrac{1}{8} + 2\dfrac{1}{6}$

**47.** $1\dfrac{5}{24} + \dfrac{5}{18}$

**48.** $6\dfrac{2}{3} + \dfrac{3}{4}$

**49.** $7\dfrac{1}{6} - 2\dfrac{1}{4}$

**50.** $7\dfrac{2}{5} - 3\dfrac{3}{4}$

**51.** $8\dfrac{5}{7} - 2\dfrac{1}{4}$

**52.** $7\dfrac{8}{15} - 2\dfrac{3}{5}$

**53.** $2\dfrac{1}{8} + 3\dfrac{2}{3}$

**54.** $3\dfrac{1}{7} + 4\dfrac{1}{3}$

**55.** $11\dfrac{1}{7} - 6\dfrac{5}{7}$

**56.** $12\dfrac{1}{3} - 5\dfrac{2}{3}$

**57.** $3\dfrac{5}{12} + 5\dfrac{7}{12}$

**58.** $9\dfrac{12}{13} + 9\dfrac{1}{13}$

## Mixed Practice

**59.** $\dfrac{7}{8} + \dfrac{1}{12}$

**60.** $\dfrac{19}{30} + \dfrac{3}{10}$

**61.** $3\dfrac{3}{16} + 4\dfrac{3}{8}$

**62.** $5\dfrac{2}{3} + 7\dfrac{2}{5}$

**63.** $\dfrac{16}{21} - \dfrac{2}{7}$

**64.** $\dfrac{15}{24} - \dfrac{3}{8}$

**65.** $5\dfrac{1}{5} - 2\dfrac{1}{2}$

**66.** $6\dfrac{1}{3} - 4\dfrac{1}{4}$

**67.** $25\dfrac{2}{3} - 6\dfrac{1}{7}$

**68.** $45\dfrac{3}{8} - 26\dfrac{1}{10}$

**69.** $1\dfrac{1}{6} + \dfrac{3}{8}$

**70.** $1\dfrac{2}{3} + \dfrac{5}{18}$

**71.** $8\dfrac{1}{4} + 3\dfrac{5}{6}$

**72.** $7\dfrac{3}{4} + 6\dfrac{2}{5}$

**73.** $36 - 2\dfrac{4}{7}$

**74.** $28 - 3\dfrac{5}{8}$

## Applications

**75.** *Inline Skating* Nancy and Sarah meet three mornings a week to skate. They skated $8\dfrac{1}{4}$ miles on Monday, $10\dfrac{2}{3}$ miles on Wednesday, and $5\dfrac{3}{4}$ miles on Friday. What was their total distance for those three days?

**76.** *Marathon Training* Paco and Eskinder are training for the Boston Marathon. Their coach gave them the following schedule: a medium run of $10\dfrac{1}{2}$ miles on Thursday, a short run of $5\dfrac{1}{4}$ miles on Friday, a rest day on Saturday, and a long run of $18\dfrac{2}{3}$ miles on Sunday. How many miles did they run over these four days?

**77. *Restaurant Management*** The manager of a Boston restaurant must have his staff replace unsafe and rusted knives and replace tables and chairs in the dining area on Monday when the restaurant is closed. He has scheduled the staff for $15\frac{1}{2}$ hours of work. He estimates it will take $3\frac{2}{3}$ hours to replace the unsafe and rusted knives. He estimates it will take $9\frac{1}{4}$ hours to replace the tables and chairs in the dining area. In the time remaining, he wants them to wash the front windows. How much time will be available for washing the front windows?

**78. *Aquariums*** Carl bought a 20-gallon aquarium. He put $17\frac{3}{4}$ gallons of water into the aquarium, but it looked too low, so he added $1\frac{1}{4}$ more gallons of water. He then put in the artificial plants and the gravel but now the water was too high, so he siphoned off $2\frac{2}{3}$ gallons of water. How many gallons of water are now in the aquarium?

## To Think About

***Carpentry*** *Carpenters use fractions in their work. The picture below is a diagram of a spice cabinet. The symbol " means inches. Use the picture to answer exercises 79 and 80.*

**79.** Before you can determine where the cabinet will fit, you need to calculate the height, $A$, and the width, $B$. Don't forget to include the $\frac{1}{2}$-inch thickness of the wood where needed.

**80.** Look at the close-up of the drawer. The width is $4\frac{9}{16}''$. In the diagram, the width of the opening for the drawer is $4\frac{5}{8}''$. What is the difference?

Why do you think the drawer is smaller than the opening?

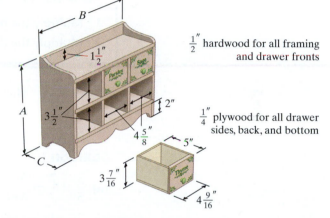

$\frac{1}{2}''$ hardwood for all framing and drawer fronts

$\frac{1}{4}''$ plywood for all drawer sides, back, and bottom

**81. *Facilities Management*** The Falmouth Country Club maintains the putting greens with a grass height of $\frac{7}{8}$ inch. The grass on the fairways is maintained at a height of $2\frac{1}{2}$ inches. How much must the mower blade be lowered by a person mowing the fairways if that person will be using the same mowing machine on the putting greens?

**82. *Facilities Management*** The director of facilities maintenance at the club in Exercise 81 discovered that due to slippage in the adjustment lever, the lawn mower actually cuts the grass $\frac{1}{16}$ of an inch too long or too short on some days. What is the maximum height that the fairway grass could be after being mowed with this machine? What is the minimum height that the putting greens could be after being mowed with this machine?

## Cumulative Review

**83. [0.1.2]** Simplify. $\dfrac{36}{44}$

**84. [0.1.3]** Change to an improper fraction. $26\dfrac{3}{5}$

---

**Quick Quiz 0.2** *Perform the operations indicated. Simplify your answers whenever possible.*

**1.** $\dfrac{3}{4} + \dfrac{1}{2} + \dfrac{5}{12}$

**2.** $2\dfrac{3}{5} + 4\dfrac{14}{15}$

**3.** $6\dfrac{1}{9} - 3\dfrac{5}{6}$

**4. Concept Check** Explain how you would find the LCD of the fractions $\frac{4}{21}$ and $\frac{5}{18}$.

# 0.3 Multiplying and Dividing Fractions ▶

**Student Learning Objectives**

After studying this section, you will be able to:

1 Multiply fractions, whole numbers, and mixed numbers. ▶

2 Divide fractions, whole numbers, and mixed numbers. ▶

## 1 Multiplying Fractions, Whole Numbers, and Mixed Numbers ▶

**Multiplying Fractions** During a recent snowstorm, the runway at Beverly Airport was plowed. However, the plow cleared only $\frac{3}{5}$ of the width and $\frac{2}{7}$ of the length. What fraction of the total runway area was cleared? To answer this question, we need to multiply $\frac{3}{5} \times \frac{2}{7}$.

The answer is that $\frac{6}{35}$ of the total runway area was cleared.

The multiplication rule for fractions states that to multiply two fractions, we multiply the two numerators and multiply the two denominators.

---

**TO MULTIPLY ANY TWO FRACTIONS**

1. Multiply the numerators.
2. Multiply the denominators.

---

**Example 1** Multiply.

(a) $\frac{3}{5} \times \frac{2}{7}$ (b) $\frac{1}{3} \times \frac{5}{4}$ (c) $\frac{7}{3} \times \frac{1}{5}$ (d) $\frac{6}{5} \times \frac{2}{3}$

**Solution**

(a) $\frac{3}{5} \times \frac{2}{7} = \frac{3 \cdot 2}{5 \cdot 7} = \frac{6}{35}$

(b) $\frac{1}{3} \times \frac{5}{4} = \frac{1 \cdot 5}{3 \cdot 4} = \frac{5}{12}$

(c) $\frac{7}{3} \times \frac{1}{5} = \frac{7 \cdot 1}{3 \cdot 5} = \frac{7}{15}$

(d) $\frac{6}{5} \times \frac{2}{3} = \frac{6 \cdot 2}{5 \cdot 3} = \frac{12}{15} = \frac{4}{5}$

Note that we must simplify this fraction. □

▶ **Student Practice 1** Multiply.

(a) $\frac{2}{7} \times \frac{5}{11}$ (b) $\frac{1}{5} \times \frac{7}{10}$ (c) $\frac{9}{5} \times \frac{1}{4}$ (d) $\frac{8}{9} \times \frac{3}{10}$

It is possible to avoid having to simplify a fraction as the last step. In many cases we can divide by a value that appears as a factor in both a numerator and a denominator. Often it is helpful to write the numbers as products of prime factors in order to do this.

**Example 2** Multiply.

(a) $\frac{3}{5} \times \frac{5}{7}$ (b) $\frac{4}{11} \times \frac{5}{2}$ (c) $\frac{15}{8} \times \frac{10}{27}$

**Solution**

(a) $\frac{3}{5} \times \frac{5}{7} = \frac{3 \cdot 5}{5 \cdot 7} = \frac{3 \cdot \overset{1}{\cancel{5}}}{7 \cdot \cancel{5}} = \frac{3}{7}$ Note that here we divided numerator and denominator by 5.

If we factor each number, we can see the common factors.

(b) $\frac{4}{11} \times \frac{5}{2} = \frac{2 \cdot \overset{1}{\cancel{2}}}{11} \times \frac{5}{\cancel{2}} = \frac{10}{11}$ (c) $\frac{15}{8} \times \frac{10}{27} = \frac{\overset{1}{\cancel{3}} \cdot 5}{2 \cdot 2 \cdot \cancel{2}} \times \frac{5 \cdot \overset{1}{\cancel{2}}}{\cancel{3} \cdot 3 \cdot 3} = \frac{25}{36}$

After dividing out common factors, the resulting multiplication problem involves smaller numbers and the answers are in simplified form. □

**Student Practice 2**   Multiply.

(a) $\dfrac{3}{5} \times \dfrac{4}{3}$         (b) $\dfrac{9}{10} \times \dfrac{5}{12}$

## Sidelight: Dividing Out Common Factors

Why does this method of dividing out a value that appears as a factor in both numerator and denominator work? Let's reexamine one of the examples we solved previously.

$$\frac{3}{5} \times \frac{5}{7} = \frac{3 \cdot 5}{5 \cdot 7} = \frac{3 \cdot \overset{1}{\cancel{5}}}{7 \cdot \underset{1}{\cancel{5}}} = \frac{3}{7}$$

Consider the following steps and reasons.

$\dfrac{3}{5} \times \dfrac{5}{7} = \dfrac{3 \cdot 5}{5 \cdot 7}$   Definition of multiplication of fractions.

$\phantom{\dfrac{3}{5} \times \dfrac{5}{7}} = \dfrac{3 \cdot 5}{7 \cdot 5}$   Change the order of the factors in the denominator, since $5 \cdot 7 = 7 \cdot 5$. This is called the commutative property of multiplication.

$\phantom{\dfrac{3}{5} \times \dfrac{5}{7}} = \dfrac{3}{7} \cdot \dfrac{5}{5}$   Definition of multiplication of fractions.

$\phantom{\dfrac{3}{5} \times \dfrac{5}{7}} = \dfrac{3}{7} \cdot 1$   Write 1 in place of $\frac{5}{5}$, since 1 is another name for $\frac{5}{5}$.

$\phantom{\dfrac{3}{5} \times \dfrac{5}{7}} = \dfrac{3}{7}$   $\frac{3}{7} \cdot 1 = \frac{3}{7}$, since any number can be multiplied by 1 without changing the value of the number.

Think about this concept. It is an important one that we will use again when we discuss rational expressions.

## Multiplying a Fraction by a Whole Number
Whole numbers can be named using fractional notation. $3$, $\frac{9}{3}$, $\frac{6}{2}$, and $\frac{3}{1}$ are ways of expressing the number *three*. Therefore,

$$3 = \frac{9}{3} = \frac{6}{2} = \frac{3}{1}.$$

When we multiply a fraction by a whole number, we merely express the whole number as a fraction whose denominator is 1 and follow the multiplication rule for fractions.

## Example 3   Multiply.

(a) $7 \times \dfrac{3}{5}$         (b) $\dfrac{3}{16} \times 4$

### Solution

(a) $7 \times \dfrac{3}{5} = \dfrac{7}{1} \times \dfrac{3}{5} = \dfrac{21}{5}$ or $4\dfrac{1}{5}$     (b) $\dfrac{3}{16} \times 4 = \dfrac{3}{16} \times \dfrac{4}{1} = \dfrac{3}{4 \cdot \cancel{4}} \times \dfrac{\cancel{4}}{1} = \dfrac{3}{4}$

Notice that in **(b)** we did not use *prime* factors to factor 16. We recognized that $16 = 4 \cdot 4$. This is a more convenient factorization of 16 for this problem. Choose the factorization that works best for each problem. If you cannot decide what is best, factor into primes.   □

 **Student Practice 3**   Multiply.

(a) $4 \times \dfrac{2}{7}$         (b) $12 \times \dfrac{3}{4}$

## Multiplying Mixed Numbers
When multiplying mixed numbers, we first change them to improper fractions and then follow the multiplication rule for fractions.

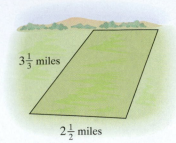

$3\frac{1}{3}$ miles

$2\frac{1}{2}$ miles

▲ **Example 4** How do we find the area of a rectangular field $3\frac{1}{3}$ miles long and $2\frac{1}{2}$ miles wide?

**Solution** To find the area, we multiply length times width.

$$3\frac{1}{3} \times 2\frac{1}{2} = \frac{10}{3} \times \frac{5}{2} = \frac{\cancel{2}\cdot 5}{3} \times \frac{5}{\cancel{2}} = \frac{25}{3} = 8\frac{1}{3}$$

The area is $8\frac{1}{3}$ square miles. ▢

▲ ▣⟹ **Student Practice 4** Delbert Robinson has a farm with a rectangular field that measures $5\frac{3}{5}$ miles long and $3\frac{3}{4}$ miles wide. What is the area of that field?

**Example 5** Multiply. $2\frac{2}{3} \times \frac{1}{4} \times 6$

**Solution**

$$2\frac{2}{3} \times \frac{1}{4} \times 6 = \frac{8}{3} \times \frac{1}{4} \times \frac{6}{1} = \frac{\cancel{4}\cdot 2}{\cancel{3}} \times \frac{1}{\cancel{4}} \times \frac{2\cdot\cancel{3}}{1} = \frac{4}{1} = 4$$ ▢

▣⟹ **Student Practice 5** Multiply.

$$3\frac{1}{2} \times \frac{1}{14} \times 4$$

## 2 Dividing Fractions, Whole Numbers, and Mixed Numbers ▶

**Dividing Fractions** To divide two fractions, we invert the second fraction (that is, the divisor) and then multiply the two fractions.

---

**TO DIVIDE TWO FRACTIONS**

1. Invert the second fraction (that is, the divisor).
2. Now multiply the two fractions.

---

**Example 6** Divide.

(a) $\frac{1}{3} \div \frac{1}{2}$　　　　(b) $\frac{2}{5} \div \frac{3}{10}$　　　　(c) $\frac{2}{3} \div \frac{7}{5}$

**Solution**

(a) $\frac{1}{3} \div \frac{1}{2} = \frac{1}{3} \times \frac{2}{1} = \frac{2}{3}$　　Note that we always invert the *second* fraction.

(b) $\frac{2}{5} \div \frac{3}{10} = \frac{2}{5} \times \frac{10}{3} = \frac{2}{\cancel{5}} \times \frac{\cancel{5}\cdot 2}{3} = \frac{4}{3}$ or $1\frac{1}{3}$　　(c) $\frac{2}{3} \div \frac{7}{5} = \frac{2}{3} \times \frac{5}{7} = \frac{10}{21}$ ▢

▣⟹ **Student Practice 6** Divide.

(a) $\frac{2}{5} \div \frac{1}{3}$　　　　　　(b) $\frac{12}{13} \div \frac{4}{3}$

**Dividing a Fraction and a Whole Number** The process of inverting the second fraction and then multiplying the two fractions should be done very carefully when one of the original values is a whole number. Remember, a whole number such as 2 is equivalent to $\frac{2}{1}$.

**Example 7** Divide.

(a) $\dfrac{1}{3} \div 2$

(b) $5 \div \dfrac{1}{3}$

**Solution**

(a) $\dfrac{1}{3} \div 2 = \dfrac{1}{3} \div \dfrac{2}{1} = \dfrac{1}{3} \times \dfrac{1}{2} = \dfrac{1}{6}$

(b) $5 \div \dfrac{1}{3} = \dfrac{5}{1} \div \dfrac{1}{3} = \dfrac{5}{1} \times \dfrac{3}{1} = \dfrac{15}{1} = 15$ ▫

**Student Practice 7** Divide.

(a) $\dfrac{3}{7} \div 6$

(b) $8 \div \dfrac{2}{3}$

**Sidelight: Number Sense**

Look at the answers to the problems in Example 7. In part (a), you will notice that $\frac{1}{6}$ is less than the original number $\frac{1}{3}$. Does this seem reasonable? Let's see. If $\frac{1}{3}$ is divided by 2, it means that $\frac{1}{3}$ will be divided into two equal parts. We would expect that each part would be less than $\frac{1}{3}$. $\frac{1}{6}$ is a reasonable answer to this division problem.

In part **(b)**, 15 is greater than the original number 5. Does this seem reasonable? Think of what $5 \div \frac{1}{3}$ means. It means that 5 will be divided into thirds. Let's think of an easier problem. What happens when we divide 1 into thirds? We get *three* thirds. We would expect, therefore, that when we divide 5 into thirds, we would get $5 \times 3$ or 15 thirds. 15 is a reasonable answer to this division problem.

**Complex Fractions** Sometimes division is written in the form of a **complex fraction** with one fraction in the numerator and one fraction in the denominator. It is best to write this in standard division notation first; then complete the problem using the rule for division.

**Example 8** Divide.

(a) $\dfrac{\frac{3}{7}}{\frac{3}{5}}$

(b) $\dfrac{\frac{2}{9}}{\frac{5}{7}}$

**Solution**

(a) $\dfrac{\frac{3}{7}}{\frac{3}{5}} = \dfrac{3}{7} \div \dfrac{3}{5} = \dfrac{\cancel{3}}{7} \times \dfrac{5}{\cancel{3}} = \dfrac{5}{7}$

(b) $\dfrac{\frac{2}{9}}{\frac{5}{7}} = \dfrac{2}{9} \div \dfrac{5}{7} = \dfrac{2}{9} \times \dfrac{7}{5} = \dfrac{14}{45}$ ▫

**Student Practice 8** Divide.

(a) $\dfrac{\frac{3}{11}}{\frac{5}{7}}$

(b) $\dfrac{\frac{12}{5}}{\frac{8}{15}}$

**Sidelight:** Invert and Multiply

Why does the method of "invert and multiply" work? The division rule really depends on the property that any number can be multiplied by 1 without changing the value of the number. Let's look carefully at an example of division of fractions:

$$\frac{2}{5} \div \frac{3}{7} = \frac{\frac{2}{5}}{\frac{3}{7}}$$

We can write the problem using a complex fraction.

$$= \frac{\frac{2}{5}}{\frac{3}{7}} \times 1$$

We can multiply by 1, since any number can be multiplied by 1 without changing the value of the number.

$$= \frac{\frac{2}{5}}{\frac{3}{7}} \times \frac{\frac{7}{3}}{\frac{7}{3}}$$

We write 1 in the form $\frac{\frac{7}{3}}{\frac{7}{3}}$, since any nonzero number divided by itself equals 1. We choose this value as a multiplier because it will help simplify the denominator.

$$= \frac{\frac{2}{5} \times \frac{7}{3}}{\frac{3}{7} \times \frac{7}{3}}$$

Definition of multiplication of fractions.

$$= \frac{\frac{2}{5} \times \frac{7}{3}}{1} = \frac{2}{5} \times \frac{7}{3}$$

The product in the denominator equals 1.

Thus we have shown that $\frac{2}{5} \div \frac{3}{7}$ is equivalent to $\frac{2}{5} \times \frac{7}{3}$ and have shown justification for the "invert and multiply rule."

**Dividing Mixed Numbers**  This method for division of fractions can be used with mixed numbers. However, we first must change the mixed numbers to improper fractions and then use the rule for dividing fractions.

**Mᴄ Example 9** Divide.

**(a)** $2\frac{1}{3} \div 3\frac{2}{3}$

**(b)** $\dfrac{2}{3\frac{1}{2}}$

**Solution**

**(a)** $2\frac{1}{3} \div 3\frac{2}{3} = \frac{7}{3} \div \frac{11}{3} = \frac{7}{\cancel{3}} \times \frac{\cancel{3}}{11} = \frac{7}{11}$

**(b)** $\dfrac{2}{3\frac{1}{2}} = 2 \div 3\frac{1}{2} = \frac{2}{1} \div \frac{7}{2} = \frac{2}{1} \times \frac{2}{7} = \frac{4}{7}$

**▶ Student Practice 9**  Divide.

**(a)** $1\frac{2}{5} \div 2\frac{1}{3}$

**(b)** $4\frac{2}{3} \div 7$

**(c)** $\dfrac{1\frac{1}{5}}{1\frac{2}{7}}$

**Example 10**  A chemist has 96 fluid ounces of a solution. She pours the solution into test tubes. Each test tube holds $\frac{3}{4}$ fluid ounce. How many test tubes can she fill?

**Solution**  We need to divide the total number of ounces, 96, by the number of ounces in each test tube, $\frac{3}{4}$.

$$96 \div \frac{3}{4} = \frac{96}{1} \div \frac{3}{4} = \frac{96}{1} \times \frac{4}{3} = \frac{\cancel{3} \cdot 32}{1} \times \frac{4}{\cancel{3}} = \frac{128}{1} = 128$$

She will be able to fill 128 test tubes.

*Check:* Pause for a moment to think about the answer. Does 128 test tubes filled with solution seem like a reasonable answer? Did you perform the correct operation? □

**Student Practice 10** A chemist has 64 fluid ounces of a solution. He wishes to fill several jars, each holding $5\frac{1}{3}$ fluid ounces. How many jars can he fill?

Sometimes when solving word problems involving fractions or mixed numbers, it is helpful to solve the problem using simpler numbers first. Once you understand what operation is involved, you can go back and solve using the original numbers in the word problem.

**Example 11** A car traveled 301 miles on $10\frac{3}{4}$ gallons of gas. How many miles per gallon did it get?

**Solution** Use simpler numbers: 300 miles on 10 gallons of gas. We want to find out how many miles the car traveled on 1 gallon of gas. You may want to draw a picture.

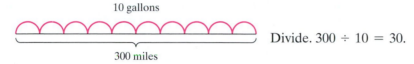

Divide. $300 \div 10 = 30$.

Now use the original numbers given in the problem.

$$301 \div 10\frac{3}{4} = \frac{301}{1} \div \frac{43}{4} = \frac{301}{1} \times \frac{4}{43} = \frac{1204}{43} = 28$$

The car got 28 miles per gallon. □

**Student Practice 11** A car traveled 126 miles on $5\frac{1}{4}$ gallons of gas. How many miles per gallon did it get?

 **STEPS TO SUCCESS** Doing Homework for Each Class Is Critical.

Many students in the class ask the question, "Is homework really that important? Do I actually have to do it?"

*You learn by doing.* It really makes a difference. Mathematics involves mastering a set of skills that you learn by practicing, not by watching someone else do it. Your instructor may make solving a mathematics problem look very easy, but for you to learn the necessary skills, you must practice them over and over.

*The key to success is practice.* Learning mathematics is like learning to play a musical instrument, to type, or to play a sport. No matter how much you watch someone else do mathematical calculations, no matter how many books you read on "how to" do it, and no matter how easy it appears to be, the key to success in mathematics is practice on each homework set.

*Do each kind of problem.* Some exercises in a homework set are more difficult than others. Some stress different concepts. Usually you need to work at least all the odd-numbered

problems in the exercise set. This allows you to cover the full range of skills in the problem set. Remember, the more exercises you do, the better you will become in your mathematical skills.

**Making it personal:** Write down your personal reason for why you think doing the homework in each section is very important for success. Which of the three points given do you find is the most convincing? ▼

## 0.3 Exercises MyMathLab®

### Verbal and Writing Skills, Exercises 1 and 2

1. Explain in your own words how to multiply two mixed numbers.

2. Explain in your own words how to divide two proper fractions.

*Multiply. Simplify your answer whenever possible.*

3. $\dfrac{28}{5} \times \dfrac{6}{35}$

4. $\dfrac{5}{7} \times \dfrac{28}{15}$

5. $\dfrac{17}{18} \times \dfrac{3}{5}$

6. $\dfrac{17}{26} \times \dfrac{13}{34}$

7. $\dfrac{4}{5} \times \dfrac{3}{10}$

8. $\dfrac{3}{11} \times \dfrac{5}{7}$

9. $\dfrac{24}{25} \times \dfrac{5}{2}$

10. $\dfrac{15}{24} \times \dfrac{8}{9}$

11. $\dfrac{7}{12} \times \dfrac{8}{28}$

12. $\dfrac{6}{21} \times \dfrac{9}{18}$

13. $\dfrac{6}{35} \times 5$

14. $\dfrac{2}{21} \times 15$

15. $9 \times \dfrac{2}{5}$

16. $\dfrac{8}{11} \times 3$

*Divide. Simplify your answer whenever possible.*

17. $\dfrac{8}{5} \div \dfrac{8}{3}$

18. $\dfrac{13}{9} \div \dfrac{13}{7}$

19. $\dfrac{3}{7} \div 3$

20. $\dfrac{7}{8} \div 4$

21. $10 \div \dfrac{5}{7}$

22. $18 \div \dfrac{2}{9}$

23. $\dfrac{6}{14} \div \dfrac{3}{8}$

24. $\dfrac{8}{12} \div \dfrac{5}{6}$

25. $\dfrac{7}{24} \div \dfrac{9}{8}$

26. $\dfrac{9}{28} \div \dfrac{4}{7}$

27. $\dfrac{\frac{7}{8}}{\frac{3}{4}}$

28. $\dfrac{\frac{5}{6}}{\frac{10}{13}}$

29. $\dfrac{\frac{5}{6}}{\frac{7}{9}}$

30. $\dfrac{\frac{3}{4}}{\frac{11}{12}}$

31. $1\dfrac{3}{7} \div 6\dfrac{1}{4}$

32. $4\dfrac{1}{2} \div 3\dfrac{3}{8}$

33. $3\dfrac{1}{3} \div 2\dfrac{1}{2}$

34. $5\dfrac{1}{2} \div 3\dfrac{3}{4}$

35. $6\dfrac{1}{2} \div \dfrac{3}{4}$

36. $\dfrac{1}{4} \div 1\dfrac{7}{8}$

37. $\dfrac{15}{2\frac{2}{5}}$

38. $\dfrac{18}{4\frac{1}{2}}$

39. $\dfrac{\frac{2}{3}}{1\frac{1}{4}}$

40. $\dfrac{\frac{5}{6}}{2\frac{1}{2}}$

### Mixed Practice *Perform the proper calculations. Simplify your answer whenever possible.*

41. $\dfrac{4}{7} \times \dfrac{21}{2}$

42. $\dfrac{12}{18} \times \dfrac{9}{2}$

43. $\dfrac{5}{14} \div \dfrac{2}{7}$

44. $\dfrac{5}{6} \div \dfrac{11}{18}$

26

**45.** $10\frac{3}{7} \times 5\frac{1}{4}$

**46.** $10\frac{2}{9} \div 2\frac{1}{3}$

**47.** $25 \div \frac{5}{8}$

**48.** $15 \div 1\frac{2}{3}$

**49.** $6 \times 4\frac{2}{3}$

**50.** $6\frac{1}{2} \times 12$

**51.** $2\frac{1}{2} \times \frac{1}{10} \times \frac{3}{4}$

**52.** $2\frac{1}{3} \times \frac{2}{3} \times \frac{3}{5}$

**53.** **(a)** $\frac{1}{15} \times \frac{25}{21}$

**(b)** $\frac{1}{15} \div \frac{25}{21}$

**54.** **(a)** $\frac{1}{6} \times \frac{24}{15}$

**(b)** $\frac{1}{6} \div \frac{24}{15}$

**55.** **(a)** $\frac{2}{3} \div \frac{12}{21}$

**(b)** $\frac{2}{3} \times \frac{12}{21}$

**56.** **(a)** $\frac{3}{7} \div \frac{21}{25}$

**(b)** $\frac{3}{7} \times \frac{21}{25}$

## Applications

**57.** ***Shirt Manufacturing*** A denim shirt at the Gap requires $2\frac{3}{4}$ yards of material. How many shirts can be made from $71\frac{1}{2}$ yards of material?

**58.** ***Pullover Manufacturing*** A fleece pullover requires $1\frac{5}{8}$ yards of material. How many fleece pullovers can be made from $29\frac{1}{4}$ yards of material?

**59.** ***Farm Management*** Jesse purchased a large rectangular field so that he could farm the land. The field measures $11\frac{1}{3}$ miles long and 12 miles wide. What is the area of his field?

**60.** ***Gardening*** Sara must find the area of her flower garden so that she can determine how much fertilizer to purchase. What is the area of her rectangular garden, which measures 15 feet long and $10\frac{1}{5}$ feet wide?

## Cumulative Review   *In exercises 61 and 62, find the missing numerator.*

**61.** [0.1.4] $\frac{11}{15} = \frac{?}{75}$

**62.** [0.1.4] $\frac{7}{9} = \frac{?}{63}$

**Quick Quiz 0.3** *Perform the operations indicated. Simplify answers whenever possible.*

**1.** $\frac{7}{15} \times \frac{25}{14}$

**2.** $3\frac{1}{4} \times 4\frac{1}{2}$

**3.** $3\frac{3}{10} \div 2\frac{1}{2}$

**4.** **Concept Check** Explain the steps you would take to perform the calculation $3\frac{1}{4} \div 2\frac{1}{2}$.

## 0.4　Using Decimals ▶

### 1　Understanding the Meaning of Decimals ▶

We can express a part of a whole as a fraction or as a decimal. A **decimal** is another way of writing a fraction whose denominator is 10, 100, 1000, and so on.

$$\frac{3}{10} = 0.3 \qquad \frac{5}{100} = 0.05 \qquad \frac{172}{1000} = 0.172 \qquad \frac{58}{10,000} = 0.0058$$

The period in decimal notation is known as the **decimal point.** The number of digits in a number to the right of the decimal point is known as the number of **decimal places** of the number. The place value of decimals is shown in the following chart.

| Hundred-thousands | Ten-thousands | Thousands | Hundreds | Tens | Ones | ← Decimal point | Tenths | Hundredths | Thousandths | Ten-thousandths | Hundred-thousandths |
|---|---|---|---|---|---|---|---|---|---|---|---|
| 100,000 | 10,000 | 1000 | 100 | 10 | 1 | • | $\frac{1}{10}$ | $\frac{1}{100}$ | $\frac{1}{1000}$ | $\frac{1}{10,000}$ | $\frac{1}{100,000}$ |

**Example 1** Write each of the following decimals as a fraction or mixed number. State the number of decimal places. Write out in words the way the number would be spoken.

(a) 0.6　　　(b) 0.29　　　(c) 0.527　　　(d) 1.38　　　(e) 0.00007

**Solution**

| Decimal Form | Fraction Form | Number of Decimal Places | The Words Used to Describe the Number |
|---|---|---|---|
| (a) 0.6 | $\frac{6}{10}$ | one | six tenths |
| (b) 0.29 | $\frac{29}{100}$ | two | twenty-nine hundredths |
| (c) 0.527 | $\frac{527}{1000}$ | three | five hundred twenty-seven thousandths |
| (d) 1.38 | $1\frac{38}{100}$ | two | one and thirty-eight hundredths |
| (e) 0.00007 | $\frac{7}{100,000}$ | five | seven hundred-thousandths |

▷ **Student Practice 1** State the number of decimal places. Write each decimal as a fraction or mixed number and in words.

(a) 0.9　　　(b) 0.09　　　(c) 0.731　　　(d) 1.371　　　(e) 0.0005

You have seen that a given fraction can be written in several different but equivalent ways. There are also several different equivalent ways of writing the decimal form of a fraction. The decimal 0.18 can be written in the following equivalent ways:

$$\text{Fractional form:} \quad \frac{18}{100} = \frac{180}{1000} = \frac{1800}{10,000} = \frac{18,000}{100,000}$$

$$\text{Decimal form:} \quad 0.18 = 0.180 = 0.1800 = 0.18000$$

Thus we see that *any number of terminal zeros may be added to the right-hand side of a decimal* without changing its value.

$$0.13 = 0.1300 \qquad 0.162 = 0.162000$$

Similarly, *any number of terminal zeros may be removed from the right-hand side of a decimal* without changing its value.

## 2 Changing a Fraction to a Decimal

A fraction can be changed to a decimal by dividing the denominator into the numerator.

**Example 2** Write each of the following fractions as a decimal.

(a) $\dfrac{3}{4}$    (b) $\dfrac{21}{20}$    (c) $\dfrac{1}{8}$    (d) $\dfrac{3}{200}$

**Solution**

(a) $\dfrac{3}{4} = 0.75$  since
$$\begin{array}{r} 0.75 \\ 4\overline{)3.00} \\ \underline{2\,8} \\ 20 \\ \underline{20} \\ 0 \end{array}$$

(b) $\dfrac{21}{20} = 1.05$  since
$$\begin{array}{r} 1.05 \\ 20\overline{)21.00} \\ \underline{20} \\ 1\,00 \\ \underline{1\,00} \\ 0 \end{array}$$

(c) $\dfrac{1}{8} = 0.125$  since
$$\begin{array}{r} 0.125 \\ 8\overline{)1.000} \\ \underline{8} \\ 20 \\ \underline{16} \\ 40 \\ \underline{40} \\ 0 \end{array}$$

(d) $\dfrac{3}{200} = 0.015$  since
$$\begin{array}{r} 0.015 \\ 200\overline{)3.000} \\ \underline{2\,00} \\ 1\,000 \\ \underline{1\,000} \\ 0 \end{array}$$

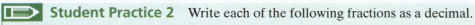

**Student Practice 2**  Write each of the following fractions as a decimal.

(a) $\dfrac{3}{8}$    (b) $\dfrac{7}{200}$    (c) $\dfrac{33}{20}$

Sometimes division yields an infinite repeating decimal. We use three dots to indicate that the pattern continues forever. For example,

$$\frac{1}{3} = 0.3333\ldots \qquad \begin{array}{r} 0.333 \\ 3\overline{)1.000} \\ \underline{9} \\ 10 \\ \underline{9} \\ 10 \\ \underline{9} \\ 1 \end{array}$$

An alternative notation is to place a bar over the repeating digit(s):

$$0.3333\ldots = 0.\overline{3} \qquad 0.575757\ldots = 0.\overline{57}$$

**Example 3** Write each fraction as a decimal.

(a) $\dfrac{2}{11}$ \qquad\qquad (b) $\dfrac{5}{6}$

**Solution**

(a) $\dfrac{2}{11} = 0.181818\ldots$ or $0.\overline{18}$ \qquad (b) $\dfrac{5}{6} = 0.8333\ldots$ or $0.8\overline{3}$

$$\begin{array}{r} 0.1818 \\ 11\overline{)2.0000} \\ \underline{1\,1} \\ 90 \\ \underline{88} \\ 20 \\ \underline{11} \\ 90 \\ \underline{88} \\ 2 \end{array}$$

$$\begin{array}{r} 0.8333 \\ 6\overline{)5.0000} \\ \underline{48} \\ 20 \\ \underline{18} \\ 20 \\ \underline{18} \\ 20 \\ \underline{18} \\ 2 \end{array}$$ Note that the 8 does not repeat. Only the digit 3 is repeating.

> **Student Practice 3** Write each fraction as a decimal.
>
> (a) $\dfrac{1}{6}$ \qquad\qquad (b) $\dfrac{5}{11}$

---

**Calculator**

**Fraction to Decimal**

You can use a calculator to change $\frac{3}{5}$ to a decimal.
Enter:

3 $\boxed{\div}$ 5 $\boxed{=}$

The display should read

$\boxed{0.6}$

Try the following.

(a) $\dfrac{17}{25}$ \qquad (b) $\dfrac{2}{9}$

(c) $\dfrac{13}{10}$ \qquad (d) $\dfrac{15}{19}$

---

Sometimes division must be carried out to many places in order to observe the repeating pattern. This is true in the following example:

$$\frac{2}{7} = 0.285714285714285714\ldots \qquad \text{This can also be written as } \frac{2}{7} = 0.\overline{285714}.$$

It can be shown that the denominator determines the maximum number of decimal places that might repeat. So $\frac{2}{7}$ must repeat in the seventh decimal place or sooner.

### 3 Changing a Decimal to a Fraction

To convert from a decimal to a fraction, merely write the decimal as a fraction with a denominator of 10, 100, 1000, 10,000, and so on, and simplify the result when possible.

**Example 4** Write each decimal as a fraction and simplify whenever possible.

(a) 0.2        (b) 0.35        (c) 0.516        (d) 0.74        (e) 0.138        (f) 0.008

**Solution**

(a) $0.2 = \dfrac{2}{10} = \dfrac{1}{5}$

(b) $0.35 = \dfrac{35}{100} = \dfrac{7}{20}$

(c) $0.516 = \dfrac{516}{1000} = \dfrac{129}{250}$

(d) $0.74 = \dfrac{74}{100} = \dfrac{37}{50}$

(e) $0.138 = \dfrac{138}{1000} = \dfrac{69}{500}$

(f) $0.008 = \dfrac{8}{1000} = \dfrac{1}{125}$ ◻

**Student Practice 4** Write each decimal as a fraction and simplify whenever possible.

(a) 0.8        (b) 0.88        (c) 0.45        (d) 0.148        (e) 0.612        (f) 0.016

All repeating decimals can also be converted to fractional form. In practice, however, repeating decimals are usually rounded to a few places. It will not be necessary, therefore, to learn how to convert $0.\overline{033}$ to $\frac{11}{333}$ for this course.

**4 Adding and Subtracting Decimals**

Last week Bob spent $19.83 on lunches purchased at the cafeteria at work. During this same period, Sally spent $24.76 on lunches. How much did the two of them spend on lunches last week?

Adding and subtracting decimals is similar to adding and subtracting whole numbers, except that it is necessary to line up decimal points. To perform the operation $19.83 + 24.76$, we line up the numbers in column form and add the digits:

$$\begin{array}{r} 19.83 \\ +\ 24.76 \\ \hline 44.59 \end{array}$$

Thus Bob and Sally spent $44.59 on lunches last week.

---

**ADDITION AND SUBTRACTION OF DECIMALS**

1. Write in column form and line up the decimal points.
2. Add or subtract the digits.

---

**Example 5** Add or subtract.

(a) $3.6 + 2.3$  (b) $127.32 - 38.48$  (c) $3.1 + 42.36 + 9.034$  (d) $5.0006 - 3.1248$

**Solution**

(a) $\begin{array}{r} 3.6 \\ +\ 2.3 \\ \hline 5.9 \end{array}$

(b) $\begin{array}{r} 127.32 \\ -\ 38.48 \\ \hline 88.84 \end{array}$

(c) $\begin{array}{r} 3.1 \\ 42.36 \\ +\ 9.034 \\ \hline 54.494 \end{array}$

(d) $\begin{array}{r} 5.0006 \\ -\ 3.1248 \\ \hline 1.8758 \end{array}$ ◻

**Student Practice 5** Add or subtract.

(a) $3.12 + 5.08$

(b) $152.003 - 136.118$

(c) $1.1 + 3.16 + 5.123$

(d) $1.0052 - 0.1234$

**Sidelight: Adding Zeros to the Right-Hand Side of the Decimal**

When we added fractions, we had to have common denominators. Since decimals are really fractions, why can we add them without having common denominators? Actually, we have to have common denominators to add any fractions, whether they are in decimal form or fraction form. However, sometimes the notation does not show this. Let's examine Example 5(c).

*Original Problem*    We are adding the three numbers:

$$3.1$$
$$42.36$$
$$+\ 9.034$$
$$\overline{54.494}$$

$$3\frac{1}{10} + 42\frac{36}{100} + 9\frac{34}{1000}$$
$$3\frac{100}{1000} + 42\frac{360}{1000} + 9\frac{34}{1000}$$
$$3.100 + 42.360 + 9.034 \quad \text{This is the new problem.}$$

*Original Problem*    *New Problem*

| | |
|---|---|
| 3.1 | 3.100 |
| 42.36 | 42.360 |
| + 9.034 | + 9.034 |
| 54.494 | 54.494 |

We notice that the results are the same. The only difference is the notation. We are using the property that any number of zeros may be added to the right-hand side of a decimal without changing its value.

This shows the convenience of adding and subtracting fractions in decimal form. Little work is needed to change the decimals so that they have a common denominator. All that is required is to add zeros to the right-hand side of the decimal (and we usually do not even write out that step except when subtracting).

As long as we line up the decimal points, we can add or subtract any decimal fractions.

In the following example we will find it useful to add zeros to the right-hand side of the decimal.

**Example 6** Perform the following operations.

(a) $1.0003 + 0.02 + 3.4$        (b) $12 - 0.057$

**Solution** We will add zeros so that each number shows the same number of decimal places.

(a)
$$1.0003$$
$$0.0200$$
$$+\ 3.4000$$
$$\overline{4.4203}$$

(b)
$$12.000$$
$$-\ 0.057$$
$$\overline{11.943}$$    □

**Student Practice 6** Perform the following operations.

(a) $0.061 + 5.0008 + 1.3$        (b) $18 - 0.126$

## 5 Multiplying Decimals

**MULTIPLICATION OF DECIMALS**

To multiply decimals, you first multiply as with whole numbers. To determine the position of the decimal point, you count the total number of decimal places in the two numbers being multiplied. This will determine the number of decimal places that should appear in the answer.

**Example 7** Multiply. $0.8 \times 0.4$

**Solution**

$$
\begin{array}{r}
0.8 \quad (\text{one decimal place}) \\
\times\, 0.4 \quad (\text{one decimal place}) \\
\hline
0.32 \quad (\text{two decimal places})
\end{array}
$$

 **Student Practice 7**  Multiply. $0.5 \times 0.3$

Note that you will often have to add zeros to the left of the digits obtained in the product so that you obtain the necessary number of decimal places.

**Example 8**  Multiply. $0.123 \times 0.5$

**Solution**

$$
\begin{array}{r}
0.123 \quad (\text{three decimal places}) \\
\times\quad 0.5 \quad (\text{one decimal place}) \\
\hline
0.0615 \quad (\text{four decimal places})
\end{array}
$$

 **Student Practice 8**   Multiply. $0.12 \times 0.4$

Here are some examples that involve more decimal places.

**Example 9**  Multiply.

**(a)** $2.56 \times 0.003$           **(b)** $0.0036 \times 0.008$

**Solution**

**(a)**
$$
\begin{array}{r}
2.56 \quad (\text{two decimal places}) \\
\times\, 0.003 \quad (\text{three decimal places}) \\
\hline
0.00768 \quad (\text{five decimal places})
\end{array}
$$

**(b)**
$$
\begin{array}{r}
0.0036 \quad (\text{four decimal places}) \\
\times\, 0.008 \quad (\text{three decimal places}) \\
\hline
0.0000288 \quad (\text{seven decimal places})
\end{array}
$$

 **Student Practice 9**   Multiply.

**(a)** $1.23 \times 0.005$           **(b)** $0.003 \times 0.00002$

**Sidelight: Counting the Number of Decimal Places**
Why do we count the number of decimal places? The rule really comes from the properties of fractions. If we write the problem in Example 8 in fraction form, we have

$$0.123 \times 0.5 = \frac{123}{1000} \times \frac{5}{10} = \frac{615}{10{,}000} = 0.0615.$$

**6 Dividing Decimals**

When discussing division of decimals, we frequently refer to the three primary parts of a division problem. Be sure you know the meaning of each term.

The **divisor** is the number you divide into another.
The **dividend** is the number to be divided.
The **quotient** is the result of dividing one number by another.

In the problem $6 \div 2 = 3$ we represent each of these terms as follows:

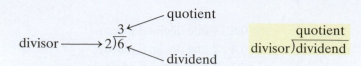

When dividing two decimals, count *the number of decimal places* in the divisor. Then *move the decimal point to the right* that *same number of places* in both *the divisor* and *the dividend.* Mark that position with a caret ($_\wedge$). Finally, perform the division. Be sure to line up the decimal point in the quotient with the position indicated by the caret.

**Example 10** Four friends went out for lunch. The total bill, including tax, was $32.68. How much did each person pay if they shared the cost equally?

**Solution** To answer this question, we must calculate $32.68 \div 4$.

$$\begin{array}{r} 8.17 \\ 4\overline{)32.68} \\ \underline{32} \phantom{.68} \\ 06 \\ \underline{4} \\ 28 \\ \underline{28} \\ 0 \end{array}$$

Since there are no decimal places in the divisor, we do not need to move the decimal point. We must be careful, however, to place the decimal point in the quotient directly above the decimal point in the dividend.

Thus $32.68 \div 4 = 8.17$, and each friend paid $8.17. ☐

**Student Practice 10** Sally Keyser purchased 6 boxes of paper for a laser printer. The cost was $31.56. There was no tax since she purchased the paper for a charitable organization. How much did she pay for each box of paper?

Note that sometimes we will need to place extra zeros in the dividend in order to move the decimal point the required number of places.

**Example 11** Divide. $16.2 \div 0.027$

**Solution**

$$0.027_\wedge \overline{)16.200_\wedge}$$

There are **three** decimal places in the divisor, so we move the decimal point **three places** to **the right** in the **divisor** and **dividend** and mark the new position by a caret. Note that we must add two zeros to 16.2 in order to do this.

three decimal places

$$\begin{array}{r} 600. \\ 0.027_\wedge \overline{)16.200_\wedge} \\ \underline{16\ 2} \phantom{00} \\ 000 \end{array}$$

Now perform the division as with whole numbers. The decimal point in the answer is directly above the caret.

Thus $16.2 \div 0.027 = 600.$ ☐

**Student Practice 11** Divide. $1800. \div 0.06$

Special care must be taken to line up the digits in the quotient. Note that sometimes we will need to place zeros in the quotient after the decimal point.

**Example 12** Divide. 0.04288 ÷ 3.2

**Solution**

$$3.2_\wedge \overline{)0.0_\wedge 4288}$$

There is **one** decimal place in the divisor, so we move the decimal point **one place** to **the right** in the **divisor** and **dividend** and mark the new position by a caret.

one decimal place

$$3.2_\wedge \overline{)0.0_\wedge 4288} \quad \begin{array}{c} 0.0134 \\ \hline \phantom{0} \end{array}$$

$$
\begin{array}{r}
0.0134 \\
3.2_\wedge \overline{)0.0_\wedge 4288} \\
\underline{32\phantom{00}} \\
108 \\
\underline{96} \\
128 \\
\underline{128} \\
0
\end{array}
$$

Now perform the division as for whole numbers. The decimal point in the answer is directly above the caret. Note the need for the initial zero after the decimal point in the answer.

Thus 0.04288 ÷ 3.2 = 0.0134.     □

 **Student Practice 12**   Divide. 0.01764 ÷ 4.9

Sidelight: **Dividing Decimals by Another Method**
Why does this method of dividing decimals work? Essentially, we are using the steps we used in Section 0.1 to change a fraction to an equivalent fraction by multiplying both the numerator and denominator by the same number. Let's reexamine Example 12.

$$0.04288 \div 3.2 = \frac{0.04288}{3.2}$$

Write the original problem using fraction notation.

$$= \frac{0.04288 \times 10}{3.2 \quad \times 10}$$

Multiply the numerator and denominator by 10. Since this is the same as multiplying by 1, we are not changing the fraction.

$$= \frac{0.4288}{32}$$

Write the result of multiplication by 10.

$$= 0.4288 \div 32$$

Rewrite the fraction as an equivalent problem with division notation.

Notice that we have obtained a new problem that is the same as the problem in Example 12 when we moved the decimal one place to the right in the divisor and dividend. We see that the reason we can move the decimal point as many places as necessary to the right in the divisor and dividend is that this is the same as multiplying the numerator and denominator of a fraction by a power of 10 to obtain an equivalent fraction.

**7** Multiplying and Dividing a Decimal by a Multiple of 10 ▶

When multiplying by 10, 100, 1000, and so on, a simple rule may be used to obtain the answer. For every zero in the multiplier, move the decimal point one place to the right.

**Example 13** Multiply.

(a) $3.24 \times 10$      (b) $15.6 \times 100$      (c) $0.0026 \times 1000$

**Solution**

(a) $3.24 \times 10 = 32.4$    One zero—move decimal point one place to the right.

(b) $15.6 \times 100 = 1560$    Two zeros—move decimal point two places to the right.

(c) $0.0026 \times 1000 = 2.6$    Three zeros—move decimal point three places to the right.

**Student Practice 13** Multiply.

(a) $0.0016 \times 100$      (b) $2.34 \times 1000$      (c) $56.75 \times 10,000$

The reverse rule is true for division. When dividing by 10, 100, 1000, 10,000, and so on, move the decimal point one place to the left for every zero in the divisor.

**Example 14** Divide.

(a) $52.6 \div 10$      (b) $0.0038 \div 100$      (c) $5936.2 \div 1000$

**Solution**

(a) $\dfrac{52.6}{10} = 5.26$    Move decimal point one place to the left.

(b) $\dfrac{0.0038}{100} = 0.000038$    Move decimal point two places to the left.

(c) $\dfrac{5936.2}{1000} = 5.9362$    Move decimal point three places to the left.

**Student Practice 14** Divide.

(a) $\dfrac{5.82}{10}$      (b) $123.4 \div 1000$      (c) $\dfrac{0.00614}{10,000}$

---

**STEPS TO SUCCESS** Do You Realize How Valuable Friendship Is?

In a math class a friend is a person of fantastic value. Robert Louis Stevenson once wrote "A friend is a gift you give yourself." This is especially true when you take a mathematics class and make a friend in the class. You will find that you enjoy sitting together and drawing support and encouragement from each other. You may want to exchange phone numbers or e-mail addresses. You may want to study together or review together before a test.

How do you get started? Try talking to the students seated around you. Ask someone for help about something you did not understand in class. Take the time to listen to them and their interests and concerns. You may discover you have a lot in common. If the first few people you talk to seem uninterested, try sitting in a different part of the room and talk to those students who are seated around you in the new location. Don't force a friendship on anyone but just look for chances to open up a good channel of communication.

**Making it personal:** What do you think is the best way to make a friend of someone in your class? Which of the given suggestions do you find the most helpful? Will you take the time to reach out to someone in your class this week and try to begin a new friendship? ▼

## 0.4 Exercises   MyMathLab®

### Verbal and Writing Skills, Exercises 1 and 4

1. A decimal is another way of writing a fraction whose denominator is _____.
2. We write 0.42 in words as _____.
3. When dividing 7432.9 by 1000 we move the decimal point _____ places to the _____.
4. When dividing 96.3 by 10,000 we move the decimal point _____ places to the _____.

*Write each fraction as a decimal.*

5. $\frac{7}{8}$    6. $\frac{18}{25}$    7. $\frac{3}{15}$    8. $\frac{9}{15}$    9. $\frac{7}{11}$    10. $\frac{1}{6}$

*Write each decimal as a fraction in simplified form.*

11. 0.8    12. 0.5    13. 0.25    14. 0.35    15. 0.625    16. 0.775

17. 0.06    18. 0.08    19. 3.4    20. 4.8    21. 5.5    22. 6.25

*Add or subtract.*

23. $1.71 + 0.38$    24. $4.64 + 0.23$    25. $2.5 + 3.42 + 4.9$    26. $6.31 + 4.2 + 8.5$

27. $46.03 + 215.1 + 0.078$    28. $33.01 + 0.38 + 175.401$    29. $147.18 - 15.39$    30. $121.52 - 79.85$

31. $6.0054 - 2.0257$    32. $5.0032 - 3.0036$    33. $125.43 - 2.8$    34. $212.54 - 3.6$

*Multiply or divide.*

35. $7.21 \times 4.2$    36. $6.12 \times 3.4$    37. $0.04 \times 0.08$    38. $6.32 \times 1.31$

39. $4.23 \times 0.025$    40. $3.84 \times 0.0017$    41. $58,200 \times 0.0015$    42. $23,000 \times 0.0042$

43. $3.616 \div 64$    44. $12.6672 \div 39$    45. $7.9728 \div 3.02$    46. $6.519 \div 2.05$

47. $0.5230 \div 0.002$    48. $0.031 \div 0.005$    49. $0.03048 \div 0.06$    50. $0.00855 \div 0.09$

*Multiply or divide by moving the decimal point.*

51. $3.45 \times 1000$    52. $1.36 \times 1000$    53. $0.76 \div 100$    54. $175,318 \div 1000$

55. $7.36 \times 10,000$    56. $0.00243 \times 100,000$    57. $73,892 \div 100,000$    58. $3.52 \div 1000$

59. $0.1498 \times 100$    60. $85.54 \times 10,000$    61. $1.931 \div 100$    62. $96.12 \div 10,000$

**Mixed Practice** *Perform the calculations indicated.*

**63.** $54.8 \times 0.15$

**64.** $8.252 \times 0.005$

**65.** $13.75 + 2.55 + 0.078$

**66.** $1.109 + 0.088 + 16.4$

**67.** $0.05724 \div 0.027$

**68.** $77.136 \div 0.003$

**69.** $0.7683 \times 1000$

**70.** $25.62 \times 10,000$

**71.** $56.37 - 4.29$

**72.** $14.3 - 0.68$

**73.** $153.7 \div 100$

**74.** $0.58 \div 1000$

**Applications**

**75.** *Measurement* While mixing solutions in her chemistry lab, Mia needed to change the measured data from pints to liters. There is 0.4732 liter in one pint and the original measurement was 5.5 pints. What is the measured data in liters?

**76.** *Mileage of Hybrid Cars* In order to minimize fuel costs, Chris Smith purchased a used Honda Civic Hybrid that averages 44 miles per gallon in the city. The gas tank holds 13.2 gallons of gas. How many miles can Chris drive the car in the city on a full tank of gas?

**77.** *Wages* Harry has a part-time job at Stop and Shop. He earns $9 an hour. He requested enough hours of work each week so that he could earn at least $185 a week. How many hours will he have to work to achieve his goal? By how much will he exceed his earning goal of $185 per week?

**78.** *Drinking Water* The EPA standard for safe drinking water is a maximum of 1.3 milligrams of copper per liter of water. A water testing firm found 6.8 milligrams of copper in a 5-liter sample drawn from Jim and Sharon LeBlanc's house. Is the water safe or not? By how much does the amount of copper exceed or fall short of the maximum allowed?

**Cumulative Review** *Perform each operation. Simplify all answers.*

**79.** **[0.3.2]** $3\frac{1}{2} \div 5\frac{1}{4}$

**80.** **[0.3.1]** $\frac{3}{8} \cdot \frac{12}{27}$

**81.** **[0.2.3]** $\frac{12}{25} + \frac{9}{20}$

**82.** **[0.2.4]** $1\frac{3}{5} - \frac{1}{2}$

**Quick Quiz 0.4**    *Perform the calculations indicated.*

**1.** $8.0567 - 2.3489$

**2.** $58.7 \times 0.06$

**3.** $4.608 \div 0.16$

**4.** **Concept Check** Explain how you would place the decimal points when performing the calculation $0.252 \div 0.0035$.

# Use Math to Save Money

## Did You Know? Managing your debts can save you money over time.

### Live Debt Free

**Understanding the Problem:**

Tracy and Max are in debt.

- Each of their three credit cards is maxed out to the limit of $8000.
- They owe $12,000 in hospital bills and $2000 for their car loan.
- After borrowing from friends, they still owe $100 to one friend and $300 to another.

Before they can save for a much-needed vacation, they must pay off their debts.

**Making a Plan:**

Tracy and Max come up with a plan for tackling their debt.

**Step 1:** They list all of their debts, ordered from smallest to largest.

**Task 1:** Complete Step 1 for Tracy and Max. Remember that there are three credit cards.

**Step 2:** They make minimum monthly payments on each debt.

- Each credit card has a minimum monthly payment of $25.
- The hospital expects a payment of $50 per month.
- The monthly car payment is $200.
- Tracy and Max arrange to pay $20 per month for each loan from friends.

**Task 2:** What is the total amount of their minimum monthly payments?

**Step 3:** They pay off their three smallest debts first.

**Task 3:** What are their three smallest debts?

**Finding a Solution:**

**Step 4:** Tracy and Max decide to eliminate any unnecessary spending until the three smallest debts are paid off. By doing this, they can pay off the two smallest debts in only two months while still making the minimum payments on the other debts.

**Task 4:** What is the total amount of the minimum monthly payments for the two smallest debts?

**Step 5:** Each month, Tracy and Max take the amount that they would have used to pay the two smallest debts and apply it toward the third smallest, all while paying the minimum monthly payments on the remaining debts.

**Task 5:** How many more months will it take Tracy and Max to pay off the third smallest debt if they follow Step 5? Round your answer to the nearest whole number when you perform division operations.

**Step 6:** After the three smallest debts are paid off, Tracy and Max take the money that they would have spent per month to pay those debts and use it on the principal of the remaining debts. To pay the debts more quickly, Tracy and Max decide to stop using credit cards for new purchases. After a few years of careful budgeting, not using credit cards, and paying more than the minimum payment on their credit card debt, they finally pay off their debts.

**Task 6:** Besides avoiding credit cards for new purchases, can you think of other ways that Tracy and Max could have budgeted their money and paid off the debt more quickly?

**Applying the Situation to Your Life:**

Debt counselors often provide this simple, practical plan for people in debt:

- Arrange debts in order.
- Pay off the smallest debt first.
- Let the consequences of paying off the smallest debt help you to pay off the rest of the debts more quickly.
- Avoid unnecessary spending and incorporate budgeting strategies into your daily life.

39

# 0.5  Percents, Rounding, and Estimating ▶

**Student Learning Objectives**

After studying this section, you will be able to:

1 Change a decimal to a percent. ▶

2 Change a percent to a decimal. ▶

3 Find the percent of a given number. ▶

4 Find the missing percent when given two numbers. ▶

5 Use rounding to estimate. ▶

## 1 Changing a Decimal to a Percent ▶

A **percent** is a fraction that has a denominator of 100. When you say "sixty-seven percent" or write 67%, you are just expressing the fraction $\frac{67}{100}$ in another way. The word *percent* is a shortened form of the Latin words *per centum*, which means "by the hundred." In everyday use, percent means per one hundred.

Russell Camp owns 100 acres of land in Montana. 49 of the acres are covered with trees. The rest of the land is open fields. We say that 49% of his land is covered with trees.

It is important to see that 49% means 49 parts out of 100 parts. It can also be written as a fraction, $\frac{49}{100}$, or as a decimal, 0.49. Understanding the meaning of the notation allows you to change from one form to another. For example,

$$49\% = 49 \text{ out of } 100 \text{ parts} = \frac{49}{100} = 0.49.$$

Similarly, you can express a fraction with denominator 100 as a percent or a decimal.

$$\frac{11}{100} \text{ means 11 parts out of 100 or } 11\%. \quad \text{So } \frac{11}{100} = 11\% = 0.11.$$

Now that we understand the concept, we can use some quick procedures to change from a decimal to a percent, and vice versa.

> **CHANGING A DECIMAL TO A PERCENT**
> 1. Move the decimal point two places to the right.
> 2. Add the % symbol.

Sometimes percents are less than 1%. When we follow the procedure above, we see that 0.01 is 1%. Thus we would expect 0.001 to be less than 1%. $0.001 = 0.1\%$ or one-tenth (0.1) of a percent.

**Example 1** Change to a percent.

(a) 0.0364            (b) 0.0008            (c) 0.4

**Solution**  We move the decimal point two places to the right and add the % symbol.

(a) $0.0364 = 3.64\%$    (b) $0.0008 = 0.08\%$    (c) $0.4 = 0.40 = 40\%$  □

 **Student Practice 1**    Change to a percent.

(a) 0.92            (b) 0.0736            (c) 0.7            (d) 0.0003

Percents can be greater than 100%. Since $1 = 1.00 = 100\%$, we would expect 1.5 to be greater than 100%. In fact, $1.5 = 150\%$.

**Example 2** Change to a percent.

(a) 1.48            (b) 2.938            (c) 4.5

**Solution**  We move the decimal point two places to the right and add the % symbol.

(a) $1.48 = 148\%$    (b) $2.938 = 293.8\%$    (c) $4.5 = 4.50 = 450\%$  □

**Student Practice 2**    Change to a percent.

(a) 3.04            (b) 5.186            (c) 2.1

**2** Changing a Percent to a Decimal

In this procedure we move the decimal point to the left and remove the % symbol.

**CHANGING A PERCENT TO A DECIMAL**

1. Move the decimal point two places to the left.
2. Remove the % symbol.

**Example 3** Change to a decimal.

(a) 4%          (b) 0.6%          (c) 254.8%

**Solution**   First we move the decimal point two places to the left. Then we remove the % symbol.

(a) $4\% = 4.\% = 0.04$

The unwritten decimal point is understood to be here.

(b) $0.6\% = 0.006$

(c) $254.8\% = 2.548$                                                               □

**Student Practice 3**   Change to a decimal.

(a) 7%          (b) 9.3%          (c) 131%          (d) 0.04%

---

**Calculator**

**Percent to Decimal**

You can use a calculator to change 52% to a decimal. If your calculator has a %  key, do the following:
Enter: 52 %
The display should read:
0.52

If your calculator does not have a %  key, divide the number by 100.
Enter:  52 ÷ 100 =
Try the following:

(a) 46%          (b) 137%

(c) 9.3%          (d) 6%

*Note:* The calculator divides by 100 when the percent key is pressed.

---

**3** Finding the Percent of a Given Number

How do we find 60% of 20? Let us relate it to problems we did in Section 0.3 involving multiplication of fractions.
Consider the following problem.

What      is      $\frac{3}{5}$      of      20?

?      =      $\frac{3}{5}$      ×      20

? $= \dfrac{3}{\overset{}{\underset{1}{\cancel{5}}}} \times \overset{4}{\cancel{20}} = 12$   The answer is 12.

Since a percent is really a fraction, a percent problem is solved similarly to the way a fraction problem is solved. Since $\frac{3}{5} = \frac{3 \cdot 20}{5 \cdot 20} = \frac{60}{100} = 60\%$, we could write the problem what is $\frac{3}{5}$ of 20? as

What      is      60%      of      20?   Replace $\frac{3}{5}$ with 60%.

?      =      60%      ×      20

?   $= 0.60 \times 20$

?   $= 12.0$   The answer is 12.

Thus we have developed the following rule.

**FINDING THE PERCENT OF A NUMBER**

To find the percent of a number, change the percent to a decimal and multiply the number by the decimal.

---

**Calculator**

**Finding the Percent of a Number**

You can use a calculator to find 12% of 48.

Enter:  12 %  × 48 =

The display should read
5.76

If your calculator does not have a %  key, do the following:
Enter:  0.12 × 48 =
What is 54% of 450?

**Example 4** Find.

(a) 10% of 36      (b) 2% of 350      (c) 182% of 12      (d) 0.3% of 42

**Solution**

(a) 10% of $36 = 0.10 \times 36 = 3.6$      (b) 2% of $350 = 0.02 \times 350 = 7$

(c) 182% of $12 = 1.82 \times 12 = 21.84$      (d) 0.3% of $42 = 0.003 \times 42 = 0.126$   □

**Student Practice 4** Find.

(a) 18% of 50            (b) 4% of 64            (c) 156% of 35

There are many real-life applications for finding the percent of a number. When you go shopping in a store, you may find sale merchandise marked 35% off. This means that the sale price is 35% off the regular price. That is, 35% of the regular price is subtracted from the regular price to get the sale price.

**Example 5** A store is having a sale of 35% off the retail price of all sofas. Melissa wants to buy a particular sofa that normally sells for $595.

(a) How much will Melissa save if she buys the sofa on sale?

(b) What will the purchase price be if Melissa buys the sofa on sale?

**Solution**

(a) To find 35% of $595 we will need to multiply $0.35 \times 595$.

$$
\begin{array}{r}
595 \\
\times\ 0.35 \\
\hline
2975 \\
1785 \\
\hline
208.25
\end{array}
$$

Thus Melissa will save $208.25 if she buys the sofa on sale.

(b) The purchase price is the difference between the original price and the amount saved.

$$
\begin{array}{r}
595.00 \\
-\ 208.25 \\
\hline
386.75
\end{array}
$$

If Melissa buys the sofa on sale, she will pay $386.75.

□

**Student Practice 5** John received a 4.2% pay raise at work this year. He had previously earned $38,000 per year.

(a) What was the amount of his pay raise in dollars?

(b) What is his new salary?

## 4 Finding the Missing Percent When Given Two Numbers

Recall that we can write $\frac{3}{4}$ as $\frac{75}{100}$ or 75%. If we were asked the question, "What percent is 3 of 4?" we would say 75%. This gives us a procedure for finding what percent one number is of a second number.

**FINDING THE MISSING PERCENT**

1. Write a fraction with the two numbers. The number *after* the word *of* is always the denominator, and the other number is the numerator.

2. Simplify the fraction (if possible).

3. Change the fraction to a decimal.

4. Express the decimal as a percent.

**Example 6**  What percent of 24 is 15?

**Solution**  This can be solved quickly as follows.

**Step 1**  $\dfrac{15}{24}$  Write the relationship as a fraction. The number after "of" is 24, so 24 is the denominator.

**Step 2**  $\dfrac{15}{24} = \dfrac{5}{8}$  Simplify the fraction (when possible).

**Step 3**  $= 0.625$  Change the fraction to a decimal.

**Step 4**  $= 62.5\%$  Change the decimal to percent.

**Student Practice 6**  What percent of 148 is 37?

The question in Example 6 can also be written as "15 is what percent of 24?" To answer the question, we begin by writing the relationship as $\frac{15}{24}$. Remember that "of 24" means 24 will be the denominator.

**Mc Example 7**

**(a)** What percent of 16 is 3.8?    **(b)** $150 is what percent of $120?

**Solution**

**(a)** What percent of 16 is 3.8?

$\dfrac{3.8}{16}$  Write the relationship as a fraction.

You can divide to change the fraction to a decimal and then change the decimal to a percent.

$$\begin{array}{r} 0.2375 \rightarrow 23.75\% \\ 16)\overline{3.8000} \end{array}$$

**(b)** $150 is what percent of $120?

$\dfrac{150}{120} = \dfrac{5}{4}$  Reduce the fraction whenever possible to make the division easier.

$$\begin{array}{r} 1.25 \rightarrow 125\% \\ 4)\overline{5.00} \end{array}$$

**Student Practice 7**

**(a)** What percent of 48 is 24?

**(b)** 4 is what percent of 25?

**Example 8**  Marcia made 29 shots on goal during the last high school field hockey season. She actually scored a goal 8 times. What percent of her total shots were goals? Round your answer to the nearest whole percent.

**Solution**  Marcia scored a goal 8 times out of 29 tries. We want to know what percent of 29 is 8.

**Step 1**  $\dfrac{8}{29}$  Express the relationship as a fraction. The number after the word *of* is 29, so 29 appears in the denominator.

**Step 2**  $\dfrac{8}{29} = \dfrac{8}{29}$  Note that this fraction cannot be reduced.

**Step 3**  $= 0.2758\ldots$  The decimal equivalent of the fraction has many digits.

**Step 4**  $= 27.58\ldots\%$  Change the decimal to a percent, rounded to the nearest whole percent.

$\approx 28\%$  (The $\approx$ symbol means approximately equal to.)

Therefore, Marcia scored a goal approximately 28% of the time she made a shot on goal.

*Continued on next page*

**Calculator**

**Finding the Percent Given Two Numbers**

You can use a calculator to find a missing percent.
What percent of 95 is 19?

**1.** Enter as a fraction.

19 $\boxed{\div}$ 95

**2.** Change to a percent.

19 $\boxed{\div}$ 95 $\boxed{\times}$ 100 $\boxed{=}$

The display should read

$\boxed{\phantom{0}20\phantom{0}}$

This means 20%.
What percent of 625 is 250?

**Student Practice 8**　Roberto scored a basket 430 times out of 1256 attempts during his high school basketball career. What percent of the time did he score a basket? Round your answer to the nearest whole percent.

## 5 Using Rounding to Estimate

Before we proceed in this section, we will take some time to be sure you understand the idea of rounding a number. You will probably recall the following simple rule from your previous mathematics courses.

**ROUNDING A NUMBER**

If the first digit to the right of the round-off place is

1. less than 5, we make no change to the digit in the round-off place.
2. 5 or more, we increase the digit in the round-off place by 1.

To illustrate, 4689 rounded to the nearest hundred is 4700. Rounding 233,987 to the nearest ten thousand, we obtain 230,000. We will now use our experience in rounding as we discuss the general area of estimation.

Estimation is the process of finding an approximate answer. It is not designed to provide an exact answer. Estimation will give you a rough idea of what the answer might be. For any given problem, you may choose to estimate in many different ways.

**ESTIMATION BY ROUNDING**

1. Round each number so that there is one nonzero digit.
2. Perform the calculation with the rounded numbers.

**Example 9**　Use estimation by rounding to find the product. $5368 \times 2864$

**Solution**

**Step 1**　Round 5368 to 5000.
Round 2864 to 3000.

**Step 2**　Multiply.

$$5000 \times 3000 = 15,000,000$$

An estimate of the product is 15,000,000.　□

**Student Practice 9**　Use estimation by rounding to find the product. $128,621 \times 378$

▲ **Example 10**　The four walls of a college classroom are $22\frac{1}{4}$ feet long and $8\frac{3}{4}$ feet high. A painter needs to know the area of these four walls in square feet. Since paint is sold in gallons, an estimate will do. Use estimation by rounding to approximate the area of the four walls.

**Solution**

**Step 1**　Round $22\frac{1}{4}$ feet to 20 feet.

Round $8\frac{3}{4}$ feet to 9 feet.

**Step 2**　Multiply $20 \times 9$ to obtain an estimate of the area of one wall.

Multiply $20 \times 9 \times 4$ to obtain an estimate of the area of all four walls.

$$20 \times 9 \times 4 = 720$$

Our estimate for the painter is 720 square feet of wall space.　□

**Student Practice 10**   Mr. and Mrs. Ramirez need to carpet two rooms of their house. One room measures $12\frac{1}{2}$ feet by $9\frac{3}{4}$ feet. The other room measures $11\frac{1}{4}$ feet by $18\frac{1}{2}$ feet. Use estimation by rounding to approximate the number of square feet (square footage) in these two rooms.

**Example 11**   Won Lin has a small compact car in Honolulu. He drove 396.8 miles in his car and used 8.4 gallons of gas.

**(a)** Estimate the number of miles he gets per gallon.

**(b)** Estimate how much he will pay for fuel to drive 2764 miles in the next month if gasoline usually costs $\$3.59\frac{9}{10}$ per gallon.

**Solution**

**(a)** Round 396.8 miles to 400 miles. Round 8.4 gallons to 8 gallons. Now divide.

$$\frac{50}{8\overline{)400}}$$

Won Lin's car gets about 50 miles per gallon.

**(b)** We will need to use the information we found in part (a) to determine how many gallons of gasoline Won Lin will use. Round 2764 miles to 3000 miles and divide 3000 miles by 50 miles per gallon.

$$\frac{60}{50\overline{)3000}}$$

Won Lin will use about 60 gallons of gas for traveling next month.

To estimate the cost, we need to ask ourselves, "What kind of an estimate are we looking for?" It may be sufficient to round $\$3.59\frac{9}{10}$ to $\$4.00$ and multiply.

$$60 \times \$4.00 = \$240.00$$

Keep in mind that this is a broad estimate. You may want an estimate that will be closer to the exact answer. In that case round $\$3.59\frac{9}{10}$ to $\$3.60$ and multiply.

$$60 \times \$3.60 = \$216.00 \qquad \square$$

**Student Practice 11**   Roberta drove 422.8 miles in her truck in Alaska and used 19.3 gallons of gas. Assume that gasoline costs $\$3.69\frac{9}{10}$ per gallon.

**(a)** Estimate the number of miles she gets per gallon.

**(b)** Estimate how much it will cost her to drive this winter if she drives 3862 miles.

  **STEPS TO SUCCESS**   What Happens When You Read a Mathematics Textbook?

Reading a math book can give you amazing insight. Always take the time at the start of a homework section by reading the section(s) assigned in your textbook.

Remember this book was written to help you become successful in this mathematics class. However, you have to read it if you want to take advantage of all its benefits.

Always read your book with a pen or a highlighter in hand. Especially watch for helpful hints and suggestions as you look over the sample explanations. Underline any step that you think is hard or not clear to you. Put question marks by words you do not understand. You may want to write down a list of these in your notebook. Ask your instructor about items you do not understand.

When you come to a sample example, make your mind work it through step by step. Underline steps that you think are especially important.

Make sure you are thinking about what you are reading. If your mind is wandering, get up and get a drink of water or walk around the room—anything to help you get your mind back on track.

**Making it personal:** Look over the given suggestions. Pick one or two ideas that you think will be the most helpful to you. Use them this week as you read the book carefully before doing homework assignments. ▼

## 0.5 Exercises  MyMathLab®

### Verbal and Writing Skills, Exercises 1 and 2

1. When you write 19%, what do you really mean? Describe the meaning in your own words.

2. When you try to solve a problem like "What percent of 80 is 30?" how do you know if you should write the fraction as $\frac{80}{30}$ or as $\frac{30}{80}$?

*Change to a percent.*

3. 0.79

4. 0.54

5. 0.568

6. 0.063

7. 0.076

8. 0.046

9. 2.39

10. 7.49

11. 3.6

12. 5.7

13. 3.672

14. 8.674

*Change to a decimal.*

15. 3%

16. 6%

17. 0.4%

18. 0.62%

19. 250%

20. 175%

21. 7.4%

22. 8.7%

23. 0.52%

24. 0.1%

25. 100%

26. 200%

*Find the following.*

27. What is 8% of 65?

28. What is 7% of 69?

29. What is 10% of 130?

30. What is 25% of 600?

31. What is 112% of 65?

32. What is 154% of 270?

33. 36 is what percent of 24?

34. 49 is what percent of 28?

35. What percent of 340 is 17?

36. What percent of 35 is 28?

37. 30 is what percent of 500?

38. 48 is what percent of 600?

39. 80 is what percent of 200?

40. 75 is what percent of 30?

### Applications

41. *Exam Grades* Dave took an exam with 80 questions. He answered 68 correctly. What was his grade for the exam? Write the grade as a percent.

42. *Assembly Line Supervisor* Ken Thompson is supervising the production line. He conducted a random inspection of 440 units as they came off the assembly line. A total of 11 of those units were defective. What percent of the units were defective?

43. *Tipping* Diana and Russ ate a meal costing $32.80 when they went out to dinner. If they want to leave the standard 15% tip for their server, how much will they tip and what will their total bill be?

44. *Startup Business Success Rate* The Knoxville Better Business Bureau examined the records of 350 new businesses that started up last year in Knoxville, TN. It discovered that there was a failure rate of 28%. If that statistic is true, how many of those 350 new businesses in Knoxville failed?

45. *Food Budget* The Gonzalez family has a combined monthly income of $1850. Their food budget is $380 per month. What percent of their monthly income is budgeted for food? (Round your answer to the nearest whole percent.)

46. *Survey* A college survey found that 137 out of 180 students had a grandparent not born in the United States. What percent of the students had a grandparent not born in the United States? (Round your answer to the nearest hundredth of a percent.)

47. *Gift Returns* Last Christmas season, the Jones Mill Outlet Store chain calculated that they sold 36,000 gift items. If they assume that there will be a 1.5% return rate after the holidays, how many gifts can they expect to be exchanged?

48. *Rent* The total cost of a downtown Boston apartment, including utilities, is $3690 per month. Sara's share is 17% of the total cost each month. What is her monthly cost?

46

**49.** *Computer Hardware Sales* Abdul sells computers for a local computer outlet. He gets paid $450 per month plus a commission of 3.8% on all the computer hardware he sells. Last year he sold $780,000 worth of computer hardware.

**(a)** What was his sales commission on the $780,000 worth of hardware he sold?

**(b)** What was his annual salary (that is, his monthly pay and commission combined)?

**50.** *Medical Supply Sales* Bruce sells medical supplies on the road. He logged 18,600 miles in his car last year. He declared 65% of his mileage as business travel.

**(a)** How many miles did he travel on business last year?

**(b)** If his company reimburses him 31 cents for each mile traveled, how much should he be paid for travel expenses?

*In exercises 51–62, use estimation by rounding to approximate the value.* **Do not find the exact value.**

**51.** $586 \times 421$

**52.** $729 \times 688$

**53.** $3547 \times 4693$

**54.** $8192 \times 5984$

**55.** $14 + 73 + 80 + 21 + 56$

**56.** $318 + 494 + 613 + 243$

**57.** $41\overline{)829,346}$

**58.** $16\overline{)5,846,213}$

**59.** $\dfrac{2714}{31,500}$

**60.** $\dfrac{53,610}{786}$

**61.** Find 17% of $21,365.85.

**62.** Find 4.9% of $9321.88.

**Applications** *In exercises 63–66, use estimation by rounding to approximate the answer.* **Do not find the exact value.**

**63.** *Checkout Sales* A typical customer at the local Piggly Wiggly supermarket spends approximately $82 at the checkout register. The store keeps four registers open, and each handles 22 customers per hour. *Estimate* the amount of money the store receives in one hour.

**64.** *Weekend Spending Money* The Westerly Credit Union has found that on Fridays, the average customer withdraws $85 from the ATM for weekend spending money. Each ATM averages 19 customers per hour. There are five machines. *Estimate* the amount of money withdrawn by customers in one hour.

**65.** *Gas Mileage* Rod's trip from Salt Lake City to his cabin in the mountains was 117.7 miles. If his car used 3.8 gallons of gas for the trip, *estimate* the number of miles his car gets to the gallon.

**66.** *Bookstore Manager* Betty is transferring her stock of books from one store to another. If her car can transport 430 books per trip and she has 11,900 books to move, *estimate* the number of trips it will take.

## Cumulative Review

**67.** [0.4.6] *Gas Mileage* Dan took a trip in his Ford Taurus. At the start of the trip his car odometer read 68,459.5 miles. At the end of the trip his car odometer read 69,229.5 miles. He used 35 gallons of gas on the trip. How many miles per gallon did his car achieve?

**68.** [0.4.4] *Rainfall* In Hilo, Hawaii, Brad spent three months working in a restaurant for a summer job. In June he observed that 4.6 inches of rain fell. In July the rainfall was 4.5 inches and in August it was 2.9 inches. What was the average monthly rainfall that summer for those three months in Hilo?

### Quick Quiz 0.5

**1.** What is 114% of 85?

**2.** 63 is what percent of 420?

**3.** Use estimation by rounding to find the quotient. $34,987\overline{)567,238}$

**4.** **Concept Check** Explain how you would change 0.0078 to a percent.

## 0.6 Using the Mathematics Blueprint for Problem Solving ▶

**Student Learning Objective**

After studying this section, you will be able to:

1 Use the Mathematics Blueprint to solve real-life problems. ▶

### 1 Using the Mathematics Blueprint to Solve Real-Life Problems ▶

When a builder constructs a new home or office building, he or she often has a blueprint. This accurate drawing shows the basic form of the building. It also shows the dimensions of the structure to be built. This blueprint serves as a useful reference throughout the construction process.

Similarly, when solving real-life problems, it is helpful to have a "mathematics blueprint." This is a simple way to organize the information provided in the word problem in a chart or in a graph. You can record the facts you need to use. You can determine what it is you are trying to find and how you can go about actually finding it. You can record other information that you think will be helpful as you work through the problem.

As we solve real-life problems, we will use three steps.

**Step 1.** *Understand the problem.* Here we will read through the problem. We draw a picture if it will help, and use the Mathematics Blueprint as a guide to assist us in thinking through the steps needed to solve the problem.

**Step 2.** *Solve and state the answer.* We will use arithmetic or algebraic procedures along with problem-solving strategies to find a solution.

**Step 3.** *Check.* We will use a variety of techniques to see if the answer in step 2 is the solution to the word problem. This will include estimating to see if the answer is reasonable, repeating our calculation, and working backward from the answer to see if we arrive at the original conditions of the problem.

▲ **Example 1** Nancy and John want to install wall-to-wall carpeting in their living room. The floor of the rectangular living room is $11\frac{2}{3}$ feet wide and $19\frac{1}{2}$ feet long. How much will it cost if the carpet is $18.00 per square yard?

#### Solution

1. *Understand the problem.* First, read the problem carefully. Drawing a sketch of the living room may help you see what is required. The carpet will cover the floor of the living room, so we need to find the area. Now we fill in the Mathematics Blueprint.

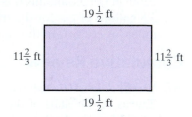

#### Mathematics Blueprint for Problem Solving

| Gather the Facts | What Am I Solving For? | What Must I Calculate? | Key Points to Remember |
|---|---|---|---|
| The living room measures $11\frac{2}{3}$ ft by $19\frac{1}{2}$ ft. The carpet costs $18.00 per square yard. | **(a)** the area of the room in square feet **(b)** the area of the room in square yards **(c)** the cost of the carpet | **(a)** Multiply $11\frac{2}{3}$ ft by $19\frac{1}{2}$ ft to get area in square feet. **(b)** Divide the number of square feet by 9 to get the number of square yards. **(c)** Multiply the number of square yards by $18.00. | There are 9 square feet, 3 feet × 3 feet, in 1 square yard; therefore, we must divide the number of square feet by 9 to obtain square yards. |

2. *Solve and state the answer.*

(a) To find the area of a rectangle, we multiply the length times the width.

$$11\frac{2}{3} \times 19\frac{1}{2} = \frac{35}{3} \times \frac{39}{2}$$

$$= \frac{455}{2} = 227\frac{1}{2}$$

A minimum of $227\frac{1}{2}$ square feet of carpet will be needed. We say a minimum because some carpet may be wasted in cutting. Carpet is sold by the square yard. We will want to know the amount of carpet needed in square yards.

**(b)** To determine the area in square yards, we divide $227\frac{1}{2}$ by 9. $(9\ \text{ft}^2 = 1\ \text{yd}^2)$

$$227\frac{1}{2} \div 9 = \frac{455}{2} \div \frac{9}{1}$$

$$= \frac{455}{2} \times \frac{1}{9} = \frac{455}{18} = 25\frac{5}{18}$$

A minimum of $25\frac{5}{18}$ square yards of carpet will be needed.

**(c)** Since the carpet costs \$18.00 per square yard, we will multiply the number of square yards needed by \$18.00.

$$25\frac{5}{18} \times 18 = \frac{455}{18} \times \frac{18}{1} = 455$$

The carpet will cost a minimum of \$455.00 for this room.

3. **Check.** We will estimate to see if our answers are reasonable.

**(a)** We will estimate by rounding each number to the nearest 10.

$$11\frac{2}{3} \times 19\frac{1}{2} \longrightarrow 10 \times 20 = 200$$

This is close to our answer of $227\frac{1}{2}\ \text{ft}^2$. Our answer is reasonable. ✓

**(b)** We will estimate by rounding to the nearest hundred and ten, respectively.

$$227\frac{1}{2} \div 9 \longrightarrow 200 \div 10 = 20$$

This is close to our answer of $25\frac{5}{18}\ \text{yd}^2$. Our answer is reasonable. ✓

**(c)** We will estimate by rounding each number to the nearest 10.

$$25\frac{5}{18} \times 18 \longrightarrow 30 \times 20 = 600$$

This is close to our answer of \$455. Our answer seems reasonable. □

*"Remember to estimate. It will save you time and money!"*

▲ ▭▶ **Student Practice 1** Jeff went to help Abby pick out wall-to-wall carpet for her new house. Her rectangular living room measures $16\frac{1}{2}$ feet by $10\frac{1}{2}$ feet. How much will it cost to carpet the room if the carpet costs \$20 per square yard?

**Mathematics Blueprint for Problem Solving**

| Gather the Facts | What Am I Solving For? | What Must I Calculate? | Key Points to Remember |
|---|---|---|---|
|  |  |  |  |

**To Think About:** *Example 1 Follow-Up* Assume that the carpet in Example 1 comes in a standard width of 12 feet. How much carpet will be wasted if it is laid out on the living room floor in one strip that is $19\frac{1}{2}$ feet long? How much carpet will be wasted if it is laid in two sections side by side that are each $11\frac{2}{3}$ feet long? Assuming you have to pay for wasted carpet, what is the minimum cost to carpet the room?

**Example 2** The following chart shows the 2015 sales of Micropower Computer Software for each of the four regions of the United States. Use the chart to answer the following questions (round all answers to the nearest whole percent):

**(a)** What percent of the sales personnel are assigned to the Northeast?

**(b)** What percent of the volume of sales is attributed to the Northeast?

**(c)** What percent of the sales personnel are assigned to the Southeast?

**(d)** What percent of the volume of sales is attributed to the Southeast?

**(e)** Which of these two regions of the country has sales personnel that appear to be more effective in terms of the volume of sales?

| Region of the U.S. | Number of Sales Personnel | Dollar Volume of Sales |
|---|---|---|
| Northeast | 12 | 1,560,000 |
| Southeast | 18 | 4,300,000 |
| Northwest | 10 | 3,660,000 |
| Southwest | 15 | 3,720,000 |
| Total | 55 | 13,240,000 |

**Solution**

1. *Understand the problem.* We will only need to deal with figures from the Northeast region and the Southeast region.

## Mathematics Blueprint for Problem Solving

| Gather the Facts | What Am I Solving For? | What Must I Calculate? | Key Points to Remember |
|---|---|---|---|
| Personnel:<br><br>12 Northeast<br><br>18 Southeast<br><br>55 total<br><br>Sales Volume:<br><br>$1,560,000 NE<br><br>$4,300,000 SE<br><br>$13,240,000 Total | **(a)** the percent of the total personnel in the Northeast<br><br>**(b)** the percent of the total sales in the Northeast<br><br>**(c)** the percent of the total personnel in the Southeast<br><br>**(d)** the percent of the total sales in the Southeast<br><br>**(e)** compare the percentages from the two regions | **(a)** 12 is what percent of 55? Divide. $12 \div 55$<br><br>**(b)** 1,560,000 is what percent of 13,240,000? $1,560,000 \div 13,240,000$<br><br>**(c)** $18 \div 55$<br><br>**(d)** $4,300,000 \div 13,240,000$ | We do not need to use the numbers for the Northwest or the Southwest. |

2. *Solve and state the answer.*

**(a)** $\dfrac{12}{55} = 0.21818\ldots$

$\approx 22\%$

**(b)** $\dfrac{1,560,000}{13,240,000} = \dfrac{156}{1324} \approx 0.1178$

$\approx 12\%$

**(c)** $\dfrac{18}{55} = 0.32727\ldots$

$\approx 33\%$

**(d)** $\dfrac{4,300,000}{13,240,000} = \dfrac{430}{1324} \approx 0.3248$

$\approx 32\%$

(e) We notice that the Northeast sales force, 22% of the total personnel, made only 12% of the sales. The percent of the sales compared to the percent of the sales force is about half (12% of 24% would be half) or 50%. The Southeast sales force, 33% of the total personnel, made 32% of the sales. The percent of sales compared to the percent of the sales force is close to 100%. We must be cautious here. *If there are no other significant factors,* it would appear that the Southeast sales force is more effective. (There may be other significant factors affecting sales, such as a recession in the Northeast, new and inexperienced sales personnel, or fewer competing companies in the Southeast.)

3. **Check.** You may want to use a calculator to check the division in step 2, or you may use estimation.

(a) $\dfrac{12}{55} \to \dfrac{10}{60} \approx 0.17$
$= 17\%$ ✓

(b) $\dfrac{1,560,000}{13,240,000} \to \dfrac{1,600,000}{13,000,000} \approx 0.12$
$= 12\%$ ✓

(c) $\dfrac{18}{55} \to \dfrac{20}{60} \approx 0.33$
$= 33\%$ ✓

(d) $\dfrac{4,300,000}{13,240,000} \to \dfrac{4,300,000}{13,000,000} \approx 0.33$
$= 33\%$ ✓

▭

**▷ Student Practice 2**   Using the chart for Example 2, answer the following questions. (Round all answers to the nearest whole percent.)

(a) What percent of the sales personnel are assigned to the Northwest?

(b) What percent of the sales volume is attributed to the Northwest?

(c) What percent of the sales personnel are assigned to the Southwest?

(d) What percent of the sales volume is attributed to the Southwest?

(e) Which of these two regions of the country has sales personnel that appear to be more effective in terms of volume of sales?

### Mathematics Blueprint for Problem Solving

| Gather the Facts | What Am I Solving For? | What Must I Calculate? | Key Points to Remember |
|---|---|---|---|
|  |  |  |  |

**To Think About:** *Example 2 Follow-Up* Suppose in 2015 the number of sales personnel (55) increased by 60%. What would the new number of sales personnel be? Suppose in 2016 the number of sales personnel decreased by 60% from the number of sales personnel in 2015. What would the new number be? Why is this number not 55, since we have increased the number by 60% and then decreased the result by 60%? Explain.

## 0.6 Exercises  MyMathLab®

**Applications** *Use the Mathematics Blueprint for Problem Solving to help you solve each of the following exercises.*

▲ 1. ***Mustang Parking Area*** Richard Penta recently restored a 1965 Mustang Coupe. He will park it on a new parking area in his backyard that uses concrete pavers. The parking area is rectangular and measures $7\frac{1}{3}$ yards by $4\frac{1}{3}$ yards . The concrete pavers each measure 1 square foot and cost $4.50 each. How much will the pavers cost that he will need to complete his parking area?

▲ 2. ***Carpentry Estimates*** Samuel is a carpenter, and he recently completed an estimate to build a new deck for Russ and Norma Camp. The deck measures $22\frac{1}{2}$ feet by $12\frac{1}{2}$ feet. He estimated the new decking would cost $8.00 per square foot. What are the estimated costs of the decking for this new deck?

▲ 3. ***Swimming Pool Estimates*** Wally and Mike install swimming pools in the summer in Manchester-by-the-Sea. They are building a swimming pool for Kevin Baird that measures $14\frac{1}{2}$ feet wide and $23\frac{1}{2}$ feet long.

(a) At the outside edge of the pool is a protective fence. How many feet of fencing would be needed to surround the pool area?

(b) Kevin discovers that he can buy a packaged box of 90 feet of fencing for $859 or he can buy cut-to-order fencing that costs $11.20 per foot. Which type of fencing should he buy? How much money does he save?

▲ 4. ***Swimming Pool Maintenance*** Wally and Mike recommend to their customers that they add the proper amount of granular chlorinator to their swimming pool each day.

(a) The swimming pool that they built in exercise 3 has an interior measurement of $12\frac{1}{2}$ feet wide by $20\frac{1}{2}$ feet long. The pool is 5 feet deep. How much water (in cubic feet) will it take to fill the pool? Round your answer to the nearest cubic foot.

(b) If each cubic foot of water is 7.5 gallons, how many gallons of water will the swimming pool hold? Round your answer to the nearest 1000 gallons.

(c) If every 5000 gallons should have 4 ounces of granular chlorinator added to the pool, how many ounces of chlorinator should be added to this pool each day?

*Exercise Training* *The following directions are posted on the wall at the gym.*

**Beginning exercise training schedule**

On day 1, each athlete will begin the morning as follows:

Jog................. $1\frac{1}{2}$ miles

Walk.............. $1\frac{3}{4}$ miles

Rest............... $2\frac{1}{2}$ minutes

Walk.............. 1 mile

5. Betty's athletic trainer told her to follow the beginning exercise training schedule on day 1. On day 2, she is to increase all distances and times by $\frac{1}{3}$ that of day 1. On day 3, she is to increase all distances and times by $\frac{1}{3}$ that of day 2. What will be her training schedule on day 3?

6. Melinda's athletic trainer told her to follow the beginning exercise training schedule on day 1. On day 2, she is to increase all distances and times by $\frac{1}{3}$ that of day 1. On day 3, she is to once again increase all distances and times by $\frac{1}{3}$ that of day 1. What will be her training schedule on day 3?

52

**To Think About**  *Refer to exercises 5 and 6 in working exercises 7–10.*

**7.** Who will have a more demanding schedule on day 3, Betty or Melinda? Why?

**8.** If Betty kept up the same type of increase day after day, how many miles would she be jogging on day 5?

**9.** If Melinda kept up the same type of increase day after day, how many miles would she be jogging on day 7?

**10.** Which athletic trainer would appear to have the best plan for training athletes if they used this plan for 14 days? Why?

**11.** *Snickers Bars*  Franklin Clarence Mars invented the Snickers Bar and introduced it in 1930. More than 15 million Snickers Bars are produced each day. Each week approximately 6,300,000 pounds of this popular bar are produced.

   **(a)** In 1 day, how many pounds of Snickers Bars are produced?

   **(b)** What percent of the total weekly pounds of Snickers Bars are produced in 1 day? Round to the nearest hundredth of a percent.

**12.** *Egg Weight*  Chicken eggs are classified by weight per dozen eggs. Large eggs weigh 24 ounces per dozen and medium eggs weigh 21 ounces per dozen.

   **(a)** If you do not include the shell, which is 12% of the total weight of an egg, how many ounces of eggs do you get from a dozen large eggs? From a dozen medium eggs?

   **(b)** At a local market, large eggs sell for $1.79 a dozen and medium eggs for $1.39 a dozen. If you do not include the shell, which is a better buy, large or medium eggs?

*Family Budgets*  *For the following exercises, use the chart below.*

Michael and Dianne have been married for two years. Michael has graduated from community college. Dianne is still attending community college. Michael works full time and Dianne part time.

| Rent | 22% | Clothing | 7% |
|---|---|---|---|
| Food | 29% | Medical | 8% |
| Utilities | 8% | Savings | 4% |
| Entertainment | 6% | Loan and credit payments | 8% |
| Cell phones | 4% | Charitable contributions | 4% |

**13.** Their combined annual income from their two jobs is $68,500.

   **(a)** If 29% of their salary is withheld for various taxes, how much money do Michael and Dianne have available for their budget?

   **(b)** How much of their take-home pay is available for food?

**14.** They want to pay off one large credit card account.

   **(a)** They determine that if they eat out less and cook more meals at home, they can save 20% of their food costs. If they achieve that goal, what is their new annual food budget in dollars?

   **(b)** If they use all that savings to pay down their credit card, how much would be available to reduce their debt?

*Paycheck Stub* *Use the following information from a paycheck stub to solve exercises 15–18.*

| TOBEY & SLATER INC. 5000 Stillwell Avenue Queens, NY 10001 | | Check Number 495885 | Payroll Period | | Pay Date 12-01-15 |
|---|---|---|---|---|---|
| | | | From Date 10-31-15 | To Date 11-30-15 | |

| Name Fred J. Gilliani | Social Security No. 012-34-5678 | I.D. Number 01 | File Number 1379 | Rate/Salary 1150.00 | Department 0100 | MS M | DEP 5 | Res NY |
|---|---|---|---|---|---|---|---|---|

| | Current | Year to Date | | Current | Year to Date |
|---|---|---|---|---|---|
| GROSS | 1,150.00 | 6,670.00 | STATE | 67.76 | 388.45 |
| FEDERAL | 138.97 | 781.07 | LOCAL | 5.18 | 30.04 |
| FICA | 87.98 | 510.28 | DIS-SUI | .00 | .00 |
| W-2 GROSS | | 6,670.00 | NET | 790.47 | 4,960.16 |

| Earnings | | | | | | Special Deductions | | |
|---|---|---|---|---|---|---|---|---|
| No. | Type | Hours | Rate | Amount | Dept/Job No. | No. | Description | Amount |
| 96 | REGULAR | | | 1,150.00 | 0100 | 82 | Retirement | 12.56 |
| | | | | | | 75 | Medical | 36.28 |
| | | | | | | 56 | Union Dues | 10.80 |

*Gross pay is the pay an employee receives for his or her services before deductions. Net pay is the pay the employee actually gets to take home. You may round each amount to the nearest whole percent for exercises 15–18.*

**15.** What percent of Fred's gross pay is deducted for federal, state, and local taxes?

**16.** What percent of Fred's gross pay is deducted for retirement and medical?

**17.** What percent of Fred's gross pay does he actually get to take home?

**18.** What percent of Fred's deductions are special deductions?

## Quick Quiz 0.6

**1.** The college courtyard has 525 square yards of space. It is being paved with squares made of Vermont granite stone. Each stone has an area of 1.5 square yards. How many stones will be required to pave the courtyard?

**2.** Melinda is paid $16,200 per year. She is also paid a sales commission of 4% of the value of her sales. Last year she sold $345,000 worth of products. What percent of her total income was her commission? Round to the nearest whole percent.

**3.** A carpenter is building a fence for a rectangular yard that measures 40 feet by 95 feet. The fence material costs $4.50 per foot. If he makes a fence large enough to surround the yard, how much will the fence cost?

**4. Concept Check** Hank knows that 1 kilometer ≈ 0.62 mile. Explain how Hank could find out how many miles he traveled on a 12-kilometer trip in Mexico.

### Background: Financial Counselor

As a college student, Sybelle had an internship with the federal government at a local Social Security Administration office. Upon graduation, she was offered a full-time position as a financial counselor. Today, she is meeting with Cora, who wants to understand how monthly retirement benefits are calculated.

### Facts

- Cora is 62 years old in 2015.
- Full retirement age is 67 years old.
- At age 66, the percent reduction in monthly benefits is about 6.7%.
- At age 65, the percent reduction in monthly benefits is about 13.3%.
- Cora's average monthly earnings are $4200.
- The guidelines used for calculating a benefit are as follows:

  90% of the first $826 of average monthly earnings
  32% of the average monthly earnings between $826 and $4980
  15% of the earnings over $4980

### Tasks

1. Sybelle must explain to Cora the calculations involved in determining the monthly benefit she'll receive at the full retirement age of 67. What does Sybelle calculate Cora's monthly benefit to be?

2. Sybelle also wants to express Cora's monthly benefit as a percent of her average monthly pre-retirement earnings. What is this percentage?

3. Finally, Sybelle wants to calculate Cora's monthly benefit if her client retires early.
   (a) By what amount would Cora's monthly benefit be reduced if she began collecting it at age 66?

   (b) By what amount would it be reduced if she began collecting it at age 65?

# Chapter 0 Organizer

| Topic and Procedure | Examples | ▶ You Try It |
|---|---|---|
| **Simplifying fractions, p. 2**<br><br>1. Write the **numerator** and **denominator** as products of prime factors.<br>2. Use the basic rule of fractions that<br>$$\frac{a \times c}{b \times c} = \frac{a}{b}$$<br>for any factor that appears in both the numerator and the denominator.<br>3. Multiply the remaining factors for the numerator and for the denominator separately. | Simplify.<br>(a) $\dfrac{15}{25} = \dfrac{\cancel{5} \cdot 3}{\cancel{5} \cdot 5} = \dfrac{3}{5}$<br>(b) $\dfrac{36}{48} = \dfrac{\cancel{2} \cdot \cancel{2} \cdot 3 \cdot \cancel{3}}{\cancel{2} \cdot \cancel{2} \cdot 2 \cdot 2 \cdot \cancel{3}} = \dfrac{3}{4}$<br>(c) $\dfrac{26}{39} = \dfrac{2 \cdot \cancel{13}}{3 \cdot \cancel{13}} = \dfrac{2}{3}$ | **1.** Simplify.<br>(a) $\dfrac{18}{27}$<br>(b) $\dfrac{45}{60}$<br>(c) $\dfrac{34}{85}$ |
| **Changing improper fractions to mixed numbers, p. 5**<br><br>1. Divide the denominator into the numerator to obtain the whole-number part of the mixed number.<br>2. The remainder from the division will be the numerator of the fraction.<br>3. The denominator remains unchanged. | Change to a mixed number.<br>(a) $\dfrac{14}{3} = 4\dfrac{2}{3}$    (b) $\dfrac{19}{8} = 2\dfrac{3}{8}$<br>since $3\overline{)14}$       since $8\overline{)19}$<br>$\quad\ \underline{12}$             $\underline{16}$<br>$\quad\ \ 2$              $3$ | **2.** Change to a mixed number.<br>(a) $\dfrac{21}{5}$<br>(b) $\dfrac{37}{7}$ |
| **Changing mixed numbers to improper fractions, p. 6**<br><br>1. Multiply the whole number by the denominator and add the result to the numerator. This will yield the new numerator.<br>2. The denominator does not change. | Change to an improper fraction.<br>(a) $4\dfrac{5}{6} = \dfrac{(4 \times 6) + 5}{6} = \dfrac{24 + 5}{6} = \dfrac{29}{6}$<br>(b) $3\dfrac{1}{7} = \dfrac{(3 \times 7) + 1}{7} = \dfrac{21 + 1}{7} = \dfrac{22}{7}$ | **3.** Change to an improper fraction.<br>(a) $2\dfrac{2}{5}$<br>(b) $6\dfrac{1}{9}$ |
| **Changing fractions to equivalent fractions with a given denominator, p. 7**<br><br>1. Divide the original denominator into the new denominator. This result is the value that we use for multiplication.<br>2. Multiply the numerator and the denominator of the original fraction by that value. | Find the missing numerator.<br>$$\frac{4}{7} = \frac{?}{21}$$<br>$7\overline{)21}^{\ 3} \leftarrow$ Use this to multiply $\dfrac{4 \times 3}{7 \times 3} = \dfrac{12}{21}$. | **4.** Find the missing numerator.<br>$\dfrac{3}{8} = \dfrac{?}{40}$ |
| **Finding the LCD (least common denominator) of two or more fractions, p. 11**<br><br>1. Write each denominator as the product of prime factors.<br>2. The LCD is a product containing each different factor.<br>3. If a factor occurs more than once in any one denominator, the LCD will contain that factor repeated the greatest number of times that it occurs in any one denominator. | (a) Find the LCD of each pair of fractions. $\dfrac{4}{15}$ and $\dfrac{3}{35}$<br>$15 = 5 \cdot 3 \quad 35 = 5 \cdot 7 \quad LCD = 3 \cdot 5 \cdot 7 = 105$<br>(b) $\dfrac{11}{18}$ and $\dfrac{7}{45}$   $18 = 3 \cdot 3 \cdot 2$ (factor 3 appears twice)<br>                   $45 = 3 \cdot 3 \cdot 5$ (factor 3 appears twice)<br>     $LCD = 2 \cdot 3 \cdot 3 \cdot 5 = 90$ | **5.** Find the LCD of each pair of fractions.<br>(a) $\dfrac{3}{10}$ and $\dfrac{7}{12}$<br>(b) $\dfrac{17}{25}$ and $\dfrac{9}{20}$ |
| **Adding and subtracting fractions that do not have a common denominator, p. 12**<br><br>1. Find the LCD.<br>2. Change each fraction to an equivalent fraction with the LCD for a denominator.<br>3. Add or subtract the fractions and simplify the answer if possible. | Perform the operation indicated.<br>(a) $\dfrac{3}{8} + \dfrac{1}{3} = \dfrac{3 \cdot 3}{8 \cdot 3} + \dfrac{1 \cdot 8}{3 \cdot 8} = \dfrac{9}{24} + \dfrac{8}{24} = \dfrac{17}{24}$<br>(b) $\dfrac{11}{12} - \dfrac{1}{4} = \dfrac{11}{12} - \dfrac{1 \cdot 3}{4 \cdot 3} = \dfrac{11}{12} - \dfrac{3}{12} = \dfrac{8}{12} = \dfrac{2}{3}$ | **6.** Perform the operation indicated.<br>(a) $\dfrac{1}{2} + \dfrac{1}{9}$<br>(b) $\dfrac{19}{24} - \dfrac{3}{8}$ |
| **Adding and subtracting mixed numbers, p. 15**<br><br>1. Change the mixed numbers to improper fractions.<br>2. Follow the rules for adding and subtracting fractions.<br>3. If necessary, change your answer to a mixed number. | (a) Add. $1\dfrac{2}{3} + 1\dfrac{3}{4} = \dfrac{5}{3} + \dfrac{7}{4} = \dfrac{5 \cdot 4}{3 \cdot 4} + \dfrac{7 \cdot 3}{4 \cdot 3}$<br>            $= \dfrac{20}{12} + \dfrac{21}{12} = \dfrac{41}{12} = 3\dfrac{5}{12}$<br>(b) Subtract. $2\dfrac{1}{4} - 1\dfrac{3}{4} = \dfrac{9}{4} - \dfrac{7}{4} = \dfrac{2}{4} = \dfrac{1}{2}$ | **7.** (a) Add. $2\dfrac{2}{3} + 3\dfrac{1}{9}$<br>(b) Subtract. $3\dfrac{5}{6} - 1\dfrac{1}{3}$ |

56

| Topic and Procedure | Examples | You Try It |
|---|---|---|
| **Multiplying fractions, p. 20**<br><br>1. If there are no common factors, multiply the numerators. Then multiply the denominators.<br>2. If possible, write the numerators and denominators as the product of prime factors. Use the basic rule of fractions to divide out any value that appears in both a numerator and a denominator. Multiply the remaining factors in the numerator. Multiply the remaining factors in the denominator. | Multiply.<br>(a) $\frac{3}{7} \times \frac{2}{13} = \frac{6}{91}$<br>(b) $\frac{6}{15} \times \frac{35}{91} = \frac{2 \cdot \cancel{3}}{\cancel{3} \cdot \cancel{5}} \times \frac{\cancel{5} \cdot \cancel{7}}{\cancel{7} \cdot 13} = \frac{2}{13}$<br>(c) $3 \times \frac{5}{8} = \frac{3}{1} \times \frac{5}{8} = \frac{15}{8}$ or $1\frac{7}{8}$ | **8.** Multiply.<br>(a) $\frac{5}{11} \times \frac{2}{3}$<br>(b) $\frac{7}{10} \times \frac{15}{21}$<br>(c) $6 \times \frac{3}{5}$ |
| **Dividing fractions, p. 22**<br><br>1. Invert the second fraction.<br>2. Multiply the fractions. | Divide.<br>(a) $\frac{4}{7} \div \frac{11}{3} = \frac{4}{7} \times \frac{3}{11} = \frac{12}{77}$<br>(b) $\frac{5}{9} \div \frac{5}{7} = \frac{\cancel{5}}{9} \times \frac{7}{\cancel{5}} = \frac{7}{9}$ | **9.** Divide.<br>(a) $\frac{3}{14} \div \frac{2}{5}$<br>(b) $\frac{7}{12} \div \frac{7}{5}$ |
| **Multiplying and dividing mixed numbers, pp. 21, 22, 24**<br><br>1. Change each mixed number to an improper fraction.<br>2. Use the rules for multiplying or dividing fractions.<br>3. If necessary, change your answer to a mixed number. | (a) Multiply.<br>$$2\frac{1}{4} \times 3\frac{3}{5} = \frac{9}{4} \times \frac{18}{5}$$<br>$$= \frac{3 \cdot 3}{2 \cdot 2} \times \frac{\cancel{2} \cdot 3 \cdot 3}{5} = \frac{81}{10} = 8\frac{1}{10}$$<br>(b) Divide. $1\frac{1}{4} \div 1\frac{1}{2} = \frac{5}{4} \div \frac{3}{2} = \frac{5}{2 \cdot \cancel{2}} \times \frac{\cancel{2}}{3} = \frac{5}{6}$ | **10.** (a) Multiply. $1\frac{5}{6} \times 2\frac{1}{4}$<br><br><br>(b) Divide. $3\frac{1}{2} \div 1\frac{3}{4}$ |
| **Changing fractional form to decimal form, p. 29**<br><br>Divide the denominator into the numerator. | Write as a decimal. $\frac{5}{8} = 0.625$ since $8\overline{)5.000}\,^{0.625}$ | **11.** Write as a decimal. $\frac{7}{8}$ |
| **Changing decimal form to fractional form, p. 30**<br><br>1. Write the decimal as a fraction with a denominator of 10, 100, 1000, and so on.<br>2. Simplify the fraction, if possible. | Write as a fraction.<br>(a) $0.37 = \frac{37}{100}$    (b) $0.375 = \frac{375}{1000} = \frac{3}{8}$ | **12.** Write as a fraction.<br>(a) 0.29<br><br>(b) 0.175 |
| **Adding and subtracting decimals, p. 31**<br><br>1. Carefully line up the decimal points as indicated for addition and subtraction. (Extra zeros may be added to the right-hand side of the decimals if desired.)<br>2. Add or subtract the appropriate digits. | (a) Add.      (b) Subtract.<br>$1.236 + 7.825$    $2 - 1.32$<br>$\quad\; 1.236$       $\quad 2.00$<br>$\underline{+\; 7.825}$      $\underline{-\; 1.32}$<br>$\quad\; 9.061$       $\quad 0.68$ | **13.** (a) Add. $2.338 + 6.195$<br>(b) Subtract. $6 - 2.54$ |
| **Multiplying decimals, p. 32**<br><br>1. First multiply the digits.<br>2. Count the total number of decimal places in the numbers being multiplied. This number determines the number of decimal places in the answer. | Multiply.<br>(a)    0.9 (one place)    (b)    0.009 (three places)<br>$\underline{\times\; 0.7}$ (one place)     $\underline{\times\; 0.07}$ (two places)<br>   0.63 (two places)     0.00063 (five places) | **14.** Multiply.<br>(a)    1.5    (b)    5.12<br>$\underline{\times\; 0.9}$      $\underline{\times\; 0.67}$ |
| **Dividing decimals, p. 33**<br><br>1. Count the number of decimal places in the divisor.<br>2. Move the decimal point to the right the same number of places in both the divisor and dividend.<br>3. Mark that position with a caret ($\wedge$).<br>4. Perform the division. Line up the decimal point in the quotient with the position indicated by the caret. | Divide. $7.5 \div 0.6$<br>Move decimal point one place to the right.<br><br>$\begin{array}{r} 12.5 \\ 0.6_{\wedge}\overline{)7.5_{\wedge}0} \\ \underline{6}\phantom{..} \\ 15 \\ \underline{12} \\ 30 \\ \underline{30} \\ 0 \end{array}$   Therefore, $7.5 \div 0.6 = 12.5$ | **15.** Divide. $9.25 \div 0.5$ |

57

| Topic and Procedure | Examples | You Try It |
|---|---|---|
| **Changing a decimal to a percent, p. 40**<br><br>1. Move the decimal point two places to the right.<br>2. Add the % symbol. | Write as a percent.<br>**(a)** $0.46 = 46\%$      **(b)** $0.002 = 0.2\%$<br>**(c)** $0.013 = 1.3\%$     **(d)** $1.59 = 159\%$<br>**(e)** $0.0007 = 0.07\%$ | **16.** Write as a percent.<br>**(a)** 0.52      **(b)** 0.008<br>**(c)** 1.86      **(d)** 0.077<br>**(e)** 0.0009 |
| **Changing a percent to a decimal, p. 41**<br><br>1. Move the decimal point two places to the left.<br>2. Remove the % symbol. | Write as a decimal.<br>**(a)** $49\% = 0.49$     **(b)** $59.8\% = 0.598$<br>**(c)** $180\% = 1.8$      **(d)** $0.13\% = 0.0013$ | **17.** Write as a decimal.<br>**(a)** 28%      **(b)** 7.42%<br>**(c)** 165%     **(d)** 0.25% |
| **Finding a percent of a number, p. 41**<br><br>1. Convert the percent to a decimal.<br>2. Multiply the decimal by the number. | Find 12% of 86.<br>$12\% = 0.12$      $0.12 \times 86 = 10.32$<br>Therefore, 12% of 86 is 10.32. | **18.** Find 15% of 92. |
| **Finding what percent one number is of another number, p. 42**<br><br>1. Place the number after the word *of* in the denominator.<br>2. Place the other number in the numerator.<br>3. If possible, simplify the fraction.<br>4. Change the fraction to a decimal.<br>5. Express the decimal as a percent. | **(a)** What percent of 8 is 7?<br><br>$$\frac{7}{8} = 0.875 = 87.5\%$$<br><br>**(b)** 42 is what percent of 12?<br><br>$$\frac{42}{12} = \frac{7}{2} = 3.5 = 350\%$$ | **19. (a)** What percent of 12 is 10? Round to the nearest tenth of a percent.<br><br>**(b)** 50 is what percent of 40? |
| **Estimation, p. 44**<br><br>1. Round each number so that there is one nonzero digit.<br>2. Perform the calculation with the rounded numbers. | Estimate the number of square feet in a room that is 22 feet long and 13 feet wide. Assume that the room is rectangular.<br>We round 22 to 20. We round 13 to 10.<br>To find the area of a rectangle, we multiply length times width.<br>$$20 \times 10 = 200.$$<br>We estimate that there are 200 square feet in the room. | **20.** Estimate the number of square feet in a rectangular game room that measures 27 feet long by 11.5 feet wide. |
| **Problem solving, p. 48**<br><br>In solving a real-life problem, you may find it helpful to complete the following steps. You will not use all of the steps all of the time. Choose the steps that best fit the conditions of the problem.<br><br>1. *Understand the problem.*<br>  **(a)** Read the problem carefully.<br>  **(b)** Draw a picture if this helps you.<br>  **(c)** Use the Mathematics Blueprint for Problem Solving.<br>2. *Solve and state the answer.*<br>3. *Check.*<br>  **(a)** Estimate to see if your answer is reasonable.<br>  **(b)** Repeat your calculation.<br>  **(c)** Work backward from your answer. Do you arrive at the original conditions of the problem? | Susan is installing wall-to-wall carpeting in her $10\frac{1}{2}$-ft-by-12-ft bedroom. How much will it cost at $20 a square yard?<br><br>1. *Understand the problem.*<br>We need to find the area of the room in square yards. Then we can find the cost.<br>2. *Solve and state the answer.*<br>Area: $10\frac{1}{2} \times 12 = \frac{21}{2} \times \frac{12}{1} = 126 \text{ ft}^2$<br>$126 \div 9 = 14 \text{ yd}^2$<br>Cost: $14 \times 20 = \$280$<br>The carpeting will cost $280.<br>3. *Check.*<br>Estimate:    $10 \times 10 = 100 \text{ ft}^2$<br>           $100 \div 10 = 10 \text{ yd}^2$<br>           $10 \times 20 = \$200$<br><br>Our answer is reasonable. ✓ | **21.** Wayne is installing new tile in his basement, which measures 15 feet by $21\frac{3}{4}$ feet. How much will the tile cost at $4.25 per square yard? Round to the nearest cent. |

58

# Chapter 0 Review Problems

## Section 0.1

*In exercises 1–4, simplify.*

1. $\dfrac{36}{48}$

2. $\dfrac{15}{50}$

3. $\dfrac{36}{82}$

4. $\dfrac{18}{30}$

5. Write $7\dfrac{1}{8}$ as an improper fraction.

6. Write $\dfrac{34}{5}$ as a mixed number.

7. Write $\dfrac{80}{3}$ as a mixed number.

*Find the missing numerator.*

8. $\dfrac{5}{8} = \dfrac{?}{24}$

9. $\dfrac{1}{7} = \dfrac{?}{35}$

10. $\dfrac{3}{5} = \dfrac{?}{75}$

11. $\dfrac{2}{5} = \dfrac{?}{55}$

## Section 0.2

*Combine.*

12. $\dfrac{3}{5} + \dfrac{1}{4}$

13. $\dfrac{7}{12} + \dfrac{5}{8}$

14. $\dfrac{7}{20} - \dfrac{1}{12}$

15. $\dfrac{7}{10} - \dfrac{4}{15}$

16. $3\dfrac{1}{6} + 2\dfrac{3}{5}$

17. $2\dfrac{7}{10} + 3\dfrac{3}{4}$

18. $6\dfrac{2}{9} - 3\dfrac{5}{12}$

19. $3\dfrac{1}{15} - 1\dfrac{3}{20}$

## Section 0.3

*Multiply.*

20. $6 \times \dfrac{5}{11}$

21. $2\dfrac{1}{3} \times 4\dfrac{1}{2}$

22. $16 \times 3\dfrac{1}{8}$

23. $\dfrac{4}{7} \times 5$

*Divide.*

24. $\dfrac{3}{8} \div 6$

25. $\dfrac{\frac{8}{3}}{\frac{5}{9}}$

26. $\dfrac{15}{16} \div 6\dfrac{1}{4}$

27. $2\dfrac{6}{7} \div \dfrac{10}{21}$

## Section 0.4

*Combine.*

28. $1.634 + 3.007 + 2.560$

29. $24.831 - 17.094$

30. $47.251 - 17.69$

31. $1.9 + 2.53 + 0.006$

*Multiply.*

32. $0.007 \times 5.35$

33. $362.341 \times 1000$

34. $2.6 \times 0.03 \times 1.02$

35. $2.51 \times 100 \times 0.5$

*Divide.*

36. $71.32 \div 1000$

37. $0.523 \div 0.4$

38. $1.35 \div 0.015$

39. $4.186 \div 2.3$

40. Write as a decimal: $\dfrac{3}{8}$.

41. Write as a fraction in simplified form: $0.36$.

## Section 0.5

*In exercises 42–45, write each percent as a decimal.*

42. $1.4\%$

43. $36.1\%$

44. $0.02\%$

45. $125.3\%$

*In exercises 46–49, write each decimal in percent form.*

46. $0.0025$

47. $0.325$

48. $0.9$

49. $0.1$

59

**50.** What is 30% of 400?

**51.** Find 7.2% of 55.

**52.** 76 is what percent of 80?

**53.** What percent of 1250 is 750?

**54.** *Cell Phones* 80% of Del Mar Community College students have a cell phone. If there are 16,850 students in the college, how many students have cell phones?

**55.** *Math Deficiency* In a given university, 720 of the 960 freshmen had a math deficiency. What percent of the class had a math deficiency?

*In exercises 56–61, estimate. Do not find an exact value.*

**56.** $234,897 \times 1,936,112$

**57.** $357 + 923 + 768 + 417$

**58.** $634,318 - 284,000$

**59.** $21\frac{1}{5} - 8\frac{4}{5} - 1\frac{2}{3}$

**60.** Find 18% of $56,297.

**61.** $12,482 \div 389$

**62.** *Salary* Estimate Carmen's salary for the week if she earns $8.35 per hour and worked 38.5 hours.

**63.** *Vacation Cost* Estimate the weekly amount owed by each of five families who are renting a resort that costs $3875 per week.

## Section 0.6

*Solve. You may use the Mathematics Blueprint for Problem Solving.*

*Gas Mileage A six-passenger Piper Cub airplane has a gas tank that holds 240 gallons. Use this information to answer exercises 64 and 65.*

**64.** When flying at cruising speed, the plane averages $7\frac{2}{3}$ miles per gallon. How far can the plane fly at cruising speed? If the pilot never plans to fly more than 80% of his maximum cruising distance, what is the longest trip he would plan to fly?

**65.** When flying at maximum speed, the plane averages $6\frac{1}{4}$ miles per gallon. How far can the plane fly at maximum speed? If the pilot never plans to fly more than 70% of his maximum flying distance when flying at full speed, what is the longest trip he would plan to fly at full speed?

**▲ 66.** *Carpeting* Mr. and Mrs. Carr are installing wall-to-wall carpeting in a room that measures $12\frac{1}{2}$ ft by $9\frac{2}{3}$ ft. How much will it cost if the carpet is $26.00 per square yard?

**67.** *Sales Commission* Mike sells sporting goods on an 8% commission. During the first week in July, he sold goods worth $5785. What was his commission for the week?

**68.** *Car Loan* Dick and Ann Wright purchased a new car. They took out a loan of $9214.50 to help pay for the car and paid the rest in cash. They paid off the loan with payments of $225 per month for four years. How much more did they pay back than the amount of the car loan? (This is the amount of interest they were charged for the car loan.)

**69.** *Wages* Kevin works as an overnight stocker at Target. He is paid $7.50 an hour for a 40-hour week. For any additional time he gets paid 1.5 times the normal rate. Last week he worked 49 hours. How much did he get paid last week?

60

# How Am I Doing? Chapter 0 Test

**MATH COACH**       **MyMathLab®**       **YouTube**

After you take this test, read through the Math Coach on pages 62–63. Math Coach videos are available via MyMathLab and YouTube. Step-by-step test solutions on the Chapter Test Prep Videos are also available via MyMathLab and YouTube. (Search "TobeyBeginningAlg" and click on "Channels.")

*In exercises 1 and 2, simplify.*

**1.** $\dfrac{16}{18}$

**2.** $\dfrac{48}{36}$

**3.** Write as an improper fraction. $6\dfrac{3}{7}$

**4.** Write as a mixed number: $\dfrac{105}{9}$

*In exercises 5–12, perform the operations indicated. Simplify answers whenever possible.*

**5.** $\dfrac{2}{3} + \dfrac{5}{6} + \dfrac{3}{8}$

**6.** $1\dfrac{1}{8} + 3\dfrac{3}{4}$

**MC 7.** $3\dfrac{2}{3} - 2\dfrac{5}{6}$

**8.** $\dfrac{5}{7} \times \dfrac{28}{15}$

**9.** $\dfrac{7}{4} \div \dfrac{1}{2}$

**MC 10.** $5\dfrac{3}{8} \div 2\dfrac{3}{4}$

**11.** $2\dfrac{1}{2} \times 3\dfrac{1}{4}$

**12.** $\dfrac{\frac{7}{8}}{\frac{1}{4}}$

*In exercises 13–18, perform the operations indicated.*

**13.** $1.6 + 3.24 + 9.8$

**14.** $7.0046 - 3.0149$

**15.** $32.8 \times 0.04$

**16.** $0.07385 \times 1000$

**MC 17.** $12.88 \div 0.056$

**18.** $26,325.9 \div 100$

**19.** Write as a percent. $0.073$

**20.** Write as a decimal. $196.5\%$

**21.** What is $3.5\%$ of $180$?

**MC 22.** $39$ is what percent of $650$?

**23.** A 4-inch stack of computer chips is on the table. Each computer chip is $\frac{2}{9}$ of an inch thick. How many computer chips are in the stack?

*In exercises 24–25, estimate. Round each number to one nonzero digit. Then calculate.*

**24.** $52,344\overline{)4,678,987}$

**25.** $285.36 + 311.85 + 113.6$

*Solve. You may use the Mathematics Blueprint for Problem Solving.*

**26.** Allison is paid $14,000 per year plus a sales commission of 3% of the value of her sales. Last year she sold $870,000 worth of products. What percent of her total income was her commission? Round to the nearest whole percent.

▲ **27.** Fred and Melinda are laying wall tile in the kitchen. Each tile covers $3\frac{1}{2}$ square inches of space. They plan to cover 210 square inches of wall space. How many tiles will they need?

1. ☐  2. ☐

3. ☐  4. ☐

5. ☐  6. ☐

7. ☐  8. ☐

9. ☐  10. ☐

11. ☐  12. ☐

13. ☐  14. ☐

15. ☐  16. ☐

17. ☐  18. ☐

19. ☐  20. ☐

21. ☐  22. ☐

23. ☐

24. ☐  25. ☐

26. ☐

27. ☐

**Total Correct:** ☐

# MATH COACH

*Mastering the skills you need to do well on the test.*

The following problems are from the Chapter 0 Test. Here are some helpful hints to keep you from making common errors on test problems.

**Chapter 0 Test, Problem 7**   Subtract. $3\frac{2}{3} - 2\frac{5}{6}$

> **HELPFUL HINT**  First change the mixed numbers to improper fractions. Next find the LCD of the two denominators. Then change the fractions to an equivalent form with the LCD as the common denominator before subtracting.

Did you change $3\frac{2}{3}$ to $\frac{11}{3}$ and $2\frac{5}{6}$ to $\frac{17}{6}$ ?

Yes _____ No _____

If you answered No, stop and change the two mixed numbers to improper fractions.

Did you find the LCD to be 6?

Yes _____ No _____

If you answered No, consider how to find the LCD of the two fractions. Once the fractions are written as equivalent

fractions with 6 as the denominator, the two like fractions can be subtracted.

If you answered Problem 7 incorrectly, go back and rework the problem using these suggestions.

---

**Chapter 0 Test, Problem 10**   Divide. $5\frac{3}{8} \div 2\frac{3}{4}$

> **HELPFUL HINT**  Be sure to change the mixed numbers to improper fractions before dividing.

Did you change $5\frac{3}{8}$ to $\frac{43}{8}$ and $2\frac{3}{4}$ to $\frac{11}{4}$ before doing any other steps?

Yes _____ No _____

If you answered No, stop and change the two mixed numbers to improper fractions.

Next did you change the division to multiplication to obtain $\frac{43}{8} \times \frac{4}{11}$ ?

Yes _____ No _____

If you answered No, stop and make this change.

Did you simplify the product?

Yes _____ No _____

If you answered No, try dividing a 4 from the second numerator and first denominator before multiplying. The product will be an improper fraction that can be converted to a mixed number.

Now go back and rework the problem using these suggestions.

Need more help? Watch the **MATH COACH** videos in MyMathLab® or on YouTube™.

62

**Chapter 0 Test, Problem 17**   Divide. $12.88 \div 0.056$

> **HELPFUL HINT**  Be careful as you move the decimal point in the divisor to the right. Make sure that the resulting divisor is an integer. Then move the decimal point in the dividend the same number of places to the right. Add zeros if necessary.

Did you move the decimal point in the divisor three places to the right to get 56?

Yes _____ No _____

If you answered No, stop and perform this step first.

Did you move the decimal point in the dividend three places to the right and add one zero to get 12880?

Yes _____ No _____

If you answered No, perform this step now. Be careful of calculation errors as you perform the division.

If you answered Problem 17 incorrectly, go back and rework the problem using these suggestions.

---

**Chapter 0 Test, Problem 22**   39 is what percent of 650?

> **HELPFUL HINT**  Write a fraction with the two numbers. The number after the word *of* is always the denominator, and the other number is the numerator.

Did you write the fraction $\dfrac{39}{650}$ ?

Yes _____ No _____

If you answered No, stop and perform this step.

Did you simplify the fraction to $\dfrac{3}{50}$ before changing the fraction to a decimal?

Yes _____ No _____

If you answered No, consider that simplifying the fraction first makes the division step a little easier. Be sure to place the decimal point correctly in your quotient.

Did you change the quotient from a decimal to a percent?

Yes _____ No _____

If you answered No, stop and perform this final step.

Now go back and rework the problem using these suggestions.

Need more help? Look for section examples marked with ᴹℂ to review.

63

# Real Numbers and Variables

## CAREER OPPORTUNITIES

## Electrical Engineer

When we think of electrical work, mathematics isn't necessarily the first thing that comes to mind. Instead, we envision electricians running cable through walls, wiring outlets, and installing appliances. In reality, though, the field is broader than this. Electrical engineers and their assistants use quite a bit of math to create, install, or repair electrical systems. They perform calculations using ratios, formulas, and geometry to perform a wide range of tasks.

To investigate how the mathematics in this chapter can help with this field, see the Career Exploration Problems on page **122**.

## 1.1 Adding Real Numbers ▶

### 1 Identifying Different Types of Numbers ▶

Let's review some of the basic terms we use to talk about numbers.

**Whole numbers** are numbers such as $0, 1, 2, 3, 4, \ldots$
**Integers** are numbers such as $\ldots, -3, -2, -1, 0, 1, 2, 3, \ldots$.
**Rational numbers** are numbers such as $\frac{3}{2}, \frac{5}{7}, -\frac{3}{8}, -\frac{4}{13}, \frac{6}{1},$ and $-\frac{8}{2}$.

Rational numbers can be written as one integer divided by another integer (as long as the denominator is not zero!). Integers can be written as fractions ($3 = \frac{3}{1}$, for example), so we can see that all integers are rational numbers. Rational numbers can be expressed in decimal form. For example, $\frac{3}{2} = 1.5$, $-\frac{3}{8} = -0.375$, and $\frac{1}{3} = 0.333\ldots$ or $0.\overline{3}$. It is important to note that rational numbers in decimal form are either terminating decimals or repeating decimals.

**Irrational numbers** are numbers that cannot be expressed as one integer divided by another integer. The numbers $\pi$, $\sqrt{2}$, and $\sqrt[3]{7}$ are irrational numbers.

Irrational numbers can be expressed in decimal form. The decimal form of an irrational number is a nonterminating, nonrepeating decimal. For example, $\sqrt{2} = 1.414213\ldots$ can be carried out to an infinite number of decimal places with no repeating pattern of digits.

Finally, **real numbers** are all the rational numbers and all the irrational numbers.

**Example 1** Classify as an integer, a rational number, an irrational number, and/or a real number.

**(a)** 5    **(b)** $-\dfrac{1}{3}$    **(c)** 2.85    **(d)** $\sqrt{2}$    **(e)** $0.777\ldots$

**Solution**   Make a table. Check off the description of the number that applies.

| | Number | Integer | Rational Number | Irrational Number | Real Number |
|---|---|---|---|---|---|
| **(a)** | 5 | ✓ | ✓ | | ✓ |
| **(b)** | $-\frac{1}{3}$ | | ✓ | | ✓ |
| **(c)** | 2.85 | | ✓ | | ✓ |
| **(d)** | $\sqrt{2}$ | | | ✓ | ✓ |
| **(e)** | $0.777\ldots$ | | ✓ | | ✓ |

▱

**▷ Student Practice 1**   Classify as an integer, a rational number, an irrational number, and/or a real number.

**(a)** $-\dfrac{2}{5}$    **(b)** $1.515151\ldots$    **(c)** $-8$    **(d)** $\pi$

Any real number can be pictured on a **number line.**

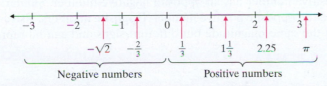

Negative numbers    Positive numbers

**Positive numbers** are to the right of 0 on a number line.

**Negative numbers** are to the left of 0 on a number line.

The **real numbers** include the positive numbers, the negative numbers, and zero.

**Student Learning Objectives**

After studying this section, you will be able to:

1 Identify different types of numbers. ▶

2 Use real numbers in real-life situations. ▶

3 Add real numbers with the same sign. ▶

4 Add real numbers with different signs. ▶

5 Use the addition properties for real numbers. ▶

## 2 Using Real Numbers in Real-Life Situations

We often encounter practical examples of number lines that include positive and negative rational numbers. For example, we can tell by reading the accompanying thermometer that the temperature is 20° below 0. From the stock market report, we see that the stock opened at 36 and closed at 34.5, so the net change for the day was $-1.5$.

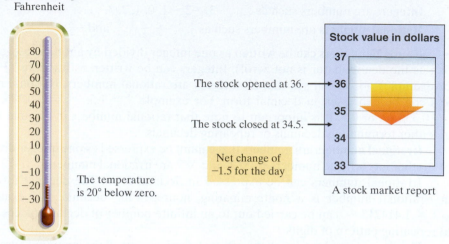

Temperature in degrees Fahrenheit

The temperature is 20° below zero.

Stock value in dollars

The stock opened at 36. → 36

The stock closed at 34.5. → 34

Net change of $-1.5$ for the day

A stock market report

In the following example we use real numbers to represent real-life situations.

**Example 2** Use a real number to represent each situation.

**(a)** A temperature of 128.6°F below zero is recorded at Vostok, Antarctica.

**(b)** The Himalayan peak K2 rises 29,064 feet above sea level.

**(c)** The Dow gains 10.24 points.

**(d)** An oil drilling platform extends 328 feet below sea level.

**Solution**    A key word can help you to decide whether a number is positive or negative.

**(a)** 128.6°F *below* zero is $-128.6$.

**(b)** 29,064 feet *above* sea level is $+29,064$.

**(c)** A *gain* of 10.24 points is $+10.24$.

**(d)** 328 feet *below* sea level is $-328$. ☐

📝 **Student Practice 2**    Use a real number to represent each situation.

**(a)** A population growth of 1259

**(b)** A depreciation of $763

**(c)** A wind-chill factor of minus 10° F

In everyday life we consider positive numbers the opposite of negative numbers. For example, a gain of 3 yards in a football game is the opposite of a loss of 3 yards; a check written for $2.16 on a checking account is the opposite of a deposit of $2.16.

Each positive number has an opposite negative number. Similarly, each negative number has an opposite positive number. **Opposite numbers,** also called **additive inverses,** have the same magnitude but different signs and can be represented on a number line.

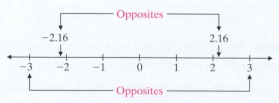

Opposites

$-2.16$                               2.16

Opposites

**Example 3** Find the additive inverse (that is, the opposite).

(a) $-7$          (b) $\dfrac{1}{4}$          (c) A temperature rise of $40°$

**Solution**

(a) The opposite of $-7$ is $+7$.

(b) The opposite of $\dfrac{1}{4}$ is $-\dfrac{1}{4}$.

(c) The opposite of $+40°$ is $-40°$.   □

**Student Practice 3** Find the additive inverse (the opposite).

(a) $\dfrac{2}{5}$          (b) $-1.92$          (c) A loss of 12 yards on a football play

## 3  Adding Real Numbers with the Same Sign

To use a real number, we need to be clear about its sign. When we write the number three as $+3$, the sign indicates that it is a positive number. The positive sign can be omitted. If someone writes three (3), it is understood that it is a positive three $(+3)$. To write a negative number such as negative three $(-3)$, we must include the sign.

A concept that will help us add and subtract real numbers is the idea of absolute value. The **absolute value** of a number is the distance between that number and zero on a number line. The absolute value of 3 is written $|3|$.

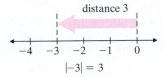 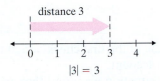

$|-3| = 3$          $|3| = 3$

Distance is always a positive number regardless of the direction of travel. This means that the absolute value of any number will be a positive value or zero. We place the symbols $|$ and $|$ around a number to mean the absolute value of the number.

The distance from 0 to 3 is 3, so $|3| = 3$. This is read "the absolute value of 3 is 3."

The distance from 0 to $-3$ is 3, so $|-3| = 3$. This is read "the absolute value of $-3$ is 3."

Some other examples are

$$|-22| = 22, \qquad |5.6| = 5.6, \qquad \text{and} \qquad |0| = 0.$$

Thus, the absolute value of a number can be thought of as the magnitude of the number, without regard to its sign.

**Example 4** Find the absolute value.

(a) $|-4.62|$          (b) $\left|\dfrac{3}{7}\right|$          (c) $\left|\dfrac{0}{5}\right|$

**Solution**

(a) $|-4.62| = 4.62$          (b) $\left|\dfrac{3}{7}\right| = \dfrac{3}{7}$          (c) $\left|\dfrac{0}{5}\right| = 0$   □

**Student Practice 4** Find the absolute value.

(a) $|-7.34|$          (b) $\left|\dfrac{5}{8}\right|$          (c) $\left|\dfrac{0}{2}\right|$

Now let's look at addition of real numbers when the two numbers have the same sign. Suppose that you are keeping track of your checking account at a local

bank. When you make a deposit of 5 dollars, you record it as +5. When you write a check for 4 dollars, you record it as −4, as a debit. (If you do not have a checking account but have a debit card, think of depositing 5 dollars in your account and then making a purchase of 4 dollars.)

**Situation 1:** *Total Deposit* You made a deposit of 20 dollars on one day and a deposit of 15 dollars the next day. You want to know the total value of your deposits.
  *Your record for situation 1.*

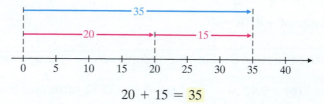

$$20 + 15 = 35$$

The amount of the deposit on the first day added to the amount of the deposit on the second day is the total of the deposits made over the two days.

**Situation 2:** *Total Debit* You write a check for 25 dollars to pay one bill and two days later write a check for 5 dollars. You want to know the total value of the debits to your account for the two checks.
  *Your record for situation 2.*

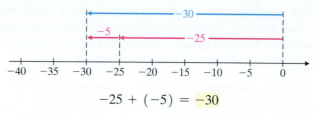

$$-25 + (-5) = -30$$

The value of the first check added to the value of the second check is the total debit to your account.

In each situation we found that we added the absolute value of each number. (That is, we added the numbers without regarding their sign.) The answer always contained the sign that was common to both numbers.

We will now state these results as a formal rule.

---

### ADDITION RULE FOR TWO NUMBERS WITH THE SAME SIGN

To add two numbers with the same sign, add the absolute values of the numbers and use the common sign in the answer.

---

**Example 5** Add.

**(a)** $14 + 16$                    **(b)** $-8 + (-7)$

**Solution**

**(a)** $14 + 16$

$14 + 16 = 30$          Add the absolute values of the numbers.

$14 + 16 = +30$          Use the common sign in the answer. Here the common sign is the + sign.

**(b)** $-8 + (-7)$

$8 + 7 = 15$          Add the absolute values of the numbers.

$-8 + (-7) = -15$          Use the common sign in the answer. Here the common sign is the − sign.    ☐

 **Student Practice 5** Add.

**(a)** $37 + 19$                    **(b)** $-23 + (-35)$

**Example 6** Add. $\dfrac{2}{3} + \dfrac{1}{7}$

**Solution**

$$\dfrac{2}{3} + \dfrac{1}{7}$$

$$\dfrac{14}{21} + \dfrac{3}{21}$$ — Change each fraction to an equivalent fraction with a common denominator of 21.

$$\dfrac{14}{21} + \dfrac{3}{21} = +\dfrac{17}{21} \text{ or } \dfrac{17}{21}$$ — Add the absolute values of the numbers. Use the common sign in the answer. Note that if no sign is written, the number is understood to be positive. ☐

 **Student Practice 6**   Add.

$$-\dfrac{3}{5} + \left(-\dfrac{4}{7}\right)$$

**Example 7** Add. $-4.2 + (-3.94)$

**Solution**

$$-4.2 + (-3.94)$$

$$4.20 + \phantom{(}3.94\phantom{)} = 8.14$$ — Add the absolute values of the numbers.

$$-4.20 + (-3.94) = -8.14$$ — Use the common sign in the answer. ☐

 **Student Practice 7**   Add. $-12.7 + (-9.38)$

The rule for adding two numbers with the same sign can be extended to more than two numbers. If we add more than two numbers with the same sign, the answer will have the sign common to all.

**Example 8** Add. $-7 + (-2) + (-5)$

**Solution**

$$-7 + (-2) + (-5)$$ — We are adding three real numbers, all with the same sign. We begin by adding the first two numbers.

$$= -9 + (-5)$$ — Add $-7 + (-2) = -9$.

$$= -14$$ — Add $-9 + (-5) = -14$.

Of course, this can be shortened by adding the three numbers without regard to sign and then using the common sign for the answer. ☐

**Student Practice 8**   Add. $-7 + (-11) + (-33)$

## 4  Adding Real Numbers with Different Signs

What if the signs of the numbers you are adding are different? Let's consider our checking account again to see how such a situation might occur.

**Situation 3:** *Net Increase* You made a deposit of 30 dollars on one day. On the next day you write a check for 25 dollars. You want to know the result of your two transactions.

*Your record for situation 3.*

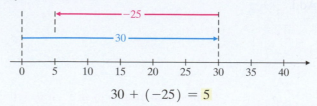

$$30 + (-25) = 5$$

A positive 30 for the deposit added to a negative 25 for the check, which is a debit, gives a net increase of 5 dollars in the account.

**Situation 4:** *Net Decrease* You made a deposit of 10 dollars on one day. The next day you write a check for 40 dollars. You want to know the result of your two transactions.

*Your record for situation 4.*

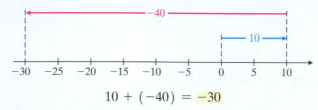

$$10 + (-40) = -30$$

A positive 10 for the deposit added to a negative 40 for the check, which is a debit, gives a net decrease of 30 dollars in the account.

The result is a negative thirty $(-30)$, because the check was larger than the deposit. If you did not have at least 30 dollars in your account at the start of *situation 4*, you have overdrawn your account.

What do we observe from *situations 3 and 4*? In each case, first we found the difference of the absolute values of the two numbers. Then the sign of the result was always the sign of the number with the greater absolute value. Thus, in *situation 3*, 30 is larger than 25. The sign of 30 is positive. The sign of the answer (5) is positive. In *situation 4*, 40 is larger than 10. The sign of 40 is negative. The sign of the answer $(-30)$ is negative.

We will now state these results as a formal rule.

> ### ADDITION RULE FOR TWO NUMBERS WITH DIFFERENT SIGNS
>
> 1. Find the difference between the larger absolute value and the smaller absolute value.
>
> 2. Give the answer the sign of the number having the larger absolute value.

**Example 9** Add. $8 + (-7)$

**Solution**

$8 + (-7)$ — We are to add two numbers with different signs.

$8 - 7 = 1$ — Find the difference between the two absolute values, which is 1.

$+8 + (-7) = +1$ or $1$ — The answer will have the sign of the number with the larger absolute value. That number is $+8$. Its sign is **positive,** so the answer will be $+1$.

 **Student Practice 9** Add. $-9 + 15$

**5** Using the Addition Properties for Real Numbers

It is useful to know the following three properties of real numbers.

1. *Addition is commutative.*
   This property states that if two numbers are added, the result is the same no matter which number is written first. The order of the numbers does not affect the result.

$$3 + 6 = 6 + 3 = 9$$
$$-7 + (-8) = (-8) + (-7) = -15$$
$$-15 + 3 = 3 + (-15) = -12$$

2. *Addition of zero to any given number will result in that given number again.*

$$0 + 5 = 5$$
$$-8 + 0 = -8$$

3. *Addition is associative.*
   This property states that if three numbers are added, it does not matter which two numbers are grouped by parentheses and added first.

$$3 + (5 + 7) = (3 + 5) + 7$$
$$3 + (12) = (8) + 7$$
$$15 = 15$$

First combine numbers inside parentheses; then combine the remaining numbers. The results are the same no matter which numbers are grouped and added first.

We can use these properties along with the rules we have for adding real numbers to add three or more numbers. We go from left to right, adding two numbers at a time.

**Example 10** Add. $\dfrac{2}{15} + \left(-\dfrac{8}{15}\right) + \dfrac{1}{15}$

**Solution**

$-\dfrac{6}{15} + \dfrac{1}{15}$    Add $\frac{2}{15} + \left(-\frac{8}{15}\right) = -\frac{6}{15}$.
The answer is negative since the larger of the two absolute values is negative.

$= -\dfrac{1}{3}$    Add $-\frac{6}{15} + \frac{1}{15} = -\frac{5}{15} = -\frac{1}{3}$.
The answer is negative since the larger of the two absolute values is negative.  □

**Student Practice 10** Add.

$$-\frac{5}{12} + \frac{7}{12} + \left(-\frac{11}{12}\right)$$

Sometimes the numbers being added have the same signs; sometimes the signs are different. When adding three or more numbers, you may encounter both situations.

**Example 11** Add. $-1.8 + 1.4 + (-2.6)$

**Solution**

$-0.4 + (-2.6)$    Add $-1.8 + 1.4$. We take the difference of 1.8 and 1.4 and use the sign of the number with the larger absolute value.

$= -3.0$    Add $-0.4 + (-2.6) = -3.0$. The signs are the same; we add the absolute values of the numbers and use the common sign.  □

**Student Practice 11** Add. $-6.3 + (-8.0) + 3.5$

If many real numbers are added, it is often easier to add numbers with the same sign in a column format. Remember that addition is commutative and associative; therefore, real numbers can be added *in any order*. You do *not* need to combine the first two numbers as your first step.

**Example 12** Add. $-8 + 3 + (-5) + (-2) + 6 + 5$

**Solution**

$$
\begin{array}{ll}
-8 & \\
-5 & \text{All the signs are the same.} \\
\underline{-2} & \text{Add the three negative} \\
-15 & \text{numbers to obtain } -15.
\end{array}
\qquad
\begin{array}{ll}
+3 & \\
+6 & \text{All the signs are the same.} \\
\underline{+5} & \text{Add the three positive} \\
+14 & \text{numbers to obtain } +14.
\end{array}
$$

Add the two results.

$$-15 + 14 = -1$$

The answer is negative because the number with the larger absolute value is negative.

**Student Practice 12** Add. $-6 + 5 + (-7) + (-2) + 5 + 3$

A word about notation: The only time we really need to show the sign of a number is when the number is negative, for example $-3$. The only time we need to show parentheses when we add real numbers is when we have two different signs preceding a number. For example, $-5 + (-6)$.

**Example 13** Add.

**(a)** $2.8 + (-1.3)$ 
**(b)** $-\dfrac{2}{5} + \left(-\dfrac{3}{4}\right)$

**Solution**

**(a)** $2.8 + (-1.3) = 1.5$

**(b)** $-\dfrac{2}{5} + \left(-\dfrac{3}{4}\right) = -\dfrac{8}{20} + \left(-\dfrac{15}{20}\right) = -\dfrac{23}{20}$ or $-1\dfrac{3}{20}$

**Student Practice 13** Add.

**(a)** $-2.9 + (-5.7)$ 
**(b)** $\dfrac{2}{3} + \left(-\dfrac{1}{4}\right)$

## 1.1 Exercises   MyMathLab®

**Verbal and Writing Skills, Exercises 1–10**

*Check off any description of the number that applies.*

| | Number | Integer | Rational Number | Irrational Number | Real Number |
|---|---|---|---|---|---|
| 1. | 23 | | | | |
| 2. | $-\frac{4}{5}$ | | | | |
| 3. | $\pi$ | | | | |
| 4. | 2.34 | | | | |
| 5. | $-6.666\ldots$ | | | | |

| | Number | Integer | Rational Number | Irrational Number | Real Number |
|---|---|---|---|---|---|
| 6. | $-\frac{7}{9}$ | | | | |
| 7. | $-2.3434\ldots$ | | | | |
| 8. | 14 | | | | |
| 9. | $\sqrt{2}$ | | | | |
| 10. | $3.232232223\ldots$ | | | | |

*Use a real number to represent each situation.*

**11.** Jules Verne wrote a book with the title *20,000 Leagues under the Sea.*

**12.** The value of the dollar is up \$0.07 with respect to the yen.

**13.** Ramona lost $37\frac{1}{2}$ pounds on Weight Watchers.

**14.** The scouts hiked from sea level to the top of a 3642-foot-high mountain.

**15.** The temperature rises 7°F.

**16.** The lowest point in Australia is Lake Eyre at 52 feet below sea level.

*Find the additive inverse (opposite).*

**17.** 8

**18.** $-\dfrac{3}{7}$

**19.** $-2.73$

**20.** 85.4

*Find the absolute value.*

**21.** $|-1.3|$

**22.** $|-5.9|$

**23.** $\left|\dfrac{5}{6}\right|$

**24.** $\left|\dfrac{3}{11}\right|$

*Add.*

**25.** $-8 + (-7)$

**26.** $-14 + (-3)$

**27.** $-20 + (-30)$

**28.** $(-17) + (-23)$

**29.** $-\dfrac{7}{20} + \dfrac{13}{20}$

**30.** $-\dfrac{3}{7} + \left(-\dfrac{2}{7}\right)$

**31.** $-\dfrac{2}{13} + \left(-\dfrac{5}{13}\right)$

**32.** $-\dfrac{3}{16} + \dfrac{5}{16}$

**33.** $-\dfrac{2}{5} + \dfrac{3}{7}$

**34.** $-\dfrac{2}{7} + \dfrac{3}{14}$

**35.** $-10.3 + (-8.9)$

**36.** $-5.4 + (-12.8)$

**37.** $0.6 + (-0.2)$

**38.** $-0.8 + 0.5$

**39.** $-5.26 + (-8.9)$

**40.** $-6.48 + (-3.7)$

**41.** $-8 + 5 + (-3)$

**42.** $5 + (-9) + (-2)$

**43.** $-2 + (-8) + 10$

**44.** $-8 + 7 + (-15)$

73

**45.** $-\dfrac{3}{10}+\dfrac{3}{4}$  **46.** $-\dfrac{3}{8}+\dfrac{11}{24}$  **47.** $-14+9+(-3)$  **48.** $-18+10+(-5)$

## Mixed Practice *Add.*

**49.** $8+(-11)$  **50.** $15+(-26)$  **51.** $-83+142$  **52.** $-114+186$

**53.** $-\dfrac{4}{9}+\dfrac{5}{6}$  **54.** $-\dfrac{3}{5}+\dfrac{2}{3}$  **55.** $-\dfrac{1}{10}+\dfrac{1}{2}$  **56.** $-\dfrac{2}{3}+\left(-\dfrac{1}{4}\right)$

**57.** $5.18+(-7.39)$  **58.** $8.33+(-14.2)$  **59.** $4+(-8)+16$

**60.** $38+(-15)+(-6)$  **61.** $26+(-19)+12+(-31)$  **62.** $-16+12+(-26)+15$

**63.** $17.85+(-2.06)+0.15$  **64.** $28.37+4.08+(-16.98)$

## Applications

**65.** *Profit/Loss* Holly paid $47 for a vase at an estate auction. She resold it to an antiques dealer for $214. What was her profit or loss?

**66.** *Temperature* When we skied at Jackson Hole, Wyoming, yesterday, the temperature at the summit was $-12°F$. Today when we called the ski report, the temperature had risen $7°F$. What is the temperature at the summit today?

**67.** *Home Equity Line of Credit* Ramon borrowed $2300 from his home equity line of credit to pay off his car loan. He then borrowed another $1500 to pay to have his kitchen repainted. Represent how much Ramon owed on his home equity line of credit as a real number.

**68.** *Time Change* During the winter, New York City is on Eastern Standard Time (EST). Melbourne, Australia, is 15 hours ahead of New York. If it is 11 P.M. in Melbourne, what time is it in New York?

**69.** *Football* During the Homecoming football game, Quentin lost 15 yards, gained 3 yards, and gained 21 yards in three successive running plays. What was his total gain or loss?

**70.** *School Fees* Wanda's financial aid account at school held $643.85. She withdrew $185.50 to buy books for the semester. Does she have enough left in her account to pay the $475.00 registration fee for the next semester? If so, how much extra money does she have? If not, how much is she short?

**71.** *Butterfly Population* The population of a particular butterfly species was 8000. Twenty years later there were 3000 fewer. Today, there are 1500 fewer. Study the graph to the right. What is the new population?

**72.** *Credit Card Balance* Aaron owes $258 to a credit card company. He makes a purchase of $32 with the card and then makes a payment of $150 on the account. How much does he still owe?

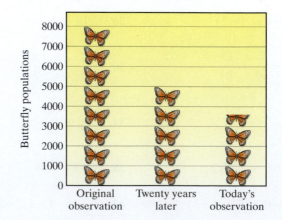

*Profit/Loss During the first five months of 2015, a regional Midwest airline posted profit and loss figures for each month of operation, as shown in the accompanying bar graph.*

**73.** For the first three months of 2015, what was the total earnings of the airline?

**74.** For the first five months of 2015, what was the total earnings for the airline?

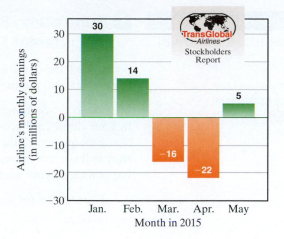

## To Think About

**75.** What number must be added to $-13$ to get 5?

**76.** What number must be added to $-18$ to get 10?

## Cumulative Review  *Perform the calculations indicated.*

**77.** **[0.2.3]** $\dfrac{15}{16} + \dfrac{1}{4}$

**78.** **[0.3.1]** $\dfrac{3}{7} \times \dfrac{14}{9}$

**79.** **[0.2.3]** $\dfrac{2}{15} - \dfrac{1}{20}$

**80.** **[0.3.2]** $2\dfrac{1}{2} \div 3\dfrac{2}{5}$

**81.** **[0.4.4]** $0.72 + 0.8$

**82.** **[0.4.4]** $1.63 - 0.98$

**83.** **[0.4.5]** $1.63 \times 0.7$

**84.** **[0.4.6]** $0.208 \div 0.8$

### Quick Quiz 1.1  *Add.*

**1.** $-18 + (-16)$

**2.** $-2.7 + 8.6 + (-5.4)$

**3.** $-\dfrac{5}{6} + \dfrac{7}{24}$

**4.** **Concept Check** Explain why when you add two negative numbers, you always obtain a negative number, but when you add one negative number and one positive number, you may obtain zero, a positive number, or a negative number.

## 1.2 Subtracting Real Numbers ▶

**Student Learning Objective**

**After studying this section, you will be able to:**

1 Subtract real numbers with the same or different signs. ▶

### 1 Subtracting Real Numbers with the Same or Different Signs ▶

So far we have developed the rules for adding real numbers. We can use these rules to subtract real numbers. Let's look at a bank account situation to see how.

**Situation 5:** *Subtract a Deposit and Add a Debit* You have a balance of 20 dollars in your debit card account. The bank calls you and says that a deposit of 5 dollars that belongs to another account was erroneously added to your account. They say they will correct the account balance to 15 dollars. You want to keep track of what's happening to your account.

*Your record for situation 5.*

$$20 - (+5) = 15$$

From your present balance *subtract* the *deposit* to get the new balance.

This equation shows what needs to be done to your account. The bank tells you that because the error happened in the past, they cannot "take it away." However, they can add to your account a debit of 5 dollars. Here is the equivalent addition.

$$20 + (-5) = 15$$

To your present balance *add* a *debit* to get the new balance.

We see that subtracting a positive 5 has the same effect as adding a negative 5.

> ### SUBTRACTION OF REAL NUMBERS
>
> To subtract real numbers, add the opposite of the second number (that is, the number you are subtracting) to the first.

The rule tells us to do three things when we subtract real numbers. First, change subtraction to addition. Second, replace the second number by its opposite. Third, add the two numbers using the rules for addition of real numbers.

**Example 1** Subtract. $6 - (-2)$

**Solution**

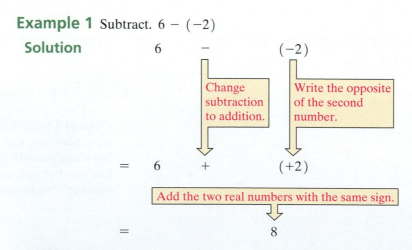

$$= 8$$

▶ **Student Practice 1** Subtract. $9 - (-3)$

**Example 2** Subtract. $-8 - (-6)$

**Solution**

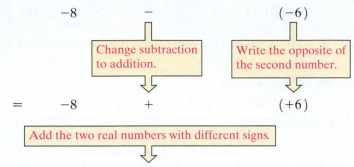

$$= -8 + (+6)$$

Add the two real numbers with different signs.

$$= -2$$    □

**Student Practice 2**  Subtract. $-12 - (-5)$

**Example 3** Subtract.

(a) $\dfrac{3}{7} - \dfrac{6}{7}$

(b) $-\dfrac{7}{18} - \left(-\dfrac{1}{9}\right)$

**Solution**

(a) $\dfrac{3}{7} - \dfrac{6}{7} = \dfrac{3}{7} + \left(-\dfrac{6}{7}\right)$    Change the subtraction problem to one of adding the opposite of the second number. We note that the problem has two fractions with the same denominator.

$$= -\dfrac{3}{7}$$    Add two numbers with different signs.

(b) $-\dfrac{7}{18} - \left(-\dfrac{1}{9}\right) = -\dfrac{7}{18} + \dfrac{1}{9}$    Change subtracting to adding the opposite.

$$= -\dfrac{7}{18} + \dfrac{2}{18}$$    Change $\frac{1}{9}$ to $\frac{2}{18}$ since LCD $= 18$.

$$= -\dfrac{5}{18}$$    Add two numbers with different signs.    □

**Student Practice 3**  Subtract.

(a) $\dfrac{5}{9} - \dfrac{7}{9}$

(b) $-\dfrac{5}{21} - \left(-\dfrac{3}{7}\right)$

**Calculator**

**More with Negative Numbers**

Subtract. Remember to use the $\boxed{+/-}$ key to change the sign of a number from $+$ to $-$ or from $-$ to $+$.

(a) $-18 - (-24)$

(b) $-6 + (-10) - (-15)$

**Ans:**

(a) $6$    (b) $-1$

**Example 4** Subtract. $-5.2 - (-5.2)$

**Solution**

$-5.2 - (-5.2) = -5.2 + 5.2$    Change the subtraction problem to one of adding the opposite of the second number.

$$= 0$$    Add two numbers with different signs.    □

**Student Practice 4**  Subtract. $-17.3 - (-17.3)$

Example 4 illustrates what is sometimes called the **additive inverse property.** When you add two real numbers that are opposites of each other, you will obtain zero. Examples of this are the following:

$$5 + (-5) = 0 \qquad -186 + 186 = 0 \qquad -\dfrac{1}{8} + \dfrac{1}{8} = 0.$$

**Example 5** Subtract.

(a) $-8 - 2$      (b) $23 - 28$      (c) $5 - (-3)$      (d) $\dfrac{1}{4} - 8$

**Solution** To subtract, we add the opposite of the second number to the first.

(a) $-8 - 2 = -8 + (-2) = -10$

In a similar fashion we have

(b) $23 - 28 = 23 + (-28) = -5$

(c) $5 - (-3) = 5 + 3 = 8$

(d) $\dfrac{1}{4} - 8 = \dfrac{1}{4} + (-8) = \dfrac{1}{4} + \left(-\dfrac{32}{4}\right) = -\dfrac{31}{4}$ or $-7\dfrac{3}{4}$

**Student Practice 5** Subtract.

(a) $-21 - 9$      (b) $17 - 36$      (c) $12 - (-15)$      (d) $\dfrac{3}{5} - 2$

**Example 6** A satellite is recording radioactive emissions from nuclear waste buried 3 miles below sea level. The satellite orbits Earth at 98 miles above sea level. How far is the satellite from the nuclear waste?

**Solution** We want to find the difference between $+98$ miles and $-3$ miles. This means we must subtract $-3$ from 98.

$$98 - (-3) = 98 + 3 = 101$$

The satellite is 101 miles from the nuclear waste.

**Student Practice 6** A helicopter is directly over a sunken vessel. The helicopter is 350 feet above sea level. The vessel lies 186 feet below sea level. How far is the helicopter from the sunken vessel?

 **STEPS TO SUCCESS** What Are the Absolute Essentials to Succeed in This Course?

Students who are successful in this course find there are six absolute essentials.
Here they are:

1. Attend every class session.
2. Read the textbook for every assigned section.
3. Take notes in class.
4. Do the assigned homework for every class.
5. Get help immediately when you need assistance.
6. Review what you are learning.

**Making it personal:** Which of the six suggestions above is the one you have the greatest trouble actually doing? Will you make a personal commitment to doing that one thing consistently for the next two weeks? You will be amazed at the results. ▼

*If you are in an online class or a nontraditional class:*

Students in an online class or a self-paced class find it really helps to spread the homework and the reading over five different days each week.

**Making it personal:** Will you try to do your homework for five days each week over the next two weeks? You will be amazed at the results. Write out a study plan for the next two weeks. ▼

Attend Class → Read Text → Take Notes → Do Homework → Get Help → Review

## 1.2 Exercises MyMathLab®

### Verbal and Writing Skills, Exercises 1 and 2

1. Explain in your own words how you would perform the necessary steps to find $-8 - (-3)$.

2. Explain in your own words how you would perform the necessary steps to find $-10 - (-15)$.

*Subtract by adding the opposite.*

3. $27 - 49$

4. $23 - 57$

5. $19 - 23$

6. $8 - 19$

7. $-14 - (-3)$

8. $-17 - (-13)$

9. $-52 - (-60)$

10. $-48 - (-80)$

11. $0 - (-5)$

12. $0 - (-7)$

13. $-18 - (-18)$

14. $-24 - (-24)$

15. $-17 - (-20)$

16. $-11 - (-19)$

17. $\frac{2}{5} - \frac{4}{5}$

18. $\frac{2}{9} - \frac{7}{9}$

19. $\frac{3}{4} - \left(-\frac{3}{5}\right)$

20. $-\frac{2}{3} - \frac{1}{4}$

21. $-\frac{3}{4} - \frac{5}{6}$

22. $-\frac{7}{10} - \frac{10}{15}$

23. $-0.6 - 0.3$

24. $-0.9 - 0.5$

25. $2.64 - (-1.83)$

26. $-0.03 - 0.06$

### Mixed Practice *Calculate.*

27. $\frac{3}{5} - 4$

28. $\frac{5}{6} - 3$

29. $-\frac{2}{3} - 4$

30. $-\frac{1}{6} - 5$

31. $34 - 87$

32. $19 - 76$

33. $-25 - 48$

34. $-74 - 11$

35. $2.3 - (-4.8)$

36. $8.4 - (-2.7)$

37. $8 - \left(-\frac{3}{4}\right)$

38. $\frac{2}{3} - (-6)$

39. $\frac{5}{6} - 7$

40. $9 - \frac{2}{3}$

41. $-\frac{2}{7} - \frac{4}{5}$

42. $-\frac{5}{6} - \frac{1}{5}$

43. $-135 - (-126.5)$

44. $-97.6 - (-146)$

45. $\frac{1}{5} - 6$

46. $\frac{2}{7} - (-3)$

47. $4.5 - (-1.56)$

48. $5.2 - (-3.88)$

49. $-3 - 2.047$

50. $-1.043 - 4$

51. Subtract $-9$ from $-2$.

52. Subtract $-12$ from $20$.

53. Subtract $13$ from $-35$.

**One Step Further** *Change each subtraction operation to "adding the opposite." Then combine the numbers.*

**54.** $9 + (-7) - 5$

**55.** $7 + (-6) - 3$

**56.** $-18 + 12 - (-6)$

**57.** $-13 + 12 - (-1)$

**58.** $-23 - (-12) - (-4) + 17$

**59.** $16 + (-20) - (-15) - 1$

**60.** $-8.3 - (-2.6) + 1.9$

**61.** $-7.8 - (-5.2) + 3.7$

## Applications

**62.** *Sea Rescue* A rescue helicopter is 300 feet above sea level. The captain has located an ailing submarine directly below it that is 126 feet below sea level. How far is the helicopter from the submarine?

**63.** *Sea Rescue* Suppose a rescue helicopter 600 feet above sea level descends 300 feet to search for debris or any other signs of a submarine in trouble. When the captain does not see anything, he ascends 200 feet and stays at that altitude while he tries to get a radio signal from the submarine. A few moments later he speaks to the captain and is informed that the submarine is 126 feet below sea level. How far is the helicopter from the submarine?

**64.** *Debit Card Account Balance* Yesterday Jackie had $156 in her debit card account. Today her account reads "balance −$37." Find the difference in these two amounts.

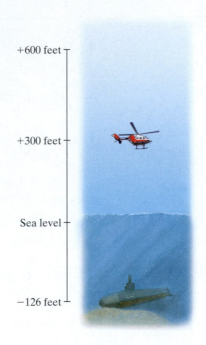

## Cumulative Review *In exercises 65–69, perform the operations indicated.*

**65.** **[1.1.4]** $-37 + 16$

**66.** **[1.1.3]** $-37 + (-14)$

**67.** **[1.1.3]** $-3 + (-6) + (-10)$

**68.** **[1.1.4]** *Temperature* On a winter morning, Alisa noticed that the outside temperature was $-5°F$. By the afternoon the temperature had risen $20°F$. What was the afternoon temperature?

**69.** **[0.3.1]** *Hiking* Sean and Khalid went hiking in the Blue Ridge Mountains. During their $8\frac{1}{3}$-mile hike, $\frac{4}{5}$ of the distance was covered with snow. How many miles were snow covered?

**Quick Quiz 1.2** *Subtract.*

**1.** $-8 - (-15)$

**2.** $-1.3 - 0.6$

**3.** $\frac{5}{8} - \left(-\frac{2}{7}\right)$

**4.** **Concept Check** Explain the different results that are possible when you start with a negative number and then subtract a negative number.

# 1.3 Multiplying and Dividing Real Numbers ▶

## 1 Multiplying Real Numbers ▶

We are familiar with the meaning of multiplication for positive numbers. For example, $5 \times 90 = 450$ might mean that you receive five weekly checks of 90 dollars each and you gain \$450. Let's look at a situation that corresponds to $5 \times (-90)$. What might that mean?

**Situation 6:** *Checking an Account Balance* You write a check for five weeks in a row to pay your weekly room rent of 90 dollars. If you do not have a checking account and instead have a debit card account, a similar result would occur if you were charged \$90 each week for five weeks. You want to know the total impact on your account balance.

*Your record for situation 6.*

| $(+5)$ | $\times$ | $(-90)$ | $=$ | $-450$ |
|---|---|---|---|---|
| The number of checks you have written | times | negative 90, the value of each check that was a debit to your account, | gives | negative 450 dollars, a net debit to your account |

Note that a multiplication symbol is not needed between the $(+5)$ and the $(-90)$ because the two sets of parentheses indicate multiplication. The multiplication $(5)(-90)$ is the same as repeated addition of five $(-90)$s. Note that 5 multiplied by $-90$ can be written as $5(-90)$ or $(5)(-90)$.

$$(-90) + (-90) + (-90) + (-90) + (-90) = -450$$

This example seems to show that a positive number multiplied by a negative number is negative.

What if the negative number is the one that is written first? If $(5)(-90) = -450$ then $(-90)(5) = -450$ by the commutative property of multiplication. This is an example showing that *when two numbers with different signs* (one positive, one negative) *are multiplied, the result is negative.*

But what if both numbers are negative? Consider the following situation.

**Situation 7:** *Renting a Room* Last year at college you rented a room at 90 dollars per week for 36 weeks, which included two semesters and summer school. This year you will not attend the summer session, so you will be renting the room for only 30 weeks. Thus the number of weekly rental checks will be six less than last year. You are making out your budget for this year. You want to know the financial impact of renting the room for six fewer weeks.

*Your record for situation 7.*

| $(-6)$ | $\times$ | $(-90)$ | $=$ | $540$ |
|---|---|---|---|---|
| The difference in the number of checks this year compared to last year is $-6$, which is negative to show a decrease, | times | $-90$, the value of each check paid out, | gives | $+540$ dollars. The product is positive, because your financial situation will be 540 dollars better this year. |

You could check that the answer is positive by calculating the total rental expenses.

|  | Dollars in rent last year | $(36)(90) =$ | $3240$ |
|---|---|---|---|
| (subtract) | Dollars in rent this year | $-(30)(90) =$ | $-2700$ |
|  | Extra dollars available this year | $=$ | $+540$ |

This agrees with our previous answer: $(-6)(-90) = +540$.

Note that $-6$ times $-90$ can be written as $-6(-90)$ or $(-6)(-90)$.

**Student Learning Objectives**

**After studying this section, you will be able to:**

1 Multiply real numbers. ▶

2 Use the multiplication properties for real numbers. ▶

3 Divide real numbers. ▶

In this situation it seems reasonable that a negative number times a negative number yields a positive answer. We already know from arithmetic that a positive number times a positive number yields a positive answer. Thus we might see the general rule that *when two numbers with the same sign* (both positive or both negative) *are multiplied, the result is positive.*

We will now state our rule.

### MULTIPLICATION OF REAL NUMBERS

To multiply two real numbers with **the same sign,** multiply the absolute values.
The sign of the result is **positive.**
To multiply two real numbers with **different signs,** multiply the absolute values.
The sign of the result is **negative.**

**Example 1** Multiply.

(a) $(3)(6)$     (b) $\left(-\dfrac{5}{7}\right)\left(-\dfrac{2}{9}\right)$     (c) $-4(8)$     (d) $\left(\dfrac{2}{7}\right)(-3)$

**Solution**

(a) $(3)(6) = 18$

> When multiplying two numbers with the same sign, the result is a positive number.

(b) $\left(-\dfrac{5}{7}\right)\left(-\dfrac{2}{9}\right) = \dfrac{10}{63}$

(c) $-4(8) = -32$

> When multiplying two numbers with different signs, the result is a negative number.

(d) $\left(\dfrac{2}{7}\right)(-3) = \left(\dfrac{2}{7}\right)\left(-\dfrac{3}{1}\right) = -\dfrac{6}{7}$

**Student Practice 1**    Multiply.

(a) $(-6)(-2)$     (b) $(7)(9)$     (c) $\left(-\dfrac{3}{5}\right)\left(\dfrac{2}{7}\right)$     (d) $\left(\dfrac{5}{6}\right)(-7)$

To multiply more than two numbers, multiply two numbers at a time.

**Example 2** Multiply. $(-4)(-3)(-2)$

**Solution**

$(-4)(-3)(-2) = (+12)(-2)$     We begin by multiplying the first two numbers, $(-4)$ and $(-3)$. The signs are the same. The answer is positive 12.

$\qquad\qquad\qquad = -24$     Now we multiply $(+12)$ and $(-2)$. The signs are different. The answer is negative 24.

**Student Practice 2**    Multiply. $(-5)(-2)(-6)$

**Example 3** Multiply.

**(a)** $-3(-1.5)$     **(b)** $\left(-\dfrac{1}{2}\right)(-1)(-4)$     **(c)** $-2(-2)(-2)(-2)$

**Solution**  Multiply two numbers at a time. See if you find a pattern.

**(a)** $-3(-1.5) = 4.5$          Be sure to place the decimal point in your answer.

**(b)** $\left(-\dfrac{1}{2}\right)(-1)(-4) = +\dfrac{1}{2}(-4) = -2$

**(c)** $-2(-2)(-2)(-2) = +4(-2)(-2) = -8(-2) = +16$ or $16$

What kind of answer would we obtain if we multiplied five negative numbers? If you guessed "negative," you probably see the pattern.    □

**Student Practice 3**  Determine the sign of the product. Then multiply to check.

**(a)** $-2(-3)$                    **(b)** $(-1)(-3)(-2)$

**(c)** $-4\left(-\dfrac{1}{4}\right)(-2)(-6)$

---

When you multiply two or more nonzero real numbers:

1. The result is always **positive** if there is an **even** number of negative signs.
2. The result is always **negative** if there is an **odd** number of negative signs.

---

## 2 Using the Multiplication Properties for Real Numbers

For convenience, we will list the properties of multiplication.

1. *Multiplication is commutative.*
   This property states that if two real numbers are multiplied, the order of the numbers does not affect the result. The result is the same no matter which number is written first.

$$(5)(7) = (7)(5) = 35, \qquad \left(\frac{1}{3}\right)\left(\frac{2}{7}\right) = \left(\frac{2}{7}\right)\left(\frac{1}{3}\right) = \frac{2}{21}$$

2. *Multiplication of any real number by zero will result in zero.*

$$(5)(0) = 0, \qquad (-5)(0) = 0, \qquad (0)\left(\frac{3}{8}\right) = 0, \qquad (0)(0) = 0$$

3. *Multiplication of any real number by 1 will result in that same number.*

$$(5)(1) = 5, \qquad (1)(-7) = -7, \qquad (1)\left(-\frac{5}{3}\right) = -\frac{5}{3}$$

4. *Multiplication is associative.*
   This property states that if three real numbers are multiplied, it does not matter which two numbers are grouped by parentheses and multiplied first.

$$2 \times (3 \times 4) = (2 \times 3) \times 4$$   First multiply the numbers in parentheses. Then multiply the remaining numbers.

$$2 \times (12) = (6) \times 4$$   The results are the same no matter which numbers are grouped and multiplied first.

$$24 = 24$$

### 3 Dividing Real Numbers

What about division? Any division problem can be rewritten as a multiplication problem.

We know that $20 \div 4 = 5$ because $4(5) = 20$.
Similarly, $-20 \div (-4) = 5$ because $-4(5) = -20$.

In both division problems the answer is positive 5. Thus we see that *when you divide two numbers with the same sign* (both positive or both negative), *the answer is positive.* What if the signs are different?

We know that $-20 \div 4 = -5$ because $4(-5) = -20$.
Similarly, $20 \div (-4) = -5$ because $-4(-5) = 20$.

In these two problems the answer is negative 5. So we have reasonable evidence to see that *when you divide two numbers with different signs* (one positive and one negative), *the answer is negative.*

We will now state our rule for division.

---

**DIVISION OF REAL NUMBERS**

To divide two real numbers with **the same sign,** divide the absolute values. The sign of the result is **positive.**

To divide two real numbers with **different signs,** divide the absolute values. The sign of the result is **negative.**

---

**Example 4** Divide.

(a) $12 \div 4$  (b) $(-25) \div (-5)$  (c) $\dfrac{-36}{18}$  (d) $\dfrac{42}{-7}$

**Solution**

(a) $12 \div 4 = 3$ ← When dividing two numbers with the same sign, the result is a positive number.

(b) $(-25) \div (-5) = 5$ ←

(c) $\dfrac{-36}{18} = -2$ ← When dividing two numbers with different signs, the result is a negative number.

(d) $\dfrac{42}{-7} = -6$ ←

**Student Practice 4** Divide.

(a) $-36 \div (-2)$  (b) $49 \div 7$  (c) $\dfrac{50}{-10}$  (d) $\dfrac{-39}{13}$

**Example 5** Divide. (a) $-36 \div 0.12$  (b) $-2.4 \div (-0.6)$

**Solution**

(a) $-36 \div 0.12$  Look at the problem to determine the sign. When dividing two numbers with different signs, the result will be a negative number.

We then divide the absolute values.

$$0.12_\wedge \overline{)36.00_\wedge} \begin{array}{r} 3\ 00. \\ \hline 36 \\ \hline 00 \end{array}$$

Thus $-36 \div 0.12 = -300$. The answer is a negative number.

**(b)** $-2.4 \div (-0.6)$  Look at the problem to determine the sign. When dividing two numbers with the same sign, the result will be positive.

We then divide the absolute values.

$$0.6_\wedge \overline{)2.4_\wedge} \quad \begin{array}{c} 4. \\ \underline{2\,4} \end{array}$$

Thus $-2.4 \div (-0.6) = 4$. The answer is a positive number.  □

 **Student Practice 5**  Divide.

**(a)** $-12.6 \div (-1.8)$ **(b)** $0.45 \div (-0.9)$

Note that the rules for multiplication and division are the same. When you **multiply** or **divide** two numbers with the **same** sign, you obtain a **positive** number. When you **multiply** or **divide** two numbers with **different** signs, you obtain a **negative** number.

**Example 6**  Divide. $-\dfrac{12}{5} \div \dfrac{2}{3}$

**Solution**

$$= \left(-\frac{12}{5}\right)\left(\frac{3}{2}\right) \quad \text{Divide two fractions. We invert the second fraction and multiply by the first fraction.}$$

$$= \left(-\frac{\overset{6}{\cancel{12}}}{5}\right)\left(\frac{3}{\underset{1}{\cancel{2}}}\right)$$

$$= -\frac{18}{5} \quad \text{or} \quad -3\frac{3}{5} \quad \text{The answer is negative since the two numbers divided have different signs.} \quad □$$

 **Student Practice 6**  Divide.

$$-\frac{5}{16} \div \left(-\frac{10}{13}\right)$$

Note that division can be indicated by the symbol $\div$ or by the fraction bar —. $\frac{2}{3}$ means $2 \div 3$.

**Example 7**  Divide. **(a)** $\dfrac{\frac{7}{8}}{-21}$ **(b)** $\dfrac{-\frac{2}{3}}{-\frac{7}{13}}$

**Solution**

**(a)** $\dfrac{\frac{7}{8}}{-21}$

$$= \frac{7}{8} \div \left(-\frac{21}{1}\right) \quad \text{Change } -21 \text{ to a fraction. } -21 = -\frac{21}{1}$$

$$= \frac{\overset{1}{\cancel{7}}}{8}\left(-\frac{1}{\underset{3}{\cancel{21}}}\right) \quad \text{Change the division to multiplication. Divide out common factor.}$$

$$= -\frac{1}{24} \quad \text{Simplify.}$$

*Continued on next page*

(b) $\dfrac{-\dfrac{2}{3}}{-\dfrac{7}{13}} = -\dfrac{2}{3} \div \left(-\dfrac{7}{13}\right) = -\dfrac{2}{3}\left(-\dfrac{13}{7}\right) = \dfrac{26}{21}$   or   $1\dfrac{5}{21}$ $\square$

**Student Practice 7**   Divide.

(a) $\dfrac{-12}{-\dfrac{4}{5}}$

(b) $\dfrac{-\dfrac{2}{9}}{\dfrac{8}{13}}$

1. *Division of 0 by any nonzero real number gives 0 as a result.*

$$0 \div 5 = 0, \qquad 0 \div \dfrac{2}{3} = 0, \qquad \dfrac{0}{5.6} = 0, \qquad \dfrac{0}{1000} = 0$$

You can divide zero by 5, $\frac{2}{3}$, 5.6, 1000, or any number (except 0).

2. *Division of any real number by 0 is* **undefined.**

$$7 \div 0 \qquad \dfrac{64}{0}$$
$$\uparrow \qquad\quad \uparrow$$

Neither of these operations is possible. **Division by zero is undefined.**

You may be wondering why division by zero is undefined. Let us think about it for a minute. We said that $7 \div 0$ is undefined. Suppose there were an answer. Let us call the answer $a$. So we assume for a minute that $7 \div 0 = a$. Then it would have to follow that $7 = 0(a)$. But this is impossible. Zero times any number is zero. So we see that if there were such a number, it would contradict known mathematical facts. Therefore there is no number $a$ such that $7 \div 0 = a$. Thus we conclude that division by zero is undefined.

When combining two numbers, it is important to be sure you know which rule applies. Think about the concepts in the following chart. See if you agree with each example.

| Operation | Two Real Numbers with the Same Sign | Two Real Numbers with Different Signs |
|---|---|---|
| Addition | Result may be positive or negative. $\quad 9 + 2 = 11$ $\quad -5 + (-6) = -11$ | Result may be positive or negative. $\quad -3 + 7 = 4$ $\quad 4 + (-12) = -8$ |
| Subtraction | Result may be positive or negative. $\quad 15 - 6 = 15 + (-6) = 9$ $\quad -12 - (-3) = -12 + 3 = -9$ | Result may be positive or negative. $\quad -12 - 3 = -12 + (-3) = -15$ $\quad 5 - (-6) = 5 + 6 = 11$ |
| Multiplication | Result is always positive. $\quad 9(3) = 27$ $\quad -8(-5) = 40$ | Result is always negative. $\quad -6(12) = -72$ $\quad 8(-3) = -24$ |
| Division | Result is always positive. $\quad 150 \div 6 = 25$ $\quad -72 \div (-2) = 36$ | Result is always negative. $\quad -60 \div 10 = -6$ $\quad 30 \div (-6) = -5$ |

**Example 8** The Hamilton-Wenham Generals recently analyzed the 48 plays their team made while in possession of the football during their last game. The bar graph illustrates the number of plays made in each category. The team statistician prepared the following chart indicating the average number of yards gained or lost during each type of play.

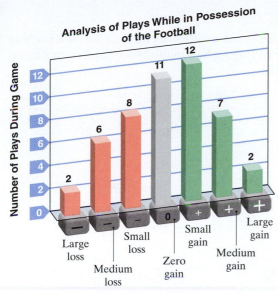

| Type of Play | Average Yards Gained or Lost for Play |
|---|---|
| Large gain | +25 |
| Medium gain | +15 |
| Small gain | +5 |
| Zero gain | 0 |
| Small loss | −5 |
| Medium loss | −10 |
| Large loss | −15 |

(a) How many yards were lost by the Generals in the plays that were considered small losses?

(b) How many yards were gained by the Generals in the plays that were considered small gains?

(c) If the total yards gained in small gains were combined with the total yards lost in small losses, what would be the result?

**Solution**

(a) We multiply the number of small losses by the average number of total yards lost on each small loss:

$$8(-5) = -40.$$

The team lost approximately 40 yards with plays that were considered small losses.

(b) We multiply the number of small gains by the average number of yards gained on each small gain:

$$12(5) = 60.$$

The team gained approximately 60 yards with plays that were considered small gains.

(c) We combine the results for (a) and (b):

$$-40 + 60 = 20.$$

A total of 20 yards was gained during the plays that were small losses and small gains.  ☐

**Student Practice 8** Using the information provided in Example 8, answer the following:

(a) How many yards were lost by the Generals in the plays that were considered medium losses?

(b) How many yards were gained by the Generals in the plays that were considered medium gains?

(c) If the total yards gained in medium gains were combined with the total yards lost in medium losses, what would be the result?

## 1.3 Exercises  MyMathLab®

### Verbal and Writing Skills, Exercises 1 and 2

1. Explain in your own words the rule for determining the correct sign when multiplying two real numbers.

2. Explain in your own words the rule for determining the correct sign when multiplying three or more real numbers.

*Multiply. Be sure to write your answer in the simplest form.*

**3.** $8(-5)$

**4.** $9(-9)$

**5.** $0(-12)$

**6.** $0(136)$

**7.** $14(3.5)$

**8.** $7.5(8)$

**9.** $(-1.32)(-0.2)$

**10.** $(-2.3)(-0.11)$

**11.** $1.8(-2.5)$

**12.** $(3.4)(-2.2)$

**13.** $\left(\frac{3}{8}\right)(-4)$

**14.** $(5)\left(-\frac{7}{10}\right)$

**15.** $\left(-\frac{3}{5}\right)\left(-\frac{15}{11}\right)$

**16.** $\left(-\frac{4}{9}\right)\left(-\frac{3}{5}\right)$

**17.** $\left(\frac{12}{13}\right)\left(\frac{-5}{24}\right)$

**18.** $\left(\frac{14}{17}\right)\left(-\frac{3}{28}\right)$

*Divide.*

**19.** $0 \div (-9)$

**20.** $0 \div (-13)$

**21.** $-48 \div (-8)$

**22.** $-64 \div 8$

**23.** $-120 \div (-8)$

**24.** $-180 \div (-4)$

**25.** $156 \div (-13)$

**26.** $-0.6 \div 0.3$

**27.** $-9.1 \div 0.07$

**28.** $8.1 \div (-0.03)$

**29.** $0.54 \div (-0.9)$

**30.** $-7.2 \div 8$

**31.** $-6.3 \div 7$

**32.** $\frac{2}{7} \div \left(-\frac{3}{5}\right)$

**33.** $\left(-\frac{1}{5}\right) \div \left(\frac{2}{3}\right)$

**34.** $\left(-\frac{5}{6}\right) \div \left(-\frac{7}{18}\right)$

**35.** $-\frac{5}{7} \div \left(-\frac{3}{28}\right)$

**36.** $\left(-\frac{4}{9}\right) \div \left(-\frac{8}{15}\right)$

**37.** $\left(-\frac{7}{12}\right) \div \left(-\frac{5}{6}\right)$

**38.** $\dfrac{12}{-\frac{2}{5}}$

**39.** $\dfrac{-6}{-\frac{3}{7}}$

**40.** $\dfrac{-\frac{3}{8}}{-\frac{2}{3}}$

**41.** $\dfrac{\frac{-2}{3}}{\frac{8}{15}}$

**42.** $\dfrac{\frac{9}{2}}{-3}$

**43.** $\dfrac{\frac{8}{3}}{-4}$

*Multiply. You may want to determine the sign of the product before you multiply.*

**44.** $-6(2)(-3)(4)$

**45.** $-1(-2)(-3)(4)$

88

**46.** $-3(2)(-1)(-2)(5)$

**47.** $-2(4)(3)(-1)(-3)$

**48.** $-6(2)(-3)(0)(-9)$

**49.** $-3(0)\left(\dfrac{1}{3}\right)(-4)(2)$

**50.** $-2(0.14)(-3)(0.5)$

**51.** $25(-0.04)(-0.3)(-1)$

**52.** $\left(\dfrac{3}{7}\right)\left(-\dfrac{2}{3}\right)\left(-\dfrac{5}{3}\right)$

**53.** $\left(-\dfrac{4}{5}\right)\left(-\dfrac{6}{7}\right)\left(-\dfrac{1}{3}\right)$

**54.** $\left(-\dfrac{2}{3}\right)\left(-\dfrac{1}{4}\right)\left(\dfrac{3}{5}\right)\left(-\dfrac{2}{7}\right)$

**55.** $\left(-\dfrac{3}{4}\right)\left(-\dfrac{7}{15}\right)\left(-\dfrac{8}{21}\right)\left(-\dfrac{5}{9}\right)$

**Mixed Practice**  *Take a minute to review the chart before Example 8. Be sure that you can remember the sign rules for each operation. Then do exercises 56–65. Perform the calculations indicated.*

**56.** $-5 - (-2)$

**57.** $-36 \div (-4)$

**58.** $-4(-8)$

**59.** $12 + (-8)$

**60.** $(-30) \div 5$

**61.** $8 - (-9)$

**62.** $-6 + (-3)$

**63.** $6(-12)$

**64.** $18 \div (-18)$

**65.** $-37 \div 37$

## Applications

**66.** *Stock Trading*  During one day in November 2015, the value of one share of Delta Air Lines stock opened at \$44.03. At the end of the day, it closed at \$42.80. If you owned 90 shares, what was your profit or loss that day?

**67.** *Equal Contributions*  Ed, Ned, Ted, and Fred went camping. They each contributed an equal share of money toward food. Fred did the shopping. When he returned from the store, he had \$17.60 left. How much money did Fred give back to each person?

**68.** *Student Loans*  Ramon will pay \$6480 on his student loan over the next three years. If \$180 is automatically deducted from his bank account each month to pay the loan off, how much does he still owe after one year?

**69.** *Car Payments*  Muriel will pay the Volkswagen dealer a total of \$14,136 to be paid in 60 equal monthly installments. What is her monthly bill?

*Football*  *The Beverly Panthers recently analyzed the 37 plays their team made while in possession of the football during their last game. The team statistician prepared the following chart indicating the number of plays in each category and the average number of yards gained or lost during each type of play. Use this chart to answer exercises 70–75.*

| Type of Play | Number of Plays | Average Yards Gained or Lost per Play |
|---|---|---|
| Large gain | 1 | +25 |
| Medium gain | 6 | +15 |
| Small gain | 4 | +5 |
| Zero gain | 5 | 0 |
| Small loss | 10 | −5 |
| Medium loss | 7 | −10 |
| Large loss | 4 | −15 |

**70.** How many yards were lost by the Panthers in the plays that were considered small losses?

**71.** How many yards were gained by the Panthers in the plays that were considered small gains?

**72.** If the total yards gained in small gains were combined with the total yards lost in small losses, what would be the result?

**73.** How many yards were lost by the Panthers in the plays that were considered medium losses?

**74.** How many yards were gained by the Panthers in the plays that were considered medium gains?

**75.** If the total yards gained in medium gains were combined with the total yards lost in medium losses, what would be the result?

## Cumulative Review

**76. [1.1.5]** $-17.4 + 8.31 + 2.40$

**77. [1.1.5]** $-\frac{3}{4} + \left(-\frac{2}{3}\right) + \left(-\frac{5}{12}\right)$

**78. [1.2.1]** $-47 - (-32)$

**79. [1.2.1]** $-37 - 51$

---

**Quick Quiz 1.3** *Perform the operations indicated.*

**1.** $\left(-\frac{3}{8}\right)(5)$

**2.** $-4(3)(-5)(-2)$

**3.** $-2.4 \div (-0.6)$

**4. Concept Check** Explain how you can determine the sign of the answer if you multiply several negative numbers.

---

 **STEPS TO SUCCESS** Getting the Greatest Value from Your Homework

***Read the textbook first before doing the homework.*** Take some time to read the text and study the examples. Try working out the Student Practice problems. You will be amazed at the amount of understanding you will obtain by studying the book before jumping into the homework exercises.

***Take your time.*** Read the directions carefully. Be sure you understand what is being asked. Check your answers with those given in the back of the textbook. If your answer is incorrect, study similar examples in the text. Then redo the problem, watching for errors.

***Make a schedule.*** You will need to allow two hours outside of class for each hour of actual class time. Make a weekly schedule of the times you have class. Now write down the times each day you will devote to doing math homework. Then write down the times you will spend doing homework for your other classes. If you have a job be sure to write down all your work hours.

**Making it personal:** Write down your own schedule of class, work, and study time. ▼

| Sunday | Monday | Tuesday | Wednesday | Thursday | Friday | Saturday |
|--------|--------|---------|-----------|----------|--------|----------|
|  |  |  |  |  |  |  |
|  |  |  |  |  |  |  |
|  |  |  |  |  |  |  |
|  |  |  |  |  |  |  |
|  |  |  |  |  |  |  |
|  |  |  |  |  |  |  |
|  |  |  |  |  |  |  |

## 1.4 Exponents ▶

### 1 Writing Numbers in Exponent Form ▶

In mathematics, we use exponents as a way to abbreviate repeated multiplication.

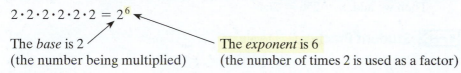

There are two parts to exponent notation: (1) the **base** and (2) the **exponent.** The **base** tells you what number is being multiplied and the **exponent** tells you how many times this number is used as a factor. (A *factor,* you recall, is a number being multiplied.)

$$2\cdot2\cdot2\cdot2\cdot2\cdot2 = 2^6$$

The *base* is 2
(the number being multiplied)

The *exponent* is 6
(the number of times 2 is used as a factor)

If the base is a *positive* real number, the exponent appears to the right and slightly above the level of the number as in, for example, $5^6$ and $8^3$. If the base is a *negative* real number, then parentheses are used around the number and the exponent appears outside the parentheses. For example, $(-2)(-2)(-2) = (-2)^3$.

In algebra, if we do not know the value of a number, we use a letter to represent the unknown number. We call the letter a **variable.** This is quite useful in the case of exponents. Suppose we do not know the value of a number, but we know the number is multiplied by itself several times. We can represent this with a variable base and a whole-number exponent. For example, when we have an unknown number, represented by the variable $x$, and this number occurs as a factor four times, we have

$$(x)(x)(x)(x) = x^4.$$

Likewise, if an unknown number, represented by the variable $w$, occurs as a factor five times, we have

$$(w)(w)(w)(w)(w) = w^5.$$

**Example 1** Write in exponent form.
**(a)** $9(9)(9)$  **(b)** $13(13)(13)(13)$  **(c)** $-7(-7)(-7)(-7)(-7)$
**(d)** $-4(-4)(-4)(-4)(-4)(-4)$  **(e)** $(x)(x)$  **(f)** $(y)(y)(y)$

**Solution**
**(a)** $9(9)(9) = 9^3$  **(b)** $13(13)(13)(13) = 13^4$
**(c)** The $-7$ is used as a factor five times. The answer must contain parentheses.
   Thus $-7(-7)(-7)(-7)(-7) = (-7)^5$.
**(d)** $-4(-4)(-4)(-4)(-4)(-4) = (-4)^6$
**(e)** $(x)(x) = x^2$  **(f)** $(y)(y)(y) = y^3$  □

**▷ Student Practice 1** Write in exponent form.
**(a)** $6(6)(6)(6)$  **(b)** $-2(-2)(-2)(-2)(-2)$
**(c)** $108(108)(108)$  **(d)** $-11(-11)(-11)(-11)(-11)(-11)$
**(e)** $(w)(w)(w)$  **(f)** $(z)(z)(z)(z)$

If the base has an exponent of 2, we say the base is **squared.**

If the base has an exponent of 3, we say the base is **cubed.**

If the base has an exponent greater than 3, we say the base is raised **to the (exponent)-th power.**

$x^2$ is read "$x$ squared."

$y^3$ is read "$y$ cubed."

$3^6$ is read "three to the sixth power" or simply "three to the sixth."

**Student Learning Objectives**

After studying this section, you will be able to:

1 Write numbers in exponent form. ▶

2 Evaluate numerical expressions that contain exponents. ▶

## 2 Evaluating Numerical Expressions That Contain Exponents

**Example 2** Evaluate.

(a) $2^5$      (b) $2^3 + 4^4$

**Solution**

(a) $2^5 = (2)(2)(2)(2)(2) = 32$

(b) First we evaluate each power.

$2^3 = 8$     $4^4 = 256$

Then we add. $8 + 256 = 264$      □

▶ **Student Practice 2**   Evaluate.

(a) $3^5$      (b) $2^2 + 3^3$

---

### Calculator

**Exponents**

You can use a calculator to evaluate $3^5$. Press the following keys:

| 3 | $y^x$ | 5 | = |

The display should read

| 243 |

Try the following.

(a) $4^6$     (b) $(0.2)^5$

(c) $18^6$     (d) $3^{12}$

**Ans:**

(a) 4096     (b) 0.00032

(c) 34,012,224    (d) 531,441

The steps needed to raise a number to a power are slightly different on some calculators.

---

If the base is negative, be especially careful in determining the sign. Notice the following:

$$(-3)^2 = (-3)(-3) = +9 \qquad (-3)^3 = (-3)(-3)(-3) = -27$$

From Section 1.3 we know that when you multiply two or more real numbers, first you multiply their absolute values.

- The result is positive if there is an even number of negative signs.
- The result is negative if there is an odd number of negative signs.

**SIGN RULE FOR EXPONENTS**

Suppose that a number is written in exponent form and the base is negative. The result is **positive** if the exponent is **even**. The result is **negative** if the exponent is **odd**.

Be careful how you read expressions with exponents and negative signs.

$$(-3)^4 \text{ means } (-3)(-3)(-3)(-3) \text{ or } +81.$$
$$-3^4 \text{ means } -(3)(3)(3)(3) \text{ or } -81.$$

**Example 3** Evaluate.

(a) $(-2)^3$     (b) $(-4)^6$     (c) $-3^6$     (d) $-(5^4)$

**Solution**

(a) $(-2)^3 = -8$     The answer is negative since the base is negative and the exponent 3 is odd.

(b) $(-4)^6 = +4096$     The answer is positive since the exponent 6 is even.

(c) $-3^6 = -729$     The negative sign is not contained in parentheses. Thus we find 3 raised to the sixth power and then take the negative of that value.

(d) $-(5^4) = -625$     The negative sign is outside the parentheses.     □

▶ **Student Practice 3**   Evaluate.

(a) $(-3)^3$     (b) $(-2)^6$     (c) $-2^4$     (d) $-(6^3)$

$\mathbb{M}_\mathbb{C}$ **Example 4**  Evaluate.

(a) $\left(\dfrac{1}{2}\right)^4$ (b) $(0.2)^4$ (c) $\left(\dfrac{2}{5}\right)^3$

(d) $(3)^3(2)^5$ (e) $2^3 - 3^4$

**Solution**

(a) $\left(\dfrac{1}{2}\right)^4 = \left(\dfrac{1}{2}\right)\left(\dfrac{1}{2}\right)\left(\dfrac{1}{2}\right)\left(\dfrac{1}{2}\right) = \dfrac{1}{16}$

(b) $(0.2)^4 = (0.2)(0.2)(0.2)(0.2) = 0.0016$

(c) $\left(\dfrac{2}{5}\right)^3 = \left(\dfrac{2}{5}\right)\left(\dfrac{2}{5}\right)\left(\dfrac{2}{5}\right) = \dfrac{8}{125}$

(d) First we evaluate each power.
$3^3 = 27 \qquad 2^5 = 32$

Then we multiply. $(27)(32) = 864$

(e) $2^3 - 3^4 = 8 - 81 = -73$

**Student Practice 4**  Evaluate.

(a) $\left(\dfrac{1}{3}\right)^3$

(b) $(0.3)^4$

(c) $\left(\dfrac{3}{2}\right)^4$

(d) $(3)^4(4)^2$

(e) $4^2 - 2^4$

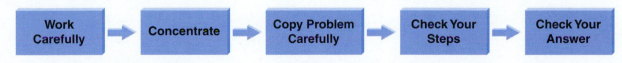

## 👣 STEPS TO SUCCESS Helping Your Accuracy

It is easy to make a mistake. But here are five ways to cut down on errors. Look over each one and think about how each suggestion can help you.

1. Work carefully and take your time. Do not rush through a problem just to get it done.
2. Concentrate on the problem. Sometimes your mind starts to wander. Then you get careless and will likely make a mistake.
3. Check your problem. Be sure you copied it correctly from the book.
4. Check your computations from step to step. Did you do each step correctly?
5. Check your final answer. Does it work? Is it reasonable?

**Making it personal:** Look over these five suggestions. Which one do you think will help you the most? Write down how you can use this suggestion to help you personally as you try to improve your accuracy. ▼

Work Carefully → Concentrate → Copy Problem Carefully → Check Your Steps → Check Your Answer

## 1.4 Exercises   MyMathLab®

**Verbal and Writing Skills, Exercises 1–6**

1. Explain in your own words how to evaluate $3^4$.

2. Explain in your own words how to evaluate $9^2$.

3. Explain how you would determine whether $(-5)^3$ is negative or positive.

4. Explain how you would determine whether $(-2)^5$ is negative or positive.

5. Explain the difference between $(-2)^4$ and $-2^4$. What answers do you obtain when you evaluate the expressions?

6. Explain the difference between $(-3)^4$ and $-3^4$. What answers do you obtain when you evaluate the expressions?

*Write in exponent form.*

7. $(6)(6)(6)(6)(6)$

8. $(8)(8)(8)(8)(8)(8)$

9. $(w)(w)$

10. $(x)(x)(x)(x)$

11. $(p)(p)(p)(p)$

12. $(r)(r)(r)(r)(r)(r)(r)$

13. $(3q)(3q)(3q)$

14. $(2w)(2w)(2w)(2w)(2w)$

*Evaluate.*

15. $3^3$

16. $9^2$

17. $3^4$

18. $7^3$

19. $6^3$

20. $12^2$

21. $(-3)^3$

22. $(-2)^3$

23. $(-4)^2$

24. $(-5)^4$

25. $-5^2$

26. $-4^2$

27. $\left(\dfrac{1}{4}\right)^2$

28. $\left(\dfrac{3}{4}\right)^2$

29. $\left(\dfrac{2}{5}\right)^3$

30. $\left(\dfrac{1}{3}\right)^3$

31. $(2.1)^2$

32. $(1.5)^2$

33. $(0.2)^5$

34. $(0.7)^3$

35. $(-16)^2$

36. $(-7)^4$

37. $-16^2$

38. $-7^4$

94

*Evaluate.*

**39.** $5^3 + 6^2$

**40.** $6^2 + 2^3$

**41.** $10^2 - 11^2$

**42.** $5^3 - 2^2$

**43.** $(-4)^2 - (12)^2$

**44.** $14^2 - (-6)^2$

**45.** $2^5 - (-3)^2$

**46.** $7^2 - (-2)^4$

**47.** $(-5)^3(-2)^3$

## Cumulative Review  *Evaluate.*

**48.** **[1.1.5]** $(-11) + (-13) + 6 + (-9) + 8$

**49.** **[1.3.3]** $\dfrac{3}{4} \div \left(-\dfrac{9}{20}\right)$

**50.** **[1.2.1]** $-17 - (-9)$

**51.** **[1.3.1]** $(-2.1)(-1.2)$

**52.** **[0.5.3]** Amanda decided to invest her summer job earnings of $1600. At the end of the year she earned 6% on her investment. How much money did Amanda have at the end of the year?

Quick Quiz 1.4  *Evaluate.*

**1.** $(-4)^4$

**2.** $(1.8)^2$

**3.** $\left(\dfrac{3}{4}\right)^3$

**4.** **Concept Check** Explain the difference between $(-2)^6$ and $-2^6$. How do you decide if the answers are positive or negative?

# 1.5 The Order of Operations ▶

**Student Learning Objective**

After studying this section, you will be able to:

1 Use the order of operations to simplify numerical expressions. ▶

## 1 Using the Order of Operations to Simplify Numerical Expressions ▶

It is important to know *when* to do certain operations as well as how to do them. For example, to simplify the expression $2 - 4 \cdot 3$, should we subtract first or multiply first?

Also remember that multiplication can be written several ways. Thus $4 \cdot 3$, $4 \times 3$, $4(3)$, and $(4)(3)$ all indicate that we are multiplying 4 times 3.

The following list will assist you. It tells which operations to do first: the correct **order of operations.** You might think of it as a *list of priorities*.

---

**ORDER OF OPERATIONS FOR NUMBERS**

Follow this order of operations:

Do first     1. Do all operations inside parentheses.

           2. Raise numbers to a power.

           3. Multiply and divide numbers from left to right.

Do last     4. Add and subtract numbers from left to right.

---

Let's return to the problem $2 - 4 \cdot 3$. There are no parentheses or numbers raised to a power, so the first thing we do is multiply. Then we subtract since this comes last on our list.

$$2 - 4 \cdot 3 = 2 - 12 \quad \text{Follow the order of operations by first multiplying } 4 \cdot 3 = 12.$$
$$= -10 \quad \text{Combine } 2 - 12 = -10.$$

**Example 1** Evaluate. $8 \div 2 \cdot 3 + 4^2$

**Solution**

$$8 \div 2 \cdot 3 + 4^2 = 8 \div 2 \cdot 3 + 16 \quad \text{Evaluate } 4^2 = 16 \text{ because the highest priority in this problem is raising to a power.}$$
$$= 4 \cdot 3 + 16 \quad \text{Next multiply and divide from left to right. So } 8 \div 2 = 4$$
$$= 12 + 16 \quad 4 \cdot 3 = 12.$$
$$= 28 \quad \text{Finally, add.} \quad \square$$

▶ **Student Practice 1** Evaluate. $25 \div 5 \cdot 6 + 2^3$

*Note:* Multiplication and division have equal priority. We do not do multiplication first. Rather, we work from left to right, doing any multiplication or division that we encounter. Similarly, addition and subtraction have equal priority.

**Example 2** Evaluate. $(-3)^3 - 2^4$

**Solution** The highest priority is to raise the expressions to the appropriate powers.

$$(-3)^3 - 2^4 = -27 - 16 \quad \text{In } (-3)^3 \text{ we are cubing the number } -3 \text{ to obtain } -27. \text{ Be careful; } -2^4 \text{ is not } (-2)^4! \text{ We raise 2 to the fourth power.}$$
$$= -43 \quad \text{The last step is to add and subtract from left to right.} \quad \square$$

▶ **Student Practice 2** Evaluate. $(-4)^3 - 2^6$

$\mathbb{M_C}$ **Example 3** Evaluate. $2 \cdot (2 - 3)^3 + 6 \div 3 + (8 - 5)^2$

**Solution**

$2 \cdot (2 - 3)^3 + 6 \div 3 + (8 - 5)^2$

$= 2 \cdot (-1)^3 + 6 \div 3 + 3^2$    Combine the numbers inside the parentheses. Note that we need parentheses for $-1$ because of the negative sign, but they are not needed for 3.

$= 2 \cdot (-1) + 6 \div 3 + 9$    Next, raise to a power.

$= -2 + 2 + 9$    Next, multiply and divide from left to right.

$= 0 + 9$

$= 9$    Finally, add and subtract from left to right. □

**Student Practice 3**    Evaluate.

$6 - (8 - 12)^2 + 8 \div 2$

**Example 4** Evaluate. $\left(-\dfrac{1}{5}\right)\left(\dfrac{1}{2}\right) - \left(\dfrac{3}{2}\right)^2$

**Solution**    The highest priority is to raise $\frac{3}{2}$ to the second power.

$$\left(\frac{3}{2}\right)^2 = \left(\frac{3}{2}\right)\left(\frac{3}{2}\right) = \frac{9}{4}$$

$$\left(-\frac{1}{5}\right)\left(\frac{1}{2}\right) - \left(\frac{3}{2}\right)^2 = \left(-\frac{1}{5}\right)\left(\frac{1}{2}\right) - \frac{9}{4}$$

$\qquad\qquad = -\dfrac{1}{10} - \dfrac{9}{4}$    Next we multiply.

$\qquad\qquad = -\dfrac{1 \cdot 2}{10 \cdot 2} - \dfrac{9 \cdot 5}{4 \cdot 5}$    We need to write each fraction as an equivalent fraction with the LCD of 20.

$\qquad\qquad = -\dfrac{2}{20} - \dfrac{45}{20}$

$\qquad\qquad = -\dfrac{47}{20}$ or $-2\dfrac{7}{20}$    Add. □

**Student Practice 4**    Evaluate.

$$\left(-\frac{1}{7}\right)\left(-\frac{14}{5}\right) + \left(-\frac{1}{2}\right) \div \left(\frac{3}{4}\right)^2$$

---

**Calculator**

**Order of Operations**

Use your calculator to evaluate $3 + 4 \cdot 5$. Enter

3 $\boxed{+}$ 4 $\boxed{\times}$ 5 $\boxed{=}$

If the display is $\boxed{\quad 23\quad}$, the correct order of operations is built in. If the display is not 23, you will need to modify the way you enter the problem. You should use

4 $\boxed{\times}$ 5 $\boxed{+}$ 3 $\boxed{=}$

Try $6 + 3 \cdot 4 - 8 \div 2$.

**Ans:**
14

## 1.5 Exercises  MyMathLab®

### Verbal and Writing Skills, Exercises 1–4

*Game Points* You have lost a game of UNO and are counting the points left in your hand. You announce that you have three fours and six fives.

1. Write this as a number expression.

2. How many points do you have in your hand?

3. What answer would you get for the number expression if you simplified it by

   (a) performing the operations from left to right?

   (b) following the order of operations?

4. Which procedure in exercise 3 gives the correct number of total points?

*Evaluate.*

5. $(7 - 9)^2 \div 2 \times 5$

6. $(5 - 8)^2 \div 3 \times 6$

7. $9 + 4(5 + 2 - 8)$

8. $13 + 2(-8 + 6 - 3)$

9. $8 - 2^3 \cdot 5 + 3$

10. $7 - 3^2 \cdot 4 + 5$

11. $4 + 42 \div 3 \cdot 2 - 8$

12. $7 + 36 \div 12 \cdot 3 - 14$

13. $3 \cdot 5 + 7 \cdot 3 - 5 \cdot 3$

14. $2 \cdot 6 + 5 \cdot 3 - 7 \cdot 4$

15. $8 - 5(2)^3 \div (-8)$

16. $11 - 3(4)^2 \div (-6)$

17. $3(5 - 7)^2 - 6(3)$

18. $-2(3 - 6)^2 - (-2)$

19. $5 \cdot 6 - (3 - 5)^2 + 8 \cdot 2$

20. $(-5)^2 + (15 \div 3) + 4 \cdot 2$

21. $\dfrac{1}{2} \div \dfrac{2}{3} + 6 \cdot \dfrac{1}{4}$

22. $\dfrac{5}{6} \div \dfrac{2}{3} - 6 \cdot \left(\dfrac{1}{2}\right)^2$

23. $0.8 + 0.3(0.6 - 0.2)^2$

24. $0.05 + 1.4 - (0.5 - 0.7)^3$

25. $\dfrac{3}{8}\left(-\dfrac{1}{6}\right) - \dfrac{7}{8} + \dfrac{1}{2}$

26. $\dfrac{1}{2} \div \dfrac{4}{5} - \dfrac{3}{4}\left(\dfrac{5}{6}\right)$

### Mixed Practice

27. $(3 - 7)^2 \div 8 + 3$

28. $(5 - 6)^2 \cdot 7 - 4$

29. $\left(\dfrac{3}{4}\right)^2(-16) + \dfrac{4}{5} \div \dfrac{-8}{25}$

30. $\left(2\dfrac{4}{7}\right) \div \left(-1\dfrac{1}{5}\right)$

98

**31.** $-2.6 - (-1.8)(2.3) + (4.1)^2$

**32.** $5.15 + 4.2 \div (-0.3) - (3.5)^2$

**33.** $\left(\dfrac{2}{3}\right)^2 + \dfrac{1}{2} - \left(\dfrac{1}{3} - \dfrac{3}{4}\right) + \left(-\dfrac{1}{2}\right)^2$

**34.** $\left(\dfrac{1}{2}\right)^3 + \dfrac{1}{4} - \left(\dfrac{2}{3} - \dfrac{1}{6}\right) + \left(-\dfrac{1}{3}\right)^2$

**Applications** *Often in golf, it becomes important to keep track of eagles, bogies, pars, and birdies. Consider the following example during a year when Tiger Woods won the Buick Invitational. The following scorecard shows the results of Round 4. There are 18 holes per round and the table shows the number of times Woods got each type of score.*

| Score on a Hole | Number of Times the Score Occurred |
|---|---|
| Eagle $(-2)$ | 1 |
| Birdie $(-1)$ | 5 |
| Par $(0)$ | 10 |
| Bogey $(+1)$ | 2 |

**35.** Write his score as the sum of eagles, birdies, pars, and bogeys.

**36.** What was his final score for the round, when compared to par?

**37.** What answer do you get if you do the arithmetic left to right?

**38.** Explain why the answers in exercises 36 and 37 do not match.

## Cumulative Review  *Simplify.*

**39. [1.4.2]** $(0.5)^3$

**40. [1.2.1]** $-\dfrac{3}{4} - \dfrac{5}{6}$

**41. [1.4.2]** $-1^{20}$

**42. [0.3.2]** $3\dfrac{3}{5} \div 6\dfrac{1}{4}$

---

**Quick Quiz 1.5**  *Evaluate.*

**1.** $7 - 3^4 + 2 - 5$

**2.** $(0.3)^2 - 4.2(-4) + 0.07$

**3.** $(7 - 9)^4 + 22 \div (-2) + 6$

**4. Concept Check** Explain in what order you would perform the calculations to evaluate the expression? $4 - (-3 + 4)^3 + 12 \div (-3)$.

# Use Math to Save Money

**Did You Know?** Having a budget can help you control how you spend and save your money.

## Time to Budget

### Understanding the Problem:

One way to improve your financial situation is to learn to manage the money you have with a budget. A budget can maximize your efforts to ensure you have enough money to cover your fixed expenses, as well as your variable expenses.

Michael is a teacher. One of Michael's goals is to go back to school to earn his master's degree in education. He knows that this will not only help him further his career but also provide better financial stability in the long run. His net monthly income is currently $2500. So, Michael knows he'll need to put himself on a budget for a period of time in order to save money to go back to school.

### Making a Plan:

Michael needs to budget his expenses to help control his spending and to save for college. He wants to see how long it will take to save up the needed money.

**Step 1:** Research shows that consumer credit counseling services recommend allocating the following percentages for each category of the monthly budget:

| Housing | 25% |
|---|---|
| Transportation | 10% |
| Savings | 5% |
| Utilities | 5% |
| Debt Payments | 20% |
| Food | 15% |
| Misc. | 20% |

**Task 1:** If Michael follows this plan, how much will he put away in savings at the end of one year?

**Task 2:** If Michael follows this plan, how much will he spend on food by the end of one year?

**Step 2:** After investigating several schools in his area, Michael chooses to attend a state college that offers a one-year program for the degree he wishes to pursue. Michael will need $3000 for tuition and fees the first semester, and $450 for textbooks.

**Task 3:** What is the total cost for books and fees, and textbooks for one semester?

### Finding a Solution:

**Step 3:** By controlling his spending Michael is able to save for his master's degree.

**Task 4:** How many months will Michael have to save to pay for one semester at this college? Round to the nearest month.

**Task 5:** If Michael can increase his savings to 10% of his monthly budget by making cuts in other areas, how long would he have to save to pay for one semester? Round to the nearest month.

**Step 4:** Once Michael earns his advanced degree, his salary will increase on the following schedule:
- Year 1 an additional $4200
- Year 2 an additional $4800
- Year 3 an additional $5200
- Year 4 an additional $5500
- Year 5 an additional $5800

**Task 6:** If Michael goes back to his original budget, how much will he be saving with his new income after five years?

**Task 7:** How much will he have available for misc. spending?

### Applying the Situation to Your Life:

Having a budget can help you control your spending.
**Task 8:** Do you have a budget?

**Task 9:** How would you adjust Michael's budget to fit your needs?

**Task 10:** You may not be planning to get a master's degree but probably you are thinking of saving up for some important purchase. Are you thinking of buying a new car that is more fuel-efficient? Are you planning a special trip? How could you adjust Michael's budget to help you with your goal?

## 1.6 Using the Distributive Property to Simplify Algebraic Expressions ▶

### 1 Using the Distributive Property to Simplify Algebraic Expressions ▶

**Student Learning Objective**

After studying this section, you will be able to:

1 Use the distributive property to simplify algebraic expressions. ▶

As we learned previously, we use letters called *variables* to represent unknown numbers. If a number is multiplied by a variable, we do not need any symbol between the number and variable. Thus, to indicate $(2)(x)$, we write $2x$. To indicate $3 \cdot y$, we write $3y$. If one variable is multiplied by another variable, we place the variables next to each other. Thus, $(a)(b)$ is written $ab$. We use exponent form if an unknown number (a variable) is used several times as a factor. Thus, $x \cdot x \cdot x = x^3$. Similarly, $(y)(y)(y)(y) = y^4$.

In algebra, we need to be familiar with several definitions. We will use them throughout the remainder of this book. Take some time to think through how each of these definitions is used.

An **algebraic expression** is a quantity that contains numbers and variables, such as $a + b$, $2x - 3$, and $5ab^2$. In this chapter we will be learning rules about adding and multiplying algebraic expressions. A **term** is a number, a variable, a product, or a quotient of numbers and variables. $17$, $x$, $5xy$, and $22xy^3$ are all examples of terms. We will refer to terms when we discuss the distributive property.

An important property of algebra is the **distributive property.** We can state it in an equation as follows:

---

**DISTRIBUTIVE PROPERTY**

For all real numbers $a$, $b$, and $c$,

$$a(b + c) = ab + ac.$$

---

A numerical example shows that it does seem reasonable.

$$5(3 + 6) = 5(3) + 5(6)$$
$$5(9) = 15 + 30$$
$$45 = 45$$

We can use the distributive property to multiply any term by the sum of two or more terms. In Section 0.1, we defined the word *factor*. Two or more algebraic expressions that are multiplied are called **factors.** Consider the following examples of multiplying algebraic expressions.

**Example 1** Multiply.

**(a)** $5(a + b)$        **(b)** $-3(3x + 2y)$

**Solution**

**(a)** $5(a + b) = 5a + 5b$     Multiply the factor $(a + b)$ by the factor 5.

**(b)** $-3(3x + 2y) = -3(3x) + (-3)(2y)$     Multiply the factor $(3x + 2y)$ by the factor $(-3)$.

           $= -9x - 6y$                       □

**▷ Student Practice 1**    Multiply.

**(a)** $3(x + 2y)$        **(b)** $-2(a + 3b)$

If the parentheses are preceded by a negative sign, we consider this to be the product of $(-1)$ and the expression inside the parentheses.

**Example 2** Multiply. $-(a - 2b)$

**Solution**

$$-(a - 2b) = (-1)(a - 2b) = (-1)(a) + (-1)(-2b) = -a + 2b \quad \square$$

**Student Practice 2** Multiply. $-(-3x + y)$

In general, we see that in all these examples we have multiplied each term of the expression in the parentheses by the expression in front of the parentheses.

**Example 3** Multiply.

**(a)** $\frac{2}{3}(x^2 - 6x + 8)$ **(b)** $1.4(a^2 + 2.5a + 1.8)$

**Solution**

**(a)** $\frac{2}{3}(x^2 - 6x + 8) = \left(\frac{2}{3}\right)(1x^2) + \left(\frac{2}{3}\right)(-6x) + \left(\frac{2}{3}\right)(8)$

$$= \frac{2}{3}x^2 + (-4x) + \frac{16}{3}$$

$$= \frac{2}{3}x^2 - 4x + \frac{16}{3}$$

**(b)** $1.4(a^2 + 2.5a + 1.8) = 1.4(1a^2) + (1.4)(2.5a) + (1.4)(1.8)$

$$= 1.4a^2 + 3.5a + 2.52 \quad \square$$

**Student Practice 3** Multiply.

**(a)** $\frac{3}{5}(a^2 - 5a + 25)$ **(b)** $2.5(x^2 - 3.5x + 1.2)$

There are times we multiply a variable by itself and use exponent notation. For example, $(x)(x) = x^2$ and $(x)(x)(x) = x^3$. In other cases there will be numbers and variables multiplied at the same time.

We will see problems like $(2x)(x) = (2)(x)(x) = 2x^2$. Some expressions will involve multiplication of more than one variable. We will see problems like $(3x)(xy) = (3)(x)(x)(y) = 3x^2y$. There will be times when we use the distributive property and all of these methods will be used. For example,

$$2x(x - 3y + 2) = 2x(x) + (2x)(-3y) + (2x)(2)$$

$$= 2x^2 + (-6)(xy) + 4(x)$$

$$= 2x^2 - 6xy + 4x.$$

We will discuss this type of multiplication of variables with exponents in more detail in Section 4.1. At that point we will expand these examples and other similar examples to develop the general rule for multiplication $(x^a)(x^b) = x^{a+b}$.

**Example 4** Multiply. $-2x(3x + y - 4)$

**Solution**

$$-2x(3x + y - 4) = -2(x)(3)(x) + (-2)(x)(y) + (-2)(x)(-4)$$

$$= -2(3)(x)(x) + (-2)(xy) + (-2)(-4)(x)$$

$$= -6x^2 - 2xy + 8x \quad \square$$

**Student Practice 4** Multiply. $-4x(x - 2y + 3)$

The distributive property can also be presented with the $a$ on the right.

$$(b + c)a = ba + ca$$

The $a$ is "distributed" over the $b$ and $c$ inside the parentheses.

**Example 5** Multiply. $(2x^2 - x)(-3)$

**Solution**

$$(2x^2 - x)(-3) = 2x^2(-3) + (-x)(-3)$$
$$= -6x^2 + 3x \qquad \square$$

 **Student Practice 5**   Multiply.

$$(3x^2 - 2x)(-4)$$

▲**Example 6**  A farmer has a rectangular field that is 300 feet wide. One portion of the field is $2x$ feet long. The other portion of the field is $3y$ feet long. Use the distributive property to find an expression for the area of this field.

**Solution**   First we draw a picture of a field that is 300 feet wide and $2x + 3y$ feet long.

To find the area of the field, we multiply the width times the length.

$$300(2x + 3y) = 300(2x) + 300(3y) = 600x + 900y$$

Thus the area of the field in square feet is $600x + 900y$. $\qquad \square$

▲ **Student Practice 6**   A farmer has a rectangular field that is 400 feet wide. One portion of the field is $6x$ feet long. The other portion of the field is $9y$ feet long. Use the distributive property to find an expression for the area of this field.

---

👣 **STEPS TO SUCCESS** Review a Little Every Day

Successful students find that review is not something you do the night before the test. Take time to review a little each day. When you are learning new material, take a little time to look over the concepts previously learned in the chapter. By this continual review you will find the pressure is reduced to prepare for a test. You need time to think about what you have learned and make sure you really understand it. This will help to tie together the different topics in the chapter. A little review of each idea and each kind of problem will enable you to feel confident. You will think more clearly and have less tension when it comes to test time.

**Making it personal:** Which of these suggestions is the one you most need to follow? Write down what you need to do to improve in this area. ▼

## 1.6 Exercises  MyMathLab®

### Verbal and Writing Skills, Exercises 1–6

*In exercises 1 and 2, complete each sentence by filling in the blank.*

**1.** A _____ is a symbol used to represent an unknown number.

**2.** When we write an expression with numbers and variables such as $7x$, it indicates that we are _____ 7 by $x$.

**3.** Explain in your own words how we multiply a problem like $(4x)(x)$.

**4.** Explain in your own words why you think the property $a(b + c) = ab + ac$ is called the distributive property. What does "distribute" mean?

**5.** Does the following distributive property work?

$$a(b - c) = ab - ac$$

Why or why not? Give an example.

**6.** Susan tried to use the distributive property and wrote

$$-5(x + 3y - 2) = -5x - 15y - 10.$$

What did she do wrong?

*Multiply. Use the distributive property.*

**7.** $5(2x - 5y)$

**8.** $6(3x - 6y)$

**9.** $-2(4a - 3b)$

**10.** $-3(2a - 4b)$

**11.** $3(3x + y)$

**12.** $8(5x + y)$

**13.** $8(-m - 3n)$

**14.** $10(-2m - n)$

**15.** $-(x - 3y)$

**16.** $-(-3y + x)$

**17.** $-9(9x - 5y + 8)$

**18.** $-3(4x + 8 - 6y)$

**19.** $2(-5x + y - 6)$

**20.** $3(2x - 6y - 5)$

**21.** $\dfrac{5}{6}(12x^2 - 24x + 18)$

**22.** $\dfrac{2}{3}(-27a^4 + 9a^2 - 21)$

**23.** $\dfrac{x}{5}(x + 10y - 4)$ $\left( Hint: \dfrac{x}{5} = \dfrac{1}{5}x \right)$

**24.** $\dfrac{y}{3}(3y - 4x - 6)$ $\left( Hint: \dfrac{y}{3} = \dfrac{1}{3}y \right)$

**25.** $5x(x + 2y + z)$

**26.** $3a(2a + b - c)$

**27.** $(-4.5x + 5)(-3)$

**28.** $(-3.2x + 5)(-4)$

**29.** $(6x + y - 1)(3x)$

**30.** $(2x - 2y + 6)(3x)$

**Mixed Practice** *Multiply. Use the distributive property.*

**31.** $(3x + 2y - 1)(-xy)$

**32.** $(5a - 3b - 1)(-ab)$

**33.** $(-a - 2b + 4)5ab$

**34.** $(2a - b - 5)3ab$

**35.** $\frac{1}{4}(8a^2 - 16a - 5)$

**36.** $\frac{1}{3}(-15a^2 - 21a + 4)$

**37.** $-0.3x(-1.2x^2 - 0.3x + 0.5)$
*Hint:* $x(x^2) = x(x)(x) = x^3$

**38.** $-0.6q(1.2q^2 + 2.5r - 0.7s)$
*Hint:* $q(q^2) = q(q)(q) = q^3$

**39.** $0.4q(-3.3q^2 - 0.7r - 10)$
*Hint:* $q(q^2) = q(q)(q) = q^3$

## Applications

▲ **40.** *Geometry* Gary Roswell owns a large rectangular field where he grows corn and wheat. The width of the field is 850 feet. The portion of the field where corn grows is $10x$ feet long and the portion where wheat grows is $7y$ feet long. Use the distributive property to find an expression for the area of this field.

▲ **41.** *Geometry* Kathy Maris has a rectangular field that is 800 feet wide. One portion of the field is $5x$ feet long. The other portion of the field is $14y$ feet long. Use the distributive property to find an expression for the area of this field.

## To Think About

▲ **42.** *Athletic Field* The athletic field at Gordon College is $2x$ feet wide. It used to be 1800 feet long. An old, rundown shed was torn down, making the field $3y$ feet longer. Use the distributive property to find an expression for the area of the new field.

▲ **43.** *Airport Runway* The runway at Beverly Airport is $4x$ feet wide. Originally, the airport was supposed to have a 3000-foot-long runway. However, some of the land was wetland, so a runway could not be built on all of it. Therefore, the length of the runway was decreased by $2y$ feet. Use the distributive property to find an expression for the area of the final runway.

**Cumulative Review** *In exercises 44–48, evaluate.*

**44.** **[1.1.5]** $-18 + (-20) + 36 + (-14)$

**45.** **[1.4.2]** $(-2)^6$

**46.** **[1.2.1]** $-27 - (-41)$

**47.** **[1.5.1]** $25 \div 5(2) + (-6)$

**48.** **[1.5.1]** $(12 - 10)^2 + (-3)(-2)$

Quick Quiz 1.6    *Multiply. Use the distributive property.*

**1.** $5(-3a - 7b)$

**2.** $-2x(x - 4y + 8)$

**3.** $-3ab(4a - 5b - 9)$

**4.** **Concept Check** Explain how you would multiply to obtain the answer for $(-\frac{3}{7})(21x^2 - 14x + 3)$.

## 1.7 Combining Like Terms

### 1 Identifying Like Terms

We can add or subtract quantities that are *like quantities*. This is called **combining like quantities.**

$$5 \text{ inches} + 6 \text{ inches} = 11 \text{ inches}$$
$$20 \text{ square inches} - 16 \text{ square inches} = 4 \text{ square inches}$$

However, we cannot combine things that are not the same.

$$16 \text{ square inches} - 4 \text{ inches} \quad (\text{Cannot be done!})$$

Similarly, in algebra we can **combine like terms.** This means to add or subtract like terms. Remember, we cannot combine terms that are not the same. Recall that a *term* is a number, a variable, a product, or a quotient of numbers and variables. **Like terms** are terms that have identical variables and exponents. In other words, like terms must have exactly the same letter parts.

**Example 1** List the like terms of each expression.

(a) $5x - 2y + 6x$

(b) $2x^2 - 3x - 5x^2 - 8x$

**Solution**

(a) $5x$ and $6x$ are like terms. These are the only like terms in this expression.

(b) $2x^2$ and $-5x^2$ are like terms.

$-3x$ and $-8x$ are like terms.

Note that $x^2$ and $x$ are not like terms.

**Student Practice 1**   List the like terms of each expression.

(a) $5a + 2b + 8a - 4b$

(b) $x^2 + y^2 + 3x - 7y^2$

Do you really understand what a term is? A term is a number, a variable, a product, or a quotient of numbers and variables. Terms are the parts of an algebraic expression separated by plus or minus signs. The sign in front of the term is considered part of the term.

### 2 Combining Like Terms

It is important to know how to combine like terms. Since

$$4 \text{ inches} + 5 \text{ inches} = 9 \text{ inches},$$

we would expect in algebra that $4x + 5x = 9x$.

Why is this true? Let's take a look at the distributive property.

Like terms may be added or subtracted by using the distributive property:
$$ab + ac = a(b + c) \quad \text{and} \quad ba + ca = (b + c)a.$$

For example,

$$-7x + 9x = (-7 + 9)x = 2x$$
$$5x^2 + 12x^2 = (5 + 12)x^2 = 17x^2.$$

**Example 2** Combine like terms.

(a) $-4x^2 + 8x^2$

(b) $5x + 3x + 2x$

**Solution**

(a) Notice that each term contains the factor $x^2$. Using the distributive property, we have

$$-4x^2 + 8x^2 = (-4 + 8)x^2 = 4x^2.$$

(b) Note that each term contains the factor $x$. Using the distributive property, we have

$$5x + 3x + 2x = (5 + 3 + 2)x = 10x. \qquad \square$$

**Student Practice 2**  Combine like terms.

(a) $16y^3 + 9y^3$

(b) $5a + 7a + 4a$

In this section, the direction *simplify* means to remove parentheses and/or combine like terms.

**Example 3** Simplify. $5a^2 - 2a^2 + 6a^2$

**Solution**

$$5a^2 - 2a^2 + 6a^2 = (5 - 2 + 6)a^2 = 9a^2 \qquad \square$$

**Student Practice 3**  Simplify. $-8y^2 - 9y^2 + 4y^2$

After doing a few problems, you will find that it is not necessary to write out the step of using the distributive property. We will omit this step for the remaining examples in this section.

**Example 4** Simplify.

(a) $5.6a + 2b + 7.3a - 6b$

(b) $3x^2y - 2xy^2 + 6x^2y$

(c) $2a^2b + 3ab^2 - 6a^2b^2 - 8ab$

**Solution**

(a) $5.6a + 2b + 7.3a - 6b = 12.9a - 4b$  We combine the $a$ terms and the $b$ terms separately.

(b) $3x^2y - 2xy^2 + 6x^2y = 9x^2y - 2xy^2$  **Note:** $x^2y$ and $xy^2$ are not like terms because of different powers.

(c) $2a^2b + 3ab^2 - 6a^2b^2 - 8ab$  These terms cannot be combined; there are no like terms in this expression. $\quad \square$

**Student Practice 4**  Simplify.

(a) $1.3x + 3a - 9.6x + 2a$

(b) $5ab - 2ab^2 - 3a^2b + 6ab$

(c) $7x^2y - 2xy^2 - 3x^2y^2 - 4xy$

The two skills in this section that a student must practice are identifying like terms and correctly adding and subtracting like terms. If a problem involves many terms, you may find it helpful to rearrange the terms so that like terms are together.

**Example 5** Simplify. $3a - 2b + 5a^2 + 6a - 8b - 12a^2$

**Solution** There are three pairs of like terms.

$$\underbrace{3a + 6a}_{a\text{ terms}} \underbrace{- 2b - 8b}_{b\text{ terms}} \underbrace{+ 5a^2 - 12a^2}_{a^2\text{ terms}}$$ You can rearrange the terms so that like terms are together, making it easier to combine them.

$$= 9a - 10b - 7a^2$$ Combine like terms.

Because of the commutative property, the order of terms in an answer to this problem is not significant. These three terms can be rearranged in a different order. $-10b + 9a - 7a^2$ and $-7a^2 + 9a - 10b$ are also correct. Later, we will learn the preferred way to write the answer.

**Student Practice 5** Simplify. $5xy - 2x^2y + 6xy^2 - xy - 3xy^2 - 7x^2y$

Use extra care with fractional values.

**Example 6** Simplify. $\dfrac{3}{4}x^2 - 5y - \dfrac{1}{8}x^2 + \dfrac{1}{3}y$

**Solution** We need the least common denominator for the $x^2$ terms, which is 8. Change $\dfrac{3}{4}$ to eighths by multiplying the numerator and denominator by 2.

$$\frac{3}{4}x^2 - \frac{1}{8}x^2 = \frac{3 \cdot 2}{4 \cdot 2}x^2 - \frac{1}{8}x^2 = \frac{6}{8}x^2 - \frac{1}{8}x^2 = \frac{5}{8}x^2$$

The least common denominator for the $y$ terms is 3. Change $-5$ to thirds.

$$-\frac{5}{1}y + \frac{1}{3}y = \frac{-5 \cdot 3}{1 \cdot 3}y + \frac{1}{3}y = \frac{-15}{3}y + \frac{1}{3}y = -\frac{14}{3}y$$

Thus, our solution is $\dfrac{5}{8}x^2 - \dfrac{14}{3}y$.

**Student Practice 6** Simplify.

$$\frac{1}{7}a^2 - \frac{5}{12}b + 2a^2 - \frac{1}{3}b$$

**Example 7** Simplify. $6(2x + 3xy) - 8x(3 - 4y)$

**Solution** First remove the parentheses; then combine like terms.

$$6(2x + 3xy) - 8x(3 - 4y) = 12x + 18xy - 24x + 32xy$$ Use the distributive property.

$$= -12x + 50xy$$ Combine like terms.

**Student Practice 7** Simplify. $5a(2 - 3b) - 4(6a + 2ab)$

## 1.7 Exercises  MyMathLab®

**Verbal and Writing Skills, Exercises 1–6**

1. Explain in your own words the mathematical meaning of the word *term*.

2. Explain in your own words the mathematical meaning of the phrase *like terms*.

3. Explain which terms are like terms in the expression $5x - 7y - 8x$.

4. Explain which terms are like terms in the expression $12a - 3b - 9a$.

5. Explain which terms are like terms in the expression $7xy - 9x^2y - 15xy^2 - 14xy$.

6. Explain which terms are like terms in the expression $-3a^2b - 12ab + 5ab^2 + 9ab$.

*Combine like terms.*

7. $-16x^2 - 15x^2$

8. $-14x^3 - 21x^3$

9. $5a^3 - 7a^2 + a^3$

10. $2b^3 + 8b^2 - 9b^3$

11. $3x + 2y - 8x - 7y$

12. $5x - 9b - 6x - 5b$

13. $1.3x - 2.6y + 5.8x - 0.9y$

14. $3.1a - 0.2b - 0.8a + 5.3b$

15. $1.6x - 2.8y - 3.6x - 5.9y$

16. $1.9x - 2.4b - 3.8x - 8.2b$

17. $3p - 4q + 2p + 3 + 5q - 21$

18. $6x - 5y - 3y + 7 - 11x - 5$

19. $2ab + 5bc - 6ac - 2ab$

20. $7ab - 3bc - 12ac + 8ab$

21. $2x^2 - 3x - 5 - 7x + 8 - x^2$

22. $5x + 7 - 6x^2 + 6 - 11x + 4x^2$

23. $2y^2 - 8y + 9 - 12y^2 - 8y + 3$

24. $3y^2 + 9y - 12 - 4y^2 - 6y + 2$

25. $\frac{1}{3}x - \frac{2}{3}y - \frac{2}{5}x + \frac{4}{7}y$

26. $\frac{2}{5}s - \frac{3}{8}t - \frac{4}{15}s - \frac{5}{12}t$

27. $\frac{3}{4}a^2 - \frac{1}{3}b - \frac{1}{5}a^2 - \frac{1}{2}b$

28. $\frac{2}{5}y - \frac{3}{4}x^2 - \frac{1}{3}y + \frac{7}{8}x^2$

29. $3rs - 8r + s - 5rs + 10r - s$

30. $-rs + 10s + 5r - rs + 6s - 2r$

31. $4xy + \frac{5}{4}x^2y + \frac{3}{4}xy + \frac{3}{4}x^2y$

32. $\frac{3}{7}ab - \frac{2}{7}a^2b + 2ab + \frac{9}{7}a^2b$

*Simplify. Use the distributive property to remove parentheses; then combine like terms.*

33. $5(2a - b) - 3(5b - 6a)$

34. $8(3x - 2y) + 4(3y - 5x)$

**109**

**35.** $-3b(5a - 3b) + 4(-3ab - 5b^2)$

**36.** $3a(2a - 3b) - 3(-5a^2 + ab)$

**37.** $6(c - 2d^2) - 2(4c - d^2)$

**38.** $-4(3cd + 2c^2) + 2c(5d - c)$

**39.** $3(4 - x) - 2(-4 - 7x)$

**40.** $5(6 - x) - 2(9 - 11x)$

## To Think About

▲ **41.** *Fencing in a Pool* Mr. Jimenez has a pool behind his house that needs to be fenced in. The backyard is an odd quadrilateral shape. The four sides are $3a$, $2b$, $4a$, and $7b$ in length. How much fencing (the length of the perimeter) would he need to enclose the pool?

▲ **42.** *Framing a Masterpiece* The new Degas masterpiece purchased by the Museum of Fine Arts in Boston needs to be reframed. If the rectangular picture measures $6x - 3$ wide by $8x - 7$ high, what is the perimeter of the painting?

▲ **43.** *Geometry* A rectangle is $5x - 10$ feet long and $2x + 6$ feet wide. What is the perimeter of the rectangle?

▲ **44.** *Geometry* A triangle has sides of length $3a + 8$ meters, $5a - b$ meters, and $12 - 2b$ meters. What is the perimeter of the triangle?

## Cumulative Review *Evaluate.*

**45.** [1.2.1] $-\dfrac{3}{4} - \dfrac{1}{3}$

**46.** [1.3.1] $\left(\dfrac{2}{3}\right)\left(-\dfrac{9}{16}\right)$

**47.** [1.1.4] $\dfrac{4}{5} + \left(-\dfrac{1}{25}\right) + \left(-\dfrac{3}{10}\right)$

**48.** [1.3.3] $\left(\dfrac{5}{7}\right) \div \left(-\dfrac{14}{3}\right)$

---

**Quick Quiz 1.7** *Combine like terms.*

**1.** $3xy - \dfrac{2}{3}x^2y - \dfrac{5}{6}xy + \dfrac{7}{3}x^2y$

**2.** $8.2a^2b + 5.5ab^2 - 7.6a^2b - 9.9ab^2$

**3.** $2(3x - 5y) - 2(-7x - 4y)$

**4.** **Concept Check** Explain how you would remove parentheses and then combine like terms to obtain the answer for $1.2(3.5x - 2.2y) - 4.5(2.0x + 1.5y)$.

# 1.8 Using Substitution to Evaluate Algebraic Expressions and Formulas ⏵

## 1 Evaluating an Algebraic Expression for a Specified Value ⏵

You will use the order of operations to **evaluate** variable expressions. Suppose we are asked to evaluate

$$6 + 3x \text{ for } x = -4.$$

In general, $x$ represents some unknown number. Here we are told $x$ has the value $-4$. We can replace $x$ with $-4$. Use parentheses around $-4$. Note that we always put replacement values in parentheses.

$$6 + 3(-4) = 6 + (-12) = -6$$

When we replace a variable by a particular value, we say we have **substituted** the value for the variable. We then evaluate the expression (that is, find a value for it).

**Student Learning Objectives**

**After studying this section, you will be able to:**

1 Evaluate an algebraic expression for a specified value. ⏵

2 Evaluate a formula by substituting values. ⏵

**Example 1** Evaluate $\frac{2}{3}x - 5$ for $x = -6$.

**Solution**

$$\frac{2}{3}x - 5 = \frac{2}{3}(-6) - 5 \quad \text{Substitute } -6 \text{ for } x. \text{ Be sure to enclose the } -6 \text{ in parentheses.}$$

$$= -4 - 5 \quad \text{Multiply } \left(\frac{2}{3}\right)\left(-\frac{6}{1}\right) = -4.$$

$$= -9 \quad \text{Combine.} \qquad \square$$

 **Student Practice 1** Evaluate $4 - \frac{1}{2}x$ for $x = -8$.

Compare parts **(a)** and **(b)** in the next example. The two parts illustrate that you must be careful what value you raise to a power. *Note:* In part **(b)** we will need parentheses within parentheses. To avoid confusion, we use brackets [ ] to represent the outside parentheses.

**Example 2** Evaluate for $x = -3$.

**(a)** $2x^2$      **(b)** $(2x)^2$

**Solution**

**(a)** Here the value $x$ is squared.

$$2x^2 = 2(-3)^2$$
$$= 2(9) \quad \text{First square } -3.$$
$$= 18 \quad \text{Then multiply.}$$

**(b)** Here the value $(2x)$ is squared.

$$(2x)^2 = [2(-3)]^2$$
$$= (-6)^2 \quad \text{First multiply the numbers inside the brackets.}$$
$$= 36 \quad \text{Then square } -6. \quad \square$$

**Student Practice 2** Evaluate for $x = -3$.

**(a)** $4x^2$      **(b)** $(4x)^2$

Carefully study the solutions to Example 2**(a)** and Example 2**(b)**. You will find that taking the time to see *how* and *why* they are different is a good investment of time.

Mc **Example 3** Evaluate $x^2 + 3x$ for $x = -4$.

**Solution**

$$x^2 + 3x = (-4)^2 + 3(-4) \quad \text{Replace each } x \text{ by } -4 \text{ in the original expression.}$$
$$= 16 + (3)(-4) \quad \text{Raise to a power.}$$
$$= 16 - 12 \quad \text{Multiply.}$$
$$= 4 \quad \text{Finally, we add the opposite of 12.} \quad \square$$

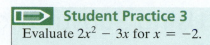

**Student Practice 3**

Evaluate $2x^2 - 3x$ for $x = -2$.

**Example 4** Evaluate $x^3 + 2xy - 3x + 1$ for $x = 2$ and $y = -\dfrac{1}{4}$.

**Solution**

$$x^3 + 2xy - 3x + 1 = (2)^3 + 2(2)\left(-\frac{1}{4}\right) - 3(2) + 1 \quad \text{Replace } x \text{ by 2 and } y \text{ by } -\frac{1}{4}.$$
$$= 8 + (-1) - 6 + 1 \quad \text{Calculate } 2^3 = 8, \text{ and multiply } 2(2)\left(-\frac{1}{4}\right) = 4\left(-\frac{1}{4}\right) = -1. \text{ Also, multiply } -3(2) = -6.$$
$$= 8 + (-1) + (-6) + 1 \quad \text{Change subtracting to adding the opposite.}$$
$$= 9 + (-7) \quad \text{Find the sum of the positive numbers and find the sum of the negative numbers.}$$
$$= 2 \quad \text{We take the difference of 9 and 7 and use the sign of the number with the larger absolute value.} \quad \square$$

**Student Practice 4**

Evaluate $6a + 4ab^2 - 5$ for $a = -\dfrac{1}{6}$ and $b = 3$.

## 2 Evaluating a Formula by Substituting Values

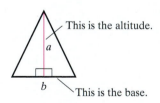

This is the altitude.

$a$

$b$   This is the base.

We can *evaluate a formula* by substituting values for the variables. For example, the area of a triangle can be found using the formula $A = \frac{1}{2}ab$, where $b$ is the length of the base of the triangle and $a$ is the altitude of the triangle (see figure). If we know values for $a$ and $b$, we can substitute those values into the formula to find the area. The units for area are *square units*.

Because some of the examples and exercises in this section involve geometry, it may be helpful to review this topic.

The following information is very important. If you have forgotten some of this material (or if you have never learned it), please take the time to learn it completely now. Throughout the entire book we will be using this information in solving applied problems.

**Perimeter** is the distance around a plane figure. Perimeter is measured in linear units (inches (in.), feet (ft), centimeters (cm), miles (mi)). **Area** is a measure of the amount of surface in a region. Area is measured in square units (square inches ($in.^2$), square feet ($ft^2$), square centimeters ($cm^2$)).

In our sketches we will show angles of 90° by using a small square (⌐). This indicates that the two lines are at right angles. All angles that measure 90° are called **right angles.** An **altitude** is perpendicular to the base of a figure. That is, the altitude forms a right angle with the base. The small corner square in a sketch helps us identify the altitude of the figure.

The following box provides a handy guide to some facts and formulas you will need to know. Use it as a reference when solving word problems involving geometric figures.

## GEOMETRIC FORMULAS: TWO-DIMENSIONAL FIGURES

A **parallelogram** is a four-sided figure with opposite sides parallel. In a parallelogram, opposite sides are equal and opposite angles are equal.

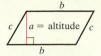

Perimeter = the sum of all four sides

Area = $ab$

A **rectangle** is a parallelogram with all interior angles measuring 90°.

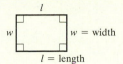

Perimeter = $2l + 2w$

Area = $lw$

A **square** is a rectangle with all four sides equal.

Perimeter = $4s$

Area = $s^2$

A **trapezoid** is a four-sided figure with two sides parallel. The parallel sides are called the *bases* of the trapezoid.

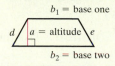

Perimeter = the sum of all four sides

Area = $\dfrac{1}{2}a(b_1 + b_2)$

A **triangle** is a closed plane figure with three sides.

Perimeter = the sum of all three sides

Area = $\dfrac{1}{2}ab$

A **circle** is a plane curve consisting of all points at an equal distance from a given point called the center. **Circumference** is the distance around a circle.

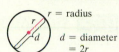

Circumference = $2\pi r$ or $\pi d$

Area = $\pi r^2$

$\pi$ (the number *pi*) is a constant associated with circles. It is an irrational number that is approximately 3.141592654. We usually use 3.14 as a sufficiently accurate approximation. Thus we write $\pi \approx 3.14$ for most of our calculations involving $\pi$.

**Example 5**  Find the area of a triangle with a base of 16 centimeters (cm) and an altitude of 12 centimeters (cm).

**Solution**  Substitute $a = 12$ cm and $b = 16$ cm in $A = \dfrac{1}{2}ab$.

$$A = \frac{1}{2}(12 \text{ cm})(16 \text{ cm})$$

$$= (6)(16)(\text{cm})^2 = 96 \text{ square centimeters}$$

The area of the triangle is 96 square centimeters or 96 cm².

**Student Practice 5**  Find the area of a triangle with an altitude of 3 meters and a base of 7 meters.

The area of a circle is given by $A = \pi r^2$. We will use 3.14 as an approximation for the *irrational number* $\pi$.

▲**Example 6**  Find the area of a circle if the radius is 2 inches.

**Solution**

$A = \pi r^2 \approx (3.14)(2 \text{ inches})^2$   Write the formula and substitute the given values for the letters.

$\approx (3.14)(4)(\text{in.})^2$   Raise to a power. Then multiply.

$\approx 12.56$ square inches or 12.56 in.²

**Student Practice 6**  Find the area of a circle if the radius is 3 meters.

The formula $C = \frac{5}{9}(F - 32)$ allows us to find the Celsius temperature if we know the Fahrenheit temperature. That is, we can substitute a value for $F$ in degrees Fahrenheit into the formula to obtain a temperature $C$ in degrees Celsius.

**Example 7** What is the Celsius temperature when the Fahrenheit temperature is $F = -22°$?

**Solution** Use the formula.

$$C = \frac{5}{9}(F - 32)$$

$$= \frac{5}{9}[(-22) - 32] \quad \text{Substitute } -22 \text{ for } F \text{ in the formula.}$$

$$= \frac{5}{9}(-54) \quad \text{Combine the numbers inside the brackets.}$$

$$= (5)(-6) \quad \text{Simplify.}$$

$$= -30 \quad \text{Multiply.}$$

The temperature is $-30°$ Celsius or $-30°C$.

**Student Practice 7** What is the Celsius temperature when the Fahrenheit temperature is $F = 68°$? Use the formula $C = \frac{5}{9}(F - 32)$.

When driving in Canada or Mexico, we must observe speed limits posted in kilometers per hour. A formula that converts $r$ (miles per hour) to $k$ (kilometers per hour) is $k \approx 1.61r$. Note that this is an approximation.

**Example 8** You are driving on a highway in Mexico. It has a posted maximum speed of 100 kilometers per hour. You are driving at 61 miles per hour. Are you exceeding the speed limit?

**Solution** Use the formula.

$$k \approx 1.61r$$

$$\approx (1.61)(61) \quad \text{Replace } r \text{ by 61.}$$

$$\approx 98.21 \quad \text{Multiply the numbers.}$$

You are driving at approximately 98 kilometers per hour. You are not exceeding the speed limit.

**Student Practice 8** You are driving behind a heavily loaded truck on a Canadian highway. The highway has a posted minimum speed of 65 kilometers per hour. When you travel at exactly the same speed as the truck ahead of you, you observe that the speedometer reads 35 miles per hour. Assuming that your speedometer is accurate, determine whether the truck is violating the minimum speed law.

## 1.8 Exercises   MyMathLab®

*Evaluate.*

1. $-3x + 5$ for $x = 4$

2. $-5x - 6$ for $x = 6$

3. $\frac{2}{5}y - 8$ for $y = -10$

4. $\frac{5}{6}y - 5$ for $y = -6$

5. $5x + 10$ for $x = \frac{1}{2}$

6. $5x + 15$ for $x = -\frac{3}{2}$

7. $2 - 4x$ for $x = 7$

8. $3 - 5x$ for $x = 8$

9. $3.5 - 2x$ for $x = 2.4$

10. $6.3 - 3x$ for $x = 2.3$

11. $9x + 13$ for $x = -\frac{3}{4}$

12. $5x + 7$ for $x = -\frac{2}{3}$

13. $x^2 - 3x$ for $x = -2$

14. $x^2 + 3x$ for $x = 4$

15. $5y^2$ for $y = -1$

16. $8y^2$ for $y = -1$

17. $-3x^3$ for $x = 2$

18. $-5x^3$ for $x = 3$

19. $-5x^2$ for $x = -2$

20. $-2x^2$ for $x = -3$

21. $2x^2 + 3x$ for $x = -3$

22. $18 + 3x^2$ for $x = -3$

23. $(2x)^2 + x$ for $x = 3$

24. $2 - x^2$ for $x = -2$

25. $2 - (-x)^2$ for $x = -2$

26. $2x - 3x^2$ for $x = -4$

27. $10a + (4a)^2$ for $a = -2$

28. $9a - (2a)^2$ for $a = -3$

29. $4x^2 - 6x$ for $x = \frac{1}{2}$

30. $5 - 9x^2$ for $x = \frac{1}{3}$

31. $x^3 - 7y + 3$ for $x = 2$ and $y = 5$

32. $a^3 + 2b - 4$ for $a = 1$ and $b = -4$

33. $\frac{1}{2}a^2 - 3b + 9$ for $a = -4$ and $b = \frac{2}{3}$

34. $\frac{1}{3}x^2 + 4y - 5$ for $x = -3$ and $y = \frac{3}{4}$

35. $2r^2 + 3s^2 - rs$ for $r = -1$ and $s = 3$

36. $-r^2 + 5rs + 4s^2$ for $r = -2$ and $s = 3$

37. $a^3 + 2abc - 3c^2$ for $a = 5, b = 9,$ and $c = -1$

38. $a^3 - 2ab + 2c^2$ for $a = 3, b = 2,$ and $c = -4$

39. $\frac{a^2 + ab}{3b}$ for $a = -1$ and $b = -2$

40. $\frac{x^2 - 2xy}{2y}$ for $x = -2$ and $y = -3$

### Applications

▲ 41. *Geometry* A sign is made in the shape of a parallelogram. The base measures 22 feet. The altitude measures 16 feet. What is the area of the sign?

▲ 42. *Geometry* A field is shaped like a parallelogram. The base measures 92 feet. The altitude measures 54 feet. What is the area of the field?

115

▲ **43.** *TV Parts* A square support unit in a television is made with a side measuring 3 centimeters. A new model being designed for next year will have a larger square with a side measuring 3.2 centimeters. By how much will the area of the square be increased?

▲ **44.** *Computer Chips* A square computer chip for last year's computer had a side measuring 23 millimeters. This year the computer chip has been reduced in size. The new square chip has a side of 20 millimeters. By how much has the area of the chip decreased?

▲ **45.** *Carpentry* A carpenter cut out a small trapezoid as a wooden support for a front step. It has an altitude of 4 inches. One base of the trapezoid measures 9 inches and the other base measures 7 inches. What is the area of this support?

▲ **46.** *Cable Television Technician* Dan is a Comcast cable technician. He works on a signal tower that has a small trapezoid frame on the top of the tower. The frame has an altitude of 9 inches. One base of the trapezoid is 20 inches and the other base measures 17 inches. What is the area of this small trapezoidal frame?

▲ **47.** *Geometry* Bradley Palmer State Park has a triangular piece of land on the border. The altitude of the triangle is 400 feet. The base of the triangle is 280 feet. What is the area of this piece of land?

▲ **48.** *Roofing* The ceiling in the Madisons' house has a leak. The roofer exposed a triangular region that needs to be sealed and then reroofed. The region has an altitude of 14 feet. The base of the region is 19 feet. What is the area of the region that needs to be reroofed?

▲ **49.** *Geometry* The radius of a circular tablecloth is 3 feet. What is the area of the tablecloth? (Use $\pi \approx 3.14$.)

**50.** *Landscape Architect Assistant* Lexi works as an assistant for a landscape architect in West Chicago, Illinois. The center of the flower garden at a new estate has a circular concrete platform for an elaborate water fountain. The circular platform has a diameter of 12 feet. What will be the area of the circular platform? (Use $\pi \approx 3.14$.)

*Temperature* *For exercises 51 and 52, use the formula* $C = \dfrac{5}{9}(F - 32)$ *to find the Celsius temperature.*

**51.** Dry ice is solid carbon dioxide. Dry ice does not melt, it goes directly from the solid state to the gaseous state. Dry ice changes from a solid to a gas at $-109.3°F$. What is this temperature in Celsius?

**52.** *Winter Jogging* Jenny went jogging in January on a trail in Minneapolis when the weather outside was $-12$ degrees Celsius. Her outside winter jogging coat is rated for $-3$ degrees Fahrenheit. Is the coat going to be warm enough? What is the outside winter jogging coat rated for in degrees Celsius?

*Solve.*

▲ **53.** *Sail Dimensions* Find the total cost of making a triangular sail that has a base dimension of 12 feet and an altitude of 20 feet if the price for making the sail is $19.50 per square foot.

**54.** *Replacement Window Technician* Caleb works for a company that installs energy-efficient replacement windows. He installed a new semicircular window in a large dining room. The owners want the window coated with a layer of sunblocking material. The semicircular window has a radius of 14 inches. The sunblock coating costs $1.15 per square inch to apply. What is the total cost of coating the window to the nearest cent? (Use $\pi \approx 3.14$.)

**55.** *Temperature Extremes on the Moon* The results of measurements made by NASA with the Lunar Orbiter show that the coldest temperature recorded on the surface of the moon was −238°C. The warmest temperature recorded on the surface of the moon was 123°C. What are the corresponding temperatures in degrees Fahrenheit? (Use the formula $F = \dfrac{9}{5}C + 32$.)

**56.** *Tour de France* A recently completed Tour de France was 3642 kilometers long and was completed in 20 stages. What was the average length of each stage in miles? Use the formula $r \approx 0.62k$, where $r$ is the number of miles and $k$ is the number of kilometers. Round to the nearest tenth of a mile.

## Cumulative Review *In exercises 57 and 58, simplify.*

**57.** **[1.5.1]** $(-2)^4 - 4 \div 2 - (-2)$

**58.** **[1.7.2]** $3(x - 2y) - (x^2 - y) - (x - y)$

---

### Quick Quiz 1.8 *Evaluate.*

**1.** $2x^2 - 4x - 14$ for $x = -2$

**2.** $5a - 6b$ for $a = \dfrac{1}{2}$ and $b = -\dfrac{1}{3}$

**3.** $x^3 + 2x^2y + 5y + 2$ for $x = -2$ and $y = 3$

**4.** **Concept Check** Explain how you would find the area of a circle if you know its diameter is 12 meters.

---

### 🌿 STEPS TO SUCCESS What Is the Best Way to Review Before a Test?

Here is what students have found.

**1.** Read over your textbook again. Make a list of any terms, rules, or formulas you need to know for the exam. Make sure you understand all of them.

**2.** Go over your notes. Look back at your homework and quizzes. Redo the problems you missed. Make sure you can get the right answer.

**3.** Practice some of each type of problem covered in the chapter(s) you are to be tested on.

**4.** At the end of the chapter are special sections you need to complete. Study each part of the Chapter Organizer and do the You Try It problems. Be sure to do the Chapter Review Problems.

**5.** When you think you are ready take the How Am I Doing? Chapter Test. Make sure you study the Math Coach notes right after the test.

**6.** Get help with concepts that are giving you difficulty. Don't be afraid to ask for help. Teachers, tutors, friends in class, and other friends are ready to help you. Don't wait. Get help now.

**Making it personal:** Which of the six steps do you most need to follow? Take some time today to start doing these things. These methods of review have helped thousands of students! They can help you—NOW! ▼

## 1.9  Grouping Symbols

**Student Learning Objective**

After studying this section, you will be able to:

1. Simplify algebraic expressions by removing grouping symbols.

### 1  Simplifying Algebraic Expressions by Removing Grouping Symbols

Many expressions in algebra use **grouping symbols** such as parentheses, brackets, and braces. Sometimes expressions are inside other expressions. Because it can be confusing to have more than one set of parentheses, brackets and braces are also used. How do we know what to do first when we see an expression like $2[5 - 4(a + b)]$?

To simplify the expression, we start with the innermost grouping symbols. Here it is a set of parentheses. We first use the distributive property to multiply.

$$2[5 - 4(a + b)] = 2[5 - 4a - 4b]$$

We use the distributive law again.

$$= 10 - 8a - 8b$$

There are no like terms, so this is our final answer.

Notice that we started with two sets of grouping symbols, but our final answer has none. So we can say we *removed* the grouping symbols. Of course, we didn't just take them away; we used the distributive property and the rules for real numbers to simplify as much as possible. Although simplifying expressions like this involves many steps, we sometimes say "remove parentheses" as a shorthand direction. Sometimes we say "simplify."

**Remember to remove the innermost grouping symbols first.** Keep working from the inside out.

### Example 1  Simplify. $3[6 - 2(x + y)]$

**Solution**   We want to remove the innermost parentheses first. Therefore, we first use the distributive property to simplify $-2(x + y)$.

$$3[6 - 2(x + y)] = 3[6 - 2x - 2y] \quad \text{Use the distributive property.}$$
$$= 18 - 6x - 6y \quad \text{Use the distributive property again.} \quad \square$$

 **Student Practice 1**   Simplify. $5[4x - 3(y - 2)]$

You recall that a negative sign in front of parentheses is equivalent to having a coefficient of negative 1. You can write the $-1$ and then multiply by $-1$ using the distributive property.

$$-(x + 2y) = -1(x + 2y) = -x - 2y$$

Notice that this has the effect of removing the parentheses. Each term in the result now has its sign changed.

Similarly, a positive sign in front of parentheses can be viewed as multiplication by $+1$.

$$+(5x - 6y) = +1(5x - 6y) = 5x - 6y$$

If a grouping symbol has a positive or negative sign in front, we mentally multiply by $+1$ or $-1$, respectively.

**Fraction bars** are also considered grouping symbols. In exercises 39 and 40 of Section 1.8, your first step was to simplify expressions above and below the fraction bars. Later in this book we will encounter further examples requiring the same first step. This type of operation will have some similarities to the operation of removing parentheses.

**Example 2** Simplify. $-2[3a - (b + 2c) + (d - 3e)]$

**Solution**

$= -2[3a - b - 2c + d - 3e]$  Remove the two innermost sets of parentheses. Since one is not inside the other, we remove both sets at once.

$= -6a + 2b + 4c - 2d + 6e$  Now we remove the brackets by multiplying each term by $-2$.  ◻

**Student Practice 2**  Simplify. $-3[2a - (3b - c) + 4d]$

**Example 3** Simplify. $2[3x - (y + w)] - 3[2x + 2(3y - 2w)]$

**Solution**

$= 2[3x - y - w] - 3[2x + 6y - 4w]$  In each set of brackets, remove the inner parentheses.

$= 6x - 2y - 2w - 6x - 18y + 12w$  Remove each set of brackets by multiplying by the appropriate number.

$= -20y + 10w$ or $10w - 20y$  Combine like terms: $6x - 6x = 0x = 0$.  ◻

**Student Practice 3**  Simplify. $3[4x - 2(1 - x)] - [3x + (x - 2)]$

You can always simplify problems with many sets of grouping symbols by the method shown. Essentially, you just keep removing one level of grouping symbols at each step. Finally, at the end you combine any like terms remaining if possible.

Sometimes it is possible to combine like terms at each step.

**Example 4** Simplify. $-3\{7x - 2[x - (2x - 1)]\}$

**Solution**

$= -3\{7x - 2[x - 2x + 1]\}$  Remove the inner parentheses by multiplying each term within the parentheses by $-1$.

$= -3\{7x - 2[-x + 1]\}$  Combine like terms by combining $+x - 2x$.

$= -3\{7x + 2x - 2\}$  Remove the brackets by multiplying each term within them by $-2$.

$= -3\{9x - 2\}$  Combine the $x$ terms.

$= -27x + 6$  Remove the braces by multiplying each term by $-3$.  ◻

**Student Practice 4**

Simplify. $-2\{5x - 3[2x - (3 - 4x)]\}$

🐾 **STEPS TO SUCCESS**  How Important Is the Quick Quiz?

At the end of each exercise set is a Quick Quiz. Please be sure to do that. You will be amazed how it helps you in proving that you have mastered the homework. Make sure you do that each time as a necessary part of your homework. It will really help you gain confidence in what you have learned. The Quiz is essential.

Here is a fun way you can use the Quick Quiz. Ask a friend in the class if he or she will "quiz you" about 5 minutes before class by asking you to work out (without using your book) the solution to one of the problems on the Quick Quiz. Tell your friend you can do the same thing for him or for her. This will

force you to "be ready for anything" when you come to class. You will be amazed how this little trick will keep you sharp and ready for class.

**Making it personal:** Which of these suggestions do you find most helpful? Use those suggestions as you do each section. ▼

## 1.9 Exercises  MyMathLab®

**Verbal and Writing Skills, Exercises 1–4**

1. Rewrite the expression $-3x - 2y$ using a negative sign and parentheses.

2. Rewrite the expression $-x + 5y$ using a negative sign and parentheses.

3. To simplify expressions with grouping symbols, we use the _____ property.

4. When an expression contains many grouping symbols, remove the _____ grouping symbols first.

*Simplify. Remove grouping symbols and combine like terms.*

5. $8x - 4(x - 3y)$

6. $-6y - 2(x - 5y)$

7. $5(c - 3d) - (3c + d)$

8. $6(2c - d) - (4c + d)$

9. $-3(x + 3y) + 2(2x + y)$

10. $-4(x + 5y) + 3(6y - 3x)$

11. $2x[4x^2 - 2(x - 3)]$

12. $4y[-3y^2 + 2(4 - y)]$

13. $2[5(x + y) - 2(3x - 4y)]$

14. $-3[2(3a + b) - 5(a - 2b)]$

15. $[10 - 4(x - 2y)] + 3(2x + y)$

16. $[-5(-x + 3y) - 12] - 4(2x - 3)$

17. $5[3a - 2a(3a + 6b) + 6a^2]$

18. $3[x - y(3x + y) + y^2]$

19. $6a(2a^2 - 3a - 4) - a(a - 2)$

20. $7b(3b^2 - 2b - 5) - 2b(4 - b)$

21. $3a^2 - 4[2b - 3b(b + 2)]$

22. $2b^2 - 3[5b + 2b(2 - b)]$

**23.** $5b + \{-[3a + 2(5a - 2b)] - 1\}$

**24.** $3a - \{-2[a - 4(3a + b)] - 2\}$

**25.** $2\{3x^2 + 5[2x - (3 - x)]\}$

**26.** $4\{4x^2 + 2[3x - (4 - x)]\}$

**27.** $-4\{3a^2 - 2[4a^2 - (b + a^2)]\}$

**28.** $-3\{x^2 - 5[x - (x - 3x^2)]\}$

## Cumulative Review

**29.** **[1.8.2]** *Melting Point of Gold* The melting point of pure gold is 1064.18°C. Use $F = 1.8C + 32$ to find the melting point of pure gold in degrees Fahrenheit. Round to the nearest hundredth of a degree.

▲ **30.** **[1.8.2]** *Geometry* Use 3.14 as an approximation for $\pi$ to compute the area covered by a circular irrigation system with radial arm of length 380 feet. Use $A = \pi r^2$.

**31.** **[1.8.2]** *Dog Weight* An average Great Dane weighs between 120 and 150 pounds. Express the range of weight for a Great Dane in kilograms. Use the formula $k = 0.45p$ (where $k =$ kilograms, $p =$ pounds).

**32.** **[1.8.2]** *Dog Weight* An average Miniature Pinscher weighs between 9 and 14 pounds. Express the range of weight for a Miniature Pinscher in kilograms. Use the formula $k = 0.45p$ (where $k =$ kilograms, $p =$ pounds).

Quick Quiz 1.9 *Simplify.*

**1.** $2[3x - 2(5x + y)]$

**2.** $3[x - 3(x + 4) + 5y]$

**3.** $-4\{2a + 2[2ab - b(1 - a)]\}$

**4.** **Concept Check** Explain how you would simplify the following expression to combine like terms whenever possible. $3\{2 - 3[4x - 2(x + 3) + 5x]\}$

### Background: Electrical Engineer

Dan Silva is employed as an electrical engineering assistant. In order to finish a bid for a new client, Dan's manager has given him a few problems to solve involving the use of math and formulas.

### Facts

- Each of two conductors carries 12A (amps) and has a voltage drop of 1.8V (volts). The formula used to calculate power loss is $P = I \times E$, where $P$ is the power loss in watts (W), $I$ is the current, and $E$ is the voltage.

- A third conductor has a voltage drop of 3% in a 240V circuit carrying a current of 24A.

- An electrical circuit supply is 120V and its resistance is 200 ohms. The formula for current flow is $I = \dfrac{E}{R}$, where $I$ is the current flow, $E$ is the voltage, and $R$ is the resistance.

- A circuit box has the dimensions 4 inches by 4 inches by $1\frac{1}{2}$ inches, and volume $(V)$ = length $(L)$ × width $(W)$ × height $(H)$.

### Tasks

1. Dan must calculate the power loss of each conductor that carries 12A (amps) and has a voltage drop of 1.8V (volts) along with the total power loss. What are these amounts?

2. Dan needs to calculate the power loss in watts for the conductor having a voltage drop of 3% in the 240V circuit carrying a current of 24A. What is this amount?

3. Dan must determine the amount of current flow in amps (A) in the circuit where the supply is 120V and the resistance is 200 ohms using the formula $I = \dfrac{E}{R}$, where $I$ is the current flow, $E$ is the voltage, and $R$ is the resistance. What is this amount?

4. Dan's final calculation involves a little geometry. He needs to determine the total space inside seven 4-by-4-by-$1\frac{1}{2}$-inch circuit boxes. What is this total volume?

# Chapter 1 Organizer

| Topic and Procedure | Examples | ▷ You Try It |
|---|---|---|
| **Absolute value, p. 67**<br><br>The absolute value of a number is the distance between that number and zero on the number line. The absolute value of any number will be positive or zero. | Find the absolute value.<br>(a) $\|3\| = 3$  (b) $\|-2\| = 2$<br>(c) $\|0\| = 0$  (d) $\left\|-\dfrac{5}{6}\right\| = \dfrac{5}{6}$<br>(e) $\|-1.38\| = 1.38$ | **1.** Find the absolute value.<br>(a) $\|5\|$  (b) $\|-1\|$<br>(c) $\|0.5\|$  (d) $\left\|-\dfrac{1}{4}\right\|$<br>(e) $\|-4.57\|$ |
| **Adding real numbers with the same sign, p. 68**<br><br>If the signs are the same, add the absolute values of the numbers. Use the common sign in the answer. | Add. $-3 + (-7) = -10$ | **2.** Add. $-10 + (-4)$ |
| **Adding real numbers with different signs, p. 70**<br><br>If the signs are different:<br>1. Find the difference between the larger and the smaller absolute values.<br>2. Give the answer the sign of the number having the larger absolute value. | Add.<br>(a) $(-7) + 13 = 6$<br>(b) $7 + (-13) = -6$ | **3.** Add.<br>(a) $(-5) + 11$<br>(b) $5 + (-11)$ |
| **Adding several real numbers, p. 71**<br><br>When adding several real numbers, separate them into two groups by sign. Find the sum of all the positive numbers and the sum of all the negative numbers. Combine these two subtotals by the method described above. | Add. $-7 + 6 + 8 + (-11) + (-13) + 22$<br><br>$\begin{array}{rr} -7 & +6 \\ -11 & +8 \\ \underline{-13} & \underline{+22} \\ -31 & +36 \end{array}$<br><br>$-31 + 36 = 5$<br>The answer is positive since 36 is positive. | **4.** Add.<br>$-4 + 1 + (-8) + 12 + (-3) + 5$ |
| **Subtracting real numbers, p. 76**<br><br>Add the opposite of the second number. | Subtract.<br>$-3 - (-13) = -3 + (+13) = 10$ | **5.** Subtract. $-8 - (-7)$ |
| **Multiplying and dividing real numbers, pp. 82, 84**<br><br>1. If the two numbers have the same sign, multiply (or divide) the absolute values. The result is positive.<br>2. If the two numbers have different signs, multiply (or divide) the absolute values. The result is negative. | Multiply or divide.<br>(a) $-5(-3) = +15$<br>(b) $-36 \div (-4) = +9$<br>(c) $28 \div (-7) = -4$<br>(d) $-6(3) = -18$ | **6.** Multiply or divide.<br>(a) $9(-6)$<br>(b) $24 \div (-3)$<br>(c) $-48 \div (-8)$<br>(d) $-3(-7)$ |
| **Exponent form, p. 91**<br><br>The base tells you what number is being multiplied. The exponent tells you how many times this number is used as a factor. | Evaluate.<br>(a) $2^5 = 2 \cdot 2 \cdot 2 \cdot 2 \cdot 2 = 32$<br>(b) $4^3 = 4 \cdot 4 \cdot 4 = 64$<br>(c) $(-3)^4 = (-3)(-3)(-3)(-3) = 81$ | **7.** Evaluate.<br>(a) $3^4$  (b) $(1.5)^2$<br>(c) $\left(\dfrac{1}{2}\right)^4$ |
| **Raising a negative number to a power, p. 92**<br><br>When the base is negative, the result is positive for even exponents and negative for odd exponents. | Evaluate.<br>(a) $(-3)^3 = -27$<br>(b) $(-2)^4 = 16$ | **8.** Evaluate.<br>(a) $(-2)^3$  (b) $(-4)^4$ |
| **Order of operations, p. 96**<br><br>Remember the proper order of operations:<br>1. Perform operations inside parentheses.<br>2. Raise to powers.<br>3. Multiply and divide from left to right.<br>4. Add and subtract from left to right. | Evaluate.<br>$3(5 + 4)^2 - 2^2 \cdot 3 \div (9 - 2^3)$<br>$= 3 \cdot 9^2 - 4 \cdot 3 \div (9 - 8)$<br>$= 3 \cdot 81 - 12 \div 1$<br>$= 243 - 12 = 231$ | **9.** Evaluate.<br>$4^2 + 2(6 - 3)^3 - (5 - 2)^2 \div 3$ |
| **Removing parentheses, p. 101**<br><br>Use the distributive property to remove parentheses. $a(b + c) = ab + ac$ | Multiply.<br>(a) $3(5x + 2) = 15x + 6$<br>(b) $-4(x - 3y) = -4x + 12y$ | **10.** Multiply.<br>(a) $4(2a - 3)$<br>(b) $-5(5x - 1)$ |
| **Combining like terms, p. 106**<br><br>Combine terms that have identical variables and exponents. | Simplify.<br>$7x^2 - 3x + 4y + 2x^2 - 8x - 9y = 9x^2 - 11x - 5y$ | **11.** Simplify.<br>$9a^2 - 10a + 3ab + 7a - 12a^2 + 5ab$ |

| Topic and Procedure | Examples | You Try It |
|---|---|---|
| **Substituting into variable expressions, p. 111**<br><br>1. Replace each variable by the numerical value given for it.<br>2. Follow the order of operations in evaluating the expression. | Evaluate. $2x^3 + 3xy + 4y^2$ for $x = -3$ and $y = 2$.<br>$2(-3)^3 + 3(-3)(2) + 4(2)^2$<br>$\qquad = 2(-27) + 3(-3)(2) + 4(4)$<br>$\qquad = -54 - 18 + 16$<br>$\qquad = -56$ | **12.** Evaluate. $6x^2 - xy + 3y^2$ for $x = 4$ and $y = -1$. |
| **Using formulas, p. 112**<br><br>1. Replace the variables in the formula by the given values.<br>2. Evaluate the expression.<br>3. Label units carefully. | Find the area of a circle with radius 4 feet. Use $A = \pi r^2$, with $\pi$ as approximately 3.14.<br>$A \approx (3.14)(4\ \text{ft})^2$<br>$\quad \approx (3.14)(16\ \text{ft}^2)$<br>$\quad \approx 50.24\ \text{ft}^2$<br>The area of the circle is approximately 50.24 square feet. | **13.** Find the area of a trapezoid whose altitude is 50 feet and whose bases are 40 feet and 60 feet. |
| **Removing grouping symbols, p. 118**<br><br>1. Remove innermost grouping symbols first.<br>2. Then remove remaining innermost grouping symbols.<br>3. Continue until all grouping symbols are removed.<br>4. Combine like terms. | Simplify.<br>$5\{3x - 2[4 + 3(x - 1)]\}$<br>$\quad = 5\{3x - 2[4 + 3x - 3]\}$<br>$\quad = 5\{3x - 8 - 6x + 6\}$<br>$\quad = 15x - 40 - 30x + 30$<br>$\quad = -15x - 10$ | **14.** Simplify.<br>$4\{9x - [2(x + 3) - 8]\}$ |

# Chapter 1 Review Problems

## Section 1.1

*Add.*

**1.** $-6 + (-2)$     **2.** $-12 + 7.8$     **3.** $5 + (-2) + (-12)$     **4.** $3.7 + (-1.8)$

**5.** $\dfrac{1}{2} + \left(-\dfrac{5}{6}\right)$     **6.** $-\dfrac{3}{11} + \left(-\dfrac{1}{22}\right)$     **7.** $\dfrac{3}{4} + \left(-\dfrac{1}{12}\right) + \left(-\dfrac{1}{2}\right)$     **8.** $\dfrac{2}{15} + \dfrac{1}{6} + \left(-\dfrac{4}{5}\right)$

## Section 1.2

*Add or subtract.*

**9.** $5 - (-3)$     **10.** $-2 - (-15)$     **11.** $-30 - (+3)$     **12.** $8 - (-1.2)$

**13.** $-\dfrac{7}{8} + \left(-\dfrac{3}{4}\right)$     **14.** $-\dfrac{3}{8} + \dfrac{5}{6}$     **15.** $-20.8 - 1.9$     **16.** $-151 - (-63)$

## Section 1.3

*Multiply or divide.*

**17.** $87 \div (-29)$     **18.** $-10.4 \div (-0.8)$     **19.** $\dfrac{-24}{-\dfrac{3}{4}}$     **20.** $-\dfrac{2}{3} \div \left(-\dfrac{4}{5}\right)$

**21.** $\dfrac{5}{7} \div \left(-\dfrac{5}{25}\right)$     **22.** $-6(3)(4)$     **23.** $-1(-4)(-3)(-5)$     **24.** $(-5)\left(-\dfrac{1}{2}\right)(4)(-3)$

## Section 1.4

*Evaluate.*

**25.** $(-3)^5$     **26.** $(-2)^6$     **27.** $(-5)^4$     **28.** $\left(-\dfrac{2}{3}\right)^3$

**29.** $-9^2$     **30.** $(0.6)^2$     **31.** $\left(\dfrac{5}{6}\right)^2$     **32.** $\left(\dfrac{3}{4}\right)^3$

## Section 1.5

*Simplify using the order of operations.*

**33.** $5(-4) + 3(-2)^3$

**34.** $8 \div 0.4 + 0.1 \times (0.2)^2$

**35.** $(3 - 6)^2 + (-12) \div (-3)(-2)$

## Section 1.6

*Use the distributive property to multiply.*

**36.** $7(-3x + y)$

**37.** $3x(6 - x + 3y)$

**38.** $-(7x^2 - 3x + 11)$

**39.** $(2xy + x - y)(-3y^2)$

## Section 1.7

*Combine like terms.*

**40.** $3a^2b - 2bc + 6bc^2 - 8a^2b - 6bc^2 + 5bc$

**41.** $9x + 11y - 12x - 15y$

**42.** $4x^2 - 13x + 7 - 9x^2 - 22x - 16$

**43.** $-x + \dfrac{1}{2} + 14x^2 - 7x - 1 - 4x^2$

## Section 1.8

*Evaluate for the given value of the variables.*

**44.** $7x - 6$ for $x = -7$

**45.** $7 - \dfrac{3}{4}x$ for $x = 8$

**46.** $x^2 + 3x - 4$ for $x = -3$

**47.** $-x^2 + 5x - 9$ for $x = 3$

**48.** $2x^3 - x^2 + 6x + 9$ for $x = -1$

**49.** $b^2 - 4ac$ for $a = -1$, $b = 5$, and $c = -2$

**50.** $\dfrac{mMG}{r^2}$ for $m = -4$, $M = 15$, $G = -1$, and $r = -2$

*Solve.*

**51.** ***Simple Interest*** Find the simple interest on a loan of $6000 at an annual interest rate of 18% per year for $\frac{3}{4}$ of a year. Use $I = prt$, where $p$ = principal, $r$ = rate per year (in decimal form), and $t$ = time in years.

**52.** ***Medication*** The label of a medication warns the user that it must be stored at a temperature between 20°C and 25°C. What is this temperature range in degrees Fahrenheit? Use the formula $F = \dfrac{9C + 160}{5}$.

▲ **53.** ***Sign Painting*** How much will it cost to paint a circular sign with a radius of 4 feet if the painter charges $1.50 per square foot? Use $A = \pi r^2$, where $\pi$ is approximately 3.14.

**54.** ***Profit*** Find the daily profit $P$ at a furniture factory if the initial cost of setting up the factory is $C = \$1200$, rent $R = \$300$, and sale price of furniture $S = \$56$. Use the profit formula $P = 180S - R - C$.

▲ **55.** ***Parking Lot Sealer*** A parking lot is in the shape of a trapezoid. The altitude of the trapezoid is 200 feet, and the bases of the trapezoid are 300 feet and 700 feet. What is the area of the parking lot? If the parking lot had a sealer applied that costs $2 per square foot, what was the cost of the amount of sealer needed for the entire parking lot?

▲ **56.** ***Signal Paint*** The Green Mountain Telephone Company has a triangular signal tester at the top of a communications tower. The altitude of the triangle is 3.8 feet and the base is 5.5 feet. What is the area of the triangular signal tester? If the signal tester was painted with a special metallic surface paint that costs $66 per square foot, what was the cost of the amount of paint needed to paint one side of the triangle?

## Section 1.9

*Simplify.*

**57.** $5x - 7(x - 6)$

**58.** $3(x - 2) - 4(5x + 3)$

**59.** $2[3 - (4 - 5x)]$

**60.** $-3x[x + 3(x - 7)]$

**61.** $2xy^3 - 6x^3y - 4x^2y^2 + 3(xy^3 - 2x^2y - 3x^2y^2)$

**62.** $-5(x + 2y - 7) + 3x(2 - 5y)$

**63.** $-(a + 3b) + 5[2a - b - 2(4a - b)]$

**64.** $-5\{2a - [5a - b(3 + 2a)]\}$

**65.** $-3\{2x - [x - 3y(x - 2y)]\}$

**66.** $2\{3x + 2[x + 2y(x - 4)]\}$

## Mixed Practice

*Simplify the following.*

**67.** $-6.3 + 4$

**68.** $4 + (-8) + 12$

**69.** $-\dfrac{2}{3} - \dfrac{4}{5}$

**70.** $-\dfrac{7}{8} - \left(-\dfrac{3}{4}\right)$

**71.** $3 - (-4) + (-8)$

**72.** $-1.1 - (-0.2) + 0.4$

**73.** $\left(-\dfrac{9}{10}\right)\left(-2\dfrac{1}{4}\right)$

**74.** $3.6 \div (-0.45)$

**75.** $-14.4 \div (-0.06)$

**76.** $(-8.2)(3.1)$

**77.** ***Jeopardy*** A Jeopardy quiz show contestant began the second round (Double Jeopardy) with $400. She buzzed in on the first two questions, answering a $1000 question correctly but then giving the incorrect answer to a $800 question. What was her score?

*Simplify the following.*

**78.** $(-0.3)^4$

**79.** $-0.5^4$

**80.** $9(5) - 5(2)^3 + 5$

**81.** $3.8x - 0.2y - 8.7x + 4.3y$

**82.** Evaluate $\dfrac{2p + q}{3q}$ for $p = -2$ and $q = 3$.

**83.** Evaluate $\dfrac{4s - 7t}{s}$ for $s = -3$ and $t = -2$.

**84.** ***Dog Body Temperature*** The normal body temperature of a dog is 38.6°C. Your dog has a temperature of 101.1°F. Does your dog have a fever? Use the formula $F = \dfrac{9}{5}C + 32$.

**85.** $-7(x - 3y^2 + 4) + 3y(4 - 6y)$

**86.** $-2\{6x - 3[7y - 2y(3 - x)]\}$

126

# How Am I Doing? Chapter 1 Test

**MATH COACH**     **MyMathLab®**     **You Tube™**

After you take this test read through the Math Coach on pages 128–129. Math Coach videos are available via MyMathLab and YouTube. Step-by-step test solutions on the Chapter Test Prep Videos are also available via MyMathLab and YouTube. (Search "TobeyBeginningAlg" and click on "Channels.")

*Simplify.*

**1.** $-2.5 + 6.3 + (-4.1)$     **2.** $-5 - (-7)$     **3.** $\left(-\dfrac{2}{3}\right)(7)$

**4.** $-5(-2)(7)(-1)$     **5.** $-12 \div (-3)$     **6.** $-1.8 \div (0.6)$

**7.** $(-4)^3$     **8.** $(1.6)^2$     **MC 9.** $\left(\dfrac{2}{3}\right)^4$

**10.** $(0.2)^2 - (2.1)(-3) + 0.46$     **MC 11.** $3(4-6)^3 + 12 \div (-4) + 2$

**12.** $-5x(x + 2y - 7)$     **13.** $-2ab^2(-3a - 2b + 7ab)$

**14.** $6ab - \dfrac{1}{2}a^2b + \dfrac{3}{2}ab + \dfrac{5}{2}a^2b$     **15.** $2.3x^2y - 8.1xy^2 + 3.4xy^2 - 4.1x^2y$

**16.** $3(2 - a) - 4(-6 - 2a)$     **17.** $5(3x - 2y) - (x + 6y)$

*In questions 18–20, evaluate for the values of the variables indicated.*

**18.** $x^3 - 3x^2y + 2y - 5$ for $x = 3$ and $y = -4$

**MC 19.** $3x^2 - 7x - 11$ for $x = -3$

**20.** $2a - 3b$ for $a = \dfrac{1}{3}$ and $b = -\dfrac{1}{2}$

**21.** If you are traveling 60 miles per hour on a highway in Canada, how fast are you traveling in kilometers per hour? (Use $k = 1.61r$, where $r$ = rate in miles per hour and $k$ = rate in kilometers per hour.)

▲ **22.** A field is in the shape of a trapezoid. The altitude of the trapezoid is 120 feet and the bases of the trapezoid are 180 feet and 200 feet. What is the area of the field?

▲ **23.** Jeff Slater's garage has a triangular roof support beam. The support beam is covered with a sheet of plywood. The altitude of the triangular region is 6.8 feet and the base is 8.5 feet. If the triangular piece of plywood was painted with paint that cost $0.80 per square foot, what was the cost of the amount of paint needed to coat one side of the triangle?

▲ **24.** You wish to apply blacktop sealer to your driveway but do not know how much to buy. If your rectangular driveway measures 60 feet long by 10 feet wide, and a can of blacktop sealer claims to cover 200 square feet, how many cans should you buy?

*Simplify.*

**25.** $3[x - 2y(x + 2y) - 3y^2]$     **MC 26.** $-3\{a + b[3a - b(1 - a)]\}$

1. _____ ☐
2. _____ ☐
3. _____ ☐
4. _____ ☐
5. _____ ☐
6. _____ ☐
7. _____ ☐
8. _____ ☐
9. _____ ☐
10. _____ ☐
11. _____ ☐
12. _____ ☐
13. _____ ☐
14. _____ ☐
15. _____ ☐
16. _____ ☐
17. _____ ☐
18. _____ ☐
19. _____ ☐
20. _____ ☐
21. _____ ☐
22. _____ ☐
23. _____ ☐
24. _____ ☐
25. _____ ☐
26. _____ ☐

**Total Correct:** ☐

# MATH COACH

*Mastering the skills you need to do well on the test.*

The following problems are from the Chapter 1 Test. Here are some helpful hints to keep you from making common errors on test problems.

**Chapter 1 Test, Problem 9**  Simplify. $\left(\frac{2}{3}\right)^4$

> **HELPFUL HINT**  Always write out the repeated multiplication.
>
> - If a number is raised to the fourth power, then we write out the multiplication with that number appearing as a factor a total of four times.
> - When the base is a fraction, this means that we must multiply the numerators and then multiply the denominators.

Did you rewrite the problem as $\left(\frac{2}{3}\right)\left(\frac{2}{3}\right)\left(\frac{2}{3}\right)\left(\frac{2}{3}\right)$?

Yes _____ No _____

If you answered No to this question, please complete this step now.

Did you multiply $2 \times 2 \times 2 \times 2$ in the numerator and $3 \times 3 \times 3 \times 3$ in the denominator?

Yes _____ No _____

If you answered No, make this correction and complete the calculations.

If you answered Problem 9 incorrectly, go back and rework the problem using these suggestions.

---

**Chapter 1 Test, Problem 11**  Simplify. $3(4 - 6)^3 + 12 \div (-4) + 2$

> **HELPFUL HINT**  First, do all operations inside parentheses. Second, raise numbers to a power. Then multiply and divide numbers from left to right. As the last step, add and subtract numbers from left to right.

Did you first combine $4 - 6$ to obtain $-2$?

Yes _____ No _____

If you answered No, perform the operation inside the parentheses first.

Next, did you calculate $(-2)^3 = -8$ *before* multiplying by 3?

Yes _____ No _____

If you answered No, go back and evaluate the exponential expression.

Did you multiply and divide *before* adding?

Yes _____ No _____

If you answered No, remember that multiplication and division must be performed before addition and subtraction once the operations inside the parentheses are done and exponents are evaluated.

Now go back and rework the problem using these suggestions.

Need more help? Watch the **MATH COACH** videos in MyMathLab® or on You Tube .

**Chapter 1 Test, Problem 19**   Evaluate $3x^2 - 7x - 11$ for $x = -3$.

> **HELPFUL HINT**  When you replace a variable by a particular value, place parentheses around that value. Then use the order of operations to evaluate the expression.

Did you first rewrite the expression as $3(-3)^2 - 7(-3) - 11$, using parentheses to complete the substitution?

Yes _____ No _____

If you answered No, please go back and perform that substitution step using parentheses around the specified value.

Did you next raise $-3$ to the second power to obtain $3(9) - 7(-3) - 11$ ?

Yes _____ No _____

If you answered No, review the order of operations and complete this step.

Remember that multiplication must be performed before addition and subtraction.

If you answered Problem 19 incorrectly, go back and rework this problem using these suggestions.

---

**Chapter 1 Test, Problem 26**   Simplify. $-3\{a + b[3a - b(1 - a)]\}$

> **HELPFUL HINT**  Work from the inside out. Remove the innermost symbols, ( ), first. Then remove the next level of innermost symbols, [ ]. Finally remove the outermost symbols, $\{\}$. Be careful to avoid sign errors.

Did you first obtain the expression $-3\{a + b[3a - b + ab]\}$ when you removed the innermost parentheses?

Yes _____ No _____

If you answered No, go back to the original problem and use the distributive property to remove the innermost grouping symbol, the parentheses.

Did you next obtain the expression $-3\{a + 3ab - b^2 + ab^2\}$ when you removed the brackets?

Yes _____ No _____

If you answered No, use the distributive property to distribute the $b$ on the outside of the bracket, [ ], to each term inside the bracket.

Finally, use the distributive property in the last step to distribute the $-3$ across all of the terms inside the outermost brackets, $\{\}$. Be careful with the $+/-$ signs.

Now go back and rework the problem using these suggestions.

Need more help? Look for section examples marked with $\mathbb{MC}$ to review.

129

# Equations and Inequalities

## CAREER OPPORTUNITIES

### Dietitian, Nutritionist

Dietitians and nutritionists design food programs that promote healthy eating habits and lifestyles. They play important roles in assisting health care professionals by evaluating patients' nutritional needs and implementing individualized dietary plans. The mathematics covered in this chapter can be applied to the development of treatment plans for maintaining healthy, balanced lifestyles.

To investigate how the mathematics in this chapter can help with this field, see the Career Exploration Problems on page **176**.

# 2.1 The Addition Principle of Equality ▶

## 1 Using the Addition Principle to Solve Equations of the Form $x + b = c$ ▶

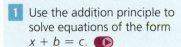

**Student Learning Objective**

After studying this section, you will be able to:

1 Use the addition principle to solve equations of the form $x + b = c$. ▶

When we use an equals sign ($=$), we are indicating that two expressions are equal in value. Such a statement is called an **equation.** For example, $x + 5 = 23$ is an equation. A **solution** of an equation is a number that when substituted for the variable makes the equation true. Thus 18 is a solution of $x + 5 = 23$ because $18 + 5 = 23$. Equations that have exactly the same solutions are called **equivalent equations.** By following certain procedures, we can often transform an equation to a simpler, equivalent one that has the form $x =$ some number. Then this number is a solution of the equation. The process of finding all solutions of an equation is called **solving the equation.**

One of the first procedures used in solving equations has an everyday application. Suppose that we place a 10-kilogram box on one side of a seesaw and a 10-kilogram stone on the other side. If the center of the box is the same distance from the balance point as the center of the stone, we would expect the seesaw to balance. The box and the stone do not look the same, but their weights are equal. If we add a 2-kilogram lead weight to the center of each object at the same time, the seesaw would still balance. The weights are still equal.

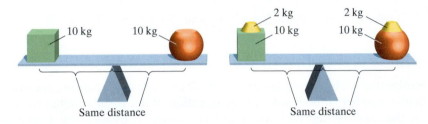

There is a similar principle in mathematics. We can state it in words as follows.

---

**THE ADDITION PRINCIPLE**

If the same number is added to both sides of an equation, the results on both sides are equal in value.

---

We can restate it in symbols this way.

For real numbers $a, b,$ and $c,$   if $a = b,$   then $a + c = b + c.$

Here is an example.

$$\text{If } 3 = \frac{6}{2}, \quad \text{then } 3 + 5 = \frac{6}{2} + 5.$$

Since we added the same amount, 5, to both sides, the sides remain equal to each other.

$$3 + 5 = \frac{6}{2} + 5$$

$$8 = \frac{6}{2} + \frac{10}{2}$$

$$8 = \frac{16}{2}$$

$$8 = 8$$

We can use the addition principle to solve certain equations.

**Example 1** Solve for $x$. $x + 16 = 20$

**Solution** $x + 16 + (-16) = 20 + (-16)$  Use the addition principle to add $-16$ to both sides.

$x + 0 = 4$  Simplify.

$x = 4$  The value of $x$ is 4.

We have just found a solution of the equation. A **solution** is a value for the variable that makes the equation true. We then say that the value 4 in our example **satisfies** the equation. We can easily verify that 4 is a solution by substituting this value into the original equation. This step is called **checking** the solution.

*Check.* $x + 16 = 20$

$4 + 16 \stackrel{?}{=} 20$

$20 = 20$ ✓

When the same value appears on both sides of the equals sign, we call the equation an **identity.** Because the two sides of the equation in our check have the same value, we know that the original equation has been solved correctly. We have found a solution, and since no other number makes the equation true, it is the only solution. ◻

 **Student Practice 1** Solve for $x$ and check your solution.

$$x + 14 = 23$$

Notice that when you are trying to solve these types of equations, you must add a particular number to both sides of the equation. What is the number to choose? Look at the number that is on the same side of the equation with $x$, that is, the number added to $x$. Then think of the number that is **opposite in sign.** This is called the **additive inverse** of the number. The additive inverse of 16 is $-16$. The additive inverse of $-3$ is 3. The number to add to both sides of the equation is precisely this additive inverse.

It does not matter which side of the equation contains the variable. The $x$-term may be on the right or left. In the next example the $x$-term will be on the right.

**Example 2** Solve for $x$. $14 = x - 3$

**Solution** $14 + 3 = x - 3 + 3$  Notice that $-3$ is being added to $x$ in the original equation. Add 3 to both sides, since 3 is the additive inverse of $-3$. This will eliminate the $-3$ on the right and isolate $x$.

$17 = x + 0$  Simplify.

$17 = x$  The value of $x$ is 17.

*Check.* $14 = x - 3$

$14 \stackrel{?}{=} 17 - 3$  Replace $x$ by 17.

$14 = 14$ ✓  Simplify. It checks. The solution is 17. ◻

**Student Practice 2** Solve for $x$ and check your solution. $17 = x - 5$

Before you add a number to both sides, you should always simplify the equation. The following example shows how combining numbers by addition—separately, on both sides of the equation—simplifies the equation.

**Example 3** Solve for $x$. $1.5 + 0.2 = 0.3 + x + 0.2$

**Solution**

$$1.5 + 0.2 = 0.3 + x + 0.2$$

$1.7 = x + 0.5$      Simplify by adding.

$1.7 + (-0.5) = x + 0.5 + (-0.5)$      Add the value $-0.5$ to both sides, since $-0.5$ is the additive inverse of $0.5$.

$1.2 = x$      Simplify. The value of $x$ is $1.2$.

*Check.*   $1.5 + 0.2 = 0.3 + x + 0.2$

$1.5 + 0.2 \overset{?}{=} 0.3 + 1.2 + 0.2$      Replace $x$ by $1.2$ in the original equation.

$1.7 = 1.7 \ \checkmark$      It checks.      □

▶ **Student Practice 3**   Solve for $x$ and check your solution.

$$0.5 - 1.2 = x - 0.3$$

In Example 3 we added $-0.5$ to each side. You could subtract $0.5$ from each side and get the same result. In Chapter 1 we discussed how subtracting a $0.5$ is the same as adding a negative $0.5$. Do you see why?

Just as it is possible to add the same number to both sides of an equation, it is also possible to subtract the same number from both sides of an equation. This is so because any subtraction problem can be rewritten as an addition problem. For example, $1.7 - 0.5 = 1.7 + (-0.5)$. Thus the addition principle tells us that we can subtract the same number from both sides of the equation.

We can determine whether a value is the solution to an equation by following the same steps used to check an answer. Substitute the value to be tested for the variable in the original equation. We will obtain an identity if the value is the solution.

**Example 4** Is $10$ the solution to the equation $-15 + 2 = x - 3$? If it is not, find the solution.

**Solution**   We substitute $10$ for $x$ in the equation and see if we obtain an identity.

$$-15 + 2 = x - 3$$

$$-15 + 2 \overset{?}{=} 10 - 3$$

$-13 \neq 7$      The values are not equal. The statement is not an identity.

Thus, $10$ is not the solution.

Now we take the original equation and solve to find the solution.

$$-15 + 2 = x - 3$$

$-13 = x - 3$      Simplify by adding.

$-13 + 3 = x - 3 + 3$      Add $3$ to both sides. $3$ is the additive inverse of $-3$.

$-10 = x$

*Check.*   Replace $x$ with $-10$ and verify that it is a solution.      □

Look at the solution above. Notice that if you make a sign error you could obtain $10$ as the solution instead of $-10$. As we saw above, $10$ is *not* a solution. We must be especially careful to write the correct sign for each number when solving equations.

▶ **Student Practice 4**   Is $-2$ the solution to the equation $x + 8 = -22 + 6$? If it is not, find the solution.

**Example 5** Find the value of $x$ that satisfies the equation.

$$\frac{1}{5} + x = -\frac{1}{10} + \frac{1}{2}$$

**Solution** To be combined, the fractions must have common denominators. The least common denominator (LCD) of the fractions is 10.

$$\frac{1 \cdot 2}{5 \cdot 2} + x = -\frac{1}{10} + \frac{1 \cdot 5}{2 \cdot 5}$$     Change each fraction to an equivalent fraction with a denominator of 10.

$$\frac{2}{10} + x = -\frac{1}{10} + \frac{5}{10}$$     This is an equivalent equation.

$$\frac{2}{10} + x = \frac{4}{10}$$     Simplify by adding.

$$\frac{2}{10} + \left(-\frac{2}{10}\right) + x = \frac{4}{10} + \left(-\frac{2}{10}\right)$$     Add the additive inverse of $\frac{2}{10}$ to each side. You could also say that you are subtracting $\frac{2}{10}$ from each side.

$$x = \frac{2}{10}$$     Add the fractions.

$$x = \frac{1}{5}$$     Simplify the answer.

*Check.* We substitute $\frac{1}{5}$ for $x$ and see if we obtain an identity.

$$\frac{1}{5} + x = -\frac{1}{10} + \frac{1}{2}$$     Write the original equation.

$$\frac{1}{5} + \frac{1}{5} \stackrel{?}{=} -\frac{1}{10} + \frac{1}{2}$$     Substitute $\frac{1}{5}$ for $x$.

$$\frac{2}{5} \stackrel{?}{=} -\frac{1}{10} + \frac{5}{10}$$

$$\frac{2}{5} \stackrel{?}{=} \frac{4}{10}$$

$$\frac{2}{5} = \frac{2}{5} \ \checkmark$$     It checks.     □

**Student Practice 5** Find the value of $x$ that satisfies the equation.

$$\frac{1}{20} - \frac{1}{2} = x + \frac{3}{5}$$

---

## ✏ STEPS TO SUCCESS Taking Good Notes in Every Class Session

***Don't copy down everything the teacher says.*** You will get overloaded with facts. Instead write down the important ideas and examples as the instructor lectures. Be sure to include any helpful hints or suggestions that your instructor gives you. You will be amazed at how easily you forget these if you do not write them down.

***Be an active listener.*** Keep your mind on what the instructor is saying. Be ready with questions whenever you do not understand something. Stay alert in class. Keep your mind on mathematics.

***Preview the lesson material.*** Before class, glance over the topics that will be covered. You can take better notes if you know the topics of the lecture ahead of time.

***Look back at your notes.*** Try to review them the same day sometime after class. You will find the content of your notes easier to understand if you read them again within a few hours of class.

**Making it personal:** Which of these suggestions are the most helpful to you? How can you improve your skills in note taking? ▼

## 2.1 Exercises MyMathLab®

**Verbal and Writing Skills, Exercises 1–6** *In exercises 1–3, fill in each blank with the appropriate word.*

1. When we use the _____ sign, we indicate two expressions are _____ in value.

2. If the _____ _____ is added to both sides of an equation, the results on each side are equal in value.

3. The _____ of an equation is a value of the variable that makes the equation true.

4. What is the additive inverse of $-20$?

5. Why do we add the additive inverse of $a$ to each side of $x + a = b$ to solve for $x$?

6. What is the additive inverse of $a$?

*Solve for x. Check your answers.*

7. $x + 14 = 21$

8. $x + 15 = 21$

9. $20 = 9 + x$

10. $23 = 8 + x$

11. $x - 3 = 14$

12. $x - 13 = 4$

13. $0 = x + 5$

14. $0 = x + 9$

15. $x - 6 = -19$

16. $x - 11 = -13$

17. $-12 + x = 50$

18. $-16 + x = 47$

19. $3 + 5 = x - 7$

20. $8 - 2 = x + 5$

21. $32 - 17 = x - 6$

22. $32 - 11 = x - 4$

23. $4 + 8 + x = 6 + 6$

24. $19 - 3 + x = 10 + 6$

25. $18 - 7 + x = 7 + 9 - 5$

26. $3 - 17 + 8 = 8 + x - 3$

27. $-12 + x - 3 = 15 - 18 + 9$

28. $-19 + x - 7 = 20 - 42 + 10$

*In exercises 29–36, determine whether the given solution is correct. If it is not, find the solution.*

29. Is $x = 5$ the solution to $-7 + x = 2$?

30. Is $x = 7$ the solution to $-13 + x = 4$?

31. Is $-6$ the solution to $-11 + 5 = x + 8$?

32. Is $-9$ the solution to $-13 - 4 = x - 8$?

33. Is $-33$ the solution to $x - 23 = -56$?

34. Is $-8$ the solution to $-39 = x - 47$?

35. Is $35$ the solution to $15 - 3 + 20 = x - 3$?

36. Is $-12$ the solution to $x + 8 = 12 - 19 + 3$?

135

*Find the value of x that satisfies each equation.*

**37.** $2.5 + x = 0.7$

**38.** $8.2 + x = 3.2$

**39.** $12.5 + x - 8.2 = 4.9$

**40.** $4.3 + x - 2.6 = 3.4$

**41.** $x - \dfrac{1}{4} = \dfrac{3}{4}$

**42.** $x + \dfrac{1}{3} = \dfrac{2}{3}$

**43.** $\dfrac{2}{3} + x = \dfrac{1}{6} + \dfrac{1}{4}$

**44.** $\dfrac{2}{5} + x = \dfrac{1}{2} - \dfrac{3}{10}$

**Mixed Practice** *Solve for x.*

**45.** $3 + x = -12 + 8$

**46.** $12 + x = -7 + 20$

**47.** $5\dfrac{1}{6} + x = 8$

**48.** $3\dfrac{3}{4} + x = 9$

**49.** $\dfrac{3}{14} - \dfrac{2}{7} = x - \dfrac{1}{2}$

**50.** $\dfrac{3}{16} - \dfrac{1}{4} = x - \dfrac{3}{8}$

**51.** $1.6 + x - 3.2 = -2 + 5.6$

**52.** $1.8 + x - 4.6 = -3 + 4.2$

**53.** $x - 18.225 = 1.975$

**54.** $x - 10.012 = -16.835$

**Cumulative Review** *Simplify by combining like terms.*

**55.** **[1.7.2]** $x + 3y - 5x - 7y + 2x$

**56.** **[1.7.2]** $y^2 + y - 12 - 3y^2 - 5y + 16$

**Quick Quiz 2.1** *Solve for the variable.*

**1.** $x - 4.7 = 9.6$

**2.** $-8.6 + x = -12.1$

**3.** $3 - 12 + 7 = 8 + x - 2$

**4.** **Concept Check** Explain how you would check to verify whether $x = 3.8$ is the solution to $-1.3 + 1.6 + 3x = -6.7 + 4x + 3.2$.

## 2.2  The Multiplication Principle of Equality ▶

### 1  Solving Equations of the Form $\frac{1}{a}x = b$ ▶

The addition principle allows us to add the same number to both sides of an equation. What would happen if we multiplied each side of an equation by the same number? For example, what would happen if we multiplied each side of an equation by 3?

To answer this question, let's return to our simple example of the box and the stone on a balanced seesaw. If we triple the weight on each side (that is, multiply the weight on each side by 3), the seesaw would still balance. The weight values of both sides remain equal.

**Student Learning Objectives**

After studying this section, you will be able to:

1 Solve equations of the form $\frac{1}{a}x = b$. ▶

2 Solve equations of the form $ax = b$. ▶

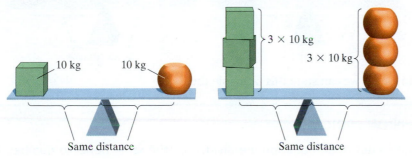

In words we can state this principle thus.

> **MULTIPLICATION PRINCIPLE**
>
> If both sides of an equation are multiplied by the same nonzero number, the results on both sides are equal in value.

In symbols we can restate the multiplication principle this way.

> For real numbers $a$, $b$, and $c$ with $c \neq 0$,   if $a = b$,   then $ca = cb$.

Let us look at an equation where it would be helpful to multiply each side by 3.

**Example 1** Solve for $x$. $\frac{1}{3}x = -15$

**Solution**  We know that $(3)\left(\frac{1}{3}\right) = 1$. We will multiply each side of the equation by 3 because we want to isolate the variable $x$.

$$3\left(\frac{1}{3}x\right) = 3(-15) \quad \text{Multiply each side of the equation by 3 since } (3)\left(\frac{1}{3}\right) = 1.$$

$$\left(\frac{3}{1}\right)\left(\frac{1}{3}\right)(x) = -45$$

$$1x = -45 \quad \text{Simplify.}$$

$$x = -45 \quad \text{The solution is } -45.$$

*Check.*  $\frac{1}{3}(-45) \overset{?}{=} -15 \quad \text{Substitute } -45 \text{ for } x \text{ in the original equation.}$

$$-15 = -15 \checkmark \quad \text{It checks.} \qquad \square$$

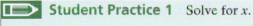 **Student Practice 1**  Solve for $x$.

$$\frac{1}{8}x = -2$$

Note that $\frac{1}{5}x$ can be written as $\frac{x}{5}$. To solve the equation $\frac{x}{5} = 3$, we could multiply each side of the equation by 5. Try it. Then check your solution.

## 2 Solving Equations of the Form $ax = b$ ▶

We can see that using the multiplication principle to multiply each side of an equation by $\frac{1}{2}$ is the same as dividing each side of the equation by 2. Thus, it would seem that the multiplication principle would allow us to divide each side of the equation by any nonzero real number. Is there a real-life example of this idea?

Let's return to our simple example of the box and the stone on a balanced seesaw. Suppose that we were to cut the two objects in half (so that the amount of weight of each was divided by 2). We then return the objects to the same places on the seesaw. The seesaw would still balance. The weight values of both sides remain equal.

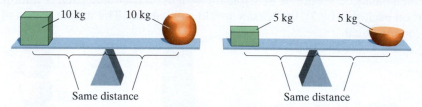

Same distance    Same distance

In words we can state this principle thus.

**DIVISION PRINCIPLE**

If both sides of an equation are divided by the same nonzero number, the results on both sides are equal in value.

*Note:* We put a restriction on the number by which we are dividing. We cannot divide by zero. We say that expressions like $\frac{2}{0}$ are not defined. Thus we restrict our divisor to *nonzero* numbers. We can restate the division principle this way.

For real numbers $a$, $b$, and $c$ where $c \neq 0$, if $a = b$, then $\dfrac{a}{c} = \dfrac{b}{c}$.

**Example 2** Solve for $x$. $5x = 125$

**Solution** $\dfrac{5x}{5} = \dfrac{125}{5}$    Divide both sides by 5.

$x = 25$    Simplify. The solution is 25.

*Check.*    $5x = 125$

$5(25) \overset{?}{=} 125$    Replace $x$ by 25.

$125 = 125$ ✓    It checks.    ☐

▣➡ **Student Practice 2**    Solve for $x$. $9x = 72$

For equations of the form $ax = b$ (a number multiplied by $x$ equals another number), we solve the equation by choosing to divide both sides by a particular number. What is the number to choose? We look at the side of the equation that contains $x$. We notice the number that is multiplied by $x$. We divide by that number. The division principle tells us that we can still have a true equation provided that we divide by that number *on both sides* of the equation.

The solution to an equation may be a proper fraction or an improper fraction.

**Example 3** Solve for $x$. $9x = 60$

**Solution** $\dfrac{9x}{9} = \dfrac{60}{9}$    Divide both sides by 9.

$x = \dfrac{20}{3}$ or $6\dfrac{2}{3}$    Simplify. The solution is $\dfrac{20}{3}$ or $6\dfrac{2}{3}$.

*Check.*     $9x = 60$

$$\overset{3}{\cancel{9}}\left(\frac{20}{\cancel{3}}\right) \overset{?}{=} 60 \qquad \text{Replace } x \text{ by } \frac{20}{3}.$$

$$60 = 60 \quad \checkmark \quad \text{It checks.} \qquad \qquad \square$$

**Student Practice 3**   Solve for $x$.

$$6x = 50$$

In Examples 2 and 3 we *divided by the number multiplied by x.* This procedure is followed regardless of whether the sign of that number is positive or negative. In equations of the form $ax = b$, $a$ is a number multiplied by $x$. The **coefficient** of $x$ is $a$. A coefficient is a multiplier.

**Sidelight**  As you work through the exercises in this book, you will notice that the solutions of equations can be integers, fractions, or decimals. Recall from page 65 that a **terminating decimal** is one that has a definite number of digits. Unless directions state that the solution should be rounded, the decimal form of a solution should be given only if it is a terminating decimal. Is the decimal form of the solution in Example 3 a terminating decimal?

**Example 4**  Solve for $x$. $-3x = 48$

**Solution**     $$\frac{-3x}{-3} = \frac{48}{-3} \qquad \text{Divide both sides by the coefficient } -3.$$

$$x = -16 \quad \text{The solution is } -16.$$

*Check.*   Can you check this solution?     $\square$

**Student Practice 4**   Solve for $x$. $-27x = 54$

The coefficient of $x$ may be 1 or $-1$. You may have to rewrite the equation so that the coefficient 1 or $-1$ is obvious. With practice you may be able to recognize the coefficient without actually rewriting the equation.

**Example 5**  Solve for $x$. $-x = -24$.

**Solution**     $$-1x = -24 \qquad \text{Rewrite the equation. } -1x \text{ is the same as } -x.$$
$$\text{Now the coefficient } -1 \text{ is obvious.}$$

$$\frac{-1x}{-1} = \frac{-24}{-1} \qquad \text{Divide both sides by } -1.$$

$$x = 24 \qquad \text{The solution is 24.}$$

*Check.*   Can you check this solution?     $\square$

**Student Practice 5**   Solve for $x$. $-x = 36$

The variable can be on either side of the equation.

**Example 6**  Solve for $x$. $-78 = 5x - 8x$

**Solution**     $$-78 = 5x - 8x \qquad \text{There are like terms on the right side.}$$
$$-78 = -3x \qquad \text{Combine like terms.}$$
$$\frac{-78}{-3} = \frac{-3x}{-3} \qquad \text{Divide both sides by } -3.$$
$$26 = x \qquad \text{The solution is 26.}$$

*Check.*   The check is left to you.     $\square$

*Continued on next page*

**Student Practice 6** Solve for $x$.
$$-51 = 3x - 9x$$

There is a mathematical concept that unites what we have learned in this section. The concept uses the idea of a multiplicative inverse. For any nonzero number $a$, the **multiplicative inverse** of $a$ is $\frac{1}{a}$. Likewise, for any nonzero number $a$, the multiplicative inverse of $\frac{1}{a}$ is $a$. So to solve an equation of the form $ax = b$, we say that we need to multiply each side by the multiplicative inverse of $a$. Thus to solve $5x = 45$, we would multiply each side of the equation by the multiplicative inverse of 5, which is $\frac{1}{5}$. In similar fashion, if we wanted to solve the equation $\frac{1}{6}x = 4$, we would multiply each side of the equation by the multiplicative inverse of $\frac{1}{6}$, which is 6. In general, all the problems we have covered so far in this section can be solved by multiplying both sides of the equation by the multiplicative inverse of the coefficient of $x$.

**Example 7** Solve for $x$. $31.2 = 6.0x - 0.8x$

**Solution**

$$31.2 = 6.0x - 0.8x \quad \text{There are like terms on the right side.}$$
$$31.2 = 5.2x \quad \text{Combine like terms.}$$
$$\frac{31.2}{5.2} = \frac{5.2x}{5.2} \quad \text{Divide both sides by 5.2 (which is the same as multiplying both sides by the multiplicative inverse of 5.2).}$$
$$6 = x \quad \text{The solution is 6.}$$

*Note:* Be sure to place the decimal point in the quotient directly above the caret ($\wedge$) when performing the division.

$$
\begin{array}{r}
6. \\
5.2_\wedge \overline{)31.2_\wedge} \\
\underline{31\ 2} \\
0
\end{array}
$$

*Check.* The check is left to you. □

**Student Practice 7** Solve for $x$. $16.2 = 5.2x - 3.4x$

---

 **STEPS TO SUCCESS** HELP! When Do I Get It? Where Do I Get It?

Getting the right kind of help at the right time can be the key ingredient to being successful in Beginning Algebra. When you have made every effort to learn the math on your own and you still need assistance, it is time to ask for help.

**When should you ask for help?** Ask as soon as you discover that you don't understand something. Don't wait until the night before a test. When you try the homework and have trouble and you are unable to clear up the difficulty in the next class period, *that is the time to seek help immediately.*

**Where do you go for help?** The best source is your instructor. Make an appointment to see your instructor and explain exactly where you are having trouble. If that is not possible, use the tutoring services at your college, visit the mathematics lab, watch the videos, use MyMathLab and let it help you, or call the 1-800 phone number for phone tutoring. You may

also want to talk with a classmate to see if you can help one another.

**Making it personal:** Look over these suggestions. Write down the one that you think would work best for you and then use it this week. Make it a priority in your life to get help immediately in this math course whenever you do not understand something. We all need a little help from time to time. ▼

## 2.2 Exercises    MyMathLab®

### Verbal and Writing Skills, Exercises 1–4

*In exercises 1–4, fill in the blank with the appropriate number.*

1. To solve the equation $6x = -24$, divide each side of the equation by _____.

2. To solve the equation $-7x = 56$, divide each side of the equation by _____.

3. To solve the equation $\frac{1}{7}x = -2$, multiply each side of the equation by _____.

4. To solve the equation $\frac{1}{9}x = 5$, multiply each side of the equation by _____.

*Solve for x. Be sure to simplify your answer. Check your solution.*

5. $\frac{1}{8}x = 6$

6. $\frac{1}{5}x = 12$

7. $\frac{1}{2}x = -15$

8. $\frac{1}{9}x = -8$

9. $\frac{x}{5} = 16$

10. $\frac{x}{12} = -7$

11. $-3 = \frac{x}{5}$

12. $\frac{x}{6} = -2$

13. $13x = 52$

14. $15x = 60$

15. $56 = 7x$

16. $46 = 2x$

17. $-16 = 6x$

18. $-35 = 21x$

19. $1.5x = 75$

20. $2x = 0.36$

21. $-15 = -x$

22. $32 = -x$

23. $-112 = 16x$

24. $-108 = -18x$

25. $0.4x = 0.08$

26. $2.5x = 0.5$

27. $-3.9x = -15.6$

28. $-4.7x = -14.1$

*Determine whether the given solution is correct. If not, find the correct solution.*

29. Is 7 the solution for $-3x = 21$?

30. Is 8 the solution for $5x = -40$?

31. Is $-15$ the solution for $-x = 15$?

32. Is $-8$ the solution for $-11x = 88$?

### Mixed Practice  *Find the value of the variable that satisfies the equation.*

33. $7y = -0.21$

34. $-6y = 2.16$

35. $-56 = -21t$

36. $26 = -39t$

37. $4.6y = -3.22$

38. $-2.8y = -3.08$

39. $4x + 3x = 21$

40. $5x + 4x = 36$

41. $2x - 7x = 20$

42. $3x - 9x = 18$

43. $\frac{1}{4}x = -9$

44. $\frac{1}{5}x = -4$

**45.** $12 - 19 = -7x$

**46.** $24 - 27 = -9x$

**47.** $8m = -14 + 30$

**48.** $8x = 26 - 50$

**49.** $\frac{3}{4}x = 63$

**50.** $\frac{5}{6}x = 40$

**51.** $-2.5133x = 26.38965$

**52.** $-5.42102x = -45.536568$

## Cumulative Review *Simplify.*

**53.** **[1.7.2]** $-3y(2x + y) + 5(3xy - y^2)$

**54.** **[1.9.1]** $-\{2(x - 3) + 3[x - (2x - 5)]\}$

**55.** **[0.6.1]** *Humpback Whales* During the summer months, a group of humpback whales gather at Stellwagen Bank, near Gloucester, Massachusetts, to feed. When they return to the Caribbean for winter, they will lose up to 25% of their body weight in blubber. If a humpback whale weighs 30 tons after feeding at Stellwagen Bank, how much will it weigh after losing 25% of its body weight?

**56.** **[0.6.1]** *Earthquakes* In an average year worldwide, there are 20 earthquakes of magnitude 7 on the Richter scale. If next year is predicted to be an exceptional year and the number of earthquakes of magnitude 7 is expected to increase by 35%, how many earthquakes of magnitude 7 can be expected?

## Quick Quiz 2.2 *Solve for the variable.*

**1.** $2.5x = -95$

**2.** $-3.9x = -54.6$

**3.** $7x - 12x = 60$

**4.** **Concept Check** Explain how you would check to verify whether $x = 36\frac{2}{3}$ is the solution to $-22 = -\frac{3}{5}x$.

## 2.3 Using the Addition and Multiplication Principles Together ▶

### 1 Solving Equations of the Form $ax + b = c$ ▶

Jenny Crawford scored several goals in field hockey during April. Her teammates scored three more than five times the number she scored for a total of 18 goals scored in April. How many did Jenny score? To solve this problem, we need to solve the equation $5x + 3 = 18$.

To solve an equation of the form $ax + b = c$, we must use both the addition principle and the multiplication or division principle.

**Student Learning Objectives**

After studying this section, you will be able to:

1 Solve equations of the form $ax + b = c$. ▶

2 Solve equations with the variable on both sides of the equation. ▶

3 Solve equations with parentheses. ▶

**Example 1** Solve for $x$ in the equation $5x + 3 = 18$ to determine how many goals Jenny scored. Check your solution.

**Solution** We first want to isolate the variable term.

$$5x + 3 = 18$$

$5x + 3 + (-3) = 18 + (-3)$    Use the addition principle to add $-3$ to both sides.

$5x = 15$    Simplify.

$\dfrac{5x}{5} = \dfrac{15}{5}$    Use the division principle to divide both sides by 5.

$x = 3$    The solution is 3. Thus Jenny scored 3 goals.

*Check.*   $5(3) + 3 \overset{?}{=} 18$

$15 + 3 \overset{?}{=} 18$

$18 = 18$ ✓   It checks.      □

▶ **Student Practice 1** Solve for $x$ and check your solution. $9x + 2 = 38$

### 2 Solving Equations with the Variable on Both Sides of the Equation ▶

In some cases the variable appears on both sides of the equation. We would like to rewrite the equation so that all the terms containing the variable appear on one side. To do this, we apply the addition principle to the variable term.

**Example 2** Solve for $x$. $9x = 6x + 15$

**Solution**

$9x + (-6x) = 6x + (-6x) + 15$    Add $-6x$ to both sides. Notice $6x + (-6x)$ eliminates the variable on the right side.

$3x = 15$    Combine like terms.

$\dfrac{3x}{3} = \dfrac{15}{3}$    Divide both sides by 3.

$x = 5$    The solution is 5.

*Check.*   The check is left to the student.      □

▶ **Student Practice 2** Solve for $x$. $13x = 2x - 66$

In many problems the variable terms and constant terms appear on both sides of the equation. You will want to get all the variable terms on one side and all the constant terms on the other side.

**Example 3** Solve for $x$ and check your solution. $9x + 3 = 7x - 2$

**Solution** First we want to isolate the variable term.

$$9x + (-7x) + 3 = 7x + (-7x) - 2 \qquad \text{Add } -7x \text{ to both sides of the equation.}$$
$$2x + 3 = -2 \qquad \text{Combine like terms.}$$
$$2x + 3 + (-3) = -2 + (-3) \qquad \text{Add } -3 \text{ to both sides.}$$
$$2x = -5 \qquad \text{Simplify.}$$
$$\frac{2x}{2} = \frac{-5}{2} \qquad \text{Divide both sides by 2.}$$
$$x = -\frac{5}{2} \text{ or } -2\frac{1}{2} \qquad \text{The solution is } -\frac{5}{2}.$$

*Check.* $9x + 3 = 7x - 2$

$$9\left(-\frac{5}{2}\right) + 3 \stackrel{?}{=} 7\left(-\frac{5}{2}\right) - 2 \qquad \text{Replace } x \text{ by } -\frac{5}{2}.$$
$$-\frac{45}{2} + 3 \stackrel{?}{=} -\frac{35}{2} - 2 \qquad \text{Simplify.}$$
$$-\frac{45}{2} + \frac{6}{2} \stackrel{?}{=} -\frac{35}{2} - \frac{4}{2} \qquad \text{Change to equivalent fractions with a common denominator.}$$
$$-\frac{39}{2} = -\frac{39}{2} \quad \checkmark \qquad \text{It checks. The solution is } -\frac{5}{2}. \qquad \square$$

> **Student Practice 3** Solve for $x$ and check your solution.
> $$3x + 2 = 5x + 2$$

In our next example we will study equations that need simplifying before any other steps are taken. Where it is possible, you should first collect like terms on one or both sides of the equation. The variable terms can be collected on the right side or the left side. In this example we will collect all the $x$-terms on the right side.

**Example 4** Solve for $y$. $5y + 26 - 6 = 9y + 12y$

**Solution** $\quad 5y + 26 - 6 = 9y + 12y$

$$5y + 20 = 21y \qquad \text{Combine like terms.}$$
$$5y + (-5y) + 20 = 21y + (-5y) \qquad \text{Add } -5y \text{ to both sides.}$$
$$20 = 16y \qquad \text{Combine like terms.}$$
$$\frac{20}{16} = \frac{16y}{16} \qquad \text{Divide both sides by 16.}$$
$$\frac{5}{4} \text{ or } 1\frac{1}{4} = y \qquad \text{Don't forget to reduce the resulting fraction.}$$

*Check.* The check is left to the student. $\qquad \square$

> **Student Practice 4** Solve for $z$.
> $$-z + 8 - z = 3z + 10 - 3$$

Do you really need all these steps? No. As you become more proficient you will be able to combine or eliminate some of these steps. However, it is best to write each step in its entirety until you are consistently obtaining the correct solution. It is much better to show every step than to take a lot of shortcuts and possibly obtain a wrong answer. This is a section of the algebra course where working neatly and accurately will help you—both now and as you progress through the course.

## 3 Solving Equations with Parentheses ▶

The equations that you just solved are simpler versions of equations that we will now discuss. These equations contain parentheses. If the parentheses are first removed, the problems then become just like those encountered previously. We use the distributive property to remove the parentheses.

**Example 5** Solve for $x$ and check your solution.

$$4(x + 1) - 3(x - 3) = 25$$

**Solution**   $4(x + 1) - 3(x - 3) = 25$     Multiply by 4 and $-3$ to remove
$$4x + 4 - 3x + 9 = 25$$     parentheses. Be careful of the signs. Remember that $(-3)(-3) = 9$.

After removing the parentheses, it is important to combine like terms on each side of the equation. Do this before going on to isolate the variable.

$$x + 13 = 25$$     Combine like terms.
$$x + 13 - 13 = 25 - 13$$     Subtract 13 from both sides to isolate the variable.
$$x = 12$$     The solution is 12.

*Check.*

$$4(12 + 1) - 3(12 - 3) \overset{?}{=} 25$$     Replace $x$ by 12.
$$4(13) - 3(9) \overset{?}{=} 25$$     Combine numbers inside parentheses.
$$52 - 27 \overset{?}{=} 25$$     Multiply.
$$25 = 25 \ \checkmark$$     Simplify. It checks.     ◻

 **Student Practice 5**   Solve for $x$ and check your solution.
$$4x - (x + 3) = 12 - 3(x - 2)$$

**Example 6** Solve for $x$. $3(-x - 7) = -2(2x + 5)$

**Solution**     $3(-x - 7) = -2(2x + 5)$
$$-3x - 21 = -4x - 10$$     Remove parentheses. Watch the signs carefully.
$$-3x + 4x - 21 = -4x + 4x - 10$$     Add $4x$ to both sides.
$$x - 21 = -10$$     Simplify.
$$x - 21 + 21 = -10 + 21$$     Add 21 to both sides.
$$x = 11$$     The solution is 11.

*Check.*   The check is left to the student.     ◻

**Student Practice 6**   Solve for $x$. $4(-2x - 3) = -5(x - 2) + 2$

In problems that involve decimals, great care should be taken. In some steps you will be multiplying decimal quantities, and in other steps you will be adding them.

**Example 7** Solve for $x$. $0.3(1.2x - 3.6) = 4.2x - 16.44$

**Solution**

$$0.3(1.2x - 3.6) = 4.2x - 16.44$$
$$0.36x - 1.08 = 4.2x - 16.44 \qquad \text{Remove parentheses.}$$
$$0.36x - 0.36x - 1.08 = 4.2x - 0.36x - 16.44 \qquad \text{Subtract } 0.36x \text{ from both sides.}$$
$$-1.08 = 3.84x - 16.44 \qquad \text{Combine like terms.}$$
$$-1.08 + 16.44 = 3.84x - 16.44 + 16.44 \qquad \text{Add 16.44 to both sides.}$$
$$15.36 = 3.84x \qquad \text{Simplify.}$$
$$\frac{15.36}{3.84} = \frac{3.84x}{3.84} \qquad \text{Divide both sides by 3.84.}$$
$$4 = x \qquad \text{The solution is 4.}$$

*Check.* The check is left to the student.

**Student Practice 7** Solve for $x$.
$$0.3x - 2(x + 0.1) = 0.4(x - 3) - 1.1$$

 **Example 8** Solve for $z$ and check.
$$2(3z - 5) + 2 = 4z - 3(2z + 8)$$

**Solution**
$$2(3z - 5) + 2 = 4z - 3(2z + 8)$$
$$6z - 10 + 2 = 4z - 6z - 24 \qquad \text{Remove parentheses.}$$
$$6z - 8 = -2z - 24 \qquad \text{Combine like terms.}$$
$$6z + 2z - 8 = -2z + 2z - 24 \qquad \text{Add } 2z \text{ to both sides.}$$
$$8z - 8 = -24 \qquad \text{Simplify.}$$
$$8z - 8 + 8 = -24 + 8 \qquad \text{Add 8 to both sides.}$$
$$8z = -16 \qquad \text{Simplify.}$$
$$\frac{8z}{8} = \frac{-16}{8} \qquad \text{Divide both sides by 8.}$$
$$z = -2 \qquad \text{Simplify. The solution is } -2.$$

*Check.*
$$2[3(-2) - 5] + 2 \overset{?}{=} 4(-2) - 3[2(-2) + 8] \qquad \text{Replace } z \text{ by } -2.$$
$$2[-6 - 5] + 2 \overset{?}{=} -8 - 3[-4 + 8] \qquad \text{Multiply.}$$
$$2[-11] + 2 \overset{?}{=} -8 - 3[4] \qquad \text{Simplify.}$$
$$-22 + 2 \overset{?}{=} -8 - 12$$
$$-20 = -20 \ \checkmark \qquad \text{It checks.}$$

**Student Practice 8** Solve for $z$ and check.
$$5(2z - 1) + 7 = 7z - 4(z + 3)$$

---

## STEPS TO SUCCESS Keep Trying! Do Not Quit!

**Math opens doors.** We live in a highly technical world that depends more and more on mathematics. You cannot afford to give up. Dropping mathematics may prevent you from entering an interesting career field. Understanding mathematics can open new doors for you.

**Practice daily.** Learning mathematics requires time and effort. You will find that regular study and daily practice are necessary. This will help your level of academic success and lead you toward a mastery of mathematics. It may open a path for you to a good paying job. Do not quit! You can do it!

**Making it personal:** Which of these statements do you find most helpful? What things can you do to help you master mathematics? Don't let yourself quit or let up on your studies. Your work now will help your financial success in the future! ▼

## 2.3 Exercises  MyMathLab®

*Find the value of the variable that satisfies each equation in exercises 1–22. Check your solution.*

**1.** $3x + 23 = 50$

**2.** $4x + 7 = 35$

**3.** $4x - 11 = 13$

**4.** $5x - 9 = 36$

**5.** $7x - 18 = -46$

**6.** $8x - 15 = -47$

**7.** $-4x + 17 = -35$

**8.** $-6x + 25 = -83$

**9.** $2x + 3.2 = 9.4$

**10.** $4x + 4.6 = 9.2$

**11.** $\frac{1}{4}x + 6 = 13$

**12.** $\frac{1}{2}x + 1 = 7$

**13.** $\frac{1}{3}x + 5 = -4$

**14.** $\frac{1}{8}x - 3 = -9$

**15.** $8x = 48 + 2x$

**16.** $5x = 22 + 3x$

**17.** $-6x = -27 + 3x$

**18.** $-7x = -26 + 6x$

**19.** $44 - 2x = 6x$

**20.** $21 - 5x = 7x$

**21.** $54 - 2x = -8x$

**22.** $72 - 4x = -12x$

*In exercises 23–26, determine whether the given solution is correct. If it is not, find the solution.*

**23.** Is 2 the solution for $2y + 3y = 12 - y$?

**24.** Is 4 the solution for $5y + 2 = 6y - 6 + y$?

**25.** Is 11 a solution for $7x + 6 - 3x = 2x - 5 + x$?

**26.** Is $-12$ a solution for $9x + 2 - 5x = -8 + 5x - 2$?

*Solve for the variable. You may move the variable terms to the right or to the left.*

**27.** $14 - 2x = -5x + 11$

**28.** $8 - 3x = 7x + 8$

**29.** $x - 6 = 8 - x$

**30.** $-x + 12 = -4 + x$

**31.** $0.6y + 0.8 = 0.1 - 0.1y$

**32.** $1.1y + 0.3 = -1.3 + 0.3y$

**33.** $5x - 9 = 3x + 23$

**34.** $9x - 5 = 7x + 43$

## To Think About, Exercises 35 and 36

*For exercises 35 and 36, first combine like terms on each side of the equation. Then solve for y by collecting all the y-terms on the left. Then solve for y by collecting all the y-terms on the right. Which approach is better?*

**35.** $-3 + 10y + 6 = 15 + 12y - 18$

**36.** $7y + 21 - 5y = 5y - 7 + y$

*Remove the parentheses and solve for the variable. Check your solution.*

**37.** $5(x + 3) = 35$

**38.** $7(x + 3) = 28$

**39.** $5(4x - 3) + 8 = -2$

**40.** $4(2x + 1) - 7 = 6 - 5$

**41.** $7x - 3(5 - x) = 10$

**42.** $8x - 2(4 - x) = 14$

**43.** $0.5x - 0.3(2 - x) = 4.6$

**44.** $0.4x - 0.2(3 - x) = 1.8$

**45.** $4(a - 3) + 2 = 2(a - 5)$

**46.** $6(a + 3) - 2 = -4(a - 4)$

**47.** $-2(x + 3) + 4 = 3(x + 4) + 2$

**48.** $-3(x + 5) + 2 = 4(x + 6) - 9$

**49.** $-3(y - 3y) + 4 = -4(3y - y) + 6 + 13y$

**50.** $2(4x - x) + 6 = 2(2x + x) + 8 - x$

**Mixed Practice** *Solve for the variable.*

**51.** $5.7x + 3 = 4.2x - 3$

**52.** $4x - 3.1 = 5.3 - 3x$

**53.** $5z + 7 - 2z = 32 - 2z$

**54.** $8 - 7z + 2z = 20 + 5z$

**55.** $-0.3a + 1.4 = -1.2 - 0.7a$

**56.** $-0.7b + 1.6 = -1.7 - 1.5b$

**57.** $6x + 8 - 3x = 11 - 12x - 13$

**58.** $4 - 7x - 13 = 8x - 3 - 5x$

**59.** $-3.5x + 1.3 = -2.7x + 1.5$

**60.** $1.4x - 0.8 = 1.2x - 0.2$

**61.** $5(4 + x) = 3(3x - 1) - 9$

**62.** $5(2x - 3) = 3(3x + 2) - 17$

**63.** $-1.7x + 4.4 + 5x = 0.3x - 0.1$

**64.** $6x - 3.7 - 1.2x = 0.8x + 1.1$

**Cumulative Review** *Evaluate using the correct order of operations. (Be careful to avoid sign errors.)*

**65. [1.5.1]** $(-6)(-8) + (-3)(2)$

**66. [1.5.1]** $(-3)^3 + (-20) \div 2$

**67. [1.5.1]** $5 + (2 - 6)^2$

**68. [0.6.1]** *Investments* On May 1, 2015, Marcella owned three different stocks: Motorola, Barnes & Noble, and CVS. Her portfolio contained the following:

35 shares of Motorola valued at $9.11 per share,

16 shares of Barnes & Noble valued at $22.70 per share, and

5 shares of CVS valued at $100.46 per share.

Find the market value of Marcella's stock holdings on May 1, 2015.

**69. [0.6.1]** *Employee Discount* Marvin works at Best Buy and gets a 10% discount on anything he buys from the store. A GPS navigation system that Marvin wishes to purchase costs $899 and is on sale at a 20% discount.

**(a)** What is the sale price if Marvin has a total discount of 30%? (Disregard sales tax.)

**(b)** What is the price if Marvin gets a 10% discount on the 20% sale price? (Disregard sales tax.)

**Quick Quiz 2.3** *Solve for the variable.*

**1.** $7x - 6 = -4x - 10$

**2.** $-3x + 6.2 = -5.8$

**3.** $2(3x - 2) = 4(5x + 3)$

**4. Concept Check** Explain how you would solve the equation $3(x - 2) + 2 = 2(x - 4)$.

# 2.4 Solving Equations with Fractions ▶

## 1 Solving Equations with Fractions ▶

Equations with fractions can be rather difficult to solve. This difficulty is simply due to the extra care we usually have to use when computing with fractions. The actual equation-solving procedures are the same, with fractions or without. To avoid unnecessary work, we transform the given equation with fractions to an equivalent equation that does not contain fractions. How do we do this? We multiply each side of the equation by the least common denominator of all the fractions contained in the equation. We then use the distributive property so that the LCD is multiplied by each term of the equation.

**Example 1** Solve for $x$. $\frac{1}{4}x - \frac{2}{3} = \frac{5}{12}x$

**Solution** First we find that the LCD = 12.

$$12\left(\frac{1}{4}x - \frac{2}{3}\right) = 12\left(\frac{5}{12}x\right) \qquad \text{Multiply both sides by 12.}$$

$$\left(\frac{12}{1}\right)\left(\frac{1}{4}\right)(x) - \left(\frac{12}{1}\right)\left(\frac{2}{3}\right) = \left(\frac{12}{1}\right)\left(\frac{5}{12}\right)(x) \qquad \text{Use the distributive property.}$$

$$3x - 8 = 5x \qquad \text{Simplify.}$$
$$3x + (-3x) - 8 = 5x + (-3x) \qquad \text{Add } -3x \text{ to both sides.}$$
$$-8 = 2x \qquad \text{Simplify.}$$
$$\frac{-8}{2} = \frac{2x}{2} \qquad \text{Divide both sides by 2.}$$
$$-4 = x \qquad \text{Simplify.}$$

*Check.* The check is left to the student. ▫

▭▶ **Student Practice 1** Solve for $x$.

$$\frac{3}{8}x - \frac{3}{2} = \frac{1}{4}x$$

In Example 1 we multiplied both sides of the equation by the LCD. However, most students prefer to go immediately to the second step and multiply each term by the LCD. This avoids having to write out a separate step using the distributive property.

**Example 2** Solve for $x$ and check your solution. $\frac{x}{3} + 3 = \frac{x}{5} - \frac{1}{3}$

**Solution** $15\left(\frac{x}{3}\right) + 15(3) = 15\left(\frac{x}{5}\right) - 15\left(\frac{1}{3}\right)$ The LCD is 15. Use the multiplication principle to multiply each term by 15.

$$5x + 45 = 3x - 5 \qquad \text{Simplify.}$$
$$5x - 3x + 45 = 3x - 3x - 5 \qquad \text{Subtract } 3x \text{ from both sides.}$$
$$2x + 45 = -5 \qquad \text{Combine like terms.}$$
$$2x + 45 - 45 = -5 - 45 \qquad \text{Subtract 45 from both sides.}$$
$$2x = -50 \qquad \text{Simplify.}$$
$$\frac{2x}{2} = \frac{-50}{2} \qquad \text{Divide both sides by 2.}$$
$$x = -25 \qquad \text{The solution is } -25.$$

*Check.* $\dfrac{-25}{3} + 3 \overset{?}{=} \dfrac{-25}{5} - \dfrac{1}{3}$

$-\dfrac{25}{3} + \dfrac{9}{3} \overset{?}{=} -\dfrac{5}{1} - \dfrac{1}{3}$

$-\dfrac{16}{3} \overset{?}{=} -\dfrac{15}{3} - \dfrac{1}{3}$

$-\dfrac{16}{3} = -\dfrac{16}{3} \quad \checkmark$ □

 **Student Practice 2** Solve for $x$ and check your solution.

$$\frac{5x}{4} - 1 = \frac{3x}{4} + \frac{1}{2}$$

**Example 3** Solve for $x$. $\dfrac{x + 2}{8} = \dfrac{x}{4} + \dfrac{1}{2}$

**Solution** $\dfrac{x}{8} + \dfrac{2}{8} = \dfrac{x}{4} + \dfrac{1}{2}$

First we rewrite the left side as two fractions. This is actually multiplying $\frac{1}{8}(x + 2) = \frac{x}{8} + \frac{2}{8}$.

$\dfrac{x}{8} + \dfrac{1}{4} = \dfrac{x}{4} + \dfrac{1}{2}$

Once we write $\frac{x+2}{8}$ as two fractions we can simplify $\frac{2}{8} = \frac{1}{4}$.

$8\left(\dfrac{x}{8}\right) + 8\left(\dfrac{1}{4}\right) = 8\left(\dfrac{x}{4}\right) + 8\left(\dfrac{1}{2}\right)$

We observe that the LCD is 8, so we multiply each term by 8.

$x + 2 = 2x + 4$ — Simplify.

$x - x + 2 = 2x - x + 4$ — Subtract $x$ from both sides.

$2 = x + 4$ — Combine like terms.

$2 - 4 = x + 4 - 4$ — Subtract 4 from both sides.

$-2 = x$ — The solution is $-2$.

**CAUTION:** We may *not* use slashes to divide out *part of an addition and subtraction* problem. We may use slashes if we are *multiplying* factors.

$\dfrac{x + \cancel{2}}{8} = \dfrac{x + 1}{4} \quad$ THIS IS WRONG!

$\dfrac{x \cdot \cancel{2}}{\cancel{8}} = \dfrac{x}{4} \quad$ THIS IS CORRECT.

*Check.* The check is left to the student. □

 **Student Practice 3** Solve for $x$.

$$\frac{x + 6}{9} = \frac{x}{6} + \frac{1}{2}$$

If a problem contains both parentheses and fractions, it is best to remove the parentheses first. Many students find it is helpful to have a written procedure to follow in solving these more-involved equations.

---

**PROCEDURE TO SOLVE EQUATIONS**

1. Remove any parentheses.
2. If fractions exist, multiply all terms on both sides by the least common denominator of all the fractions.
3. Combine like terms if possible.
4. Add or subtract terms on both sides of the equation to get all terms with the variable on one side of the equation.
5. Add or subtract a constant value on both sides of the equation to get all terms not containing the variable on the other side of the equation.
6. Divide both sides of the equation by the coefficient of the variable.
7. Simplify the solution (if possible).
8. Check your solution.

$\mathbb{M}_\mathbb{C}$**Example 4**   Solve for $x$ and check your solution.

$$\frac{1}{3}(x - 2) = \frac{1}{5}(x + 4) + 2$$

**Solution**

**Step 1**          $\dfrac{x}{3} - \dfrac{2}{3} = \dfrac{x}{5} + \dfrac{4}{5} + 2$   Remove parentheses.

**Step 2**   Multiply by the LCD, 15.

$$15\left(\frac{x}{3}\right) - 15\left(\frac{2}{3}\right) = 15\left(\frac{x}{5}\right) + 15\left(\frac{4}{5}\right) + 15(2)$$

$5x - 10 = 3x + 12 + 30$   Simplify.

**Step 3**          $5x - 10 = 3x + 42$   Combine like terms.

**Step 4**     $5x - 3x - 10 = 3x - 3x + 42$   Subtract $3x$ from both sides.

$2x - 10 = 42$   Simplify.

**Step 5**     $2x - 10 + 10 = 42 + 10$   Add 10 to both sides.

$2x = 52$   Simplify.

**Step 6**          $\dfrac{2x}{2} = \dfrac{52}{2}$   Divide both sides by 2.

**Step 7**          $x = 26$   Simplify the solution.

**Step 8**   *Check.*  $\dfrac{1}{3}(26 - 2) \overset{?}{=} \dfrac{1}{5}(26 + 4) + 2$   Replace $x$ by 26.

$\dfrac{1}{3}(24) \overset{?}{=} \dfrac{1}{5}(30) + 2$   Combine values within parentheses.

$8 \overset{?}{=} 6 + 2$   Simplify.

$8 = 8 \checkmark$   The solution is 26. ◻

**Student Practice 4**   Solve for $x$ and check your solution.

$$\frac{1}{2}(x + 5) = \frac{1}{5}(x - 2) + \frac{1}{2}$$

Remember that not every step will be needed in each problem. You can combine some steps as well, *as long as you are consistently obtaining the correct solution.* However, you are encouraged to write out every step as a way of helping you to avoid careless errors.

It is important to remember that when we write decimals, these numbers are really fractions written in a special way. Thus, $0.3 = \frac{3}{10}$ and $0.07 = \frac{7}{100}$. It is possible to take an equation containing decimals and to multiply each term by the appropriate value to obtain integer coefficients.

$\mathbb{M}_\mathbb{C}$**Example 5**   Solve for $x$. $0.2(1 - 8x) + 1.1 = -5(0.4x - 0.3)$

**Solution**

$0.2 - 1.6x + 1.1 = -2.0x + 1.5$   Remove parentheses.

Next, we multiply each term by 10 to move the decimal point one place to the right.

$$10(0.2) - 10(1.6x) + 10(1.1) = 10(-2.0x) + 10(1.5)$$
$$2 - 16x + 11 = -20x + 15$$
$$-16x + 13 = -20x + 15$$   Simplify.

**Student Practice 5**   Solve for $x$.

$$2.8 = 0.3(x - 2) + 2(0.1x - 0.3)$$

$$-16x + 20x + 13 = -20x + 20x + 15 \quad \text{Add } 20x \text{ to both sides.}$$
$$4x + 13 = 15 \quad \text{Simplify.}$$
$$4x + 13 - 13 = 15 - 13 \quad \text{Subtract 13 from both sides.}$$
$$4x = 2 \quad \text{Simplify.}$$
$$\frac{4x}{4} = \frac{2}{4} \quad \text{Divide both sides by 4.}$$
$$x = \frac{1}{2} \quad \text{or} \quad 0.5 \quad \text{Simplify.}$$

Earlier we stated that the decimal form of a solution should only be given if it is a terminating decimal. You can decide which form of the answer you want to use in the check. Here we use 0.5.

*Check.*

$$0.2[1 - 8(0.5)] + 1.1 \overset{?}{=} -5[0.4(0.5) - 0.3]$$
$$0.2[1 - 4] + 1.1 \overset{?}{=} -5[0.2 - 0.3]$$
$$0.2[-3] + 1.1 \overset{?}{=} -5[-0.1]$$
$$-0.6 + 1.1 \overset{?}{=} 0.5$$
$$0.5 = 0.5 \ \checkmark$$

**To Think About:** *Does Every Equation Have One Solution?* Actually, no. There are some rare cases where an equation has no solution at all. Suppose we try to solve the equation

$$5(x + 3) = 2x - 8 + 3x.$$

If we remove the parentheses and combine like terms, we have

$$5x + 15 = 5x - 8.$$

If we add $-5x$ to each side, we obtain

$$15 = -8.$$

Clearly this is impossible. There is no value of $x$ for which these two numbers are equal. We would say this equation has **no solution.**

One additional surprise may happen. An equation may have an infinite number of solutions. Suppose we try to solve the equation

$$9x - 8x - 7 = 3 + x - 10.$$

If we combine like terms on each side, we have the equation

$$x - 7 = x - 7.$$

If we add $-x$ to each side, we obtain

$$-7 = -7.$$

Now this statement is always true, no matter what the value of $x$. We would say this equation has **an infinite number of solutions.**

In the To Think About exercises in this section, we will encounter some equations that have no solution or an infinite number of solutions.

## 2.4 Exercises   MyMathLab®

*In exercises 1–16, solve for the variable and check your answer.*

1. $\frac{1}{6}x + \frac{2}{3} = -\frac{1}{2}$

2. $\frac{1}{3}x + \frac{5}{6} = \frac{1}{2}$

3. $\frac{2}{3}x = \frac{1}{15}x + \frac{3}{5}$

4. $\frac{4}{15}x + \frac{1}{5} = \frac{2}{3}x$

5. $\frac{x}{2} + \frac{x}{5} = \frac{7}{10}$

6. $\frac{x}{8} + \frac{x}{4} = -\frac{3}{4}$

7. $5 - \frac{1}{3}x = \frac{1}{12}x$

8. $15 - \frac{1}{2}x = \frac{1}{4}x$

9. $2 + \frac{y}{2} = \frac{3y}{4} - 3$

10. $\frac{x}{3} + 3 = \frac{5x}{6} + 2$

11. $\frac{x-3}{5} = 1 - \frac{x}{3}$

12. $\frac{y-5}{4} = 1 - \frac{y}{5}$

13. $\frac{x+3}{4} = \frac{x}{2} + \frac{1}{6}$

14. $\frac{x-2}{3} = \frac{x}{12} + \frac{5}{4}$

15. $0.6x + 5.9 = 3.8$

16. $-3.2x - 5.1 = 2.9$

*Answer Yes or No for exercises 17–20.*

17. Is 4 a solution to $\frac{1}{2}(y - 2) + 2 = \frac{3}{8}(3y - 4)$?

18. Is 2 a solution to $\frac{1}{5}(y + 2) = \frac{1}{10}y + \frac{3}{5}$?

19. Is $\frac{5}{8}$ a solution to $\frac{1}{2}\left(y - \frac{1}{5}\right) = \frac{1}{5}(y + 2)$?

20. Is $\frac{1}{2}$ a solution to $\frac{1}{3}\left(x - \frac{1}{4}\right) = \frac{1}{8} + \frac{1}{3}x$?

*Remove parentheses first. Then combine like terms. Solve for the variable.*

21. $\frac{3}{4}(3x + 1) = 2(3 - 2x) + 1$

22. $\frac{1}{4}(3x + 1) = 2(2x - 4) - 8$

23. $2(x - 2) = \frac{2}{5}(3x + 1) + 2$

24. $2(x - 4) = \frac{5}{6}(x + 6) - 6$

25. $0.3x - 0.2(3 - 5x) = -0.5(x - 6)$

26. $0.2(x + 1) + 0.5x = -0.3(x - 4)$

27. $-8(0.1x + 0.4) - 0.9 = -0.1$

28. $0.6x + 1.5 = 0.3x - 0.6(2x + 5)$

**Mixed Practice** *Solve.*

**29.** $\frac{1}{3}(y + 2) = 3y - 5(y - 2)$

**30.** $\frac{1}{4}(y + 6) = 2y - 3(y - 3)$

**31.** $\frac{1 + 2x}{5} + \frac{4 - x}{3} = \frac{1}{15}$

**32.** $\frac{1 + 3x}{2} + \frac{2 - x}{3} = \frac{5}{6}$

**33.** $\frac{3}{4}(x - 2) + \frac{3}{5} = \frac{1}{5}(x + 1)$

**34.** $\frac{2}{3}(x + 4) = 6 - \frac{1}{4}(3x - 2) - 1$

**35.** $\frac{1}{3}(x - 2) = 3x - 2(x - 1) + \frac{16}{3}$

**36.** $\frac{1}{4}(x + 5) = 3x - 2(3 - x) - 7$

**37.** $\frac{4}{5}x - \frac{2}{3} = \frac{3x + 1}{2}$

**38.** $\frac{5}{12}x + \frac{1}{3} = \frac{2x - 3}{4}$

**39.** $0.3x - 0.2(5x - 1) = -0.4(x + 2)$

**40.** $0.7(x + 3) = 0.2(x - 5) + 0.1$

**To Think About** *Solve. Be careful to examine your work to see if the equation has a solution, no solution, or an infinite number of solutions.*

**41.** $-1 + 5(x - 2) = 12x + 3 - 7x$

**42.** $x + 3x - 2 + 3x = -11 + 7(x + 2)$

**43.** $9(x + 3) - 6 = 24 - 2x - 3 + 11x$

**44.** $7(x + 4) - 10 = 3x + 20 + 4x - 2$

**45.** $-3(4x - 1) = 5(2x - 1) + 8$

**46.** $11x - 8 = -4(x + 3) + 4$

**47.** $3(4x + 1) - 2x = 2(5x - 3)$

**48.** $5(-3 + 4x) = 4(2x + 4) + 12x$

## Cumulative Review

**49.** **[0.3.1]** Multiply. $\left(-3\frac{1}{4}\right)\left(5\frac{1}{3}\right)$

**50.** **[0.3.2]** Divide. $5\frac{1}{2} \div 1\frac{1}{4}$

**51.** **[0.6.1]** *Peregrine Falcons* Peregrine falcons are the fastest birds on record, reaching horizontal speeds of 40–55 miles per hour. Also, they are one of the few animals in which the females are larger than the males. If female peregrine falcons are 30% larger than the males and males measure 440–750 grams, what is the weight range for females?

**52.** **[1.8.2]** *Auditorium Seating* The seating area of an auditorium is shaped like a trapezoid, with front and back sides parallel. The front of the auditorium measures 88 feet across, the back of the auditorium measures 150 feet across, and the auditorium is 200 feet from front to back. If each seat requires a space that is 2.5 feet wide by 3 feet deep, how many seats will the auditorium hold? (This will only be an approximation because of the angled side walls. Round off to the nearest whole number.)

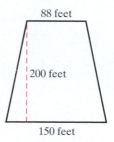

88 feet

200 feet

150 feet

---

**Quick Quiz 2.4** *Solve for the variable.*

**1.** $\dfrac{3}{4}x + \dfrac{5}{12} = \dfrac{1}{3}x - \dfrac{1}{6}$

**2.** $\dfrac{2}{3}x - \dfrac{3}{5} + \dfrac{7}{5}x + \dfrac{1}{3} = 1$

**3.** $\dfrac{2}{3}(x + 2) + \dfrac{1}{4} = \dfrac{1}{2}(5 - 3x)$

**4.** **Concept Check** Explain how you would solve the equation $\dfrac{x + 5}{6} = \dfrac{x}{2} + \dfrac{3}{4}$.

# Use Math to Save Money

## Did You Know?   You may be paying more than you think.

### See Through the Hidden Fees

#### Understanding the Problem:

Sam lives in California. Before taking a trip one weekend in June 2015 he needed to put gas in his car.

- On his street is a Shell gas station, which charged $4.55 per gallon for gas.
- There is also an ARCO gas station, which charged $4.43 per gallon for gas.
- The ARCO station also charged a single "ATM Transaction Fee" of $0.45.

Sam needed to decide which gas station to go to in order to pay the least amount for gas.

#### Making a Plan:

Sam needed to know how much gas would cost at each station to make a choice that would save him money.

**Step 1:**  Sam needed to find out how much filling up his car at the gas station really cost after any hidden fees.

**Task 1:**  Determine how much it cost Sam to buy one gallon of gas at each station.

**Task 2:**  Determine how much it cost Sam to buy three gallons of gas at each station.

**Task 3:**  Determine how much it cost Sam to buy four gallons of gas at each station.

**Task 4:**  Determine how much it cost Sam to buy 10 gallons of gas at each station.

**Step 2:**  Notice that when Sam bought more gas, it was less expensive to buy at the ARCO station than at the Shell station. The price at Shell rose faster than the price at ARCO.

**Task 5:**  Find the number of gallons for which the cost was the same.

#### Making a Decision:

**Step 3:**  Sam needed to take into consideration the price of gas and any fees involved when he made his decision about where to buy gas.

**Task 6:**  Which station was less expensive if Sam only needed a small amount of gas, say less than four gallons?

**Task 7:**  Which station was less expensive if Sam needed more than four gallons of gas?

#### Applying the Situation to Your Life:

Some gas stations have different prices depending on whether you pay with cash or a credit card.

**Task 8:**  Does the station where you normally buy gas charge the same price for cash or credit?

**Task 9:**  Do you know if the gas station charges an ATM transaction fee?

**Task 10:**  Have the increases in gas prices caused you to change your driving habits? If so, please explain.

There is a simple plan to help you save money by avoiding hidden fees:

- Always check for any hidden fees.
- Find the real total cost for your purchases, including fees, taxes, shipping costs, and tips.
- Watch for different prices for cash or credit.

**2.5**    # Formulas

**Student Learning Objective**

After studying this section, you will be able to:

 Solve a formula for a specified variable.

## 1 Solving a Formula for a Specified Variable ◉

Formulas are equations with one or more variables that are used to describe real-life situations. The formula describes the relationship that exists among the variables. For example, in the formula $d = rt$, distance ($d$) is equal to the rate of speed ($r$) multiplied by the time ($t$). We can use this formula to find distance if we know the rate and time. Sometimes, however, we are given the distance and the rate, and we are asked to find the time.

**Example 1** American Airlines scheduled a nonstop flight in a Boeing 777 jet from Chicago to London. The approximate air distance traveled on the flight was 3975 miles. The average speed of the aircraft on the trip was 530 miles per hour. How many hours did it take the Boeing 777 to fly this trip? (*Source:* www.aa.com)

**Solution**

$$d = rt \qquad \text{Use the distance formula.}$$
$$3975 = 530t \qquad \text{Substitute the known values for the variables.}$$
$$\frac{3975}{530} = \frac{530t}{530} \qquad \text{Divide both sides of the equation by 530 to solve for } t.$$
$$7.5 = t \qquad \text{Simplify.}$$

It took the Boeing 777 about 7.5 hours to fly this trip from Chicago to London.    □

**Student Practice 1**    The airlines are planning a nonstop flight from Chicago to Prague. This distance is approximately 4565 miles and the time of the flight is 8.3 hours. Find the average rate of speed for the flight.

If we have many problems that ask us to find the time given the distance and rate, it may be worthwhile to rewrite the formula in terms of time.

**Example 2** Solve for $t$. $d = rt$

**Solution**

$$\frac{d}{r} = \frac{rt}{r} \qquad \text{We want to isolate } t. \text{ Therefore we divide both sides of the equation by the coefficient of } t, \text{ which is } r.$$
$$\frac{d}{r} = t \qquad \text{We have solved for the variable indicated.} \qquad \square$$

**Student Practice 2**    Einstein's equation relating energy $E$ to mass $m$ and the speed of light $c$ is $E = mc^2$. Solve it for $m$.

A straight line can be described by an equation of the form $Ax + By = C$, where $A$, $B$, and $C$ are real numbers and $A$ and $B$ are not both zero. We will study this in later chapters. Often it is useful to solve such an equation for the variable $y$ in order to make graphing the equation easier.

**Example 3**  Solve for $y$. $3x - 2y = 6$

**Solution**

$$-2y = 6 - 3x$$  We want to isolate the term containing $y$, so we subtract $3x$ from both sides.

$$\frac{-2y}{-2} = \frac{6 - 3x}{-2}$$  Divide both sides by the coefficient of $y$.

$$y = \frac{6}{-2} + \frac{-3x}{-2}$$  Rewrite the fraction on the right side as two fractions.

$$y = \frac{3}{2}x - 3$$  Simplify and reorder the terms on the right.

This is known as the slope–intercept form of the equation of a line.  □

**Student Practice 3**  Solve for $y$.

$$8 - 2y + 3x = 0$$

Our procedure for solving an equation can be rewritten to give us a procedure for solving a formula for a specified variable.

**PROCEDURE TO SOLVE A FORMULA FOR A SPECIFIED VARIABLE**

1. Remove any parentheses.
2. If fractions exist, multiply all terms on both sides by the LCD of all the fractions.
3. Combine like terms on each side if possible.
4. Add or subtract terms on both sides of the equation to get all terms with the desired variable on one side of the equation.
5. Add or subtract the appropriate quantities to get all terms that do *not* have the desired variable on the other side of the equation.
6. Divide both sides of the equation by the coefficient of the desired variable.
7. Simplify if possible.

▲ **Example 4**  A trapezoid is a four-sided figure with two parallel sides. In some houses, two of the sides of the roof are in the shape of a trapezoid. If the parallel sides are $a$ and $b$ and the altitude is $h$, the area is given by

$$A = \frac{h}{2}(a + b).$$

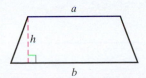

Solve this equation for $a$.

*Continued on next page*

**Solution**

$$A = \frac{h}{2}(a + b)$$

$$A = \frac{ha}{2} + \frac{hb}{2} \qquad \text{Remove the parentheses.}$$

$$2(A) = 2\left(\frac{ha}{2}\right) + 2\left(\frac{hb}{2}\right) \qquad \text{Multiply all terms by the LCD of 2.}$$

$$2A = ha + hb \qquad \text{Simplify.}$$

$$2A - hb = ha \qquad \text{We want to isolate the term containing } a.$$
$$\text{Therefore, we subtract } hb \text{ from both sides.}$$

$$\frac{2A - hb}{h} = \frac{ha}{h} \qquad \text{Divide both sides by } h \text{ (the coefficient of } a\text{).}$$

$$\frac{2A - hb}{h} = a \qquad \text{The solution is obtained.}$$

*Note:* Although the solution is in simple form, it could be written in an alternative way. Since

$$\frac{2A - hb}{h} = \frac{2A}{h} - \frac{hb}{h} = \frac{2A}{h} - b,$$

we could also have written $\dfrac{2A}{h} - b = a$.

 **Student Practice 4** Solve for $x$.

$$H = \frac{3}{4}(2y + x)$$

---

**STEPS TO SUCCESS** Evaluating Your Study Habits

*Take a close look at your study habits.* You will be surprised how you can make a difference in your performance by **learning how to learn**. Practice the key techniques discussed in the Steps to Success and improve your performance in all your classes as well as in your place of employment.

**Making it personal:**

1. The chart on the inside front cover shows where you can find each Steps to Success section. Reread these techniques for success. Which of the suggestions do you think will help you the most? Make a list of these suggestions.

2. Based on this list, write out a study plan that you will follow for the remainder of the semester.

3. Discuss this plan with your instructor or counselor to see if they have any other suggestions.

Now follow this plan for the remainder of the semester! ▼

## 2.5 Exercises   MyMathLab®

### Verbal and Writing Skills, Exercises 1 and 2

1. **Temperature** The formula for calculating the temperature in degrees Fahrenheit when you know the temperature in degrees Celsius is $F = \frac{9}{5}C + 32$. Explain in your own words how you would solve this equation for $C$.

▲ 2. **Geometry** The formula for finding the area of a trapezoid with an altitude of 9 meters and bases of $b$ meters and $c$ meters is given by the equation $A = \frac{9}{2}(b + c)$. Explain in your own words how you would solve this equation for $b$.

### Applications

▲ 3. **Geometry** The formula for the area of a triangle is $A = \frac{1}{2}ab$, where $b$ is the *base* of the triangle and $a$ is the *altitude* of the triangle.

   (a) Use this formula to find the base of a triangle that has an area of 60 square meters and an altitude of 12 meters.

   (b) Use this formula to find the altitude of a triangle that has an area of 88 square meters and a base of 11 meters.

4. **Simple Interest** The formula for calculating simple interest is $I = Prt$, where $P$ is the *principal* (amount of money invested), $r$ is the *rate* (in decimal form) at which the money is invested, and $t$ is the *time*.

   (a) Use this formula to find how long it would take to earn $720 in interest on an investment of $3000 at the rate of 6%.

   (b) Use this formula to find the rate of interest if $5000 earns $400 interest in 2 years.

   (c) Use this formula to find the amount of money invested if the interest earned was $120 and the rate of interest was 5% over 3 years.

5. The equation $2x + 9y = 18$ describes a line and is written in standard form.

   (a) Solve for the variable $y$.

   (b) Use this result to find $y$ when $x = -9$.

6. The equation $-3x + 8y = 24$ describes a line and is written in standard form.

   (a) Solve for the variable $x$.

   (b) Use this result to find $x$ when $y = 6$.

*In each formula or equation, solve for the variable indicated.*

**Area of a triangle**

▲ 7. $A = \frac{1}{2}bh$   Solve for $b$.

▲ 8. $A = \frac{1}{2}bh$   Solve for $h$.

**Simple interest formula**

9. $I = Prt$   Solve for $P$.

10. $I = Prt$   Solve for $t$.

**Slope–intercept form of a line**

**11.** $y = mx + b$   Solve for $m$.

**12.** $y = mx + b$   Solve for $b$.

**Standard form of a line**

**13.** $5x - 3y = -30$   Solve for $y$.

**14.** $2x - 7y = 14$   Solve for $y$.

**Slope–intercept form of a line**

**15.** $y = -\dfrac{3}{5}x + 6$   Solve for $x$.

**16.** $y = -\dfrac{5}{6}x + 10$   Solve for $x$.

**Standard form of a line**

**17.** $ax + by = c$   Solve for $y$.

**18.** $ax + by = c$   Solve for $x$.

**Area of a circle**

▲ **19.** $A = \pi r^2$   Solve for $r^2$.

**Surface area of a sphere**

▲ **20.** $s = 4\pi r^2$   Solve for $r^2$.

**Gravitational acceleration**

**21.** $g = \dfrac{GM}{r^2}$   Solve for $r^2$.

**Volume of sphere**

▲ **22.** $V = \dfrac{4}{3}\pi r^3$   Solve for $r^3$.

**Simple interest formula**

**23.** $A = P(1 + rt)$   Solve for $t$.

**Area of a trapezoid**

▲ **24.** $A = \dfrac{1}{2}a(b_1 + b_2)$   Solve for $b_1$.

**Surface area of a right circular cylinder**

▲ **25.** $S = 2\pi rh + 2\pi r^2$   Solve for $h$.

**26.** $H = 5as + 10a^2$   Solve for $s$.

**Kinetic energy**

**27.** $K = \dfrac{1}{2}mv^2$   Solve for $m$.

**28.** $K = \dfrac{1}{2}mv^2$   Solve for $v^2$.

**Volume of a rectangular prism**

▲ **29.** $V = LWH$   Solve for $L$.

▲ **30.** $V = LWH$   Solve for $H$.

*Volume of a cone*

▲ **31.** $V = \dfrac{1}{3}\pi r^2 h$　Solve for $r^2$.

▲ **32.** $V = \dfrac{1}{3}\pi r^2 h$　Solve for $h$.

*Perimeter of a rectangle*

▲ **33.** $P = 2L + 2W$　Solve for $W$.

*Finding the nth term of an arithmetic sequence*

**34.** $N = F + d(n - 1)$　Solve for $d$.

*Pythagorean Theorem*

▲ **35.** $c^2 = a^2 + b^2$　Solve for $a^2$.

▲ **36.** $c^2 = a^2 + b^2$　Solve for $b^2$.

*Temperature conversion formulas*

**37.** $F = \dfrac{9}{5}C + 32$　Solve for $C$.

**38.** $C = \dfrac{5}{9}(F - 32)$　Solve for $F$.

*Ohm's law*

**39.** $P = \dfrac{E^2}{R}$　Solve for $R$.

*Density formula (density, mass, and volume)*

**40.** $d = \dfrac{m}{v}$　Solve for $v$.

*Area of a sector of a circle*

▲ **41.** $A = \dfrac{\pi r^2 S}{360}$　Solve for $S$.

▲ **42.** $A = \dfrac{\pi r^2 S}{360}$　Solve for $r^2$.

## Applications

▲ **43.** *Geometry* Use the result you obtained in exercise 33 to solve the following problem. A farmer has a rectangular field with a perimeter of 5.8 miles and a length of 2.1 miles. Find the width of the field.

**44.** Use the result you obtained in exercise 34 to solve the following problem. If you are given an arithmetic sequence where the first term ($F$) is 6, 3 is the value of $n$, and the third ($n$th) term is 24 (so $N$ is 24), what is the difference ($d$)?

▲ **45.** *Geometry* Use the result you obtained in exercise 29 to solve the following problem. The foundation of a house is in the shape of a rectangular solid. The volume of the foundation is 5940 cubic feet. The height of the foundation is 9 feet and the width is 22 feet. What is the length of the foundation?

▲ **46.** *Geometry* Use the result you obtained in exercise 30 to solve the following problem. The fish tank at the Mandarin Danvers Restaurant is in the shape of a rectangular solid. The volume of the tank is 3024 cubic inches. The length of the tank is 18 inches while the width of the tank is 14 inches. What is the height of the tank?

**47.** ***Tourism*** The number of overseas visitors in thousands ($V$) to the United States in any given year can be predicted by the equation $V = 3780x + 55,440$, where $x$ is the number of years since 2009. For example, if $x = 5$ (this would be the year 2014), the predicted number of overseas visitors in thousands would be $3780(5) + 55,440 = 74,340$. Thus we would predict that in 2014, a total of 74,340,000 overseas visitors came to the United States. (*Source:* United States Department of Commerce, National Travel and Tourism Office)

**(a)** Solve this equation for $x$.

**(b)** Use the result of your answer in (a) to find the year in which the number of overseas visitors is predicted to be 89,460,000. (*Hint:* Let $V = 89,460$ in your answer for (a).)

**48.** ***Tourism*** The number of North American visitors (visitors from Mexico and Canada) in thousands ($V$) to the United States in any given year can be predicted by the equation $V = 1709x + 31,560$, where $x$ is the number of years since 2009. For example, if $x = 6$ (this would be the year 2015), the predicted number of North American visitors in thousands would be $1709(6) + 31,560 = 41,814$. Thus we would predict that in 2015, a total of 41,814,000 North American visitors came to the United States. (*Source:* United States Department of Commerce, National Travel and Tourism Office)

**(a)** Solve this equation for $x$.

**(b)** Use the result of your answer in (a) to find the year in which the number of North American visitors is predicted to be 50,359,000. (*Hint:* Let $V = 50,359$ in your answer for (a).)

## To Think About

▲ **49.** In the formula $A = \frac{1}{2}ab$, if $b$ doubles, what is the effect on $A$?

▲ **50.** In the formula $A = \frac{1}{2}ab$, if both $a$ and $b$ double, what is the effect on $A$?

▲ **51.** In $A = \pi r^2$, if $r$ doubles, what is the effect on $A$?

▲ **52.** In $A = \pi r^2$, if $r$ is halved, what is the effect on $A$?

## Cumulative Review

**53.** **[0.5.3]** Find 20% of $80.

**54.** **[0.5.3]** What is 0.5% of 200?

**55.** **[0.3.1]** ***Electronic Games*** A very popular handheld electronic game requires $3\frac{1}{4}$ square feet of a certain type of durable plastic in the manufacturing process. How many square feet of durable plastic does this company need to make 12,000 handheld games?

**56.** **[0.2.4]** ***Spotlight Rental*** The Superstar Lighting Company rents out giant spotlights that shine up into the sky to mark the location of special events, such as the opening of a movie, a major sports play-off game, or a huge sales event at an auto dealership. A giant spotlight was used for $4\frac{1}{3}$ hours on Saturday, $2\frac{3}{4}$ hours on Tuesday, and $3\frac{1}{2}$ hours on Wednesday. What was the total number of hours that the spotlight was in use?

## Quick Quiz 2.5

**1.** Solve for $x$.

$$A = 3x + 2w$$

**2.** Solve for $a$.

$$A = \frac{1}{3}h(a + b)$$

**3.** Solve for $y$.

$$2ax = 2axy - 5$$

**4.** **Concept Check** Explain how you would solve for $x$ in the equation $y = \frac{3}{8}x - 9$.

# 2.6 Solving Inequalities in One Variable ▶

## 1 Interpreting Inequality Statements ▶

We frequently speak of one value being greater than or less than another value. We say that "5 is less than 7" or "9 is greater than 4." These relationships are called **inequalities.** We can write inequalities in mathematics by using symbols. We use the symbol $<$ to represent the words "**is less than.**" We use the symbol $>$ to represent the words "**is greater than.**"

| *Statement in Words* | *Statement in Algebra* |
|---|---|
| 5 is less than 7. | $5 < 7$ |
| 9 is greater than 4. | $9 > 4$ |

*Note:* "5 is less than 7" and "7 is greater than 5" have the same meaning. Similarly, $5 < 7$ and $7 > 5$ have the same meaning. They represent two equivalent ways of describing the same relationship between the two numbers 5 and 7.

We can better understand the concept of inequality if we examine a number line.

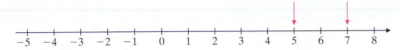

We say that one number is greater than another if it is to the right of the other on the number line. Thus $7 > 5$, since 7 is to the right of 5.

What about negative numbers? We can say "$-1$ is greater than $-3$" and write it in symbols as $-1 > -3$ because we know that $-1$ lies to the right of $-3$ on the number line.

**Example 1** In each statement, replace the question mark with the symbol $<$ or $>$.

**(a)** $3 \, ? \, {-1}$    **(b)** $-2 \, ? \, 1$    **(c)** $-3 \, ? \, {-4}$    **(d)** $0 \, ? \, 3$    **(e)** $-3 \, ? \, 0$

**Solution**

**(a)** $3 > -1$    Use $>$, since 3 is to the right of $-1$ on the number line.

**(b)** $-2 < 1$    Use $<$, since $-2$ is to the left of 1. (Or equivalently, we could say that 1 is to the right of $-2$.)

**(c)** $-3 > -4$   Note that $-3$ is to the right of $-4$.

**(d)** $0 < 3$    Use $<$, since 0 is to the left of 3.

**(e)** $-3 < 0$    Note that $-3$ is to the left of 0.    ◻

**▶ Student Practice 1** In each statement, replace the question mark with the symbol $<$ or $>$.

**(a)** $7 \, ? \, 2$ **(b)** $-2 \, ? \, {-4}$ **(c)** $-1 \, ? \, 2$ **(d)** $-8 \, ? \, {-5}$ **(e)** $0 \, ? \, {-2}$ **(f)** $5 \, ? \, {-3}$

## 2 Graphing an Inequality on a Number Line ▶

Sometimes we will use an inequality to express the relationship between a variable and a number. $x > 3$ means that $x$ could have the value of *any number* greater than 3.

Any number that makes an inequality true is called a **solution** of the inequality. The set of all numbers that make the inequality true is called the **solution set.** A picture that represents all of the solutions of an inequality is called a **graph** of the inequality.

The inequality $x > 3$ can be graphed on a number line as follows:

**Case 1**

Note that all of the points to the right of 3 are shaded. The open circle at 3 indicates that we do not include the point for the number 3.

Similarly, we can graph $x < -2$ as follows:

**Case 2**

Note that all of the points to the left of $-2$ are shaded.

Sometimes a variable will be either greater than or equal to a certain number. In the statement "$x$ is greater than or equal to 3," we are implying that $x$ could have the value of 3 or any number greater than 3. We write this as $x \geq 3$. We graph it as follows:

**Case 3**

Note that the closed circle at 3 indicates that we *do* include the point for the number 3.

Similarly, we can graph $x \leq -1$ as follows:

**Case 4**

*Note*: Be careful you do not confuse ○——→ with ●——→. It is important to decide if you need an open circle or a closed one. Case 1 and Case 2 use open circles. Case 3 and Case 4 use closed circles.

**Example 2** State each mathematical relationship in words and then graph it.

**(a)** $x < -1$      **(b)** $x \geq -2$      **(c)** $-3 < x$      **(d)** $x \leq -\dfrac{1}{2}$

**Solution**

**(a)** We state that "$x$ is less than $-1$."

$$x < -1$$

**(b)** We state that "$x$ is greater than or equal to $-2$."

$$x \geq -2$$

**(c)** We can state that "$-3$ is less than $x$" or, equivalently, that "$x$ is greater than $-3$." Be sure you see that $-3 < x$ is equivalent to $x > -3$. Although both statements are correct, we *usually write the variable first* in a simple inequality containing a variable and a numerical value.

$$x > -3$$

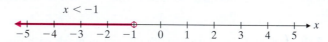

**(d)** We state that "$x$ is less than or equal to $-\frac{1}{2}$."

$$x \leq -\dfrac{1}{2}$$

**▶ Student Practice 2** State each mathematical relationship in words and then graph it on the given number line.

**(a)** $x > 5$

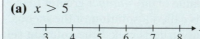

**(b)** $x \leq -2$

**(c)** $3 > x$

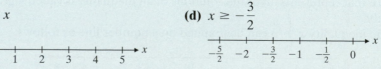

**(d)** $x \geq -\dfrac{3}{2}$

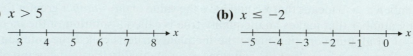

## 3 Translating English Phrases into Algebraic Statements ▶

We can translate many everyday situations into algebraic statements with an unknown value and an inequality symbol. This is the first step in solving word problems using inequalities.

**Example 3** Translate each English statement into an algebraic statement.

(a) The police on the scene said that the car was traveling more than 80 miles per hour. (Use the variable $s$ for speed.)

(b) The owner of the trucking company said that the payload of a truck must never exceed 4500 pounds. (Use the variable $p$ for payload.)

### Solution

(a) Since the speed must be greater than 80, we have $s > 80$.

(b) If the payload of the truck can never exceed 4500 pounds, then the payload must be always less than or equal to 4500 pounds. Thus we write $p \leq 4500$.    □

**Student Practice 3** Translate each English statement into an algebraic statement.

(a) During the drying cycle, the temperature inside the clothes dryer must never exceed 180° Fahrenheit. (Use the variable $t$ for temperature.)

(b) The bank loan officer said that the total consumer debt incurred by Wally and Mary must be less than $15,000 if they want to qualify for a mortgage to buy their first home. (Use the variable $d$ for debt.)

## 4 Solving and Graphing an Inequality

When we **solve an inequality,** we are finding *all* the values that make it true. To solve an inequality, we simplify it to the point where we can clearly see all possible values for the variable. We've solved equations by adding, subtracting, multiplying by, and dividing by a particular value on both sides of the equation. Here we perform similar operations with inequalities with one important exception. We'll show some examples so that you can see how these operations can be used with inequalities just as with equations.

We will first examine the pattern that occurs when we perform these operations *with a positive value* on both sides of an inequality.

| Original Inequality | Operations with a Positive Number | New Inequality |
|---|---|---|
| $4 < 6$ | Add 2 to both sides. ⟶ | $6 < 8$ |
| | Subtract 2 from both sides. ⟶ | $2 < 4$ |
| | Multiply both sides by 2. ⟶ | $8 < 12$ |
| | Divide both sides by 2. ⟶ | $2 < 3$ |

Notice that the inequality symbol remains the same when these operations are performed with a positive value.

Now let us examine what happens when we perform these operations *with a negative value.*

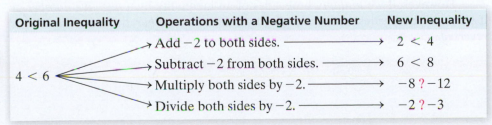

| Original Inequality | Operations with a Negative Number | New Inequality |
|---|---|---|
| $4 < 6$ | Add $-2$ to both sides. ⟶ | $2 < 4$ |
| | Subtract $-2$ from both sides. ⟶ | $6 < 8$ |
| | Multiply both sides by $-2$. ⟶ | $-8 \,?\, -12$ |
| | Divide both sides by $-2$. ⟶ | $-2 \,?\, -3$ |

What happens to the inequality sign when we multiply both sides by a negative number? Since $-8$ is to the right of $-12$ on a number line, we know that the new inequality should be $-8 > -12$ if we want the statement to remain true. Notice how we reverse the direction of the inequality from $<$ (less than) to $>$ (greater than) when we multiply by a negative value. Thus we have the following.

$$4 < 6 \longrightarrow \text{Multiply both sides by } -2. \longrightarrow -8 > -12$$

The same thing happens when we divide by a negative number. The inequality is reversed from $<$ to $>$. We know this since $-2$ is to the right of $-3$ on a number line.

$$4 < 6 \longrightarrow \text{Divide both sides by } -2. \longrightarrow -2 > -3$$

Similar reversals take place in the next example.

**Example 4** Perform the given operations and write the new inequalities.

| Original Inequality | | New Inequality |
|---|---|---|
| (a) $-2 < -1$ | $\longrightarrow$ Multiply both sides by $-3$. | $\longrightarrow$ $6 > 3$ |
| (b) $0 > -4$ | $\longrightarrow$ Divide both sides by $-2$. | $\longrightarrow$ $0 < 2$ |
| (c) $8 \geq 4$ | $\longrightarrow$ Divide both sides by $-4$. | $\longrightarrow$ $-2 \leq -1$ |

**Solution** Notice that we perform the arithmetic with signed numbers just as we always do. But the new inequality signs are reversed (from those of the original inequalities).

*Whenever both sides of an inequality are multiplied or divided by a negative quantity, the direction of the inequality is reversed.* □

**Student Practice 4** Perform the given operations and write the new inequalities.

(a) $7 > 2$        Multiply each side by $-2$.

(b) $-3 < -1$     Multiply each side by $-1$.

(c) $-10 \geq -20$    Divide each side by $-10$.

(d) $-15 \leq -5$     Divide each side by $-5$.

---

**PROCEDURE FOR SOLVING INEQUALITIES**

You may use the same procedures to solve inequalities that you used to solve equations *except* that the direction of an inequality is *reversed* if you *multiply* or *divide* both sides *by a negative number.*

---

It may be helpful to think over quickly what we have discussed here. The inequalities remain the same when we add a number to both sides or subtract a number from both sides of the inequality. The inequalities remain the same when we multiply both sides by a positive number or divide both sides by a positive number.

However, if we *multiply* both sides of an inequality by a *negative number* or if we *divide* both sides of an inequality by a *negative number*, then *the inequality is reversed.*

**Example 5** Solve and graph. $3x + 7 \geq 13$

**Solution**

$$3x + 7 \; -7 \; \geq 13 \; -7 \qquad \text{Subtract 7 from both sides.}$$

$$3x \geq 6 \qquad \text{Simplify.}$$

$$\frac{3x}{3} \geq \frac{6}{3} \qquad \text{Divide both sides by 3.}$$

$$x \geq 2 \qquad \text{Simplify. Note that the direction of the inequality is not changed, since we have divided by a positive number.}$$

The graph is as follows:

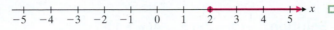

**Student Practice 5** Solve and graph. $8x - 2 < 3$

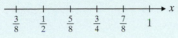

$\mathbb{M}_{\mathbb{C}}$ **Example 6** Solve and graph. $5 - 3x > 7$

**Solution**

$$5 \; -5 \; - 3x > 7 \; -5 \qquad \text{Subtract 5 from both sides.}$$

$$-3x > 2 \qquad \text{Simplify.}$$

$$\frac{-3x}{-3} < \frac{2}{-3} \qquad \text{Divide by } -3 \text{ and \textbf{reverse the inequality} since we are dividing by a negative number.}$$

$$x < -\frac{2}{3} \qquad \text{Note the direction of the inequality.}$$

The graph is as follows:

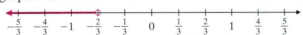

**Student Practice 6** Solve and graph.

$$4 - 5x > 7$$

Just like equations, some inequalities contain parentheses and fractions. The initial steps to solve these inequalities will be the same as those used to solve equations with parentheses and fractions. When the variable appears on both sides of the inequality, it is advisable to collect the $x$-terms on the left side of the inequality symbol.

**Example 7** Solve and graph. $-\dfrac{13x}{2} \leq \dfrac{x}{2} - \dfrac{15}{8}$

**Solution**

$$8\left(\frac{-13x}{2}\right) \leq 8\left(\frac{x}{2}\right) - 8\left(\frac{15}{8}\right) \qquad \text{Multiply all terms by LCD} = 8. \text{ We do \textbf{not} reverse the direction of the inequality symbol since we are multiplying by a positive number.}$$

$$-52x \leq 4x - 15 \qquad \text{Simplify.}$$

$$-52x - 4x \leq 4x - 15 - 4x \qquad \text{Subtract } 4x \text{ from both sides.}$$

$$-56x \leq -15 \qquad \text{Combine like terms.}$$

$$\frac{-56x}{-56} \geq \frac{-15}{-56} \qquad \text{Divide both sides by } -56. \text{ We \textbf{reverse} the direction of the inequality when we divide both sides by a negative number.}$$

$$x \geq \frac{15}{56}$$

The graph is as follows:

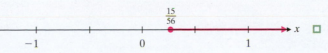

**Student Practice 7**    Solve and graph.

$$\frac{1}{2}x + 3 < \frac{2}{3}x$$

16  17  18  19  20  21    $x$

**Example 8**  Solve and graph. $\frac{1}{3}(3 - 2x) \le -4(x + 1)$

**Solution**    $\frac{1}{3}(3 - 2x) \le -4(x + 1)$

$$1 - \frac{2x}{3} \le -4x - 4 \qquad \text{Remove parentheses. } \frac{3}{3} = 1.$$

$$3(1) - 3\left(\frac{2x}{3}\right) \le 3(-4x) - 3(4) \qquad \text{Multiply all terms by LCD} = 3.$$

$$3 - 2x \le -12x - 12 \qquad \text{Simplify.}$$

$$3 - 2x + 12x \le -12x + 12x - 12 \qquad \text{Add } 12x \text{ to both sides.}$$

$$3 + 10x \le -12 \qquad \text{Combine like terms.}$$

$$3 - 3 + 10x \le -12 - 3 \qquad \text{Subtract 3 from both sides.}$$

$$10x \le -15 \quad \text{Simplify.}$$

$$\frac{10x}{10} \le \frac{-15}{10} \qquad \text{Divide both sides by 10. Since we are dividing by a \textbf{positive} number, the inequality is \textbf{not} reversed.}$$

$$x \le -\frac{3}{2}$$

The graph is as follows:

$-\frac{9}{2}$  $-4$  $-\frac{7}{2}$  $-3$  $-\frac{5}{2}$  $-2$  $-\frac{3}{2}$  $-1$  $-\frac{1}{2}$  $0$  $\frac{1}{2}$   $x$

**Student Practice 8**    Solve and graph.

$$\frac{1}{2}(3 - x) \le 2x + 5$$

$-2$  $-\frac{9}{5}$  $-\frac{8}{5}$  $-\frac{7}{5}$  $-\frac{6}{5}$  $-1$    $x$

**CAUTION:** The most common error students make when solving inequalities is forgetting to reverse the direction of the inequality symbol when multiplying or dividing both sides of the inequality by a negative number.

Normally when you solve inequalities, you solve for $x$ by putting the variables on the left side. If you solve by placing the variables on the right side, you will end up with statements like $3 > x$. This is equivalent to $x < 3$. It is wise to express your answer with the variables on the left side.

**Example 9** A hospital director has determined that the costs of operating one floor of the hospital for an eight-hour shift must never exceed $3000. An expression for the cost of operating one floor of the hospital is $130n + 1200$, where $n$ is the number of nurses. This expression is based on an estimate of $1200 in fixed costs and a cost of $200 per nurse for an eight-hour shift. Solve the inequality $130n + 1200 \leq 3000$ to determine the number of nurses that may be on duty on this floor during an eight-hour shift if the director's cost control measure is followed.

**Solution**

$$200n + 1200 \leq 3000 \qquad \text{The inequality we must solve.}$$

$$200n + 1200 - 1200 \leq 3000 - 1200 \qquad \text{Subtract 1200 from each side.}$$

$$200n \leq 1800 \qquad \text{Simplify.}$$

$$\frac{200n}{200} \leq \frac{1800}{200} \qquad \text{Divide each side by 130.}$$

$$n \leq 9$$

The number of nurses on duty on this floor during an eight-hour shift must always be less than or equal to nine. □

 **Student Practice 9** The company president of Staywell, Inc., wants the monthly profits never to be less than $2,500,000. He has determined that an expression for monthly profit is $2000n - 700,000$. In the expression, $n$ is the number of exercise machines manufactured each month. The profit on each machine is $2000, and the $-$700,000 in the expression represents the fixed costs of running the company.

Solve the inequality $2000n - 700,000 \geq 2,500,000$ to find how many machines must be made and sold each month to satisfy these financial goals.

---

## 👣 STEPS TO SUCCESS  Why Do We Need to Study Mathematics?

Students often question the value of studying math. They see little real use for it in their everyday lives. Knowing math can help in various ways.

*Get a good job.* Mathematics is often the key that opens the door to a better-paying job or just to get a job if you are unemployed. In our technological world, people use mathematics daily. Many vocational and professional areas—such as business, statistics, economics, psychology, finance, computer science, chemistry, physics, engineering, electronics, nuclear energy, banking, quality control, nursing, medical technician, and teaching—require a certain level of expertise in mathematics. Those who want to work in these fields must be able to function at a given mathematical level. Those who cannot will not be able to enter these job areas.

*Save money.* These are challenging financial times. We are all looking for ways to save money. The more mathematics you learn, the more you will be able to find ways to save money. Several suggestions for saving money are given in this book.

Be sure to read over each one and think how it might apply to your life.

*Make decisions.* Should I buy a car or lease one? Should I buy a house or rent? Should I drive to work or take public transportation? What career field should I pick if I want to increase my chances of getting a good job? Mathematics will help you to think more clearly and make better decisions.

Making it personal: Which of these three paragraphs is most relevant to you? Write down what you think is the most important reason why you should study mathematics. ▼

**Verbal and Writing Skills, Exercises 1 and 2**

1. Is the statement $5 > -6$ equivalent to the statement $-6 < 5$? Why?

2. Is the statement $-8 < -3$ equivalent to the statement $-3 > -8$? Why?

*Replace the ? by $<$ or $>$.*

3. $8 \; ? \; -6$

4. $-10 \; ? \; 6$

5. $0 \; ? \; -8$

6. $-8 \; ? \; 0$

7. $-4 \; ? \; -2$

8. $-5 \; ? \; -8$

9. (a) $-7 \; ? \; 2$
   (b) $2 \; ? \; -7$

10. (a) $-5 \; ? \; 11$
    (b) $11 \; ? \; -5$

11. (a) $15 \; ? \; -15$
    (b) $-15 \; ? \; 15$

12. (a) $-17 \; ? \; 17$
    (b) $17 \; ? \; -17$

13. $\dfrac{1}{3} \; ? \; \dfrac{9}{10}$

14. $\dfrac{4}{6} \; ? \; \dfrac{7}{9}$

15. $\dfrac{7}{8} \; ? \; \dfrac{25}{31}$

16. $\dfrac{9}{11} \; ? \; \dfrac{41}{53}$

17. $-6.6 \; ? \; -8.9$

18. $-4.2 \; ? \; -7.3$

19. $-4.2 \; ? \; 3.5$

20. $-3.7 \; ? \; 3.7$

21. $-\dfrac{10}{3} \; ? \; -3$

22. $-5 \; ? \; -\dfrac{29}{4}$

23. $-\dfrac{5}{8} \; ? \; -\dfrac{3}{5}$

24. $-\dfrac{2}{3} \; ? \; -\dfrac{1}{2}$

*Graph each inequality on the number line.*

25. $x > 7$

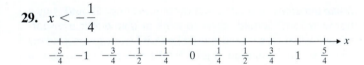

26. $x < 1$

27. $x \geq -6$

28. $x \leq -2$

29. $x < -\dfrac{1}{4}$

30. $x \leq -\dfrac{3}{2}$

31. $x \leq -5.3$

32. $x > -3.5$

33. $25 < x$

34. $35 \geq x$

*Translate each graph to an inequality using the variable x.*

35.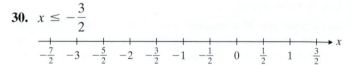

36.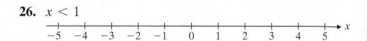

37.

38.

**39.**

**40.** 

*Translate each English statement into an inequality.*

**41.** *Full-Time Student* At Normandale Community College the number of credits a student takes per semester must not be less than 12 to be considered full time. (Use the variable $c$ for credits.)

**42.** *BMI Category* A person is considered underweight if his or her BMI (body mass index) measurement is smaller than 18.5. (Use the variable $B$ for BMI.)

**43.** *Height Limit* In order to ride the roller coaster at the theme park, your height must be at least 48 inches. (Use $h$ for height.)

**44.** *Boxing Category* To box in the featherweight category, your weight must not exceed 126 pounds. (Use $w$ for weight.)

## To Think About, Exercises 45 and 46

**45.** Suppose that the variable $x$ must satisfy *all* of these conditions.

$$x \leq 2, \quad x > -3, \quad x < \frac{5}{2}, \quad x \geq -\frac{5}{2}$$

Graph on a number line the region that satisfies all of the conditions.

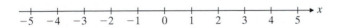

**46.** Suppose that the variable $x$ must satisfy *all* of these conditions.

$$x < 4, \quad x > -4, \quad x \leq \frac{7}{2}, \quad x \geq -\frac{9}{2}$$

Graph on a number line the region that satisfies all of the conditions.

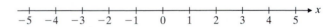

*Solve and graph the result.*

**47.** $x + 7 \leq 4$

**48.** $x - 5 < -3$

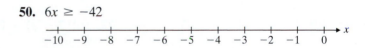

**49.** $5x \leq 25$

**50.** $6x \geq -42$

**51.** $-2x < 18$

**52.** $-7x < 28$

**53.** $\frac{1}{2}x \geq 4$

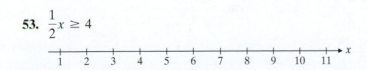

**54.** $\frac{1}{3}x \leq 2$

**55.** $-\frac{1}{4}x > 3$

**56.** $-\frac{1}{5}x < 10$

**57.** $8 - 5x > 13$

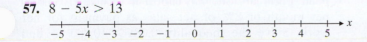

**58.** $9 - 4x \leq 21$

**59.** $-4 + 5x < -3x + 8$

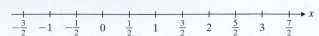

**60.** $-6 - 4x < 1 - 6x$

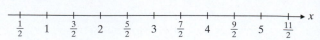

**61.** $\dfrac{5x}{6} - 5 > \dfrac{x}{6} - 9$

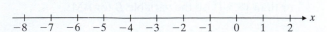

**62.** $\dfrac{x}{4} - 2 < \dfrac{3x}{4} + 5$

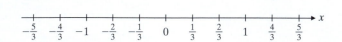

**63.** $2(3x + 4) > 3(x + 3)$

**64.** $5(x - 3) \le 2(x - 3)$

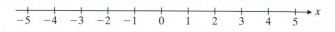

**Verbal and Writing Skills** *In exercises 65 and 66, answer the questions in your own words.*

**65.** Add $-2$ to both sides of the inequality $5 > 3$. What is the result? Why is the direction of the inequality not reversed?

**66.** Divide $-3$ into both sides of the inequality $-21 > -29$. What is the result? Why is the direction of the inequality reversed?

**Mixed Practice** *Solve. Collect the variable terms on the left side of the inequality.*

**67.** $5x + 2 > 8x - 7$

**68.** $7x + 8 < 12x - 2$

**69.** $6x - 2 \ge 4x + 6$

**70.** $9x - 8 \le 7x + 4$

**71.** $0.3(x - 1) < 0.1x - 0.5$

**72.** $0.4(2 - x) + 0.6 > 0.2(x - 2)$

**73.** $3 + 5(2 - x) \ge -3(x + 5)$

**74.** $9 - 3(2x - 1) \le 4(x + 2)$

**75.** $\dfrac{x + 6}{7} - \dfrac{3}{7} > \dfrac{x + 3}{2}$

**76.** $\dfrac{3x + 5}{4} - \dfrac{7}{12} > -\dfrac{x}{6}$

**Applications**

**77.** *Course Average* To pass a course with a B grade, a student must have an average of 80 or greater. A student's grades on three tests are 75, 83, and 86. Solve the inequality $\frac{75 + 83 + 86 + x}{4} \ge 80$ to find what score the student must get on the next test to get a B average or better.

**78.** *Payment Options* Sharon sells very expensive European sports cars. She may choose to receive $10,000.00 per month or 8% of her sales as payment for her work. Solve the inequality $0.08x > 10,000$ to find how much she needs to sell to make the 8% offer a better deal.

79. *Elephant Weight* The average African elephant weighs 268 pounds at birth. During its first three weeks, a baby elephant will usually gain about 4 pounds per day. Assuming that growth rate, solve the inequality $268 + 4x \geq 300$ to find how many days it will be until a baby elephant weighs at least 300 pounds.

80. *Car Loan* Rennie is buying a used car that costs $4500. The deal called for a $600 down payment, and payments of $260 monthly. He wants to know whether he can pay off the car within a year. Solve the inequality $600 + 260x \geq 4500$ to find out the minimum number of months it will take to pay off the car.

## Cumulative Review

81. **[0.5.3]** Find 16% of 38.

82. **[0.5.4]** 18 is what percent of 120?

83. **[0.5.4]** *Percent Accepted* For the most coveted graduate study positions, only 16 out of 800 students are accepted. What percent are accepted?

84. **[0.5.4]** Write the fraction $\frac{3}{8}$ as a percent.

### Quick Quiz 2.6

1. Graph $x \leq -3.5$ on the given number line.

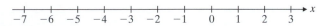

*Solve and graph the result for each of the following.*

2. $-12 + 4x \leq 2x$

3. $\dfrac{x}{2} - 1 < \dfrac{3}{2}x + 4$

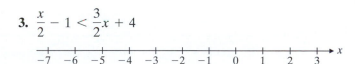

4. **Concept Check** Explain the difference between $12 < x$ and $x > 12$. Would the graphs of these inequalities be the same or different?

### Background: Dietitian

Jules has decided to pursue a career as a registered dietitian. She finds the prospect of educating and helping people to care for themselves by promoting healthy, nutritious eating habits to be exciting.

Part of her education and training involves working at a local hospital's walk-in clinic under the guidance of a registered dietitian. Jules often consults with overweight and diabetic patients about their treatment plans and diets. It's common for her to use equations to identify a number of measurements, such as patients' energy requirements, resting metabolic rates, and ideal body weights. She finds herself using equations often because her evaluations of even a single patient are an ongoing process of monitoring his or her condition or weight.

### Facts

- The Mifflin-St. Jeor equation is used to calculate the amount of energy (in calories) that a body needs to function while at rest for 24 hours. This amount of energy is called basic metabolic rate (BMR), and for men, the equation takes the form:

$$BMR = 10W + 6.25H - 5A + 5,$$

  where $W$ is weight in kilograms, $H$ is height in centimeters, and $A$ is age in years.

- Another commonly used equation is one for finding ideal body weight (IBW). For males, this equation is:

$$IBW = 106 \text{ lb} + 6(\text{the number of inches over 5 feet tall}).$$

  For females, the equation is:

$$IBW = 100 \text{ lb} + 5(\text{the number of inches over 5 feet tall}).$$

### Tasks

1. Jules wants to solve the Mifflin-St. Jeor equation for both weight ($W$) and height ($H$). What are her solutions?

2. Jules wants to determine the height of a male whose ideal weight is 190 pounds. What should she do to determine this?

3. She now wants to identify her own ideal weight, her IBW. She is 5 feet 5 inches tall. What is her ideal weight?

# Chapter 2 Organizer

| Topic and Procedure | Examples | ➡ You Try It |
|---|---|---|
| **Solving equations without parentheses or fractions, p. 143**<br><br>1. On both sides of the equation, combine like terms if possible.<br>2. Add or subtract terms on both sides of the equation in order to get all terms with the variable on one side of the equation.<br>3. Add or subtract a value on both sides of the equation to get all terms not containing the variable on the other side of the equation.<br>4. Divide both sides of the equation by the coefficient of the variable.<br>5. If possible, simplify the solution.<br>6. Check your solution by substituting the obtained value into the original equation. | Solve for $x$. $\quad 5x + 2 + 2x = -10 + 4x + 3$<br>$7x + 2 = -7 + 4x$<br>$7x - 4x + 2 = -7 + 4x - 4x$<br>$3x + 2 = -7$<br>$3x + 2 - 2 = -7 - 2$<br>$3x = -9$<br>$\dfrac{3x}{3} = \dfrac{-9}{3}$<br>$x = -3$<br>*Check:*<br>Is $-3$ the solution of $5x + 2 + 2x = -10 + 4x + 3$?<br>$5(-3) + 2 + 2(-3) \overset{?}{=} -10 + 4(-3) + 3$<br>$-15 + 2 + (-6) \overset{?}{=} -10 + (-12) + 3$<br>$-13 - 6 \overset{?}{=} -22 + 3$<br>$-19 = -19 \ ✓$ | **1.** Solve for $x$.<br>$-8x - 1 + x = 13 - 6x - 2$ |
| **Solving equations with parentheses and/ or fractions, pp. 145, 150**<br><br>1. Remove any parentheses.<br>2. Simplify, if possible.<br>3. If fractions exist, multiply all terms on both sides by the least common denominator of all the fractions.<br>4. Now follow the steps for solving an equation without parentheses or fractions. | Solve for $y$. $\quad 5(3y - 4) = \dfrac{1}{4}(6y + 4) - 48$<br>$15y - 20 = \dfrac{3}{2}y + 1 - 48$<br>$15y - 20 = \dfrac{3}{2}y - 47$<br>$2(15y) - 2(20) = 2\left(\dfrac{3}{2}y\right) - 2(47)$<br>$30y - 40 = 3y - 94$<br>$30y - 3y - 40 = 3y - 3y - 94$<br>$27y - 40 = -94$<br>$27y - 40 + 40 = -94 + 40$<br>$27y = -54$<br>$\dfrac{27y}{27} = \dfrac{-54}{27}$<br>$y = -2$ | **2.** Solve for $y$.<br>$\dfrac{1}{3}(y + 5) = \dfrac{1}{4}(5y - 8)$ |
| **Solving formulas for a specified variable, p. 158**<br><br>1. Remove any parentheses.<br>2. If fractions exist, multiply all terms on both sides by the LCD, which may be a variable.<br>3. Add or subtract terms on both sides of the equation in order to get all terms containing the *desired variable* on one side of the equation.<br>4. Add or subtract terms on both sides of the equation in order to get all other terms on the opposite side of the equation.<br>5. Divide both sides of the equation by the coefficient of the desired variable. This division may involve other variables.<br>6. Simplify, if possible.<br>7. (Optional) Check your solution by substituting the obtained expression into the original equation. | Solve for $z$. $\quad B = \dfrac{1}{3}(hx + hz)$<br>First we remove parentheses.<br>$B = \dfrac{1}{3}hx + \dfrac{1}{3}hz$<br>Now we multiply each term by 3.<br>$3(B) = 3\left(\dfrac{1}{3}hx\right) + 3\left(\dfrac{1}{3}hz\right)$<br>$3B = hx + hz$<br>$3B - hx = hx - hx + hz$<br>$3B - hx = hz$<br>The coefficient of $z$ is $h$, so we divide each side by $h$.<br>$\dfrac{3B - hx}{h} = z$ | **3.** Solve for $a$.<br>$H = \dfrac{1}{4}(ca + b)$ |

177

| Topic and Procedure | Examples |  You Try It |
|---|---|---|
| **Solving inequalities, p. 167**<br>1. Follow the steps for solving an equation up to the multiplication or division step.<br>2. If you divide or multiply both sides of the inequality by a *positive number*, the **direction** of the inequality **is not reversed.**<br>3. If you divide or multiply both sides of the inequality by a *negative number*, the **direction** of the inequality **is reversed.** | Solve for $x$ and graph your solution.<br><br>$\frac{1}{2}(3x - 2) \leq -5 + 5x - 3$ — First remove parentheses and simplify.<br><br>$\frac{3}{2}x - 1 \leq -8 + 5x$<br><br>$2\left(\frac{3}{2}x\right) - 2(1) \leq 2(-8) + 2(5x)$ — Now multiply each term by 2.<br><br>$3x - 2 \leq -16 + 10x$<br><br>$3x - 10x - 2 \leq -16 + 10x - 10x$<br><br>$-7x - 2 \leq -16$<br><br>$-7x - 2 + 2 \leq -16 + 2$<br><br>$-7x \leq -14$<br><br>$\frac{-7x}{-7} \geq \frac{-14}{-7}$ — When we divide both sides by a negative number, the inequality is reversed.<br><br>$x \geq 2$<br><br>Graph of the solution: | **4.** Solve for $x$ and graph your solution.<br><br>$4 + 3x - 5 \geq \frac{1}{3}(10x + 1)$ |

# Chapter 2 Review Problems

## Sections 2.1–2.3

*Solve for the variable.*

**1.** $3x + 2x = -35$

**2.** $x - 19 = -29 + 7$

**3.** $18 - 10x = 63 + 5x$

**4.** $x - (0.5x + 2.6) = 17.6$

**5.** $3(x - 2) = -4(5 + x)$

**6.** $12 - 5x = -7x - 2$

**7.** $2(3 - x) = 1 - (x - 2)$

**8.** $4(x + 5) - 7 = 2(x + 3)$

**9.** $3 = 2x + 5 - 3(x - 1)$

**10.** $2(5x - 1) - 7 = 3(x - 1) + 5 - 4x$

## Section 2.4

*Solve for the variable.*

**11.** $\frac{3}{4}x - 3 = \frac{1}{2}x + 2$

**12.** $1 = \frac{5x}{6} + \frac{2x}{3}$

**13.** $\frac{7x}{5} = 5 + \frac{2x}{5}$

**14.** $\frac{7x - 3}{2} - 4 = \frac{5x + 1}{3}$

**15.** $\dfrac{3x-2}{2} + \dfrac{x}{4} = 2 + x$

**16.** $\dfrac{-3}{2}(x+5) = 1 - x$

**17.** $-0.2(x+1) = 0.3(x+11)$

**18.** $1.2x - 0.8 = 0.8x + 0.4$

**19.** $3.2 - 0.6x = 0.4(x-2)$

**20.** $\dfrac{1}{3}(x-2) = \dfrac{x}{4} + 2$

**21.** $\dfrac{3}{4} - \dfrac{2}{3}x = \dfrac{1}{3}x + \dfrac{3}{4}$

**22.** $-\dfrac{8}{3}x - 8 + 2x - 5 = -\dfrac{5}{3}$

**23.** $\dfrac{1}{6} + \dfrac{1}{3}(x-3) = \dfrac{1}{2}(x+9)$

**24.** $\dfrac{1}{7}(x+5) - \dfrac{3}{7} = \dfrac{1}{2}(x+3)$

## Section 2.5

*Solve for the variable indicated.*

**25.** Solve for $y$. $3x - y = 10$

**26.** Solve for $y$. $5x + 2y + 7 = 0$

**27.** Solve for $r$. $A = P(1 + rt)$

**28.** Solve for $h$. $A = 4\pi r^2 + 2\pi rh$

**29.** Solve for $p$. $H = \dfrac{1}{3}(a + 2p + 3)$

**30.** Solve for $y$. $ax + by = c$

**31.** **(a)** Solve for $A$. $x = \dfrac{ABC}{10}$

   **(b)** Use your result to find $A$ if $x = 6$, $B = -1$, and $C = 1.5$.

**32.** **(a)** Solve for $w$. $2l + 2w = P$

   **(b)** Use your result to find $w$ if $P = 34$ and $l = 10.5$.

**33.** **(a)** Solve for $h$. $V = lwh$

   **(b)** Use your result to find $h$ when $V = 48$, $l = 2$, and $w = 4$.

## Section 2.6

*Solve each inequality and graph the result.*

**34.** $9 + 2x \leq 6 - x$

**35.** $2x - 3 + x > 5(x + 1)$

**36.** $-x + 4 < 3x + 16$

**37.** $8 - \dfrac{1}{3}x \leq x$

**38.** $7 - \dfrac{3}{5}x > 4$

**39.** $-4x - 14 < 4 - 2(3x - 1)$

**40.** $3(x - 2) + 8 < 7x + 14$

*Use an inequality to solve.*

**41.** *Wages* Julian earns \$15 per hour as a plasterer's assistant. His employer determines that the current job allows him to pay \$480 in wages to Julian. What are the maximum number of hours that Julian can work on this job? (*Hint:* Use $15h \leq 480$.)

**42.** *Hiring a Substitute* The cost of hiring a substitute elementary teacher for a day is \$110. The school's budget for substitute teachers is \$2420 per month. What is the maximum number of times a substitute teacher may be hired during a month? (*Hint:* Use $110n \leq 2420$.)

## Mixed Practice

*Solve for the variable.*

**43.** $10(2x + 4) - 13 = 8(x + 7) - 3$

**44.** $-9x + 15 - 2x = 4 - 3x$

**45.** $-2(x - 3) = -4x + 3(3x + 2)$

**46.** $\dfrac{1}{2} + \dfrac{5}{4}x = \dfrac{2}{5}x - \dfrac{1}{10} + 4$

*Solve each inequality and graph the result.*

**47.** $5 - \dfrac{1}{2}x > 4$

**48.** $2(x - 1) \geq 3(2 + x)$

**49.** $\dfrac{1}{3}(x + 2) \leq \dfrac{1}{2}(3x - 5)$

**50.** $4(2 - x) - (-5x + 1) \geq -8$

180

# How Am I Doing? Chapter 2 Test

**MATH COACH**  **MyMathLab®**  **You Tube**

After you take this test read through the Math Coach on pages 182–183. Math Coach videos are available via MyMathLab and YouTube. Step-by-step test solutions on the Chapter Test Prep Videos are also available via MyMathLab and YouTube. (Search "TobeyBeginningAlg" and click on "Channels.")

*Solve for the variable.*

**1.** $3x + 5.6 = 11.6$

**2.** $9x - 8 = -6x - 3$

**3.** $2(2y - 3) = 4(2y + 2)$

**4.** $\frac{1}{7}y + 3 = \frac{1}{2}y$

**5.** $4(7 - 4x) = 3(6 - 2x)$

**MC 6.** $0.8x + 0.18 - 0.4x = 0.3(x + 0.2)$

**7.** $\frac{2y}{3} + \frac{1}{5} - \frac{3y}{5} + \frac{1}{3} = 1$

**8.** $3 - 2y = 2(3y - 2) - 5y$

**9.** $5(20 - x) + 10x = 165$

**10.** $5(x + 40) - 6x = 9x$

**11.** $-2(2 - 3x) = 76 - 2x$

**MC 12.** $20 - (2x + 6) = 5(2 - x) + 2x$

*In questions 13–17, solve for x.*

**13.** $2x - 3 = 12 - 6x + 3(2x + 3)$

**14.** $\frac{1}{3}x - \frac{3}{4}x = \frac{1}{12}$

**15.** $\frac{3}{5}x + \frac{7}{10} = \frac{1}{3}x + \frac{3}{2}$

**16.** $\frac{15x - 2}{28} = \frac{5x - 3}{7}$

**MC 17.** $\frac{2}{3}(x + 8) + \frac{3}{5} = \frac{1}{5}(11 - 6x)$

**18.** Solve for $w$. $A = 3w + 2P$

**19.** Solve for $w$. $\frac{2w}{3} = 4 - \frac{1}{2}(x + 6)$

**20.** Solve for $a$. $A = \frac{1}{2}h(a + b)$

**21.** Solve for $y$. $5ax(2 - y) = 3axy + 5$

**22.** Solve for $B$. $V = \frac{1}{3}Bh$

▲ **23.** Use your result from question 22 to find the area of the base ($B$) of a pyramid if the volume ($V$) is 140 cubic inches and the height ($h$) is 14 inches.

*Solve and graph the inequality.*

**24.** $3(x - 2) \geq 5x$

$\xrightarrow{\;-5\quad-4\quad-3\quad-2\quad-1\quad 0\;} x$

**MC 25.** $2 - 7(x + 1) - 5(x + 2) < 0$

$\xrightarrow{\;-\frac{3}{2}\quad-\frac{5}{4}\quad-1\quad-\frac{3}{4}\quad-\frac{1}{2}\quad-\frac{1}{4}\;} x$

**26.** $5 + 8x - 4 < 2x + 13$

$\xrightarrow{\;0\quad 1\quad 2\quad 3\quad 4\quad 5\;} x$

**27.** $\frac{1}{4}x + \frac{1}{16} \leq \frac{1}{8}(7x - 2)$

$\xrightarrow{\;-1\quad-\frac{1}{2}\quad 0\quad\frac{1}{2}\quad 1\quad\frac{3}{2}\;} x$

1. _____ ☐
2. _____ ☐
3. _____ ☐
4. _____ ☐
5. _____ ☐
6. _____ ☐
7. _____ ☐
8. _____ ☐
9. _____ ☐
10. _____ ☐
11. _____ ☐
12. _____ ☐
13. _____ ☐
14. _____ ☐
15. _____ ☐
16. _____ ☐
17. _____ ☐
18. _____ ☐
19. _____ ☐
20. _____ ☐
21. _____ ☐
22. _____ ☐
23. _____ ☐
24. _____ ☐
25. _____ ☐
26. _____ ☐
27. _____ ☐

**Total Correct:** ☐

# MATH COACH

*Mastering the skills you need to do well on the test.*

The following problems are from the Chapter 2 Test. Here are some helpful hints to keep you from making common errors on test problems.

**Chapter 2 Test, Problem 6**   Solve for the variable. $0.8x + 0.18 - 0.4x = 0.3(x + 0.2)$

> **HELPFUL HINT**  After removing parentheses, it might be most helpful for you to multiply both sides of the equation by 100 in order to obtain a simpler, equivalent equation without decimals. Check to make sure that you did not make any errors in calculations before solving the equation.

Did you remove the parentheses to get the equation $0.8x + 0.18 - 0.4x = 0.3x + 0.06$?

Yes _____ No _____

If you answered No, go back and use the distributive property to remove the parentheses. Be careful to place the decimal point in the correct location when multiplying 0.3 and 0.2 together.

Did you multiply each term of the equation by 100 to move the decimal point two places to the right to get the equivalent equation $80x + 18 - 40x = 30x + 6$?

Yes _____ No _____

If you answered No, stop and carefully complete this step before solving the equation. Remember that you may need to add a 0 to the end of a term in order to move the decimal point two places to the right.

If you answered Problem 6 incorrectly, go back and rework the problem using these suggestions.

---

**Chapter 2 Test, Problem 12**   Solve for the variable. $20 - (2x + 6) = 5(2 - x) + 2x$

> **HELPFUL HINT**  Slowly complete the necessary steps to remove each set of parentheses before doing any other steps. Be careful to avoid sign errors.

Did you obtain the equation $20 - 2x - 6 = 10 - 5x + 2x$ after removing each set of parentheses?

Yes _____ No _____

If you answered No, go back and carefully use the distributive property to remove each set of parentheses. Locate any mistakes you have made and make a note of the type of error discovered.

Did you combine like terms to get the equation $14 - 2x = 10 - 3x$?

Yes _____ No _____

If you answered No, stop and perform that step correctly.

Now go back and rework the problem using these suggestions.

Need more help? Watch the **MATH COACH** videos in MyMathLab® or on You Tube™.

182

**Chapter 2 Test, Problem 17**   Solve for $x$. $\dfrac{2}{3}(x + 8) + \dfrac{3}{5} = \dfrac{1}{5}(11 - 6x)$

> **HELPFUL HINT**  Remove the parentheses first. This is the most likely place to make a mistake. Next, carefully show every step of your work as you multiply each fraction by the LCD. Be sure to check your work.

Did you remove each set of parentheses to obtain the equation $\dfrac{2}{3}x + \dfrac{16}{3} + \dfrac{3}{5} = \dfrac{11}{5} - \dfrac{6}{5}x$?

Yes _____ No _____

If you answered No, stop and carefully redo your steps of multiplication, showing every part of your work.

Did you identify the LCD as 15 and then multiply each term by 15 to get $10x + 80 + 9 = 33 - 18x$?

Yes _____ No _____

If you answered No, stop and write out your steps slowly.

If you answered Problem 17 incorrectly, go back and rework the problem using these suggestions.

---

**Chapter 2 Test, Problem 25**   Solve and graph the inequality. $2 - 7(x + 1) - 5(x + 2) < 0$

> **HELPFUL HINT**  Be sure to remove parentheses and combine any like terms on each side of the inequality before solving for the variable. Always verify the following:
>  1. Did you multiply or divide by a negative number? If so, did you reverse the inequality symbol?
>  2. In the graph, is your choice of an open circle or closed circle correct?

Did you remove parentheses to get $2 - 7x - 7 - 5x - 10 < 0$? Did you combine like terms to obtain the inequality $-15 - 12x < 0$? Next, did you add 15 to both sides of the inequality?

Yes _____ No _____

If you answered No to any of these questions, stop now and perform those steps.

Did you remember to reverse the inequality symbol in the last step?

Yes _____ No _____

If you answered No, please review the rules for when to reverse the inequality symbol and then go back and perform this step.

Did you use an open circle in your number line graph? Is your arrow pointing to the right?

Yes _____ No _____

If you answered No to either question, please review the rules for how to graph an inequality involving the $<$ or $>$ inequality symbols.

Now go back and rework the problem using these suggestions.

Need more help? Look for section examples marked with ᴹℂ to review.

CHAPTER 3

# Solving Applied Problems

## CAREER OPPORTUNITIES

### Landscape Design Architect

Landscape design architects not only work on planting greenery, but they also help to design, create, and install outdoor living and work spaces. Whether the purpose is to improve the overall appearance of a landscape by matching the natural environment with existing structures or to create an exterior space that meets functional and aesthetic needs, landscape designers use the basic math, algebra, and geometry that are studied in this chapter.

To investigate how the mathematics in this chapter can help with this field, see the Career Exploration Problems on page **233**.

# 3.1 Translating English Phrases into Algebraic Expressions ▶

## 1 Translating English Phrases into Algebraic Expressions ▶

One of the most useful applications of algebra is solving word problems. One of the first steps in solving a word problem is translating the conditions of the problem into algebra. In this section we show you how to translate common English phrases into algebraic expressions. This process is similar to translating between languages like Spanish and French.

Several English phrases describe the operation of addition. If we represent an unknown number by the variable $x$, all of the following phrases can be translated into algebra as $x + 3$.

| *English Phrases Describing Addition* | *Algebraic Expression* | *Diagram* |
|---|---|---|
| Three *more than* a number | | |
| The *sum of* a number and three | | |
| A number *increased by* three | $x + 3$ | |
| Three is *added to* a number. | | |
| Three *greater than* a number | | |
| A number *plus* three | | |

In a similar way we can use algebra to express English phrases that describe the operations of subtraction, multiplication, and division.

**CAUTION:** Since subtraction is not commutative, the order is essential. A number decreased by five is $x - 5$. It is not correct to say $5 - x$. Use extra care as you study each example. Make sure you understand the proper order.

| *English Phrases Describing Subtraction* | *Algebraic Expression* | *Diagram* |
|---|---|---|
| A number *decreased by* four | | |
| Four *less than* a number | | |
| Four is *subtracted from* a number. | | |
| Four *smaller than* a number | $x - 4$ | |
| Four *fewer than* a number | | |
| A number *diminished by* four | | |
| A number *minus* 4 | | |
| The *difference between* a number and four | | |

| *English Phrases Describing Multiplication* | *Algebraic Expression* | *Diagram* |
|---|---|---|
| *Double* a number | | |
| *Twice* a number | | |
| The *product of* two and a number | $2x$ | |
| Two *of* a number | | |
| Two *times* a number | | |

Since division is not commutative, the order is essential. A number divided by 3 is $\frac{x}{3}$. It is not correct to say $\frac{3}{x}$. Use extra care as you study each example.

**Student Learning Objectives**

After studying this section, you will be able to:

1 Translate English phrases into algebraic expressions. ▶

2 Write an algebraic expression to compare two or more quantities. ▶

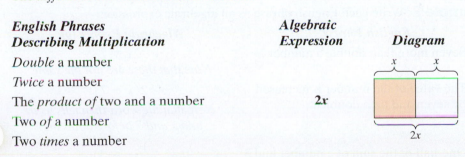

| *English Phrases Describing Division* | *Algebraic Expression* | *Diagram* |
|---|---|---|
| A number *divided by* five | | |
| *One-fifth of* a number | $\dfrac{x}{5}$ | |
| The *quotient of* a number and five | | |

Often other words are used in English instead of the word *number*. We can use a variable, such as $x$, here also.

**Example 1** Write each English phrase as an algebraic expression.

| *English Phrase* | *Algebraic Expression* |
|---|---|
| **(a)** A *quantity* is increased by five. | $x + 5$ |
| **(b)** Double the *value* | $2x$ |
| **(c)** One-third of the *weight* | $\dfrac{x}{3}$ or $\dfrac{1}{3}x$ |
| **(d)** Twelve more than an *unknown number* | $x + 12$ |
| **(e)** Seven less than a *number* | $x - 7$ |

Note that the algebraic expression for "seven less than a number" does not follow the order of the words in the English phrase. The variable or expression that follows the words *less than* always comes first.

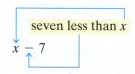

The variable or expression that follows the words *more than* technically comes before the plus sign. However, since addition is commutative, it also can be written after the plus sign. ◻

**Student Practice 1** Write each English phrase as an algebraic expression.

**(a)** Four more than a number **(b)** Triple a value

**(c)** Eight less than a number **(d)** One-fourth of a height

More than one operation can be described in an English phrase. Sometimes parentheses must be used to make clear which operation is done first.

**Example 2** Write each English phrase as an algebraic expression.

| *English Phrase* | *Algebraic Expression* |
|---|---|
| **(a)** Seven more than double a number | $2x + 7$ |
| | *Note that these are **not** the same.* |
| **(b)** The value of the number is increased by seven and then doubled. | $2(x + 7)$ |
| | *Note that the word **then** tells us to add x and 7 before doubling.* |
| **(c)** One-half of the sum of a number and 3 | $\dfrac{1}{2}(x + 3)$ |

◻

**Student Practice 2**   Write each English phrase as an algebraic expression.

(a) Eight more than triple a number

(b) A number is increased by eight and then it is tripled.

(c) One-third of the sum of a number and 4

**2** Writing an Algebraic Expression to Compare Two or More Quantities

Often in a word problem two or more quantities are described in terms of another. We will want to use a variable to represent one quantity and then write an algebraic expression using *the same variable* to represent the other quantity. Which quantity should we let the variable represent? We usually let the variable represent the quantity that is the basis of comparison: the quantity that the others are being *compared to*.

**Example 3**  Use a variable and an algebraic expression to describe the two quantities in the English sentence "Mike's salary is $2000 more than Fred's salary."

**Solution**   The two quantities that are being compared are Mike's salary and Fred's salary. Since Mike's salary is being *compared to* Fred's salary, we let the variable represent Fred's salary. The choice of the letter $f$ helps us to remember that the variable represents Fred's salary.

$$\text{Let } f = \text{Fred's salary.}$$

Then $f + \$2000 = $ Mike's salary, since Mike's salary is $2000 *more than* Fred's.  □

 **Student Practice 3**   Use a variable and an algebraic expression to describe the two quantities in the English sentence "Marie works 17 hours per week less than Ann."

**Example 4** The length of a rectangle is 3 meters shorter than twice the width. Use a variable and an algebraic expression to describe the length and the width. Draw a picture of the rectangle and label the length and width.

**Solution**   The length of the rectangle is being *compared to* the width. Use the letter $w$ for width.

$$\text{Let } w = \text{ the width.}$$

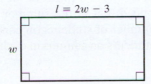

$$\text{Then } 2w - 3 = \text{ the length.}$$

A picture of the rectangle is shown.

$$l = 2w - 3$$

□

 **Student Practice 4**   The length of a rectangle is 5 meters longer than double the width. Use a variable and an algebraic expression to describe the length and the width. Draw a picture of the rectangle and label the length and width.

**Example 5** The measure of the first angle of a triangle is triple the measure of the second angle. The measure of the third angle of the triangle is 12° more than the measure of the second angle. Describe each angle measure algebraically. Draw a diagram of the triangle and label its parts.

**Solution** Since the first and third angle measures are described in terms of the second angle, we let the variable represent the number of degrees in the second angle.

Let $s =$ the number of degrees in the second angle.

Then $3s =$ the number of degrees in the first angle.

And $s + 12 =$ the number of degrees in the third angle.

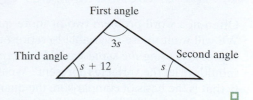

**Student Practice 5** The measure of the first angle of a triangle is 16° less than the measure of the second angle. The measure of the third angle is double the measure of the second angle. Describe each angle measure algebraically. Draw a diagram of the triangle and label its parts.

Some comparisons will involve fractions.

**Example 6** A theater manager was examining the records of attendance for last year. The number of people attending the theater in January was one-half of the number of people attending the theater in February. The number of people attending the theater in March was three-fifths of the number of people attending the theater in February. Use algebra to describe the attendance each month.

**Solution** What are we looking for? The *number of people* who attended the theater *each month.* The basis of comparison is February. That is where we begin.

Let $f =$ the number of people who attended in February.

Then $\frac{1}{2}f =$ the number of people who attended in January.

And $\frac{3}{5}f =$ the number of people who attended in March.

**Student Practice 6** The college dean noticed that in the spring the number of students on campus was two-thirds of the number of students on campus in the fall. She also noticed that in the summer the number of students on campus was one-fifth the number of students on campus in the fall. Use algebra to describe the number of students on campus in each of these three time periods.

## 3.1 Exercises   MyMathLab®

**Verbal and Writing Skills** *Write an algebraic expression for each quantity. Let x represent the unknown value.*

1. eleven more than a number

2. the sum of a number and five

3. twelve less than a number

4. seven subtracted from a number

5. one-eighth of a quantity

6. one-sixth of a quantity

7. twice a quantity

8. triple a number

9. three more than half of a number

10. five more than one-third of a number

11. double a quantity increased by nine

12. ten times a number increased by 1

13. one-third of the sum of a number and seven

14. one-fourth of the sum of a number and 5

15. one-third of a number reduced by twice the same number

16. one-fifth of a number reduced by double the same number

17. five times a quantity decreased by eleven

18. four less than seven times a number

*Write an algebraic expression for each of the quantities being compared.*

19. **Stock Value** The value of a share of IBM stock one day was $74.50 more than the value of a share of AT&T stock.

20. **Investments** One day in June 2015, the value of a share of Twitter stock was $47.49 less than the value of a share of Target stock.

▲ 21. **Geometry** The length of a rectangle is 7 inches more than double the width.

▲ 22. **Geometry** The length of a rectangle is 3 meters more than triple the width.

23. **Cookie Sales** The number of boxes of cookies sold by Sarah was 43 fewer than the number of boxes of cookies sold by Keiko. The number of boxes of cookies sold by Imelda was 53 more than the number sold by Keiko.

24. **April Rainfall** The average April rainfall in Savannah, Georgia, is about 13 inches more than that of Burlington, Vermont. The average rainfall in Phoenix, Arizona, is about 28 inches less than that of Burlington.

▲ 25. **Geometry** The measure of the first angle of a triangle is 25 degrees less than the measure of the second angle. The measure of the third angle of the triangle is triple the measure of the second angle.

▲ 26. **Geometry** The measure of the second angle of a triangle is double the measure of the first angle. The measure of the third angle of the triangle is 30 degrees more than the measure of the first angle.

27. **Exports** The value of the exports of Japan was twice the value of the exports of Canada.

28. **Text Messages** The number of text messages Marisol received was three times the number received by her brother.

189

29. ***Concert Tickets*** The price of the All Star Concert tickets was one-half the price of the Summer on the Beach Concert tickets.

30. ***Technology Convention*** The attendance at the Innovative Technology convention last year was one-third the attendance at the convention this year.

## To Think About

***Kayak Rentals*** *The following bar graph depicts the number of people who rented sea kayaks at Essex Boat Rental during the summer of 2015. Use the bar graph to answer exercises 31 and 32.*

31. Write an expression for the number of men who rented sea kayaks at Essex Boat Rental in each age category. Start by using $x$ for the number of men aged 16 to 24 who rented kayaks.

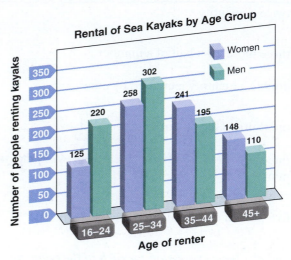

32. Write an expression for the number of women who rented sea kayaks at Essex Boat Rental. Start by using $y$ for the number of women aged 25 to 34 who rented kayaks.

## Cumulative Review  *Solve for the variable.*

33. **[2.4.1]** $x + \dfrac{1}{2}(x - 3) = 9$

34. **[2.4.1]** $\dfrac{3}{5}x - 3(x - 1) = 9$

---

## Quick Quiz 3.1  *Write an algebraic expression for each quantity. Let x represent the unknown value.*

1. Ten greater than a number

2. Five less than double a number

3. The measure of the first angle of a triangle is 15 degrees more than the measure of the second angle. The measure of the third angle of the triangle is five times the measure of the second angle. Write an algebraic expression for the measure of each of the three angles.

4. **Concept Check** Explain how you would decide whether to use $\frac{1}{3}(x + 7)$ or $\frac{1}{3}x + 7$ as the algebraic expression for the phrase "one-third of the sum of a number and seven."

# 3.2 Using Equations to Solve Word Problems

In Section 0.6 we introduced a simple three-step procedure to solve applied problems. You have had an opportunity to use that approach to solve word problems in Exercises 0.6 and in the Applications exercises in Chapters 1 and 2. Now we are going to focus our attention on solving applied problems that require the use of variables, translating English phrases into algebraic expressions, and setting up equations. The process is a little more involved. Some students find the following outline a helpful way to keep organized while solving such problems.

**Student Learning Objectives**

After studying this section, you will be able to:

1 Solve number problems.

2 Use the Mathematics Blueprint to solve applied word problems.

3 Use formulas to solve word problems.

1. **Understand the problem.**
   (a) Read the word problem carefully to get an overview.
   (b) Determine what information you will need to solve the problem.
   (c) Draw a sketch. Label it with the known information. Determine what needs to be found.
   (d) Choose a variable to represent one unknown quantity.
   (e) If necessary, represent other unknown quantities in terms of that very same variable.

2. **Write an equation.**
   (a) Look for key words to help you to translate the words into algebraic symbols and expressions.
   (b) Use a given relationship in the problem or an appropriate formula to write an equation.

3. **Solve and state the answer.**

4. **Check.**
   (a) Check the solution in the original equation. Is the answer reasonable?
   (b) Be sure the solution to the equation answers the question in the word problem. You may need to do some additional calculations if it does not.

## 1 Solving Number Problems

**Example 1** Two-thirds of a number is eighty-four. What is the number?

**Solution**

1. **Understand the problem.** Draw a sketch.

   Let $x = $ the unknown number.

2. **Write an equation.**

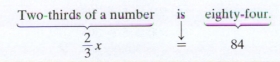

*Continued on next page*

3. *Solve and state the answer.*

$$\frac{2}{3}x = 84$$

$$3\left(\frac{2}{3}x\right) = 3(84) \quad \text{Multiply both sides of the equation by 3.}$$

$$2x = 252 \quad \text{Simplify.}$$

$$\frac{2x}{2} = \frac{252}{2} \quad \text{Divide both sides by 2.}$$

$$x = 126$$

The number is 126.

4. *Check.* Is two-thirds of 126 eighty-four?

$$\frac{2}{3}(126) \overset{?}{=} 84$$

$$84 = 84 \quad \checkmark$$

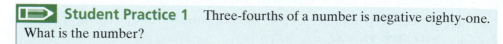 **Student Practice 1** Three-fourths of a number is negative eighty-one. What is the number?

Learning to solve problems like Examples 1 and 2 is a very useful skill. You will find that learning the rest of the material in Chapter 3 will be much easier if you can master the procedure used in these two examples.

**Example 2** Five more than six times a quantity is three hundred five. Find the number.

**Solution**

1. *Understand the problem.* Read the problem carefully. You may not need to draw a sketch.

   Let $x =$ the unknown quantity.

2. *Write an equation.*

   $$\underbrace{\text{Five more than}}_{5\ +} \quad \underbrace{\text{six times a quantity}}_{6x} \quad \underbrace{\text{is}}_{=} \quad \underbrace{\text{three hundred five.}}_{305}$$

3. *Solve and state the answer.* You may want to rewrite the equation to make it easier to solve.

$$6x + 5 = 305$$

$$6x + 5 - 5 = 305 - 5 \quad \text{Subtract 5 from both sides.}$$

$$6x = 300 \quad \text{Simplify.}$$

$$\frac{6x}{6} = \frac{300}{6} \quad \text{Divide both sides by 6.}$$

$$x = 50$$

The quantity, or number, is 50.

4. *Check.* Is five more than six times 50 three hundred five?

$$6(50) + 5 \overset{?}{=} 305$$

$$300 + 5 \overset{?}{=} 305$$

$$305 = 305 \quad \checkmark$$

**Student Practice 2** Two less than triple a number is forty-nine. Find the number.

**Example 3** The larger of two numbers is three more than twice the smaller. The sum of the numbers is thirty-nine. Find each number.

### Solution

1. **Understand the problem.** The problem refers to *two* numbers. We must write an algebraic expression for *each number* before writing the equation. If we are comparing one quantity to others, we usually let the variable represent the quantity that the others are being *compared* to. In this case, the larger number is being compared to the smaller number. We want to use *one variable* to describe each number.

$$\text{Let } s = \text{the smaller number.}$$
$$\text{Then } \underbrace{2s + 3}_{\text{three more than twice the smaller number}} = \text{the larger number.}$$

2. **Write an equation.** The sum of the numbers    is    thirty-nine.

$$s + (2s + 3) \qquad = \qquad 39$$

3. **Solve.**

$$s + (2s + 3) = 39$$
$$3s + 3 = 39 \quad \text{Combine like terms.}$$
$$3s = 36 \quad \text{Subtract 3 from each side.}$$
$$s = 12 \quad \text{Divide both sides by 3.}$$

4. **Check.**

$$12 + [2(12) + 3] \overset{?}{=} 39$$
$$39 = 39 \quad \checkmark$$

The solution checks, but have we solved the word problem? We need to find *each* number. 12 is the smaller number. Substitute 12 into the expression $2s + 3$ to find the larger number.

$$2s + 3 = 2(12) + 3 = 27$$

The smaller number is 12. The larger number is 27. □

> **Student Practice 3** Consider two numbers. The second number is twelve less than triple the first number. The sum of the two numbers is twenty-four. Find each number.

## 2 Using the Mathematics Blueprint to Solve Applied Word Problems ▶

To facilitate understanding more involved word problems, we will use a Mathematics Blueprint similar to the one we used in Section 0.6. This format is a simple way to organize facts, determine what to set variables equal to, and select a method or approach that will assist you in finding the desired quantity. You will find using this form helpful, particularly in those cases when you read through a word problem and mentally say to yourself, "Now where do I begin?" You begin by responding to the headings of the blueprint. Soon a procedure for solving the problem will emerge.

### Mathematics Blueprint for Problem Solving

| Gather the Facts | Assign the Variable | Basic Formula or Equation | Key Points to Remember |
|---|---|---|---|
|  |  |  |  |
|  |  |  |  |

**Example 4** The mean annual snowfall in Juneau, Alaska, is 105.8 inches. This is 20.2 inches less than three times the annual snowfall in Boston. What is the annual snowfall in Boston?

### Solution

*Understand the problem and write an equation.*

Mathematics Blueprint for Problem Solving

| Gather the Facts | Assign the Variable | Basic Formula or Equation | Key Points to Remember |
|---|---|---|---|
| Snowfall in Juneau is 105.8 inches.

This is 20.2 inches less than three times the snowfall in Boston. | We do not know the snowfall in Boston.

Let $b$ = annual snowfall in Boston. Then $3b - 20.2$ = annual snowfall in Juneau. | Set $3b - 20.2$ equal to 105.8, which is the snowfall in Juneau. | All measurements of snowfall are recorded in inches. |

Juneau's snowfall is 20.2 less than three times Boston's snowfall.

$$105.8 = 3b - 20.2$$

*Solve and state the answer.* You may want to rewrite the equation to make it easier to solve.

$$3b - 20.2 = 105.8$$
$$3b = 126 \qquad \text{Add 20.2 to both sides.}$$
$$b = 42 \qquad \text{Divide both sides by 3.}$$

The annual snowfall in Boston is 42 inches.

*Check.*   Reread the word problem. Work backward.

Three times 42 is 126.

126 less 20.2 is 105.8.

Is this the annual snowfall in Juneau? Yes. ✓                                      ◻

**Student Practice 4**   The maximum recorded rainfall for a 24-hour period in the United States occurred in Alvin, Texas, on July 25–26, 1979. The maximum recorded rainfall for a 24-hour period in Canada occurred in Ucluelet Brynnor Mines, British Columbia, on October 6, 1967. Alvin, Texas, received 43 inches of rain in that period. The amount recorded in Texas was 14 inches less than three times the amount recorded in Canada. How much rainfall was recorded at Ucluelet Brynnor Mines in Canada? (*Source:* National Oceanic and Atmospheric Administration)

Some word problems require a simple translation of the facts. Others require a little more detective work. You will not always need to use the Mathematics Blueprint to solve every word problem. As you gain confidence in problem solving, you will no doubt leave out some of the steps. We suggest that you use the procedure when you find yourself on unfamiliar ground. It is a powerful organizational tool.

### 3 Using Formulas to Solve Word Problems

Sometimes the relationship between two quantities is so well understood that we have developed a formula to describe that relationship. We have already done some work with formulas in Section 2.5. The following examples show how you can use a formula to solve a word problem.

**Example 5** Two people traveled in separate cars. They each traveled a distance of 330 miles on an interstate highway. To maximize fuel economy, Fred traveled at exactly 50 mph. Sam traveled at exactly 55 mph. How much time did the trip take each person? (Use the formula distance = rate · time or $d = rt$.)

Sam's speed was 55 mph
Fred's speed was 50 mph
330 miles

## Solution

### Mathematics Blueprint for Problem Solving

| Gather the Facts | Assign the Variable | Basic Formula or Equation | Key Points to Remember |
|---|---|---|---|
| Each person drove 330 miles. | Time is the unknown quantity for each driver. | distance = (rate)(time) or $d = rt$ | The time is expressed in hours. |
| Fred drove at 50 mph. <br> Sam drove at 55 mph. | Use subscripts to denote different values of $t$. <br> $t_f$ = Fred's time <br> $t_s$ = Sam's time | | |

We must find $t$ in the formula $d = rt$. To simplify calculations we can solve for $t$ before we substitute values:

$$\frac{d}{r} = \frac{rt}{r} \rightarrow \frac{d}{r} = t$$

Substitute the known values into the formula and solve for $t$.

$$\frac{d}{r} = t \qquad\qquad \frac{d}{r} = t$$

$$\frac{330}{50} = t_f \qquad\qquad \frac{330}{55} = t_s$$

$$6.6 = t_f \qquad\qquad 6 = t_s$$

It took Fred 6.6 hours to drive 330 miles.   It took Sam 6 hours to drive 330 miles.

*Check.* Is this reasonable? Yes, you would expect Fred to take longer to drive the same distance because Fred was driving at a slower speed.

  *Note:* You may wish to express 6.6 hours in hours and minutes. To change 0.6 hour to minutes, proceed as follows:

$$0.6 \text{ hour} \cdot \frac{60 \text{ minutes}}{1 \text{ hour}} = (0.6)(60) \text{ minutes} = 36 \text{ minutes}$$

Thus, Fred drove for 6 hours and 36 minutes.  ◻

**Student Practice 5**   Sarah left the city to visit her aunt and uncle, who live in a rural area north of the city. She traveled the 220-mile trip in 4 hours. On her way home she took a slightly longer route, which measured 225 miles on the car odometer. The return trip took 4.5 hours.

**(a)** What was her average speed on the trip leaving the city?

**(b)** What was her average speed on the return trip?

**(c)** On which trip did she travel faster and by how much?

**Example 6** A teacher told Melinda that she had a course average of 78 based on her six math tests. When she got home, Melinda found five of her tests. She had scores of 87, 63, 79, 71, and 96 on the five tests. She could not find her sixth test. What score did she obtain on that test? (Use the formula that an average = the sum of scores ÷ the number of scores.)

**Solution**

Mathematics Blueprint for Problem Solving

| Gather the Facts | Assign the Variable | Basic Formula or Equation | Key Points to Remember |
|---|---|---|---|
| Her five known test scores are 87, 63, 79, 71, and 96. | We do not know the score Melinda received on her sixth test. | average = $\dfrac{\text{sum of scores}}{\text{number of scores}}$ | Since there are six test scores, we will need to divide the sum by 6. |
| Her course average is 78. | Let $x$ = the score on the sixth test. | | |

When you average anything, you total up the sum of all the values and then divide it by the number of values.

We now write the equation for the average of six items. This involves adding the scores for all the tests and dividing by 6.

$$\frac{87 + 63 + 79 + 71 + 96 + x}{6} = 78$$

$$\frac{396 + x}{6} = 78 \qquad \text{Add the numbers in the numerator.}$$

$$6\left(\frac{396 + x}{6}\right) = 6(78) \qquad \text{Multiply both sides of the equation by 6 to remove the fraction.}$$

$$396 + x = 468 \qquad \text{Simplify.}$$

$$x = 72 \qquad \text{Subtract 396 from both sides to find } x.$$

Melinda's score on the sixth test was 72.

*Check.* To verify that this is correct, we check that the average of the six tests is 78.

$$\frac{87 + 63 + 79 + 71 + 96 + 72}{6} \stackrel{?}{=} 78$$

$$\frac{468}{6} \stackrel{?}{=} 78$$

$$78 = 78 \quad \checkmark$$

The problem checks. We know that the score on the sixth test was 72. ☐

**Student Practice 6** Barbara's math course has four tests and one final exam. The final exam counts as much as two tests. Barbara has test scores of 78, 80, 100, and 96 on her four tests. What grade does she need on the final exam if she wants to have a 90 average for the course? (Use the formula that an average = the sum of scores ÷ the number of scores.)

## 3.2 Exercises    MyMathLab®

*Solve. Check your solution.*

1. What number minus 543 gives 718?

2. What number added to 74 gives 265?

3. A number divided by eight is 296. What is the number?

4. A number divided by nine is 189. What is the number?

5. Seventeen greater than a number is 199. Find the number.

6. Three times a number is one. What is the number?

7. A number is doubled and then increased by seven. The result is ninety-three. What is the original number?

8. Eight times a number is decreased by thirty-two. The result is one hundred twenty. Find the original number.

9. When eighteen is reduced by two-thirds of a number, the result is 12. Find the number.

10. Twice a number is increased by one-third the same number. The result is 42. Find the number.

11. Eight less than triple a number is the same as five times the same number. Find the number.

12. Ten less than double a number is the same as seven times the number. Find the number.

13. A number, half of that number, and one-third of that number are added. The result is 22. What is the number?

14. A number, twice that number, and one-third of that number are added. The result is 20. What is the number?

**Applications**  *Solve. Check to see if your answer is reasonable.*

15. ***Tablet Cases*** Lester orders supplies for the Tech Warehouse and noticed that the number of black tablet cases sold this year was three times the number of red cases. If he orders 120 black tablet cases and expects the same sales pattern, how many red cases should he order?

16. ***Inventory*** The college store maintains an inventory of baseball T-shirts and sweatshirts that have the team logo. Due to demand, the number of baseball T-shirts kept in inventory is four times the number of baseball sweatshirts. If the store has 164 T-shirts, how many sweatshirts are in inventory at the college store?

17. ***Electric Boat Rentals*** Island Rentals charges a $50 base fee plus $25 per hour to rent an electric boat that can carry 6 people. Thomas and his friends put their money together to rent an electric boat for a day. For how many hours can they rent the boat if they have $225?

18. ***On-Demand Movies*** Joseph just changed his cable television plan to include on-demand movies for $2.95 each. The charge for the basic cable package is $24.95 per month. If his bill one month was $39.70, how many on-demand movies did he watch?

19. ***Facility Rental*** Nicole Tran is hosting a large bridal shower for her friend and has $280 in her budget to rent a facility. The facility she chose for the party requires an $80 deposit plus $60 per hour. If there is no damage to the facility, one-half of the deposit is refunded. If Nicole expects to get the deposit refund, for how many hours can she rent the facility for the bridal shower?

20. ***iPod Nano*** Kerry just bought a new Apple iPod Nano for $129. On the same day, she bought several albums from iTunes for $11.99 each. If she spent $224.92 total, how many albums did she buy?

**21.** ***One-Day Sale*** Raquelle went to Weller's Department Store's one-day sale. She bought two blouses for $38 each and a pair of shoes for $49. She also wanted to buy some jewelry. Each item of jewelry was bargain priced at $11.50 each. If she brought $171 with her, how many pieces of jewelry could she buy?

**22.** ***Sales Clerk Bonus*** Samuel is a sales clerk in a major department store and earns $11 per hour. Last week the store had a credit card promotion offering a bonus of $2.50 for each store credit card application the sales clerks obtained from customers. Samuel worked 30 hours during the promotional week, and his paycheck for that week was $377.50 before deductions. How many credit card applications did Samuel obtain during the promotional week?

**23.** ***Gravity*** It has been shown that the force of gravity on a planet varies with the mass of the planet. The force of gravity on Jupiter, for example, is about two and a half times that of Earth. Using this information, approximately how much would a 220-lb astronaut weigh on Jupiter?

**24.** ***Sunday Comics*** Charles Schulz, the creator of *Peanuts,* once estimated that he had drawn close to 2600 Sunday comics over his career. At that time, about how many years had he been drawing *Peanuts*?

**25.** ***In-Line Skating*** Two in-line skaters, Nell and Kristin, start from the same point and skate in the same direction. Nell skates at 12 miles per hour and Kristin skates at 14 miles per hour. If they can keep up that pace for 2.5 hours, how far apart will they be at the end of that time?

**26.** ***Truck Drivers*** Allan and Tiana are transporting cars on their trucks from the same pick-up point and taking them to the same dealership. Tiana is traveling at 55 mph, while Allan's speed is a little slower, 45 mph, since he has a larger load. If they both continue at the same speed, how far apart will they be after traveling for 3.5 hours?

**27.** ***Travel Routes*** Nella drove from Albuquerque, New Mexico, to the Garden of the Gods rock formation in Colorado Springs. It took her six hours to travel 312 miles over the mountain road. She came home on the highway. On the highway she took five hours to travel 320 miles. How fast did she travel using the mountain route? How much faster (in miles per hour) did she travel using the highway route?

**28.** ***Road Trip*** Ester and her roommate MaryAnn decided to visit a friend at her beach home. MaryAnn had to return a few days earlier than Ester so they took separate cars. They both left their apartment at the same time for the 420-mile trip to the beach house. MaryAnn's car is not fuel efficient so she traveled at 50 mph to save money on gas, while Ester drove 60 mph. How much time did the trip take each person? How much longer did it take MaryAnn to travel to the beach house?

**29.** *Grading System* Danielle's chemistry professor has a complicated grading system. Each lab counts once and each test counts as two labs. At the end of the semester there is a final lab that counts as three regular labs. Danielle received scores of 84, 81, and 93 on her labs and scores of 89 and 94 on her tests. What score must she get on the final lab to receive an A (an A is 90)?

**30.** *Life Expectancy* Just before Sara's presentation on the top seven countries' life expectancies, she lost a few note cards for her speech, and therefore she did not have the data for Spain and Sweden. She remembered they both had the exact same age for life expectancy. Sara used the following data from her note cards to calculate the missing information. Japan: 82.6 years; Hong Kong: 82.2 years; Iceland: 81.8 years; Switzerland: 81.7 years; Australia 81.2 years; the average age for all 7 countries: 81.6 years. What age did Sara calculate as the life expectancy for Spain and Sweden? Round your answer to the nearest tenth.

## To Think About

**31.** *Cricket Chirps* In warmer climates, approximate temperature predictions can be made by counting the number of chirps of a cricket during a minute. The Fahrenheit temperature decreased by forty is equivalent to one-fourth of the number of cricket chirps.

  **(a)** Write an equation for this relationship.

  **(b)** Approximately how many chirps per minute should be recorded if the temperature is 90°F?

  **(c)** If a person recorded 148 cricket chirps in a minute, what would be the Fahrenheit temperature according to this formula?

## Cumulative Review *Simplify.*

**32.** **[1.6.1]** $5x(2x^2 - 6x - 3)$

**33.** **[1.6.1]** $-2a(ab - 3b + 5a)$

**34.** **[1.7.2]** $7x - 3y - 12x - 8y + 5y$

**35.** **[1.7.2]** $5x^2y - 7xy^2 - 8xy - 9x^2y$

## Quick Quiz 3.2

**1.** A number is tripled and then decreased by 15. The result is 36. What is the number?

**2.** The sum of one-half of a number, one-sixth of the number, and one-eighth of the number is thirty-eight. Find the number.

**3.** James scored 84, 89, 73, and 80 on four tests. What must he score on the next test to have an 80 average in the course?

**4.** **Concept Check** Explain how you would set up an equation to solve the following problem.

Phil purchased two shirts for $23 each and then purchased several pairs of socks. The socks were priced at $0.75 per pair. How many pairs of socks did he purchase if the total cost was $60.25?

# 3.3 Solving Word Problems: Comparisons

## 1 Solving Word Problems Involving Comparisons

Many real-life problems involve comparisons. We often compare quantities such as length, height, or income. Sometimes not all the information is known about the quantities that are being compared. You need to identify each quantity and write an algebraic expression that describes the situation in the word problem.

**Example 1** The Center City Animal Hospital treated a total of 18,360 dogs and cats last year. The hospital treated 1376 more dogs than cats. How many dogs were treated last year? How many cats were treated last year?

### Solution

1. **Understand the problem.**

   What information is given?   The combined number of dogs and cats is 18,360.
   What is being compared?   There were 1376 more dogs than cats.

   If you compare one quantity to another, usually the second quantity is represented by the variable. Since we are comparing the number of dogs to the number of cats, we start with the number of cats.

   Let $c$ = the number of cats treated at the hospital.
   Then $c + 1376$ = the number of dogs treated at the hospital.

2. **Write an equation.**

   The number of cats   plus   the number of dogs   is   18,360.
   $$c \qquad + \qquad (c + 1376) \qquad = \qquad 18{,}360$$

3. **Solve and state the answer.**

   $$c + c + 1376 = 18{,}360$$
   $$2c + 1376 = 18{,}360 \qquad \text{Combine like terms.}$$
   $$2c + 1376 - 1376 = 18{,}360 - 1376 \qquad \text{Subtract 1376 from both sides.}$$
   $$2c = 16{,}984 \qquad \text{Simplify.}$$
   $$\frac{2c}{2} = \frac{16{,}984}{2} \qquad \text{Divide both sides by 2.}$$
   $$c = 8492 \qquad \text{The number of cats treated is 8492.}$$
   $$c + 1376 = 8492 + 1376 = 9868 \qquad \text{The number of dogs treated is 9868.}$$

4. **Check.**   The number of dogs treated plus the number of cats treated should total 18,360.

   $$9868 + 8492 \overset{?}{=} 18{,}360$$
   $$18{,}360 = 18{,}360 \checkmark$$

**Student Practice 1**   A deck hand on a fishing boat is working with a rope that measures 89 feet. He needs to cut it into two pieces. The long piece must be 17 feet longer than the short piece. Find the length of each piece of rope.

If the word problem contains three unknown quantities, determine the basis of comparison for two of the quantities.

**Example 2** An airport filed a report showing the number of plane departures from the airport during each month last year. The number of departures in March was 50 more than the number of departures in January. In July, the number of departures was 150 less than triple the number of departures in January. In those three months, the airport had 2250 departures. How many departures were recorded for each month?

### Solution

1. **Understand the problem.** What is the basis of comparison?
   *The number of departures in March is compared to the number in January.*
   *The number of departures in July is compared to the number in January.*
   Express this algebraically. It may help to underline the key phrases.

   Let $j$ = the departures in January.

   March had 50 more than January

   Then $j + 50$ = the departures in March.

   July had 150 less than triple January

   So $3j - 150$ = the departures in July.

2. **Write an equation.**

   | number of departures in January | number of departures in March | number of departures in July | three months' total departures |
   |---|---|---|---|
   | $j$ | $+ (j + 50)$ | $+ (3j - 150) =$ | $2250$ |

3. **Solve and state the answer.**

   $$j + (j + 50) + (3j - 150) = 2250$$
   $$5j - 100 = 2250 \quad \text{Combine like terms.}$$
   $$5j = 2350 \quad \text{Add 100 to each side.}$$
   $$j = 470 \quad \text{Divide both sides by 5.}$$

   Now, if $j = 470$, then $j + 50 = 470 + 50 = 520$
   and $3j - 150 = 3(470) - 150 = 1410 - 150 = 1260$.

   The number of departures in January was 470; the number of departures in March was 520; the number of departures in July was 1260.

4. **Check.** Is the number of departures in March 50 more than those in January?

   $$520 \overset{?}{=} 50 + 470$$
   $$520 = 520 \ \checkmark$$

   Is the number of departures in July 150 less than triple those in January?

   $$1260 \overset{?}{=} 3(470) - 150$$
   $$1260 \overset{?}{=} 1410 - 150$$
   $$1260 = 1260 \ \checkmark$$

   Is the total number of departures in the three months equal to 2250?

   $$470 + 520 + 1260 \overset{?}{=} 2250$$
   $$2250 = 2250 \ \checkmark$$

**Student Practice 2** A social services worker was comparing the cost incurred by three families in heating their homes for the year. The first family had an annual heating bill that was $360 more than that of the second family. The third family had a heating bill that was $200 less than double the heating bill of the second family. The total annual heating bill for the three families was $3960. What was the annual heating bill for each family?

$\mathbb{M}_{\mathbb{C}}$ **Example 3** A small plot of land is in the shape of a rectangle. The length is 7 meters longer than the width. The perimeter of the rectangle is 86 meters. Find the dimensions of the rectangle.

**Solution**

1. *Understand the problem.* Read the problem:

   What information is given?

   *The perimeter of the rectangle is 86 meters.*

   What is being compared?

   *The length is being compared to the width.*

   Express this algebraically and draw a picture.

   Let $w$ = the width.
   Then $w + 7$ = the length.

   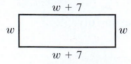

   Reread the problem: What are you being asked to do?

   *Find the dimensions of the rectangle. The dimensions of the rectangle are the length and the width of the rectangle.*

2. *Write an equation.* The perimeter is the total distance around the rectangle.

   $$w + (w + 7) + w + (w + 7) = 86$$

3. *Solve and state the answer.*

   $$w + (w + 7) + w + (w + 7) = 86$$
   $4w + 14 = 86$   Combine like terms.
   $4w = 72$   Subtract 14 from both sides.
   $w = 18$   Divide both sides by 4.

   The width of the rectangle is 18 meters. What is the length?

   $w + 7$ = the length
   $18 + 7 = 25$

   The length of the rectangle is 25 meters.

4. *Check.* Put the actual dimensions in your drawing and add the lengths of the sides. (See the figure below.) Is the sum 86 meters? ✓

   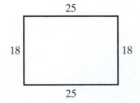

**Student Practice 3** A farmer needs 720 meters of wire fencing to enclose a pasture. The pasture is in the shape of a triangle. The first side of the triangle is 30 meters less than the second side. The third side is one-half as long as the second side. Find the dimensions of the triangle.

## 3.3 Exercises  MyMathLab®

### Applications

*Solve. Check to see if your answer is reasonable. Have you answered the question that was asked?*

1. **Plumber** Eduardo is a plumber and is preparing to install copper pipes in a home. He must cut a 17.5-foot piece of copper pipe into two different-size pieces. The shorter piece must be 4.5 feet shorter than the long piece of pipe. Find the length of each piece.

2. **Calorie Intake** Leslie is watching her calorie intake. She drinks a vegetable drink for breakfast and will limit the total calories for lunch and dinner to 1100 calories. The plan is for her lunch to be 300 calories less than dinner. How many calories can Leslie have for dinner? For her lunch?

3. **Salaries** James and Lin got married and got new jobs in Chicago. Lin earns $3400 more per year than James. Together they earn $82,300. How much do each of them earn per year?

4. **Horse Sales** Johnny breeds horses and then sells them. This month he plans to sell two quarter horses, one that he named Max and another named Banjo. In order to make a profit, Johnny must earn $18,750 from the sale of both horses. Max is a better breed, so Johnny will sell Max for two times the price that he sells Banjo. How much will he earn from the sale of Max? From the sale of Banjo?

5. **Volunteering** Three siblings—Dave, Kate, and Sarah—spent 100 hours helping Habitat for Humanity build new housing. Sarah worked 15 hours more than Dave. Kate worked 5 hours fewer than Dave. How many hours did each of these siblings volunteer?

6. **Beach Rentals** Isaac works for a rental company located near the beach. The company rents kayaks, paddleboards, and wetsuits. This month he rented 38 more paddleboards than kayaks and 20 fewer wetsuits than kayaks. The total rentals for the month were 171. How many of each item did he rent?

7. **College Professor** Professor Fullerton was demonstrating the ease of making a bar graph using information about students' home states. He was amazed that his class of 32 represented only 3 states. There were 8 more students born in Texas than were born in Oklahoma. The number of students born in Arizona was 9 fewer than the number born in Oklahoma. How many were born in each state?

8. **Multiplex** An 800-seat multiplex is divided into three theaters. Theater One is the smallest. Theater Two has 180 more seats than Theater One. Theater Three has 60 seats fewer than twice the number in Theater One. How large is each theater?

▲ 9. **Bridge Support** The length of a rectangular piece of steel in a bridge is 3 meters less than double the width. The perimeter of the piece of steel is 42 meters. Find the length of the piece of steel. Find the width of the piece of steel.

▲ 10. **State Park** A state park in Colorado has a perimeter of 92 miles. The park is in the shape of a rectangle. The length of the park is 30 miles less than triple the width of the park. Find the length of the park. Find the width of the park.

203

▲ **11.** *Artist* Rashau made a rectangular frame for her latest oil painting. The length is 20 centimeters more than triple the width. The perimeter of the frame is 96 centimeters. Find the length and the width of the frame.

**12.** *Carpenter* Beatrice is making a large rectangular hope chest. In order for the chest to fit in the space chosen by her customer, it must have a perimeter around the base of 156 inches and a length that is 18 inches more than the width. Find the length and width of the hope chest.

▲ **13.** *Jewelry Box* A solid-gold jewelry box was found in the underwater palace of Cleopatra just off the shore of Alexandria. The length of the rectangular box is 35 centimeters less than triple the width. The perimeter of the box is 190 centimeters. Find the length and width of Cleopatra's solid-gold jewelry box.

**14.** *Livestock Pen* Andy owns a farm in Nebraska and is building a small, rectangular livestock pen near his barn. He has 40 feet of material that he will use to place a fence around the perimeter of the pen. Andy's plan for constructing the pen indicates that the length of the livestock pen will be 2 feet more than double the width. Find the length and width of the livestock pen.

**15.** *Running Speed* The top running speed of a cheetah is double the top running speed of a jackal. The top running speed of an elk is 10 miles per hour faster than that of a jackal. If each of these three animals could run at top speed for an hour (which, of course, is not possible), they could run a combined distance of 150 miles. What is the top running speed of each of these three animals? (*Source:* American Museum of Natural History)

**16.** *River Length* The Missouri River is 149 miles longer than double the length of the Snake River. The Potomac River is 796 miles shorter than the Snake River. The combined lengths of these three rivers is 3685 miles. What is the length of each of the three rivers? (*Source:* U.S. Department of the Interior)

▲ **17.** *Kite Design* Mr. Shen wished to enter the Orange County Kite Competition with an innovative kite of his own design. He did tests that showed an irregular quadrilateral design was best. (See diagram.) The longest side (A) was twice the length of the shortest side (C). Side B was one and one-half times the shortest side (C). Side D needed to be 3" (3 inches) longer than side C. If the competition requires that the perimeter of the kite must be 58", what are the dimensions of Mr. Shen's kite?

▲ **18.** *Pennant Design* The Tri-County schools are participating in a Renaissance Faire. The team from Appleton High School has designed a pennant to fly over its team tent. The pennant is in the shape of a triangle. The longest side is 5" (5 inches) less than twice the middle length side. The shortest side is 3" longer than half the middle length side. If the team captain bought 72" of satin edging for the banner, and there were 4" left over after the pennant was made, what are the dimensions of the pennant?

## To Think About

▲ **19.** *Geometry* A small square is constructed. Then a new square is made by increasing each side by 2 meters. The perimeter of the new square is 3 meters shorter than five times the length of one side of the original square. Find the dimensions of the original square.

▲ **20.** *Geometry* A rectangle is constructed. The length of this rectangle is double the width. Then a new rectangle is made by increasing each side by 3 meters. The perimeter of the new rectangle is 2 meters greater than four times the length of the old rectangle. Find the dimensions of the original rectangle.

## Cumulative Review

*Simplify.*

**21. [0.1.2]** Simplify. $\dfrac{30}{54}$

**22. [2.4.1]** Solve for $x$. $\dfrac{2}{3}x + 6 = 4(x - 11)$

**23. [1.7.2]** $-7x + 10y - 12x - 8y - 2$

**24. [1.7.2]** $3x^2y - 6xy^2 + 7xy + 6x^2y$

### Quick Quiz 3.3

**1.** The length of a rectangular yard is 80 yards longer than double the width. The perimeter of the yard is 610 yards. Find the width and the length of the yard.

**2.** There are three times as many people in Jim's psychology class as in his art history class. There are 14 more people in Jim's algebra class than in his art history class. The total enrollment of the three classes is 79 students. How many students are enrolled in each class?

**3.** A triangular support beam of the roof of a large warehouse has a perimeter of 73 meters. The length of the first side is two-thirds the length of the second side. The length of the third side is 15 meters shorter than the length of the the second side. What are the lengths of the three sides of the triangular support beam?

**4. Concept Check** Explain how you would set up an equation to solve the following problem.

When Don and Laurie got married, Laurie earned $2600 more per year than Don. Together they earned $71,200 per year. How much did each of them earn for the year?

# Use Math to Save Money

**Did You Know?** You can save money each semester by choosing the right meal plan.

## Choosing a Meal Plan

### Understanding the Problem:

Jake is a college student who lives in a college dormitory. He is going over the meal plans and calculating his food budget for the semester. He is interested in choosing a meal plan that will cost him the least while still meeting his needs.

### Making a Plan:

First he needs to compare the prices and options for the meal plans. Then he needs to determine how many meals he will eat in the dining halls so he can determine which plan is right for him.

**Step 1:** There are three dining plans that cover the 15-week semester.

- The Gold plan costs $2205 a semester and covers 21 meals a week.
- The Silver plan costs $1764 a semester and covers 16 meals a week, with extra meals costing $10.
- The Bronze plan costs $1386 a semester and covers 12 meals a week, with extra meals costing $10.

Which plan is best for the following situations:

**Task 1:** Jake plans on eating 20 meals per week in the dining halls?

**Task 2:** Jake plans on eating 18 meals per week in the dining halls?

**Task 3:** Jake plans on eating 16 meals per week in the dining halls?

**Task 4:** Jake plans on eating 14 meals per week in the dining halls?

**Step 2:** Jake also has the option of buying "dining dollars" to use at other eating establishments on campus. With dining dollars Jake will get 10% off the cost of all food purchases.

How much will Jake save over the course of the semester by using dining dollars in the following situations:

**Task 5:** if he spends $20 a week at these establishments?

**Task 6:** if he spends $40 a week at these establishments?

**Task 7:** if he spends $60 a week at these establishments?

### Finding a Solution:

**Step 3:** Jake plans on eating 17 meals a week in the dining halls and spending $30 a week at other establishments.

**Task 8:** Which plan should he buy?

**Task 9:** How much will Jake spend on food for the semester?

### Applying the Situation to Your Life:

You can do these same calculations for your own circumstances if your college has a meal plan.

- See what plans are offered.
- Determine how often you will eat in the dining halls.
- Decide which plan would be right for you.
- Calculate how much you will need to spend on food for the semester.

206

## 3.4 Solving Word Problems: The Value of Money and Percents

The problems we now present are frequently encountered in business. They deal with money: buying, selling, and renting items; earning and borrowing money; and the value of collections of stamps or coins. Many applications require an understanding of the use of percents and decimals. Review Sections 0.4 and 0.5 if you are weak in these skills.

### 1 Solving Problems Involving Periodic Rate Charges

**Example 1** A business executive rented a car. The Supreme Car Rental Agency charged $39 per day and $0.28 per mile. The executive rented the car for two days and the total rental cost was computed to be $176. How many miles did the executive drive the rented car?

**Student Learning Objectives**

After studying this section, you will be able to:

1 Solve problems involving periodic rate charges.

2 Solve percent problems.

3 Solve investment problems involving simple interest.

4 Solve coin problems.

#### Solution

1. *Understand the problem.* How do you calculate the cost of renting a car?

   *total cost = per-day cost + mileage cost*

   What is known?

   *It cost $176 to rent the car for two days, and the mileage cost is $0.28 per mile.*

   What do you need to find?

   *The number of miles the car was driven.*

   Choose a variable:

   Let $m$ = the number of miles driven in the rented car.

2. *Write an equation.* Use the relationship for calculating the total cost.

$$\text{per-day cost} + \text{mileage cost} = \text{total cost}$$
$$(39)(2) + (0.28)m = 176$$

3. *Solve and state the answer.*

$$(39)(2) + (0.28)(m) = 176$$
$$78 + 0.28m = 176 \quad \text{Simplify the equation.}$$
$$0.28m = 98 \quad \text{Subtract 78 from both sides.}$$
$$\frac{0.28m}{0.28} = \frac{98}{0.28} \quad \text{Divide both sides by 0.28.}$$
$$m = 350 \quad \text{Simplify.}$$

   The executive drove 350 miles.

4. *Check.* Does this seem reasonable? If he drove the car 350 miles in two days, would it cost $176?

$$(\text{cost of \$39 per day} + (\text{cost of \$0.28 per mile} \overset{?}{=} \text{total cost of \$176}$$
$$\text{for 2 days}) \qquad \text{for 350 miles})$$
$$(\$39)(2) + (350)(\$0.28) \overset{?}{=} \$176$$
$$\$78 + \$98 \overset{?}{=} \$176$$
$$\$176 = \$176 \ \checkmark$$

> **Student Practice 1**　Alfredo wants to rent a truck to move to Florida. He has determined that the cheapest rental rates for a truck of the correct size are from a local company that will charge him $25 per day and $0.20 per mile. He has not yet completed an estimate of the mileage of the trip, but he knows that he will need the truck for three days. He has allowed $350 in his moving budget for the truck. How far can he travel for a rental cost of exactly $350?

## 2 Solving Percent Problems

Many applied situations require finding a percent of an unknown number. If we want to find 23% of $400, we multiply 0.23 by 400: $0.23(400) = 92$. If we want to find 23% of an unknown number, we can express this using algebra by writing $0.23n$, where $n$ represents the unknown number.

**Example 2**　A sofa was marked with the following sign: "The price of this sofa has been reduced by 23%. You can save $138 if you buy now." What was the original price of the sofa?

### Solution

1. **Understand the problem.**

$$\text{Let } s = \text{ the original price of the sofa.}$$
$$\text{Then } 0.23s = \text{ the amount of the price reduction, which is \$138.}$$

2. **Write an equation and solve.**

$$0.23s = 138 \qquad \text{Write the equation.}$$
$$\frac{0.23s}{0.23} = \frac{138}{0.23} \qquad \text{Divide each side of the equation by 0.23.}$$
$$s = 600 \qquad \text{Simplify.}$$

The original price of the sofa was $600.

3. **Check.**　Is $600 a reasonable answer?　✓　Does 23% of $600 = $138?　✓　□

> **Student Practice 2**　John earns a commission of 38% of the cost of every set of chef's knives that he sells. Last year he earned $17,100 in commissions. What was the cost of the chef's knives that he sold last year?

**Example 3**　Hector received a pay raise this year. The raise was 6% of last year's salary. This year he will earn $15,900. What was his salary last year before the raise?

### Solution

1. **Understand the problem.**　What do we need to find?

   *Hector's salary last year.*

   What do we know?

   *Hector received a 6% pay raise and now earns $15,900.*

   What does this mean?

   Reword the problem:　*This year's salary of $15,900 is 6% more than last year's salary.*

   Choose a variable:

$$\text{Let } x = \text{ Hector's salary last year.}$$
$$\text{Then } 0.06x = \text{ the amount of the raise.}$$

**2. *Write an equation and solve.***

$$\begin{array}{ccccc} \text{Last year's} & & \text{the amount} & & \text{this year's} \\ \text{salary} & + & \text{of his raise} & = & \text{salary} \end{array}$$

| | | | | | |
|---|---|---|---|---|---|
| $x$ | $+$ | $0.06x$ | $=$ | $15{,}900$ | Write the equation. |
| $1.00x$ | $+$ | $0.06x$ | $=$ | $15{,}900$ | Rewrite $x$ as $1.00x$. |
| | | $1.06x$ | $=$ | $15{,}900$ | Combine like terms. |
| | | $\dfrac{1.06x}{1.06}$ | $=$ | $\dfrac{15{,}900}{1.06}$ | Divide by $1.06$. |
| | | $x$ | $=$ | $15{,}000$ | Simplify. |

Thus Hector's salary was \$15,000 last year before the raise.

**3. *Check.*** Does it seem reasonable that Hector's salary last year was \$15,000? The check is up to you. ☐

> **Student Practice 3** The price of Betsy's new car is 7% more than the price of a similar model last year. She paid \$19,795 for her car this year. What would a similar model have cost last year?

## 3 Solving Investment Problems Involving Simple Interest ▶

**Interest** is a charge for borrowing money or an income from investing money. Interest rates affect our lives. They affect the national economy and they affect a consumer's ability to borrow money for big purchases. For these reasons, a student of mathematics should be able to solve problems involving interest.

There are two basic types of interest: simple and compound. **Simple interest** is computed by multiplying the amount of money borrowed or invested (which is called the *principal*) times the rate of interest times the period of time over which it is borrowed or invested (usually measured in years unless otherwise stated).

> Simple interest = principal × rate × time
>
> $I = prt$

You often hear of banks offering a certain interest rate *compounded* quarterly, monthly, weekly, or daily. In **compound interest** the amount of interest is added to the amount of the original principal at the end of each time period, so future interest is based on the sum of both principal and previous interest. Most financial institutions use compound interest in their transactions.

*However, all examples and exercises in this chapter will involve* **simple interest.**

**Example 4** Find the interest on \$3000 borrowed at a simple interest rate of 18% for one year.

### Solution

| | |
|---|---|
| $I = prt$ | The simple interest formula. |
| $I = (3000)(0.18)(1)$ | Substitute the values of the variables: principal $= 3000$, the rate $= 18\% = 0.18$, the time $=$ one year. |
| $I = 540$ | Simplify. |

Thus the interest charge for borrowing \$3000 for one year at a simple interest rate of 18% is \$540. ☐

> **Student Practice 4** Find the interest on \$7000 borrowed at a simple interest rate of 12% for one year.

---

### Calculator

#### Interest

You can use a calculator to find simple interest. Find the interest on \$450 invested at an annual rate of 6.5% for 15 months. Notice that the time is in months. Since the interest formula $I = prt$ is in years, you need to change 15 months to years by dividing 15 by 12.

Enter: 15 ÷ 12 =

Display: 1.25

Leave this on the display and multiply as follows:

1.25 × 450 ×
6.5 % =

The display should read

36.5625

which would round to \$36.56.
Try finding the simple interest in the following.

**(a)** \$9516 invested at 12% for 30 months

**(b)** \$593 borrowed at 8% for 5 months

**Ans:**

**(a)** \$2854.80

**(b)** \$19.77

Now we apply this concept to a word problem about investments.

**Mc** **Example 5** A woman invested an amount of money in two accounts for one year. She invested some at 8% simple interest and the rest at 6% simple interest. Her total amount invested was $1250. At the end of the year she had earned $86 in interest. How much money had she invested in each account?

**Student Practice 5** A woman invested her savings of $8000 in two accounts that each calculate interest only once per year. She placed one amount in a special notice account that yields 9% annual interest. The remainder she placed in a tax-free All-Savers account that yields 7% annual interest. At the end of the year, she had earned $630 in interest from the two accounts together. How much had she invested in each account?

## Solution

### Mathematics Blueprint for Problem Solving

| Gather the Facts | Assign the Variable | Basic Formula or Equation | Key Points to Remember |
|---|---|---|---|
| $1250 is invested: part at 8% simple interest, part at 6% simple interest.<br><br>The total interest for the year was $86. | $x$ = amount invested at 8%.<br><br>$1250 - x$ = amount invested at 6%.<br><br>$0.08x$ = amount of interest for $x$ dollars at 8%.<br><br>$0.06(1250 - x)$ = amount of interest for $(1250 - x)$ dollars at 6%. | Interest earned at 8% + interest earned at 6% = total interest earned during the year, which is $86. | Be careful to write $1250 - x$ for the amount of money invested at 6%.<br><br>The order of the amounts is $1250 - x$. Do not use $x - 1250$. |

$$\underset{\text{at 8\%}}{\underset{\text{earned}}{\text{interest}}} + \underset{\text{at 6\%}}{\underset{\text{earned}}{\text{interest}}} = \underset{\text{the year}}{\underset{\text{earned during}}{\text{total interest}}}$$

$$0.08x \quad + \quad 0.06(1250 - x) \quad = \quad 86$$

*Note:* Be sure you write $(1250 - x)$ for the amount of money invested at 6%. Students often write it backwards by mistake. It is *not* correct to use $(x - 1250)$ instead of $(1250 - x)$. The order of the terms is very important.

*Solve and state the answer.*

$$
\begin{aligned}
0.08x + 75 - 0.06x &= 86 &&\text{Remove parentheses.}\\
0.02x + 75 &= 86 &&\text{Combine like terms.}\\
0.02x &= 11 &&\text{Subtract 75 from both sides.}\\
\frac{0.02x}{0.02} &= \frac{11}{0.02} &&\text{Divide both sides by 0.02.}\\
x &= 550 &&\text{The amount invested at 8\% simple}\\
&&&\text{interest is \$550.}
\end{aligned}
$$

$1250 - x = 1250 - 550 = 700$    The amount invested at 6% simple interest is $700.

*Check.* Are these values reasonable? Yes. Do the amounts equal $1250?

$$\$550 + \$700 \overset{?}{=} \$1250$$
$$\$1250 = \$1250 \quad \checkmark$$

Would these amounts earn $86 interest in one year invested at the specified rates?

$$0.08(\$550) + 0.06(\$700) \overset{?}{=} \$86$$
$$\$44 + \$42 \overset{?}{=} \$86$$
$$\$86 = \$86 \quad \checkmark$$

## 4 Solving Coin Problems ▶

Coin problems provide an unmatched opportunity to use the concept of *value*. We must make a distinction between *how many coins* there are and the *value* of the coins.

Consider the next example. Here we know the *value* of some coins, but do not know *how many* we have.

**Example 6** When Bob got out of math class, he wanted to buy a coffee at the snack bar. He had exactly enough dimes and quarters to buy a coffee that would cost $2.55. He had one fewer quarter than he had dimes. How many coins of each type did he have?

### Solution

1. ***Understand the problem.***

$$\text{Let } d = \text{the number of dimes.}$$
$$\text{Then } d - 1 = \text{the number of quarters.}$$

The total value of the coins was $2.55. How can we represent the value of the dimes and the value of the quarters? Think.

Each dime is worth $0.10.      Each quarter is worth $0.25.

5 dimes are worth $(5)(0.10) = 0.50$.     8 quarters are worth $(8)(0.25) = 2.00$.

$d$ dimes are woth $(d)(0.10) = 0.10d$.     $(d - 1)$ quarters are worth:
$$(d - 1)(0.25) = 0.25(d - 1).$$

2. ***Write an equation and solve.***    Now we can write an equation for the total value.

$$(\text{value of dimes}) + (\text{value of quarters}) = \$2.55$$

$$0.10d + 0.25(d - 1) = 2.55$$

$0.10d + 0.25d - 0.25 = 2.55$    Remove parentheses.

$0.35d - 0.25 = 2.55$    Combine like terms.

$0.35d = 2.80$    Add 0.25 to both sides.

$\dfrac{0.35d}{0.35} = \dfrac{2.80}{0.35}$    Divide both sides by 0.35.

$d = 8$    Simplify.

$d - 1 = 8 - 1 = 7$

Thus Bob had eight dimes and seven quarters.

3. ***Check.***    Is this answer reasonable? Yes. Does Bob have one fewer quarter than he has dimes?

$$8 - 7 \overset{?}{=} 1$$
$$1 = 1 \ \checkmark$$

Are eight dimes and seven quarters worth $2.55?

$$8(\$0.10) + 7(\$0.25) \overset{?}{=} \$2.55$$
$$\$0.80 + \$1.75 \overset{?}{=} \$2.55$$
$$\$2.55 = \$2.55 \ \checkmark \qquad \qquad \square$$

▶ **Student Practice 6**    Ginger has five more quarters than dimes. She has $5.10 in change. If she has only quarters and dimes, how many coins of each type does she have?

**Example 7** Michele and her two children returned from the grocery store with $2.80 in change. She had twice as many quarters as nickels. She had two more dimes than nickels. How many nickels, dimes, and quarters did she have?

**Solution**

**Mathematics Blueprint for Problem Solving**

| Gather the Facts | Assign the Variable | Basic Formula or Equation | Key Points to Remember |
|---|---|---|---|
| Michele had $2.80 in change. She had twice as many quarters as nickels. She had two more dimes than nickels. | $x$ = number of nickels. $2x$ = number of quarters. $x + 2$ = number of dimes. $0.05x$ = value of the nickels. $0.25(2x)$ = value of the quarters. $0.10(x + 2)$ = value of the dimes. | The value of the nickels + the value of the dimes + the value of the quarters = $2.80. | Don't add the number of coins to get $2.80. You must add the values of the coins! |

$$(\text{value of nickels}) + (\text{value of dimes}) + (\text{value of quarters}) = \$2.80$$
$$0.05x \quad + \quad 0.10(x + 2) \quad + \quad 0.25(2x) \quad = \quad 2.80$$

*Solve.*

$$0.05x + 0.10x + 0.20 + 0.50x = 2.80 \qquad \text{Remove parentheses.}$$
$$0.65x + 0.20 = 2.80 \qquad \text{Combine like terms.}$$
$$0.65x = 2.60 \qquad \text{Subtract 0.20 from both sides.}$$
$$\frac{0.65x}{0.65} = \frac{2.60}{0.65} \qquad \text{Divide both sides by 0.65.}$$
$$x = 4 \qquad \text{Simplify. Michele had four nickels.}$$
$$2x = 2(4) = 8 \qquad \text{She had eight quarters.}$$
$$x + 2 = 4 + 2 = 6 \qquad \text{She had six dimes.}$$

When Michele left the grocery store she had four nickels, eight quarters, and six dimes.

*Check.* Is the answer reasonable? Yes. Did Michele have twice as many quarters as nickels?

$$(4)(2) \stackrel{?}{=} 8 \qquad 8 = 8 \checkmark$$

Did she have two more dimes than nickels?

$$4 + 2 \stackrel{?}{=} 6 \qquad 6 = 6 \checkmark$$

Do four nickels, eight quarters, and six dimes have a value of $2.80?

$$4(\$0.05) + 8(\$0.25) + 6(\$0.10) \stackrel{?}{=} \$2.80$$
$$\$0.20 + \$2.00 + \$0.60 \stackrel{?}{=} \$2.80$$
$$\$2.80 = \$2.80 \checkmark \qquad \square$$

**Student Practice 7** A young boy told his friend that he had twice as many nickels as dimes in his pocket. He also said that he had four more quarters than dimes. He said that he had $2.35 in change in his pocket. Can you determine how many nickels, dimes, and quarters he had?

STEPS TO SUCCESS Brainstorming with Yourself to Find Greater Success When You Take a Test

Do you sometimes get a test back and feel that you could have done a lot better? Most of us have had that happen. Here are some great ideas to improve your test performance.

When you get a graded test back, take some time within the next four hours to look it over carefully and then quickly write some suggestions to yourself. Do some brainstorming. What do you think you could have done to improve your test score?

1. Did you study too much the night before the test and not adequately prepare as you were going through the chapter?

2. Did you neglect to do some of the homework assignments?

3. Did you fail to get help with some of the problems that you did not understand or could not do?

4. Did you neglect to review the Chapter Organizer or work the Chapter Review problems in the textbook?

**Making it personal:** Think about how you did on the last test you took in this course. Which of the four questions listed here is the most relevant to the way you studied? What can you do to make sure your performance is better on the next test? ▼

## 3.4 Exercises MyMathLab®

### Applications

*Solve. All problems involving interest refer to simple interest.*

1. **Assistant Manager** Clyde works as a part-time assistant manager for the Barkley Coffeehouse to earn money to help pay for his college expenses. He makes $11.50 per hour. The coffeehouse sells whole bean coffee, mugs, and other coffee products. Clyde earns $0.50 for each of these items he sells. Last week he worked 30 hours and earned $364. How many coffee products did he sell last week?

2. **Waitress** Kayleigh must earn $385 a week in order to cover her expenses while attending nursing school. Due to her past experience, she was hired at a high-end restaurant and scheduled weekends, which resulted in higher tip earnings. The manager told her the pay is $8.75 per hour plus tips, which average $8 per table. If Kayleigh works 12 hours a week, how many tables would she have to serve in order to make $385 per week?

3. **Tax Accountant** Eli Moran is an accountant at J & R Tax Prep, and during tax season he works overtime to meet the demand. He is paid $32 per hour for the first 40 hours and $48 per hour for each hour in the week worked over 40 hours. To save for a nice vacation, Eli's goal is to earn $1520 per week. How many hours overtime per week will he need to achieve his goal?

4. **Construction Apprentice** Anna is an apprentice sheet metal worker for a construction company. When the company falls behind schedule on a job, she works overtime. She is paid $20 per hour for the first 40 hours and $30 per hour for each hour in the week worked over 40 hours. Anna is saving money for a new truck, so she would like to work enough overtime to earn $980 per week. How many hours overtime per week will she need to achieve this goal?

5. **Layaway** Mrs. Peterson's triplets all want to attend the local private academy, where they are required to wear uniforms. Mrs. Peterson knows that the only way she can afford these uniforms is to shop early, put the uniforms on layaway, and pay a little every week. The total uniform cost for the three girls is $1817.75. If she put down a deposit of $600.00 and she could afford $105.00 per week, how long would it take her to pay off the uniforms? (Round to nearest whole number.)

6. **Party Costs** The Swedish Chef Catering Company charges $50.00 for setup, $85.00 for cleanup, and the special menu that Judy Carter ordered cost $23.50 per person. If Judy Carter's bill came to $558.00, how many people came to the party?

7. **Camera Sale** The camera Melissa wanted for her birthday is on sale at 28% off the usual price. The amount of the discount is $100.80. What was the original price of the camera?

8. **Work Force** The number of women working full-time in Springfield has risen 12% this year. This means 216 more women have full-time jobs. What was the number of women working full-time last year?

9. **Salary** The cost of living last year went up 3%. Fortunately, Alice Swanson got a 3% raise in her salary from last year. This year she is earning $22,660. How much did she make last year?

10. **Stock Profit** A speculator bought stocks and later sold them for $5136, making a profit of 7%. How much did the stocks cost him?

11. **Investments** Katerina Lubov inherited some money and invested it at 6% simple interest. At the end of the year, the total amount of her original principal and the interest was $12,720. How much did she originally invest?

12. **Investments** Robert Campbell invested some money at 8% simple interest. At the end of the year, the total amount of his original principal and the interest was $7560. How much did he originally invest?

13. ***Trust Fund*** Mr. and Mrs. Wright set up a trust fund for their children last year. Part of it is earning 7% simple interest per year while the rest of it is earning 5% simple interest per year. They placed $5000 in the trust fund. In one year the trust fund has earned $310. How much did they invest at each interest rate?

14. ***Savings Bonds*** Anne and Michael invested $8000 last year in tax-free bonds. Some of the bonds earned 8% simple interest while the rest earned 6% simple interest. At the end of the year, they had earned $580 in interest. How much did they invest at each interest rate?

15. ***Mutual Funds*** Plymouth Rock Bank invested $400,000 last year in mutual funds. The conservative fund earned 8% simple interest. The growth fund earned 12% simple interest. At the end of the year, the bank had earned $38,000 from these mutual funds. How much did it invest in each fund?

16. ***Mutual Funds*** Millennium Securities last year invested $600,000 in mutual funds. The international fund earned 11% simple interest. The high-tech fund earned 7% simple interest. At the end of the year, the company had earned $50,000 in interest. How much did it invest in each fund?

17. ***Investments*** Dave Horn invested half of his money at 5%, one-third of his money at 4%, and the rest of his money at 3.5%. If his total annual investment income was $530, how much had he invested?

18. ***Investments*** When Karin Schumann won the lottery, she decided to invest half her winnings in a mutual fund paying 4% interest, one-third of it in a credit union paying 4.5% interest, and the rest in a bank CD paying 3% interest. If her annual investment income was $2400, how much money had she invested?

19. ***Coin Bank*** Little Melinda has nickels and quarters in her bank. She has four fewer nickels than quarters. She has $3.70 in the bank. How many coins of each type does she have?

20. ***Coin Change*** Reggie's younger brother had several coins when he returned from his paper route. He had a total of $5.35 in dimes and quarters. He had six more quarters than he had dimes. How many of each coin did he have?

21. ***Coin Change*** A newspaper carrier has $3.75 in change. He has three more quarters than dimes but twice as many nickels as quarters. How many coins of each type does he have?

22. ***Coin Change*** Tim Whitman has $4.50 in change in his desk drawer to use on the vending machines downstairs. He has four more quarters than dimes. He has three times as many nickels as dimes. How many coins of each type does he have?

23. ***Office Manager*** Huy is an office manager for a small law firm. One of his responsibilities is to order supplies. He noticed that in the past, the number of boxes of white paper ordered was double the number of boxes of beige paper. He must spend no more than $112.50 on paper so that when tax and shipping are added he stays within the office budget. If a box of white paper cost $3.50 and a box of beige paper cost $4.25, how many boxes of each can Huy order and stay within his $112.50 budget?

24. ***Ticket Donation*** Mario's Marionettes donated free tickets for its show to the local Boys & Girls Club. It claimed that the ticket value was $176.75. A child's ticket cost $5.50 and an adult's ticket cost $8.75. If the number of children's tickets was three times the number of adults' tickets, how many adults and children got to attend the show for free?

**25.** *Paper Currency* Jim and Amy sold many pieces of furniture and made $1380 during their garage sale. They had sixteen more $10 bills than $50 bills. They had one more than three times as many $20 bills as $50 bills. How many of each denomination did they have?

**26.** *Paper Currency* Roberta Burgess came home with $325 in tips from two nights on her job as a waitress. She had $20 bills, $10 bills, and $5 bills. She discovered that she had three times as many $5 bills as she had $10 bills. She also found that she had 4 fewer $20 bills than she had $10 bills. How many of each denomination did she have?

**27.** *Salary* Madelyn Logan is an office furniture dealer who earns an $18,000 base salary. She also earns a 4% commission on sales. How much must she sell to earn a total of $55,000?

**28.** *Animal Shelter* The local animal shelter accepts abandoned cats and dogs. They usually receive twice as many cats as dogs. They estimate that 80% of the cats and 60% of the dogs that come in need some kind of medical treatment. If they treated 286 animals last year, how many cats and dogs did they take in?

## Cumulative Review   *Perform the operations in the proper order.*

**29.** **[1.5.1]** $5(3) + 6 \div (-2)$

**30.** **[1.5.1]** $5(-3) - 2(12 - 15)^2 \div 9$

*Evaluate for $a = -1$ and $b = 4$.*

**31.** **[1.8.1]** $a^2 - 2ab + b^2$

**32.** **[1.8.1]** $a^3 + ab^2 - b - 5$

## Quick Quiz 3.4

**1.** Charles is renting a large copy machine for his company and has been given a budget of $1250. The machine he wants to rent will require a $224 installation fee and a $114-per-month rental contract. How many months will Charles be able to rent the copy machine with his current budget?

**2.** Last year the cost of Marcia's health plan went up 7%. The cost for coverage for Marcia and her family this year is $12,412. What did it cost last year before the increase went into effect?

**3.** Walter and Barbara invested $5000 in two CDs last year. One of the CDs paid 4% interest. The other one paid 5% interest. At the end of one year they earned $228 in simple interest. How much did they invest at each interest rate?

**4.** **Concept Check** Explain how you would set up an equation to solve the following problem.

Robert has $2.55 in change consisting of nickels, dimes, and quarters. He has twice as many dimes as quarters. He has one more nickel than he has quarters. How many of each coin does he have?

# 3.5 Solving Word Problems Using Geometric Formulas

## 1 Finding Area, Perimeter, and Unknown Angles

In Section 1.8 we reviewed a number of area and perimeter formulas. We repeat this list of formulas here for convenience.

**Student Learning Objectives**

**After studying this section, you will be able to:**

**1** Find the area and the perimeter of two-dimensional objects. Find the unknown angle of a triangle.

**2** Find the volume and the surface area of three-dimensional objects.

**3** Solve more complicated geometric problems.

### AREA AND PERIMETER FORMULAS

A **parallelogram** is a four-sided figure with opposite sides parallel. In a parallelogram, opposite sides are equal and opposite angles are equal.

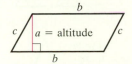

Perimeter = the sum of all four sides
Area = $ab$

A **rectangle** is a parallelogram with all interior angles measuring 90°.

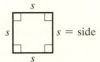

Perimeter = $2l + 2w$
Area = $lw$

A **square** is a rectangle with all four sides equal.

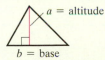

Perimeter = $4s$
Area = $s^2$

A **trapezoid** is a four-sided figure with two sides parallel. The parallel sides are called the bases of the trapezoid.

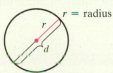

Perimeter = the sum of all four sides
Area = $\frac{1}{2}a(b_1 + b_2)$

A **triangle** is a closed plane figure with three sides.

Perimeter = the sum of the three sides
Area = $\frac{1}{2}ab$

A **circle** is a plane curve consisting of all points at an equal distance from a given point called the center.

**Circumference** is the distance around a circle. A **radius** is a line segment from the center of the circle to a point on the circle. A **diameter** is a line segment across the circle that passes through the center with endpoints on the circle.

Circumference = $2\pi r$
Area = $\pi r^2$

$d$ = diameter = $2r$

$\pi$ is a constant associated with circles. It is an irrational number that is approximately 3.141592654. We usually use 3.14 as a sufficiently accurate approximation. Thus we write $\pi \approx 3.14$ for most of our calculations involving $\pi$.

We frequently encounter triangles in word problems. There are four important facts about triangles, which we list for convenient reference.

**TRIANGLE FACTS**

1. The sum of the interior angles of any triangle is 180°. That is,

   measure of $\angle A$ + measure of $\angle B$ + measure of $\angle C$ = 180°.

2. An **equilateral** triangle is a triangle with three sides equal in length and three angles that measure 60° each.

   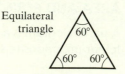

3. An **isosceles** triangle is a triangle with two equal sides. The two angles opposite the equal sides are also equal.

   measure of $\angle A$ = measure of $\angle B$

   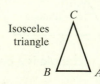

4. A **right** triangle is a triangle with one angle that measures 90°.

   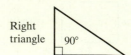

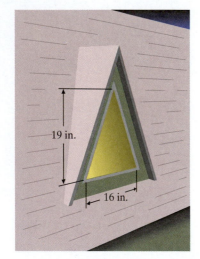

▲ **Example 1** Find the area of a triangular window whose base is 16 inches and whose altitude is 19 inches.

**Solution** We can use the formula for the area of a triangle. Substitute the known values in the formula and solve for the unknown.

$$A = \frac{1}{2}ab \qquad \text{Write the formula for the area of a triangle.}$$

$$= \frac{1}{2}(19 \text{ in.})(16 \text{ in.}) \qquad \text{Substitute the known values in the formula.}$$

$$= \frac{1}{2}(19)(16)(\text{in.})(\text{in.}) \qquad \text{Simplify.}$$

$$A = 152 \text{ in.}^2 \quad \text{or} \quad 152 \text{ square inches}$$

The area of the triangle is 152 square inches. ☐

**Student Practice 1** Find the area of a triangle whose base is 14 inches and whose altitude is 20 inches.

**Example 2** The area of an NBA basketball court is 4700 square feet. Find the width of the court if the length is 94 feet.

**Solution** Draw a diagram.

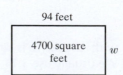

Write the formula for the area of a rectangle and solve for the unknown value.

$$A = lw$$

$$4700 \text{ (ft)}^2 = (94 \text{ ft})(w) \quad \text{Substitute the known values into the formula.}$$

$$\frac{4700 \text{ (ft)}(\text{ft})}{94 \text{ ft}} = \frac{94 \text{ ft}}{94 \text{ ft}} w \quad \text{Divide both sides by 94 feet.}$$

$$50 \text{ ft} = w$$

The width of the basketball court is 50 feet. ☐

▲ ▭➡ **Student Practice 2**  The area of a rectangular field is 120 square yards. If the width of the field is 8 yards, what is the length?

▲ **Example 3** The area of a trapezoid is 400 square inches. The altitude is 20 inches and one of the bases is 15 inches. Find the length of the other base.

**Solution**

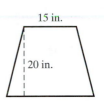

$$A = \frac{1}{2}a \, (b_1 + b_2) \qquad \text{Write the formula for the area of a trapezoid.}$$

$$400(\text{in.})^2 = \frac{1}{2}(20 \text{ in.})(15 \text{ in.} + b_2) \qquad \text{Substitute the known values.}$$

$$400(\text{in.})(\text{in.}) = 10(\text{in.})(15 \text{ in.} + b_2) \qquad \text{Simplify.}$$

$$400(\text{in.})(\text{in.}) = 150(\text{in.})(\text{in.}) + (10 \text{ in.})b_2 \qquad \text{Remove parentheses.}$$

$$250(\text{in.})(\text{in.}) = (10 \text{ in.})b_2 \qquad \text{Subtract } 150 \text{ in.}^2 \text{ from both sides.}$$

$$25 \text{ in.} = b_2 \qquad \text{Divide both sides by 10 in.}$$

The other base is 25 inches long. ☐

▲ ▭➡ **Student Practice 3**  The area of a trapezoid is 256 square feet. The bases are 12 feet and 20 feet. Find the altitude.

▲ **Example 4** Find the area of a circular sign whose diameter is 14 inches. (Use 3.14 for $\pi$.) Round to the nearest square inch.

**Solution**  $A = \pi r^2$  Write the formula for the area of a circle.

Note that the length of the diameter is given in the word problem. We need to know the length of the radius to use the formula. Since $d = 2r$, then 14 in. $= 2r$ and $r = 7$ in.

$$A = \pi(7 \text{ in.})^2 \qquad \text{Substitute known values into the formula.}$$

$$= (3.14)(7 \text{ in.})(7 \text{ in.})$$

$$= (3.14)(49 \text{ in.}^2)$$

$$\approx 153.86 \text{ in.}^2$$

Rounded to the nearest square inch, the area of the circle is approximately 154 square inches.

In this example, you were asked to find the area of the circle. Do not confuse the formula for area with the formula for circumference. $A = \pi r^2$, while $C = 2\pi r$. Remember that area involves square units, so it is only natural that in the area formula you would square the radius. ☐

▲ ▭➡ **Student Practice 4**  Find the circumference of a circle whose radius is 15 meters. (Use 3.14 for $\pi$.) Round your answer to the nearest meter.

▲ **Example 5** Find the perimeter of a parallelogram whose longer sides are 4 feet and whose shorter sides are 2.6 feet.

**Solution** Draw a picture.

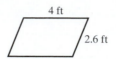

The perimeter is the distance around a figure. To find the perimeter, add the lengths of the sides. Since the opposite sides of a parallelogram are equal, we can write the following.

$$P = 2(4 \text{ feet}) + 2(2.6 \text{ feet})$$
$$= 8 \text{ feet} + 5.2 \text{ feet}$$
$$= 13.2 \text{ feet}$$

The perimeter of the parallelogram is 13.2 feet. ☐

▲ **Student Practice 5** Find the perimeter of an equilateral triangle with a side that measures 15 centimeters.

**Example 6** The smallest angle of an isosceles triangle measures 24°. The other two angles are larger. What are the measurements of the other two angles?

**Solution** We know that in an isosceles triangle the measures of two angles are equal. We know that the sum of the measures of all three angles is 180°. Both of the larger angles must be different from 24°, therefore these two larger angles must be equal.

Let $x =$ the measure in degrees of each of the larger angles.

Then we can write

$$24° + x + x = 180°$$
$$24° + 2x = 180° \quad \text{Add like terms.}$$
$$2x = 156° \quad \text{Subtract 24 from each side.}$$
$$x = 78° \quad \text{Divide each side by 2.}$$

Thus the measure of each of the other two angles of the triangle must be 78°. ☐

**Student Practice 6** The largest angle of an isosceles triangle measures 132°. The other two angles are smaller. What are the measurements of the other two angles?

## 2 Finding Volume and Surface Area of Three-Dimensional Objects ▶

Now let's examine three-dimensional figures. **Surface area** is the total area of the faces of a figure. You can find surface area by calculating the area of each face and then finding the sum. **Volume** is the measure of the amount of space inside a figure. Some formulas for the surface area and the volume of regular figures can be found in the following table.

### GEOMETRIC FORMULAS: THREE-DIMENSIONAL FIGURES

Rectangular prism

Note that all of the faces are rectangles.

$h$ = height
$l$ = length
$w$ = width

Surface area = $2lw + 2wh + 2lh$
Volume = $lwh$

**GEOMETRIC FORMULAS: THREE-DIMENSIONAL FIGURES (*continued*)**

Sphere

$r$ = radius

Surface area $= 4\pi r^2$

Volume $= \dfrac{4}{3}\pi r^3$

Right circular cylinder

$r$ = radius
$h$ = height

Surface area $= 2\pi rh + 2\pi r^2$
Volume $= \pi r^2 h$

▲ **Example 7** Find the volume of a sphere with radius 4 centimeters. (Use 3.14 for $\pi$.) Round your answer to the nearest cubic centimeter.

**Solution**

$$V = \frac{4}{3}\pi r^3 \qquad \text{Write the formula for the volume of a sphere.}$$

$$= \frac{4}{3}(3.14)(4 \text{ cm})^3 \quad \text{Substitute the known values into the formula.}$$

$$= \frac{4}{3}(3.14)(64) \text{ cm}^3 \approx 267.946667 \text{ cm}^3$$

Rounded to the nearest whole number, the volume of the sphere is approximately 268 cubic centimeters. ◻

▲ **Student Practice 7** Find the surface area of a sphere with radius 5 meters. (Use 3.14 for $\pi$.)

▲ **Example 8** A can is made of aluminum. It has a flat top and a flat bottom. The height is 5 inches and the radius is 2 inches. How much aluminum is needed to make the can? How much aluminum is needed to make 10,000 cans?

**Solution**

*Understand the problem.* What do we need to find? Reword the problem.

*We need to find the total surface area of a cylinder.*

Side of the cylinder    5

Circumference
$2\pi r = 4\pi$

Top
2

$A = \pi r^2$
$A = 4\pi$

Bottom
2

$A = \pi r^2$
$A = 4\pi$

You may calculate the area of each piece and find the sum or you may use the formula. We will use 3.14 to approximate $\pi$.

$$\text{Surface area} = 2\pi rh + 2\pi r^2$$

$$= 2(3.14)(2 \text{ in.})(5 \text{ in.}) + 2(3.14)(2 \text{ in.})^2$$

$$= 62.8 \text{ in.}^2 + 25.12 \text{ in.}^2 = 87.92 \text{ in.}^2$$

Approximately 87.92 square inches of aluminum are needed to make each can.

*Continued on next page*

Have we answered all the questions in the word problem? Reread the problem. How much aluminum is needed to make 10,000 cans?

$$(\text{aluminum for 1 can})(10{,}000) = (87.92 \text{ in.}^2)(10{,}000) = 879{,}200 \text{ square inches}$$

It would take approximately 879,200 square inches of aluminum to make 10,000 cans.

▲ ▶ **Student Practice 8** Sand is stored in a cylindrical drum that is 4 feet high and has a radius of 3 feet. How much sand can be stored in the drum? Round your answer to the nearest cubic foot.

## 3 Solving More Complicated Geometric Problems ▶

**Example 9** A quarter-circle (of radius 1.5 yards) is connected to two rectangles with dimensions as labeled on the sketch below. You need to lay carpet in your house according to this sketch. How many square yards of carpeting will be needed? (Use 3.14 for $\pi$. Round your final answer to the nearest tenth.

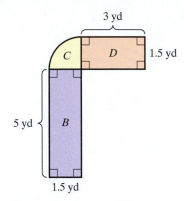

**Solution** The desired area is the sum of three areas, which we will call $B$, $C$, and $D$. Area $B$ and area $D$ are rectangular in shape. It is relatively easy to find the areas of these shapes.

$$A_B = (5 \text{ yd})(1.5 \text{ yd}) = 7.5 \text{ yd}^2 \quad A_D = (3 \text{ yd})(1.5 \text{ yd}) = 4.5 \text{ yd}^2$$

Area $C$ is one-fourth of a circle. The radius of the circle is 1.5 yd.

$$A_C = \frac{\pi r^2}{4} = \frac{(3.14)(1.5 \text{ yd})^2}{4} = \frac{(3.14)(2.25) \text{ yd}^2}{4}$$
$$= 1.76625 \text{ yd}^2 \approx 1.8 \text{ yd}^2$$

The total area is $A_B + A_D + A_C \approx 7.5 \text{ yd}^2 + 4.5 \text{ yd}^2 + 1.8 \text{ yd}^2 \approx 13.8 \text{ yd}^2$.

Approximately 13.8 square yards of carpeting will be needed to cover the floor.

▲ ▶ **Student Practice 9** John has a swimming pool that measures 8 feet by 12 feet. He plans to make a concrete walkway around the pool that is 3 feet wide. What will be the total cost of the walkway at $12 per square foot?

## 3.5 Exercises　MyMathLab®

### Verbal and Writing Skills Exercises 1–6

*Geometry* *Fill in the blank to complete each sentence.*

▲ 1. Perimeter is the _____ a plane figure.

▲ 2. _____ is the distance around a circle.

▲ 3. Area is a measure of the amount of _____ in a region.

▲ 4. A _____ is a four-sided figure with exactly two sides parallel.

▲ 5. The sum of the interior angles of any triangle is _____.

▲ 6. An equilateral triangle is a triangle in which each angle measures _____.

▲ 7. *Geometry* A large triangle has an altitude of 18 feet and a base of 42 feet. Find the area of the triangle.

▲ 8. *Geometry* Find the area of a triangle whose altitude is 12 inches and whose base is 6.5 inches.

▲ 9. *Geometry* Find the area of a parallelogram whose altitude is 14 inches and whose base is 7 inches.

▲ 10. *Geometry* The altitude of a parallelogram is 8 meters and the base is 5 meters. Find the area of the parallelogram.

▲ 11. *Geometry* The perimeter of a given parallelogram is 46 inches. If the length of one side is 14 inches, what is the length of a side adjacent to it?

▲ 12. *Geometry* The length of one side of a parallelogram is 20 cm. What is the length of the adjacent side to it if the perimeter is 60 cm?

▲ 13. *Geometry* Find the area of a circular sign whose radius is 7.00 feet. (Use 3.14 to approximate $\pi$.)

▲ 14. *Flower Bed* Find the area of a circular flower bed whose diameter is 6 meters. (Use 3.14 to approximate $\pi$.)

▲ 15. *Mercury* The diameter of the planet Mercury is approximately 3032 miles. Find the distance around its equator. (Use 3.14 to approximate $\pi$.)

▲ 16. *Olympic Medals* The medals made for the 2014 Sochi Olympics measure 100 mm, or about 3.94 inches, in diameter. Find the area of the medals in square inches. (Use 3.14 to approximate $\pi$.) Round your answer to the nearest hundredth.

### Applications

▲ 17. *Streets of Manhattan* In midtown Manhattan, the street blocks have a uniform size of 80 meters north-south by 280 meters east-west. If a typical New York City neighborhood is two blocks east-west by three blocks north-south, how much area does it cover?

▲ 18. *Computer Cases* A computer manufacturing plant makes its computer cases from plastic. Two sides of a triangular scrap of plastic measure 7 cm and 25 cm.

(a) If the perimeter of the plastic piece is 56 cm, how long is the third side?

(b) If each side of the triangular piece is decreased by 0.1 cm, what would the new perimeter be?

223

▲ 19. *Geometry* The circumference of a circle is 31.4 centimeters. Find the radius. (Use 3.14 to approximate $\pi$.)

▲ 20. *Geometry* The perimeter of an equilateral triangle is 27 inches. Find the length of each side of the triangle.

▲ 21. *Tires* An automobile tire has a diameter of 64 cm. What is the circumference of the tire? (Use 3.14 to approximate $\pi$.)

▲ 22. *Golden Ratio* The ancient Greeks discovered that rectangles where the length-to-width ratio was 8 to 5 (known as the golden ratio) were most pleasing to the eye. Kayla wants to create a painting that will be framed by some antique molding that she found. If she calculates that the molding can be used to make a frame of 104 inches in perimeter, what dimensions should her painting be so that it is visually appealing?

▲ 23. *Geometry* Each of the equal angles of an isosceles triangle is twice as large as the third angle. What is the measure of each angle?

▲ 24. *Geometry* The measure of the first angle in a triangle is triple the measure of the second angle. The measure of the third angle is 10 degrees more than the second angle. What is the measure of each angle?

▲ 25. *Playhouse Construction* Roger constructs children's playhouses. At the top of each house he places an isosceles triangle to form the roof. The largest angle measures 120°. What are the measurements of the other two angles in this triangle?

▲ 26. *TV Antenna* The largest angle of an isosceles triangle used in holding a cable TV antenna measures 146°. The other two angles are smaller. What are the measurements of the other two angles in the triangular piece of the cable TV antenna?

▲ 27. *Garden Border* Stella wants to put a small fence in the shape of a circle around a tree in her backyard. She wants the radius of the circle to be 3 feet, leaving room to plant flowers around the tree. How many feet of fencing does she need to buy if it is sold by the foot? (Use 3.14 to approximate $\pi$.)

▲ 28. *Rock Garden* Nate and Jenny want to order large rocks to use as a border around three rectangular gardens. Each garden measures 4 feet by 9 feet, and each rock will each measure about 6 inches across. How many rocks do they need to order?

▲ 29. *Geometry* What is the volume of a cylinder whose height is 8 inches and whose radius is 10 inches? (Use 3.14 to approximate $\pi$.)

▲ 30. *Gas Cylinder* A cylinder holds propane gas. The volume of the cylinder is 235.5 cubic feet. Find the height if the radius is 5 feet. (Use 3.14 to approximate $\pi$.)

▲ **31.** *Weather Balloon* A spherical weather balloon needs to hold at least 175 cubic feet of helium to be buoyant enough to lift the meteorological instruments. Will a helium-filled balloon with a diameter of 8 feet stay aloft?

▲ **32.** *Storage Freezer* Nathaniel is considering purchasing a new freezer. He has space for a freezer that measures 5 feet long and 2.5 feet high. He also needs to have 25 cubic feet of storage space. How wide can the freezer be? (Disregard the thickness of the freezer walls.)

▲ **33.** *Milk Container* A plastic cylinder made to hold milk is constructed with a solid top and bottom. The radius is 6 centimeters and the height is 4 centimeters. **(a)** Find the volume of the cylinder. **(b)** Find the total surface area of the cylinder. (Use 3.14 to approximate $\pi$.)

▲ **34.** *Pyrex Sphere* A Pyrex glass sphere is made to hold liquids in a science lab. The radius of the sphere is 3 centimeters. **(a)** Find the volume of the sphere. **(b)** Find the total surface area of the sphere. (Use 3.14 to approximate $\pi$.)

▲ **35.** *Goat Tether* Dan Perkin's goat is attached to the corner of the 8-foot-by-4-foot pump house by an 8-foot length of rope. The goat eats the grass around the pump house as shown in the diagram. This pattern consists of three-fourths of a circle of radius 8 feet and one-fourth of a circle of radius 4 feet. What area of grass can Dan Perkin's goat eat? (Use 3.14 to approximate $\pi$.)

▲ **36.** *Seven-Layer Torte* For a party, Josiah is making a seven-layer torte, which consists of seven 9-inch-diameter cake layers with a cream topping on each. He knows that one recipe of cream topping will cover 65 square inches of cake. How many batches of the topping must he make to complete his torte? (Use 3.14 to approximate $\pi$.)

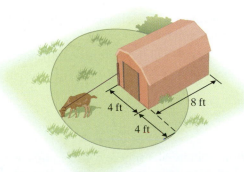

▲ **37.** *Tile Installation* Jordon is building a circular spa for a client. He will place tile around the rim. The outer radius is 8.5 feet and the inner radius is 7 feet. **(a)** How many square feet of tile will he need? (Use 3.14 to approximate $\pi$.) Round your final answer to the nearest tenth. **(b)** If the tile costs $2.50 per square foot, how much will the tile cost?

▲ **38.** *Granite Counter Installation* Sal is installing a granite kitchen countertop with a semicircular area for eating. The counter consists of two rectangles with dimensions as shown in the diagram. The edge of one of the counters is half of a circle with diameter 4 feet. (Use 3.14 to approximate $\pi$.) **(a)** How many square feet will the countertop be? **(b)** If the granite sells for $65 per square foot, how much will the granite for the countertop cost?

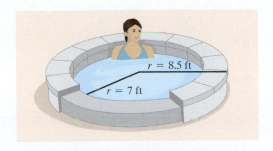

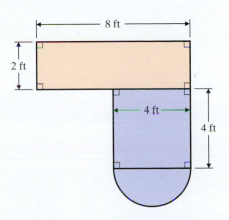

**To Think About**

▲ **39.** *Moon* Assume that a very long rope supported by poles that are 3 feet tall is stretched around the moon. (Neglect gravitational pull and assume that the rope takes the shape of a large circle.) The radius of the moon is approximately 1080 miles. How much longer would you need to make the rope if you wanted the rope to be supported by poles that are 4 feet tall? (Use 3.14 to approximate $\pi$.)

## Cumulative Review

**40.** **[1.8.2]** Find the area of a circle with a diameter of 10 centimeters. Use $\pi \approx 3.14$

**41.** **[2.4.1]** Solve for $x$. $\frac{1}{2}x - 3 = \frac{1}{4}(3x + 3)$

**42.** **[0.6.1]** Lester's gross pay for one week is $950. He noticed that he had $47.50 deducted for medical insurance and $66.50 deducted for his retirement plan.

(a) What percent of his gross pay is deducted for medical insurance?

(b) What percent of his gross pay is deducted for his retirement plan?

**43.** **[0.6.1]** Jessica is setting up a budget so she can save to go to college. Her monthly take-home pay is $2600, and she will have 15% deducted and placed in a college savings account. Her parents agreed to match her monthly savings by contributing one-quarter of the amount she saves each month.

(a) How much will Jessica have deducted each month for her college account?

(b) How much will her parents contribute to her college savings each month?

## Quick Quiz 3.5

**1.** Find the area of a trapezoid if the two bases are 12 feet and 17 feet and the altitude is 8 feet.

**2.** Find the volume of a sphere with a radius of 3 inches. Use 3.14 as an approximation for $\pi$. Round your answer to the nearest cubic inch.

**3.** The gym hallway is a square that measures 5 yards on a side. The square is connected to a rectangle that measures 5 yards by 14 yards. How much would it cost to carpet this hallway with all-weather carpeting that costs $24 per square yard?

**4.** **Concept Check** Explain how you would set up an equation to solve the following problem.

Two angles of a triangle measure 135° and 11°. What is the measure of the third angle?

# Using Inequalities to Solve Word Problems

## 1 Translating English Sentences into Inequalities

Real-life situations often involve inequalities. For example, if a student wishes to maintain a certain grade point average (GPA), you may hear him or her say, "I have to get a grade of at least an 87 on the final exam."

What does "at least an 87" mean? What grade does the student need to achieve? 87 would fit the condition. What about 88? The student would be happy with an 88. In fact the student would be happy with any grade that *is greater than or equal to* an 87. We can write this as

$$\text{grade} \geq 87.$$

Notice that the English phrase "at least an 87" translates mathematically as "greater than or equal to an 87."

Be careful when translating English phrases that involve inequalities into algebra. If you are not sure which symbol to use, try a few numbers to see if they fit the condition of the problem. Choose the mathematical symbol accordingly.

Here is a list that may help you.

| *English Phrases Describing Inequality* | *Algebraic Symbol* |
|---|---|
| **Greater than** | |
| Is more than | |
| Is greater than | $>$ |
| **Less than** | |
| Is less than | |
| Is smaller than | $<$ |
| **Greater than or equal to** | |
| Is greater than or equal to | |
| Is not less than | $\geq$ |
| Is at least | |
| Cannot be less than | |
| **Less than or equal to** | |
| Is less than or equal to | |
| Is not greater than | $\leq$ |
| Is at most | |
| Is not more than | |
| Does not exceed | |

**Example 1** Translate each English sentence into an inequality.

(a) The perimeter of the rectangle must be no more than 70 inches.

(b) The average number of errors must be less than two mistakes per page.

(c) His income must be at least $950 per week.

(d) The new car will cost at most $19,070.

### Solution

(a) Perimeter must be *no more than* 70.

$$\text{Perimeter} \leq 70$$

Try 80. Does 80 satisfy this requirement? (That is, is 80 more than 70?) No, it does not. Do 60 and 50 satisfy this requirement? Yes. Use the "less than or equal to" symbol.

(b) Average number must be *less than* 2.

$$\text{Average} < 2$$

*Continued on next page*

**Student Learning Objectives**

After studying this section, you will be able to:

1 Translate English sentences into inequalities.

2 Solve applied problems using inequalities.

**(c)** Income must be *at least* $950.

$$\text{Income} \geq 950$$

"At least" means "not less than." The income must be $950 or more. Use $\geq$.

**(d)** The new car will cost *at most* $19,070.

$$\text{Car} \leq 19{,}070$$

"At most $19,070" means it is the top price for the car. The cost of the car cannot be more than $19,070. Use $\leq$. ▫

**Student Practice 1** Translate each English sentence into an inequality.

**(a)** The height of the person must be no greater than 6 feet.

**(b)** The speed of the car must have been greater than 65 miles per hour.

**(c)** The area of the room must be greater than or equal to 560 square feet.

**(d)** The profit margin cannot be less than $50 per television set.

## 2 Solving Applied Problems Using Inequalities

We can use the same problem-solving techniques we used previously to solve word problems involving inequalities.

**Example 2** A manufacturing company makes 60-watt fluorescent light bulbs. For every 1000 bulbs manufactured, a random sample of five bulbs is selected and tested. To meet quality control standards, the average number of hours these five bulbs last must be at least 900 hours. The test engineer tested five bulbs. Four of them lasted 850 hours, 1050 hours, 1000 hours, and 950 hours, respectively. How many hours must the fifth bulb last in order for the batch to meet quality control standards?

### Solution

**1. Understand the problem.**

Let $n =$ the number of hours the fifth bulb must last

The average that the five bulbs last *must be at least* 900 hours. We could write

$$\text{Average} \geq 900 \text{ hours.}$$

**2. Write an inequality.**

$$\text{Average of five bulbs} \geq 900 \text{ hours.}$$

$$\frac{850 + 1050 + 1000 + 950 + n}{5} \geq 900$$

**3. Solve and state the answer.**

$$\frac{850 + 1050 + 1000 + 950 + n}{5} \geq 900$$

$$\frac{3850 + n}{5} \geq 900 \qquad \text{Simplify the numerator.}$$

$$3850 + n \geq 4500 \qquad \text{Multiply both sides by 5.}$$

$$n \geq 650 \qquad \text{Subtract 3850 from each side.}$$

If the fifth bulb burns for 650 hours or more, the quality control standards will be met.

**4. Check.** Any value of 650 or larger should satisfy the original inequality. Suppose we pick $n = 650$ and substitute it into the inequality.

$$\frac{850 + 1050 + 1000 + 950 + 650}{5} \overset{?}{\geq} 900$$

$$\frac{4500}{5} \overset{?}{\geq} 900$$

$$900 \geq 900 \quad \checkmark$$

If we try $n = 700$, the left side of the inequality will be 910. $910 \geq 900$. In fact, if we try any number greater than 650 for $n$, the left side of the inequality will be greater than 900. The inequality will be valid for $n = 650$ or any larger value. ☐

**Student Practice 2** A manufacturing company makes coils of $\frac{1}{2}$-inch maritime rope. For every 300 coils of rope made, a random sample of four coils of rope is selected and tested. To meet quality control standards, the average number of pounds that each coil will hold must be at least 1100 pounds. The test engineer tested four coils of rope. The first three held 1050 pounds, 1250 pounds, and 950 pounds, respectively. How many pounds must the last coil hold if the sample is to meet quality control standards?

**Example 3** Juan is selling supplies for restaurants in the Southwest. He gets paid $700 per month plus 8% of the value of all the supplies he sells to restaurants. If he wants to earn more than $3100 per month, what value of products will he need to sell?

**Student Practice 3** Rita is selling commercial fire and theft alarm systems to local companies. She gets paid $1400 per month plus 2% of the cost of all the systems she sells in one month. She wants to earn more than $2200 per month. What value of products will she need to sell each month?

**Solution**

1. **Understand the problem.** Juan earns a fixed salary and a salary that depends on how much he sells (commission salary) each month.

   Let $x$ = the value of the supplies he sells each month.

   Then $0.08x$ = the amount of the commission salary he earns each month.

   He wants his total income to be more than $3100 per month; thus

   $$\text{total income} > 3100.$$

2. **Write the inequality.**

| The fixed income | added to | the commission income | must be greater than | 3100. |
|:---:|:---:|:---:|:---:|:---:|
| ↓ | ↓ | ↓ | ↓ | ↓ |
| 700 | + | $0.8x$ | > | 3100 |

3. **Solve and state the answer.**

   $700 + 0.08x > 3100$

   $0.08x > 2400$    Subtract 700 from each side.

   $x > 30,000$    Divide each side by 0.08.

   Juan must sell more than $30,000 worth of supplies each month.

4. **Check.** Any value greater than 30,000 should satisfy the original inequality.

   Suppose we pick $x = 30,001$ and substitute it into the inequality.

   $$700 + 0.08(30,001) \overset{?}{>} 3100$$
   $$700 + 2400.08 \overset{?}{>} 3100$$
   $$3100.08 > 3100 \quad \checkmark$$    ☐

## 3.6 Exercises MyMathLab®

**Verbal and Writing Skills** *Translate to an inequality using a variable.*

1. The cost is greater than $67,000.

2. Anna wants her time in a 5-kilometer race to be less than 21 minutes.

3. Elliot would like his weight to be less than 175 pounds.

4. The tax increase is more than $7890.

5. The height of stratocumulus clouds is not more than 1500 feet.

6. The temperature in the children's wading pools should not drop below 82 degrees.

7. To earn an A in algebra, Ramon cannot get less than a 93 average.

8. The camp-ins at the Museum of Science in Boston cannot accommodate more than 600 people.

**Applications** *Solve using an inequality.*

▲ 9. *Window Design* A triangular window on Ross Camp's new Searay boat has two sides that measure 87 centimeters and 64 centimeters, respectively. The perimeter of the triangle must not exceed 291 centimeters. What are the possible values for the length of the third side of the window?

▲ 10. *Package Size* The U.S. Postal Service provides guidelines for package size. One of the guidelines specifies the dimensions of the package: The length plus the girth (circumference around the widest part) cannot exceed 130 inches. If you are preparing a package that is 33 inches wide and 28 inches high, how long is the package permitted to be?

11. *Flower Delivery* Flora's Flowers uses the Speedy Delivery Service. Flora estimates that if she averages at least 24 deliveries per week, she will save money by hiring a student to deliver her flowers after school. Her contract with Speedy Delivery is up for renewal in 6 weeks. Flora tracks her deliveries for 5 weeks: 18, 40, 21, 7, and 36. At least how many deliveries must she average in week 6 to convince her to drop the delivery service?

12. *Debate Competition Scores* Syed is competing in a debate competition, and so far his scores have been 20, 23, 17, 20, and 22. He has one more debate left and must obtain a total average score of 20 or more to make it to the semifinals. What possible values can he obtain on his last debate in order to make it to the semifinals?

▲ 13. *Garden Design* Joel plans to plant a flower garden in the front of his house, where he can fit a garden that is 8 feet wide. He wishes to use the can of wildflower mix that he got for his birthday. The can of seeds claims that it will luxuriously carpet an area of no more than 60 square feet. What are the possible dimensions for the depth of Joel's garden, so that it will be covered with wildflowers?

14. *Class Size* The Lee County Elementary School District noted that there was an increase of 1785 pupils during the fall semester. How many new teachers must they add to keep their average class size under 28? (Assume one teacher for each class.)

15. *Truck Rental* Abby and Frank decided that to save money, they would move themselves from their apartment to their new house. They had to rent a truck to do this. Because their budget was very strict, they could afford to spend no more than $100. The rental truck costs $29.95 per day plus $0.49 per mile. If they needed the truck for two days, what was the limit on the mileage they could put on the truck?

16. *Software Sales* Marci just graduated from college, and her field of study was technology related to business. She was hired by a company that sells software technology to businesses, and she is paid a monthly salary of $3000 plus 6% commission. Marci wants to earn at least $4200 per month to meet her monthly bills and save for a home and retirement. What is the amount of sales she must make monthly to meet her goal?

17. *Chemistry Lab Technician* Malinni is a technician for a large research company. For an experiment she is doing involving chemicals, the compound may never reach a temperature less than 20 degrees Celsius. What would be the corresponding temperature range in degrees Fahrenheit? (Use $C = \frac{5}{9}(F - 32)$ in your inequality.)

18. *IQ* A person's IQ is found by multiplying mental age ($m$), as indicated by standard tests, by 100 and dividing this result by the chronological age ($c$). This gives the formula IQ $= 100\frac{m}{c}$. If a group of 14-year-old children has an IQ of not less than 80, what is the mental age range of this group?

19. *Assistant Manager* Marc is an assistant manager at a manufacturing plant. One of his duties is to oversee the company's budget. Marc's cost to produce a new item must not exceed $160,700 for the first quarter of the year. The cost ($C$) to produce this particular item is given by the equation $C = 32n + 700$, where $n$ is the number of items produced. Use this equation to determine how many new items Marc could produce in the first quarter without exceeding the budget.

20. *Farmer* Clyde is making an enclosed outdoor pen for some of the animals on his farm. He has enough fencing material to make a fence that is no more than 1400 feet. The length of the pen is fixed at 380 feet due to the constraints of the outdoor area he is using. What are the possible values for the width of the fence for the pen? (Use the formula $P = 2L + 2W$ in your inequality.)

21. *Commission* Frank makes $16,000 per year plus a 6% commission on all sales over $30,000. He wants to make more than $22,000 this year. How many dollars in sales must he achieve to make this salary goal?

22. *Commission* Judy makes $18,000 per year plus a 7% commission on all sales over $40,000. She wants to make more than $26,400 this year. How many dollars in sales must she achieve to make this salary goal?

## To Think About

**23.** *CD Manufacturing* A compact disc recording company's weekly costs can be described by the equation cost $= 5000 + 7n$, where $n$ is the number of compact discs manufactured in one week. The equation for the amount of income produced from selling these $n$ discs is income $= 18n$. At least how many discs need to be manufactured and sold in one week for the income to be greater than the cost?

**24.** *DVD Player Manufacturing* A company that manufactures DVD players has weekly costs described by the equation cost $= 15{,}000 + 120n$, where $n$ is the number of DVD players manufactured in one week. The equation for the amount of income produced from selling these $n$ DVD players is income $= 250n$. At least how many DVD players need to be manufactured and sold in one week for the income to be greater than the cost?

## Cumulative Review

**25.** **[2.5.1]** Solve for $q$. $3(q + 1) = 8q - p$

**26.** **[2.5.1]** Solve for $t$. $I = Prt$

*Solve.*

**27.** **[2.6.4]** $30 - 2(x + 1) \leq 4x$

**28.** **[2.6.4]** $2(x + 3) - 22 < 4(x - 2)$

## Quick Quiz 3.6

**1.** Jane wants to plant a rectangular garden that is 8 feet wide. She wants to have a garden that is no more than 79.2 square feet in area. In order to comply with that plan, what should be the length of Jane's garden?

**2.** The student activities office has four students who work part-time on a work-study program. Three of the students work 15 hours per week, 17 hours per week, and 12 hours per week. The office policy is that the average number of hours students work per week in the student activities office is less than 15.5 hours. What is the restriction for the number of hours that the fourth student can work?

**3.** Chris makes $17,000 per year plus a 4% commission on all sales over $10,000. He wants to make more than $28,000 this year. How many dollars in sales must he achieve to make his salary goal?

**4.** *Concept Check* Explain how you would set up an inequality to solve the following problem.

Caleb wants to spend no more than $150 to rent a truck to move things for his dad's company. The rental company charges $37.50 per day to rent the truck, and 19.5 cents per mile that the truck is driven. If he rented the truck for two days, how many miles could he drive the truck?

### Background: Landscape Design Architect

Jerry is pursuing a degree in horticulture with the goal of becoming a landscape design architect, hoping someday to start his own business. He works at Greener Pastures Landscape Design with designers who include him in all aspects of a job.

His current task involves redesigning the landscape features of a newly purchased home for its owners. Decisions have been made to install maintenance-free fencing on three sides of the owner's backyard and install a row of hedges forming a natural fence on the fourth side.

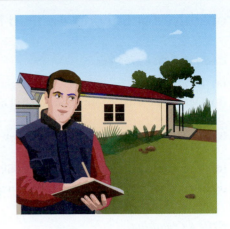

### Facts

Jerry has been given the following information about the backyard lot. The layout is rectangular with a perimeter of 426 feet. The width is 15 feet longer than half the length. Three of the sides, one length and two widths, are to be enclosed with the maintenance-free fencing. A natural fence of arborvitaes, planted 6 feet apart, will border the fourth side.

Fencing and installation costs $11 per foot for materials and $6 per foot for installation.

The arborvitaes are to be planted 6 feet apart. Their cost including materials and installation should be no more than $1130. Labor to plant each arborvitae is known to be $35 more than the cost of the arborvitae, and the number of man-hours needed for the installation of the entire row is considered to be 16 hours.

### Tasks

1. Jerry must determine the backyard's dimensions. What is the length and the width of the yard?

2. Jerry can now compute the total cost of the fence and its installation. What is the total cost?

3. Jerry must determine the maximum price that should be spent on each arborvitae along with the total cost of this aspect of the project. Recalling an algebra course he took and the concept of inequalities, the following is what he comes up with, knowing that the hedges must be planted 6 feet apart.

   Let the cost of a single arborvitae $= x$

   Labor cost $= 16(x + 35)$

   Number of arborvitaes needed $= \dfrac{l}{6}$

   Based on his calculations, how much can Jerry pay per plant while staying within the given budget for this portion of the project?

# Chapter 3 Organizer

## Procedure for Solving Applied Problems

**1. *Understand the problem.***

  **(a)** Read the word problem carefully to get an overview.

  **(b)** Determine what information you will need to solve the problem.

  **(c)** Draw a sketch. Label it with the known information. Determine what needs to be found.

  **(d)** Choose a variable to represent one unknown quantity.

  **(e)** If necessary, represent other unknown quantities in terms of that same variable.

**2. *Write an equation.***

  **(a)** Look for key words to help you to translate the words into algebraic symbols and expressions.

  **(b)** Use a given relationship in the problem or an appropriate formula in order to write an equation.

**3. *Solve and state the answer.***

**4. *Check.***

  **(a)** Check the solution in the original equation. Is the answer reasonable?

  **(b)** Be sure the solution to the equation answers the question in the word problem. You may need to do some additional calculations if it does not.

| Topic and Procedure | Examples | You Try It |
|---|---|---|
| **Comparisons, p. 200**<br><br>Select as the variable the quantity that is being compared to. | The perimeter of a rectangle is 126 meters. The length of the rectangle is 6 meters less than double the width. Find the dimensions of the rectangle.<br><br>**1. *Understand the problem.*** We want to find the length and the width of a rectangle whose perimeter is 126 meters.<br><br>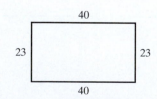<br><br>The length is *compared to* the width, so we start with the width.  Let $w$ = width.<br><br>The length is 6 meters less than double the width.  Then $l = 2w - 6$.<br><br>**2. *Write an equation.*** The perimeter of a rectangle is $P = 2l + 2w$.  $126 = 2(2w - 6) + 2w$<br><br>**3. *Solve.***<br>$$126 = 4w - 12 + 2w$$<br>$$126 = 6w - 12$$<br>$$138 = 6w$$<br>$$23 = w \quad \text{The width is 23 meters.}$$<br>$$2w - 6 = 2(23) - 6 =$$<br>$$46 - 6 = 40$$<br><br>The length is 40 meters.<br><br>**4. *Check.*** Is this reasonable? Yes. A rectangle 23 meters wide and 40 meters long seems to be about right for the perimeter to be 126 meters. Is the perimeter exactly 126 meters? Is the length exactly 6 meters less than double the width?<br><br>$$2(23) + 2(40) \stackrel{?}{=} 126$$<br>$$46 + 80 \stackrel{?}{=} 126$$<br>$$126 = 126 \ \checkmark$$<br>$$40 \stackrel{?}{=} 2(23) - 6$$<br>$$40 \stackrel{?}{=} 46 - 6$$<br>$$40 = 40 \ \checkmark$$ | **1.** The perimeter of a rectangle is 112 meters. The length of the rectangle is 8 meters longer than double the width. Find the dimensions of the rectangle. |

| Topic and Procedure | Examples | You Try It |
|---|---|---|
| **Money and percents, p. 207**<br><br>Let $x =$ one amount and the total $- x =$ the other amount. Find the percent of each amount. | Gina saved some money for college. She invested $2400 for one year and earned $225 in simple interest. She invested part of it at 12% and the rest of it at 9%. How much did she invest at each rate?<br><br>**1. Understand the problem.**<br>We want to find each amount that was invested.<br><br>$$\text{interest earned at 12%} + \text{interest earned at 9%} = \text{total interest of \$225}$$<br><br>We let $x$ represent one quantity of money.<br>Let $x =$ amount of money invested at 12%.<br><br>We started with $2400. If we invest $x$ at 12%, we still have $(2400 - x)$ left.<br>Then $2400 - x =$ the amount of money invested at 9%.<br><br>Interest $= prt$<br><br>Interest at 12%     $I_1 = 0.12x$<br><br>Interest at 9%     $I_2 = 0.09(2400 - x)$<br><br>**2. Write an equation.**<br>$$I_1 + I_2 = 225.$$<br>$$0.12x + 0.09(2400 - x) = 225$$<br><br>**3. Solve.**<br>$$0.12x + 216 - 0.09x = 225$$<br>$$0.03x + 216 = 225$$<br>$$0.03x = 9$$<br>$$\frac{0.03x}{0.03} = \frac{9}{0.03}$$<br>$$x = 300$$<br>$300 was invested at 12%.<br>$2400 - x = 2400 - 300 = 2100$<br>$2100 was invested at 9%.<br><br>**4. Check.**<br>Is this reasonable? Yes.<br>Are the conditions of the problem satisfied?<br>Does the total amount invested equal $2400?<br>Will $2100 at 9% and $300 at 12% yield $225 in interest?<br><br>$$2100 + 300 \overset{?}{=} 2400$$<br>$$2400 = 2400 \checkmark$$<br>$$0.09(2100) + 0.12(300) \overset{?}{=} 225$$<br>$$189 + 36 \overset{?}{=} 225$$<br>$$225 = 225 \checkmark$$ | **2.** Caleb saved some money from his summer job. He invested $3600 for one year and earned $132 in simple interest. He invested part of it at 2% and the rest of it at 4%. How much did he invest at each rate? |
| **Geometric formulas, pp. 217–221**<br><br>Know the definitions of and formulas related to basic geometric shapes such as the triangle, rectangle, parallelogram, trapezoid, and circle. Review these formulas from the beginning of Section 3.5 as needed. Formulas for the volume and total surface area of three solids are also given there. | The volume of a right circular cylinder found in a four-cylinder car engine is determined by the formula $V = \pi r^2 h$. Find $V$ if $r = 6$ centimeters and $h = 5$ centimeters. Use $\pi \approx 3.14$.<br><br>$$V = 3.14(6 \text{ cm})^2(5 \text{ cm})$$<br>$$= 3.14(36)(5) \text{ cm}^3$$<br>$$\approx 565.2 \text{ cm}^3 \text{ (or cc)}$$<br><br>Four such cylinders would make this a 2260.8-cc engine. | **3.** Find the volume of a right circular cylinder that has a radius of 10 inches and a height of 12 inches. Use $\pi \approx 3.14$. |
| **Using inequalities to solve word problems, p. 227**<br><br>Be sure that the inequality you use fits the conditions of the word problem.<br><br>**1.** *At least 87* means $\geq 87$.<br>**2.** *No less than $150* means $\geq \$150$.<br>**3.** *At most $500* means $\leq \$500$.<br>**4.** *No more than 70* means $\leq 70$.<br>**5.** *Less than 2* means $< 2$.<br>**6.** *More than 10* means $> 10$. | Julie earns 2% on her sales over $1000. How much must Julie make in sales to earn at least $300?<br>Let $x =$ amount of sales.<br><br>$$(2\%)(x - \$1000) \geq \$300$$<br>$$0.02(x - 1000) \geq 300$$<br>$$x - 1000 \geq 15{,}000$$<br>$$x \geq 16{,}000$$<br><br>Julie must make sales of $16,000 or more (at least $16,000). | **4.** Lexi earns 3% on sales over $2000. How much must Lexi make in sales to earn at least $840? |

# Chapter 3 Review Problems

## Section 3.1

*Write an algebraic expression. Use the variable x to represent the unknown value.*

**1.** 19 more than a number

**2.** two-thirds of a number

**3.** half a number

**4.** 18 less than a number

**5.** triple the sum of a number and 4

**6.** twice a number decreased by three

*Write an expression for each of the quantities compared. Use the letter specified.*

**7.** *Workforce* The number of working people is four times the number of retired people. The number of unemployed people is one-half the number of retired people. (Use the letter $r$.)

▲ **8.** *Geometry* The length of a rectangle is 5 meters more than triple the width. (Use the letter $w$.)

▲ **9.** *Geometry* A triangle has three angles $A$, $B$, and $C$. The number of degrees in angle $A$ is double the number of degrees in angle $B$. The number of degrees in angle $C$ is 17 degrees less than the number of degrees in angle $B$. (Use the letter $b$.)

**10.** *Class Size* There are 29 more students in biology class than in algebra. There are one-half as many students in geology class as in algebra. (Use the letter $a$.)

## Section 3.2

*Solve.*

**11.** Fourteen is taken away from triple a number. The result is $-5$. What is the number?

**12.** When twice a number is reduced by seven, the result is $-21$. What is the number?

**13.** *Table Set* Six chairs and a table cost $450. If the table costs $210, how much does one chair cost?

**14.** *Age* Jon is twice as old as his cousin David. If Jon is 32, how old is David?

**15.** *Speed Rates* Two cars left San Francisco and drove to San Diego, 800 miles away. One car was driven at 60 miles an hour. The other car was driven at 65 miles an hour. How long did it take each car to make the trip? Round your answers to the nearest tenth.

**16.** *Biology Score* Zach took four biology tests and got scores of 83, 86, 91, and 77. He needs to complete one more test. If he wants to get an average of 85 on all five tests, what score does he need on his last test?

## Section 3.3

▲ **17.** *Geometry* An isosceles triangle has a perimeter of 85 feet. If one of the two equal sides measures 31 feet, what is the measure of the third side?

▲ **18.** *Geometry* The measure of the second angle of a triangle is three times the measure of the first angle. The measure of the third angle is 12 degrees less than twice the measure of the first. Find the measure of each of the three angles.

**19.** *Rope* A piece of rope 50 yards long is cut into two pieces. One piece is three-fifths as long as the other. Find the length of each piece.

**20.** *Income* George and Heather have a combined income of $65,000. Heather earns $2000 more than half of what George earns. How much does each earn?

## Section 3.4

21. *Electric Bill* The electric bill at Jane's house this month was $71.50. The charge is based on a flat rate of $25 per month plus a charge of $0.15 per kilowatt-hour of electricity used. How many kilowatt-hours of electricity were used?

22. *Car Rental* Abel rented a car from Sunshine Car Rentals. He was charged $39 per day and $0.25 per mile. He rented the car for three days and paid a rental cost of $187. How many miles did he drive the car?

23. *Simple Interest* Boxell Associates has $7400 invested at 5.5% simple interest. After money was withdrawn, $242 was earned for one year on the remaining funds. How much was withdrawn?

24. *CD Player* Jamie bought a new CD player for her car. When she bought it, she found that the price had been decreased by 18%. She was able to buy the CD player for $36 less than the original price. What was the original price?

25. *Investments* Peter and Shelly invested $9000 for one year. Part of it was invested at 12% and the remainder was invested at 8%. At the end of one year the couple had earned exactly $1000 in simple interest. How much did they invest at each rate?

26. *Investments* A man invested $5000 in a savings bank. He placed part of it in a checking account that earns 4.5% and the rest in a regular savings account that earns 6%. His total annual income from this investment in simple interest was $270. How much was invested in each account?

27. *Coin Change* Mary has $3.75 in nickels, dimes, and quarters. She has three more quarters than dimes. She has twice as many nickels as quarters. How many of each coin does she have?

28. *Tip Jar* After the morning rush at the neighborhood coffee shop, the tip jar contained $9.80 in coins. There were two more quarters than nickels. There were three fewer dimes than nickels. How many coins of each type were there?

## Section 3.5

*Solve.*

▲ 29. *Geometry* Find the surface area of a tennis ball that has a diameter of 2.6 inches. (Use $\pi \approx 3.14.$) Round your answer to the nearest hundredth.

▲ 30. *Trench Length* Wilfredo and Tina bought a new screen house for camping. It was hexagonal (6 sides) in shape and 4 ft long on a side. If they wanted to dig a trench around their house so that rainwater would flow away, how long would the trench be? (You are looking for the perimeter of the screen house.)

▲ 31. *Geometry* Find the third angle of a triangle if one angle measures 62 degrees and the second angle measures 47 degrees.

▲ 32. *Geometry* Find the area of a triangle whose base is 8 miles and whose altitude is 10.5 miles.

▲ 33. *Geometry* Find the volume of a storage room that has a width of 6 feet, a length of 11 feet, and a height of 7 feet.

▲ 34. *Geometry* Find the volume of a basketball whose radius is 4.5 inches. (Use $\pi \approx 3.14.$)

▲ 35. *Geometry* The Clock Tower (also known as Big Ben) in London has four clock faces measuring 23 feet in diameter. Find the area of each clock face. (Use $\pi \approx 3.14.$) Round your answer to the nearest hundredth.

▲ 36. *Geometry* The perimeter of an isosceles triangle is 46 inches. The two equal sides are each 17 inches long. How long is the third side?

237

▲ **37.** *House Painting* Jim wants to have the exterior siding on his house painted. Two sides of the house have a height of 18 feet and a length of 22 feet, and the other two sides of the house have a height of 18 feet and a length of 19 feet. The painter has quoted a price of $2.50 per square foot. The windows and doors of the house take up 100 square feet. How much will the painter charge to paint the siding?

▲ **38.** *Aluminum Cylinder* Find the cost to build a cylinder out of aluminum if the radius is 5 meters, the height is 14 meters, and the cylinder will have a flat top and flat bottom. The cost factor has been determined to be $40.00 per square meter. (*Hint:* First find the surface area.) (Use $\pi \approx 3.14$.)

## Section 3.6

*Solve.*

**39.** *Truck Rental* Brian rents a truck for $20 a day and $0.25 a mile. How far can he drive if he wants to rent the truck for three days and does not want to spend more than $100?

**40.** *Income* Greta earns $25,000 per year in salary and 5% commission on her sales. How much were her sales if her annual income was more than $38,000?

**41.** *Transcontinental Railroad* In 1862, two companies were granted the rights to build the transcontinental railroad from Omaha, Nebraska, to Sacramento, California. The Central Pacific railroad began in 1863 from Sacramento heading east. The Union Pacific began 24 months later, leaving from Omaha and heading west. The Central Pacific averaged 8.75 miles of track each month, while the Union Pacific averaged 20 miles of track per month. The two companies met at Promontory, Utah, after completing 1590 miles of track. How much track did each build?

**42.** *Clothes Shopping* Michael plans to spend less than $70 on shirts and ties. He bought three shirts for $17.95 each. How much can he spend on ties?

## Mixed Practice

*Solve.*

**43.** *Electricity Use* Fred and Nancy Sullivan are planning to purchase a new refrigerator. The most efficient model costs $870, but the electricity to operate it would cost only $41 per year. The less-efficient model is the same size. It costs $744, but the electricity to operate it would cost $83 per year. How long will it be before the total cost is greater for the less-efficient model?

**44.** *Buying vs. Renting* Custom Computer Works of Gloucester rents its office building and warehouse for $1800 per month. The owner is thinking of buying the building. He could purchase it with a down payment of $42,000. He would then make monthly mortgage payments of $1100. How many months would it take for his total costs to be less to purchase the building rather than rent it?

**45.** *Basketball* In a recent basketball game the Boston Celtics scored 99 points. They scored three times as many field goals as free throws. They also made 12 three-point baskets. How many field goals did they make? How many free throws did they make? (A field goal is worth two points and a free throw is worth one point.)

**46.** *Investments* Sarah Tanguay has $2000 more invested at 8.5% than she does at 9.75%. Both investments are earning simple interest. If the annual return from each investment is the same, how much is invested at each rate?

**47.** *Jet Travel* A jet plane travels 32 miles in three minutes. How far can the plane travel in one hour if it continues at that speed? How many minutes would it take the jet to travel from Boston to Denver (approximately 1800 miles) at that same speed?

**48.** *Football* A running back in a Wheaton College football game catches a pass and runs for the goal line 80 yards away. He can run a 50-yard dash in six seconds. If he runs at 80% of his speed in the 50-yard dash, how long will it take him to reach the goal line?

# How Am I Doing? Chapter 3 Test

**MATH COACH**   MyMathLab®   You Tube

After you take this test read through the Math Coach on pages 241–242. Math Coach videos are available via MyMathLab and YouTube. Step-by-step test solutions on the Chapter Test Prep Videos are also available via MyMathLab and YouTube. (Search "TobeyBeginningAlg" and click on "Channels.")

*Solve.*

1. A number is doubled and then decreased by 11. The result is 59. What is the number?

2. The sum of one-half of a number, one-ninth of the number, and one-twelfth of the number is twenty-five. Find the number.

3. Double the sum of a number and 5 is the same as fourteen more than triple the same number. Find the number.

▲ 4. A triangular region has a perimeter of 66 meters. The first side is two-thirds of the second side. The third side is 14 meters shorter than the second side. What are the lengths of the three sides of the triangular region?

▲ 5. A rectangle has a length 7 meters longer than double the width. The perimeter is 134 meters. Find the dimensions of the rectangle.

6. Three harmful pollutants were measured by a consumer group in the city. The total sample contained 15 parts per million (ppm) of three harmful pollutants. The amount of the first pollutant was double the amount of the second. The amount of the third pollutant was 75% of the amount of the second. How many parts per million of each pollutant were found?

7. Raymond has a budget of $1940 to rent a computer for his company office. The computer company he wants to rent from charges $200 for installation and service as a one-time fee. Then they charge $116 per month rental for the computer. How many months will Raymond be able to rent a computer with this budget?

8. Last year the yearly tuition at Westmont College went up 8%. This year's charge for tuition for the year is $34,560. What was it last year before the increase went into effect?

MC 9. Franco invested $4000 in money market funds. Part was invested at 14% interest, the rest at 11% interest. At the end of each year the fund company pays interest. After one year he earned $482 in simple interest. How much was invested at each interest rate?

10. Mary has $3.50 in change. She has twice as many nickels as quarters. She has one fewer dimes than she has quarters. How many of each coin does she have?

1. _____ ☐

2. _____ ☐

3. _____ ☐

4. _____ ☐

5. _____ ☐

6. _____ ☐

7. _____ ☐

8. _____ ☐

9. _____ ☐

10. _____ ☐

239

11. _____ □

▲ 11. Find the circumference of a circle with radius 34 inches. Use $\pi \approx 3.14$ as an approximation.

12. _____ □

▲ 12. Find the area of a trapezoid if the two bases are 10 inches and 14 inches and the altitude is 16 inches.

13. _____ □

▲ 13. Find the volume of a sphere with radius 10 inches. Use $\pi \approx 3.14$ as an approximation and round your answer to the nearest cubic inch.

14. _____ □

▲ 14. Find the area of a parallelogram with a base of 12 centimeters and an altitude of 8 centimeters.

15. _____ □

▲ 15. $\mathbb{M_C}$ How much would it cost to carpet the area shown in the figure if carpeting costs $12 per square yard?

12 yd

1.5 yd

9 yd

2 yd

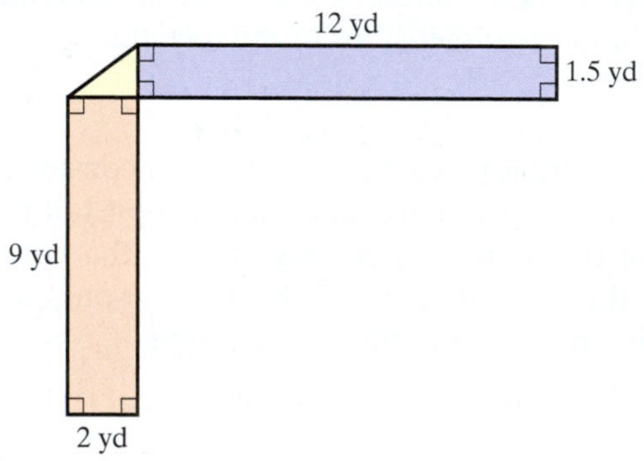

16. _____ □

16. Cathy scored 76, 84, and 78 on three tests. How much must she score on the next test to have at least an 80 average?

17. _____ □

$\mathbb{M_C}$ 17. Carol earns $15,000 plus 5% commission on sales over $10,000. How much must she make in sales to earn more than $20,000?

18. _____ □

**Total Correct:** _____

▲ 18. Andrew is putting down tile for a new gymnasium. He can lay 25 square feet of tile in an hour. How long will it take him to do a gymnasium 150 feet long and 100 feet wide?

240

# MATH COACH

*Mastering the skills you need to do well on the test.*

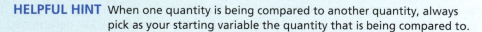

The following problems are from the Chapter 3 test. Here are some helpful hints to keep you from making common errors on test problems.

**Chapter 3 Test, Problem 5**  A rectangle has a length 7 meters longer than double the width. The perimeter is 134 meters. Find the dimensions of the rectangle.

> **HELPFUL HINT**  When one quantity is being compared to another quantity, always pick as your starting variable the quantity that is being compared to.

Did you first let $w$ = the width since the length is compared to the width? Did you then write $2w + 7$ for the length?

Yes _____ No _____

If you answered No, remember that the rectangle has a length 7 meters longer than double the width. We need to start with the width. Since the length is 7 meters longer than double the width, we write $2w + 7$ as the length.

Did you then write the equation for the perimeter? $P = 2l + 2w$. Thus we would write $2(2w + 7) + 2w = 134$. This becomes $4w + 14 + 2w = 134$. Finally we obtain $6w + 14 = 134$. Were you able to write that equation?

Yes _____ No _____

If you answered No, remember that we must have double the width plus double the length equals the perimeter of 134.

If you answered Problem 5 incorrectly, rework it now using these suggestions.

---

**Chapter 3 Test, Problem 9**  Franco invested $4000 in money market funds. Part was invested at 14% interest, the rest at 11% interest. At the end of each year the fund company pays interest. After one year he earned $482 in simple interest. How much was invested at each interest rate?

> **HELPFUL HINT**  Let $x$ = the amount of one investment and then the $(\text{total} - x)$ = the amount of the other investment. Then multiply each investment by the respective interest rate.

Let us suppose that you picked $x$ as the amount of money that was invested at 14% interest. Did you realize that you should use $(4000 - x)$ to represent the amount of money invested at 11%?

Yes _____ No _____

If you answered No, remember that we want the $(\text{total} - x)$ to represent the second amount of money. The two amounts together have to add up to $4000. That is why we use $x$ for one amount and $(4000 - x)$ to represent the other amount.

Were you able to write the equation $0.14x + 0.11(4000 - x) = 482$? Can you simplify that equation to obtain $0.14x + 440 - 0.11x = 482$? Do you realize that this becomes the equation $0.03x + 440 = 482$?

Yes _____ No _____

If you answered No, remember that we must multiply $x$ by 14%, which is $0.14x$. We must multiply $(4000 - x)$ by 11%, which is $0.11(4000 - x)$. We must be very careful when we remove the parentheses to get the simplified equation.

If you answered Problem 9 incorrectly, rework it now using these suggestions.

Need more help? Watch the MATH COACH videos in MyMathLab® or on You Tube .

241

**Chapter 3 Test, Problem 15** How much would it cost to carpet the area shown in the figure if carpeting costs $12 per square yard?

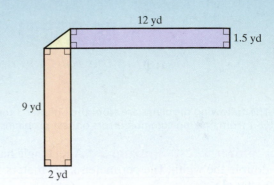

> **HELPFUL HINT** Consider that the figure consists of three separate parts. Two are rectangles and one is a triangle. Find the area of each separate part. Then add these smaller areas in order to obtain the total area.

Did you find the area of the top rectangle to be $12 \times 1.5 = 18$ square yards? Did you find the area of the bottom rectangle to be $9 \times 2 = 18$ square yards?

Yes _____ No _____

If you answered No, please go back and look at each rectangle separately. For each one you must multiply the length times the width. Be sure to label your answer as square yards.

Did you realize that the triangle has an altitude of 1.5 yards and a base of 2 yards?

Did you multiply $\left(\frac{1}{2}\right)(1.5)(2)$ and obtain an area for the triangle of 1.5 square yards?

Did you add up the three areas to obtain 37.5 square yards?

Yes _____ No _____

If you answered No, please go back and examine the dimensions of the triangle carefully. Although they are not labeled on the figure, they can be quickly determined by looking at the other side of each rectangle. Then use the area formula $A = \left(\frac{1}{2}\right)(ab)$. Then find the total of the three areas.

If you answered Problem 15 incorrectly, rework it now using these suggestions.

---

**Chapter 3 Test, Problem 17** Carol earns $15,000 plus 5% commission on sales over $10,000. How much must she make in sales to earn more than $20,000?

> **HELPFUL HINT** The commission income is based on how much above $10,000 she obtains in sales. So if we let $x$ = the amount of sales she needs, then she is only earning commission on $(x - 10,000)$ dollars.

Did you realize that the expression for how much she earns from commission sales is $0.05(x - 10,000)$, where $x$ is the amount of sales in dollars?

Yes _____ No _____

If you answered No, remember that we must find 5% of the amount of money that she obtains in sales that is above $10,000. Therefore we have the expression $0.05(x - 10,000)$.

Did you obtain the final inequality of her earnings to be $15,000 + 0.05(x - 10,000) > 20,000$ since she wants to earn more than $20,000?

Can you simplify this inequality to obtain $15,000 + 0.05x - 500 > 20,000$? Can you further simplify this to obtain $14,500 + 0.05x > 20,000$?

Yes _____ No _____

If you answered No, remember that we must add her yearly salary plus her commission salary and that this must be greater than $20,000.

If you answered Problem 17 incorrectly, rework it now using these suggestions.

Need more help? Look for section examples marked with MC to review.

# Cumulative Test for Chapters 0–3

*This test provides a comprehensive review of the key objectives for Chapters 0–3.*

1. Divide. $2.4\overline{)8.856}$

2. Add. $\dfrac{3}{8} + \dfrac{5}{12} + \dfrac{1}{2}$

3. Multiply. $\left(-\dfrac{2}{3}\right)\left(\dfrac{9}{14}\right)$

4. Divide. $4\dfrac{3}{8} \div \dfrac{1}{2}$

5. Simplify. $\dfrac{24}{42}$

6. Write $\dfrac{3}{5}$ as a decimal.

7. Simplify. $4(4x - y + 5) - 3(6x - 2y)$

8. Simplify. $3[x - 2y(x + 2y) - 3y^2]$

9. Evaluate $3x^2 - 7x - 11$ for $x = -3$.

10. Evaluate. $12 - 3(2 - 4) + 12 \div 4$

11. Solve for $b$. $H = \dfrac{1}{2}(3a + 5b)$

12. Solve for $x$ and graph your solution. $5x - 3 \le 2(4x + 1) + 4$

13. Solve for $y$. $\dfrac{2y}{3} - \dfrac{1}{4} = \dfrac{1}{6} + \dfrac{y}{4}$

14. The number of students in a literature class is 12 fewer than the number of students in sociology. The total enrollment for the two classes is 96 students. How many students are in each class?

▲ 15. A rectangle has a perimeter of 78 centimeters. The length of the rectangle is 11 centimeters longer than triple the width. Find the dimensions of the rectangle.

16. This year the sales for Mayflower Medical are 15% higher than last year. This year the sales were $265,000. What were last year's sales? Round your answer to the nearest dollar.

17. Hassan invested $7000 for one year. He invested part of it in a high-risk fund that pays 15% interest. He placed the rest in a safe, low-risk fund that pays 7% interest. At the end of one year he earned $730 in simple interest. How much did he invest in each of the two funds?

▲ 18. Find the volume of a sphere that has a radius of 3.00 inches. How much will the contents of the sphere weigh if it is filled with a liquid that weighs 1.50 pounds per cubic inch? Use $\pi \approx 3.14$ as an approximation.

1. _____

2. _____

3. _____

4. _____

5. _____

6. _____

7. _____

8. _____

9. _____

10. _____

11. _____

12. _____

13. _____

14. _____

15. _____

16. _____

17. _____

18. _____

243

# CHAPTER 4

# Exponents and Polynomials

## CAREER OPPORTUNITIES

### Biotechnology Professionals

The biotechnology industry features a wide range of careers. Using the most recent advances made in biology and technology, such jobs deal with important issues in the areas of pharmaceuticals, farming, food production, genetics, disease prevention, and the environment, to name just a few. The mathematics seen in this chapter is used by biotechnology professionals in the research, development, and problem solving that lead to better, safer products and services for the entire world.

To investigate how the mathematics in this chapter can help with this field, see the Career Exploration Problems on page **288**.

# 4.1  The Rules of Exponents ▶

## 1  Using the Product Rule to Multiply Exponential Expressions with Like Bases ▶

Recall that $x^2$ means $x \cdot x$. That is, $x$ appears as a factor two times. The 2 is called the **exponent.** The **base** is the variable $x$. The expression $x^2$ is called an **exponential expression.** What happens when we multiply $x^2 \cdot x^2$? Is there a pattern that will help us form a general rule?

$$\overbrace{(2^2)(2^3) = (2 \cdot 2)(2 \cdot 2 \cdot 2)}^{5 \text{ twos}} = 2^5$$

The exponent means 2 occurs 5 times as a factor.

$$\overbrace{(3^3)(3^4) = (3 \cdot 3 \cdot 3)(3 \cdot 3 \cdot 3 \cdot 3)}^{7 \text{ threes}} = 3^7$$

Notice that $3 + 4 = 7$.

$$\overbrace{(x^3)(x^5) = (x \cdot x \cdot x)(x \cdot x \cdot x \cdot x \cdot x)}^{8 \ x\text{'s}} = x^8$$

The sum of the exponents is $3 + 5 = 8$.

$$\overbrace{(y^4)(y^2) = (y \cdot y \cdot y \cdot y)(y \cdot y)}^{6 \ y\text{'s}} = y^6$$

The sum of the exponents is $4 + 2 = 6$.

We can state the pattern in words and then use variables.

---

**THE PRODUCT RULE**

To multiply two exponential expressions that have the same base, keep the base and *add the exponents.*

$$x^a \cdot x^b = x^{a+b}$$

---

Be sure to notice that this rule applies only to expressions that have the *same base.* Here $x$ represents the base, while the letters $a$ and $b$ represent the exponents that are added.

It is important that you apply this rule even when an exponent is 1. Every variable that does not have a written exponent is understood to have an exponent of 1. Thus $x = x^1$, $y = y^1$, and so on.

**Example 1** Multiply.   **(a)** $x^3 \cdot x^6$   **(b)** $m \cdot m^5$

**Solution**

**(a)** $x^3 \cdot x^6 = x^{3+6} = x^9$

**(b)** $m \cdot m^5 = m^{1+5} = m^6$   Note that the exponent of the first $x$ is 1.  ☐

 **Student Practice 1**  Multiply.   **(a)** $a^7 \cdot a^5$   **(b)** $w^{10} \cdot w$

**Example 2** Simplify, if possible.   **(a)** $y^5 \cdot y^{11}$   **(b)** $2^3 \cdot 2^5$   **(c)** $x^6 \cdot y^8$

**Solution**

**(a)** $y^5 \cdot y^{11} = y^{5+11} = y^{16}$

**(b)** $2^3 \cdot 2^5 = 2^{3+5} = 2^8$   Note that the base does not change! Only the exponent changes.

**(c)** $x^6 \cdot y^8$   The rule for multiplying exponential expressions does not apply since the bases are not the same. This cannot be simplified.  ☐

**Student Practice 2**  Simplify, if possible.

**(a)** $x^3 \cdot x^9$   **(b)** $3^7 \cdot 3^4$   **(c)** $a^3 \cdot b^2$

We can now look at multiplying expressions such as $(2x^5)(3x^6)$.

The number 2 in $2x^5$ is called the **numerical coefficient.** Recall that a numerical coefficient is a number that is multiplied by a variable. When we multiply two expressions such as $2x^5$ and $3x^6$, we first multiply the numerical coefficients; we multiply the variables with exponents separately.

As you do the following problems, keep in mind the rule for multiplying expressions with exponents and the rules for multiplying signed numbers.

**Example 3** Multiply.

**(a)** $(2x^5)(3x^6)$      **(b)** $(-5x^3)(x^6)$      **(c)** $(-6x)(-4x^5)$

**Solution**

**(a)** $(2x^5)(3x^6) = (2 \cdot 3)(x^5 \cdot x^6)$

$\qquad\qquad\quad = 6(x^5 \cdot x^6)$     Multiply the numerical coefficients.

$\qquad\qquad\quad = 6x^{11}$     Use the rule for multiplying expressions with exponents. Add the exponents.

**(b)** Every variable that does not have a visible numerical coefficient is understood to have a numerical coefficient of 1. Thus $x^6$ has a numerical coefficient of 1.

$$(-5x^3)(x^6) = (-5x^3)(1x^6) = (-5 \cdot 1)(x^3 \cdot x^6) = -5x^9$$

**(c)** $(-6x)(-4x^5) = (-6)(-4)(x^1 \cdot x^5) = 24x^6$    Remember that $x$ has an exponent of 1.

 **Student Practice 3**   Multiply.

**(a)** $(7a^8)(a^4)$      **(b)** $(3y^2)(-2y^3)$      **(c)** $(-4x^3)(-5x^2)$

Problems of this type may involve more than one variable or more than two factors.

**Example 4** Multiply. $(5ab)\left(-\frac{1}{3}a\right)(9b^2)$

**Solution**   $(5ab)\left(-\frac{1}{3}a\right)(9b^2) = (5)\left(-\frac{1}{3}\right)(9)(a \cdot a)(b \cdot b^2)$

$\qquad\qquad\qquad\qquad\qquad\quad = -15a^2b^3$

 **Student Practice 4**   Multiply.

$$(2xy)\left(-\tfrac{1}{4}x^2y\right)(6xy^3)$$

### 2   Using the Quotient Rule to Divide Exponential Expressions with Like Bases ▶

Frequently, we must divide exponential expressions. Because division by zero is undefined, in all problems in this chapter we assume that the denominator of any variable expression is not zero. We'll look at division in three separate parts.

Suppose that we want to simplify $x^5 \div x^2$. We could do the division the long way.

$$\frac{x^5}{x^2} = \frac{(x)(x)(x)\cancel{(x)}\cancel{(x)}}{\cancel{(x)}\cancel{(x)}} = x^3$$

Here we are using the arithmetical property of reducing fractions (see Section 0.1). When the same factor appears in both numerator and denominator, that factor can be removed.

A simpler way is to *subtract the exponents.* Notice that the base remains the same.

**THE QUOTIENT RULE (PART 1)**

$\dfrac{x^a}{x^b} = x^{a-b}$ Use this form if the larger exponent is in the numerator and $x \neq 0$.

**Example 5** Divide.

(a) $\dfrac{2^{16}}{2^{11}}$ 
(b) $\dfrac{x^5}{x^3}$ 
(c) $\dfrac{y^{16}}{y^7}$

**Solution**

(a) $\dfrac{2^{16}}{2^{11}} = 2^{16-11} = 2^5$ Note that the base does *not* change.

(b) $\dfrac{x^5}{x^3} = x^{5-3} = x^2$ 
(c) $\dfrac{y^{16}}{y^7} = y^{16-7} = y^9$  □

▶ **Student Practice 5** Divide.

(a) $\dfrac{10^{13}}{10^7}$ 
(b) $\dfrac{x^{11}}{x}$ 
(c) $\dfrac{y^{18}}{y^8}$

Now we consider the situation where the larger exponent is in the denominator. Suppose that we want to simplify $x^2 \div x^5$.

$$\frac{x^2}{x^5} = \frac{\cancel{(x)}\cancel{(x)}}{\cancel{(x)}\cancel{(x)}(x)(x)(x)} = \frac{1}{x^3}$$

**THE QUOTIENT RULE (PART 2)**

$\dfrac{x^a}{x^b} = \dfrac{1}{x^{b-a}}$ Use this form if the larger exponent is in the denominator and $x \neq 0$.

**Example 6** Divide.

(a) $\dfrac{12^{17}}{12^{20}}$ 
(b) $\dfrac{b^7}{b^9}$ 
(c) $\dfrac{x^{20}}{x^{24}}$

**Solution**

(a) $\dfrac{12^{17}}{12^{20}} = \dfrac{1}{12^{20-17}} = \dfrac{1}{12^3}$ Note that the base does *not* change.

(b) $\dfrac{b^7}{b^9} = \dfrac{1}{b^{9-7}} = \dfrac{1}{b^2}$ 
(c) $\dfrac{x^{20}}{x^{24}} = \dfrac{1}{x^{24-20}} = \dfrac{1}{x^4}$  □

▶ **Student Practice 6** Divide.

(a) $\dfrac{c^3}{c^4}$ 
(b) $\dfrac{10^{31}}{10^{56}}$ 
(c) $\dfrac{z^{15}}{z^{21}}$

When there are numerical coefficients, use the rules for dividing signed numbers to reduce fractions to lowest terms.

**Example 7** Divide.

(a) $\dfrac{5x^5}{25x^7}$ 　　　　(b) $\dfrac{-12y^8}{4y^3}$ 　　　　(c) $\dfrac{-16x^7}{-24x^8}$

**Solution**

(a) $\dfrac{5x^5}{25x^7} = \dfrac{1}{5x^{7-5}} = \dfrac{1}{5x^2}$ 　　　　(b) $\dfrac{-12y^8}{4y^3} = -3y^{8-3} = -3y^5$

(c) $\dfrac{-16x^7}{-24x^8} = \dfrac{2}{3x^{8-7}} = \dfrac{2}{3x}$

**Student Practice 7** Divide.

(a) $\dfrac{-7x^7}{-21x^9}$ 　　　　(b) $\dfrac{15x^{11}}{-3x^4}$ 　　　　(c) $\dfrac{23b^8}{46b^9}$

You have to work very carefully if two or more variables are involved. Treat the coefficients and each variable separately.

**Example 8** Divide.

(a) $\dfrac{a^3b^2}{5ab^6}$ 　　　　(b) $\dfrac{-3x^2y^5}{12x^6y^8}$

**Solution**

(a) $\dfrac{a^3b^2}{5ab^6} = \dfrac{a^2}{5b^4}$ 　　　　(b) $\dfrac{-3x^2y^5}{12x^6y^8} = -\dfrac{1}{4x^4y^3}$

**Student Practice 8** Divide.

(a) $\dfrac{r^7s^9}{s^{10}}$ 　　　　(b) $\dfrac{12x^5y^6}{-24x^3y^8}$

Suppose that a given base appears with the same exponent in the numerator and denominator of a fraction. In this case we can use the fact that *any nonzero number divided by itself is* 1.

**Example 9** Divide.

(a) $\dfrac{7^6}{7^6}$ 　　　　(b) $\dfrac{3x^5}{x^5}$

**Solution**

(a) $\dfrac{7^6}{7^6} = 1$ 　　　　(b) $\dfrac{3x^5}{x^5} = 3\left(\dfrac{x^5}{x^5}\right) = 3(1) = 3$

**Student Practice 9** Divide.

(a) $\dfrac{10^7}{10^7}$ 　　　　(b) $\dfrac{12a^4}{2a^4}$

Do you see that if we had subtracted exponents when simplifying $\dfrac{x^6}{x^6}$ we would have obtained $x^0$ in Example 9? So we can surmise that any number (except 0) to the 0 power equals 1. We can write this fact as a separate rule.

**THE QUOTIENT RULE (PART 3)**

$$\frac{x^a}{x^a} = x^0 = 1 \quad \text{if } x \neq 0 \quad (0^0 \text{ remains undefined}).$$

**To Think About:** *What Is 0 to the 0 Power?* What about $0^0$? Why is it undefined? $0^0 = 0^{1-1}$. If we use the quotient rule, $0^{1-1} = \frac{0}{0}$. Since division by zero is undefined, we must agree that $0^0$ is undefined.

**Example 10** Divide.

(a) $\dfrac{4x^0y^2}{8^0y^5z^3}$ 　　　　　　　　　　　(b) $\dfrac{5b^2c}{10b^2c^3}$

**Solution**

(a) $\dfrac{4x^0y^2}{8^0y^5z^3} = \dfrac{4(1)y^2}{(1)y^5z^3} = \dfrac{4y^2}{y^5z^3} = \dfrac{4}{y^3z^3}$ 　　(b) $\dfrac{5b^2c}{10b^2c^3} = \dfrac{1b^0}{2c^2} = \dfrac{(1)(1)}{2c^2} = \dfrac{1}{2c^2}$ □

**⇨ Student Practice 10** Divide.

(a) $\dfrac{-20a^3b^8c^4}{28a^3b^7c^5}$ 　　　　　　　　　(b) $\dfrac{5x^0y^6}{10x^4y^8}$

We can combine all three parts of the quotient rule we have developed.

**THE QUOTIENT RULE**

$\dfrac{x^a}{x^b} = x^{a-b}$ 　　Use this form if the larger exponent is in the numerator and $x \neq 0$.

$\dfrac{x^a}{x^b} = \dfrac{1}{x^{b-a}}$ 　　Use this form if the larger exponent is in the denominator and $x \neq 0$.

$\dfrac{x^a}{x^a} = x^0 = 1$ 　　if $x \neq 0$.

We can combine the product rule and the quotient rule to simplify algebraic expressions that involve both multiplication and division.

**Example 11** Simplify. $\dfrac{(8x^2y)(-3x^3y^2)}{-6x^4y^3}$

**Solution**

$$\frac{(8x^2y)(-3x^3y^2)}{-6x^4y^3} = \frac{-24x^5y^3}{-6x^4y^3} = 4x \qquad □$$

**⇨ Student Practice 11** Simplify.

$$\frac{(-6ab^5)(3a^2b^4)}{16a^5b^7}$$

## 3 Raising Exponential Expressions to a Power

How do we simplify an expression such as $(x^4)^3$? $(x^4)^3$ is $x^4$ raised to the third power. For this type of problem we say that we are raising a power to a power. A problem such as $(x^4)^3$ could be done by writing the following.

$$(x^4)^3 = x^4 \cdot x^4 \cdot x^4 \quad \text{By definition}$$
$$= x^{12} \quad \text{By adding exponents}$$

Notice that when we add the exponents we get $4 + 4 + 4 = 12$. This is the same as multiplying 4 by 3. That is, $4 \cdot 3 = 12$. This process can be summarized by the following rule.

---

### RAISING A POWER TO A POWER

To raise a power to a power, keep the same base and multiply the exponents.

$$(x^a)^b = x^{ab}$$

---

Recall what happens when you raise a negative number to a power. $(-1)^2 = 1$. $(-1)^3 = -1$. In general,

$$(-1)^n = \begin{cases} +1 & \text{if } n \text{ is even} \\ -1 & \text{if } n \text{ is odd.} \end{cases}$$

**Example 12** Simplify.

**(a)** $(x^3)^5$ **(b)** $(2^7)^3$ **(c)** $(-1)^8$

**Solution**

**(a)** $(x^3)^5 = x^{3 \cdot 5} = x^{15}$ **(b)** $(2^7)^3 = 2^{7 \cdot 3} = 2^{21}$ **(c)** $(-1)^8 = +1$

Note that in both parts **(a)** and **(b)** the base does not change. ☐

**Student Practice 12** Simplify.

**(a)** $(a^4)^3$ **(b)** $(10^5)^2$ **(c)** $(-1)^{15}$

Here are two rules involving products and quotients that are very useful: the product raised to a power rule and the quotient raised to a power rule. We'll illustrate each with an example.

If a product in parentheses is raised to a power, the parentheses indicate that *each factor* must be raised to that power.

$$(xy)^2 = x^2 y^2 \qquad (xy)^3 = x^3 y^3$$

---

### PRODUCT RAISED TO A POWER

$$(xy)^a = x^a y^a$$

---

**Example 13** Simplify.

**(a)** $(ab)^8$  **(b)** $(3x)^4$  **(c)** $(-2x^2)^3$

**Solution**

**(a)** $(ab)^8 = a^8 b^8$  **(b)** $(3x)^4 = 3^4 x^4 = 81x^4$

**(c)** $(-2x^2)^3 = (-2)^3 \cdot (x^2)^3 = -8x^6$ ☐

**Student Practice 13**  Simplify.

**(a)** $(3xy)^3$  **(b)** $(yz)^{37}$  **(c)** $(-3x^3)^2$

If a fractional expression within parentheses is raised to a power, the parentheses indicate that both numerator and denominator must be raised to that power.

$$\left(\frac{x}{y}\right)^5 = \frac{x^5}{y^5} \qquad \left(\frac{x}{y}\right)^2 = \frac{x^2}{y^2} \qquad \text{if } y \neq 0$$

**QUOTIENT RAISED TO A POWER**

$$\left(\frac{x}{y}\right)^a = \frac{x^a}{y^a} \qquad \text{if } y \neq 0$$

**Example 14** Simplify.

**(a)** $\left(\dfrac{x}{y}\right)^5$  **(b)** $\dfrac{(7w^2)^3}{(2w)^4}$

**Solution**

**(a)** $\left(\dfrac{x}{y}\right)^5 = \dfrac{x^5}{y^5}$  **(b)** $\dfrac{(7w^2)^3}{(2w)^4} = \dfrac{7^3 w^6}{2^4 w^4} = \dfrac{343 w^6}{16 w^4} = \dfrac{343 w^2}{16}$ ☐

**Student Practice 14**  Simplify.

**(a)** $\left(\dfrac{x}{5}\right)^3$  **(b)** $\dfrac{(4a)^2}{(ab)^6}$

Many expressions can be simplified by using the previous rules involving exponents. Be sure to take particular care to determine the correct sign, especially if there is a negative numerical coefficient.

**Example 15** Simplify. $\left(\dfrac{-3x^2 z^0}{y^3}\right)^4$

**Solution**

$$\left(\frac{-3x^2 z^0}{y^3}\right)^4 = \left(\frac{-3x^2}{y^3}\right)^4 \qquad \text{Simplify inside the parentheses first. Note that } z^0 = 1.$$

$$= \frac{(-3)^4 x^8}{y^{12}} \qquad \text{Apply the rules for raising a power to a power. Notice that we wrote } (-3)^4 \text{ and not } -3^4. \text{ We are raising } -3 \text{ to the fourth power.}$$

$$= \frac{81 x^8}{y^{12}} \qquad \text{Simplify the coefficient: } (-3)^4 = +81. \qquad ☐$$

**Student Practice 15**  Simplify.

$$\left(\frac{-2x^3 y^0 z}{4xz^2}\right)^5$$

We list here the rules of exponents we have discussed in Section 4.1.

$$x^a \cdot x^b = x^{a+b}$$

$$\frac{x^a}{x^b} = \begin{cases} x^{a-b} & \text{if} \quad a > b \quad x \neq 0 \\ \dfrac{1}{x^{b-a}} & \text{if} \quad b > a \quad x \neq 0 \\ x^0 = 1 & \text{if} \quad a = b \quad x \neq 0 \end{cases}$$

$$(x^a)^b = x^{ab}$$

$$(xy)^a = x^a y^a$$

$$\left(\frac{x}{y}\right)^a = \frac{x^a}{y^a} \quad y \neq 0$$

## STEPS TO SUCCESS  Why Does Reviewing Make Such a Big Difference?

Students are often amazed that reviewing makes such a huge difference in helping them to learn. It is one of the most powerful tools that a math student can use.

Mathematics involves learning concepts one step at a time. Then the concepts are put together in a chapter. At the end of the chapter you need to know each of these concepts. Therefore, to succeed in each chapter you need to be able to put together all the pieces of each chapter. Reviewing each section and reviewing at the end of each chapter are the amazing tools that help you to master the mathematical concepts.

As you review, if you find you cannot work out a problem, be sure to study the examples and the Student Practice problems very carefully. Then things will become more clear.

**Making it personal:** Start with this section. Do each Cumulative Review Problem at the end of the chapter. Check your answer for each one in the back of the book. If you miss any problem, then go back to the appropriate section of the book for help.

For example if you miss problem 97 in Section 4.1, notice that it is coded [1.3.3]. This means to go back to Section 1.3 of the book and look at Objective 3. There you will find similar examples that explain this kind of problem. Which of these suggestions do you find most helpful? Write out your plan and begin using it today. ▼

## (4.1) Exercises MyMathLab®

**Verbal and Writing Skills, Exercises 1–6**

1. Write in your own words the product rule for exponents.

2. To be able to use the rules of exponents, what must be true of the bases?

3. If the larger exponent is in the denominator, the quotient rule states that $\dfrac{x^a}{x^b} = \dfrac{1}{x^{b-a}}$ if $x \neq 0$. Provide an example to show why this is true.

*In exercises 4 and 5, identify the numerical coefficient, the base(s), and the exponent(s).*

4. $-5xy^3$

5. $6x^{11}y$

6. Evaluate **(a)** $3x^0$ and **(b)** $(3x)^0$. **(c)** Why are the results different?

*Write in simplest exponent form.*

7. $2 \cdot 2 \cdot a \cdot a \cdot a \cdot b$

8. $6 \cdot x \cdot x \cdot x \cdot x \cdot y$

9. $(-5)(x)(y)(z)(y)(x)(x)(z)$

10. $(-4)(b)(b)(a)(b)(c)(c)(b)(c)$

*Multiply. Leave your answer in exponent form.*

11. $(7^4)(7^6)$

12. $(3^2)(3^3)$

13. $(8^9)(8^{12})$

14. $(3^7)(3^8)$

15. $x^4 \cdot x^8$

16. $a^8 \cdot a^{12}$

17. $t^{15} \cdot t$

18. $w^{18} \cdot w$

*Multiply.*

19. $-5x^4(4x^2)$

20. $6x^2(-9x^3)$

21. $(5x)(10x^2)$

22. $(-2x^3)(-5x)$

23. $(2xy^3)(9x^2y^5)$

24. $(-3a^2b)(7ab^4)$

25. $\left(\dfrac{2}{5}xy^3\right)\left(\dfrac{1}{3}x^2y^2\right)$

26. $\left(\dfrac{4}{5}x^5y\right)\left(\dfrac{15}{16}x^2y^4\right)$

27. $(1.1x^2z)(-2.5xy)$

28. $(2.3wx^4)(-3.5xy^4)$

29. $(8a)(2a^3b)(0)$

30. $(5ab)(2a^2)(0)$

31. $(-16x^2y^4)(-5xy^3)$

32. $(-12x^4y)(-7x^5y^3)$

33. $(-8x^3y^2)(3xy^5)$

34. $(9x^2y^6)(-11x^3y^3)$

**35.** $(-2x^3y^2)(0)(-3x^4y)$ **36.** $(-4x^8y^2)(13y^3)(0)$ **37.** $(8a^4b^3)(-3x^2y^5)$ **38.** $(-4wz^4)(-9x^2y^3)$

**39.** $(2x^2y)(-3y^3z^2)(5xz^4)$ **40.** $(3ab)(5a^2c)(-2b^2c^3)$

*Divide. Leave your answer in exponent form. Assume that all variables in any denominator are nonzero.*

**41.** $\dfrac{y^{12}}{y^5}$ **42.** $\dfrac{x^{11}}{x^4}$ **43.** $\dfrac{y^5}{y^8}$ **44.** $\dfrac{b^{20}}{b^{23}}$

**45.** $\dfrac{11^{18}}{11^{30}}$ **46.** $\dfrac{8^9}{8^{12}}$ **47.** $\dfrac{2^{17}}{2^{10}}$ **48.** $\dfrac{7^{18}}{7^9}$

**49.** $\dfrac{a^{13}}{4a^5}$ **50.** $\dfrac{b^{16}}{5b^{13}}$ **51.** $\dfrac{x^7}{y^9}$ **52.** $\dfrac{x^5}{y^4}$

**53.** $\dfrac{48x^5y^3}{24xy^3}$ **54.** $\dfrac{45a^4b^3}{15a^4b^2}$ **55.** $\dfrac{16x^5y}{-32x^2y^3}$ **56.** $\dfrac{-36x^3y^7}{72x^5y}$

**57.** $\dfrac{1.8f^4g^3}{54f^2g^8}$ **58.** $\dfrac{3.1s^5t^3}{62s^8t}$ **59.** $\dfrac{(-17x^5y^4)(5y^6)}{-5xy^7}$ **60.** $\dfrac{(-6xy)(10x^5y^2)}{-4x^6y}$

**61.** $\dfrac{8^0x^2y^3}{16x^5y}$ **62.** $\dfrac{2^3x^5y^3}{2^0x^3y^7}$ **63.** $\dfrac{18a^6b^3c^0}{24a^5b^3}$ **64.** $\dfrac{12a^7b^8}{16a^3b^8c^0}$

*Simplify.*

**65.** $(x^2)^6$ **66.** $(w^5)^8$ **67.** $(x^3y)^5$ **68.** $(ab^3)^4$

**69.** $(rs^2)^6$ **70.** $(m^3n^2)^5$ **71.** $(3a^3b^2c)^3$ **72.** $(2x^4yz^3)^2$

**73.** $(-3a^4)^2$ **74.** $(-2a^5)^4$ **75.** $\left(\dfrac{x}{2m^4}\right)^7$ **76.** $\left(\dfrac{p^5}{6x}\right)^5$

**77.** $\left(\dfrac{5x}{7y^2}\right)^2$

**78.** $\left(\dfrac{3b^3}{2a^4}\right)^3$

**79.** $(-3a^2b^3c^0)^4$

**80.** $(-a^3b^0c^4)^5$

**81.** $(-2x^3y^0z)^3$

**82.** $(-4xy^0z^4)^3$

**83.** $\dfrac{(3x)^5}{(3x^2)^3}$

**84.** $\dfrac{(4y)^4}{(4y^5)^2}$

## Mixed Practice

**85.** $(-5a^2b^3)^2(ab)$

**86.** $(-2ab^3)^5(a^2b)$

**87.** $\left(\dfrac{7}{a^5}\right)^2$

**88.** $\left(\dfrac{4}{x^6}\right)^3$

**89.** $\left(\dfrac{2x}{y^3}\right)^4$

**90.** $\dfrac{18a^3}{(3a)(2b)}$

**91.** $\dfrac{(10ac^3)(7a)}{40b}$

**92.** $\dfrac{7x^3y^6}{35x^4y^8}$

**93.** $\dfrac{11x^7y^2}{33x^8y^3}$

## Cumulative Review *Simplify.*

**94.** **[1.2.1]** $-3-8$

**95.** **[1.1.5]** $-17+(-32)+(-24)+27$

**96.** **[1.3.1]** $\left(-\dfrac{3}{5}\right)\left(-\dfrac{2}{15}\right)$

**97.** **[1.3.3]** $-\dfrac{5}{4}\div\dfrac{5}{16}$

*Amazon Rainforest In 2012, the size of the Amazon rainforest was 7,760,000 km². Over half the rainforest, 4,966,400 km², lies in Brazil. (Source: rainforests.mongabay.com) Round your answers to the nearest tenth of a percent.*

**98.** **[0.5.3]** What percent of the Amazon rainforest lay in Brazil in 2012?

**99.** **[0.5.3]** In 2013, 7500 km² of the Amazon rainforest in Brazil were lost. This number would have been much higher if conservation efforts were not in place to slow the deforestation rate. Through these efforts, the goal is to lose at most 5250 km² of the rainforest for each of the years 2014, 2015, 2016, and 2017. What percent of the Amazon rainforest in Brazil will be lost from 2013 through 2017?

Quick Quiz 4.1 *Simplify the following.*

**1.** $(2x^2y^3)(-5xy^4)$

**2.** $\dfrac{-28x^6y^6}{35x^3y^8}$

**3.** $(-3x^3y^5)^4$

**4.** **Concept Check** Explain the steps you would need to follow to simplify the expression.

$$\dfrac{(4x^3)^2}{(2x^4)^3}$$

# 4.2 Negative Exponents and Scientific Notation

**Student Learning Objectives**

After studying this section, you will be able to:

1 Use negative exponents.

2 Use scientific notation.

## 1 Using Negative Exponents

If $n$ is an integer and $x \neq 0$, then $x^{-n}$ is defined as follows:

> **DEFINITION OF A NEGATIVE EXPONENT**
>
> $$x^{-n} = \frac{1}{x^n} \quad \text{where } x \neq 0$$

**Example 1** Write with positive exponents.

(a) $y^{-3}$     (b) $z^{-6}$     (c) $w^{-1}$

**Solution**

(a) $y^{-3} = \dfrac{1}{y^3}$     (b) $z^{-6} = \dfrac{1}{z^6}$     (c) $w^{-1} = \dfrac{1}{w^1} = \dfrac{1}{w}$

**Student Practice 1** Write with positive exponents.

(a) $x^{-12}$     (b) $w^{-5}$     (c) $z^{-2}$

To evaluate a numerical expression with a negative exponent, first write the expression with a positive exponent. Then simplify.

**Example 2** Evaluate.

(a) $3^{-2}$     (b) $2^{-5}$

**Solution**

(a) $3^{-2} = \dfrac{1}{3^2} = \dfrac{1}{9}$     (b) $2^{-5} = \dfrac{1}{2^5} = \dfrac{1}{32}$

**Student Practice 2** Evaluate.

(a) $4^{-3}$     (b) $2^{-4}$

All the previously studied laws of exponents are true for any integer exponent. These laws are summarized in the following box.

> **LAWS OF EXPONENTS WHERE X, Y, $\neq$ 0**
>
> **The Product Rule**
> $$x^a \cdot x^b = x^{a+b}$$
>
> **The Quotient Rule**
> $$\frac{x^a}{x^b} = x^{a-b} \quad \text{Use if } a > b. \qquad \frac{x^a}{x^b} = \frac{1}{x^{b-a}} \quad \text{Use if } a < b.$$
>
> **Power Rules**
> $$(xy)^a = x^a y^a, \qquad (x^a)^b = x^{ab}, \qquad \left(\frac{x}{y}\right)^a = \frac{x^a}{y^a}$$

By using the definition of a negative exponent and the properties of fractions, we can derive two more helpful properties of exponents.

---

**PROPERTIES OF NEGATIVE EXPONENTS WHERE $X, Y, \neq 0$**

$$\frac{1}{x^{-n}} = x^n \qquad \frac{x^{-m}}{y^{-n}} = \frac{y^n}{x^m}$$

---

**Example 3** Simplify. Write the expression with no negative exponents.

(a) $\dfrac{1}{x^{-6}}$  (b) $\dfrac{x^{-3}y^{-2}}{z^{-4}}$  (c) $x^{-2}y^3$

**Solution**

(a) $\dfrac{1}{x^{-6}} = x^6$  (b) $\dfrac{x^{-3}y^{-2}}{z^{-4}} = \dfrac{z^4}{x^3 y^2}$  (c) $x^{-2}y^3 = \dfrac{y^3}{x^2}$  □

---

**Student Practice 3** Simplify. Write the expression with no negative exponents.

(a) $\dfrac{3}{w^{-4}}$  (b) $\dfrac{x^{-6}y^4}{z^{-2}}$  (c) $x^{-6}y^{-5}$

---

**Example 4** Simplify. Write the expression with no negative exponents.

(a) $(3x^{-4}y^2)^{-3}$  (b) $\dfrac{x^2 y^{-4}}{x^{-5}y^3}$

**Solution**

(a) $(3x^{-4}y^2)^{-3} = 3^{-3}x^{12}y^{-6}$

We use the power to a power rule:
$3^{1(-3)} = 3^{-3}; x^{-4(-3)} = x^{12}; y^{2(-3)} = y^{-6}.$

$= \dfrac{x^{12}}{3^3 y^6} = \dfrac{x^{12}}{27y^6}$

We rewrite the expression so that only positive exponents appear, then simplify.

(b) $\dfrac{x^2 y^{-4}}{x^{-5}y^3} = \dfrac{x^2 x^5}{y^4 y^3} = \dfrac{x^7}{y^7}$

First rewrite the expression so that only positive exponents appear. Then simplify using the product rule.  □

**Student Practice 4** Simplify. Write the expression with no negative exponents.

(a) $(2x^4 y^{-5})^{-2}$  (b) $\dfrac{y^{-3}z^{-4}}{y^2 z^{-6}}$

---

**2 Using Scientific Notation** ▶

One common use of negative exponents is in writing numbers in scientific notation. Scientific notation is most useful in expressing very large and very small numbers.

---

**SCIENTIFIC NOTATION**

A positive number is written in **scientific notation** if it is in the form $a \times 10^n$, where $1 \le a < 10$ and $n$ is an integer.

---

**Example 5** Write in scientific notation.

(a) 4567 　　　　　　　　　　(b) 157,000,000

**Solution**

(a) $4567 = 4.567 \times 1000$ 　To change 4567 to a number that is greater than or equal to 1 but less than 10, we move the decimal point *three* places to the *left*. We must then multiply the number by a power of 10 so that we do not change the value of the number. Use 1000.

$$= 4.567 \times 10^3$$

Notice we moved the decimal point **3** places to the left, so we must multiply by $10^3$.

(b) $157,000,000 = 1.57000000 \times 100000000$

　　　　　　　　　　　　　8 places　　　　8 zeros

$$= 1.57 \times 10^8$$

 **Student Practice 5** Write in scientific notation.

(a) 78,200 　　　　　　　　　　(b) 4,786,000

Numbers that are smaller than **1** will have a *negative power* of 10 if they are written in scientific notation.

**Example 6** Write in scientific notation.

(a) 0.061 　　　　　　　　　(b) 0.000052

**Solution**

(a) We need to write 0.061 as a number that is greater than or equal to 1 but less than 10. In which direction do we move the decimal point?

$0.061 = 6.1 \times 10^{-2}$ 　Move the decimal point 2 places to the *right*.

(b) $0.000052 = 5.2 \times 10^{-5}$ 　Why?

**Student Practice 6** Write in scientific notation.

(a) 0.98 　　　　　　　　　(b) 0.000092

The reverse procedure transforms scientific notation into ordinary decimal notation.

**Example 7** Write in decimal notation.

(a) $1.568 \times 10^2$ 　　　　　　　(b) $7.432 \times 10^{-3}$

**Solution**

(a) $1.568 \times 10^2 = 1.568 \times 100 = 156.8$

**Alternative Method**

$1.568 \times 10^2 = 156.8$ 　The exponent 2 tells us to move the decimal point 2 places to the right.

(b) $7.432 \times 10^{-3} = 7.432 \times \dfrac{1}{1000} = 0.007432$

**Alternative Method**

$7.432 \times 10^{-3} = 0.007432$ 　The exponent $-3$ tells us to move the decimal point 3 places to the left.

**Student Practice 7** Write in decimal notation.

(a) $1.93 \times 10^6$ 　　　　　　　(b) $8.562 \times 10^{-5}$

---

## Calculator

### Scientific Notation

Most scientific calculators can display only eight digits at one time. Numbers with more than eight digits are usually shown in scientific notation. 1.12 E 08 or 1.12 8 means $1.12 \times 10^8$. You can use a calculator to compute with large numbers by entering the numbers using scientific notation. For example,

$$(7.48 \times 10^{24}) \times (3.5 \times 10^8)$$

is entered as follows.

7.48 $\boxed{\text{EXP}}$ 24 $\boxed{\times}$

3.5 $\boxed{\text{EXP}}$ 8 $\boxed{=}$

Display: $\boxed{\text{2.618 E 33}}$

or $\boxed{\text{2.618　33}}$

Note: Some calculators have an $\boxed{\text{EE}}$ key instead of $\boxed{\text{EXP}}$

The distance light travels in one year is called a *light-year*. A light-year is a convenient unit of measure to use when investigating the distances between stars.

**Example 8** A light-year is a distance of 9,460,000,000,000,000 meters. Write this number in scientific notation.

**Solution** 9,460,000,000,000,000 meters $= 9.46 \times 10^{15}$ meters                    □

**Student Practice 8** Astronomers measure distances to faraway galaxies in parsecs. A parsec is a distance of 30,900,000,000,000,000 meters. Write this number in scientific notation.

To perform a calculation involving very large or very small numbers, it is usually helpful to write the numbers in scientific notation and then use the laws of exponents to do the calculation.

**Example 9** Use scientific notation and the laws of exponents to find the following. Leave your answer in scientific notation.

**(a)** $(32,000,000)(1,500,000,000,000)$          **(b)** $\dfrac{0.00063}{0.021}$

**Solution**

**(a)** $(32,000,000)(3,500,000,000,000)$

$= (3.2 \times 10^7)(3.5 \times 10^{12})$     Write each number in scientific notation.

$= 3.2 \times 3.5 \times 10^7 \times 10^{12}$     Rearrange the order. Remember that multiplication is commutative.

$= 11.2 \times 10^{19}$     Multiply $3.2 \times 3.5$. Multiply $10^7 \times 10^{12}$.

$= 1.12 \times 10^{20}$     Rewrite 11.2 in scientific notation and combine powers of 10.

**(b)** $\dfrac{0.00063}{0.021} = \dfrac{6.3 \times 10^{-4}}{2.1 \times 10^{-2}}$     Write each number in scientific notation.

$= \dfrac{6.3}{2.1} \times \dfrac{10^{-4}}{10^{-2}}$     Rearrange the order. We are actually using the definition of multiplication of fractions.

$= \dfrac{6.3}{2.1} \times \dfrac{10^2}{10^4}$     Rewrite with positive exponents.

$= 3.0 \times 10^{-2}$                    □

**Student Practice 9** Use scientific notation and the laws of exponents to find the following. Leave your answer in scientific notation.

**(a)** $(56,000)(1,400,000,000)$          **(b)** $\dfrac{0.000111}{0.00000037}$

When we use scientific notation, we are often writing approximate numbers. We must include some zeros so that the decimal point can be properly located. However, all other digits except for these zeros are considered **significant digits.** The number 34.56 has four significant digits. The number 0.0049 has two significant digits. The zeros are considered placeholders. The number 634,000 has three significant digits (unless we have specific knowledge to the contrary). The zeros are considered placeholders. We sometimes round numbers to a specific number of significant digits. For example, 0.08746 rounded to two significant digits is 0.087. When we round 1,348,593 to three significant digits, we obtain 1,350,000.

**Example 10** The approximate distance from Earth to the star Polaris is 208 parsecs. A parsec is a distance of approximately $3.09 \times 10^{13}$ kilometers. How long would it take a space probe traveling at 40,000 kilometers per hour to reach the star? Round to three significant digits.

**Solution**

1. **Understand the problem.** Recall that the distance formula is

$$\text{distance} = \text{rate} \times \text{time}.$$

We are given the distance and the rate. We need to find the time.

Let's take a look at the distance. The distance is given in parsecs, but the rate is given in kilometers per hour. We need to change the distance to kilometers. We are told that a parsec is approximately $3.09 \times 10^{13}$ kilometers. That is, there are $3.09 \times 10^{13}$ kilometers per parsec. We use this information to change 208 parsecs to kilometers.

$$208 \text{ parsecs} = \frac{(208 \text{ parsecs})(3.09 \times 10^{13} \text{ kilometers})}{1 \text{ parsec}} = 642.72 \times 10^{13} \text{ kilometers}$$

2. **Write an equation.** Use the distance formula.

$$d = r \times t$$

3. **Solve the equation and state the answer.** Substitute the known values into the formula and solve for the unknown, time.

$$642.72 \times 10^{13} \text{ km} = \frac{40,000 \text{ km}}{1 \text{ hr}} \times t$$

$$6.4272 \times 10^{15} \text{ km} = \frac{4 \times 10^4 \text{ km}}{1 \text{ hr}} \times t \qquad \textcolor{red}{\text{Change the numbers to scientific notation.}}$$

Next, multiply both sides by the reciprocal of $\dfrac{4 \times 10^4 \text{ km}}{1 \text{ hr}}$.

$$6.4272 \times 10^{15} \text{ km} \times \frac{1 \text{ hr}}{4 \times 10^4 \text{ km}} = \frac{4 \times 10^4 \text{ km}}{1 \text{ hr}} \times \frac{1 \text{ hr}}{4 \times 10^4 \text{ km}} \times t$$

$$\frac{(6.4272 \times 10^{15} \text{ km})(1 \text{ hr})}{4 \times 10^4 \text{ km}} = t \qquad \textcolor{red}{\text{Simplify.}}$$

$$1.6068 \times 10^{11} \text{ hr} = t$$

Rounding to three significant digits, we have

$$1.6068 \times 10^{11} \text{ hr} \approx 1.61 \times 10^{11} \text{ hr}.$$

4. **Check.** Unless you have had a great deal of experience working in astronomy, it would be difficult to determine whether this is a reasonable answer. You may wish to reread your analysis and redo your calculations as a check. □

**Student Practice 10** The average distance from Earth to the distant star Betelgeuse is 159 parsecs. How many hours would it take a space probe to travel from Earth to Betelgeuse at a speed of 50,000 kilometers per hour? Round to three significant digits.

## 4.2 Exercises MyMathLab®

*Simplify. Express your answer with positive exponents. Assume that all variables are nonzero.*

1. $x^{-4}$

2. $y^{-5}$

3. $3^{-4}$

4. $2^{-4}$

5. $\dfrac{1}{y^{-8}}$

6. $\dfrac{1}{z^{-10}}$

7. $\dfrac{x^{-4}y^{-5}}{z^{-6}}$

8. $\dfrac{x^{-6}y^{-2}}{z^{-5}}$

9. $a^3b^{-2}$

10. $a^5b^{-8}$

11. $(2x^{-3})^{-3}$

12. $(4x^{-4})^{-2}$

13. $3x^{-2}$

14. $5y^{-7}$

15. $(3xy^2)^{-2}$

16. $(5x^3y)^{-3}$

### Mixed Practice, Exercises 17–28

17. $\dfrac{3xy^{-2}}{z^{-3}}$

18. $\dfrac{4x^{-2}y^{-3}}{y^4}$

19. $\dfrac{(4xy)^{-1}}{(4xy)^{-2}}$

20. $\dfrac{(3a^2b)^{-2}}{(3a^2b)^{-3}}$

21. $a^{-1}b^3c^{-4}d$

22. $x^{-5}y^{-2}z^3$

23. $(8^{-2})(2^3)$

24. $(3^{-3})(9^2)$

25. $\left(\dfrac{3x^0y^2}{z^4}\right)^{-2}$

26. $\left(\dfrac{2a^3b^0}{c^2}\right)^{-3}$

27. $\dfrac{x^{-2}y^{-3}}{x^4y^{-2}}$

28. $\dfrac{a^3b^{-1}}{a^{-5}b^{-4}}$

*Write in scientific notation.*

29. 123,780

30. 5,786,100

31. 0.063

32. 0.0000871

33. 889,610,000,000

34. 7,652,000,000

35. 0.00000342

36. 0.00783

*In exercises 37–42, write in decimal notation.*

37. $3.02 \times 10^5$

38. $8.137 \times 10^7$

39. $4.7 \times 10^{-4}$

40. $5.36 \times 10^{-2}$

41. $9.83 \times 10^5$

42. $3.5 \times 10^{-8}$

43. **Bamboo Growth** The growth rate of some species of bamboo is 0.0000237 miles per hour. Write this in scientific notation.

44. **Neptune** Neptune is $2.793 \times 10^9$ miles from the sun. Write this in decimal notation.

45. **Astronomical Unit** The astronomical unit (AU) is a unit of length approximately equal to $1.496 \times 10^8$ km. Write this in decimal notation.

46. **Gold Atom** The average volume of an atom of gold is 0.0000000000000000000001695 cubic centimeters. Write this in scientific notation.

*Evaluate by using scientific notation and the laws of exponents. Leave your answer in scientific notation.*

47. $(42,000,000)(150,000,000)$

48. $(55,000,000,000)(16,000,000)$

49. $\dfrac{(5,000,000)(16,000)}{8,000,000,000}$

50. $(0.0075)(0.0000002)(0.001)$

51. $(0.003)^4$

52. $(500,000)^4$

53. $(150,000,000)(0.00005)(0.002)(30,000)$

54. $\dfrac{(160,000)(0.0003)}{1600}$

## Applications

*National Debt* In May 2015, the national debt was about $1.8 \times 10^{13}$ dollars. (Source: www.factfinder.census.gov)

**55.** The Census Bureau estimates that in May 2015, the population of the United States was $3.21 \times 10^8$ people. If the national debt were evenly divided among every person in the country, how much debt would be assigned to each individual? Round to three significant digits.

**56.** The Census Bureau estimates that in May 2015, the number of people in the United States who were 18 years or older was approximately $2.82 \times 10^8$. If the national debt were evenly divided among every person 18 years or older in the country, how much would be assigned to each individual? Round to three significant digits.

**57.** *Watch Hand* The tip of a $\frac{1}{3}$-inch-long hour hand on a watch travels at a speed of 0.00000275 miles per hour. How far has it traveled in a day?

**58.** *Neutron Mass* The mass of a neutron is approximately $1.675 \times 10^{-27}$ kilogram. Find the mass of 150,000 neutrons.

**59.** *Mission to Pluto* In January 2006, a spacecraft called *New Horizons* began its journey to Pluto. The trip was $3.5 \times 10^9$ miles long and took $9\frac{1}{2}$ years. How many miles did *New Horizons* travel per year? (Source: www.pluto.jhuapl.edu)

**60.** *Molecules per Mole* Avogadro's number says that there are approximately $6.02 \times 10^{23}$ molecules/mole. How many molecules can one expect in 0.00483 mole?

**61.** *Construction Costs* In March 2015, the cost for construction of new private buildings was estimated at $\$5.46 \times 10^{10}$. In March 2005, the estimated cost for construction of new private buildings was $\$6.29 \times 10^{10}$. What was the percent decrease from March 2005 to March 2015?

**62.** *Construction Costs* In March 2015, the cost for construction of new public buildings was estimated at $\$1.81 \times 10^{11}$. In March 2005, the estimated cost for construction of new public buildings was $\$1.65 \times 10^{11}$. What was the percent increase from March 2005 to March 2015?

## Cumulative Review *Simplify.*

**63.** **[1.2.1]** $-2.7 - (-1.9)$

**64.** **[1.4.2]** $(-1)^{33}$

**65.** **[1.1.4]** $-\dfrac{3}{4} + \dfrac{5}{7}$

---

**Quick Quiz 4.2** *Simplify and write your answer with only positive exponents.*

**1.** $3x^{-3}y^2z^{-4}$

**2.** $\dfrac{4a^3b^{-4}}{8a^{-5}b^{-3}}$

**3.** Write in scientific notation. 0.00876

**4.** **Concept Check** Explain how you would simplify a problem like the following so that your answer had only positive exponents.

$$(4x^{-3}y^4)^{-3}$$

## 4.3 Fundamental Polynomial Operations

### 1 Recognizing Polynomials and Determining Their Degrees

A **polynomial** in $x$ is the sum of a finite number of terms of the form $ax^n$, where $a$ is any real number and $n$ is a whole number. Usually these polynomials are written in descending powers of the variable, as in

$$5x^3 + 3x^2 - 2x - 5 \quad \text{and} \quad 3.2x^2 - 1.4x + 5.6.$$

A **multivariable polynomial** is a polynomial with more than one variable. The following are multivariable polynomials:

$$5xy + 8, \quad 2x^2 - 7xy + 9y^2, \quad 17x^3y^9$$

The **degree of a term** is the sum of the exponents of all of the variables in the term. For example, the degree of $7x^3$ is three. The degree of $4xy$ is two. The degree of $10x^4y^2$ is six.

The **degree of a polynomial** is the highest degree of all of the terms in the polynomial. For example, the degree of $5x^3 + 8x^2 - 20x - 2$ is three. The degree of $6xy - 4x^2y + 2xy^3$ is four. A polynomial consisting of a constant only is said to have degree 0.

There are special names for polynomials with one, two, or three terms.

A **monomial** has *one* term:

$$5a, \quad 3x^3yz^4, \quad 12xy$$

A **binomial** has *two* terms:

$$7x + 9y, \quad -6x - 4, \quad 5x^4 + 2xy^2$$

A **trinomial** has *three* terms:

$$8x^2 - 7x + 4, \quad 2ab^3 - 6ab^2 - 15ab, \quad 2 + 5y + y^4$$

**Example 1** State the degree of the polynomial and whether it is a monomial, a binomial, or a trinomial.

**(a)** $5xy + 3x^3$ **(b)** $-7a^5b^2$ **(c)** $8x^4 - 9x - 15$

### Solution

**(a)** This polynomial is of degree 3. It has two terms, so it is a binomial.

**(b)** The sum of the exponents is $5 + 2 = 7$. Therefore this polynomial is of degree 7. It has one term, so it is a monomial.

**(c)** This polynomial is of degree 4. It has three terms, so it is a trinomial. □

**▷ Student Practice 1** State the degree of the polynomial and whether it is a monomial, a binomial, or a trinomial.

**(a)** $-7x^5 - 3xy$ **(b)** $22a^3b^4$ **(c)** $-3x^3 + 3x^2 - 6x$

### 2 Adding Polynomials

We usually write a polynomial in $x$ so that the exponents on $x$ decrease from left to right. For example, the polynomial

$$5x^2 - 6x + 2$$

is said to be written in **decreasing order** since each exponent is decreasing as we move from left to right.

You can add, subtract, multiply, and divide polynomials. Let us take a look at addition. To add two polynomials, we add their like terms.

**Example 2** Add. $(5x^2 - 6x - 12) + (-3x^2 - 9x + 5)$

**Solution**

$$
\begin{aligned}
(5x^2 - 6x - 12) + (-3x^2 - 9x + 5) &= [5x^2 + (-3x^2)] + [-6x + (-9x)] + [-12 + 5] \\
&= [(5 - 3)x^2] + [(-6 - 9)x] + [-12 + 5] \\
&= 2x^2 + (-15x) + (-7) \\
&= 2x^2 - 15x - 7
\end{aligned}
$$

**Student Practice 2** Add. $(-8x^3 + 3x^2 + 6) + (2x^3 - 7x^2 - 3)$

The numerical coefficients of polynomials may be any real number. Thus polynomials may have numerical coefficients that are decimals or fractions.

**Example 3** Add. $\left(\frac{1}{2}x^2 - 6x + \frac{1}{3}\right) + \left(\frac{1}{5}x^2 - 2x - \frac{1}{2}\right)$

**Solution**

$$
\begin{aligned}
\left(\tfrac{1}{2}x^2 - 6x + \tfrac{1}{3}\right) + \left(\tfrac{1}{5}x^2 - 2x - \tfrac{1}{2}\right) &= \left[\tfrac{1}{2}x^2 + \tfrac{1}{5}x^2\right] + [-6x + (-2x)] + \left[\tfrac{1}{3} + \left(-\tfrac{1}{2}\right)\right] \\
&= \left[\left(\tfrac{1}{2} + \tfrac{1}{5}\right)x^2\right] + [(-6 - 2)x] + \left[\tfrac{1}{3} + \left(-\tfrac{1}{2}\right)\right] \\
&= \left[\left(\tfrac{5}{10} + \tfrac{2}{10}\right)x^2\right] + [-8x] + \left[\tfrac{2}{6} - \tfrac{3}{6}\right] \\
&= \tfrac{7}{10}x^2 - 8x - \tfrac{1}{6}
\end{aligned}
$$

**Student Practice 3** Add.

$$
\left(-\tfrac{1}{3}x^2 - 6x - \tfrac{1}{12}\right) + \left(\tfrac{1}{4}x^2 + 5x - \tfrac{1}{3}\right)
$$

**Example 4** Add. $(1.2x^3 - 5.6x^2 + 5) + (-3.4x^3 - 1.2x^2 + 4.5x - 7)$

**Solution** Group like terms.

$$
\begin{aligned}
(1.2x^3 - 5.6x^2 + 5) + (-3.4x^3 - 1.2x^2 + 4.5x - 7) &= (1.2 - 3.4)x^3 + (-5.6 - 1.2)x^2 + 4.5x + (5 - 7) \\
&= -2.2x^3 - 6.8x^2 + 4.5x - 2
\end{aligned}
$$

**Student Practice 4** Add.

$$
(3.5x^3 - 0.02x^2 + 1.56x - 3.5) + (-0.08x^2 - 1.98x + 4)
$$

## 3 Subtracting Polynomials ▶

Recall that subtraction of real numbers can be defined as adding the opposite of the second number. Thus $a - b = a + (-b)$. That is, $3 - 5 = 3 + (-5)$. A similar method is used to subtract two polynomials.

To subtract two polynomials, change the sign of each term in the second polynomial and then add.

**Example 5** Subtract. $(7x^2 - 6x + 3) - (5x^2 - 8x - 12)$

**Solution**   We change the sign of each term in the second polynomial and then add.

$$(7x^2 - 6x + 3) - (5x^2 - 8x - 12) = (7x^2 - 6x + 3) + (-5x^2 + 8x + 12)$$
$$= (7 - 5)x^2 + (-6 + 8)x + (3 + 12)$$
$$= 2x^2 + 2x + 15 \qquad \square$$

**Student Practice 5**   Subtract.
$$(5x^3 - 15x^2 + 6x - 3) - (-4x^3 - 10x^2 + 5x + 13)$$

As mentioned previously, polynomials may involve more than one variable. When subtracting polynomials in two variables, you will need to use extra care in determining which terms are like terms. For example, $6x^2y$ and $5x^2y$ are like terms. In a similar fashion, $3xy$ and $8xy$ are like terms. However, $7xy^2$ and $15x^2y^2$ are not like terms. Every exponent of every variable in the two terms must be the same if the terms are to be like terms. You will use this concept in Example 6.

**Example 6** Subtract.
$$(-6x^2y - 3xy + 7xy^2) - (5x^2y - 8xy - 15x^2y^2)$$

**Solution**   Change the sign of each term in the second polynomial and add. Look for like terms.

$$(-6x^2y - 3xy + 7xy^2) + (-5x^2y + 8xy + 15x^2y^2)$$
$$= (-6 - 5)x^2y + (-3 + 8)xy + 7xy^2 + 15x^2y^2$$
$$= -11x^2y + 5xy + 7xy^2 + 15x^2y^2$$

Nothing further can be done to combine these four terms. $\qquad \square$

**Student Practice 6**   Subtract.
$$(x^3 - 7x^2y + 3xy^2 - 2y^3) - (2x^3 + 4xy - 6y^3)$$

## 4 Evaluating Polynomials to Predict a Value

Sometimes polynomials are used to predict values. In such cases we need to **evaluate** the polynomial. We do this by substituting a known value for the variable and determining the value of the polynomial.

**Example 7** Passenger cars sold in the United States have become more fuel efficient over the years due to regulations from Congress. The number of miles per gallon obtained by the average passenger car in the United States can be described by the polynomial

$$0.3x + 27.6,$$

where $x$ is the number of years since 1985. (*Source:* www.rita.dot.gov.) Use this polynomial to estimate the number of miles per gallon obtained by the average passenger car in

**(a)** 1988 **(b)** 2020

**Solution**

**(a)** The year 1988 is three years later than 1985, so $x = 3$.

Thus the number of miles per gallon obtained by the average passenger car in 1988 can be estimated by evaluating $0.3x + 27.6$ when $x = 3$.

$$0.3(3) + 27.6 = 0.9 + 27.6$$
$$= 28.5$$

We estimate that the average passenger car in 1988 obtained 28.5 miles per gallon.

**(b)** The year 2020 is 35 years after 1985, so $x = 35$.

Thus the estimated number of miles per gallon obtained by the average passenger car in 2020 can be predicted by evaluating $0.3x + 27.6$ when $x = 35$.

$$0.3(35) + 27.6 = 10.5 + 27.6$$
$$= 38.1$$

We therefore predict that the average passenger car in 2020 will obtain 38.1 miles per gallon. □

**Student Practice 7** The number of miles per gallon obtained by the average light truck (less than 8500 lb) in the United States can be described by the polynomial $0.16x + 20.7$, where $x$ is the number of years since 1985. (*Source:* www.rita.dot.gov.) Use this polynomial to estimate the number of miles per gallon obtained by the average truck in

**(a)** 1990 **(b)** 2025

## 4.3 Exercises  MyMathLab®

### Verbal and Writing Skills, Exercises 1–4

1. State in your own words a definition for a polynomial in $x$ and give an example.

2. State in your own words a definition for a multivariable polynomial and give an example.

3. State in your own words how to determine the degree of a polynomial in $x$.

4. State in your own words how to determine the degree of a multivariable polynomial.

*State the degree of the polynomial and whether it is a monomial, a binomial, or a trinomial.*

5. $6x^3y$

6. $-9x^2y^3$

7. $20x^5 + 6x^3 - 7x$

8. $9x^3 - 10x^2 + 5$

9. $4x^2y^3 - 7x^3y^3$

10. $-7x^4y + 12x^5y^2$

*Add.*

11. $(6x - 11) + (-9x - 4)$

12. $(-2x + 19) + (5x - 6)$

13. $(6x^2 + 5x - 6) + (-8x^2 - 3x + 5)$

14. $(x^2 - 4x - 8) + (-x^2 - x + 1)$

15. $\left(\frac{1}{2}x^2 + \frac{1}{3}x - 4\right) + \left(\frac{1}{3}x^2 + \frac{1}{6}x - 5\right)$

16. $\left(\frac{1}{4}x^2 - \frac{2}{3}x - 10\right) + \left(-\frac{1}{3}x^2 + \frac{1}{9}x + 2\right)$

17. $(3.4x^3 - 7.1x + 3.4) + (2.2x^2 - 6.1x - 8.8)$

18. $(-4.6x^3 + 5.6x - 0.3) + (9.8x^2 + 4.5x - 1.7)$

*Subtract.*

19. $(2x - 19) - (-3x + 5)$

20. $(5x - 5) - (6x - 3)$

21. $\left(\frac{2}{5}x^2 - \frac{1}{2}x + 5\right) - \left(\frac{1}{3}x^2 - \frac{3}{7}x - 6\right)$

22. $\left(\frac{3}{8}x^2 - \frac{2}{3}x - 7\right) - \left(\frac{2}{3}x^2 - \frac{1}{2}x + 2\right)$

23. $(4x^3 + 3x) - (x^3 + x^2 - 5x)$

24. $(2x^3 - x^2 + 5) - (6x^3 + x^2 - x + 1)$

25. $(0.5x^4 - 0.7x^2 + 8.3) - (5.2x^4 + 1.6x + 7.9)$

26. $(1.3x^4 - 3.1x^3 + 6.3x) - (x^4 - 5.2x^2 + 6.5x)$

*Perform the operations indicated.*

27. $(8x + 2) + (x - 7) - (3x + 1)$

28. $(x - 5) - (3x + 8) - (5x - 2)$

**29.** $(-4x^2y^2 + 9xy - 3) + (8x^2y^2 - 5xy - 7)$

**30.** $(12x^2y - xy^2 + 5) + (-2x^2y + 5xy^2 - 8)$

**31.** $(3x^4 - 4x^2 - 18) - (2x^4 + 3x^3 + 6)$

**32.** $(3b^3 - 5b^2 - 7b) - (2b^3 + 3b - 5)$

*Prisons* *The number of prisoners held in federal and state prisons, measured in thousands, can be described by the polynomial* $-2.06x^2 + 77.82x + 743$. *The variable x represents the number of years since 1990. (Source:* www.ojp.usdoj.gov)

**33.** Estimate the number of prisoners in 1990.

**34.** Estimate the number of prisoners in 2005.

**35.** According to the polynomial, by how much did the prison population increase from 2002 to 2007?

**36.** According to the polynomial, by how much did the prison population decrease from 2008 to 2012?

## Applications

▲ **37.** *Geometry* The lengths and the widths of the following three rectangles are labeled. Create a polynomial that describes the sum of the *area* of these three rectangles.

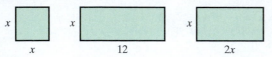

▲ **38.** *Geometry* The dimensions of the sides of the following figure are labeled. Create a polynomial that describes the *perimeter* of this figure.

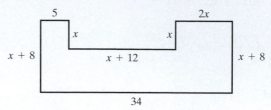

## Cumulative Review

**39.** **[2.5.1]** Solve for $y$. $3y - 8x = 2$

**40.** **[2.6.4]** Solve and graph. $\dfrac{5x}{7} - 4 > \dfrac{2x}{7} - 1$

**41.** **[2.3.3]** Solve for $x$. $-2(x - 5) + 6 = 2^2 - 9 + x$

**42.** **[2.4.1]** Solve for $x$. $\dfrac{x}{6} + \dfrac{x}{2} = \dfrac{4}{3}$

---

**Quick Quiz 4.3** *Combine.*

**1.** $(3x^2 - 5x + 8) + (-7x^2 - 6x - 3)$

**2.** $(2x^2 - 3x - 7) - (-4x^2 + 6x + 9)$

**3.** $(5x - 3) - (2x - 4) + (-6x + 7)$

**4.** **Concept Check** Explain how you would determine the degree of the following polynomial and how you would decide if it is a monomial, a binomial, or a trinomial.

$$2xy^2 - 5x^3y^4$$

# Use Math to Save Money

**Did You Know?** With some simple changes in your shopping habits, you can save a lot on your grocery bill.

## Saving at the Grocery Store

### Understanding the Problem:

A rise in the cost of gas affects our budget with higher gas prices; it also affects our budget indirectly by causing the price of groceries to increase due to an increased cost in shipping and production. Jenny is a single mom with two children. Concerned with rising costs, she decided to track her grocery expenses for three months. In the first month she spent $450, in the second month she spent $425, and in the third month she spent $460.

### Making a Plan:

By using coupons and buying store brand items, it is possible to reduce the amount of money spent at the grocery store. Jenny wants to incorporate these strategies into her shopping so she can save money.

**Step 1:** Jenny determines her monthly average spending on groceries so she'll have a baseline to use to calculate her savings.

**Task 1:** What are Jenny's average monthly grocery expenses?

**Step 2:** To help bring the monthly costs down, Jenny starts clipping coupons and cutting back on expensive treats. In the fourth month she saved her family 6% of their average expenses by using coupons and an additional $30 by cutting out treats.

**Task 2:** How much was the grocery bill the fourth month?

**Step 3:** Jenny is pleased with the results but knows she can save even more. The next month, she decides to choose store brand products when possible. She is happy to see that it saved her 20% of what she spent in the fourth month.

**Task 3:** How much was the grocery bill the fifth month?

**Task 4:** Compare Jenny's original average costs to her costs in the fifth month and then calculate the savings. Show it as a

dollar amount and as a percentage of original costs, rounded to the nearest tenth of a percent.

### Finding a Solution:

Jenny decides to make these changes permanent. These are substantial savings and will keep adding up as long as she sticks to her new shopping strategy.

**Step 4:** To give her family encouragement to accept the changes she's made in her shopping habits, Jenny calculates how much she will save in a year.

**Task 5:** If she continues to save the same amount every month, how much will she have saved at the end of one year?

### Applying the Situation to Your Life:

Grocery shopping is an expense that everyone has. Incorporating these savings strategies into your own shopping will help you save money as well. Calculate how much you spend a month, and then calculate what you could save a month using the percentage of savings you solved for above. That figure should give you encouragement to implement these strategies in your own life. Even if you only use one of these methods, you will see savings. Some additional ways to save on groceries are to stock up on items when they are on sale and to buy less convenience food, instead making more meals from scratch.

## 4.4 Multiplying Polynomials ▶

**Student Learning Objectives**

After studying this section, you will be able to:

1 Multiply a monomial by a polynomial. ▶

2 Multiply two binomials. ▶

### 1 Multiplying a Monomial by a Polynomial ▶

We use the distributive property to multiply a monomial by a polynomial. Remember, the distributive property states that for real numbers $a$, $b$, and $c$,

$$a(b + c) = ab + ac.$$

**Example 1** Multiply. $3x^2(5x - 2)$

**Solution**

$$3x^2(5x - 2) = 3x^2(5x) + 3x^2(-2) \qquad \text{Use the distributive property.}$$
$$= (3 \cdot 5)(x^2 \cdot x) + (3)(-2)x^2$$
$$= 15x^3 - 6x^2$$

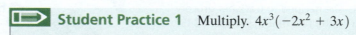 **Student Practice 1** Multiply. $4x^3(-2x^2 + 3x)$

Try to do as much of the multiplication as you can mentally.

**Example 2** Multiply.

**(a)** $2x(x^2 + 3x - 1)$ **(b)** $-2xy^2(x^2 - 2xy - 3y^2)$

**Solution**

**(a)** $2x(x^2 + 3x - 1) = 2x^3 + 6x^2 - 2x$

**(b)** $-2xy^2(x^2 - 2xy - 3y^2) = -2x^3y^2 + 4x^2y^3 + 6xy^4$

Notice in part **(b)** that you are multiplying each term by the negative expression $-2xy^2$. This will change the sign of each term in the product.

**Student Practice 2** Multiply.

**(a)** $-3x(x^2 + 2x - 4)$ **(b)** $6xy(x^3 + 2x^2y - y^2)$

When we multiply by a monomial, the monomial may be on the right side.

**Example 3** Multiply. $(x^2 - 2x + 6)(-2xy)$

**Solution** $(x^2 - 2x + 6)(-2xy) = -2x^3y + 4x^2y - 12xy$

**Student Practice 3** Multiply.

$$(-6x^3 + 4x^2 - 2x)(-3xy)$$

### 2 Multiplying Two Binomials ▶

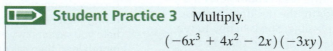

We can build on our knowledge of the distributive property and our experience with multiplying monomials to learn how to multiply two binomials. Let's suppose that we want to multiply $(x + 2)(3x + 1)$. We can use the distributive property. Since $a$ can represent any quantity, let $a = x + 2$. Then let $b = 3x$ and $c = 1$. We now have the following.

$$a(b + c) = ab + ac$$
$$(x + 2)(3x + 1) = (x + 2)(3x) + (x + 2)(1)$$
$$= 3x^2 + 6x + x + 2$$
$$= 3x^2 + 7x + 2$$

Let's take another look at the original problem, $(x + 2)(3x + 1)$. This time we will assign a letter to each term in the binomials. That is, let $a = x$, $b = 2$, $c = 3x$, and $d = 1$. Using substitution, we have the following.

$$\begin{aligned}
(x + 2)(3x + 1) &= (a + b)(c + d) \\
&= (a + b)c + (a + b)d \\
&= ac + bc + ad + bd \\
&= (x)(3x) + (2)(3x) + (x)(1) + (2)(1) \quad \text{By substitution} \\
&= 3x^2 + 6x + x + 2 \\
&= 3x^2 + 7x + 2
\end{aligned}$$

How does this compare with the preceding result?

The distributive property shows us *how* the problem can be done and *why* it can be done. In actual practice there is a memory device to help students remember the steps involved. It is often referred to as FOIL. The letters FOIL stand for the following.

F multiply the *First* terms

O multiply the *Outer* terms

I multiply the *Inner* terms

L multiply the *Last* terms

The FOIL letters are simply a way to remember the four terms in the final product and how they are obtained. Let's return to our original problem.

$(x + 2)(3x + 1)$    F    Multiply the *first* terms to obtain $3x^2$.

$(x + 2)(3x + 1)$    O    Multiply the *outer* terms to obtain $x$.

$(x + 2)(3x + 1)$    I    Multiply the *inner* terms to obtain $6x$.

$(x + 2)(3x + 1)$    L    Multiply the *last* terms to obtain $2$.

The result so far is $3x^2 + x + 6x + 2$. These four terms are the same four terms that we obtained when we multiplied using the distributive property. We can combine the like terms to obtain the final answer: $3x^2 + 7x + 2$. Now let's study the FOIL method in a few examples.

**Example 4** Multiply. $(2x - 1)(3x + 2)$

**Solution**

First      Last        First $+$ Outer $+$ Inner $+$ Last

$(2x - 1)(3x + 2)$

$-3x$ Inner

$+4x$ Outer

         F      O      I      L

$= 6x^2 + 4x - 3x - 2$

$= 6x^2 + x - 2$    Combine like terms.

Notice that we combine the inner and outer terms to obtain the middle term.   ◻

**Student Practice 4**   Multiply. $(5x - 1)(x - 2)$

**Example 5** Multiply. $(3a - 2b)(4a - b)$

**Solution**

$$(3a - 2b)(4a - b) = 12a^2 - 3ab - 8ab + 2b^2$$
$$= 12a^2 - 11ab + 2b^2$$

First — Last — Inner — Outer

**Student Practice 5** Multiply. $(8a - 5b)(3a - b)$

After you have done several problems, you may be able to combine the outer and inner products mentally.

In some problems the inner and outer products cannot be combined.

**Example 6** Multiply. $(3x + 2y)(5x - 3z)$

**Solution**

$$(3x + 2y)(5x - 3z) = 15x^2 - 9xz + 10xy - 6yz$$

First — Last — Inner — Outer

Since there are no like terms, we cannot combine any terms.

**Student Practice 6** Multiply. $(3a + 2b)(2a - 3c)$

**Example 7** Multiply. $(7x - 2y)^2$

**Solution**

$(7x - 2y)(7x - 2y)$   When we square a binomial, it is the same as multiplying the binomial by itself.

$$(7x - 2y)(7x - 2y) = 49x^2 - 14xy - 14xy + 4y^2$$
$$= 49x^2 - 28xy + 4y^2$$

First — Last — Inner — Outer

**Student Practice 7** Multiply. $(3x - 2y)^2$

We can multiply binomials containing exponents that are greater than 1. That is, we can multiply binomials containing $x^2$ or $y^3$, and so on.

**Example 8** Multiply. $(3x^2 + 4y^3)(2x^2 + 5y^3)$

**Solution** $(3x^2 + 4y^3)(2x^2 + 5y^3) = 6x^4 + 15x^2y^3 + 8x^2y^3 + 20y^6$
$$= 6x^4 + 23x^2y^3 + 20y^6$$

**Student Practice 8** Multiply. $(2x^2 + 3y^2)(5x^2 + 6y^2)$

▲ **Example 9** The width of a living room is $(x + 4)$ feet. The length of the room is $(3x + 5)$ feet. What is the area of the room in square feet?

$3x + 5$  $x + 4$

**Solution** $A = (\text{length})(\text{width}) = (3x + 5)(x + 4)$
$$= 3x^2 + 12x + 5x + 20$$
$$= 3x^2 + 17x + 20$$

There are $(3x^2 + 17x + 20)$ square feet in the room.

▲ **Student Practice 9** What is the area in square feet of a room that is $(2x - 1)$ feet wide and $(7x + 3)$ feet long?

---

👣 **STEPS TO SUCCESS** Keeping Yourself on Schedule

*The key to success is to keep on schedule.* In a class where you determine your own pace, you will need to commit yourself to follow the suggested pace provided in your course materials. Check off each assignment as you do it so you can see your progress.

*Make sure all your class materials are organized.* Keep all course schedules and assignments where you can quickly find them. Review them often to be sure you are doing everything that you should.

*Discipline yourself to follow the detailed schedule.* Professor Tobey and Professor Slater have both taught online classes for several years. They have found that students usually succeed in the course if they do *every* suggested activity. This will give you a guideline for determining what you need to be successful.

**Making it personal:** Write all your assignment due dates, quiz and exam schedules in the space below. Then plan when you will complete each assignment and study for quizzes and exams. Then write this in a daily planner. ▼

## (4.4) Exercises  MyMathLab®

*Multiply.*

**1.** $-2x(6x^3 - x)$

**2.** $5x(-3x^4 + 4x)$

**3.** $4x^2(6x - 1)$

**4.** $-4x^2(2x - 9)$

**5.** $2x^3(-2x^3 + 5x - 1)$

**6.** $5x^3(-x^2 + 3x + 2)$

**7.** $\frac{1}{2}(2x + 3x^2 + 5x^3)$

**8.** $\frac{2}{3}(4x + 6x^2 - 2x^3)$

**9.** $(2x^3 - 4x^2 + 5x)(-x^2y)$

**10.** $(3b^3 + 3b^2 - ab)(-5a^2)$

**11.** $(3x^3 + x^2 - 8x)(3xy)$

**12.** $(x^4 + 2x^3 - 8x)(xy)$

**13.** $(x^3 - 3x^2 + 5x - 2)(3x)$

**14.** $(-3x^3 + 2x^2 - 6x + 5)(5x)$

**15.** $(x^2y^2 - 6xy + 8)(-2xy)$

**16.** $(x^2y^2 + 5xy - 9)(-3xy)$

**17.** $(-7x^3 + 3x^2 + 2x - 1)(4x^2y)$

**18.** $(-4x^3 + 8x^2 - 9x - 2)(xy^2)$

**19.** $(3d^4 - 4d^2 + 6)(-2c^2d)$

**20.** $(-4x^3 + 6x^2 - 5x)(-7xy^2)$

**21.** $6x^3(2x^4 - x^2 + 3x + 9)$

**22.** $8x^3(-2x^4 + 3x^2 - 5x - 14)$

**23.** $-2x^3(8x^3 - 5x^2 + 6x)$

**24.** $-4x^6(x^2 - 3x + 5)$

*Multiply. Try to do most of the exercises mentally without writing down intermediate steps.*

**25.** $(x + 5)(x + 7)$

**26.** $(x + 6)(x + 3)$

**27.** $(x + 6)(x + 2)$

**28.** $(x + 9)(x + 3)$

**29.** $(x - 8)(x + 2)$

**30.** $(x + 3)(x - 6)$

**31.** $(x - 5)(x - 4)$

**32.** $(x - 6)(x - 5)$

**33.** $(5x - 2)(-4x - 3)$

**34.** $(7x + 1)(-2x - 3)$

**35.** $(2x - 5)(x + 3y)$

**36.** $(x - 3)(2x + 3y)$

**37.** $(5x + 2)(3x - y)$

**38.** $(3x + 4)(5x - y)$

**39.** $(4y + 1)(5y - 3)$

**40.** $(5y + 1)(6y - 5)$

**41.** $(5x^2 + 4y^3)(2x^2 + 3y^3)$

**42.** $(3a^2 + 4b^4)(2a^2 + b^4)$

### To Think About

**43.** What is wrong with this multiplication?
$(x - 2)(-3) = 3x - 6$

**44.** What is wrong with this answer?
$-(3x - 7) = -3x - 7$

**45.** What is the missing term?
$(5x + 2)(5x + 2) = 25x^2 + $ _____ $ + 4$

**46.** Multiply the binomials and write a brief description of what is special about the result. $(5x - 1)(5x + 1)$

**274**

**Mixed Practice** *Multiply.*

**47.** $(4x - 3y)(5x - 2y)$

**48.** $(3b - 5c)(2b - 7c)$

**49.** $(7x - 2)^2$

**50.** $(3x - 7)^2$

**51.** $(4a + 2b)^2$

**52.** $(5a + 3b)^2$

**53.** $(0.2x + 3)(4x - 0.3)$

**54.** $(0.5x - 2)(6x - 0.2)$

**55.** $\left(\frac{1}{2}x + \frac{1}{3}\right)\left(\frac{1}{2}x - \frac{1}{4}\right)$

**56.** $\left(\frac{1}{3}x + \frac{1}{5}\right)\left(\frac{1}{3}x - \frac{1}{2}\right)$

**57.** $(2x^2 + 4y^3)(3x^2 + 2y^3)$

**58.** $(4x^3 + 2y^2)(2x^3 + 3y^2)$

*Find the area of the rectangle.*

▲ **59.**

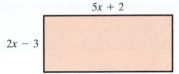

$5x + 2$

$2x - 3$

▲ **60.**

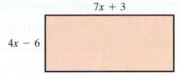

$7x + 3$

$4x - 6$

## Cumulative Review

**61.** **[2.3.3]** Solve for $x$. $3(x - 6) = -2(x + 4) + 6x$

**62.** **[2.3.3]** Solve for $w$. $3(w - 7) - (4 - w) = 11w$

**63.** **[3.4.4]** *Paper Currency* Heather returned from the bank with $375. She had one more $20 bill than she had $10 bills. The number of $5 bills she had was one less than triple the number of $10 bills. How many of each denomination did she have?

*Social Security* *Every year, the U.S. government disburses billions of dollars in veterans benefits checks. The polynomial 2.23x + 25 can be used to estimate the total value of veterans benefits checks sent in a year in billions of dollars, where x is the number of years since 2000. Round your answers to the nearest tenth. (Source: washingtontimes.com)*

**64.** **[4.3.4]** Estimate the total value of veterans benefits checks sent in 2000.

**65.** **[4.3.4]** Estimate the total value of veterans benefits checks sent in 2005.

**66.** **[4.3.4]** Estimate the total value of veterans benefits checks sent in 2015.

**67.** **[4.3.4]** Predict the total value of veterans benefits checks that will be sent in 2019.

## Quick Quiz 4.4 *Multiply.*

**1.** $(2x^2y^2 - 3xy + 4)(4xy^2)$

**2.** $(2x + 3)(3x - 5)$

**3.** $(6a - 4b)(2a - 3b)$

**4.** **Concept Check** Explain how you would multiply $(7x - 3)^2$.

## 4.5 Multiplication: Special Cases

### 1 Multiplying Binomials of the Type $(a + b)(a - b)$

The case when you multiply $(x + y)(x - y)$ is interesting and deserves special consideration. Using the FOIL method, we find

$$(x + y)(x - y) = x^2 - xy + xy - y^2 = x^2 - y^2.$$

Notice that the sum of the inner product and the outer product is zero. We see that

$$(x + y)(x - y) = x^2 - y^2.$$

This works in all cases when the binomials are the sum and difference of the same two terms. That is, in one factor the terms are added, while in the other factor the same two terms are subtracted.

$$(5a + 2b)(5a - 2b) = 25a^2 - 10ab + 10ab - 4b^2$$
$$= 25a^2 - 4b^2$$

The product is the difference of the squares of the terms. That is, $(5a)^2 - (2b)^2$ or $25a^2 - 4b^2$.

Many students find it helpful to memorize this equation.

> **MULTIPLYING BINOMIALS: A SUM AND A DIFFERENCE**
>
> $$(a + b)(a - b) = a^2 - b^2$$

You may use this relationship to find the product quickly in cases where it applies. The terms must be the same and there must be a sum and a difference.

**Example 1** Multiply. $(7x + 2)(7x - 2)$

**Solution**

$$(7x + 2)(7x - 2) = (7x)^2 - (2)^2 = 49x^2 - 4$$

*Check.* Multiply the binomials using FOIL to verify that the sum of the inner and outer products is zero. ☐

**Student Practice 1** Multiply. $(6x + 7)(6x - 7)$

**Example 2** Multiply. $(5x - 8y)(5x + 8y)$

**Solution**

$$(5x - 8y)(5x + 8y) = (5x)^2 - (8y)^2 = 25x^2 - 64y^2$$ ☐

**Student Practice 2** Multiply. $(3x - 5y)(3x + 5y)$

### 2 Multiplying Binomials of the Type $(a + b)^2$ and $(a - b)^2$

A second case that is worth special consideration is a binomial that is squared. Consider the following problem.

$$(3x + 2)^2 = (3x + 2)(3x + 2)$$
$$= 9x^2 + 6x + 6x + 4$$
$$= 9x^2 + 12x + 4$$

If you complete enough problems of this type, you will notice a pattern. The answer always contains the square of the first term added to double the product of the first and last terms added to the square of the last term.

| $3x$ is the first term | $2$ is the last term | Square the first term: $(3x)^2$ | Double the product of the first and last terms: $2(3x)(2)$ | Square the last term: $(2)^2$ |
|:---:|:---:|:---:|:---:|:---:|
| ↓ | ↓ | ↓ | ↓ | ↓ |
| $(3x$ | $+\quad 2)^2 \quad =$ | $9x^2$ | $+\qquad 12x$ | $+\quad 4$ |

We can show the same steps using variables instead of words.

$$(a + b)^2 = a^2 + 2ab + b^2$$

There is a similar formula for the square of a difference:

$$(a - b)^2 = a^2 - 2ab + b^2$$

We can use this formula to simplify $(2x - 3)^2$.

$$(2x - 3)^2 = (2x)^2 - 2(2x)(3) + (3)^2$$
$$= 4x^2 - 12x + 9$$

You may wish to multiply this product using FOIL to verify.

These two types of products, the square of a sum and the square of a difference, can be summarized as follows.

---

**A BINOMIAL SQUARED**

$$(a + b)^2 = a^2 + 2ab + b^2$$
$$(a - b)^2 = a^2 - 2ab + b^2$$

---

**Example 3** Multiply.

**(a)** $(5y - 2)^2$　　　　　　　　　　**(b)** $(8x + 9y)^2$

**Solution**

**(a)** $(5y - 2)^2 = (5y)^2 - (2)(5y)(2) + (2)^2$
$$= 25y^2 - 20y + 4$$

**(b)** $(8x + 9y)^2 = (8x)^2 + (2)(8x)(9y) + (9y)^2$
$$= 64x^2 + 144xy + 81y^2$$ ◻

**Student Practice 3** Multiply.

**(a)** $(4a - 9b)^2$　　　　　　　　　　**(b)** $(5x + 4)^2$

**CAUTION:** $(a + b)^2 \neq a^2 + b^2$! The two sides are not equal! Squaring the sum $(a + b)$ does not give $a^2 + b^2$! Beginning algebra students often make this error. Make sure you remember that when you square a binomial, there is always a *middle term*.

$$(a + b)^2 = a^2 + 2ab + b^2$$

Sometimes a numerical example helps you to see this.

$$(3 + 4)^2 \neq 3^2 + 4^2$$
$$7^2 \neq 9 + 16$$
$$49 \neq 25$$

Notice that what is missing on the right is $2ab = 2 \cdot 3 \cdot 4 = 24$.

### 3 Multiplying Polynomials with More Than Two Terms

We used the distributive property to multiply two binomials $(a + b)(c + d)$, and we obtained $ac + ad + bc + bd$. We could also use the distributive property to multiply the polynomials $(a + b)$ and $(c + d + e)$, and we would then obtain $ac + ad + ae + bc + bd + be$. Let us see if we can find a direct way to multiply products such as $(3x - 2)(x^2 - 2x + 3)$. It can be done quickly using an approach similar to that used in arithmetic for multiplying whole numbers. Consider the following arithmetic problem.

$$
\begin{array}{r}
128 \\
\times\ 43 \\
\hline
384 \\
512\phantom{0} \\
\hline
5504
\end{array}
$$

$384 \leftarrow$ The product of 128 and 3
$512 \leftarrow$ The product of 128 and 4 moved one space to the left
$5504 \leftarrow$ The sum of the two partial products

Let us use a similar format to multiply the two polynomials. For example, multiply $(x^2 - 2x + 3)$ and $(3x - 2)$.

$$
\begin{array}{r}
x^2 - 2x + 3 \\
3x - 2 \\
\hline
-2x^2 + 4x - 6 \\
3x^3 - 6x^2 + 9x\phantom{- 6} \\
\hline
3x^3 - 8x^2 + 13x - 6
\end{array}
$$

This is often called **vertical multiplication.**
$-2x^2 + 4x - 6 \leftarrow$ The product $(x^2 - 2x + 3)(-2)$
$3x^3 - 6x^2 + 9x \leftarrow$ The product $(x^2 - 2x + 3)(3x)$ moved one space to the left so that like terms are underneath each other
$3x^3 - 8x^2 + 13x - 6 \leftarrow$ The sum of the two partial products

**Example 4** Multiply vertically. $(3x^3 + 2x^2 + x)(x^2 - 2x - 4)$

**Solution**

$$
\begin{array}{r}
3x^3 + 2x^2 + x \\
x^2 - 2x - 4 \\
\hline
-12x^3 - 8x^2 - 4x \\
-6x^4 - 4x^3 - 2x^2 \\
3x^5 + 2x^4 + x^3 \\
\hline
3x^5 - 4x^4 - 15x^3 - 10x^2 - 4x
\end{array}
$$

We place one polynomial over the other.
$-12x^3 - 8x^2 - 4x \leftarrow$ The product $(3x^3 + 2x^2 + x)(-4)$
$-6x^4 - 4x^3 - 2x^2 \leftarrow$ The product $(3x^3 + 2x^2 + x)(-2x)$
$3x^5 + 2x^4 + x^3 \leftarrow$ The product $(3x^3 + 2x^2 + x)(x^2)$
$3x^5 - 4x^4 - 15x^3 - 10x^2 - 4x \leftarrow$ The sum of the three partial products

Note that the answers for each partial product are placed so that like terms are underneath each other. □

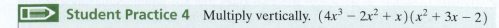
**Student Practice 4** Multiply vertically. $(4x^3 - 2x^2 + x)(x^2 + 3x - 2)$

### Alternative Method: FOIL Horizontal Multiplication
Some students prefer to do this type of multiplication using a horizontal format similar to the FOIL method. The following example illustrates this approach.

**Example 5** Multiply horizontally. $(x^2 + 3x + 5)(x^2 - 2x - 6)$

**Solution**  We will use the distributive property repeatedly.

$$(x^2 + 3x + 5)(x^2 - 2x - 6) = x^2(x^2 - 2x - 6) + 3x(x^2 - 2x - 6) + 5(x^2 - 2x - 6)$$

$$= \boxed{x^4 - 2x^3 - 6x^2} + \boxed{3x^3 - 6x^2 - 18x} + \boxed{5x^2 - 10x - 30}$$

$$= x^4 + x^3 - 7x^2 - 28x - 30 \qquad \square$$

**Student Practice 5**  Multiply horizontally. $(2x^2 + 5x + 3)(x^2 - 3x - 4)$

Some problems may need to be done in two or more separate steps.

**Example 6** Multiply. $(2x - 3)(x + 2)(x + 1)$

**Solution**  We first need to multiply any two of the binomials. Let us select the first pair.

$$\underbrace{(2x - 3)(x + 2)}(x + 1)$$

Find this product first.

$$(2x - 3)(x + 2) = 2x^2 + 4x - 3x - 6$$
$$= 2x^2 + x - 6$$

Now we replace the first two factors with their resulting product.

$$\underbrace{(2x^2 + x - 6)}(x + 1)$$

First product

We then multiply again.

$$(2x^2 + x - 6)(x + 1) = (2x^2 + x - 6)x + (2x^2 + x - 6)1$$
$$= 2x^3 + x^2 - 6x + 2x^2 + x - 6$$
$$= 2x^3 + 3x^2 - 5x - 6$$

The vertical format of Example 4 is an alternative method for this type of problem.

$$
\begin{array}{r}
2x^2 + x - 6 \\
x + 1 \\
\hline
2x^2 + x - 6 \\
2x^3 + x^2 - 6x \\
\hline
2x^3 + 3x^2 - 5x - 6
\end{array}
$$

Thus we have

$$(2x - 3)(x + 2)(x + 1) = 2x^3 + 3x^2 - 5x - 6.$$

Note that it does not matter which two binomials are multiplied first. For example, you could first multiply $(2x - 3)(x + 1)$ to obtain $2x^2 - x - 3$ and then multiply that product by $(x + 2)$ to obtain the same result. $\qquad \square$

**Student Practice 6**  Multiply.

$$(3x - 2)(2x + 3)(3x + 2)$$

(*Hint:* Rearrange the factors.)

Sometimes we encounter a binomial raised to a third power. In such cases we would write out the binomial three times as a product and then multiply. So to evaluate $(3x + 4)^3$ we would first write $(3x + 4)(3x + 4)(3x + 4)$ and then follow the method of Example 6.

## 4.5 Exercises  MyMathLab®

**Verbal and Writing Skills, Exercises 1–4**

**1.** In the special case of $(a + b)(a - b)$, a binomial times a binomial is a _____.

**2.** Identify which of the following could be the answer to a problem using the formula for $(a + b)(a - b)$. Why?

    **(a)** $9x^2 - 16$          **(b)** $4x^2 + 25$

    **(c)** $9x^2 + 12x + 4$      **(d)** $x^4 - 1$

**3.** A student evaluated $(4x - 7)^2$ as $16x^2 + 49$. What is missing? State the correct answer.

**4.** The square of a binomial, $(a - b)^2$, always produces which of the following?

    **(a)** binomial          **(b)** trinomial

    **(c)** four-term polynomial

*Use the formula $(a + b)(a - b) = a^2 - b^2$ to multiply.*

**5.** $(y - 7)(y + 7)$

**6.** $(x + 3)(x - 3)$

**7.** $(x - 9)(x + 9)$

**8.** $(x + 10)(x - 10)$

**9.** $(6x - 5)(6x + 5)$

**10.** $(5x + 2)(5x - 2)$

**11.** $(2x - 7)(2x + 7)$

**12.** $(4x + 1)(4x - 1)$

**13.** $(5x - 3y)(5x + 3y)$

**14.** $(6a + 5b)(6a - 5b)$

**15.** $(0.6x + 3)(0.6x - 3)$

**16.** $(3x - 0.8)(3x + 0.8)$

*Use the formula for a binomial squared to multiply.*

**17.** $(2y + 5)^2$

**18.** $(3x - 1)^2$

**19.** $(5x - 4)^2$

**20.** $(6x + 5)^2$

**21.** $(7x + 3)^2$

**22.** $(8x - 3)^2$

**23.** $(3x - 7)^2$

**24.** $(3x + 5y)^2$

**25.** $\left(\dfrac{2}{3}x + \dfrac{1}{4}\right)^2$

**26.** $\left(\dfrac{3}{4}x + \dfrac{1}{2}\right)^2$

**27.** $(9xy + 4z)^2$

**28.** $(7y - 3xz)^2$

**Mixed Practice, Exercises 29–36** *Multiply. Use the special formula that applies.*

**29.** $(7x + 3y)(7x - 3y)$

**30.** $(12x - 5y)(12x + 5y)$

**31.** $(3c - 5d)^2$

**32.** $(6c - d)^2$

**33.** $(9a - 10b)(9a + 10b)$    **34.** $(11a + 6b)(11a - 6b)$    **35.** $(5x + 9y)^2$    **36.** $(4x + 8y)^2$

*Use the distributive property to multiply.*

**37.** $(x^2 - x + 5)(x - 3)$    **38.** $(x^2 + 4x + 2)(x - 5)$    **39.** $(2x + 1)(x^3 + 3x^2 - x + 4)$

**40.** $(3x - 1)(x^3 + x^2 - 4x - 2)$    **41.** $(a^2 - 3a + 2)(a^2 + 4a - 3)$    **42.** $(b^2 + 5b - 1)(b^2 - 4b + 1)$

**43.** $(x + 3)(x - 1)(3x - 8)$    **44.** $(x - 7)(x + 4)(2x - 5)$    **45.** $(2x - 5)(x - 1)(x + 3)$

**46.** $(3x - 2)(x + 6)(x - 3)$    **47.** $(a - 5)(2a + 3)(a + 5)$    **48.** $(b - 3)(b + 3)(2b - 1)$

## To Think About

▲ **49.** Find the volume of this object.

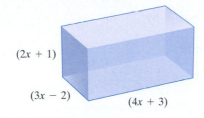

▲ **50.** *Geometry* The formula for the volume of a pyramid is $V = \dfrac{1}{3}Bh$, where $B$ is the area of the base and $h$ is the height. Find the volume of the following pyramid.

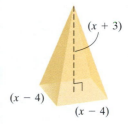

## Cumulative Review

**51.** **[3.2.1]** One number is three more than twice a second number. The sum of the two numbers is 60. Find both numbers.

▲ **52.** **[3.3.1]** *Room Dimensions* The perimeter of a rectangular room measures 34 meters. The width is 2 meters more than half the length. Find the dimensions of the room.

**Quick Quiz 4.5** *Multiply.*

**1.** $(7x - 12y)(7x + 12y)$

**2.** $(2x + 3)(x - 2)(3x + 1)$

**3.** $(3x - 2)(5x^3 - 2x^2 - 4x + 3)$

**4.** **Concept Check** Using the formula $(a + b)^2 = a^2 + 2ab + b^2$, explain how to multiply $(6x - 9y)^2$.

# 4.6 Dividing Polynomials ▶

**Student Learning Objectives**

After studying this section, you will be able to:

1. Divide a polynomial by a monomial. ▶

2. Divide a polynomial by a binomial. ▶

## 1 Dividing a Polynomial by a Monomial ▶

To divide a polynomial by a monomial, divide each term of the numerator by the denominator; then write the sum of the results. We are using the property of fractions that states that

$$\frac{a+b}{c} = \frac{a}{c} + \frac{b}{c}$$

**DIVIDING A POLYNOMIAL BY A MONOMIAL**

1. Divide each term of the polynomial by the monomial.

2. When dividing variables, use the property $\frac{x^a}{x^b} = x^{a-b}$.

**Example 1** Divide. $\dfrac{8y^6 - 8y^4 + 24y^2}{8y^2}$

**Solution** $\dfrac{8y^6 - 8y^4 + 24y^2}{8y^2} = \dfrac{8y^6}{8y^2} - \dfrac{8y^4}{8y^2} + \dfrac{24y^2}{8y^2} = y^4 - y^2 + 3$ ☐

**Student Practice 1** Divide.
$$\frac{15y^4 - 27y^3 - 21y^2}{3y^2}$$

## 2 Dividing a Polynomial by a Binomial ▶

Division of a polynomial by a binomial is similar to long division in arithmetic. Notice the similarity in the following division problems.

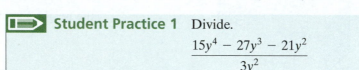

*Division of a three-digit number by a two-digit number*

```
      32
21)672
      63
      42
      42
       0
```

*Division of a polynomial by a binomial*

```
              3x + 2
2x + 1)6x² + 7x + 2
         6x² + 3x
                4x + 2
                4x + 2
                     0
```

**DIVIDING A POLYNOMIAL BY A BINOMIAL**

1. Place the terms of the polynomial and binomial in descending order. Insert a 0 for any missing term.

2. Divide the first term of the polynomial by the first term of the binomial. The result is the first term of the answer.

3. Multiply the first term of the answer by the binomial and subtract the result from the first two terms of the polynomial. Bring down the next term to obtain a new polynomial.

4. Divide the new polynomial by the binomial using the process described in step 2.

5. Continue dividing, multiplying, and subtracting until the degree of the remainder is less than the degree of the binomial divisor.

6. Write the remainder as the numerator of a fraction that has the binomial divisor as its denominator.

**Example 2** Divide. $(x^3 + 5x^2 + 11x + 4) \div (x + 2)$

**Solution**

**Step 1** The terms are arranged in descending order. No terms are missing.

**Step 2** Divide the first term of the polynomial by the first term of the binomial. In this case, divide $x^3$ by $x$ to get $x^2$.

$$x + 2 \overline{) x^3 + 5x^2 + 11x + 4} \quad \text{with } x^2 \text{ above}$$

**Step 3** Multiply $x^2$ by $x + 2$ and subtract the result from the first two terms of the polynomial, $x^3 + 5x^2$ in this case.

$$
\begin{array}{r}
x^2 \\
x + 2 \overline{) x^3 + 5x^2 + 11x + 4} \\
\underline{x^3 + 2x^2} \qquad \downarrow \\
3x^2 + 11x
\end{array}
$$

Subtract: $5x^2 - 2x^2 = 3x^2$
Bring down the next term.

**Step 4** Continue to use the step 2 process. Divide $3x^2$ by $x$. Write the resulting $3x$ as the next term of the answer.

$$
\begin{array}{r}
x^2 + 3x \\
x + 2 \overline{) x^3 + 5x^2 + 11x + 4} \\
\underline{x^3 + 2x^2} \\
3x^2 + 11x
\end{array}
$$

**Step 5** Continue multiplying, dividing, and subtracting until the degree of the remainder is less than the degree of the divisor. In this case, we stop when the remainder does not have an $x$.

$$
\begin{array}{r}
x^2 + 3x + 5 \\
x + 2 \overline{) x^3 + 5x^2 + 11x + 4} \\
\underline{x^3 + 2x^2} \\
3x^2 + 11x \\
\underline{3x^2 + 6x} \\
5x + 4 \\
\underline{5x + 10} \\
-6
\end{array}
$$

$3x(x + 2) = 3x^2 + 6x$
Bring down 4.
Subtract $4 - 10 = -6$.
The remainder is $-6$.

**Step 6** The answer is $x^2 + 3x + 5 + \dfrac{-6}{x + 2}$.

*Check.* To check the answer, we multiply $(x + 2)(x^2 + 3x + 5)$ and add the remainder, $-6$.

$$(x + 2)(x^2 + 3x + 5) + (-6) = x^3 + 5x^2 + 11x + 10 - 6$$
$$= x^3 + 5x^2 + 11x + 4$$

This is the original polynomial. It checks. □

**Student Practice 2** Divide.

$$(x^3 + 10x^2 + 31x + 25) \div (x + 4)$$

Take great care with the subtraction step when negative numbers are involved.

**Example 3** Divide. $(5x^3 - 24x^2 + 9) \div (5x + 1)$

**Student Practice 3** Divide.

$(2x^3 - x^2 + 1) \div (x - 1)$

**Solution** We must first insert $0x$ to represent the missing $x$-term. Then we divide $5x^3$ by $5x$.

$$
\begin{array}{r}
x^2 \phantom{-24x^2 + 0x + 9} \\
5x + 1 \overline{)5x^3 - 24x^2 + 0x + 9} \\
\underline{5x^3 + \phantom{2}x^2} \\
-25x^2
\end{array}
$$

Note that we are subtracting:
$-24x^2 - (+1x^2) = -24x^2 - 1x^2$
$= -25x^2$

Next we divide $-25x^2$ by $5x$.

$$
\begin{array}{r}
x^2 - 5x \phantom{+ 9} \\
5x + 1 \overline{)5x^3 - 24x^2 + 0x + 9} \\
\underline{5x^3 + \phantom{2}x^2} \\
-25x^2 + 0x \\
\underline{-25x^2 - 5x} \\
5x
\end{array}
$$

Note that we are subtracting:
$0x - (-5x) = 0x + 5x = 5x$

Finally, we divide $5x$ by $5x$.

$$
\begin{array}{r}
x^2 - 5x + 1 \\
5x + 1 \overline{)5x^3 - 24x^2 + 0x + 9} \\
\underline{5x^3 + \phantom{2}x^2} \\
-25x^2 + 0x \\
\underline{-25x^2 - 5x} \\
5x + 9 \\
\underline{5x + 1} \\
8
\end{array}
$$

$\longleftarrow$ The remainder is 8.

The answer is $x^2 - 5x + 1 + \dfrac{8}{5x + 1}$.

*Check.* To check, multiply $(5x + 1)(x^2 - 5x + 1)$ and add the remainder, 8.

$$(5x + 1)(x^2 - 5x + 1) + 8 = 5x^3 - 24x^2 + 1 + 8$$
$$= 5x^3 - 24x^2 + 9$$

This is the original polynomial. Our answer is correct. □

Now we will perform the division by writing a minimum of steps. See if you can follow each step.

**Example 4** Divide and check. $(12x^3 - 11x^2 + 8x - 4) \div (3x - 2)$

**Solution**

$$
\begin{array}{r}
4x^2 - \phantom{1}x + 2 \\
3x - 2 \overline{\smash{)}12x^3 - 11x^2 + 8x - 4} \\
\underline{12x^3 - \phantom{1}8x^2} \phantom{+ 8x - 4} \\
-3x^2 + 8x \phantom{- 4} \\
\underline{-3x^2 + 2x} \phantom{- 4} \\
6x - 4 \\
\underline{6x - 4} \\
0
\end{array}
$$

*Check.* $(3x - 2)(4x^2 - x + 2) = 12x^3 - 3x^2 + 6x - 8x^2 + 2x - 4$
$$= 12x^3 - 11x^2 + 8x - 4 \quad \text{Our answer is correct.}$$

 **Student Practice 4**  Divide and check.

$$(20x^3 - 11x^2 - 11x + 6) \div (4x - 3)$$

 **STEPS TO SUCCESS** Key Points to Help You While You Are Taking an Exam

You will be amazed to see how your performance on exams can improve just by using a good strategy while taking an exam. Look over the following suggestions, then apply them while taking your next exam. You will really see a difference!

**Before You Begin the Exam**

*Arrive early.* This will give you time to collect your thoughts and ready yourself.

*Relax.* After you get your exam, take two or three moderately deep breaths. You will feel your entire body begin to relax.

*Look over the entire exam quickly.* This will help you pace yourself and use your time wisely. If point values are given, plan to spend more time on questions that are worth more points. Write down formulas in a blank area of the test.

**During the Test**

*Read directions carefully.* Be sure to answer all questions clearly and ask your instructor about anything that is not clear.

*Answer questions that are easiest first.* This will help you build your confidence.

*Keep track of time.* Do not get bogged down with one question. Return to it after you complete the rest of the questions.

*Check your work.* Keep your work neat and organized so that you are less likely to make errors and can check your steps easily.

*Stay calm.* Stop and take a deep breath periodically. If others leave before you, don't worry. You are entitled to use the full amount of allotted time.

**Making it personal:** Which of these suggestions do you find most helpful? Can you list any other suggestions? Use these strategies on your next exam. ▼

## 4.6 Exercises    MyMathLab®

*Divide.*

1. $\dfrac{25x^4 - 15x^2 + 20x}{5x}$

2. $\dfrac{20b^4 - 4b^3 + 16b^2}{4b}$

3. $\dfrac{3y^5 + 21y^3 - 9y^2}{3y^2}$

4. $\dfrac{10y^4 - 35y^3 + 5y^2}{5y^2}$

5. $\dfrac{81x^7 - 36x^5 - 63x^3}{9x^3}$

6. $\dfrac{28x^8 - 14x^6 + 35x^4}{7x^3}$

7. $(48x^7 - 54x^4 + 36x^3) \div 6x^3$

8. $(72x^8 - 45x^6 - 36x^3) \div 9x^3$

*Divide. Check your answers for exercises 9–16 by multiplication.*

9. $\dfrac{6x^2 + 13x + 5}{2x + 1}$

10. $\dfrac{12x^2 + 19x + 5}{3x + 1}$

11. $\dfrac{x^2 - 8x - 17}{x - 5}$

12. $\dfrac{x^2 - 9x - 5}{x - 3}$

13. $\dfrac{3x^3 - x^2 + 4x - 2}{x + 1}$

14. $\dfrac{2x^3 - 3x^2 - 3x + 6}{x + 1}$

15. $\dfrac{4x^3 + 4x^2 - 19x - 15}{2x + 5}$

16. $\dfrac{6x^3 + 11x^2 - 8x + 5}{2x + 5}$

17. $\dfrac{10x^3 + 11x^2 - 11x + 2}{5x - 2}$

18. $\dfrac{5x^3 - 28x^2 - 20x + 21}{5x - 3}$

19. $\dfrac{4x^3 + 3x + 5}{2x - 3}$

20. $\dfrac{8x^3 + 8x + 5}{2x - 1}$

21. $(y^3 - y^2 - 13y - 12) \div (y + 3)$

**22.** $(y^3 - 2y^2 - 26y - 4) \div (y + 4)$  **23.** $(y^4 - 9y^2 - 5) \div (y - 2)$  **24.** $(2y^4 + 3y^2 - 5) \div (y - 2)$

## Cumulative Review

**25.** [0.5.3] *Milk Prices* In January 2008, the average price for a gallon of whole milk was $3.87. By January 2010, the price had decreased by 16%. In January 2015, the price increased 16% from the 2010 price. What was the average price of a gallon of whole milk in January 2010? What was the price in January 2015? (*Source:* www.bls.gov)

**26.** [3.2.1] *Page Numbers* Thomas was assigned to read a special two-page case study in his psychology book. He can't remember the page numbers, but does remember that the pages are consecutive and the two numbers add to 341. What are the page numbers?

**27.** [3.4.2] *Hurricane Pattern* The National Hurricane Center has noticed an interesting pattern in the number of Atlantic hurricanes each year. Examine the bar graph, then answer the following questions. Round all answers to the nearest tenth.

(a) What was the mean number of hurricanes per year during the years 2003 to 2006?

(b) What was the mean number of hurricanes per year during the years 2007 to 2010?

(c) What was the mean number of hurricanes per year during the years 2011 to 2014?

(d) What was the percent decrease of the mean from the four-year period 2003 to 2006 to the four-year period 2007 to 2010?

(e) What was the percent decrease of the mean from the four-year period 2007 to 2010 to the four-year period 2011 to 2014?

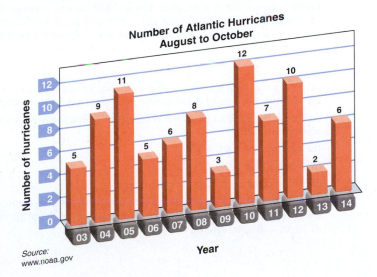

Source: www.noaa.gov

Quick Quiz 4.6 *Divide.*

**1.** $\dfrac{20x^5 - 64x^4 - 8x^3}{4x^2}$

**2.** $(8x^3 + 2x^2 - 19x - 6) \div (2x + 3)$

**3.** $(x^3 + 4x - 3) \div (x - 2)$

**4.** **Concept Check** Explain how you would check your answer to problem 3. Perform the check. Does your answer check?

## Background: Biological Technician

Jeannie is a biological technician working at a major biotechnology company. Jeannie's responsibilities are varied in her role assisting a team of scientists in the protocols of their experiments. She spends a good deal of time observing, recording, and calculating results from these experiments.

## Facts

- One of the projects she's working on involves the study of how to best modify bacterial cells, and she needs to calculate a transformation efficiency value, a measure of the cell's ability to be transformed. Typically these values are calculated and expressed in scientific notation. After measuring and recording a first set of data, her final calculation appears as:

$$\text{Transformation efficiency} = 140 \text{ transformants} \div \left[ \frac{250}{2,500,000} \right]$$

A second set of data generates the following calculation:

$$\text{Transformation efficiency} = 164 \text{ transformants} \times \left[ \frac{1}{5 \times 10^{-3}} \right]$$

In her lab reports, Jeannie's measurements for cell density, absorption, and number of bacteria must be converted from scientific notation to decimal notation.

- Another project Jeannie is working on involves genetics. In it, she's writing polynomial expressions that describe the possible genetic makeup for the offspring of two parents. Each parent has half dominant genes, symbolized by $X$, and half recessive genes, symbolized by $y$. The binomial describing each parent's makeup is $(0.5X + 0.5y)$.

## Tasks

1. What are the final transformation efficiency values expressed in scientific notation for the two sets of data Jeannie has measured and recorded?

2. What are the decimal equivalents of the following measurements expressed in scientific notation?

   (a) Cell density: $5.01 \times 10^7$

   (b) Absorption: $3.02 \times 10^{-3}$

   (c) Number of bacteria: $237 \times 10 \times 10^5$

3. Given that the binomial describing each parent's genetic makeup is $(0.5X + 0.5y)$, the polynomial that describes the genetic makeup of the offspring is the product of $(0.5X + 0.5y)^2$. What is this product?

# Chapter 4 Organizer

| Topic and Procedure | Examples | You Try It |
|---|---|---|
| **Multiplying monomials, p. 245**<br><br>$$x^a \cdot x^b = x^{a+b}$$<br><br>1. Multiply the numerical coefficients.<br>2. Add the exponents of a given base. | (a) $3^{12} \cdot 3^{15} = 3^{27}$<br>(b) $(-3x^2)(6x^3) = -18x^5$<br>(c) $(2ab)(4a^2b^3) = 8a^3b^4$ | **1.** Multiply.<br>(a) $2^9 \cdot 2^{14}$<br>(b) $(-8a^3)(-2a^5)$<br>(c) $(-ab^2)(3a^4b^2)$ |
| **Dividing monomials, p. 247** $(x \neq 0)$<br><br>$$\frac{x^a}{x^b} = \begin{cases} x^{a-b} & \text{Use if } a \text{ is greater than } b. \\ \dfrac{1}{x^{b-a}} & \text{Use if } b \text{ is greater than } a. \end{cases}$$<br><br>1. Divide or reduce the fraction created by the quotient of the numerical coefficients.<br>2. Subtract the exponents of a given base. | (a) $\dfrac{16x^7}{8x^3} = 2x^4$<br>(b) $\dfrac{5x^3}{25x^5} = \dfrac{1}{5x^2}$<br>(c) $\dfrac{-12x^5y^7}{18x^3y^{10}} = -\dfrac{2x^2}{3y^3}$ | **2.** Divide.<br>(a) $\dfrac{21x^5}{3x^2}$<br>(b) $\dfrac{-3x}{9x^2}$<br>(c) $\dfrac{14ab^7}{28a^3b}$ |
| **Exponent of zero, p. 249**<br><br>$$x^0 = 1 \quad \text{if } x \neq 0$$ | (a) $5^0 = 1$     (b) $\dfrac{x^6}{x^6} = 1$<br>(c) $w^0 = 1$     (d) $3x^0y = 3y$ | **3.** Simplify.<br>(a) $9^0$     (b) $m^0$<br>(c) $\dfrac{a^5}{a^5}$     (d) $6ab^0$ |
| **Raising a power, product, or quotient to a power, pp. 250–251**<br><br>$$(x^a)^b = x^{ab}$$<br>$$(xy)^a = x^a y^a$$<br>$$\left(\dfrac{x}{y}\right)^a = \dfrac{x^a}{y^a} \quad (y \neq 0)$$<br><br>1. Raise the numerical coefficient to the power outside the parentheses.<br>2. Multiply the exponent outside the parentheses times all exponents inside the parentheses. | (a) $(x^9)^3 = x^{27}$<br>(b) $(3x^2)^3 = 27x^6$<br>(c) $\left(\dfrac{2x^2}{y^3}\right)^3 = \dfrac{8x^6}{y^9}$<br>(d) $(-3x^4y^5)^4 = 81x^{16}y^{20}$<br>(e) $(-5ab)^3 = -125a^3b^3$ | **4.** Simplify.<br>(a) $(a^4)^5$<br>(b) $(2n^3)^2$<br>(c) $\left(\dfrac{3x^3}{y}\right)^3$<br>(d) $(-5s^2t^5)^2$<br>(e) $(-a^2b)^5$ |
| **Negative exponents, p. 256**<br><br>If $x \neq 0$ and $y \neq 0$, then<br><br>$$x^{-n} = \dfrac{1}{x^n}, \quad \dfrac{1}{x^{-n}} = x^n, \quad \dfrac{x^{-m}}{y^{-n}} = \dfrac{y^n}{x^m}$$ | Write with positive exponents.<br>(a) $x^{-6} = \dfrac{1}{x^6}$    (b) $\dfrac{1}{w^{-3}} = w^3$<br>(c) $\dfrac{w^{-12}}{z^{-5}} = \dfrac{z^5}{w^{12}}$    (d) $2^{-2} = \dfrac{1}{2^2} = \dfrac{1}{4}$ | **5.** Rewrite with positive exponents.<br>(a) $a^{-3}$    (b) $\dfrac{1}{x^{-1}}$<br>(c) $\dfrac{m^{-9}}{n^{-6}}$    (d) $3^{-2}$ |
| **Scientific notation, p. 257**<br><br>A positive number is written in scientific notation if it is in the form $a \times 10^n$, where $1 \le a < 10$ and $n$ is an integer. | (a) $2{,}568{,}000 = 2.568 \times 10^6$<br>(b) $0.0000034 = 3.4 \times 10^{-6}$ | **6.** Write in scientific notation.<br>(a) $386{,}400$<br>(b) $0.000052$ |
| **Performing calculations with numbers written in scientific notation, p. 259**<br><br>Rearrange the order when multiplying.<br><br>Use the definition of multiplication of fractions when dividing. | Multiply or divide.<br>(a) $(5.2 \times 10^7)(1.8 \times 10^5)$<br>$\quad = 5.2 \times 1.8 \times 10^7 \times 10^5$<br>$\quad = 9.36 \times 10^{12}$<br>(b) $\dfrac{4.5 \times 10^6}{6 \times 10^3}$<br>$\quad = \dfrac{4.5}{6} \times \dfrac{10^6}{10^3}$<br>$\quad = 0.75 \times 10^3$<br>$\quad = 7.5 \times 10^2$ | **7.** Multiply or divide.<br>(a) $(3.1 \times 10^6)(2.5 \times 10^4)$<br>(b) $\dfrac{3.8 \times 10^9}{1.25 \times 10^5}$ |
| **Add polynomials, pp. 263–264**<br><br>To add two polynomials, we add their like terms. | $(-7x^3 + 2x^2 + 5) + (x^3 + 3x^2 + x)$<br>$= -6x^3 + 5x^2 + x + 5$ | **8.** Add.<br>$(x^4 - 5x^3 + 2x^2) + (-7x^4 + x^3 - x^2)$ |
| **Subtracting polynomials, p. 265**<br><br>To subtract the polynomials, change all signs of the second polynomial and add the result to the first polynomial.<br><br>$$a - b = a + (-b)$$ | $(5x^2 - 6) - (-3x^2 + 2) = (5x^2 - 6) + (3x^2 - 2)$<br>$\qquad\qquad\qquad = 8x^2 - 8$ | **9.** Subtract. $(8 - x^2) - (5 + 2x^2)$ |

289

| Topic and Procedure | Examples | ▷ You Try It |
|---|---|---|
| **Multiplying a monomial by a polynomial, p. 270**<br><br>Use the distributive property.<br>$a(b + c) = ab + ac$<br>$(b + c)a = ba + ca$ | Multiply.<br>**(a)** $-5x(2x^2 + 3x - 4) = -10x^3 - 15x^2 + 20x$<br>**(b)** $(6x^3 - 5xy - 2y^2)(3xy) = 18x^4y - 15x^2y^2 - 6xy^3$ | **10.** Multiply.<br>**(a)** $-2a(3a^2 - 5a + 1)$<br><br>**(b)** $(-x^2 + 3xy - 3y^2)(4xy)$ |
| **Multiplying two binomials, pp. 270–272, 276**<br><br>1. The product of the sum and difference of the same two terms:<br>$(a + b)(a - b) = a^2 - b^2$<br>2. The square of a binomial:<br>$(a + b)^2 = a^2 + 2ab + b^2$<br>$(a - b)^2 = a^2 - 2ab + b^2$<br>3. Use FOIL for other binomial multiplication. The middle terms can often be combined, giving a trinomial. | 1. $(3x + 7y)(3x - 7y) = 9x^2 - 49y^2$<br><br>2. $(3x + 7y)^2 = 9x^2 + 42xy + 49y^2$<br>$(3x - 7y)^2 = 9x^2 - 42xy + 49y^2$<br><br>3. $(3x - 5)(2x + 7) = 6x^2 + 21x - 10x - 35$<br>$= 6x^2 + 11x - 35$ | **11.** Multiply.<br>**(a)** $(2a + 5b)(2a - 5b)$<br>**(b)** $(2a + 5b)^2$<br>**(c)** $(2a - 5b)^2$<br>**(d)** $(2a - b)(3a + 5b)$ |
| **Multiplying two polynomials, p. 278**<br><br>To multiply two polynomials, multiply each term of one by each term of the other. This method is similar to the multiplication of many-digit numbers. | Vertical method:<br>$$\begin{array}{r} 3x^2 - 7x + 4 \\ \times \quad 3x - 1 \\ \hline -3x^2 + 7x - 4 \\ 9x^3 - 21x^2 + 12x \quad\quad \\ \hline 9x^3 - 24x^2 + 19x - 4 \end{array}$$<br>Horizontal method: $(5x + 2)(2x^2 - x + 3)$<br>$= 10x^3 - 5x^2 + 15x + 4x^2 - 2x + 6$<br>$= 10x^3 - x^2 + 13x + 6$ | **12.** Multiply.<br>**(a)** $$\begin{array}{r} 6x^2 - 5x + 3 \\ 2x + 1 \\ \hline \end{array}$$<br>**(b)** $(x - 5)(3x^2 - 2x + 1)$ |
| **Multiplying three or more polynomials, p. 279**<br><br>1. Multiply any two polynomials.<br>2. Multiply the result by any remaining polynomials. | $(2x + 1)(x - 3)(x + 4) = (2x^2 - 5x - 3)(x + 4)$<br>$= 2x^3 + 3x^2 - 23x - 12$ | **13.** Multiply. $(x + 5)(x - 1)(3x + 2)$ |
| **Dividing a polynomial by a monomial, p. 282**<br><br>1. Divide each term of the polynomial by the monomial.<br>2. When dividing variables, use the property $\frac{x^a}{x^b} = x^{a-b}$. | Divide. $(15x^3 + 20x^2 - 30x) \div (5x)$<br>$= \frac{15x^3}{5x} + \frac{20x^2}{5x} - \frac{30x}{5x}$<br>$= 3x^2 + 4x - 6$ | **14.** Divide. $(18a^3 - 9a^2 + 3a) \div (3a)$ |
| **Dividing a polynomial by a binomial, p. 282**<br><br>1. Place the terms of the polynomial and binomial in descending order. Insert a 0 for any missing term.<br>2. Divide the first term of the polynomial by the first term of the binomial.<br>3. Multiply the first term of the answer by the binomial, and subtract the result from the first two terms of the polynomial. Bring down the next term to obtain a new polynomial.<br>4. Divide the new polynomial by the binomial using the process described in step 2.<br>5. Continue dividing, multiplying, and subtracting until the degree of the remainder is less than the degree of the binomial divisor.<br>6. Write the remainder as the numerator of a fraction that has the binomial divisor as its denominator. | Divide.<br>$(8x^3 + 2x^2 - 13x + 7) \div (4x - 1)$<br>$$\begin{array}{r} 2x^2 + x - 3 \\ 4x - 1 \overline{)8x^3 + 2x^2 - 13x + 7} \\ \underline{8x^3 - 2x^2} \\ 4x^2 - 13x \\ \underline{4x^2 - x} \\ -12x + 7 \\ \underline{-12x + 3} \\ 4 \end{array}$$<br>The answer is<br>$2x^2 + x - 3 + \dfrac{4}{4x - 1}.$ | **15.** Divide.<br>$(3x^3 + 13x^2 - 13x + 2) \div (3x - 2)$ |

# Chapter 4 Review Problems

## Section 4.1

*Simplify. In problems 1–12, leave your answer in exponent form.*

1. $(-6a^2)(3a^5)$

2. $(5^{10})(5^{13})$

3. $(3xy^2)(2x^3y^4)$

4. $(2x^3y^4)(-7xy^5)$

5. $\dfrac{7^{15}}{7^{27}}$

6. $\dfrac{x^{12}}{x^{17}}$

7. $\dfrac{y^{30}}{y^{16}}$

8. $\dfrac{9^{13}}{9^{24}}$

9. $\dfrac{-15xy^2}{25x^6y^6}$

10. $\dfrac{-12a^3b^6}{18a^2b^{12}}$

11. $(x^3)^8$

12. $\dfrac{(2b^2)^4}{(5b^3)^6}$

13. $(-3a^3b^2)^2$

14. $(3x^3y)^4$

15. $\left(\dfrac{5ab^2}{c^3}\right)^2$

16. $\left(\dfrac{x^0y^3}{4w^5z^2}\right)^3$

## Section 4.2

*Simplify. Write with positive exponents.*

17. $a^{-3}b^5$

18. $m^8p^{-5}$

19. $\dfrac{2x^{-6}}{y^{-3}}$

20. $(2x^{-5}y)^{-3}$

21. $(6a^4b^5)^{-2}$

22. $\dfrac{3x^{-3}}{y^{-2}}$

23. $\dfrac{4x^{-5}y^{-6}}{w^{-2}z^8}$

24. $\dfrac{3^{-3}a^{-2}b^5}{c^{-3}d^{-4}}$

*Write in scientific notation.*

25. 156,340,200,000

26. 179,632

27. 0.00092

28. 0.00000174

*Write in decimal notation.*

29. $1.2 \times 10^5$

30. $6.034 \times 10^6$

31. $2.5 \times 10^{-1}$

32. $4.32 \times 10^{-5}$

*Perform the calculation indicated. Leave your answer in scientific notation.*

33. $\dfrac{(28{,}000{,}000)(5{,}000{,}000{,}000)}{7000}$

34. $(3.12 \times 10^5)(2.0 \times 10^6)(1.5 \times 10^8)$

35. $\dfrac{(0.00078)(0.000005)(0.00004)}{0.002}$

36. **Mission to Pluto** The *New Horizons* spacecraft began its $3.5 \times 10^9$-mile journey to Pluto in January 2006. If the cost of this mission is $0.20 per mile, find the total cost. (*Source*: www.pluto.jhuapl.edu)

37. **Atomic Clock** An atomic clock is based on the fact that cesium emits 9,192,631,770 cycles of radiation in one second. How many of these cycles occur in one day? Round to three significant digits.

38. **Computer Speed** Today's fastest modern computers can perform one operation in $1 \times 10^{-11}$ second. How many operations can such a computer perform in 1 minute?

## Section 4.3

*Combine.*

39. $(2.8x^2 - 1.5x + 3.4) + (2.7x^2 + 0.5x - 5.7)$

40. $(4x^3 - x^2 - x + 3) - (-3x^3 + 2x^2 + 5x - 1)$

41. $\left(\dfrac{3}{5}x^2y - \dfrac{1}{3}x + \dfrac{3}{4}\right) - \left(\dfrac{1}{2}x^2y + \dfrac{2}{7}x + \dfrac{1}{3}\right)$

42. $\dfrac{1}{2}x^2 - \dfrac{3}{4}x + \dfrac{1}{5} - \left(\dfrac{1}{4}x^2 - \dfrac{1}{2}x + \dfrac{1}{10}\right)$

43. $(x^2 - 9) - (4x^2 + 5x) + (5x - 6)$

291

## Section 4.4

*Multiply.*

**44.** $(3x + 1)(5x - 1)$      **45.** $(7x - 2)(4x - 3)$      **46.** $(2x + 3)(10x + 9)$      **47.** $5x(2x^2 - 6x + 3)$

**48.** $(xy^2 + 5xy - 6)(-4xy^2)$      **49.** $(5a + 7b)(a - 3b)$      **50.** $(2x^2 - 3)(4x^2 - 5y)$

## Section 4.5

*Multiply.*

**51.** $(4x + 3)^2$      **52.** $(a + 5b)(a - 5b)$      **53.** $(7x + 6y)(7x - 6y)$      **54.** $(5a - 2b)^2$

**55.** $(x^2 + 7x + 3)(4x - 1)$      **56.** $(x - 6)(2x - 3)(x + 4)$

## Section 4.6

*Divide.*

**57.** $(12y^3 + 18y^2 + 24y) \div (6y)$      **58.** $(30x^5 + 35x^4 - 90x^3) \div (5x^2)$      **59.** $(16x^3 - 24x^2 + 32x) \div (4x)$

**60.** $(15x^2 + 11x - 14) \div (5x + 7)$      **61.** $(12x^2 - x - 63) \div (4x + 9)$      **62.** $(2x^3 - x^2 + 3x - 1) \div (x + 2)$

**63.** $(6x^2 + x - 9) \div (2x + 3)$      **64.** $(x^3 - x - 24) \div (x - 3)$      **65.** $(2x^3 - 3x + 1) \div (x - 2)$

## Applications

*Solve. Express your answer in scientific notation.*

**66.** *Fighting Obesity* President Obama signed into law the Healthy Hunger-Free Kids Act of 2010, which includes $4.5 \times 10^9$ dollars to provide children with healthy food in schools and fight childhood obesity. If the population of the United States in 2010 was $3.1 \times 10^8$ people, how much per person did the United States spend to fight childhood obesity? Write your answer in dollars and cents. (*Source:* www.whitehouse.gov)

**67.** *Population* The population of Africa in 2020 is projected to be $1.25 \times 10^9$. The projected population of North America in 2020 is $5.97 \times 10^8$. What is the total population of the two countries projected to be? (*Hint:* First write $5.97 \times 10^8$ as $0.597 \times 10^9$ before you do the calculations.)

**68.** *Electron Mass* The mass of an electron is approximately $9.11 \times 10^{-28}$ gram. Find the mass of 30,000 electrons.

**69.** *Gray Whales* During feeding season, gray whales eat $3.4 \times 10^5$ pounds of food per day. If the feeding period lasts for 140 days, how many pounds of food total will a gray whale consume?

## To Think About

*Find a polynomial that describes the shaded area.*

▲ **70.**

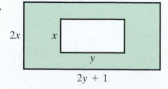

▲ **71.**

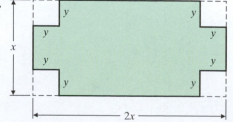

292

# How Am I Doing? Chapter 4 Test

MATH COACH          MyMathLab®          You Tube

After you take this test read through the Math Coach on pages 294–295. Math Coach videos are available via MyMathLab and YouTube. Step-by-step test solutions on the Chapter Test Prep Videos are also available via MyMathLab and YouTube. (Search "TobeyBeginningAlg" and click on "Channels.")

*Simplify. Leave your answer in exponent form.*

**1.** $(3^{10})(3^{24})$

**2.** $\dfrac{25^{18}}{25^{34}}$

**3.** $(8^4)^6$

*In questions 4–8, simplify.*

**4.** $(-3xy^4)(-4x^3y^6)$

**5.** $\dfrac{-35x^8y^{10}}{25x^5y^{10}}$

**6.** $(-5xy^6)^3$

**7.** $\left(\dfrac{7a^7b^2}{3c^0}\right)^2$

**MC 8.** $\dfrac{(3x^2)^3}{(6x)^2}$

**9.** Evaluate. $4^{-3}$

*In questions 10 and 11, simplify and write with only positive exponents.*

**10.** $6a^{-4}b^{-3}c^5$

**MC 11.** $\dfrac{3x^{-3}y^2}{x^{-4}y^{-5}}$

**12.** Write in scientific notation. $0.0005482$

**13.** Write in decimal notation. $5.82 \times 10^8$

**14.** Multiply. Leave your answer in scientific notation.
$(4.0 \times 10^{-3})(3.0 \times 10^{-8})(2.0 \times 10^4)$

*Combine.*

**15.** $(2x^2 - 3x - 6) + (-4x^2 + 8x + 6)$

**16.** $(3x^3 - 4x^2 + 3) - (14x^3 - 7x + 11)$

*Multiply.*

**17.** $-7x^2(3x^3 - 4x^2 + 6x - 2)$

**18.** $(5x^2y^2 - 6xy + 2)(3x^2y)$

**19.** $(5a - 4b)(2a + 3b)$

**MC 20.** $(3x + 2)(2x + 1)(x - 3)$

**21.** $(7x^2 + 2y^2)^2$

**22.** $(5s - 11t)(5s + 11t)$

**23.** $(3x - 2)(4x^3 - 2x^2 + 7x - 5)$

**24.** $(3x^2 - 5xy)(x^2 + 3xy)$

*Divide.*

**25.** $(15x^6 - 5x^4 + 25x^3) \div 5x^3$

**26.** $(8x^3 - 22x^2 - 5x + 12) \div (4x + 3)$

**MC 27.** $(2x^3 - 6x - 36) \div (x - 3)$

*Solve. Express your answer in scientific notation. Round to the nearest hundredth.*

**28.** At the end of 2014, Saudi Arabia was estimated to have $2.6 \times 10^{11}$ barrels of oil reserves. By one estimate, they have 69 years of reserves remaining. If they disbursed the oil in an equal amount each year, how many barrels of oil would they pump each year? (*Source:* saudiembassy.net)

**29.** A space probe is traveling from Earth to Pluto at a speed of $2.49 \times 10^4$ miles per hour. How far would this space probe travel in one week?

1. _____
2. _____
3. _____
4. _____
5. _____
6. _____
7. _____
8. _____
9. _____
10. _____
11. _____
12. _____
13. _____
14. _____
15. _____
16. _____
17. _____
18. _____
19. _____
20. _____
21. _____
22. _____
23. _____
24. _____
25. _____
26. _____
27. _____
28. _____
29. _____

**Total Correct:** _____

# MATH COACH

*Mastering the skills you need to do well on the test.*

The following problems are from the Chapter 4 Test. Here are some helpful hints to keep you from making common errors on test problems.

**Chapter 4 Test, Problem 8**   Simplify. $\dfrac{(3x^2)^3}{(6x)^2}$

> **HELPFUL HINT** Do the problem in three stages. First, use the power to a power rule to raise the numerator to the third power. Second, raise the denominator to the second power. Then divide the monomials using the rules of exponents. Be sure to simplify any fractions.

Did you use the power to a power rule to raise both $3^1$ and $x^2$ to the third power in the numerator and both $6^1$ and $x^1$ to the second power in the denominator?

Yes _____ No _____

If you answered No, stop and review the power to a power rule before completing these steps again.

Did you remember to simplify the fraction $\dfrac{27}{36}$?

Yes _____ No _____

Finally, did you remember to use the quotient rule to subtract the exponents in the $x$ terms?

Yes _____ No _____

If you answered No to either of these questions, go back and examine your work carefully before completing these steps again.

If you answered Problem 8 incorrectly, go back and rework the problem using these suggestions.

---

**Chapter 4 Test, Problem 11**   Simplify and write with only positive exponents. $\dfrac{3x^{-3}y^2}{x^{-4}y^{-5}}$

> **HELPFUL HINT** First, use the definition of a negative exponent to rewrite the expression using only positive exponents. Then use the rules for exponents to simplify the resulting expression.

Did you remove the negative exponents by rewriting the expression as $\dfrac{3x^4y^2y^5}{x^3}$?

Yes _____ No _____

If you answered No, review the definition of negative exponents in Section 4.2 and complete this step again.

Did you use the quotient rule to simplify the $x$ terms and the product rule to simplify the $y$ terms?

Yes _____ No _____

If you answered No, review the rules for exponents in Sections 4.1 and 4.2 and simplify the expression again.

Now go back and rework the problem using these suggestions.

Need more help? Watch the **MATH COACH** videos in MyMathLab® or on You Tube™.

294

**Chapter 4 Test, Problem 20**  Multiply.  $(3x + 2)(2x + 1)(x - 3)$

> **HELPFUL HINT** A good approach is to start by multiplying the first two binomials. Then multiply that result by the third binomial. Be careful to avoid sign errors when multiplying, and be careful to write down the correct exponent for each term.

Did you use the FOIL method to multiply the first two binomials and obtain $6x^2 + 7x + 2$?

Yes _____ No _____

If you answered No, stop and complete this step.

Did you multiply the result above by $(x - 3)$?

Yes _____ No _____

Did you multiply *each term* of $6x^2 + 7x + 2$ by $x$?

Yes _____ No _____

Did you multiply *each term* of $6x^2 + 7x + 2$ by $-3$?

Yes _____ No _____

If you answered No to any of these questions, go back and examine each step of the multiplication carefully. Be sure to write the correct exponent each time that you multiply. Be careful to avoid sign errors when multiplying by $-3$. Then combine like terms before writing your final answer.

If you answered Problem 20 incorrectly, go back and rework the problem using these suggestions.

---

**Chapter 4 Test, Problem 27**  Divide.  $(2x^3 - 6x - 36) \div (x - 3)$

> **HELPFUL HINT** Review the procedure for dividing a polynomial by a binomial in Section 4.6. Make sure you understand each step. Be sure you understand where the expression $0x^2$ came from in the dividend. Be careful with subtraction. Write out the subtraction steps to avoid sign errors.

Did you write the division problem in the form $x - 3 \overline{)2x^3 + 0x^2 - 6x - 36}$?

Yes _____ No _____

If you answered No, remember that every power must be represented. We must use $0x^2$ as a placeholder so that we can perform our division.

When you carried out the first step of division, did you obtain $2x^2$ as the first part of your answer?

Yes _____ No _____

When you multiplied $x - 3$ by $2x^2$ and then subtracted, did you get the result $6x^2$?

Yes _____ No _____

If you answered No to these questions, stop and examine your first division step carefully. Make sure that you subtracted carefully too. Remember to write out the subtraction steps: $0x^2 - (-6x^2) = 6x^2$.

Next, did you bring down $-6x$ from the dividend to obtain $6x^2 - 6x$?

Yes _____ No _____

If you answered No, go back and look at the dividend again and see how to obtain this result.

Now go back and rework the problem using these suggestions.

Need more help? Look for section examples marked with $^{M}C$ to review.

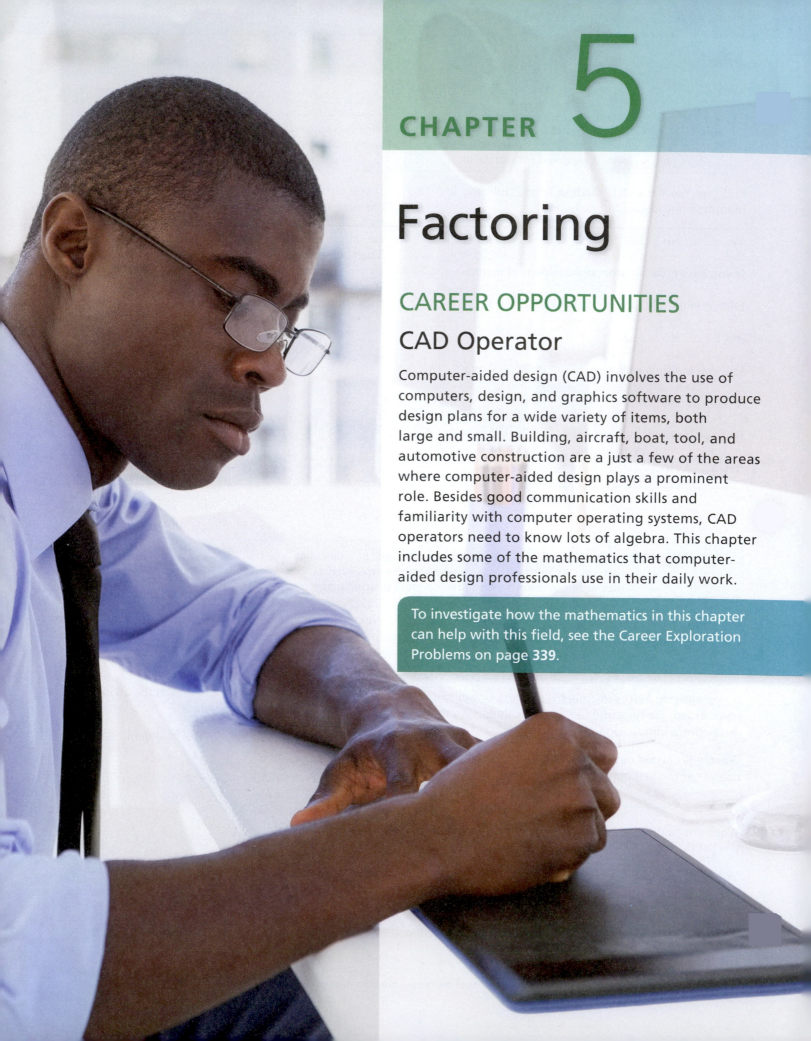

# Factoring

## CAREER OPPORTUNITIES

### CAD Operator

Computer-aided design (CAD) involves the use of computers, design, and graphics software to produce design plans for a wide variety of items, both large and small. Building, aircraft, boat, tool, and automotive construction are a just a few of the areas where computer-aided design plays a prominent role. Besides good communication skills and familiarity with computer operating systems, CAD operators need to know lots of algebra. This chapter includes some of the mathematics that computer-aided design professionals use in their daily work.

To investigate how the mathematics in this chapter can help with this field, see the Career Exploration Problems on page **339**.

## 5.1  Removing a Common Factor ▶

### 1  Factoring Polynomials Whose Terms Contain a Common Factor ▶

**Student Learning Objective**

After studying this section, you will be able to:

**1** Factor polynomials whose terms contain a common factor. ▶

Recall that when two or more numbers, variables, or algebraic expressions are multiplied, each is called a **factor.**

$$\underbrace{3 \cdot 2}_{\text{factor factor}} \qquad \underbrace{3x^2 \cdot 5x^3}_{\text{factor factor}} \qquad \underbrace{(2x - 3)(x + 4)}_{\text{factor factor}}$$

When you are asked **to factor** a number or an algebraic expression, you are being asked, "What factors, when multiplied, will give that number or expression?"

For example, you can factor 6 as $3 \cdot 2$ since $3 \cdot 2 = 6$. You can factor $15x^5$ as $3x^2 \cdot 5x^3$ since $3x^2 \cdot 5x^3 = 15x^5$. Factoring is simply the reverse of multiplying. 6 and $15x^5$ are simple expressions to factor and can be factored in different ways.

The factors of the polynomial $2x^2 + 5x - 12$ are not so easy to recognize. Factoring a polynomial changes an addition and/or subtraction problem into a multiplication problem. In this chapter we will be learning techniques for finding the factors of a polynomial. We will begin with **common factors.**

**Example 1**  Factor.  **(a)** $3x - 6y$    **(b)** $9x + 2xy$

**Solution**  Begin by looking for a common factor, a factor that both terms have in common. Then rewrite the expression as a product.

**(a)** $3x - 6y = 3(x - 2y)$    This is true because $3(x - 2y) = 3x - 6y$.

**(b)** $9x + 2xy = x(9 + 2y)$    This is true because $x(9 + 2y) = 9x + 2xy$.

Some people find it helpful to think of factoring as the distributive property in reverse. When we write $3x - 6y = 3(x - 2y)$, we are doing the reverse of distributing the 3. This is the main point of this section. We are doing problems of the form $ca + cb = c(a + b)$. The common factor $c$ becomes the first factor of our answer.  □

▶ **Student Practice 1**  Factor.

**(a)** $21a - 7b$    **(b)** $5xy + 8x$

When we factor, we begin by looking for the **greatest common factor.** For example, in the polynomial $48x - 16y$, a common factor is 2. We could factor $48x - 16y$ as $2(24x - 8y)$. *However, this is not complete.* To factor $48x - 16y$ completely, we look for the greatest common factor of 48 and of 16.

$$48x - 16y = 16(3x - y)$$

**Example 2**  Factor $24xy + 12x^2 + 36x^3$. Remember to remove the greatest common factor.

**Solution**  We start by finding the greatest common factor of 24, 12, and 36. You may want to factor each number, or you may notice that 12 is a common factor. 12 is the greatest numerical common factor.

Notice also that $x$ is a factor of each term. Thus, $12x$ is the greatest common factor.

$$24xy + 12x^2 + 36x^3 = 12x(2y + x + 3x^2)$$  □

▶ **Student Practice 2**  Factor $12a^2 + 16ab^2 - 12a^2b$. Be careful to remove the greatest common factor.

**FACTORING A POLYNOMIAL WITH COMMON FACTORS**

1. Determine the greatest common numerical factor by asking, "What is the largest integer that will divide into the coefficients of all the terms?"
2. Determine the greatest common variable factor by first asking, "What variables are common to all the terms?" Then, for each variable that is common to all the terms, ask, "What is the largest exponent of the variable that is common to all the terms?"
3. The common factor(s) found in steps 1 and 2 are the first part of the answer.
4. After removing the common factor(s), what remains is placed in parentheses as the second factor.

**Example 3** Factor. **(a)** $12x^2 - 18y^2$ **(b)** $x^2y^2 + 3xy^2 + y^3$

**Solution**

**(a)** Note that the largest integer that is common to both terms is 6 (not 3 or 2).

$$12x^2 - 18y^2 = 6(2x^2 - 3y^2)$$

**(b)** Although $y$ is common to all of the terms, we factor out $y^2$ since 2 is the largest exponent of $y$ that is common to all terms. We do not factor out $x$, since $x$ is not common to all of the terms.

$$x^2y^2 + 3xy^2 + y^3 = y^2(x^2 + 3x + y)$$ □

 **Student Practice 3** Factor.

**(a)** $16a^3 - 24b^3$ **(b)** $r^3s^2 - 4r^4s + 7r^5$

*Checking Is Very Important!* You can check any factoring problem by multiplying the factors you obtain. The result must be the same as the original polynomial.

**Example 4** Factor. $8x^3y + 16x^2y^2 - 24x^3y^3$

**Solution** We see that 8 is the largest integer that will divide evenly into the three numerical coefficients. We can factor $x^2$ out of each term. We can also factor $y$ out of each term.

$$8x^3y + 16x^2y^2 - 24x^3y^3 = 8x^2y(x + 2y - 3xy^2)$$

*Check.*

$$8x^2y(x + 2y - 3xy^2) = 8x^3y + 16x^2y^2 - 24x^3y^3 \checkmark$$ □

 **Student Practice 4** Factor. $18a^3b^2c - 27ab^3c^2 - 45a^2b^2c^2$

**Example 5** Factor. $9a^3b^2 + 9a^2b^2$

**Solution** We observe that both terms contain a common factor of 9. We can also factor $a^2$ and $b^2$ out of each term. Thus the greatest common factor is $9a^2b^2$.

$$9a^3b^2 + 9a^2b^2 = 9a^2b^2(a + 1)$$

**CAUTION:** Don't forget to include the 1 inside the parentheses in Example 5. The solution is wrong without it. You will see why if you try to check a result written without the 1. □

**Student Practice 5** Factor and check. $30x^3y^2 - 24x^2y^2 + 6xy^2$

**Example 6** Factor. $3x(x - 4y) + 2(x - 4y)$

**Solution** Be sure you understand what are *terms* and what are *factors* of the polynomial in this example. There are two terms. The expression $3x(x - 4y)$ is the first term. The expression $2(x - 4y)$ is the second term. Each term is made up of two factors.

Observe that the binomial $(x - 4y)$ is a common factor of the terms. A common factor may be any type of polynomial. Thus we can factor out the common factor $(x - 4y)$.

$$3x(x - 4y) + 2(x - 4y) = (x - 4y)(3x + 2) \qquad \square$$

**Student Practice 6** Factor. $3(a + 5b) + x(a + 5b)$

**Example 7** Factor. $7x^2(2x - 3y) - (2x - 3y)$

**Solution** The common factor of the terms is $(2x - 3y)$. What happens when we factor out $(2x - 3y)$? What are we left with in the second term?

Recall that $(2x - 3y) = 1(2x - 3y)$.

Thus $7x^2(2x - 3y) - (2x - 3y)$

$= 7x^2(2x - 3y) - \mathbf{1}(2x - 3y)$ $\qquad$ Rewrite the original expression with 1 as a factor.

$= (2x - 3y)(7x^2 - 1)$ $\qquad$ Factor out $(2x - 3y)$. $\qquad \square$

**Student Practice 7** Factor. $8y(9y^2 - 2) - (9y^2 - 2)$

▲**Example 8** A computer programmer is writing a program to find the total area of 4 circles. She uses the formula $A = \pi r^2$. The radii of the circles are $a$, $b$, $c$, and $d$, respectively. She wants the final answer to be in factored form with the value of $\pi$ occurring only once, in order to minimize the rounding error. Write the total area of the 4 circles with a formula that has $\pi$ occurring only once.

**Solution** For each circle, $A = \pi r^2$, where $r = a, b, c,$ or $d$.

We add the area of each of the 4 circles.

The total area is $\pi a^2 + \pi b^2 + \pi c^2 + \pi d^2$.

In factored form the total area $= \pi(a^2 + b^2 + c^2 + d^2)$. $\qquad \square$

▲ **Student Practice 8** Use $A = \pi r^2$ to find the shaded area. The radius of the larger circle is $b$. The radius of the smaller circle is $a$. Write the shaded-area formula in factored form so that $\pi$ appears only once.

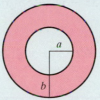

## 5.1 Exercises  MyMathLab®

**Verbal and Writing Skills, Exercises 1–4** *In exercises 1 and 2, write a word or words to complete each sentence.*

**1.** In the expression $3x^2 \cdot 5x^3$, $3x^2$ and $5x^3$ are called _____.

**2.** In the expression $3x^2 + 5x^3$, $3x^2$ and $5x^3$ are called _____.

**3.** We can factor $30a^4 + 15a^3 - 45a^2$ as $5a(6a^3 + 3a^2 - 9a)$. Is the factoring complete? Why or why not?

**4.** We can factor $4x^3 - 8x^2 + 20x$ as $4(x^3 - 2x^2 + 5x)$. Is the factoring complete?

*Remove the greatest common factor. Check your answers for exercises 5–28 by multiplication.*

**5.** $8a^2 + 8a$

**6.** $7b^2 + 7b$

**7.** $21ab - 14ab^2$

**8.** $24ab - 18ab^2$

**9.** $2\pi rh + 2\pi r^2$

**10.** $5a^2b^2 - 35ab$

**11.** $5x^3 + 25x^2 - 15x$

**12.** $8x^3 - 10x^2 - 14x$

**13.** $12ab - 28bc + 20ac$

**14.** $14xy + 21yz - 42xz$

**15.** $16x^5 + 24x^3 - 32x^2$

**16.** $36x^6 + 45x^4 - 18x^2$

**17.** $14x^2y - 35xy - 63x$

**18.** $40a^2 - 16ab - 24a$

**19.** $54x^2 - 45xy + 18x$

**20.** $48xy - 24y^2 + 40y$

**21.** $3xy^2 - 2ay + 5xy - 2y$

**22.** $2ab^3 + 3xb^2 - 5b^4 + 2b^2$

**23.** $24x^2y - 40xy^2$

**24.** $35abc^2 - 49ab^2c$

**25.** $7x^3y^2 + 21x^2y^2$

**26.** $8x^3y^2 + 32xy^2$

**27.** $16x^4y^2 - 24x^2y^2 - 8x^2y$

**28.** $18x^2y^2 - 12xy^3 + 6xy$

*Hint: In exercises 29–42, refer to Examples 6 and 7.*

**29.** $7a(x + 2y) - b(x + 2y)$

**30.** $6(3a + b) - z(3a + b)$

**31.** $3x(x - 4) - 2(x - 4)$

**32.** $5x(x - 7) + 3(x - 7)$

**33.** $6b(2a - 3c) - 5d(2a - 3c)$

**34.** $7x(3y + 5z) - 6t(3y + 5z)$

**35.** $7c(b - a^2) - 5d(b - a^2) + 2f(b - a^2)$

**36.** $2a(x^2 - y) - 5b(x^2 - y) + c(x^2 - y)$

**37.** $3a(ab - 4) - 5(ab - 4) - b(ab - 4)$

**38.** $4b(3ab - 2) + 2(3ab - 2) - 5a(3ab - 2)$

**39.** $4a^3(a - 3b) + (a - 3b)$

**40.** $2b^2(2a - 5b) + (2a - 5b)$

**41.** $(a + 2) - x(a + 2)$

**42.** $y(3x - 2) - (3x - 2)$

*Use the figure below for exercises 43 and 44.*

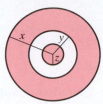

▲ **43.** Use $C = 2\pi r$ to find the circumference of each circle. The radii of the circles are $x$, $y$, and $z$, respectively. Write the sum of the circumferences of all three circles in factored form so $\pi$ appears only once.

▲ **44.** Use $A = \pi r^2$ to find the shaded area. The radii of the circles are $x$, $y$, and $z$, respectively. Write the shaded-area formula in factored form so $\pi$ appears only once.

## Cumulative Review

*Coffee Production In a recent year, the top six coffee-producing countries of the world produced 6,400,000 metric tons of coffee beans. The percent of this 6.40 million metric tons produced by each of these countries is shown in the following graph. Use the graph to answer the following questions. (Source: worldatlas.com)*

**Percent of Coffee Produced by the Top Six Coffee-Producing Countries of the World**

Ethiopia 6% — India 5%
Indonesia 8% —
Brazil 43%
Colombia 12% —
Vietnam 26% —

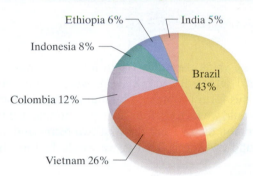

**45.** [0.5.3] How many metric tons of coffee were produced in Vietnam?

**46.** [0.5.3] How many metric tons of coffee were produced in Brazil?

**47.** [0.6.1] The population of Vietnam in 2015 was approximately 89,700,000 people. How many pounds of coffee were produced in Vietnam for each person? (Round your answer to the nearest whole number. One metric ton is about 2205 pounds.)

**48.** [0.6.1] The population of Brazil in 2015 was approximately 200,400,000 people. How many pounds of coffee were produced in Brazil for each person? (Round your answer to the nearest whole number. One metric ton is about 2205 pounds.)

**Quick Quiz 5.1** *Remove the greatest common factor.*

**1.** $3x - 4x^2 + 2xy$

**2.** $20x^3 - 25x^2 - 5x$

**3.** $8a(a + 3b) - 7b(a + 3b)$

**4. Concept Check** Explain how you would remove the greatest common factor from the following polynomial.
$$36a^3b^2 - 72a^2b^3$$

# 5.2 Factoring by Grouping ▶

**Student Learning Objective**

After studying this section, you will be able to:

1. Factor expressions with four terms by grouping. ▶

## 1 Factoring Expressions with Four Terms by Grouping ▶

A common factor of a polynomial can be a number, a variable, or an algebraic expression. Sometimes the polynomial is written so that it is easy to recognize the common factor. This is especially true when the common factor is enclosed by parentheses.

**Example 1** Factor. $x(x - 3) + 2(x - 3)$

**Solution** Observe each term:

$$\underbrace{x(x - 3)}_{\substack{\text{first} \\ \text{term}}} + \underbrace{2(x - 3)}_{\substack{\text{second} \\ \text{term}}}$$

The common factor of the first and second terms is the quantity $(x - 3)$, so we have

$$x(x - 3) + 2(x - 3) = (x - 3)(x + 2). \qquad \square$$

 **Student Practice 1** Factor. $3y(2x - 7) - 8(2x - 7)$

　　　Now let us face a new challenge. Think carefully. Try to follow this new idea. Suppose the polynomial in Example 1, $x(x - 3) + 2(x - 3)$, were written in the form $x^2 - 3x + 2x - 6$. (Note that this form is obtained by multiplying the factors of the first and second terms.) How would we factor a four-term polynomial like this?

　　　In such cases we remove a common factor from the first two terms and a different common factor from the second two terms. That is, we would factor $x$ from $x^2 - 3x$ and 2 from $2x - 6$.

$$x^2 - 3x + 2x - 6 = x(x - 3) + 2(x - 3)$$

Because the resulting terms have a common factor (the binomial enclosed by the parentheses), we would then proceed as we did in Example 1 to obtain the factored form $(x - 3)(x + 2)$. This procedure for factoring is often called **factoring by grouping.**

**Example 2** Factor. $2x^2 + 3x + 6x + 9$

**Solution**

$$2x^2 + 3x \qquad + \qquad 6x + 9$$

| Factor out a common factor of $x$ from the first two terms. | Factor out a common factor of 3 from the second two terms. |

$$\underbrace{x(2x + 3)}_{\Big\uparrow} \qquad\qquad \underbrace{3(2x + 3)}_{\Big\uparrow}$$

　　　Note that the sets of parentheses in the two terms contain the same expression at this step.

　　　The expression in parentheses is now a common factor of the terms. Now we finish the factoring.

$$2x^2 + 3x + 6x + 9 = x(2x + 3) + 3(2x + 3)$$
$$= (2x + 3)(x + 3) \qquad \square$$

 **Student Practice 2** Factor. $6x^2 - 15x + 4x - 10$

**Example 3** Factor by grouping. $4x + 8y + ax + 2ay$

**Solution**

Factor out a common
factor of 4 from
the first two terms.

$$\overbrace{4x + 8y} + \underbrace{ax + 2ay} = \overbrace{4(x + 2y)} + \underbrace{a(x + 2y)}$$

Factor out a common
factor of $a$ from
the second two terms.

$4(x + 2y) + a(x + 2y) = (x + 2y)(4 + a)$  The common factor of the terms
is the expression in parentheses,
$x + 2y$.

▭ **Student Practice 3**  Factor by grouping. $ax + 2a + 4bx + 8b$

In some problems the terms are out of order. In this case, we have to rearrange
the order of the terms first so that the first two terms have a common factor.

**Example 4** Factor. $bx + 4y + 4b + xy$

**Solution**

$bx + 4y + 4b + xy = bx + 4b + xy + 4y$  Rearrange the terms so that the first
two terms have a common factor.

$\qquad\qquad\qquad = b(x + 4) + y(x + 4)$  Factor out the common factor $b$ from
the first two terms and the common
factor $y$ from the second two terms.

$\qquad\qquad\qquad = (x + 4)(b + y)$  ▭

▭ **Student Practice 4**  Factor. $6a^2 + 5bc + 10ab + 3ac$

Sometimes you will need to *factor out a negative common factor* from the second two terms to obtain two terms that contain the same parenthetical expression.

**Example 5** Factor. $2x^2 + 5x - 4x - 10$

**Solution**

$2x^2 + 5x - 4x - 10 = x(2x + 5) - 4x - 10$  Factor out the common factor $x$
from the first two terms.

$\qquad\qquad\qquad\qquad = x(2x + 5) - 2(2x + 5)$  Factor out the common factor $-2$
from the second two terms.

$\qquad\qquad\qquad\qquad = (2x + 5)(x - 2)$

**CAUTION:** If you factored out the common factor $+2$ in the second step, the two
resulting terms would not contain the same parenthetical expression. If the expressions inside the two sets of parentheses are *not exactly the same*, you cannot express
the polynomial as a product of two factors!  ▭

▭ **Student Practice 5**  Factor. $6xy + 14x - 15y - 35$

$\mathbb{M}_C$ **Example 6** Factor. $2ax - a - 2bx + b$

**Solution**

$2ax - a - 2bx + b$
$= a(2x - 1) - b(2x - 1)$

Factor out the common factor $a$ from the first two terms. Factor out the common factor $-b$ from the second two terms.

$= (2x - 1)(a - b)$

Since the two resulting terms contain the same parenthetical expression, we can complete the factoring. □

 **Student Practice 6**   Factor.

$$3x + 6y - 5ax - 10ay$$

**CAUTION:** Many students find that they make a factoring error in the first step of problems like Example 6. When factoring out $-b$, be sure to check your signs carefully: $-2bx + b = -b(2x - 1)$.

**Example 7**  Factor and check your answer. $8ad + 21bc - 6bd - 28ac$

**Solution**   We observe that the first two terms do not have a common factor.

$8ad + 21bc - 6bd - 28ac$
$= 8ad - 6bd - 28ac + 21bc$

Rearrange the order using the commutative property of addition.

$= 2d(4a - 3b) - 7c(4a - 3b)$

Factor out the common factor $2d$ from the first two terms and the common factor $-7c$ from the last two terms.

$= (4a - 3b)(2d - 7c)$

Factor out the common factor $(4a - 3b)$.

To check, we multiply the two binomials using the FOIL procedure.

$(4a - 3b)(2d - 7c) = 8ad - 28ac - 6bd + 21bc$
$= 8ad + 21bc - 6bd - 28ac$ ✓

Rearrange the order of the terms. This is the original problem. Thus it checks. □

 **Student Practice 7**   Factor and check your answer.

$$10ad + 27bc - 6bd - 45ac$$

## 5.2 Exercises  MyMathLab®

**Verbal and Writing Skills, Exercises 1 and 2**

1. To factor $3x^2 - 6xy + 5x - 10y$, we must first remove a common factor of $3x$ from the first two terms. What do we do with the last two terms? What should we get for the answer?

2. To factor $5x^2 + 15xy - 2x - 6y$, we must first remove a common factor of $5x$ from the first two terms. What do we do with the last two terms? What should we get for the answer?

*Factor by grouping. Check your answers for exercises 3–26.*

3. $ab - 4a + 6b - 24$

4. $xy - 2x + 7y - 14$

5. $x^3 - 4x^2 + 3x - 12$

6. $x^3 - 8x^2 + 3x - 24$

7. $2ax + 6bx - ay - 3by$

8. $4y - 12x - 3yw + 9xw$

9. $3ax + bx - 6a - 2b$

10. $ad + 3a - d^2 - 3d$

11. $5a + 12bc + 10b + 6ac$

12. $4u^2 + v + 4uv + u$

13. $6c - 12d + cx - 2dx$

14. $xy + 5x - 5y - 25$

15. $y^2 - 2y - 3y + 6$

16. $ax - 3a - 2bx + 6b$

17. $54 - 6y + 9y - y^2$

18. $35 - 5a + 7a - a^2$

19. $6ax - y + 2ay - 3x$

20. $6tx + r - 3t - 2rx$

21. $2x^2 + 8x - 3x - 12$

22. $3y^2 - y + 9y - 3$

**23.** $t^3 - t^2 + t - 1$

**24.** $x^2 - 2x - xy + 2y$

**25.** $6x^2 + 15xy^2 + 8xw + 20y^2w$

**26.** $10xw + 14x^2 + 25wy^2 + 35xy^2$

## To Think About

**27.** Although $6a^2 - 12bd - 8ad + 9ab = 6(a^2 - 2bd) - a(8d - 9b)$ is true, it is not the correct solution to the problem "Factor $6a^2 - 12bd - 8ad + 9ab$." Explain. Can this expression be factored?

**28.** Tim was trying to factor $5x^2 - 3xy - 10x + 6y$. In his first step he wrote down $x(5x - 3y) + 2(-5x + 3y)$. Was he doing the problem correctly? What is the answer?

## Cumulative Review

**29.** [1.3.3] Divide. $\dfrac{6}{7} \div \left(-\dfrac{2}{5}\right)$

**30.** [1.1.5] Add. $-\dfrac{2}{3} + \dfrac{4}{5}$

**31.** [4.1.2] Simplify. $\dfrac{-5a^2b^8}{25ab^{10}}$

**32.** [4.5.2] Multiply. $(2x - 5)^2$

**33.** [3.3.1] *Salaries in the Pharmacy Industry* In 2015, the average annual salary of a pharmacist in the United States was $22,000 more than triple the average annual salary of a pharmacy technician. The two average annual salaries total $162,000. Find the average salary of a pharmacist and the average salary of a pharmacy technician. (*Source:* www.pharmacist.com)

**34.** [3.4.2] *Oil Production in North America* In 2011, in North America 17,000 thousand barrels of oil were produced each day. By 2013, this figure had increased by 13% in North America. By 2015, this figure had increased by 11% from the amount produced in 2013. How many thousand barrels of oil were produced each day in North America in 2015?

---

### Quick Quiz 5.2 *Factor by grouping.*

**1.** $7ax + 12a - 14x - 24$

**2.** $2xy^2 - 15 + 6x - 5y^2$

**3.** $10xy - 3x + 40by - 12b$

**4.** **Concept Check** Explain how you would factor the following polynomial.

$$10ax + b^2 + 2bx + 5ab$$

# 5.3 Factoring Trinomials of the Form $x^2 + bx + c$ ▶

## 1 Factoring Polynomials of the Form $x^2 + bx + c$ ▶

Suppose that you wanted to factor $x^2 + 5x + 6$. After some trial and error you *might* obtain $(x + 2)(x + 3)$, or you might get discouraged and not get an answer. If you did get these factors, you could check this answer by the FOIL method.

$$(x + 2)(x + 3) = x^2 + 3x + 2x + 6$$
$$= x^2 + 5x + 6$$

But trial and error can be a long process. There is another way. Let's look at the preceding equation again.

$$
\begin{array}{c}
\text{F} \quad\text{O}\quad\text{I}\quad\text{L} \\
(x + 2)(x + 3) = x^2 + \underbrace{3x + 2x} + 6 \\
= x^2 \quad + 5x \quad + 6
\end{array}
$$

The first thing to notice is that the product of the first terms in the factors gives the first term of the polynomial. That is, $x \cdot x = x^2$.

→ The first term is the product of these terms.

$x^2 + 5x + 6 \qquad = \qquad (x + 2)(x + 3)$

The next thing to notice is that the sum of the products of the outer and inner terms in the factors produces the middle term of the polynomial. That is, $(x \cdot 3) + (2 \cdot x) = 3x + 2x = 5x$. Thus we see that the sum of the second terms in the factors, $2 + 3$, gives the coefficient of the middle term, 5.

Finally, note that the product of the last terms of the factors gives the last term of the polynomial. That is, $2 \cdot 3 = 6$.

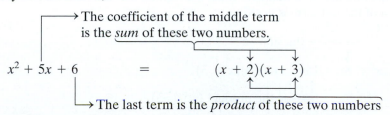

→ The coefficient of the middle term is the *sum* of these two numbers.

$x^2 + 5x + 6 \qquad = \qquad (x + 2)(x + 3)$

→ The last term is the *product* of these two numbers

Let's summarize our observations in general terms and then try a few examples.

> ### FACTORING TRINOMIALS OF THE FORM $x^2 + bx + c$
>
> 1. The answer will be of the form $(x + m)(x + n)$.
> 2. $m$ and $n$ are numbers such that:
>    (a) When you multiply them, you get the last term, which is $c$.
>    (b) When you add them, you get the coefficient of the middle term, which is $b$.

**Example 1** Factor. $x^2 + 7x + 12$

**Solution** The answer will be of the form $(x + m)(x + n)$. We want to find the two numbers, $m$ and $n$, that you can multiply to get 12 and add to get 7. The numbers are 3 and 4.

$$x^2 + 7x + 12 = (x + 3)(x + 4) \qquad \square$$

▶ **Student Practice 1** Factor. $x^2 + 8x + 12$

### Student Learning Objectives

After studying this section, you will be able to:

1 Factor polynomials of the form $x^2 + bx + c$. ▶

2 Factor polynomials that have a common factor and a factor of the form $x^2 + bx + c$. ▶

**Example 2** Factor. $x^2 + 12x + 20$

**Solution** We want two numbers that have a product of 20 and a sum of 12. The numbers are 10 and 2.

$$x^2 + 12x + 20 = (x + \underline{10})(x + \underline{2})$$

*Note:* If you cannot think of the numbers in your head, write down the possible factors whose product is 20.

| Product | Sum |
|---|---|
| $1 \cdot 20 = 20$ | $1 + 20 = 21$ |
| $2 \cdot 10 = 20$ | $2 + 10 = 12 \leftarrow$ |
| $4 \cdot 5 = 20$ | $4 + 5 = 9$ |

Then select the pair whose sum is 12.                    Select this pair.

**Student Practice 2** Factor. $x^2 + 17x + 30$

You may find that it is helpful to first list all the factors whose product is 30.

So far we have factored only trinomials of the form $x^2 + bx + c$, where $b$ and $c$ are positive numbers. The same procedure applies if $b$ is a negative number and $c$ is positive. Because $m$ and $n$ have a positive product and a negative sum, they must both be negative.

**Example 3** Factor. $x^2 - 8x + 15$

**Solution** We want two numbers that have a product of $+15$ and a sum of $-8$. They must be negative numbers since the sign of the middle term is negative and the sign of the last term is positive.

the sum $-5 + (-3)$

$$x^2 - 8x + 15 = (x - 5)(x - 3)$$

the product $(-5)(-3)$

*Think:* $(-5)(-3) = +15$
*and* $-5 + (-3) = -8.$

Multiply using FOIL to check.

**Student Practice 3** Factor. $x^2 - 11x + 18$

**Example 4** Factor. $x^2 - 9x + 14$

**Solution** We want two numbers whose product is 14 and whose sum is $-9$. The numbers are $-7$ and $-2$. So

$$x^2 - 9x + 14 = (x - 7)(x - 2) \text{ or } (x - 2)(x - 7).$$

**Student Practice 4** Factor. $x^2 - 11x + 24$

All the examples so far have had a positive last term. What happens when the last term is negative? If the last term is a negative number, one of the numbers $m$ or $n$ must be a positive number and the other must be a negative number. Why? The product of a positive number and a negative number is negative.

**Example 5** Factor. $x^2 - 3x - 10$

**Solution**   We want two numbers whose product is $-10$ and whose sum is $-3$. The two numbers are $-5$ and $+2$.

$$x^2 - 3x - 10 = (x - 5)(x + 2)$$     □

**Student Practice 5**   Factor. $x^2 - 5x - 24$

What if we made a sign error and *incorrectly* factored the trinomial $x^2 - 3x - 10$ as $(x + 5)(x - 2)$? We could detect the error immediately since the sum of $+5$ and $-2$ is 3. We need a sum of $-3$!

**Example 6** Factor and check your answer. $y^2 + 10y - 24$

**Solution**   The two numbers whose product is $-24$ and whose sum is $+10$ are $+12$ and $-2$.

$$y^2 + 10y - 24 = (y + 12)(y - 2)$$

**CAUTION:** It is very easy to make a sign error in these problems. Make sure that you mentally multiply your answer by FOIL to obtain the original expression. Check each sign carefully.

*Check.*   $(y + 12)(y - 2) = y^2 - 2y + 12y - 24 = y^2 + 10y - 24$ ✓     □

**Student Practice 6**   Factor $y^2 + 17y - 60$.
Multiply your answer to check.

**Example 7**   Factor. $x^2 - 16x - 36$

**Solution**   We want two numbers whose product is $-36$ and whose sum is $-16$.

List all the possible factors of 36 (without regard to sign). Find the pair that has a difference of 16. We are looking for a difference because the signs of the factors are different.

| *Factors of 36* | *The Difference Between the Factors* |
|---|---|
| 36 and 1 | 35 |
| 18 and 2 | 16 ← This is the value we want. |
| 12 and 3 | 9 |
| 9 and 4 | 5 |
| 6 and 6 | 0 |

Once we have picked the pair of numbers (18 and 2), it is not difficult to find the signs. For the coefficient of the middle term to be $-16$, we will have to add the numbers $-18$ and $+2$.

$$x^2 - 16x - 36 = (x - 18)(x + 2)$$     □

**Student Practice 7**   Factor. $x^2 - 7x - 60$
You may find it helpful to list the pairs of numbers whose product is 60.

At this point you should work several problems to develop your factoring skill. This is one section where you really need to drill by doing many problems.

Feel a little confused about the signs? If you do, you may find these facts helpful.

---

**FACTS ABOUT FACTORING TRINOMIALS OF THE FORM $x^2 + bx + c$**

The *two numbers m* and *n* will have the *same sign* if the last term of the polynomial is *positive*.

1. They will both be *positive* if the coefficient of the *middle* term is *positive*.

$$x^2 + bx + c = (x \quad m)(x \quad n)$$
$$x^2 + 5x + 6 = (x + 2)(x + 3)$$

2. They will both be *negative* if the coefficient of the *middle* term is *negative*.

$$x^2 - 5x + 6 = (x - 2)(x - 3)$$

The two numbers *m* and *n* will have *opposite signs* if the last term is *negative*.

1. The *larger* of the absolute values of the two numbers will be given a plus sign if the coefficient of the *middle term* is *positive*.

$$x^2 + 6x - 7 = (x + 7)(x - 1)$$

2. The larger of the absolute values of the two numbers will be given a negative sign if the coefficient of the *middle term* is *negative*.

$$x^2 - 6x - 7 = (x - 7)(x + 1)$$

---

Do not memorize these facts; rather, try to understand the pattern.

Sometimes the exponent of the first term of the polynomial will be greater than 2. If the exponent is an even power, it is a square. For example, $x^4 = (x^2)(x^2)$. Likewise, $x^6 = (x^3)(x^3)$.

**Example 8** Factor. $y^4 - 2y^2 - 35$

**Solution** Think: $y^4 = (y^2)(y^2)$ This will be the first term of each set of parentheses.

$$(y^2 \quad )(y^2 \quad )$$

$$(y^2 + \ )(y^2 - \ ) \qquad \text{The last term of the polynomial is negative.}$$
Thus the signs of *m* and *n* will be different.

$$(y^2 + 5)(y^2 - 7) \qquad \text{Now think of factors of 35 whose difference is 2.}$$

Multiply using FOIL to check. ☐

**Student Practice 8** Factor. $a^4 + a^2 - 42$

**2** Factoring Polynomials That Have a Common Factor and a Factor of the Form $x^2 + bx + c$ ▶

Some factoring problems require two steps. Often we must first factor out a common factor from each term of the polynomial. Once this is done, we may find that the other factor is a trinomial that can be factored using the methods previously discussed in this section.

**Example 9** Factor. $2x^2 + 36x + 160$

**Solution**

$2x^2 + 36x + 160 = 2(x^2 + 18x + 80)$    First factor out the common factor 2 from each term of the polynomial.

$= 2(x + 8)(x + 10)$    Then factor the remaining polynomial.

The final answer is $2(x + 8)(x + 10)$. *Be sure to list all parts of the answer.*

*Check.* $2(x + 8)(x + 10) = 2(x^2 + 18x + 80) = 2x^2 + 36x + 160$ ✓

Thus we are sure that the answer is $2(x + 8)(x + 10)$.  ☐

**Student Practice 9** Factor. $3x^2 + 45x + 150$

**Example 10** Factor. $3x^2 + 9x - 162$

**Solution**

$3x^2 + 9x - 162 = 3(x^2 + 3x - 54)$    First factor out the common factor 3 from each term of the polynomial.

$= 3(x - 6)(x + 9)$    Then factor the remaining polynomial.

The final answer is $3(x - 6)(x + 9)$. *Be sure you include the 3.*

*Check.* $3(x - 6)(x + 9) = 3(x^2 + 3x - 54) = 3x^2 + 9x - 162$ ✓

Thus we are sure that the answer is $3(x - 6)(x + 9)$.  ☐

**Student Practice 10**

Factor. $4x^2 - 8x - 140$

**CAUTION:** Don't forget the common factor!

It is quite easy to forget to look for a greatest common factor as the first step of factoring a trinomial. Therefore, it is a good idea to examine your final answer in any factoring problem and ask yourself, "Can I factor out a common factor from any binomial contained inside a set of parentheses?" Often you will be able to see a common factor at that point if you missed it in the first step of the problem.

▲ **Example 11** Find a polynomial in factored form for the shaded area in the figure.

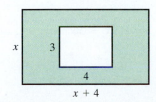

**Solution**    To obtain the shaded area, we find the area of the larger rectangle and subtract from it the area of the smaller rectangle. Thus we have the following:

shaded area $= x(x + 4) - (4)(3)$

$= x^2 + 4x - 12$

Now we factor this polynomial to obtain the shaded area $= (x + 6)(x - 2)$.  ☐

▲ **Student Practice 11**    Find a polynomial in factored form for the shaded area in the figure.

**Student Practice 11**

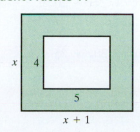

## (5.3) Exercises MyMathLab®

**Verbal and Writing Skills, Exercises 1 and 2**

*Fill in the blanks.*

1. To factor $x^2 + 5x + 6$, find two numbers whose _____ is 6 and whose _____ is 5.
2. To factor $x^2 + 5x - 6$, find two numbers whose _____ is $-6$ and whose _____ is 5.

*Factor.*

3. $x^2 + 8x + 16$
4. $x^2 + 13x + 42$
5. $x^2 + 12x + 35$
6. $x^2 + 10x + 21$

7. $x^2 - 4x + 3$
8. $x^2 - 8x + 15$
9. $x^2 - 11x + 28$
10. $x^2 - 13x + 12$

11. $x^2 + 5x - 24$
12. $x^2 + 5x - 36$
13. $x^2 - 13x - 14$
14. $x^2 - 6x - 16$

15. $x^2 + 2x - 35$
16. $x^2 + x - 12$
17. $x^2 - 2x - 24$
18. $x^2 - 11x - 26$

19. $x^2 + 15x + 36$
20. $x^2 + 15x + 44$
21. $x^2 - 10x + 24$
22. $x^2 - 13x + 42$

23. $x^2 + 13x + 30$
24. $x^2 + 9x + 20$
25. $x^2 - 6x + 5$
26. $x^2 - 15x + 54$

**Mixed Practice** *Look over your answers to exercises 3–26 carefully. Be sure that you are clear on your sign rules. Exercises 27–42 contain a mixture of all the types of problems in this section. Make sure you can do them all. Check your answers by multiplication.*

27. $a^2 + 6a - 16$
28. $a^2 - 10a + 24$
29. $x^2 - 12x + 32$
30. $x^2 - 6x - 27$

31. $x^2 + 4x - 21$
32. $x^2 - 9x + 18$
33. $x^2 + 15x + 56$
34. $x^2 + 20x + 99$

35. $y^2 + 4y - 45$
36. $x^2 + 12x - 45$
37. $x^2 + 9x - 36$
38. $x^2 - 13x + 36$

39. $x^2 - 2xy - 15y^2$
40. $x^2 - 3xy - 18y^2$
41. $x^2 - 16xy + 63y^2$
42. $x^2 + 19xy + 48y^2$

*In exercises 43–54, first factor out the greatest common factor from each term. Then factor the remaining polynomial. Refer to Examples 9 and 10.*

43. $4x^2 + 24x + 20$
44. $4x^2 + 28x + 40$
45. $6x^2 + 18x + 12$
46. $6x^2 + 24x + 18$

47. $5x^2 - 30x + 25$
48. $2x^2 - 20x + 32$
49. $3x^2 - 6x - 72$
50. $3x^2 - 18x - 48$

51. $7x^2 + 21x - 70$
52. $4x^2 - 8x - 60$
53. $5x^2 - 35x + 30$
54. $7x^2 - 35x + 42$

▲ **55.** *Geometry* Find a polynomial in factored form for the shaded area. Both figures are rectangles with dimensions as labeled.

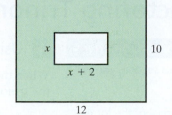

▲ **56.** *Geometry* How much larger is the perimeter of the large rectangle than the perimeter of the small rectangle?

## Cumulative Review

**57.** [4.1.1] Multiply. $(9ab^3)(2a^5b^6c^0)$

**58.** [4.1.3] Simplify. $(-5y^6)^2$

**59.** [4.2.1] Simplify. Write the expression with no negative exponents. $\dfrac{x^4y^{-3}}{x^{-2}y^5}$

**60.** [4.4.2] Multiply. $(2x + 3y)(4x - 2y)$

**61.** [3.2.3] *Travel Speed* A new car that maintains a constant speed travels from Watch Hill, Rhode Island, to Greenwich, Connecticut, in 2 hours. A train, traveling 20 mph faster, makes the trip in $1\frac{1}{2}$ hours. How far is it from Watch Hill to Greenwich? (*Hint:* Let $c =$ the car's speed. First find the speed of the car and the speed of the train. Then be sure to answer the question.)

**62.** [0.6.1] *Salary* Carla works in an electronics store. She is paid $600 per month plus 4% commission on her sales. If Carla's sales for the month are $80,000, how much will she earn for the month?

*Average Temperature* *The equation $T = 19 + 2M$ has been used by some meteorologists to predict the monthly average temperature for the small island of Menorca off the coast of Spain during the first 6 months of the year. The variable T represents the average monthly temperature measured in degrees Celsius. The variable M represents the number of months since January.*

**63.** [1.8.2] What is the average temperature of Menorca during the month of April?

**64.** [1.8.2] During what month is the average temperature 29°C?

**Quick Quiz 5.3** *Factor completely.*

**1.** $x^2 + 17x + 70$

**2.** $x^2 - 14x + 48$

**3.** $2x^2 - 4x - 96$

**4.** **Concept Check** Explain how you would completely factor $4x^2 - 4x - 120$.

# 5.4 Factoring Trinomials of the Form $ax^2 + bx + c$

**Student Learning Objectives**

After studying this section, you will be able to:

1 Factor a trinomial of the form $ax^2 + bx + c$ by the trial-and-error method.

2 Factor a trinomial of the form $ax^2 + bx + c$ by the grouping method.

3 Factor a trinomial of the form $ax^2 + bx + c$ after a common factor has been factored out of each term.

## 1 Using the Trial-and-Error Method

When the coefficient of the $x^2$-term in a trinomial of the form $ax^2 + bx + c$ is not 1, the trinomial is more difficult to factor. Several possibilities must be considered.

### Example 1 Factor. $2x^2 + 5x + 3$

**Solution** In order for the coefficient of the $x^2$-term of the polynomial to be 2, the coefficients of the $x$-terms in the factors must be 2 and 1.

$$\text{Thus } 2x^2 + 5x + 3 = (2x \quad)(x \quad).$$

In order for the last term of the polynomial to be 3, the constants in the factors must be 3 and 1.

Since all signs in the polynomial are positive, we know that each factor in parentheses will contain only positive signs. However, we still have two possibilities. They are as follows:

$$(2x + 1)(x + 3)$$
$$(2x + 3)(x + 1)$$

We check them by multiplying by the FOIL method.

$$(2x + 1)(x + 3) = 2x^2 + 7x + 3 \quad \text{Wrong middle term}$$
$$(2x + 3)(x + 1) = 2x^2 + 5x + 3 \quad \text{Correct middle term}$$

Thus the correct answer is

$$(2x + 3)(x + 1) \quad \text{or} \quad (x + 1)(2x + 3). \qquad \square$$

**Student Practice 1** Factor. $2x^2 + 7x + 5$

Some problems have many more possibilities.

### Example 2 Factor. $4x^2 - 13x + 3$

**Solution**

| *The Different Factorizations of 4 Are:* | *The Factorization of 3 Is:* |
|---|---|
| $(2)(2)$ | $(1)(3)$ |
| $(1)(4)$ | |

Let us list the possible factoring combinations and compute the middle term by the FOIL method. Note that the signs of the constants in both factors will be negative. Why?

| *Possible Factors* | *Middle Term* | *Correct?* |
|---|---|---|
| $(2x - 3)(2x - 1)$ | $-8x$ | No |
| $(4x - 3)(x - 1)$ | $-7x$ | No |
| $(4x - 1)(x - 3)$ | $-13x$ | Yes |

The correct answer is $(4x - 1)(x - 3)$ or $(x - 3)(4x - 1)$.
This method is called the **trial-and-error method.** $\qquad \square$

**Student Practice 2** Factor. $9x^2 - 64x + 7$

**Example 3** Factor. $3x^2 - 2x - 8$

**Solution**

| ***Factorization of 3*** | ***Factorizations of 8*** |
|---|---|
| $(3)(1)$ | $(8)(1)$ |
| | $(4)(2)$ |

Let us list only one-half of the possibilities. We'll let the constant in the first factor of each product be positive.

| ***Possible Factors*** | ***Middle Term*** | ***Correct Factors?*** |
|---|---|---|
| $(x + 8)(3x - 1)$ | $+23x$ | No |
| $(x + 1)(3x - 8)$ | $-5x$ | No |
| $(x + 4)(3x - 2)$ | $+10x$ | No |
| $(x + 2)(3x - 4)$ | $+2x$ | No (but only because the sign is wrong) |

So we just *reverse* the signs of the constants in the factors.

| | ***Middle Term*** | ***Correct Factors?*** |
|---|---|---|
| $(x - 2)(3x + 4)$ | $-2x$ | Yes |

The correct answer is $(x - 2)(3x + 4)$   or   $(3x + 4)(x - 2)$. □

 **Student Practice 3**   Factor. $3x^2 - x - 14$

**Example 4** Factor. $6x^2 + 5x - 4$

**Solution**

| ***Factorizations of 6*** | ***Factorizations of 4*** |
|---|---|
| $(6)(1)$ | $(4)(1)$ |
| $(2)(3)$ | $(2)(2)$ |

We list a few of the possibilities (without regard to the sign) starting with $(2x)(3x)$ for the first term of each product.

$$(2x \quad 4)(3x \quad 1) \quad (2x \quad 1)(3x \quad 4) \quad (2x \quad 2)(3x \quad 2)$$

Look at the options below. Using the FOIL method, which option would yield a middle term $+5x$ when adding the outer and inner products?

$$(2x \quad 4)(3x \quad 1) \quad (2x \quad 1)(3x \quad 4) \quad (2x \quad 2)(3x \quad 2)$$

$$12x \qquad 3x \qquad 6x$$
$$2x \qquad 8x \qquad 4x$$

No possible sum of $+5x$ — $5x$ ✓ $+5x$ is possible with $-3x$ and $+8x$ — No possible sum of $+5x$

The middle possibility yields $+5x$ if we make the $3x$ negative. Once we have picked the pair of numbers, it is not difficult to find the signs and write the factors $(2x - 1)(3x + 4)$.

Now check your answer using FOIL to be sure you obtain the original expression. Pay particular attention to the last term; it must be $-4$.

*Check.* $(2x - 1)(3x + 4) = 6x^2 + 5x - 4$  ✓

*Note*: If you cannot find the factors using $(2)(3)$ for the coefficient of $x^2$, then try the same method using $(6)(1)$. □

**Student Practice 4**   Factor. $10x^2 + 7x - 6$

It takes a good deal of practice to readily factor problems of this type. The more problems you do, the more proficient you will become. The following method may help you factor more quickly.

## 2 Using the Grouping Method

One way to factor a trinomial of the form $ax^2 + bx + c$ is to write it with four terms and factor by grouping, as we did in Section 5.2. For example, the trinomial $2x^2 + 13x + 20$ can be written as $2x^2 + 5x + 8x + 20$. Using the methods of Section 5.2, we factor it as follows.

$$2x^2 + 5x + 8x + 20 = x(2x + 5) + 4(2x + 5)$$
$$= (2x + 5)(x + 4)$$

We can factor all factorable trinomials of the form $ax^2 + bx + c$ in this way. We will use the following procedure.

---

**GROUPING METHOD FOR FACTORING TRINOMIALS OF THE FORM $ax^2 + bx + c$**

1. Obtain the grouping number $ac$.
2. Find the two numbers whose product is the grouping number and whose sum is $b$.
3. Use those numbers to write $bx$ as the sum of two terms.
4. Factor by grouping.
5. Multiply to check.

---

**Example 5** Factor by grouping. $3x^2 - 2x - 8$

**Solution**

1. The grouping number is $(3)(-8) = -24$.
2. We want two numbers whose product is $-24$ and whose sum is $-2$. They are $-6$ and $4$.
3. We write $-2x$ as the sum $-6x + 4x$.
4. Factor by grouping.

$$3x^2 - 6x + 4x - 8 = 3x(x - 2) + 4(x - 2)$$
$$= (x - 2)(3x + 4) \qquad \square$$

**Student Practice 5**  Factor by grouping. $3x^2 + 4x - 4$

**Example 6** Factor by grouping. $4x^2 - 13x + 3$

**Solution**

1. The grouping number is $(4)(3) = 12$.
2. The factors of 12 are $(12)(1)$ or $(4)(3)$ or $(6)(2)$. Note that the middle term of the polynomial is negative. Thus we choose the numbers $-12$ and $-1$ because their product is still 12 and their sum is $-13$.
3. We write $-13x$ as the sum $-12x + (-1x)$ or $-12x - 1x$.
4. Factor by grouping.

$$4x^2 - 13x + 3 = 4x^2 - 12x - 1x + 3$$
$$= 4x(x - 3) - 1(x - 3) \quad \text{Remember to factor out } -1 \text{ from the last two terms so that both sets of parentheses contain the same expression.}$$
$$= (x - 3)(4x - 1) \qquad \square$$

**Student Practice 6**    Factor by grouping. $9x^2 - 64x + 7$

To factor polynomials of the form $ax^2 + bx + c$, use the method, either trial-and-error or grouping, that works best for you.

### 3 Factoring Out a Common Factor

Some problems require first factoring out the greatest common factor and then factoring the trinomial by one of the two methods of this section.

**Example 7** Factor. $9x^2 + 3x - 30$

**Solution**

$$9x^2 + 3x - 30 = 3(3x^2 + 1x - 10) \quad \text{We first factor out the common factor 3 from each term of the trinomial.}$$
$$= 3(3x - 5)(x + 2) \quad \text{We then factor the trinomial by the grouping method or by the trial-and-error method.}$$

**Student Practice 7**    Factor. $8x^2 + 8x - 6$

Be sure to remove the greatest common factor as the very first step.

**Example 8** Factor. $32x^2 - 40x + 12$

**Solution**

$$32x^2 - 40x + 12 = 4(8x^2 - 10x + 3) \quad \text{We first factor out the greatest common factor 4 from each term of the trinomial.}$$
$$= 4(2x - 1)(4x - 3) \quad \text{We then factor the trinomial by the grouping method or by the trial-and-error method.}$$

**Student Practice 8**    Factor. $24x^2 - 38x + 10$

---

 **STEPS TO SUCCESS** Look Ahead to See What Is Coming

You will find that learning new material is much easier if you know what is coming. Take a few minutes at the end of your study time to glance over the next section of the book. If you quickly look over the topics and ideas in this new section, it will help you get your bearings when the instructor presents new material. Students find that when they preview new material, it enables them to see what is coming. It helps them to be able to grasp new ideas much more quickly.

**Making it personal:** Do this right now. Look ahead to the next section of the book. Glance over the ideas and concepts. Write down a couple of facts about the next section. ▼

## 5.4 Exercises MyMathLab®

*Factor by the trial-and-error method. Check your answers using FOIL.*

1. $4x^2 + 21x + 5$
2. $3x^2 + 13x + 12$
3. $5x^2 + 7x + 2$
4. $4x^2 + 5x + 1$

5. $4x^2 + 5x - 6$
6. $5x^2 + 9x - 2$
7. $2x^2 - 5x - 3$
8. $2x^2 - x - 6$

*Factor by the grouping method. Check your answers by using FOIL.*

9. $9x^2 + 9x + 2$
10. $4x^2 + 11x + 6$
11. $15x^2 - 34x + 15$
12. $10x^2 - 29x + 10$

13. $2x^2 + 3x - 20$
14. $6x^2 + 11x - 10$
15. $8x^2 + 10x - 3$
16. $5x^2 - 34x - 7$

*Factor by any method.*

17. $6x^2 - 5x - 6$
18. $3x^2 - 13x - 10$
19. $10x^2 + 3x - 1$
20. $6x^2 + x - 5$

21. $7x^2 - 5x - 18$
22. $9x^2 - 22x - 15$
23. $9y^2 - 13y + 4$
24. $5y^2 - 11y + 2$

25. $5a^2 - 13a - 6$
26. $2a^2 + 5a - 12$
27. $14x^2 + 17x - 6$
28. $32x^2 + 36x - 5$

29. $15x^2 + 4x - 4$
30. $8x^2 - 11x + 3$
31. $12x^2 + 28x + 15$
32. $24x^2 + 17x + 3$

33. $12x^2 - 16x - 3$
34. $12x^2 + x - 6$
35. $3x^4 - 14x^2 - 5$
36. $4x^4 + 8x^2 - 5$

37. $2x^2 + 11xy + 15y^2$
38. $15x^2 + 28xy + 5y^2$
39. $5x^2 + 16xy - 16y^2$
40. $12x^2 + 11xy - 5y^2$

*Factor by first factoring out the greatest common factor. See Examples 7 and 8.*

41. $10x^2 - 25x - 15$
42. $20x^2 - 25x - 30$
43. $6x^3 + 9x^2 - 60x$
44. $6x^3 - 16x^2 - 6x$

### Mixed Practice *Factor.*

45. $5x^2 + 3x - 2$
46. $6x^2 + x - 2$
47. $12x^2 - 38x + 20$
48. $30x^2 - 26x + 4$

49. $12x^3 - 20x^2 + 3x$
50. $12x^3 - 13x^2 + 3x$
51. $8x^2 + 24x - 14$
52. $12x^2 + 26x - 10$

## Cumulative Review

**53. [0.5.4]** *Baseball* In the regular 2014 baseball season, the New York Yankees won 84 out of 162 games. What percentage of the games did they win? Round your answer to the nearest tenth of a percent.

**54. [2.4.1]** Solve. $\dfrac{x}{3} - \dfrac{x}{5} = \dfrac{7}{15}$

*Overseas Travelers* *The double-bar graph shows the top states visited by foreign travelers (excluding travelers from Canada and Mexico) for last year and five years ago.*

Top States Visited by Foreign Travelers

*Source:* www.census.gov

**55. [0.6.1]** **(a)** How many overseas travelers visited these five states last year?

**(b)** What percent of these travelers visited Hawaii (HI)? Round to the nearest tenth.

**56. [0.6.1]** Of the overseas visitors to these five states five years ago, what percent traveled to California (CA)? Round to the nearest tenth.

**57. [0.6.1]** **(a)** Find the difference in the number of visitors to California (CA) last year versus five years ago.

**(b)** This decrease is what percent of the visitors from five years ago? Round to the nearest tenth.

**58. [0.6.1]** **(a)** Find the difference in the number of visitors to New York (NY) last year versus five years ago.

**(b)** This increase is what percent of the visitors from five years ago? Round to the nearest tenth.

---

### Quick Quiz 5.4    *Factor by any method.*

**1.** $12x^2 + 16x - 3$

**2.** $10x^2 - 21x + 9$

**3.** $6x^3 - 3x^2 - 30x$

**4. Concept Check** Explain how you would factor $10x^3 + 18x^2y - 4xy^2$.

# Use Math to Save Money

**Did You Know?**  Even if you always pay on time, having a high debt-to-credit ratio will hurt your credit score.

## Get Out of Debt and Improve Your Credit Score

### Understanding the Problem:

Megan would like to purchase a house in a few years, but currently her credit score is not very high. This will affect her ability to get a mortgage. She needs to find a way to improve her credit score. Part of her problem is that she has almost maxed out her credit cards.

- Credit card 1 has a current balance of $5100 and a credit limit of $5500.
- Credit card 2 has a current balance of $3800 and a credit limit of $4000.
- Credit card 3 has a current balance of $3200 and a credit limit of $3500.

### Making a Plan:

Megan talks to a credit counselor and is told that even though she has always paid her bills on time, she has a high debt-to-credit ratio, and this is lowering her credit score. (To calculate this ratio, first total the current balances on all the credit cards. Next, total the credit limits on all the credit cards. Then divide the total current balance by the total credit limit. Finally, convert this value to a percentage.) Lowering this percentage will increase her credit score. Getting this percentage down to less than 50% would be good. It would be even better for her credit score if she could get this percentage down to less than 33%.

**Step 1:** Megan and her counselor total up her balances and come up with a plan for Megan to reduce her debt over the next couple of years. Reducing her balances without reducing her credit limits will lower her debt-to-credit ratio.

**Task 1:** What is the total current balance (or total debt) on Megan's cards?

**Task 2:** What is her total credit limit?

**Step 2:** Now they can calculate her debt-to-credit ratio.

**Task 3:** Calculate Megan's debt-to-credit ratio as a percentage. Round to the nearest percent.

### Finding a Solution:

They decide that Megan should try to reduce her debt-to-credit ratio to 50% over the next two years.

**Step 3:** They calculate how much she needs to pay off to achieve this goal.

**Task 4:** What amount of debt would be 50% of her total credit limit?

**Task 5:** How much does she need to pay off to reach that amount?

**Step 4:** Megan is going to have to cut some spending to reach this goal in two years, but she knows it is going to be worth her efforts. As she pays down her debt, a greater percentage of her payments will go toward the principal, so she will be paying down her balances faster as time goes on. She wonders if she will be able to reduce her debt-to-credit ratio to 33% by the end of the third year.

**Task 6:** How much would she need to pay off in the third year to reach 33%?

### Applying the Situation to Your Life:

Most people know that paying their bills on time is very important and that it has a positive impact on their credit score. What many people don't know is that having a high debt-to-credit ratio will lower their score. Paying down the debt is a great way to fix this problem. Some people think they should cancel their credit cards after they pay them off, but keeping them open increases the total credit limit. This reduces the debt-to-credit ratio, which will boost the credit score. You should close your accounts only if you are having trouble controlling your spending.

## 5.5 Special Cases of Factoring

As we proceed in this section you will be able to reduce the time it takes you to factor polynomials by quickly recognizing and factoring two special types of polynomials: the difference of two squares and perfect-square trinomials.

### 1 Factoring the Difference of Two Squares

Recall the formula from Section 4.5:

$$(a + b)(a - b) = a^2 - b^2.$$

In reverse form we can use it for factoring.

> **DIFFERENCE OF TWO SQUARES**
> $$a^2 - b^2 = (a + b)(a - b)$$

**Student Learning Objectives**

After studying this section, you will be able to:

1 Recognize and factor expressions of the type $a^2 - b^2$ (difference of two squares).

2 Recognize and factor expressions of the type $a^2 + 2ab + b^2$ (perfect-square trinomial).

3 Recognize and factor expressions that require factoring out a common factor and then using a special-case formula.

We can state it in words in this way:

"The difference of two squares can be factored into the sum and difference of those values that were squared."

**Example 1** Factor. $9x^2 - 1$

**Solution** We see that the polynomial is in the form of the difference of two squares. $9x^2$ is a square and 1 is a square. So using the formula we can write the following.

$$9x^2 - 1 = (3x + 1)(3x - 1) \quad \text{Because } 9x^2 = (3x)^2 \text{ and } 1 = (1)^2 \qquad \square$$

 **Student Practice 1** Factor. $64x^2 - 1$

**MC Example 2** Factor. $25x^2 - 16$

**Solution** Again we use the formula for the difference of squares.

$$25x^2 - 16 = (5x + 4)(5x - 4) \quad \text{Because } 25x^2 = (5x)^2 \text{ and } 16 = (4)^2 \qquad \square$$

**Student Practice 2** Factor. $36x^2 - 49$

Sometimes the polynomial contains two variables.

**Example 3** Factor. $4x^2 - 49y^2$

**Solution** We see that

$$4x^2 - 49y^2 = (2x + 7y)(2x - 7y). \qquad \square$$

 **Student Practice 3** Factor. $100x^2 - 81y^2$

**CAUTION:** Please note that the difference of two squares formula only works if the last term is negative. So if Example 3 had been to factor $4x^2 + 49y^2$, we would **not** have been able to factor the problem. We will examine this in more detail in Section 5.6.

Some problems may involve more than one step.

**Example 4** Factor. $81x^4 - 1$

**Solution** We see that

$$81x^4 - 1 = (9x^2 + 1)(9x^2 - 1). \quad \text{Because } 81x^4 = (9x^2)^2 \text{ and } 1 = (1)^2$$

Is the factoring complete? No. We can factor $9x^2 - 1$.

$$81x^4 - 1 = (9x^2 + 1)(3x + 1)(3x - 1) \quad \text{Because } (9x^2 - 1) = (3x + 1)(3x - 1)$$

 **Student Practice 4** Factor. $x^8 - 1$

## 2 Factoring Perfect-Square Trinomials

There is a formula that will help us to very quickly factor certain trinomials called **perfect-square trinomials.** Recall from Section 4.5 the formulas for binomials squared.

$$(a + b)^2 = a^2 + 2ab + b^2$$
$$(a - b)^2 = a^2 - 2ab + b^2$$

We can use these two equations in reverse form for factoring.

---

**PERFECT-SQUARE TRINOMIALS**

$$a^2 + 2ab + b^2 = (a + b)^2$$
$$a^2 - 2ab + b^2 = (a - b)^2$$

---

A perfect-square trinomial is a trinomial that is the result of squaring a binomial. How can we recognize a perfect-square trinomial?

1. The first and last terms are *perfect squares*.
2. The middle term is twice the product of the values whose squares are the first and last terms.

**Example 5** Factor. $x^2 + 6x + 9$

**Solution** This is a perfect-square trinomial.

1. The first and last terms are perfect squares because $x^2 = (x)^2$ and $9 = (3)^2$.
2. The middle term, $6x$, is twice the product of $x$ and 3.

Since $x^2 + 6x + 9$ is a perfect-square trinomial, we can use the formula

$$a^2 + 2ab + b^2 = (a + b)^2$$

with $a = x$ and $b = 3$. So we have

$$x^2 + 6x + 9 = (x + 3)^2.$$

 **Student Practice 5** Factor. $x^2 + 10x + 25$

Mc **Example 6** Factor. $4x^2 - 20x + 25$

**Solution** This is a perfect-square trinomial.

Note that $20x = 2(2x \cdot 5)$. Also note the negative sign.

Thus we have the following.

$$4x^2 - 20x + 25 = (2x - 5)^2 \quad \text{Since } a^2 - 2ab + b^2 = (a - b)^2$$

**Student Practice 6**

Factor. $25x^2 - 30x + 9$

A polynomial may have more than one variable and its exponents may be higher than 2. The same principles apply.

**Example 7** Factor.

(a) $49x^2 + 42xy + 9y^2$

(b) $36x^4 - 12x^2 + 1$

**Solution**

(a) This is a perfect-square trinomial. Why?

$49x^2 + 42xy + 9y^2 = (7x + 3y)^2$   Because $49x^2 = (7x)^2$, $9y^2 = (3y)^2$, and $42xy = 2(7x \cdot 3y)$

(b) This is a perfect-square trinomial. Why?

$36x^4 - 12x^2 + 1 = (6x^2 - 1)^2$   Because $36x^4 = (6x^2)^2$, $1 = (1)^2$, and $12x^2 = 2(6x^2 \cdot 1)$    □

 **Student Practice 7**   Factor.

(a) $25x^2 + 60xy + 36y^2$

(b) $64x^6 - 48x^3 + 9$

Be *careful*. Some polynomials appear to be perfect-square trinomials but are not. They were factored in other ways in Section 5.4.

**Example 8** Factor. $49x^2 + 35x + 4$

**Solution**   This is *not* a perfect-square trinomial! Although the first and last terms are perfect squares since $(7x)^2 = 49x^2$ and $(2)^2 = 4$, the middle term, $35x$, is not double the product of 2 and $7x$. $35x \neq 28x$! So we must factor by trial and error or by grouping to obtain

$$49x^2 + 35x + 4 = (7x + 4)(7x + 1).$$    □

**Student Practice 8**   Factor. $9x^2 + 15x + 4$

## 3 Factoring Out a Common Factor Before Using a Special-Case Formula

For some polynomials, we first need to factor out the greatest common factor. Then we will find an opportunity to use the difference of two squares formula or one of the perfect-square trinomial formulas.

**Example 9** Factor. $12x^2 - 48$

**Solution**   We see that the greatest common factor is 12.

$12x^2 - 48 = 12(x^2 - 4)$    First we factor out 12.

$= 12(x + 2)(x - 2)$    Then we use the difference of two squares formula, $a^2 - b^2 = (a + b)(a - b)$.    □

**Student Practice 9**   Factor. $20x^2 - 45$

Look carefully at Example 10. Can you identify the greatest common factor?

**Example 10** Factor. $24x^2 - 72x + 54$

**Solution**

$24x^2 - 72x + 54 = 6(4x^2 - 12x + 9)$    First we factor out the greatest common factor, 6.

$= 6(2x - 3)^2$    Then we use the perfect-square trinomial formula, $a^2 - 2ab + b^2 = (a - b)^2$.    □

**Student Practice 10**   Factor. $75x^2 - 60x + 12$

## 5.5 Exercises  MyMathLab®

*Factor by using the difference of two squares formula.*

1. $100x^2 - 1$

2. $36x^2 - 1$

3. $81x^2 - 16$

4. $49x^2 - 25$

5. $x^2 - 49$

6. $9x^2 - 64$

7. $25x^2 - 81$

8. $49x^2 - 4$

9. $x^2 - 25$

10. $x^2 - 36$

11. $1 - 16x^2$

12. $1 - 64x^2$

13. $16x^2 - 49y^2$

14. $25x^2 - 81y^2$

15. $36x^2 - 169y^2$

16. $4x^2 - 9y^2$

17. $100x^2 - 81$

18. $16a^2 - 25$

19. $25a^2 - 81b^2$

20. $9x^2 - 49y^2$

*Factor by using the perfect-square trinomial formula.*

21. $9x^2 + 6x + 1$

22. $25x^2 + 10x + 1$

23. $y^2 - 10y + 25$

24. $y^2 - 12y + 36$

25. $36x^2 - 60x + 25$

26. $9x^2 - 42x + 49$

27. $49x^2 + 28x + 4$

28. $25x^2 + 30x + 9$

29. $x^2 + 14x + 49$

30. $x^2 + 8x + 16$

31. $25x^2 - 40x + 16$

32. $64x^2 - 16x + 1$

33. $81x^2 + 36xy + 4y^2$

34. $36x^2 + 60xy + 25y^2$

35. $9x^2 - 30xy + 25y^2$

36. $49x^2 - 28xy + 4y^2$

**Mixed Practice**  *Factor by using either the difference of two squares or the perfect-square trinomial formula.*

37. $16a^2 + 72ab + 81b^2$

38. $169a^2 + 26ab + b^2$

39. $49x^2 - 42xy + 9y^2$

40. $9x^2 - 30xy + 25y^2$

41. $64x^2 + 80x + 25$

42. $16x^2 + 56x + 49$

43. $144x^2 - 1$

44. $16x^2 - 121$

45. $x^4 - 16$

46. $x^4 - 1$

47. $9x^4 - 24x^2 + 16$

48. $64x^4 - 48x^2 + 9$

### To Think About, Exercises 49–52

49. In Example 4, first we factored $81x^4 - 1$: $(9x^2 + 1)(9x^2 - 1)$. Then we factored $9x^2 - 1$: $(3x + 1)(3x - 1)$. Show why you cannot factor $9x^2 + 1$.

50. What two numbers could replace the $b$ in $36x^2 + bx + 49$ so that the resulting trinomial would be a perfect square? (*Hint*: One number is negative.)

51. What value could you give to $c$ so that $16y^2 - 24y + c$ would become a perfect-square trinomial? Is there only one answer or more than one?

52. Jerome says that he can find two values of $b$ so that $100x^2 + bx - 9$ will be a perfect square. Kesha says there is only one that fits, and Larry says there are none. Who is correct and why?

*Factor by first looking for a greatest common factor. See Examples 9 and 10.*

**53.** $16x^2 - 36$

**54.** $27x^2 - 75$

**55.** $147x^2 - 3y^2$

**56.** $16y^2 - 100x^2$

**57.** $16x^2 - 16x + 4$

**58.** $45x^2 - 60x + 20$

**59.** $98x^2 + 84x + 18$

**60.** $50x^2 + 80x + 32$

**Mixed Practice** *Factor. Be sure to look for common factors first.*

**61.** $x^2 + 16x + 63$

**62.** $x^2 - 3x - 40$

**63.** $2x^2 + 5x - 3$

**64.** $15x^2 - 11x + 2$

**65.** $12x^2 - 27$

**66.** $16x^2 - 36$

**67.** $9x^2 + 42x + 49$

**68.** $9x^2 + 30x + 25$

**69.** $36x^2 - 36x + 9$

**70.** $6x^2 + 60x + 150$

**71.** $2x^2 - 32x + 126$

**72.** $2x^2 - 34x + 140$

## Cumulative Review

**73.** **[4.6.2]** Divide. $(x^3 + x^2 - 2x - 11) \div (x - 2)$

**74.** **[4.6.2]** Divide. $(6x^3 + 11x^2 - 11x - 20) \div (3x + 4)$

*Iguana Diet The green iguana can reach a length of 6 feet and weigh up to 18 pounds. Of the basic diet of the iguana, 40% should consist of greens such as lettuce, spinach, and parsley; 35% should consist of bulk vegetables such as broccoli, zucchini, and carrots; and 25% should consist of fruit.*

**75.** **[0.6.1]** If a certain iguana weighing 150 ounces has a daily diet equal to 2% of its body weight, compose a diet for it in ounces that will meet the iguana's one-day requirement for nutrition.

**76.** **[0.6.1]** If another iguana weighing 120 ounces has a daily diet equal to 3% of its body weight, compose a diet for it in ounces that will meet the iguana's one-day requirement for nutrition.

**Quick Quiz 5.5** *Factor the following.*

**1.** $49x^2 - 81y^2$

**2.** $9x^2 - 48x + 64$

**3.** $162x^2 - 200$

**4.** **Concept Check** Explain how to factor the polynomial $24x^2 + 120x + 150$.

## 5.6 A Brief Review of Factoring ▶

### 1 Identifying and Factoring Polynomials ▶

Often the various types of factoring problems are all mixed together. We need to be able to identify each type of polynomial quickly. The following table summarizes the information we have learned about factoring.

Many polynomials require more than one factoring method. When you are asked to factor a polynomial, it is expected that you will factor it completely. Usually, the first step is factoring out a common factor; then the next step will become apparent.

Carefully go through each example in the following **Factoring Organizer.** Be sure you understand each step that is involved.

### Factoring Organizer

| Number of Terms in the Polynomial | Identifying Name and/or Formula | Example |
|---|---|---|
| A. Any number of terms | **Common factor** <br> The terms have a common factor consisting of a number, a variable, or both. | $2x^2 - 16x = 2x(x - 8)$ <br> $3x^2 + 9y - 12 = 3(x^2 + 3y - 4)$ <br> $4x^2y + 2xy^2 - wxy + xyz = xy(4x + 2y - w + z)$ |
| B. Two terms | **Difference of two squares** <br> First and last terms are perfect squares. <br> $a^2 - b^2 = (a + b)(a - b)$ | $16x^2 - 1 = (4x + 1)(4x - 1)$ <br> $25y^2 - 9x^2 = (5y + 3x)(5y - 3x)$ |
| C. Three terms | **Perfect-square trinomial** <br> First and last terms are perfect squares. <br> $a^2 + 2ab + b^2 = (a + b)^2$ <br> $a^2 - 2ab + b^2 = (a - b)^2$ | $25x^2 - 10x + 1 = (5x - 1)^2$ <br> $16x^2 + 24x + 9 = (4x + 3)^2$ |
| D. Three terms | **Trinomial of the form $x^2 + bx + c$** <br> It starts with $x^2$. The constants of the two factors are numbers whose product is $c$ and whose sum is $b$. | $x^2 - 7x + 12 = (x - 3)(x - 4)$ <br> $x^2 + 11x - 26 = (x + 13)(x - 2)$ <br> $x^2 - 8x - 20 = (x - 10)(x + 2)$ |
| E. Three terms | **Trinomial of the form $ax^2 + bx + c$** <br> It starts with $ax^2$, where $a$ is any number but 1. | Use trial-and-error or the grouping method to factor $12x^2 - 5x - 2$. <br> **1.** The grouping number is $-24$. <br> **2.** The two numbers whose product is $-24$ and whose sum is $-5$ are 3 and $-8$. <br> **3.** $12x^2 - 5x - 2 = 12x^2 + 3x - 8x - 2$ <br> $\quad = 3x(4x + 1) - 2(4x + 1)$ <br> $\quad = (4x + 1)(3x - 2)$ |
| F. Four terms | **Factor by grouping** <br> Rearrange the order if the first two terms do not have a common factor. | $wx - 6yz + 2wy - 3xz = wx + 2wy - 3xz - 6yz$ <br> $\quad = w(x + 2y) - 3z(x + 2y)$ <br> $\quad = (x + 2y)(w - 3z)$ |

**Example 1** Factor. These example problems are mixed.

(a) $25x^3 - 10x^2 + x$      (b) $20x^2y^2 - 45y^2$      (c) $15x^2 - 3x^3 + 18x$

**Solution**

(a) $25x^3 - 10x^2 + x = x(25x^2 - 10x + 1)$    Factor out the common factor $x$.

$\qquad\qquad\qquad\quad = x(5x - 1)^2$    The other factor is a perfect-square trinomial.

**(b)** $20x^2y^2 - 45y^2 = 5y^2(4x^2 - 9)$      Factor out the common factor $5y^2$.

$= 5y^2(2x + 3)(2x - 3)$      The other factor is a difference of squares.

**(c)** $15x^2 - 3x^3 + 18x = -3x^3 + 15x^2 + 18x$      Rearrange the terms in descending order of powers of $x$.

$= -3x(x^2 - 5x - 6)$      Factor out the common factor $-3x$.

$= -3x(x - 6)(x + 1)$      Factor the trinomial.      ☐

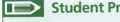 **Student Practice 1**    Factor. Be careful. These practice problems are mixed.

**(a)** $3x^2 - 36x + 108$      **(b)** $9x^4y^2 - 9y^2$

**(c)** $12x - 9 - 4x^2$

**Example 2** Factor. $ax^2 - 9a + 2x^2 - 18$

**Solution** We factor by grouping since there are four terms.

$ax^2 - 9a + 2x^2 - 18 = a(x^2 - 9) + 2(x^2 - 9)$      Factor out the common factor $a$ from the first two terms and 2 from the second two terms.

$= (a + 2)(x^2 - 9)$      Factor out the common factor $(x^2 - 9)$.

$= (a + 2)(x - 3)(x + 3)$      Factor $x^2 - 9$ using the difference of two squares formula.      ☐

 **Student Practice 2**    Factor. $5x^3 - 20x + 2x^2 - 8$

**2** **Determining Whether a Polynomial Is Prime** ▶

*Not all polynomials can be factored using the methods in this chapter.* If we cannot factor a polynomial by elementary methods, we will identify it as a **prime** polynomial. If, after you have mastered the factoring techniques in this chapter, you encounter a polynomial that you cannot factor with these methods, you should feel comfortable enough to say, "The polynomial cannot be factored with the methods in this chapter, so it is prime," rather than "I can't do it—I give up!"

**Example 3** Factor, if possible.

**(a)** $x^2 + 6x + 12$      **(b)** $25x^2 + 4$

**Solution**

**(a)** The factors of 12 are

$$(1)(12) \text{ or } (2)(6) \text{ or } (3)(4).$$

None of these pairs add up to 6, the coefficient of the middle term. Thus the problem *cannot be factored* by the methods of this chapter. It is prime.

**(b)** We have a formula to factor the difference of two squares. There is no way to factor the sum of two squares. That is, $a^2 + b^2$ *cannot be factored.*

Thus   $25x^2 + 4$ is prime.      ☐

 **Student Practice 3**    Factor, if possible.

**(a)** $x^2 - 9x - 8$      **(b)** $25x^2 + 82x + 4$

## 5.6 Exercises MyMathLab®

*Review the six basic types of factoring in the Factoring Organizer on page 326. Each of the six types is included in exercises 1–12. Factor if possible. Check your answer by multiplying.*

1. $3x^2 - 6xy + 5x$

2. $8x^2 + 5xy - 7x$

3. $16x^2 - 25y^2$

4. $36x^2 - 49y^2$

5. $x^2 + 64$

6. $25x^2 + 80xy + 64y^2$

7. $x^2 + 8x + 15$

8. $x^2 + 15x + 54$

9. $15x^2 + 7x - 2$

10. $6x^2 + 13x - 5$

11. $ax - 3cx + 3ay - 9cy$

12. $bx - 2dx + 5by - 10dy$

**Mixed Practice** *Factor, if possible. Be sure to factor completely. Always factor out the greatest common factor first, if one exists.*

13. $y^2 + 14y + 49$

14. $y^2 + 16y + 64$

15. $4x^2 - 12x + 9$

16. $16x^2 + 25$

17. $2x^2 - 11x + 12$

18. $3x^2 - 10x + 8$

19. $x^2 - 3xy - 70y^2$

20. $x^2 - 6xy - 16y^2$

21. $ax - 5a + 3x - 15$

22. $by + 7b - 6y - 42$

23. $16x - 4x^3$

24. $8y^2 + 10y - 12$

25. $2x^3 + 3x^2 - 36x$

26. $7x^2 + 3x - 2$

27. $3xyz^2 - 6xyz - 9xy$

28. $-16x^2 - 2x - 32x^3$

29. $3x^2 + 6x - 105$

30. $7x^2 + 21x - 70$

31. $5x^3y^3 - 10x^2y^3 + 5xy^3$

32. $2x^4y - 12x^3y + 18x^2y$

33. $7x^2 - 2x^4 + 4$

34. $14x^2 - x^3 + 32x$

**35.** $6x^2 - 3x + 2$

**36.** $4x^3 + 8x^2 - 60x$

**37.** $5x^4 - 5x^2 + 10x^3y - 10xy$

**38.** $2a^4b - 8a^2b + 3a^2b - 12b$

*Remove the greatest common factor first. Then continue to factor, if possible.*

**39.** $5x^2 + 10xy - 30y$

**40.** $6x^2 + 30x - 54y$

**41.** $30x^3 + 3x^2y - 6xy^2$

**42.** $48x^3 + 20x^2y - 8xy^2$

**43.** $30x^2 - 38x + 12$

**44.** $15x^2 - 35x + 10$

## To Think About

**45.** A polynomial that cannot be factored by the methods of this chapter is called _____.

**46.** A binomial of the form $x^2 - d$ can be quickly factored or identified as prime. If it can be factored, what is true of the number $d$?

## Cumulative Review

**47.** **[3.4.2]** *Independent Contractor* When Dave Barry decided to leave the company and work as an independent contractor, he took a pay cut of 14%. He earned $24,080 this year. What did he earn in his previous job?

**48.** **[0.6.1]** *Antiviral Drug* A major pharmaceutical company is testing a new, powerful antiviral drug. It kills 13 strains of virus every hour. If there are presently 294 live strains of virus in the test container, how many live strains were there 6 hours ago?

**49.** **[1.8.1]** Evaluate $3a^2 + ab - 4b^2$ for $a = 5$ and $b = -1$.

**50.** **[2.5.1]** Solve for $d$. $8(x + d) = 3(y - d)$

**51.** **[3.2.1]** Solve. A number is doubled and then reduced by 17. The result is 51. What is the number?

**52.** **[2.6.4]** Solve for $x$.
$$\frac{1}{2}x - 3 \le \frac{1}{4}(3x + 3)$$

**Quick Quiz 5.6** *Completely factor the following, if possible.*

**1.** $6x^2 - 17x + 12$

**2.** $60x^2 - 9x - 6$

**3.** $25x^2 + 49$

**4.** **Concept Check** Explain how to completely factor $2x^2 + 6xw - 5x - 15w$.

# 5.7 Solving Quadratic Equations by Factoring ▶

## 1 Solving Quadratic Equations by Factoring ▶

In Chapter 2, we learned how to solve linear equations such as $3x + 5 = 0$ by finding the root (or value of $x$) that satisfied the equation. Now we turn to the question of how to solve equations like $3x^2 + 5x + 2 = 0$. Such equations are called **quadratic equations.** A quadratic equation is a polynomial equation in one variable that contains a variable term of degree 2 and no terms of higher degree.

> The *standard form* of a quadratic equation is $ax^2 + bx + c = 0$, where $a$, $b$, and $c$ are real numbers and $a \neq 0$.

In this section, we will study quadratic equations in standard form, where $a$, $b$, and $c$ are integers.

Many quadratic equations have two real number solutions (also called **real roots**). But how can we find them? The most direct approach is the factoring method. This method depends on a very powerful property.

> **ZERO FACTOR PROPERTY**
>
> If $a \cdot b = 0$, then $a = 0$ or $b = 0$.

Notice the word *or* in the zero factor property. When we make a statement in mathematics using this word, we intend it to mean *one or the other or both.* Therefore, the zero factor property states that if the product $a \cdot b$ is zero, then $a$ can equal zero or $b$ can equal zero or *both $a$ and $b$ can equal zero.* We can use this principle to solve quadratic equations. Before you start, make sure that the equation is in standard form.

> **SOLVING A QUADRATIC EQUATION**
>
> 1. Make sure the equation is set equal to zero.
> 2. Factor, if possible, the quadratic expression that equals zero.
> 3. Set each factor containing a variable equal to zero.
> 4. Solve the resulting equations to find each root.
> 5. Check each root.

**Example 1** Solve the equation to find the two roots and check. $3x^2 + 5x + 2 = 0$

**Solution**

$$3x^2 + 5x + 2 = 0 \qquad \text{The equation is in standard form.}$$
$$(3x + 2)(x + 1) = 0 \qquad \text{Factor the quadratic expression.}$$
$$3x + 2 = 0 \qquad x + 1 = 0 \qquad \text{Set each factor equal to 0.}$$
$$3x = -2 \qquad x = -1 \qquad \text{Solve the equations to find the two roots.}$$
$$x = -\frac{2}{3}$$

The two roots (that is, solutions) are $-\frac{2}{3}$ and $-1$.

*Check.* We can determine if the two numbers $-\frac{2}{3}$ and $-1$ are solutions to the equation. Substitute $-\frac{2}{3}$ for $x$ in the *original equation*. If an identity results, $-\frac{2}{3}$ is a solution. Do the same for $-1$.

$$3x^2 + 5x + 2 = 0 \qquad\qquad 3x^2 + 5x + 2 = 0$$

$$3\left(-\frac{2}{3}\right)^2 + 5\left(-\frac{2}{3}\right) + 2 \overset{?}{=} 0 \qquad 3(-1)^2 + 5(-1) + 2 \overset{?}{=} 0$$

$$3\left(\frac{4}{9}\right) + 5\left(-\frac{2}{3}\right) + 2 \overset{?}{=} 0 \qquad 3(1) + 5(-1) + 2 \overset{?}{=} 0$$

$$\frac{4}{3} - \frac{10}{3} + 2 \overset{?}{=} 0 \qquad\qquad 3 - 5 + 2 \overset{?}{=} 0$$

$$\qquad\qquad\qquad -2 + 2 \overset{?}{=} 0$$

$$\frac{4}{3} - \frac{10}{3} + \frac{6}{3} \overset{?}{=} 0 \qquad\qquad 0 = 0 \ \checkmark$$

$$0 = 0 \ \checkmark$$

Thus $-\frac{2}{3}$ and $-1$ are both roots of the equation $3x^2 + 5x + 2 = 0$. □

**Student Practice 1** Solve the equation by factoring to find the two roots and check. $10x^2 - x - 2 = 0$

**Example 2** Solve the equation to find the two roots. $2x^2 + 13x - 7 = 0$

**Solution**

$$2x^2 + 13x - 7 = 0 \qquad \text{The equation is in standard form.}$$

$$(2x - 1)(x + 7) = 0 \qquad \text{Factor.}$$

$$2x - 1 = 0 \qquad x + 7 = 0 \qquad \text{Set each factor equal to 0.}$$

$$2x = 1 \qquad\quad x = -7 \qquad \text{Solve the equations to find the two roots.}$$

$$x = \frac{1}{2}$$

The two roots are $\frac{1}{2}$ and $-7$.

*Check.* If $x = \frac{1}{2}$, then we have the following.

$$2\left(\frac{1}{2}\right)^2 + 13\left(\frac{1}{2}\right) - 7 = 2\left(\frac{1}{4}\right) + 13\left(\frac{1}{2}\right) - 7$$

$$= \frac{1}{2} + \frac{13}{2} - \frac{14}{2} = 0 \ \checkmark$$

If $x = -7$, then we have the following.

$$2(-7)^2 + 13(-7) - 7 = 2(49) + 13(-7) - 7$$

$$= 98 - 91 - 7 = 0 \ \checkmark$$

Thus $\frac{1}{2}$ and $-7$ are both roots of the equation $2x^2 + 13x - 7 = 0$. □

**Student Practice 2** Solve the equation to find the two roots.

$$3x^2 + 11x - 4 = 0$$

If the quadratic equation $ax^2 + bx + c = 0$ has no visible constant term, then $c = 0$. All such quadratic equations can be solved by factoring out a common factor and then using the zero factor property to obtain two solutions that are real numbers.

**Example 3** Solve the equation to find the two roots. $7x^2 - 3x = 0$

**Solution**

$$7x^2 - 3x = 0 \qquad \text{The equation is in standard form. Here } c = 0.$$

$$x(7x - 3) = 0 \qquad \text{Factor out the common factor.}$$

$$x = 0 \qquad 7x - 3 = 0 \qquad \text{Set each factor equal to 0 by the zero factor property.}$$

$$7x = 3 \qquad \text{Solve the equations to find the two roots.}$$

$$x = \frac{3}{7}$$

The two roots are 0 and $\frac{3}{7}$.

*Check.* Verify that 0 and $\frac{3}{7}$ are the roots of $7x^2 - 3x = 0$. ☐

**Student Practice 3** Solve the equation to find the two roots.

$$7x^2 + 11x = 0$$

If the quadratic equation is not in standard form, we use the same basic algebraic methods we studied in Sections 2.1–2.3 to place the terms on one side and zero on the other so that we can use the zero factor property.

**Example 4** Solve. $x^2 = 12 - x$

**Solution**

$$x^2 = 12 - x \qquad \text{The equation is not in standard form.}$$

$$x^2 + x - 12 = 0 \qquad \text{Add } x \text{ and } -12 \text{ to both sides of the equation so that the right side is equal to zero; we can now factor.}$$

$$(x - 3)(x + 4) = 0 \qquad \text{Factor.}$$

$$x - 3 = 0 \quad x + 4 = 0 \qquad \text{Set each factor equal to 0 by the zero factor property.}$$

$$x = 3 \qquad x = -4 \qquad \text{Solve the equations for } x.$$

*Check.* If $x = 3$: $(3)^2 \overset{?}{=} 12 - 3 \qquad$ If $x = -4$: $(-4)^2 \overset{?}{=} 12 - (-4)$

$$9 \overset{?}{=} 12 - 3 \qquad\qquad\qquad\qquad 16 \overset{?}{=} 12 + 4$$

$$9 = 9 \checkmark \qquad\qquad\qquad\qquad\qquad 16 = 16 \checkmark$$

Both roots check. ☐

**Student Practice 4** Solve. $x^2 - 6x + 4 = -8 + x$

**Example 5** Solve. $\dfrac{x^2 - x}{2} = 6$

**Solution** We must first clear the fractions from the equation.

$$2\left(\frac{x^2 - x}{2}\right) = 2(6) \qquad \text{Multiply each side by 2.}$$

$$x^2 - x = 12 \qquad \text{Simplify.}$$

$$x^2 - x - 12 = 0 \qquad \text{Place in standard form.}$$

$$(x - 4)(x + 3) = 0 \qquad \text{Factor.}$$

$$x - 4 = 0 \quad x + 3 = 0 \qquad \text{Set each factor equal to zero.}$$

$$x = 4 \qquad x = -3 \qquad \text{Solve the equations for } x.$$

The check is left to the student. ☐

**Student Practice 5** Solve.

$$\frac{2x^2 - 7x}{3} = 5$$

**2** **Using Quadratic Equations to Solve Applied Problems** ▶

Certain types of word problems—for example, some geometry applications—lead to quadratic equations. We'll show how to solve such word problems in this section.

It is particularly important to check the apparent solutions to the quadratic equation with conditions stated in the word problem. Often a particular solution to the quadratic equation will be eliminated by the conditions of the word problem.

▲ **Example 6** Carlos lives in Mexico City. He has a rectangular brick walkway in front of his house. The length of the walkway is 3 meters longer than twice the width. The area of the walkway is 44 square meters. Find the length and width of the rectangular walkway.

**Solution**

1. **Understand the problem.**
   Draw a picture.

   Let $w$ = the width in meters.

   Then $2w + 3$ = the length in meters.

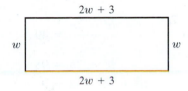

2. **Write an equation.**

$$\text{area} = (\text{width})(\text{length})$$
$$44 = w(2w + 3)$$

3. **Solve and state the answer.**

| | |
|---|---|
| $44 = w(2w + 3)$ | |
| $44 = 2w^2 + 3w$ | Remove parentheses. |
| $0 = 2w^2 + 3w - 44$ | Put in standard form. |
| $0 = (2w + 11)(w - 4)$ | Factor. |
| $2w + 11 = 0 \quad w - 4 = 0$ | Set each factor equal to 0. |
| $2w = -11 \quad w = 4$ | Simplify and solve. |
| $w = -5\frac{1}{2}$ | Although $-5\frac{1}{2}$ is a solution to the quadratic equation, it is not a valid solution to the word problem. It would not make sense to have a rectangle with a negative number as a width. |

Since $w = 4$, the width of the walkway is 4 meters. The length is $2w + 3$, so we have $2(4) + 3 = 8 + 3 = 11$.

Thus the length of the walkway is 11 meters.

4. **Check.** Is the length 3 meters more than twice the width?

$$11 \stackrel{?}{=} 3 + 2(4) \qquad 11 = 3 + 8 \ \checkmark$$

Is the area of the rectangle 44 square meters?

$$4 \times 11 \stackrel{?}{=} 44 \qquad 44 = 44 \ \checkmark \qquad \square$$

▲ ▶ **Student Practice 6** The length of a rectangle is 2 meters longer than triple the width. The area of the rectangle is 85 square meters. Find the length and width of the rectangle.

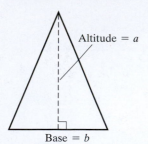

Altitude = $a$

Base = $b$

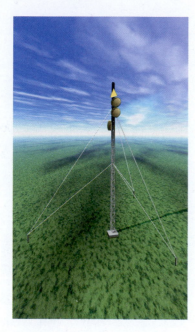

▲ **Example 7** The top of a local cable television tower has several small triangular reflectors. The area of each triangle is 49 square centimeters. The altitude of each triangle is 7 centimeters longer than the base. Find the altitude and the base of one of the triangles.

**Solution**

Let $b$ = the length of the base in centimeters

$b + 7$ = the length of the altitude in centimeters

To find the area of a triangle, we use

$$\text{area} = \frac{1}{2}(\text{altitude})(\text{base}) = \frac{1}{2}ab = \frac{ab}{2}.$$

| | |
|---|---|
| $\dfrac{ab}{2} = 49$ | Write an equation. |
| $\dfrac{(b+7)(b)}{2} = 49$ | Substitute the expressions for altitude and base. |
| $\dfrac{b^2 + 7b}{2} = 49$ | Simplify. |
| $b^2 + 7b = 98$ | Multiply each side of the equation by 2. |
| $b^2 + 7b - 98 = 0$ | Place the quadratic equation in standard form. |
| $(b - 7)(b + 14) = 0$ | Factor. |
| $b - 7 = 0 \qquad b + 14 = 0$ | Set each factor equal to zero. |
| $b = 7 \qquad\qquad b = -14$ | Solve the equations for $b$. |

We cannot have a base of $-14$ centimeters, so we reject the negative answer. The only possible solution is 7. So the base is 7 centimeters. The altitude is $b + 7 = 7 + 7 = 14$. The altitude is 14 centimeters.

The triangular reflector has a base of 7 centimeters and an altitude of 14 centimeters.

*Check.* When you do the check, answer the following two questions.

**1.** Is the altitude 7 centimeters longer than the base?

**2.** Is the area of a triangle with a base of 7 centimeters and an altitude of 14 centimeters actually 49 square centimeters? ▫

▲ ⬛➡ **Student Practice 7** A triangle has an area of 35 square centimeters. The altitude of the triangle is 3 centimeters shorter than the base. Find the altitude and the base of the triangle.

Many problems in the sciences require the use of quadratic equations. You will study these in more detail if you take a course in physics or calculus in college. Often a quadratic equation is given as part of the problem.

When an object is thrown upward, its height ($S$) in meters is given, approximately, by the quadratic equation

$$S = -5t^2 + vt + h.$$

The letter $h$ represents the initial height in meters. The letter $v$ represents the initial velocity of the object thrown. The letter $t$ represents the time in seconds starting from the time the object is thrown.

**Example 8**  A tennis ball is thrown upward with an initial velocity of 8 meters/second. Suppose that the initial height above the ground is 4 meters. At what time $t$ will the ball hit the ground?

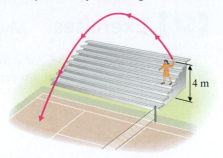

**Solution**   In this case $S = 0$ since the ball will hit the ground. The initial upward velocity is $v = 8$ meters/second. The initial height is 4 meters, so $h = 4$.

$$S = -5t^2 + vt + h \quad \text{Write an equation.}$$

$$0 = -5t^2 + 8t + 4 \quad \text{Substitute all values into the equation.}$$

$$5t^2 - 8t - 4 = 0 \quad \text{Isolate the terms on the left side. (Most students can factor more readily if the squared variable is positive.)}$$

$$(5t + 2)(t - 2) = 0 \quad \text{Factor.}$$

$$5t + 2 = 0 \qquad t - 2 = 0 \quad \text{Set each factor equal to 0.}$$

$$5t = -2 \qquad t = 2 \quad \text{Solve the equations for } t.$$

$$t = -\frac{2}{5}$$

We want a positive time $t$ in seconds; thus we do not use $t = -\frac{2}{5}$.

Therefore, the ball will strike the ground 2 seconds after it is thrown.

*Check.* Verify the solution.

 **Student Practice 8**   A Mexican cliff diver does a dive from a cliff 45 meters above the ocean. This constitutes free fall, so the initial velocity is $v = 0$, and since there is no upward velocity (no springboard), then $h = 45$ meters. How long will it be until he breaks the water's surface?

## 👣 STEPS TO SUCCESS   Be Involved.

*If you are in a traditional class:*

Don't just sit on the sidelines of the class and watch. Take part in the classroom discussion. People learn mathematics best through active participation. Whenever you are not clear about something, ask a question. Usually your questions will be helpful to other students in the room. When the teacher asks for suggestions, be sure to contribute your own ideas. Sit near the front where you can see and hear well. This will help you to focus on the material being covered.

**Making it personal:** Which of the suggestions above is the one you most need to follow? Write down what you need to do to improve in this area. ▼

*If you are in an online class or a nontraditional class:*

Be sure to e-mail the teacher. Talk to the tutor on duty. Ask questions. Think about concepts. Make your mind interact with the textbook. Be mentally involved. This active mental interaction is the key to your success.

**Making it personal:** Which of the suggestions above is the one you most need to follow? Write down what you need to do to improve in this area. ▼

## 5.7 Exercises   MyMathLab®

*Using the factoring method, solve for the roots of each quadratic equation. Be sure to place the equation in standard form before factoring. Check your answers.*

**1.** $x^2 - 4x - 12 = 0$

**2.** $x^2 - x - 12 = 0$

**3.** $x^2 + 14x + 24 = 0$

**4.** $x^2 - 6x - 40 = 0$

**5.** $2x^2 - 7x + 6 = 0$

**6.** $3x^2 - 17x + 10 = 0$

**7.** $6x^2 - 13x = -6$

**8.** $10x^2 + 19x = 15$

**9.** $x^2 + 13x = 0$

**10.** $6x^2 - x = 0$

**11.** $8x^2 = 72$

**12.** $9x^2 = 81$

**13.** $5x^2 + 3x = 8x$

**14.** $3x^2 - x = 4x$

**15.** $6x^2 = 16x - 8$

**16.** $24x^2 = -10x + 4$

**17.** $(x - 5)(x + 2) = -4(x + 1)$

**18.** $(x - 5)(x + 4) = 2(x - 5)$

**19.** $9x^2 + x + 1 = -5x$

**20.** $4x^2 - 5x + 25 = 15x$

**21.** $\dfrac{x^2}{4} + \dfrac{5x}{4} + 2 = 2$

**22.** $\dfrac{x^2}{6} + \dfrac{2x}{3} + 1 = 1$

**23.** $\dfrac{x^2 + 10x}{8} = -2$

**24.** $\dfrac{x^2 + 5x}{4} = 9$

**25.** $\dfrac{10x^2 - 25x}{12} = 5$

**26.** $\dfrac{12x^2 - 4x}{5} = 8$

**336**

## To Think About

**27.** Why can an equation in standard form with $c = 0$ (that is, an equation of the form $ax^2 + bx = 0$) always be solved?

**28.** Martha solved $(x + 3)(x - 2) = 14$ as follows:
$$x + 3 = 14 \quad \text{or} \quad x - 2 = 14$$
$$x = 11 \quad \text{or} \quad x = 16$$
Josette said this had to be wrong because these values do not check. Explain what is wrong with Martha's method.

## Applications

▲ **29.** *Geometry* The area of a rectangular garden is 140 square meters. The width is 3 meters longer than one-half of the length. Find the length and the width of the garden.

▲ **30.** *Geometry* The area of a triangular sign is 33 square meters. The base of the triangle is 1 meter less than double the altitude. Find the altitude and the base of the sign.

*Forming Groups* Suppose the number of students in a mathematics class is x. The teacher insists that each student participate in group work each class. The number of possible groups is:

$$G = \frac{x^2 - 3x + 2}{2}$$

**31.** The class has 13 students. How many possible groups are there?

**32.** Four students withdraw from the class in exercise 31. How many fewer groups can be formed?

**33.** A teacher claims each student could be in 36 different groups. How many students are there?

**34.** The teacher wants each student to participate in a different group each day. There are 45 class days in the semester. How many students must be in the class?

*Falling Object* Use the following information for exercises 35 and 36. When an object is thrown upward, its height (S), in meters, is given (approximately) by the quadratic equation

$$S = -5t^2 + vt + h,$$
where $v =$ the upward initial velocity in meters/second,
$t =$ the time of flight in seconds, and
$h =$ the height above level ground from which the object is thrown.

**35.** Johnny is standing on a platform 6 meters high and throws a ball straight up as high as he can at a velocity of 13 meters/second. At what time $t$ will the ball hit the ground? How far from the ground is the ball after 2 seconds have elapsed from the time of the throw? (Assume that the ball is 6 meters from the ground when it leaves Johnny's hand.)

**36.** You are standing on the edge of a cliff near Acapulco, overlooking the ocean. The place where you stand is 180 meters from the ocean. You drop a pebble into the water. ("Dropping" the pebble implies that there is no initial velocity, so $v = 0$.) How many seconds will it take to hit the water? How far has the pebble dropped after 3 seconds?

*Internal Phone Calls The technology and communication office of a local company has set up a new telephone system so that each employee has a separate telephone and extension number. They discovered that if there are x people in the office and each person talks to everyone else in the office by phone, the number of different calls (T) that take place is described by the equation $T = 0.5(x^2 - x)$, where x is the number of people in the office. Use this information to answer exercises 37 and 38.*

**37.** If 70 people are presently employed at the office, how many different telephone calls can be made between these 70 people?

**38.** On the day after Thanksgiving, only a small number of employees were working at the office. It has been determined that on that day, a total of 105 different phone calls could have been made from people working in the office to other people working in the office. How many people worked on the day after Thanksgiving?

*The same equation given above is used in the well-known "handshake problem." If there are n people at a party and each person shakes hands with every other person once, the number of handshakes that take place among the n people is $H = 0.5(n^2 - n)$.*

**39.** Barry is hosting a holiday party and has invited 16 friends. If all his friends attend the party and everyone, including Barry, shakes hands with everyone else at the party, how many handshakes will take place?

**40.** At a mathematics conference, 55 different handshakes took place among the organizers during the opening meeting. How many organizers were there?

## Cumulative Review *Simplify.*

**41.** **[4.1.1]** $(2x^2y^3)(-5x^3y)$

**42.** **[4.1.1]** $(3a^4b^5)(4a^6b^8)$

**43.** **[4.1.2]** $\dfrac{21a^5b^{10}}{-14ab^{12}}$

**44.** **[4.1.2]** $\dfrac{18x^3y^6}{54x^8y^{10}}$

---

### Quick Quiz 5.7 *Solve.*

**1.** $15x^2 - 8x + 1 = 0$

**2.** $4 + x(x - 2) = 7$

**3.** $4x^2 = 9x + 9$

**4.** **Concept Check** Explain how you would solve the following problem: A rectangle has an area of 65 square feet. The length of the rectangle is 3 feet longer than double the width. Find the length and the width of the rectangle.

## Background: CAD Operator

Mauricio is a skilled CAD operator who has worked with boats since he was a child. As a recent college graduate with a degree in computer-aided design technology, he's accepted a job with a nationally known sail maker and marine supply manufacturer. He feels incredibly lucky to be putting his skills to use in a career he loves.

As he begins his workweek, Mauricio is presented with two new job orders. The first is to create a construction design drawing with dimensions for a boat cover based on general specifications. The second project is to create a design for a racing sailboat's mainsail.

## Facts

- The boat storage cover has two pieces, one rectangular and the other triangular. Mauricio examines the following information: For the rectangular piece, 126 square feet of material is needed, and its length has to be 3 feet shorter than 3 times its width. For the triangular piece, 10 square feet of material is needed. The height of the triangular section is 1 foot less than the length of its base. Mauricio begins to determine the dimensions of both pieces, writing a program that he'll be able to use for similar orders.

- The second order Mauricio needs to complete is a construction design drawing with the specifications of the racing boat's mainsail. He knows that the material used for these sails is very expensive and is measured in square yards, so ultimately, he'll have to convert to those units of measurement. The general information given is that the total area for the constructed sail is 132 square feet. Regulations require that the sail's height, when fully assembled, be 2 feet less than double its base.

## Tasks

1. Mauricio must write a program to calculate dimensions for new orders of boat storage covers. What are the dimensions of the rectangular piece (length and width) and the triangular piece (base and height) for this initial storage cover? (Note: A linear dimension can only be a positive number, so eliminate any negative solutions.)

2. Mauricio must complete a construction design drawing with the specifications for the racing boat's mainsail. What are the dimensions (base and height) of the mainsail?

3. Because the material used for sail construction is measured in square yards, Mauricio must convert the measurements for the sail's area. What is the sail's area measured in square yards?

# Chapter 5 Organizer

| Topic and Procedure | Examples | ▶ You Try It |
|---|---|---|
| **Common factor, p. 297**<br><br>Factor out the largest common factor from each term. | Factor.<br>**(a)** $2x^2 - 2x = 2x(x - 1)$<br>**(b)** $3a^2 + 3ab - 12a = 3a(a + b - 4)$<br>**(c)** $8x^4y - 24x^3 = 8x^3(xy - 3)$ | **1.** Factor.<br>**(a)** $5a^2 - 15a$<br>**(b)** $4x^2 - 8xy + 4x$<br>**(c)** $6x^4 - 18x^2$ |
| **Four terms. Factor by grouping, p. 302**<br><br>Rearrange the terms if necessary so that the first two terms have a common factor. Then factor out the common factors.<br>$ax + ay - bx - by = a(x + y) - b(x + y)$<br>$\qquad = (x + y)(a - b)$ | Factor by grouping.<br>$2ax^2 + 21 + 14x^2 + 3a$<br>$= 2ax^2 + 14x^2 + 3a + 21$<br>$= 2x^2(a + 7) + 3(a + 7)$<br>$= (a + 7)(2x^2 + 3)$ | **2.** Factor by grouping.<br>$3ax^2 - 12x^2 - 8 + 2a$ |
| **Trinomials of the form** $x^2 + bx + c$**, p. 307**<br><br>Factor trinomials of the form $x^2 + bx + c$ by finding two numbers that have a product of $c$ and a sum of $b$.<br><br>If each term of the trinomial has a common factor, factor it out as the first step. | Factor.<br>**(a)** $x^2 - 18x + 77 = (x - 7)(x - 11)$<br>**(b)** $x^2 + 7x - 18 = (x + 9)(x - 2)$<br>**(c)** $5x^2 - 10x - 40 = 5(x^2 - 2x - 8)$<br>$\qquad\qquad\qquad = 5(x - 4)(x + 2)$ | **3.** Factor.<br>**(a)** $x^2 + 9x + 18$<br>**(b)** $x^2 + 2x - 35$<br>**(c)** $3x^2 - 9x - 12$ |
| **Trinomials of the form** $ax^2 + bx + c$**, where** $a \neq 1$**, p. 314**<br><br>Factor trinomials of the form $ax^2 + bx + c$ by the grouping method or by the trial-and-error method. | Factor.<br>$6x^2 + 11x - 10$<br>Grouping number $= -60$<br>Two numbers whose product is $-60$ and whose sum is $+11$ are $+15$ and $-4$.<br>$6x^2 + 15x - 4x - 10 = 3x(2x + 5) - 2(2x + 5)$<br>$\qquad\qquad\qquad = (2x + 5)(3x - 2)$ | **4.** Factor. $8x^2 + 6x - 9$ |
| **Special cases**<br>**Difference of two squares, p. 321**<br>**Perfect-square trinomials, p. 322**<br><br>If you recognize the special cases, you will be able to factor quickly.<br>$a^2 - b^2 = (a + b)(a - b)$<br>$a^2 + 2ab + b^2 = (a + b)^2$<br>$a^2 - 2ab + b^2 = (a - b)^2$ | Factor.<br>**(a)** $25x^2 - 36y^2 = (5x + 6y)(5x - 6y)$<br>**(b)** $16x^4 - 1 = (4x^2 + 1)(2x + 1)(2x - 1)$<br>**(c)** $25x^2 + 10x + 1 = (5x + 1)^2$<br>**(d)** $49x^2 - 42xy + 9y^2 = (7x - 3y)^2$ | **5.** Factor.<br>**(a)** $9x^2 - 16y^2$<br>**(b)** $81x^4 - 1$<br>**(c)** $16a^2 + 24a + 9$<br>**(d)** $4x^2 - 20xy + 25y^2$ |
| **Multistep factoring, p. 326**<br><br>Many problems require two or three steps of factoring. Always try to factor out the greatest common factor as the first step. | Factor completely.<br>**(a)** $3x^2 - 21x + 36 = 3(x^2 - 7x + 12)$<br>$\qquad\qquad\qquad = 3(x - 4)(x - 3)$<br>**(b)** $2x^3 - x^2 - 6x = x(2x^2 - x - 6)$<br>$\qquad\qquad\qquad = x(2x + 3)(x - 2)$<br>**(c)** $25x^3 - 49x = x(25x^2 - 49)$<br>$\qquad\qquad\qquad = x(5x + 7)(5x - 7)$<br>**(d)** $8x^2 - 24x + 18 = 2(4x^2 - 12x + 9)$<br>$\qquad\qquad\qquad = 2(2x - 3)^2$ | **6.** Factor completely.<br>**(a)** $4x^2 + 4x - 24$<br>**(b)** $3x^3 + 7x^2 + 2x$<br>**(c)** $9x^3 - 64x$<br>**(d)** $48x^2 - 24x + 3$ |
| **Prime polynomials, p. 327**<br><br>A polynomial that is not factorable is called prime. | **(a)** $x^2 + y^2$ is prime.<br><br>**(b)** $x^2 + 5x + 7$ is prime. | **7.** Explain why each polynomial is prime.<br>**(a)** $x^2 + 4$<br>**(b)** $x^2 + x + 2$ |
| **Solving quadratic equations by factoring, p. 330**<br><br>**1.** Write as $ax^2 + bx + c = 0$.<br>**2.** Factor.<br>**3.** Set each factor equal to 0.<br>**4.** Solve the resulting equations. | Solve. $3x^2 + 5x = 2$<br>$3x^2 + 5x - 2 = 0$<br>$(3x - 1)(x + 2) = 0$<br>$3x - 1 = 0 \qquad x + 2 = 0$<br>$x = \dfrac{1}{3} \qquad x = -2$ | **8.** Solve. $2x^2 - x = 3$ |

340

| Topic and Procedure | Examples | You Try It |
|---|---|---|
| *Using quadratic equations to solve applied problems, p. 333*<br><br>Some word problems, like those involving the product of two numbers, area, and formulas with a squared variable, can be solved using the factoring methods we have shown. | The length of a rectangle is 4 less than three times the width. Find the length and width if the area is 55 square inches.<br><br>$\quad$ Let $w$ = width.<br>Then $3w - 4$ = length.<br>$$55 = w(3w - 4)$$<br>$$55 = 3w^2 - 4w$$<br>$$0 = 3w^2 - 4w - 55$$<br>$$0 = (3w + 11)(w - 5)$$<br>$$w = -\tfrac{11}{3} \quad \text{or} \quad w = 5$$<br>$-\tfrac{11}{3}$ is not a valid solution. Thus width = 5 inches and length = 11 inches. | 9. The length of a rectangle is 3 more than twice the width. Find the length and width if the area is 90 square feet. |

# Chapter 5 Review Problems

## Section 5.1

*Factor out the greatest common factor.*

1. $12x^3 - 20x^2y$

2. $10x^3 - 35x^3y$

3. $24x^3y - 8x^2y^2 - 16x^3y^3$

4. $3a^3 + 6a^2 - 9ab + 12a$

5. $2a(a + 3b) - 5(a + 3b)$

6. $15x^3y + 6xy^2 + 3xy$

## Section 5.2

*Factor by grouping.*

7. $2ax + 5a - 8x - 20$

8. $a^2 - 4ab + 7a - 28b$

9. $x^2y + 3y - 2x^2 - 6$

10. $30ax - 15ay + 42x - 21y$

11. $15x^2 - 3x + 10x - 2$

12. $30w^2 - 18w + 5wz - 3z$

## Section 5.3

*Factor completely. Be sure to factor out the greatest common factor as your first step.*

13. $x^2 + 6x - 27$

14. $x^2 + 9x - 10$

15. $x^2 + 14x + 48$

16. $x^2 + 8xy + 15y^2$

17. $x^4 + 13x^2 + 42$

18. $x^4 - 2x^2 - 35$

19. $6x^2 + 30x + 36$

20. $2x^2 - 28x + 96$

341

## Section 5.4

*Factor completely. Be sure to factor out the greatest common factor as your first step.*

**21.** $4x^2 + 7x - 15$

**22.** $12x^2 + 11x - 5$

**23.** $2x^2 - x - 3$

**24.** $3x^2 + 2x - 8$

**25.** $20x^2 + 48x - 5$

**26.** $20x^2 + 21x - 5$

**27.** $6x^2 + 4x - 10$

**28.** $6x^2 - 4x - 10$

**29.** $4x^2 - 26x + 30$

**30.** $4x^2 - 20x - 144$

**31.** $12x^2 + xy - 6y^2$

**32.** $6x^2 + 5xy - 25y^2$

## Section 5.5

*Factor these special cases. Be sure to factor out the greatest common factor.*

**33.** $49x^2 - y^2$

**34.** $16x^2 - 36y^2$

**35.** $y^2 - 36x^2$

**36.** $9y^2 - 25x^2$

**37.** $36x^2 + 12x + 1$

**38.** $25x^2 - 20x + 4$

**39.** $16x^2 - 24xy + 9y^2$

**40.** $49x^2 - 28xy + 4y^2$

**41.** $2x^2 - 32$

**42.** $3x^2 - 27$

**43.** $28x^2 + 140x + 175$

**44.** $72x^2 - 192x + 128$

## Section 5.6

*If possible, factor each polynomial completely. If a polynomial cannot be factored, state that it is prime.*

**45.** $4x^2 - 9y^2$

**46.** $x^2 + 13x - 30$

**47.** $9x^2 - 9x - 4$

**48.** $50x^3y^2 + 30x^2y^2 - 10x^2y^2$

**49.** $3x^2 - 18x + 27$

**50.** $25x^3 - 60x^2 + 36x$

**51.** $4x^2 - 13x - 12$

**52.** $3x^3a^3 - 11x^4a^2 - 20x^5a$

**53.** $12a^2 + 14ab - 10b^2$

**54.** $121a^2 + 66ab + 9b^2$

**55.** $7a - 7 - ab + b$

**56.** $3x^3 - 3x + 5yx^2 - 5y$

## Mixed Practice

*If possible, factor each polynomial completely. If a polynomial cannot be factored, state that it is prime.*

**57.** $18b - 42 + 3bc - 7c$

**58.** $10b + 16 - 24x - 15bx$

**59.** $5xb - 35x + 4by - 28y$

**60.** $x^4 - 81y^{12}$

**61.** $6x^4 - x^2 - 15$

**62.** $28yz - 16xyz + x^2yz$

**63.** $12x^3 + 17x^2 + 6x$

**64.** $12w^2 - 12w + 3$

**65.** $4y^3 + 10y^2 - 6y$

**66.** $9x^4 - 144$

**67.** $x^2 - 6x + 12$

**68.** $8x^2 - 19x - 6$

**69.** $8y^5 + 4y^3 - 60y$

**70.** $16x^4y^2 - 56x^2y + 49$

**71.** $2ax + 5a - 10b - 4bx$

**72.** $2x^3 - 9 + x^2 - 18x$

## Section 5.7

*Solve the following equations by factoring.*

**73.** $x^2 + x - 20 = 0$

**74.** $2x^2 + 11x - 6 = 0$

**75.** $7x^2 = 15x + x^2$

**76.** $5x^2 - x = 4x^2 + 12$

**77.** $2x^2 + 9x - 5 = 0$

**78.** $x^2 + 11x + 24 = 0$

**79.** $x^2 + 14x + 45 = 0$

**80.** $5x^2 = 7x + 6$

**81.** $3x^2 + 6x = 2x^2 - 9$

**82.** $4x^2 + 9x - 9 = 0$

**83.** $5x^2 - 11x + 2 = 0$

*Solve.*

▲ **84.** *Geometry* The area of a triangle is 25 square inches. The altitude is 5 inches longer than the base. Find the length of the base and the altitude.

▲ **85.** *Geometry* The area of a rectangle is 30 square feet. The length of the rectangle is 4 feet shorter than double the width. Find the length and width of the rectangle.

**86.** *Rocket Height* The height in feet that a model rocket attains is given by $h = -16t^2 + 80t + 96$, where $t$ is the time measured in seconds. How many seconds will it take until the rocket finally reaches the ground? (*Hint:* At ground level $h = 0$.)

**87.** *Output Power* An electronic technician is working with a 100-volt electric generator. The output power of the generator is given by the equation $p = -5x^2 + 100x$, where $x$ is the amount of current measured in amperes and $p$ is measured in watts. The technician wants to find the value for $x$ when the power is 480 watts. Can you find the two answers?

# How Am I Doing? Chapter 5 Test

**MATH COACH**   **MyMathLab®**   **You Tube**

After you take this test read through the Math Coach on pages 345–346. Math Coach videos are available via MyMathLab and YouTube. Step-by-step test solutions on the Chapter Test Prep Videos are also available via MyMathLab and YouTube. (Search "TobeyBeginningAlg" and click on "Channels.")

*If possible, factor each polynomial completely. If a polynomial cannot be factored, state that it is prime.*

**1.** $x^2 + 12x - 28$

**MC 2.** $16x^2 - 81$

**3.** $10x^2 + 27x + 5$

**MC 4.** $9a^2 - 30a + 25$

**5.** $7x - 9x^2 + 14xy$

**MC 6.** $10xy + 15by - 8x - 12b$

**7.** $6x^3 - 20x^2 + 16x$

**8.** $5a^2c - 11ac + 2c$

**9.** $81x^2 - 100$

**10.** $9x^2 - 15x + 4$

**11.** $20x^2 - 45$

**12.** $36x^2 + 1$

**13.** $3x^3 + 11x^2 + 10x$

**14.** $60xy^2 - 20x^2y - 45y^3$

**15.** $81x^2 - 1$

**16.** $81y^4 - 1$

**17.** $2ax + 6a - 5x - 15$

**18.** $aw^2 - 8b + 2bw^2 - 4a$

**MC 19.** $3x^2 - 3x - 90$

**20.** $2x^3 - x^2 - 15x$

*Solve.*

**21.** $x^2 + 14x + 45 = 0$

**22.** $14 + 3x(x + 2) = -7x$

**23.** $2x^2 + x - 10 = 0$

**24.** $x^2 - 3x - 28 = 0$

*Solve using a quadratic equation.*

▲ **25.** The park service is studying a rectangular piece of land that has an area of 91 square miles. The length of this piece of land is 1 mile shorter than double the width. Find the length and width of this rectangular piece of land.

1. _____
2. _____
3. _____
4. _____
5. _____
6. _____
7. _____
8. _____
9. _____
10. _____
11. _____
12. _____
13. _____
14. _____
15. _____
16. _____
17. _____
18. _____
19. _____
20. _____
21. _____
22. _____
23. _____
24. _____
25. _____

**Total Correct:** _____

# MATH COACH

*Mastering the skills you need to do well on the test.*

The following problems are from the Chapter 5 Test. Here are some helpful hints to keep you from making common errors on test problems.

**Chapter 5 Test, Problem 2** Factor completely. $16x^2 - 81$

> **HELPFUL HINT** It is important to learn the difference-of-two-squares formula: $a^2 - b^2 = (a + b)(a - b)$. Remember that the numerical values in both terms will be perfect squares. The first ten perfect squares are 1, 4, 9, 16, 25, 36, 49, 64, 81, and 100.

Did you remember that $(4x)^2 = 16x^2$?

Yes _____ No _____

Did you remember that $9^2 = 81$?

Yes _____ No _____

If you answered No to these questions, stop and review the list of the first ten perfect squares. Consider that $4^2 = 16$ and $x \cdot x = x^2$.

Do you see how $16x^2 - 81$ can be factored using the formula $a^2 - b^2$?

Yes _____ No _____

If you answered No, stop and review the difference-of-two-squares formula again. Make sure that one set of parentheses contains a + sign and the other set of parentheses contains a − sign.

If you answered Problem 2 incorrectly, go back and rework the problem using these suggestions.

---

**Chapter 5 Test, Problem 4** Factor completely. $9a^2 - 30a + 25$

> **HELPFUL HINT** Remember the perfect-square-trinomial formula: $a^2 - 2ab + b^2 = (a - b)^2$. You must verify two things to determine if you can use this formula:
> (1) The numerical values in the first term and the last term must be perfect squares.
> (2) The middle term must equal "twice the product of the values whose squares are the first and last terms."

Did you remember that $(3a)^2 = 9a^2$?

Yes _____ No _____

Did you remember that $5^2 = 25$?

Yes _____ No _____

If you answered No to these questions, stop and review the first ten perfect squares. Consider that $3^2 = 9$ and $a \cdot a = a^2$.

Do you see how $9a^2 - 30a + 25$ can be factored using the formula $(a - b)^2$?

Yes _____ No _____

If you answered No, check to see if the middle term, $30a$, equals twice the product of $3a$ and 5.

Now go back and rework the problem using these suggestions.

Need more help? Watch the **MATH COACH** videos in MyMathLab® or on YouTube.

**Chapter 5 Test, Problem 6**  Factor completely. $10xy + 15by - 8x - 12b$

> **HELPFUL HINT** Look for common factors first. We can find the greatest common factor of the first two terms and factor. Then we can find the greatest common factor of the second two terms and factor. Make sure that you obtain the *same binomial factor* for each step. Be careful with $+/-$ signs.

Did you identify $5y$ as the greatest common factor of the first two terms: $10xy + 15by$?

Yes _____ No _____

Did you identify 4 as the greatest common factor of the second two terms: $-8x - 12b$?

Yes _____ No _____

If you answered Yes to these questions, then you obtained $(2x + 3b)$ in the first term and $(-2x - 3b)$ in the second term. These are *not* the same binomial factor. Stop and consider how to get the same binomial factor of $(2x + 3b)$.

In your final answer, is the binomial factor of $(2x + 3b)$ listed once?

Yes _____ No _____

If you answered No, remember that to factor completely, we must remove all common factors from both terms. Stop now and complete this step.

If you answered Problem 6 incorrectly, go back and rework the problem using these suggestions.

---

**Chapter 5 Test, Problem 19**  Factor completely. $3x^2 - 3x - 90$

> **HELPFUL HINT** Look for the greatest common factor of all three terms as your *first step*. Don't forget to include this common factor as part of your answer. Always check your final product to make sure that it matches the original polynomial. Do this by multiplying.

Did you obtain $(3x - 18)(x + 5)$ or $(3x + 15)(x - 6)$ as your answer?

Yes _____ No _____

If you answered Yes, then you forgot to factor out the greatest common factor 3 as your first step.

Do you see how to factor $x^2 - x - 30$?

Yes _____ No _____

If you answered No, remember that we are looking for two numbers with a product of $-30$ and a sum of $-1$.

Be sure to double-check your final answer to be sure there are no common factors and include 3 in your final answer.

Now go back and rework the problem using these suggestions.

Need more help? Look for section examples marked with $\mathbb{MC}$ to review.

CHAPTER 6

# Rational Expressions and Equations

## CAREER OPPORTUNITIES

### Operations Analyst

The effective management of the logistics and daily operations of a business is vital to its success. Teams of operations management and logistics professionals are the individuals who accomplish this. From predicting average costs to identifying the time required for manufacturing goods, these professionals use math every day to simulate the best practices needed for creating efficient operations.

To investigate how the mathematics in this chapter can help with this field, see the Career Exploration Problems on page **387**.

## 6.1 Simplifying Rational Expressions ▶

**Student Learning Objective**

After studying this section, you will be able to:

1 Simplify rational expressions by factoring. ▶

Recall that a rational number is a number that can be written as one integer divided by another integer, such as $3 \div 4$ or $\frac{3}{4}$. We usually use the word *fraction* to mean $\frac{3}{4}$. We can extend this idea to algebraic expressions. A **rational expression** is a polynomial divided by another polynomial, such as

$$(3x + 2) \div (x + 4) \quad \text{or} \quad \frac{3x + 2}{x + 4}.$$

The last fraction is sometimes also called a **fractional algebraic expression.** There is a special restriction for all fractions, including fractional algebraic expressions: The denominator of the fraction cannot be 0. For example, in the expression

$$\frac{3x + 2}{x + 4},$$

the denominator cannot be 0. Therefore, the value of $x$ cannot be $-4$. The following important restriction will apply throughout this chapter. We state it here to avoid having to mention it repeatedly throughout this chapter.

> **RESTRICTION**
>
> The denominator of a rational expression cannot be zero. Any value of the variable that would make the denominator zero is not allowed.

We have discovered that fractions can be simplified (or reduced) in the following way.

$$\frac{15}{25} = \frac{3 \cdot \cancel{5}}{5 \cdot \cancel{5}} = \frac{3}{5}$$

This is sometimes referred to as the **basic rule of fractions** and can be stated as follows.

> **BASIC RULE OF FRACTIONS**
>
> For any rational expression $\frac{a}{b}$ and any polynomials $a$, $b$, and $c$ (where $b \neq 0$ and $c \neq 0$),
>
> $$\frac{ac}{bc} = \frac{a}{b}.$$

We will examine several examples where $a$, $b$, and $c$ are real numbers, as well as more involved examples where $a$, $b$, and $c$ are polynomials. In either case we shall make extensive use of our factoring skills in this section.

One essential property is revealed by the basic rule of fractions: If the numerator and denominator of a given fraction are multiplied by the same nonzero quantity, an equivalent fraction is obtained. The rule can be used two ways. You can start with $\frac{ac}{bc}$ and end with the equivalent fraction $\frac{a}{b}$. Or, you can start with $\frac{a}{b}$ and end with the equivalent fraction $\frac{ac}{bc}$. In this section we focus on the first process.

**Example 1** Reduce. $\dfrac{21}{39}$

**Solution** $\dfrac{21}{39} = \dfrac{7 \cdot \cancel{3}}{13 \cdot \cancel{3}} = \dfrac{7}{13}$   Use the rule $\frac{ac}{bc} = \frac{a}{b}$. Let $c = 3$ because 3 is the greatest common factor of 21 and 39. □

▷ **Student Practice 1**   Reduce.

$$\frac{28}{63}$$

## 1 Simplifying Rational Expressions by Factoring ▶

The process of reducing fractions shown in Example 1 is sometimes called *dividing out* common factors. When you do this, you are **simplifying the fraction.** Remember, only **factors** of both the numerator and the denominator can be divided out. To apply the basic rule of fractions, it is usually helpful if the numerator and denominator of the fraction are completely factored. You will need to use your factoring skills from Chapter 5 to accomplish this step.

**Example 2** Simplify. $\dfrac{4x + 12}{5x + 15}$

**Solution**

$$\dfrac{4x + 12}{5x + 15} = \dfrac{4(x + 3)}{5(x + 3)} \qquad \text{Factor 4 from the numerator.}$$
$$\text{Factor 5 from the denominator.}$$
$$= \dfrac{4\,\cancel{(x + 3)}}{5\,\cancel{(x + 3)}} \qquad \text{Apply the basic rule of fractions.}$$
$$= \dfrac{4}{5} \qquad\qquad\qquad \square$$

**Student Practice 2** Simplify.

$$\dfrac{12x - 6}{14x - 7}$$

**Example 3** Simplify. $\dfrac{x^2 + 9x + 14}{x^2 - 4}$

**Solution**

$$= \dfrac{(x + 7)(x + 2)}{(x - 2)(x + 2)} \qquad \text{Factor the numerator.}$$
$$\text{Factor the denominator.}$$
$$= \dfrac{(x + 7)\,\cancel{(x + 2)}}{(x - 2)\,\cancel{(x + 2)}} \qquad \text{Apply the basic rule of fractions.}$$
$$= \dfrac{x + 7}{x - 2} \qquad\qquad \square$$

**CAUTION:** Do not try to remove terms that are added. In Example 3, do not try to remove the $x^2$ in the top and the $x^2$ in the bottom of the fraction $\frac{x^2 + 9x + 14}{x^2 - 4}$. The basic rule of fractions applies only to quantities that are factors of both numerator and denominator.

**Student Practice 3** Simplify.

$$\dfrac{4x^2 - 9}{2x^2 - x - 3}$$

Some problems may involve more than one step of factoring. Always remember to factor out the greatest common factor as the first step, if it is possible to do so.

**Example 4** Simplify. $\dfrac{x^3 - 9x}{x^3 + x^2 - 6x}$

**Solution**

$$= \dfrac{x(x^2 - 9)}{x(x^2 + x - 6)} \qquad \text{Factor out the greatest common factor from the polynomials in the numerator and the denominator.}$$
$$= \dfrac{\cancel{x}\cancel{(x + 3)}(x - 3)}{\cancel{x}\cancel{(x + 3)}(x - 2)} \qquad \text{Factor each polynomial and apply the basic rule of fractions.}$$
$$= \dfrac{x - 3}{x - 2} \qquad\qquad \square$$

*Continued on next page*

**▶ Student Practice 4** Simplify.

$$\frac{x^3 - 16x}{x^3 - 2x^2 - 8x}$$

When you are simplifying, be on the lookout for the special situation where *a factor in the denominator is the opposite of a factor in the numerator*. In such a case you should factor a negative number from one of the factors so that it becomes equivalent to the other factor and can be divided out. Look carefully at the following two examples.

**Example 5** Simplify. $\dfrac{5x - 15}{6 - 2x}$

**Solution** Notice that the variable term in the numerator, $5x$, and the variable term in the denominator, $-2x$, *are opposite in sign*. Likewise, the numerical terms $-15$ and $6$ *are opposite in sign*. Factor out a negative number from the denominator.

$$\frac{5x - 15}{6 - 2x} = \frac{5(x - 3)}{-2(-3 + x)} \quad \begin{array}{l} \text{Factor 5 from the numerator.} \\ \text{Factor } -2 \text{ from the denominator.} \\ \text{Note that } (x - 3) \text{ and } (-3 + x) \text{ are} \\ \text{equivalent since } (+x - 3) = (-3 + x). \end{array}$$

$$= \frac{5\cancel{(x - 3)}}{-2\cancel{(-3 + x)}} \quad \text{Apply the basic rule of fractions.}$$

$$= -\frac{5}{2}$$

Note that $\frac{5}{-2}$ is not considered to be in simplest form. We usually avoid leaving a negative number in the denominator. Therefore, to simplify, give the result as $-\frac{5}{2}$ or $\frac{-5}{2}$. □

**▶ Student Practice 5** Simplify.

$$\frac{8x - 20}{15 - 6x}$$

**Example 6** Simplify. $\dfrac{2x^2 - 11x + 12}{16 - x^2}$

**Solution**

$$= \frac{(x - 4)(2x - 3)}{(4 - x)(4 + x)} \quad \begin{array}{l} \text{Factor the numerator and the denominator.} \\ \text{Observe that } (x - 4) \text{ and } (4 - x) \text{ are opposites.} \end{array}$$

$$= \frac{(x - 4)(2x - 3)}{-1(-4 + x)(4 + x)} \quad \text{Factor } -1 \text{ out of } (+4 - x) \text{ to obtain } -1(-4 + x).$$

$$= \frac{\cancel{(x - 4)}(2x - 3)}{-1\cancel{(-4 + x)}(4 + x)} \quad \begin{array}{l} \text{Apply the basic rule of fractions since } (x - 4) \\ \text{and } (-4 + x) \text{ are equivalent.} \end{array}$$

$$= \frac{2x - 3}{-1(4 + x)}$$

$$= -\frac{2x - 3}{4 + x} \qquad \qquad □$$

**▶ Student Practice 6** Simplify.

$$\frac{4x^2 + 3x - 10}{25 - 16x^2}$$

After doing Examples 5 and 6, you will notice a pattern. Whenever the factor in the numerator and the factor in the denominator are opposites, the value $-1$ results. We could actually make this a property.

For all monomials $A$ and $B$ where $A \neq B$, it is true that
$$\frac{A - B}{B - A} = -1.$$

You may use this property when reducing fractions if it is helpful to you. Otherwise, you may use the factoring method shown in Examples 5 and 6.

Some problems will involve two or more variables. In such cases, you will need to factor carefully and make sure that each set of parentheses contains the correct terms.

**Example 7** Simplify. $\dfrac{x^2 - 7xy + 12y^2}{2x^2 - 7xy - 4y^2}$

**Solution**

$$= \frac{(x - 4y)(x - 3y)}{(2x + y)(x - 4y)} \qquad \text{Factor the numerator.}$$
Factor the denominator.

$$= \frac{\cancel{(x - 4y)}(x - 3y)}{(2x + y)\cancel{(x - 4y)}} \qquad \text{Apply the basic rule of fractions.}$$

$$= \frac{x - 3y}{2x + y} \qquad \qquad \square$$

**Student Practice 7** Simplify.
$$\frac{x^2 - 8xy + 15y^2}{2x^2 - 11xy + 5y^2}$$

**Example 8** Simplify. $\dfrac{6a^2 + ab - 7b^2}{36a^2 - 49b^2}$

**Solution**

$$= \frac{(6a + 7b)(a - b)}{(6a + 7b)(6a - 7b)} \qquad \text{Factor the numerator.}$$
Factor the denominator.

$$= \frac{\cancel{(6a + 7b)}(a - b)}{\cancel{(6a + 7b)}(6a - 7b)} \qquad \text{Apply the basic rule of fractions.}$$

$$= \frac{a - b}{6a - 7b} \qquad \qquad \square$$

**Student Practice 8** Simplify.
$$\frac{25a^2 - 16b^2}{10a^2 + 3ab - 4b^2}$$

## 6.1 Exercises MyMathLab®

*Simplify.*

1. $\dfrac{4x - 24y}{x - 6y}$

2. $\dfrac{3x + 4y}{9x + 12y}$

3. $\dfrac{6x + 18}{x^2 + 3x}$

4. $\dfrac{25x - 30}{5x^3 - 6x^2}$

5. $\dfrac{9x^2 + 6x + 1}{1 - 9x^2}$

6. $\dfrac{16x^2 + 8x + 1}{1 - 16x^2}$

7. $\dfrac{3a^2b(a - 2b)}{6ab^2}$

8. $\dfrac{6a^2b}{9a^2b^2(a + 3b)}$

9. $\dfrac{x^2 + x - 2}{x^2 - x}$

10. $\dfrac{x^2 + x - 12}{x^2 - 3x}$

11. $\dfrac{x^2 - 3x - 10}{3x^2 + 5x - 2}$

12. $\dfrac{4x^2 - 10x + 6}{2x^2 + x - 3}$

13. $\dfrac{x^2 + 4x - 21}{x^3 - 49x}$

14. $\dfrac{x^3 - 3x^2 - 40x}{x^2 + 10x + 25}$

15. $\dfrac{3x^2 - 11x - 4}{x^2 + x - 20}$

16. $\dfrac{x^2 - 9x + 18}{2x^2 - 9x + 9}$

17. $\dfrac{3x^2 - 8x + 5}{4x^2 - 5x + 1}$

18. $\dfrac{3y^2 - 8y - 3}{3y^2 - 10y + 3}$

19. $\dfrac{5x^2 - 27x + 10}{5x^2 + 3x - 2}$

20. $\dfrac{2x^2 + 4x - 6}{x^2 + 3x - 4}$

**Mixed Practice** *Take some time to review exercises 1–20 before you proceed with exercises 21–30.*

21. $\dfrac{12 - 3x}{5x^2 - 20x}$

22. $\dfrac{20 - 4ab}{a^2b - 5a}$

23. $\dfrac{2x^2 - 7x - 15}{25 - x^2}$

24. $\dfrac{49 - x^2}{2x^2 - 9x - 35}$

25. $\dfrac{(3x + 4)^2}{9x^2 + 9x - 4}$

26. $\dfrac{12x^2 - 11x - 5}{(4x - 5)^2}$

27. $\dfrac{3x^2 + 13x - 10}{20 - x - x^2}$

28. $\dfrac{2x^2 + 15x + 18}{42 + x - x^2}$

29. $\dfrac{a^2 + ab - 6b^2}{3a^2 + 8ab - 3b^2}$

30. $\dfrac{a^2 + 2ab - 15b^2}{4a^2 - 13ab + 3b^2}$

## Cumulative Review  *Multiply.*

31. **[4.5.2]** $(3x - 7)^2$

32. **[4.5.1]** $(7x + 6y)(7x - 6y)$

33. **[4.4.2]** $(2x + 3)(x - 4)$

34. **[4.5.3]** $(2x + 3)(x - 4)(x - 2)$

35. **[1.7.2]** Simplify. $\dfrac{2a^2}{7} + \dfrac{3b}{2} + 3a^2 - \dfrac{3b}{4}$

36. **[1.3.3]** Divide. $\dfrac{-35}{12} \div \dfrac{5}{14}$

37. **[0.3.2]** *Dividing Acreage* David and Connie Swensen wish to divide $4\frac{7}{8}$ acres of farmland into three equal-size house lots. What will be the acreage of each lot?

38. **[0.3.1]** *Roasting a Turkey* Ron and Mary Larson are planning to cook a $17\frac{1}{2}$-pound turkey. The directions suggest a cooking time of 22 minutes per pound for turkeys that weigh between 16 and 20 pounds. How many hours and minutes should they allow for an approximate cooking time?

### Quick Quiz 6.1  *Simplify.*

1. $\dfrac{x^3 + 3x^2}{x^3 - 2x^2 - 15x}$

2. $\dfrac{6 - 2ab}{ab^2 - 3b}$

3. $\dfrac{8x^2 + 6x - 5}{16x^2 + 40x + 25}$

4. **Concept Check** Explain why it is important to completely factor both the numerator and the denominator when simplifying

$$\dfrac{x^2y - y^3}{x^2y + xy^2 - 2y^3}$$

# 6.2 Multiplying and Dividing Rational Expressions

**Student Learning Objectives**

After studying this section, you will be able to:

1 Multiply rational expressions.

2 Divide rational expressions.

## 1 Multiplying Rational Expressions

To multiply two rational expressions, we multiply the numerators and multiply the denominators. As before, the denominators cannot equal zero.

For any two rational expressions $\frac{a}{b}$ and $\frac{c}{d}$ where $b \neq 0$ and $d \neq 0$,

$$\frac{a}{b} \cdot \frac{c}{d} = \frac{ac}{bd}.$$

Simplifying or reducing fractions *prior to multiplying them* usually makes the computations easier to do. Leaving the reducing step until the end makes the simplifying process longer and increases the chance for error. This long approach should be avoided.

As an example, let's do the same problem two ways to see which one is easier. Let's simplify the following problem by multiplying first and then reducing the result.

$$\frac{5}{7} \times \frac{49}{125}$$

$$\frac{5}{7} \times \frac{49}{125} = \frac{245}{875} \qquad \text{Multiply the numerators and multiply the denominators.}$$

$$= \frac{7}{25} \qquad \text{Reduce the fraction. (\textit{Note:} It can take a bit of trial and error to discover how to reduce it.)}$$

Compare this with the following method, where we reduce the fractions prior to multiplying them.

$$\frac{5}{7} \times \frac{49}{125}$$

$$\frac{5}{7} \times \frac{7 \cdot 7}{5 \cdot 5 \cdot 5} \qquad \text{\textbf{Step 1.} It is easier to factor first. We factor the numerator and denominator of the second fraction.}$$

$$= \frac{\cancel{5}}{\cancel{7}} \times \frac{\cancel{7} \cdot 7}{\cancel{5} \cdot 5 \cdot 5} = \frac{7}{25} \qquad \text{\textbf{Step 2.} Then we apply the basic rule of fractions to divide out the common factors 5 and 7 that appear in both a numerator and in a denominator.}$$

A similar approach can be used with the multiplication of rational expressions. We first factor the numerator and denominator of each fraction. Then we divide out any factor that is common to a numerator and a denominator. Finally, we multiply the remaining numerators and the remaining denominators.

**Example 1** Multiply. $\dfrac{x^2 - x - 12}{x^2 - 16} \cdot \dfrac{2x^2 + 7x - 4}{x^2 - 4x - 21}$

**Solution**

$$\frac{(x-4)(x+3)}{(x-4)(x+4)} \cdot \frac{(x+4)(2x-1)}{(x+3)(x-7)} \qquad \text{Factoring is always the first step.}$$

$$= \frac{\cancel{(x-4)}\cancel{(x+3)}}{\cancel{(x-4)}\cancel{(x+4)}} \cdot \frac{\cancel{(x+4)}(2x-1)}{\cancel{(x+3)}(x-7)} \qquad \text{Apply the basic rule of fractions. (Three pairs of factors divide out.)}$$

$$= \frac{2x-1}{x-7} \qquad \text{The final answer.} \qquad \square$$

▶ **Student Practice 1** Multiply.

$$\frac{6x^2 + 7x + 2}{x^2 - 7x + 10} \cdot \frac{x^2 + 3x - 10}{2x^2 + 11x + 5}$$

In some cases, it may take several steps to factor a given numerator or denominator. You should always check for the *greatest common factor* as your first step.

**Example 2** Multiply. $\dfrac{x^4 - 16}{x^3 + 4x} \cdot \dfrac{2x^2 - 8x}{4x^2 + 2x - 12}$

**Solution**

$$= \frac{(x^2 + 4)(x^2 - 4)}{x(x^2 + 4)} \cdot \frac{2x(x - 4)}{2(2x^2 + x - 6)}$$ Factor each numerator and denominator. Factoring out the greatest common factor first is very important.

$$= \frac{(x^2 + 4)(x + 2)(x - 2)}{x(x^2 + 4)} \cdot \frac{2x(x - 4)}{2(x + 2)(2x - 3)}$$ Factor again where possible.

$$= \frac{(x^2 + 4)(x + 2)(x - 2)}{x(x^2 + 4)} \cdot \frac{2x(x - 4)}{2(x + 2)(2x - 3)}$$ Divide out factors that appear in both a numerator and a denominator. (There are four such pairs of factors.)

$$= \frac{(x - 2)(x - 4)}{2x - 3} \quad \text{or} \quad \frac{x^2 - 6x + 8}{2x - 3}$$ Write the answer as one fraction. (Usually, if there is more than one factor in a numerator or denominator, the answer is left in factored form.)

 **Student Practice 2** Multiply.

$$\frac{2y^2 - 6y - 8}{y^2 - y - 2} \cdot \frac{y^2 - 5y + 6}{2y^2 - 32}$$

## 2 Dividing Rational Expressions

For any two fractions $\frac{a}{b}$ and $\frac{c}{d}$, the operation of division can be performed by inverting the second fraction and multiplying it by the first fraction. When we invert a fraction, we are finding its *reciprocal*. Two numbers are **reciprocals** of each other if their product is 1. The reciprocal of $\frac{3}{5}$ is $\frac{5}{3}$. The reciprocal of 7 is $\frac{1}{7}$. The reciprocal of $\frac{a}{b}$ is $\frac{b}{a}$. Sometimes people state the rule for dividing fractions this way: "To divide two fractions, keep the first fraction unchanged and multiply by the reciprocal of the second fraction."

> The definition for division of fractions is
> $$\frac{a}{b} \div \frac{c}{d} = \frac{a}{b} \cdot \frac{d}{c}.$$

This property holds whether $a$, $b$, $c$, and $d$ are polynomials or numerical values. (It is assumed, of course, that no denominator is zero.)

In the first step for dividing two rational expressions, invert the second fraction and rewrite the quotient as a product. Then follow the procedure for multiplying rational expressions.

 **Example 3** Divide. $\dfrac{6x + 12y}{2x - 6y} \div \dfrac{9x^2 - 36y^2}{4x^2 - 36y^2}$

**Solution**

$$= \frac{6x + 12y}{2x - 6y} \cdot \frac{4x^2 - 36y^2}{9x^2 - 36y^2}$$ Invert the second fraction and write the problem as the product of two fractions.

$$= \frac{6(x + 2y)}{2(x - 3y)} \cdot \frac{4(x^2 - 9y^2)}{9(x^2 - 4y^2)}$$ Factor each numerator and denominator.

$$= \frac{(3)(2)(x + 2y)}{2(x - 3y)} \cdot \frac{(2)(2)(x + 3y)(x - 3y)}{(3)(3)(x + 2y)(x - 2y)}$$ Factor again where possible.

 **Student Practice 3** Divide.

$$\frac{x^2 + 5x + 6}{x^2 + 8x} \div \frac{2x^2 + 5x + 2}{2x^2 + x}$$

*Continued on next page*

$$= \frac{\cancel{(3)(2)}(x + 2y)}{\cancel{2}(x - 3y)} \cdot \frac{(2)(2)(x + 3y)\cancel{(x - 3y)}}{\cancel{(3)}(3)\cancel{(x + 2y)}(x - 2y)}$$

Divide out factors that appear in both a numerator and a denominator.

$$= \frac{(2)(2)(x + 3y)}{3(x - 2y)}$$

Write the result as one fraction.

$$= \frac{4(x + 3y)}{3(x - 2y)}$$

Simplify.

Although it is correct to write this answer as $\dfrac{4x + 12y}{3x - 6y}$, it is customary to leave the answer in factored form to ensure that the final answer is simplified.

A polynomial that is not in fraction form can be written as a fraction if you give it a denominator of 1.

**Example 4** Divide. $\dfrac{15 - 3x}{x + 6} \div (x^2 - 9x + 20)$

**Solution**

Note that $x^2 - 9x + 20$ can be written as $\dfrac{x^2 - 9x + 20}{1}$.

$$= \frac{15 - 3x}{x + 6} \cdot \frac{1}{x^2 - 9x + 20}$$

Invert and multiply.

$$= \frac{-3(-5 + x)}{x + 6} \cdot \frac{1}{(x - 5)(x - 4)}$$

Factor where possible. Note that we had to factor $-3$ from the first numerator so that it would have a factor in common with the second denominator.

$$= \frac{-3\cancel{(-5 + x)}}{x + 6} \cdot \frac{1}{\cancel{(x - 5)}(x - 4)}$$

Divide out the common factor. $(-5 + x)$ is equivalent to $(x - 5)$.

$$= \frac{-3}{(x + 6)(x - 4)}$$

The final answer. Note that the answer can be written in several equivalent forms.

or $\quad -\dfrac{3}{(x + 6)(x - 4)} \quad$ or $\quad \dfrac{3}{(x + 6)(4 - x)}$

**Student Practice 4** Divide.

$$\frac{x + 3}{x - 3} \div (9 - x^2)$$

**CAUTION:** It is logical to assume that the exercises in Section 6.2 have at least one common factor that can be divided out. Therefore, if after factoring, you do not observe any common factors, you should be somewhat suspicious. In such cases, it would be wise to double-check your factoring steps for errors.

## 6.2 Exercises   MyMathLab®

### Verbal and Writing Skills, Exercises 1–2

1. Before multiplying rational expressions, we should always first try to

_____

_____ .

2. Division of two rational expressions is done by keeping the first fraction unchanged and then

_____

_____ .

*Multiply.*

3. $\dfrac{4x + 12}{x - 4} \cdot \dfrac{x^2 + x - 20}{x^2 + 6x + 9}$

4. $\dfrac{3x - 6}{x + 5} \cdot \dfrac{x^2 + 6x + 5}{2x^2 - 8}$

5. $\dfrac{24x^3}{4x^2 - 36} \cdot \dfrac{2x^2 + 6x}{16x^2}$

6. $\dfrac{2x^2}{6x + 15} \cdot \dfrac{4x^3 + 26x^2 + 40x}{10x^3}$

7. $\dfrac{x^2 + 3x - 10}{x^2 + x - 20} \cdot \dfrac{x^2 - 3x - 4}{x^2 + 4x + 3}$

8. $\dfrac{x^2 - 2x - 35}{x^2 - 5x - 6} \cdot \dfrac{x^2 - 2x - 24}{x^2 - 3x - 28}$

*Divide.*

9. $\dfrac{x + 6}{x - 8} \div \dfrac{x + 5}{x^2 - 6x - 16}$

10. $\dfrac{x - 4}{x + 9} \div \dfrac{x - 7}{x^2 + 13x + 36}$

11. $(5x + 4) \div \dfrac{25x^2 - 16}{5x^2 + 11x - 12}$

12. $\dfrac{4x^2 - 25}{4x^2 - 20x + 25} \div (4x + 10)$

13. $\dfrac{3x^2 + 12xy + 12y^2}{x^2 + 4xy + 3y^2} \div \dfrac{4x + 8y}{x + y}$

14. $\dfrac{5x^2 + 10xy + 5y^2}{x^2 + 5xy + 6y^2} \div \dfrac{3x + 3y}{x + 2y}$

### Mixed Practice  *Perform the operation indicated.*

15. $\dfrac{(x + 5)^2}{3x^2 - 7x + 2} \cdot \dfrac{x^2 - 4x + 4}{x + 5}$

16. $\dfrac{3x^2 - 10x - 8}{(4x + 5)^2} \cdot \dfrac{4x + 5}{(x - 4)^2}$

17. $\dfrac{x^2 + x - 30}{10 - 2x} \div \dfrac{x^2 + 4x - 12}{5x + 15}$

18. $\dfrac{x^2 + 4x - 12}{16 - 4x^2} \div \dfrac{x^2 - 3x - 54}{x^2 + 10x + 16}$

19. $\dfrac{y^2 + 4y - 12}{y^2 + 2y - 24} \cdot \dfrac{y^2 - 16}{y^2 + 2y - 8}$

20. $\dfrac{4y^2 + 13y + 3}{12y^2 - y - 1} \cdot \dfrac{6y^2 + y - 1}{2y^2 + 7y + 3}$

21. $\dfrac{x^2 + 10x + 24}{2x^2 + 13x + 6} \cdot \dfrac{2x^2 - 5x - 3}{x^2 + 5x - 24}$

22. $\dfrac{x^2 + x - 12}{3x^2 - 8x - 3} \cdot \dfrac{3x^2 - 14x - 5}{x^2 + 2x - 8}$

## Cumulative Review

23. [2.3.2] Solve. $6x^2 + 3x - 18 = 5x - 2 + 6x^2$

24. [1.2.1] Perform the operations indicated. $\dfrac{3}{4} + \dfrac{1}{2} - \dfrac{4}{7}$

▲ 25. [3.5.3] *Golden Gate Bridge* The Golden Gate Bridge has a total length (including approaches) of 8981 feet and a road width of 90 feet. The width of the sidewalk is 10.5 feet. (The sidewalk spans the entire length of the bridge.) Assume it would cost $\$x$ per square foot to resurface the road or the sidewalk. Write an expression for how much more it would cost to resurface the road than the sidewalk.

▲ 26. [3.5.3] *Garden Design* Harold Rafton planted a square garden bed. George Avis also planted a garden that was 2 feet less in width but 3 feet longer in length than Harold's garden. If the area of George's garden was 36 square feet, find the dimensions of each garden.

## Quick Quiz 6.2

1. Multiply. $\dfrac{2x - 10}{x - 4} \cdot \dfrac{x^2 + 5x + 4}{x^2 - 4x - 5}$

2. Multiply. $\dfrac{3x^2 - 13x - 10}{3x^2 + 2x} \cdot \dfrac{x^2 - 25x}{x^2 - 25}$

3. Divide. $\dfrac{2x^2 - 18}{3x^2 + 3x} \div \dfrac{x^2 + 6x + 9}{x^2 + 4x + 3}$

4. **Concept Check** Explain how you would divide $\dfrac{21x - 7}{9x^2 - 1} \div \dfrac{1}{3x + 1}$.

## 6.3 Adding and Subtracting Rational Expressions ▶

### 1 Adding and Subtracting Rational Expressions with a Common Denominator ▶

If rational expressions have the same denominator, they can be combined in the same way as arithmetic fractions. The numerators are added or subtracted and the denominator remains the same.

**ADDING RATIONAL EXPRESSIONS**

For any rational expressions $\dfrac{a}{b}$ and $\dfrac{c}{b}$,

$$\frac{a}{b} + \frac{c}{b} = \frac{a+c}{b} \qquad \text{where } b \neq 0.$$

**Example 1** Add. $\dfrac{5a}{a+2b} + \dfrac{6a}{a+2b}$

**Solution**

$$\frac{5a}{a+2b} + \frac{6a}{a+2b} = \frac{5a+6a}{a+2b} = \frac{11a}{a+2b}$$

Note that the denominators are the same. Only add the numerators. Keep the same denominator.
Do not change the denominator. ☐

 **Student Practice 1** Add.

$$\frac{2s+t}{2s-t} + \frac{s-t}{2s-t}$$

**SUBTRACTING RATIONAL EXPRESSIONS**

For any rational expressions $\dfrac{a}{b}$ and $\dfrac{c}{b}$,

$$\frac{a}{b} - \frac{c}{b} = \frac{a-c}{b} \qquad \text{where } b \neq 0.$$

**Example 2** Subtract. $\dfrac{3x}{(x+y)(x-2y)} - \dfrac{8x}{(x+y)(x-2y)}$

**Solution**

$$\frac{3x}{(x+y)(x-2y)} - \frac{8x}{(x+y)(x-2y)} = \frac{3x-8x}{(x+y)(x-2y)}$$

Write as one fraction.

$$= \frac{-5x}{(x+y)(x-2y)}$$

Simplify. ☐

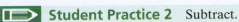 **Student Practice 2** Subtract.

$$\frac{b}{(a-2b)(a+b)} - \frac{2b}{(a-2b)(a+b)}$$

### 2 Determining the LCD for Rational Expressions with Different Denominators ▶

How do we add or subtract rational expressions when the denominators are not the same? First we must find the **least common denominator** (LCD). You need to be clear on how to find the least common denominator and how to add and subtract fractions from arithmetic before you attempt this section. Review Sections 0.1 and 0.2 if you have any questions about this topic.

> **HOW TO FIND THE LCD OF TWO OR MORE RATIONAL EXPRESSIONS**
>
> 1. Factor each denominator completely.
> 2. The LCD is a product containing each *different factor*.
> 3. If a factor occurs more than once in any one denominator, the LCD will contain that factor repeated the greatest number of times that it occurs in any one denominator.

**Example 3** Find the LCD. $\dfrac{5}{2x - 4}, \dfrac{6}{3x - 6}$

**Solution** Factor each denominator.

$$2x - 4 = 2(x - 2) \qquad 3x - 6 = 3(x - 2)$$

The different factors are 2, 3, and $(x - 2)$. Since no factor appears more than once in any one denominator, the LCD is the product of these three factors.

$$LCD = (2)(3)(x - 2) = 6(x - 2)$$

▷ **Student Practice 3** Find the LCD.

$$\frac{7}{6x + 21}, \frac{13}{10x + 35}$$

**Example 4** Find the LCD.

(a) $\dfrac{5}{12ab^2c}, \dfrac{13}{18a^3bc^4}$ 

(b) $\dfrac{8}{x^2 - 5x + 4}, \dfrac{12}{x^2 + 2x - 3}$

**Solution** If a factor occurs more than once in any one denominator, the LCD will contain that factor repeated the greatest number of times that it occurs in any one denominator.

(a) $12ab^2c = 2 \cdot 2 \cdot 3 \cdot \quad a \cdot \quad b \cdot b \cdot c$
$18a^3bc^4 = \quad 2 \cdot 3 \cdot 3 \cdot a \cdot a \cdot a \cdot b \cdot \quad c \cdot c \cdot c \cdot c$

$$LCD = 2 \cdot 2 \cdot 3 \cdot 3 \cdot a \cdot a \cdot a \cdot b \cdot b \cdot c \cdot c \cdot c \cdot c$$
$$LCD = 2^2 \cdot 3^2 \cdot a^3 \cdot b^2 \cdot c^4 = 36a^3b^2c^4$$

(b) $x^2 - 5x + 4 = (x - 4)(x - 1)$
$x^2 + 2x - 3 = \quad (x - 1)(x + 3)$

$$LCD = (x - 4)(x - 1)(x + 3)$$

▷ **Student Practice 4** Find the LCD.

(a) $\dfrac{3}{50xy^2z}, \dfrac{19}{40x^3yz}$ 

(b) $\dfrac{2}{x^2 + 5x + 6}, \dfrac{6}{3x^2 + 5x - 2}$

**3** **Adding and Subtracting Rational Expressions with Different Denominators** ▶

If two rational expressions have different denominators, we first change them to equivalent rational expressions with the least common denominator. Then we add or subtract the numerators and keep the common denominator.

**Example 5** Add. $\dfrac{5}{xy} + \dfrac{2}{y}$

**Solution**   The denominators are different. We must find the LCD. The two factors are $x$ and $y$. We observe that the LCD is $xy$.

$$\frac{5}{xy} + \frac{2}{y} = \frac{5}{xy} + \frac{2}{y} \cdot \frac{x}{x} \qquad \text{Multiply the second fraction by } \frac{x}{x}.$$

$$= \frac{5}{xy} + \frac{2x}{xy} \qquad \text{Now each fraction has a common denominator of } xy.$$

$$= \frac{5 + 2x}{xy} \qquad \text{Write the sum as one fraction.} \qquad \square$$

 **Student Practice 5**   Add.

$$\frac{7}{a} + \frac{3}{abc}$$

**Example 6** Add. $\dfrac{3x}{x^2 - y^2} + \dfrac{5}{x + y}$

**Solution**   We factor the first denominator so that $x^2 - y^2 = (x + y)(x - y)$. Thus, the factors of the denominators are $(x + y)$ and $(x - y)$. We observe that the LCD $= (x + y)(x - y)$.

$$\frac{3x}{(x + y)(x - y)} + \frac{5}{(x + y)} \cdot \frac{x - y}{x - y} \qquad \text{Multiply the second fraction by } \frac{x - y}{x - y}.$$

$$= \frac{3x}{(x + y)(x - y)} + \frac{5x - 5y}{(x + y)(x - y)} \qquad \begin{array}{l}\text{Now each fraction has a common} \\ \text{denominator of } (x + y)(x - y).\end{array}$$

$$= \frac{3x + 5x - 5y}{(x + y)(x - y)} \qquad \begin{array}{l}\text{Write the sum of the numerators over} \\ \text{the common denominator.}\end{array}$$

$$= \frac{8x - 5y}{(x + y)(x - y)} \qquad \text{Combine like terms.} \qquad \square$$

**Student Practice 6**   Add.

$$\frac{2a - b}{a^2 - 4b^2} + \frac{2}{a + 2b}$$

It is important to remember that the LCD is the smallest algebraic expression into which each denominator can be divided. For rational expressions the LCD must contain *each factor* that appears in any denominator. If the factor is repeated, the LCD must contain that factor the greatest number of times that it appears in any one denominator.

In many cases, the denominators in an addition or subtraction problem are not in factored form. You must factor each denominator to determine the LCD. Combine like terms in the numerator; then determine whether that final numerator can be factored. If so, you may be able to simplify the fraction.

**Example 7** Add. $\dfrac{5}{x^2 - y^2} + \dfrac{3x}{x^3 + x^2 y}$

**Solution**

$$\dfrac{5}{x^2 - y^2} + \dfrac{3x}{x^3 + x^2 y}$$

$$= \dfrac{5}{(x+y)(x-y)} + \dfrac{3x}{x^2(x+y)}$$

Factor the two denominators. Observe that the LCD is $x^2(x+y)(x-y)$.

$$= \dfrac{5}{(x+y)(x-y)} \cdot \dfrac{x^2}{x^2} + \dfrac{3x}{x^2(x+y)} \cdot \dfrac{x-y}{x-y}$$

Multiply each fraction by the appropriate expression to obtain a common denominator of $x^2(x+y)(x-y)$.

$$= \dfrac{5x^2}{x^2(x+y)(x-y)} + \dfrac{3x^2 - 3xy}{x^2(x+y)(x-y)}$$

$$= \dfrac{5x^2 + 3x^2 - 3xy}{x^2(x+y)(x-y)}$$

Write the sum of the numerators over the common denominator.

$$= \dfrac{8x^2 - 3xy}{x^2(x+y)(x-y)}$$

Combine like terms.

$$= \dfrac{x(8x - 3y)}{x^2(x+y)(x-y)}$$

Divide out the common factor $x$ in the numerator and denominator and simplify.

$$= \dfrac{8x - 3y}{x(x+y)(x-y)}$$

□

**Student Practice 7** Add.

$$\dfrac{7a}{a^2 + 2ab + b^2} + \dfrac{4}{a^2 + ab}$$

It is very easy to make a sign mistake when subtracting two fractions. You will find it helpful to place parentheses around the numerator of the second fraction so that you will not forget to subtract the entire numerator.

$\mathbb{M}_{\mathbb{C}}$ **Example 8** Subtract. $\dfrac{3x + 4}{x - 2} - \dfrac{x - 3}{2x - 4}$

**Solution** Factor the second denominator.

$$= \dfrac{3x + 4}{x - 2} - \dfrac{x - 3}{2(x - 2)}$$

Observe that the LCD is $2(x - 2)$.

$$= \dfrac{2}{2} \cdot \dfrac{3x + 4}{x - 2} - \dfrac{x - 3}{2(x - 2)}$$

Multiply the first fraction by $\frac{2}{2}$ so that the resulting fraction will have the common denominator.

$$= \dfrac{2(3x + 4) - (x - 3)}{2(x - 2)}$$

Write the indicated subtraction as one fraction. Note the parentheses around $x - 3$.

$$= \dfrac{6x + 8 - x + 3}{2(x - 2)}$$

Remove the parentheses in the numerator.

$$= \dfrac{5x + 11}{2(x - 2)}$$

Combine like terms. □

**Student Practice 8** Subtract.

$$\dfrac{x + 7}{3x - 9} - \dfrac{x - 6}{x - 3}$$

## Sidelight: Alternate Method

To avoid making errors when subtracting two fractions, some students prefer to change subtraction to addition of the opposite of the second fraction. In other words, we use the property that $\dfrac{a}{b} - \dfrac{c}{b} = \dfrac{a}{b} + \dfrac{-c}{b}$.

Let's revisit Example 8 to see how to use this method.

$$\dfrac{3x + 4}{x - 2} - \dfrac{(x - 3)}{2(x - 2)} \qquad \text{Insert parentheses around } x - 3.$$

$$= \dfrac{2}{2} \cdot \dfrac{3x + 4}{x - 2} + \dfrac{-(x - 3)}{2(x - 2)} \qquad \text{Change subtraction to addition of the opposite, and}$$
multiply the first fraction by $\dfrac{2}{2}$.

$$= \dfrac{2(3x + 4) + -(x - 3)}{2(x - 2)}$$

$$= \dfrac{6x + 8 + (-x) + 3}{2(x - 2)} \qquad \text{Apply distributive property to simplify the numerator.}$$

$$= \dfrac{5x + 11}{2(x - 2)} \qquad \text{Combine like terms.}$$

Try each method and see which one helps you avoid making errors.

**Example 9** Subtract and simplify. $\dfrac{8x}{x^2 - 16} - \dfrac{4}{x - 4}$

### Solution

$$\dfrac{8x}{x^2 - 16} - \dfrac{4}{x - 4}$$

$$= \dfrac{8x}{(x + 4)(x - 4)} + \dfrac{-4}{x - 4} \qquad \text{Factor the first denominator. Use the property}$$
that $\dfrac{a}{b} - \dfrac{c}{b} = \dfrac{a}{b} + \dfrac{-c}{b}$.

$$= \dfrac{8x}{(x + 4)(x - 4)} + \dfrac{-4}{x - 4} \cdot \dfrac{x + 4}{x + 4} \qquad \text{Multiply the second fraction by } \dfrac{x + 4}{x + 4}.$$

$$= \dfrac{8x + (-4)(x + 4)}{(x + 4)(x - 4)} \qquad \text{Write the sum of the numerators over the common denominator.}$$

$$= \dfrac{8x - 4x - 16}{(x + 4)(x - 4)} \qquad \text{Remove parentheses in the numerator.}$$

$$= \dfrac{4x - 16}{(x + 4)(x - 4)} \qquad \text{Combine like terms. Note that the numerator can be factored.}$$

$$= \dfrac{4(x - 4)}{(x + 4)(x - 4)} \qquad \text{Since } (x - 4) \text{ is a } \textit{factor} \text{ of the numerator } \textit{and} \text{ the denominator, we may divide out the common factor.}$$

$$= \dfrac{4}{x + 4} \qquad \square$$

**Student Practice 9** Subtract and simplify.

$$\dfrac{x - 2}{x^2 - 4} - \dfrac{x + 1}{2x^2 + 4x}$$

## 6.3 Exercises   MyMathLab®

**Verbal and Writing Skills, Exercises 1 and 2**

1. Suppose two rational expressions have denominators of $(x + 3)(x + 5)$ and $(x + 3)^2$. Explain how you would determine the LCD.

2. Suppose two rational expressions have denominators of $(x - 4)^2(x + 7)$ and $(x - 4)^3$. Explain how you would determine the LCD.

*Perform the operation indicated. Be sure to simplify.*

3. $\dfrac{3x + 2}{5 + 2x} + \dfrac{x}{2x + 5}$

4. $\dfrac{x + 6}{3x + 8} + \dfrac{4}{8 + 3x}$

5. $\dfrac{3x}{x + 3} - \dfrac{x + 5}{x + 3}$

6. $\dfrac{4x + 8}{x - 6} - \dfrac{x + 6}{x - 6}$

7. $\dfrac{8x + 3}{5x + 7} - \dfrac{6x + 10}{5x + 7}$

8. $\dfrac{7x + 6}{5x + 2} - \dfrac{2x + 3}{5x + 2}$

*Find the LCD. Do not combine fractions.*

9. $\dfrac{10}{3a^2b^3}, \dfrac{8}{ab^2}$

10. $\dfrac{12}{5a^2}, \dfrac{9}{a^3}$

11. $\dfrac{5}{18x^2y^5}, \dfrac{7}{30x^3y^3}$

12. $\dfrac{9}{14xy^3}, \dfrac{14}{35x^4y^2}$

13. $\dfrac{9}{2x - 6}, \dfrac{5}{9x - 27}$

14. $\dfrac{12}{3x + 12}, \dfrac{5}{5x + 20}$

15. $\dfrac{8}{x + 3}, \dfrac{15}{x^2 - 9}$

16. $\dfrac{5}{x^2 - 4}, \dfrac{3}{x + 2}$

17. $\dfrac{7}{3x^2 + 14x - 5}, \dfrac{4}{9x^2 - 6x + 1}$

18. $\dfrac{4}{2x^2 - 9x - 35}, \dfrac{3}{4x^2 + 20x + 25}$

*Add.*

19. $\dfrac{7}{ab} + \dfrac{3}{b}$

20. $\dfrac{8}{cd} + \dfrac{9}{d}$

21. $\dfrac{3}{x + 7} + \dfrac{8}{x^2 - 49}$

22. $\dfrac{5}{x^2 - 2x + 1} + \dfrac{3}{x - 1}$

23. $\dfrac{4y}{y + 1} + \dfrac{y}{y - 1}$

24. $\dfrac{5}{y - 3} + \dfrac{2}{y + 3}$

25. $\dfrac{6}{5a} + \dfrac{5}{3a + 2}$

26. $\dfrac{2}{3a - 5} + \dfrac{3}{4a}$

27. $\dfrac{2}{3xy} + \dfrac{1}{6yz}$

28. $\dfrac{5}{4xy} + \dfrac{5}{12yz}$

*Subtract.*

**29.** $\dfrac{5x + 6}{x - 3} - \dfrac{x - 2}{2x - 6}$

**30.** $\dfrac{7x + 3}{x - 4} - \dfrac{x - 3}{2x - 8}$

**31.** $\dfrac{3x}{x^2 - 25} - \dfrac{2}{x + 5}$

**32.** $\dfrac{7x}{x^2 - 9} - \dfrac{6}{x + 3}$

**33.** $\dfrac{a + 3b}{2} - \dfrac{a - b}{5}$

**34.** $\dfrac{2b - a}{4} - \dfrac{a + 3b}{3}$

**35.** $\dfrac{8}{2x - 3} - \dfrac{6}{x + 2}$

**36.** $\dfrac{5}{3x + 2} - \dfrac{2}{x - 4}$

**37.** $\dfrac{x}{x^2 + 2x - 3} - \dfrac{x}{x^2 - 5x + 4}$

**38.** $\dfrac{1}{x^2 - 2x} - \dfrac{5}{x^2 - 4x + 4}$

**39.** $\dfrac{3}{x^2 + 9x + 20} + \dfrac{1}{x^2 + 10x + 24}$

**40.** $\dfrac{2}{x^2 - 3x - 10} + \dfrac{5}{x^2 - 2x - 15}$

**41.** $\dfrac{3x - 8}{x^2 - 5x + 6} + \dfrac{x + 2}{x^2 - 6x + 8}$

**42.** $\dfrac{3x + 5}{x^2 + 4x + 3} + \dfrac{-x + 5}{x^2 + 2x - 3}$

**Mixed Practice** *Add or subtract.*

**43.** $\dfrac{6x}{y - 2x} - \dfrac{5x}{2x - y}$

**44.** $\dfrac{8b}{2b - a} - \dfrac{3b}{a - 2b}$

**45.** $\dfrac{3y}{8y^2 + 2y - 1} - \dfrac{5y}{2y^2 - 9y - 5}$

**46.** $\dfrac{3x}{x^2 - 5x - 6} - \dfrac{x + 5}{x^2 + 3x + 2}$

**47.** $\dfrac{2x}{2x^2 - 5x - 3} + \dfrac{3}{x - 3}$

**48.** $\dfrac{4x}{3x^2 - 5x - 2} + \dfrac{5}{x - 2}$

## Cumulative Review

**49.** **[2.4.1]** Solve. $\dfrac{1}{3}(x - 2) + \dfrac{1}{2}(x + 3) = \dfrac{1}{4}(3x + 1)$

**50.** **[2.3.3]** Solve. $4.8 - 0.6x = 0.8(x - 1)$

**51.** **[2.6.4]** Solve for $x$. $x - \dfrac{1}{5}x > \dfrac{1}{2} + \dfrac{1}{10}x$

**52.** **[4.1.3]** Simplify. $(3x^3y^4)^4$

**53.** **[3.6.2]** *Commuting Costs* A single subway fare costs $2.75. A monthly unlimited-ride subway pass costs $90. How many days per month would you have to use the subway to go to work (assume one subway fare to get to work and one subway fare to get back home) in order for it to be cheaper to buy a monthly subway pass?

**54.** **[0.5.3]** *Languages in Finland* In Finland, 91.5% of the population speaks Finnish. The other official language is Swedish, spoken by 5.3% of the population. One of the minority languages is Sámi, spoken by 0.04% of the population. (*Source:* www.stat.fi) In 2015 there were 5,475,000 people in Finland. How many more people spoke Swedish than Sámi?

---

**Quick Quiz 6.3** *Perform the operation indicated. Simplify.*

**1.** $\dfrac{3}{x^2 - 2x - 8} + \dfrac{2}{x - 4}$

**2.** $\dfrac{2x + y}{xy} - \dfrac{b - y}{by}$

**3.** $\dfrac{2}{x^2 - 9} + \dfrac{3}{x^2 + 7x + 12}$

**4.** **Concept Check** Explain how to find the LCD of the fractions $\dfrac{3}{10xy^2z}$, $\dfrac{6}{25x^2yz^3}$.

# Use Math to Save Money

**Did You Know?** A lower credit score could cost you hundreds of dollars a year in higher finance fees.

## Credit Card Follies

### Understanding the Problem:

Stores tempt customers with the promise of "15% off all purchases made today" if the customer opens and puts the purchases on a store credit card. Adam is thinking about applying for one. He has a lot of shopping to do, and saving 15% on his $500 purchase is hard to resist. He hesitates, though, because his friend Matt is telling him that in some cases, people lose money on the deal.

### Making a Plan:

Opening up a store credit card can cost you in two different ways. The first way is through the high interest rates store cards usually charge. The second way is not as straightforward. Opening a store credit account can lower your credit score because it increases your debt liability. If you need to apply for credit elsewhere, you will pay a higher interest rate due to your lower score. You need to weigh these costs against the benefit of the one-time savings.

**Step 1:** Adam calculates how much he will save today if he opens a store credit account.
**Task 1:** How much will Adam save today?

**Step 2:** The card has an annual interest rate of 25%. Using simple interest, Adam calculates that he would pay about $62.50 in finance fees if he carried the balance for six months. (*Note:* $500 \times \dfrac{0.25}{2} = 62.50$)
**Task 2:** How much would Adam pay in interest if he carried the balance for a year?

**Step 3:** Adam is surprised at how much he could end up paying in finance charges. Matt then tells him other ways it could cost him. His credit score would go down due to his opening a new account. Also, since the store card has a $1000 limit, the $500 purchase would be 50% of the available credit. The higher the percentage of available credit you are using, the lower your credit score. Matt tells Adam that if he needs to put the purchase on a card, he should use his existing card, which has a $5000 limit.

**Task 3:** What percentage of available credit would Adam be using if he put the purchase on his existing card?

**Step 4:** Even though Adam isn't planning on applying for a large loan next year, it is still a possibility. What if his car broke down and he decided to finance a new one? An increase in the interest rate due to a lower credit score could cost him hundreds of dollars over the life of the loan.
**Task 4:** If he paid $9 more a month because of a higher interest rate, how much more would he pay over the life of a four-year auto loan?

### Finding a Solution:

Adam decides that the risk of higher future fees is not worth the one-time savings on his purchase. He declines the offer and puts the purchase on his existing card.

### Applying the Situation to Your Life:

The next time you are tempted to sign up for a store credit card, you will have a better idea of the costs associated with that decision. That way, you can weigh the choices and make an informed decision.

## 6.4 Simplifying Complex Rational Expressions ⏺

### 1 Simplifying Complex Rational Expressions by Adding or Subtracting in the Numerator and Denominator ⏺

A **complex rational expression** (also called a **complex fraction**) has a fraction in the numerator or in the denominator or both.

$$\frac{3 + \dfrac{2}{x}}{\dfrac{x}{7} + 2} \qquad \frac{\dfrac{x}{y} + 1}{2} \qquad \frac{\dfrac{a + b}{3}}{\dfrac{x - 2y}{4}}$$

The bar in a complex rational expression is both a grouping symbol and a symbol for division.

$$\frac{\dfrac{a + b}{3}}{\dfrac{x - 2y}{4}} \qquad \text{is equivalent to} \qquad \left(\frac{a + b}{3}\right) \div \left(\frac{x - 2y}{4}\right)$$

We need a procedure for simplifying complex rational expressions.

> **PROCEDURE TO SIMPLIFY A COMPLEX RATIONAL EXPRESSION: ADDING AND SUBTRACTING**
>
> 1. Add or subtract so that you have a single fraction in the numerator and in the denominator.
> 2. Divide the fraction in the numerator by the fraction in the denominator. This is done by inverting the fraction in the denominator and multiplying it by the numerator.

**Example 1** Simplify. $\dfrac{\dfrac{1}{x}}{\dfrac{2}{y^2} + \dfrac{1}{y}}$

**Solution**

**Step 1** Add the two fractions in the denominator.

$$\frac{\dfrac{1}{x}}{\dfrac{2}{y^2} + \dfrac{1}{y} \cdot \dfrac{y}{y}} = \frac{\dfrac{1}{x}}{\dfrac{2 + y}{y^2}}$$

**Step 2** Divide the fraction in the numerator by the fraction in the denominator.

$$\frac{1}{x} \div \frac{2 + y}{y^2} = \frac{1}{x} \cdot \frac{y^2}{2 + y} = \frac{y^2}{x(2 + y)} \qquad \square$$

▭➡ **Student Practice 1**

Simplify. $\dfrac{\dfrac{1}{a} + \dfrac{1}{a^2}}{\dfrac{2}{b^2}}$

A complex rational expression may contain two or more fractions in the numerator and the denominator.

**Example 2** Simplify. $\dfrac{\dfrac{1}{x} + \dfrac{1}{y}}{\dfrac{3}{a} - \dfrac{2}{b}}$

**Student Practice 2**

Simplify. $\dfrac{\dfrac{1}{a} + \dfrac{1}{b}}{\dfrac{x}{2} - \dfrac{5}{y}}$

**Solution** We observe that the LCD of the fractions in the numerator is $xy$. The LCD of the fractions in the denominator is $ab$.

$$= \dfrac{\dfrac{1}{x}\cdot\dfrac{y}{y} + \dfrac{1}{y}\cdot\dfrac{x}{x}}{\dfrac{3}{a}\cdot\dfrac{b}{b} - \dfrac{2}{b}\cdot\dfrac{a}{a}}$$ Multiply each fraction by the appropriate value to obtain common denominators.

$$= \dfrac{\dfrac{y+x}{xy}}{\dfrac{3b-2a}{ab}}$$ Add the two fractions in the numerator.

Subtract the two fractions in the denominator.

$$= \dfrac{y+x}{xy} \cdot \dfrac{ab}{3b-2a}$$ Invert the fraction in the denominator and multiply it by the numerator.

$$= \dfrac{ab(y+x)}{xy(3b-2a)}$$ Write the answer as one fraction. □

For some complex rational expressions, factoring may be necessary to determine the LCD and to combine fractions.

**Example 3** Simplify. $\dfrac{\dfrac{1}{x^2-1} + \dfrac{2}{x+1}}{x}$

**Solution** We need to factor $x^2 - 1$.

$$= \dfrac{\dfrac{1}{(x+1)(x-1)} + \dfrac{2}{x+1}\cdot\dfrac{x-1}{x-1}}{x}$$ The LCD for the fractions in the numerator is $(x+1)(x-1)$.

$$= \dfrac{\dfrac{1+2x-2}{(x+1)(x-1)}}{x}$$ Add the two fractions in the numerator.

$$= \dfrac{2x-1}{(x+1)(x-1)}\cdot\dfrac{1}{x}$$ Simplify the numerator. Invert the fraction in the denominator and multiply.

$$= \dfrac{2x-1}{x(x+1)(x-1)}$$ Write the answer as one fraction. □

**Student Practice 3**

Simplify. $\dfrac{\dfrac{x}{x^2+4x+3} + \dfrac{2}{x+1}}{x+1}$

When simplifying complex rational expressions, always check to see if the final fraction can be reduced or simplified.

**Example 4** Simplify. $\dfrac{\dfrac{3}{a+b} - \dfrac{3}{a-b}}{\dfrac{5}{a^2 - b^2}}$

**Solution** The LCD of the two fractions in the numerator is $(a+b)(a-b)$.

$$= \frac{\dfrac{3}{a+b} \cdot \dfrac{a-b}{a-b} - \dfrac{3}{a-b} \cdot \dfrac{a+b}{a+b}}{\dfrac{5}{a^2 - b^2}}$$

$$= \frac{\dfrac{3a - 3b}{(a+b)(a-b)} - \dfrac{3a + 3b}{(a+b)(a-b)}}{\dfrac{5}{a^2 - b^2}}$$

Study carefully how we combine the two fractions in the numerator. Do you see how we obtain $-6b$?

$$= \frac{\dfrac{-6b}{(a+b)(a-b)}}{\dfrac{5}{(a+b)(a-b)}} \qquad \text{Factor } a^2 - b^2 \text{ as } (a+b)(a-b).$$

$$= \frac{-6b}{(a+b)(a-b)} \cdot \frac{(a+b)(a-b)}{5} \qquad \begin{array}{l} \text{Since } (a+b)(a-b) \text{ are factors in} \\ \text{both a numerator and a denominator,} \\ \text{they may be divided out.} \end{array}$$

$$= \frac{-6b}{5} \quad \text{or} \quad -\frac{6b}{5} \qquad \qquad \Box$$

**Student Practice 4** Simplify.

$$\frac{\dfrac{6}{x^2 - y^2}}{\dfrac{1}{x-y} + \dfrac{3}{x+y}}$$

## 2 Simplifying Complex Rational Expressions Using the LCD

There is another way to simplify complex rational expressions: Multiply the numerator and denominator of the complex fraction by the least common denominator of all the denominators appearing in the complex fraction.

**PROCEDURE TO SIMPLIFY A COMPLEX RATIONAL EXPRESSION: MULTIPLYING BY THE LCD**

1. Determine the LCD of all individual denominators occurring in the numerator and denominator of the complex rational expression.

2. Multiply both the numerator and the denominator of the complex rational expression by the LCD.

3. Simplify, if possible.

**Example 5** Simplify by multiplying by the LCD. $\dfrac{\dfrac{5}{ab^2} - \dfrac{2}{ab}}{3 - \dfrac{5}{2a^2b}}$

**Solution** The LCD of all the denominators in the complex rational expression is $2a^2b^2$.

$$= \frac{2a^2b^2\left(\dfrac{5}{ab^2} - \dfrac{2}{ab}\right)}{2a^2b^2\left(3 - \dfrac{5}{2a^2b}\right)} \qquad \text{Multiply the numerator and denominator by the LCD.}$$

$$= \frac{2a^2b^2\left(\dfrac{5}{ab^2}\right) - 2a^2b^2\left(\dfrac{2}{ab}\right)}{2a^2b^2(3) - 2a^2b^2\left(\dfrac{5}{2a^2b}\right)} \qquad \text{Multiply each term by } 2a^2b^2.$$

$$= \frac{10a - 4ab}{6a^2b^2 - 5b} \qquad \text{Simplify.} \qquad \square$$

**Student Practice 5** Simplify by multiplying by the LCD.

$$\frac{\dfrac{2}{3x^2} - \dfrac{3}{y}}{\dfrac{5}{xy} - 4}$$

So that you can compare the two methods, we will redo Example 4 by multiplying by the LCD.

**Example 6** Simplify by multiplying by the LCD. $\dfrac{\dfrac{3}{a + b} - \dfrac{3}{a - b}}{\dfrac{5}{a^2 - b^2}}$

**Solution** The LCD of all individual fractions contained in the complex fraction is $(a + b)(a - b)$.

$$= \frac{(a + b)(a - b)\left(\dfrac{3}{a + b}\right) - (a + b)(a - b)\left(\dfrac{3}{a - b}\right)}{(a + b)(a - b)\left(\dfrac{5}{(a + b)(a - b)}\right)} \qquad \text{Multiply each term by the LCD.}$$

$$= \frac{3(a - b) - 3(a + b)}{5} \qquad \text{Simplify.}$$

$$= \frac{3a - 3b - 3a - 3b}{5} \qquad \text{Remove parentheses.}$$

$$= -\frac{6b}{5} \qquad \text{Simplify.} \qquad \square$$

**Student Practice 6** Simplify by multiplying by the LCD.

$$\frac{\dfrac{6}{x^2 - y^2}}{\dfrac{7}{x - y} + \dfrac{3}{x + y}}$$

## 6.4 Exercises    MyMathLab®

*Simplify.*

1. $\dfrac{\dfrac{5}{x}}{\dfrac{4}{x} + \dfrac{3}{x^2}}$

2. $\dfrac{\dfrac{8}{x}}{\dfrac{5}{x^2} + \dfrac{6}{x}}$

3. $\dfrac{\dfrac{4}{a} + \dfrac{1}{b}}{\dfrac{5}{ab}}$

4. $\dfrac{\dfrac{3}{a} + \dfrac{5}{b}}{\dfrac{4}{ab}}$

5. $\dfrac{\dfrac{x}{6} - \dfrac{1}{3}}{\dfrac{2}{3x} + \dfrac{5}{6}}$

6. $\dfrac{\dfrac{x}{6} - \dfrac{1}{3}}{\dfrac{5}{6x} + \dfrac{1}{2}}$

7. $\dfrac{\dfrac{7}{5x} - \dfrac{1}{x}}{\dfrac{3}{5} + \dfrac{2}{x}}$

8. $\dfrac{\dfrac{8}{x} - \dfrac{2}{3x}}{\dfrac{2}{3} + \dfrac{5}{x}}$

9. $\dfrac{\dfrac{5}{x} + \dfrac{3}{y}}{3x + 5y}$

10. $\dfrac{\dfrac{1}{x} + \dfrac{1}{y}}{x + y}$

11. $\dfrac{4 - \dfrac{1}{x^2}}{2 + \dfrac{1}{x}}$

12. $\dfrac{1 - \dfrac{36}{x^2}}{1 - \dfrac{6}{x}}$

13. $\dfrac{\dfrac{2}{x + 6}}{\dfrac{2}{x - 6} - \dfrac{2}{x^2 - 36}}$

14. $\dfrac{\dfrac{9}{x^2 - 1}}{\dfrac{4}{x - 1} - \dfrac{2}{x + 1}}$

15. $\dfrac{a + \dfrac{3}{a}}{\dfrac{a^2 + 2}{3a}}$

16. $\dfrac{x + \dfrac{4}{x}}{\dfrac{x^2 + 3}{4x}}$

17. $\dfrac{\dfrac{3}{x - 3}}{\dfrac{1}{x^2 - 9} + \dfrac{2}{x + 3}}$

18. $\dfrac{\dfrac{4}{x + 5}}{\dfrac{2}{x - 5} - \dfrac{1}{x^2 - 25}}$

19. $\dfrac{\dfrac{3}{x - 1} + 4}{\dfrac{3}{x - 1} - 4}$

20. $\dfrac{\dfrac{2x}{x - 2} + 3}{\dfrac{3x}{x - 2} + 4}$

**To Think About, Exercises 21 and 22**

**21.** Consider the complex fraction $\dfrac{\dfrac{4}{x+3}}{\dfrac{5}{x}-1}$. What values are not allowable replacements for the variable $x$?

**22.** Consider the complex fraction $\dfrac{\dfrac{5}{x-2}}{\dfrac{6}{x}+1}$. What values are not allowable replacements for the variable $x$?

*Simplify.*

**23.** $\dfrac{x+5y}{x-6y} \div \left( \dfrac{1}{5y} - \dfrac{1}{x+5y} \right)$

**24.** $\left( \dfrac{1}{x+2y} - \dfrac{1}{x-y} \right) \div \dfrac{2x-4y}{x^2-3xy+2y^2}$

## Cumulative Review

**25.** **[2.5.1]** Solve for $y$. $5x + 6y = 8$

**26.** **[2.6.4]** Solve and graph. $7 + x < 11 + 5x$

**27.** **[3.2.1]** When nine is subtracted from double a number, the result is the same as one-half of the same number. What is the number?

**28.** **[3.4.2]** Isabella received a pay raise this year. The raise was 5% of last year's salary. This year she will earn $25,200. What was her salary last year before the raise?

---

**Quick Quiz 6.4** *Simplify.*

**1.** $\dfrac{\dfrac{a}{4b} - \dfrac{1}{3}}{\dfrac{5}{4b} - \dfrac{4}{a}}$

**2.** $\dfrac{a+b}{\dfrac{1}{a} + \dfrac{1}{b}}$

**3.** $\dfrac{\dfrac{10}{x^2-25}}{\dfrac{3}{x+5} + \dfrac{2}{x-5}}$

**4.** **Concept Check** To simplify the following complex fraction, explain how you would add the two fractions in the numerator.

$$\dfrac{\dfrac{7}{x-3} + \dfrac{15}{2x-6}}{\dfrac{2}{x+5}}$$

## 6.5 Solving Equations Involving Rational Expressions

### 1 Solving Equations Involving Rational Expressions That Have Solutions

In Section 2.4 we developed procedures to solve linear equations containing fractions whose denominators are numbers. In this section we use a similar approach to solve equations containing fractions whose denominators are polynomials. It would be wise for you to review Section 2.4 briefly *before you begin this section*. It will be especially helpful to carefully study Examples 1 and 2 in that section.

> **TO SOLVE AN EQUATION CONTAINING RATIONAL EXPRESSIONS**
>
> 1. Determine the LCD of all the denominators.
> 2. Multiply each term of the equation by the LCD.
> 3. Solve the resulting equation.
> 4. Check your solution. Exclude from your solution any value that would make the LCD equal to zero.

**Example 1** Solve for $x$ and check your solution. $\dfrac{5}{x} + \dfrac{2}{3} = -\dfrac{3}{x}$

**Solution**

$$3x\left(\frac{5}{x}\right) + 3x\left(\frac{2}{3}\right) = 3x\left(-\frac{3}{x}\right) \qquad \text{Observe that the LCD is } 3x. \text{ Multiply each term by } 3x.$$

$$15 + 2x = -9$$
$$2x = -9 - 15 \qquad \text{Subtract 15 from both sides.}$$
$$2x = -24$$
$$x = -12 \qquad \text{Divide both sides by 2.}$$

*Check.*

$$\frac{5}{-12} + \frac{2}{3} \overset{?}{=} -\frac{3}{-12} \qquad \text{Replace each } x \text{ by } -12.$$

$$-\frac{5}{12} + \frac{8}{12} \overset{?}{=} \frac{3}{12}$$

$$\frac{3}{12} = \frac{3}{12} \quad \checkmark \quad \text{It checks.} \qquad \square$$

**Student Practice 1** Solve for $x$ and check your solution.

$$\frac{3}{x} + \frac{4}{5} = -\frac{2}{x}$$

**Example 2** Solve and check. $\dfrac{6}{x + 3} = \dfrac{3}{x}$

**Solution**

Observe that the LCD $= x(x + 3)$.

$$x(x + 3)\left(\frac{6}{x + 3}\right) = x(x + 3)\left(\frac{3}{x}\right) \qquad \text{Multiply both sides by } x(x + 3).$$

$$6x = 3(x + 3) \qquad \text{Simplify. Do you see how this is done?}$$
$$6x = 3x + 9 \qquad \text{Remove parentheses.}$$
$$3x = 9 \qquad \text{Subtract } 3x \text{ from both sides.}$$
$$x = 3 \qquad \text{Divide both sides by 3.}$$

*Check.*  $\dfrac{6}{3+3} \overset{?}{=} \dfrac{3}{3}$   Replace each $x$ by 3.

$\dfrac{6}{6} = \dfrac{3}{3}$   It checks. ✓

**Student Practice 2**   Solve and check.

$$\frac{6}{2x+1} = \frac{2}{x+2}$$

It is sometimes necessary to factor denominators before the correct LCD can be determined.

$\mathbb{M}_{\mathbb{C}}$ **Example 3** Solve and check.  $\dfrac{3}{x+5} - 1 = \dfrac{4-x}{2x+10}$

**Solution**

$$\frac{3}{x+5} - 1 = \frac{4-x}{2(x+5)} \quad \text{Factor } 2x+10. \text{ We determine that the LCD is } 2(x+5).$$

$$2(x+5)\left(\frac{3}{x+5}\right) - 2(x+5)(1) = 2(x+5)\left[\frac{4-x}{2(x+5)}\right] \quad \text{Multiply each term by the LCD.}$$

$2(3) - 2(x+5) = 4 - x$   Simplify.

$6 - 2x - 10 = 4 - x$   Remove parentheses.

$-2x - 4 = 4 - x$   Combine like terms.

$-4 = 4 + x$   Add $2x$ to both sides.

$-8 = x$   Subtract 4 from both sides.

*Check.*  $\dfrac{3}{-8+5} - 1 \overset{?}{=} \dfrac{4-(-8)}{2(-8)+10}$   Replace each $x$ in the original equation by $-8$.

$\dfrac{3}{-3} - 1 \overset{?}{=} \dfrac{4+8}{-16+10}$

$-1 - 1 \overset{?}{=} \dfrac{12}{-6}$

$-2 = -2$ ✓   It checks. The solution is $-8$.

**Student Practice 3**

Solve and check.

$$\frac{x-1}{x^2-4} = \frac{2}{x+2} + \frac{4}{x-2}$$

**2  Determining Whether an Equation Involving Rational Expressions Has No Solution** ▶

Equations containing rational expressions sometimes appear to have solutions when in fact they do not. By this we mean that the "solutions" we get by using completely correct methods are, in actuality, not solutions.

In the case where a value makes a denominator in the equation equal to zero, we say it is not a solution to the equation. Such a value is called an **extraneous solution.** An extraneous solution is an apparent solution that does *not* satisfy the original equation. If all of the apparent solutions of an equation are extraneous solutions, we say that the equation has **no solution.** It is important that you check all apparent solutions in the original equation.

**Example 4** Solve and check. $\dfrac{y}{y-2} - 4 = \dfrac{2}{y-2}$

**Solution**

Observe that the LCD is $y - 2$.

$$(y-2)\left(\frac{y}{y-2}\right) - (y-2)(4) = (y-2)\left(\frac{2}{y-2}\right) \quad \text{Multiply each term by } (y-2).$$

$$y - 4(y-2) = 2 \qquad \text{Simplify. Do you see how this is done?}$$
$$y - 4y + 8 = 2 \qquad \text{Remove parentheses.}$$
$$-3y + 8 = 2 \qquad \text{Combine like terms.}$$
$$-3y = -6 \qquad \text{Subtract 8 from both sides.}$$
$$\frac{-3y}{-3} = \frac{-6}{-3} \qquad \text{Divide both sides by } -3.$$
$$y = 2 \qquad \text{2 is only an apparent solution.}$$

**This equation has no solution.**

Why? We can see immediately that $y = 2$ is not a solution of the original equation. When we substitute 2 for $y$ in a denominator, the denominator is equal to zero and the expression is undefined.

*Check.* $\qquad \dfrac{y}{y-2} - 4 = \dfrac{2}{y-2} \qquad$ Suppose that you try to check the apparent solution by substituting 2 for $y$.

$$\frac{2}{2-2} - 4 \overset{?}{=} \frac{2}{2-2}$$

$$\frac{2}{0} - 4 = \frac{2}{0} \qquad \text{This does not check since you do not obtain a real number when you divide by zero.}$$

These expressions are not defined.

There is no such number as $2 \div 0$. We see that 2 does *not* check. This equation has **no solution.** ☐

 **Student Practice 4** Solve and check.

$$\frac{2x}{x+1} = \frac{-2}{x+1} + 1$$

## 6.5 Exercises  MyMathLab®

*Solve and check exercises 1–16.*

1. $\dfrac{7}{x} + \dfrac{3}{4} = \dfrac{-2}{x}$

2. $\dfrac{8}{x} + \dfrac{2}{5} = \dfrac{-2}{x}$

3. $\dfrac{3}{x} - \dfrac{5}{4} = \dfrac{1}{2x}$

4. $\dfrac{1}{3x} + \dfrac{5}{6} = \dfrac{2}{x}$

5. $\dfrac{5x + 3}{3x} = \dfrac{7}{3} - \dfrac{9}{x}$

6. $\dfrac{2x + 3}{4x} - \dfrac{1}{x} = \dfrac{3}{2}$

7. $\dfrac{x + 5}{3x} = \dfrac{1}{2}$

8. $\dfrac{x - 4}{5x} = \dfrac{3}{10}$

9. $\dfrac{6}{3x - 5} = \dfrac{3}{2x}$

10. $\dfrac{3}{x + 4} = \dfrac{2}{x}$

11. $\dfrac{2}{2x + 5} = \dfrac{4}{x - 4}$

12. $\dfrac{3}{x + 5} = \dfrac{3}{3x - 2}$

13. $\dfrac{2}{x} + \dfrac{x}{x + 1} = 1$

14. $\dfrac{5}{2} = 3 + \dfrac{2x + 7}{x + 6}$

15. $\dfrac{85 - 4x}{x} = 7 - \dfrac{3}{x}$

16. $\dfrac{63 - 2x}{x} = 2 - \dfrac{5}{x}$

**Mixed Practice**  *Solve and check. If there is no solution, say so.*

17. $\dfrac{1}{x + 4} - 2 = \dfrac{3x - 2}{x + 4}$

18. $\dfrac{2}{x + 5} - 1 = \dfrac{3x - 4}{x + 5}$

19. $\dfrac{2}{x - 6} - 5 = \dfrac{2(x - 5)}{x - 6}$

20. $5 - \dfrac{x}{x + 3} = \dfrac{3}{3 + x}$

21. $\dfrac{2}{x + 1} - \dfrac{1}{x - 1} = \dfrac{2x}{x^2 - 1}$

22. $\dfrac{8x}{4x^2 - 1} = \dfrac{3}{2x + 1} + \dfrac{3}{2x - 1}$

23. $\dfrac{x + 2}{x^2 - x - 12} = \dfrac{1}{x + 3} - \dfrac{1}{x - 4}$

24. $\dfrac{x + 3}{x^2 - 3x - 10} = \dfrac{2}{x - 5} - \dfrac{2}{x + 2}$

25. $\dfrac{2x}{x + 4} - \dfrac{8}{x - 4} = \dfrac{2x^2 + 32}{x^2 - 16}$

26. $\dfrac{4x}{x + 3} - \dfrac{12}{x - 3} = \dfrac{4x^2 + 36}{x^2 - 9}$

**27.** $\dfrac{4}{x^2 - 1} + \dfrac{7}{x + 1} = \dfrac{5}{x - 1}$

**28.** $\dfrac{9}{9x^2 - 1} + \dfrac{1}{3x + 1} = \dfrac{2}{3x - 1}$

**29.** $\dfrac{x + 11}{x^2 - 5x + 4} + \dfrac{3}{x - 1} = \dfrac{5}{x - 4}$

**30.** $\dfrac{6}{x - 3} = \dfrac{-5}{x - 2} + \dfrac{-5}{x^2 - 5x + 6}$

**To Think About** *In each of the following equations, what values are not allowable replacements for the variable x? Do not solve the equation.*

**31.** $\dfrac{5x}{x + 6} - \dfrac{2x}{3x + 1} = \dfrac{2}{3x^2 + 19x + 6}$

**32.** $\dfrac{4x}{x - 3} - \dfrac{3x}{2x - 1} = \dfrac{6}{2x^2 - 5x - 3}$

## Cumulative Review

**33.** **[5.4.1]** Factor. $8x^2 - 2x - 1$

**34.** **[2.3.3]** Solve. $5(x - 2) = 8 - (3 + x)$

▲ **35.** **[3.3.1]** *Geometry* The perimeter of a rectangular computer monitor is 44 inches. The length is 8 inches less than twice the width. Find the dimensions of the monitor.

**36.** **[0.6.1]** *Mortgage Payments* Wally and Adele Panzas plan to purchase a new home and borrow $115,000. They plan to take out a 25-year mortgage at an annual interest rate of 8.75%. The bank will charge them $8.23 per month for each $1000 of mortgage. What will their monthly payments be?

**Quick Quiz 6.5** *Solve. If there is no solution, say so.*

**1.** $\dfrac{3}{4x} - \dfrac{5}{6x} = 2 - \dfrac{1}{2x}$

**2.** $\dfrac{x}{x - 1} - \dfrac{2}{x} = \dfrac{1}{x - 1}$

**3.** $\dfrac{6}{x^2 - 2x - 8} + \dfrac{5}{x + 2} = \dfrac{1}{x - 4}$

**4.** **Concept Check** Explain how to find the LCD for the following equation. Do not solve the equation.
$$\dfrac{x}{x^2 - 9} + \dfrac{2}{3x - 9} = \dfrac{5}{2x + 6} + \dfrac{3}{2x^2 - 18}$$

# 6.6  Ratio, Proportion, and Other Applied Problems

## 1 Solving Problems Involving Ratio and Proportion

A **ratio** is a comparison of two quantities. You may be familiar with ratios that compare miles to hours or miles to gallons. A ratio is often written as a quotient in the form of a fraction. For example, the ratio of 7 to 9 can be written as $\frac{7}{9}$.

A **proportion** is an equation that states that two ratios are equal. For example,

$$\frac{7}{9} = \frac{21}{27}, \quad \frac{2}{3} = \frac{10}{15}, \quad \text{and} \quad \frac{a}{b} = \frac{c}{d} \quad \text{are proportions.}$$

Let's take a closer look at the last proportion. We can see that the LCD of the fractional equation is $bd$.

$$(bd)\frac{a}{b} = (bd)\frac{c}{d} \quad \text{Multiply each side by the LCD and simplify.}$$

$$da = bc$$

$$ad = bc \qquad \text{Since multiplication is commutative, } da = ad.$$

Thus we have proved the following.

**Student Learning Objectives**

After studying this section, you will be able to:

1 Solve problems involving ratio and proportion.

2 Solve problems involving similar triangles.

3 Solve distance problems involving rational expressions.

4 Solve work problems.

---

**THE PROPORTION EQUATION**

$$\text{If} \quad \frac{a}{b} = \frac{c}{d},$$

$$\text{then} \quad ad = bc$$

for all real numbers $a$, $b$, $c$, and $d$, where $b \neq 0$ and $d \neq 0$.

---

This is sometimes called **cross multiplying.** It can be applied only if you have *one* fraction and nothing else on each side of the equation.

**Example 1** Michael took 5 hours to drive 245 miles on the turnpike. At the same rate, how many hours will it take him to drive a distance of 392 miles?

**Solution**

1. **Understand the problem.** Let $x =$ the number of hours it will take to drive 392 miles. If 5 hours are needed to drive 245 miles, then $x$ hours are needed to drive 392 miles.

2. **Write an equation.** We can write this as a proportion. Compare time to distance in each ratio.

$$\begin{array}{rcl} \text{Time} \longrightarrow & \dfrac{5 \text{ hours}}{245 \text{ miles}} = \dfrac{x \text{ hours}}{392 \text{ miles}} & \longleftarrow \text{Time} \\ \text{Distance} \longrightarrow & & \longleftarrow \text{Distance} \end{array}$$

3. **Solve and state the answer.**

$$5(392) = 245x \quad \text{Cross-multiply.}$$

$$\frac{1960}{245} = x \qquad \text{Divide both sides by 245.}$$

$$8 = x$$

It will take Michael 8 hours to drive 392 miles.

4. **Check.** Is $\frac{5}{245} = \frac{8}{392}$? Do the computation and see. □

**Student Practice 1** It took Brenda 8 hours to drive 420 miles. At the same rate, how long would it take her to drive 315 miles?

**Example 2** If $\frac{3}{4}$ inch on a map represents an actual distance of 20 miles, how long is the distance represented by $4\frac{1}{8}$ inches on the same map?

**Solution** Let $x$ = the distance represented by $4\frac{1}{8}$ inches.

Initial measurement on map $\longrightarrow$ $\dfrac{3}{4}$   $4\dfrac{1}{8}$ $\longleftarrow$ Second measurement on the map

Initial distance $\longrightarrow$ $\dfrac{3}{20} = \dfrac{4\frac{1}{8}}{x}$ $\longleftarrow$ Second distance

$$\left(\frac{3}{4}\right)(x) = (20)\left(4\frac{1}{8}\right) \qquad \text{Cross-multiply.}$$

$$\left(\frac{3}{4}\right)(x) = (\overset{5}{\cancel{20}})\left(\frac{33}{\underset{2}{\cancel{8}}}\right) \qquad \text{Write } 4\frac{1}{8} \text{ as } \frac{33}{8} \text{ and simplify.}$$

$$\frac{3x}{4} = \frac{165}{2} \qquad \text{Multiply the fractions.}$$

$$4\left(\frac{3x}{4}\right) = \overset{2}{\cancel{4}}\left(\frac{165}{\cancel{2}}\right) \qquad \text{Multiply each side by 4.}$$

$$3x = 330 \qquad \text{Simplify.}$$

$$x = 110 \qquad \text{Divide both sides by 3.}$$

$4\frac{1}{8}$ inches on the map represents an actual distance of 110 miles. ◻

**Student Practice 2** If $\frac{5}{8}$ inch on a map represents an actual distance of 30 miles, how long is the distance represented by $2\frac{1}{2}$ inches on the same map?

## 2 Solving Problems Involving Similar Triangles ▶

Similar triangles are triangles that have the same shape but may be different sizes. For example, if you draw a triangle on a sheet of paper, place the paper in a photocopy machine, and make a copy that is reduced by 25%, you would create a triangle that is similar to the original triangle. The two triangles will have the same shape. The corresponding sides of the triangles will be proportional. The corresponding angles of the triangles will be equal.

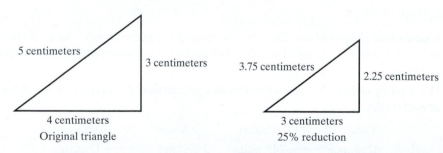

5 centimeters    3 centimeters         3.75 centimeters    2.25 centimeters

4 centimeters         3 centimeters
Original triangle       25% reduction

You can use the proportion equation to show that the corresponding sides of the triangles above are proportional. In fact, you can use the proportion equation to find an unknown length of a side of one of two similar triangles.

▲ **Example 3** A ramp is 32 meters long and rises up 15 meters. A ramp at the same angle is 9 meters long. How high does the second ramp rise?

**Solution** To answer this question, we find the length of side $x$ in the following two similar triangles.

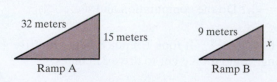

32 meters    15 meters         9 meters    $x$

Ramp A            Ramp B

Length of ramp A $\longrightarrow \dfrac{32}{9} = \dfrac{15}{x} \longleftarrow$ Rise of ramp A
Length of ramp B $\longrightarrow$ $\phantom{\dfrac{32}{9}}$ $\longleftarrow$ Rise of ramp B

$$32x = (9)(15) \quad \textcolor{red}{\text{Cross-multiply.}}$$

$$32x = 135 \quad \textcolor{red}{\text{Simplify.}}$$

$$x = \frac{135}{32} \quad \textcolor{red}{\text{Divide both sides by 32.}}$$

$$\text{or} \quad x = 4\frac{7}{32}$$

The ramp rises $4\frac{7}{32}$ meters high.  □

**Student Practice 3**   Triangle C is similar to Triangle D. Find the length of side $x$. Express your answer as a mixed number.

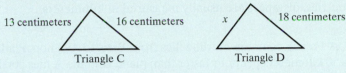

13 centimeters    16 centimeters    $x$    18 centimeters

Triangle C          Triangle D

We can also use similar triangles for indirect measurement—for instance, to find the height of an object that is too tall to measure using standard measuring devices. When the sun shines on two vertical objects at the same time, the shadows and the objects form similar triangles.

▲ **Example 4**  A woman who is 5 feet tall casts a shadow that is 8 feet long. At the same time of day, a building casts a shadow that is 72 feet long. How tall is the building?

**Solution**

1. *Understand the problem.*  First we draw a sketch. We do not know the height of the building, so we call it $x$.

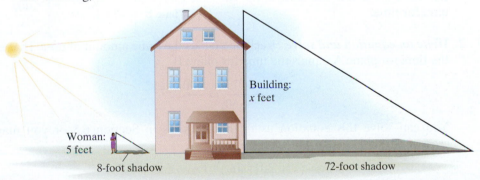

Building:
$x$ feet

Woman:
5 feet

8-foot shadow          72-foot shadow

2. *Write an equation and solve.*

Height of woman $\longrightarrow \dfrac{5}{8} = \dfrac{x}{72} \longleftarrow$ Height of building
Length of woman's shadow $\longrightarrow$ $\phantom{\dfrac{5}{8}}$ $\longleftarrow$ Length of building's shadow

$$(5)(72) = 8x \quad \textcolor{red}{\text{Cross-multiply.}}$$

$$360 = 8x$$

$$45 = x$$

The height of the building is 45 feet.  □

**Student Practice 4**   A man who is 6 feet tall casts a shadow that is 7 feet long. At the same time of day, a large flagpole casts a shadow that is 38.5 feet long. How tall is the flagpole?

In problems such as Example 4, we are assuming that the building and the person are standing exactly perpendicular to the ground. In other words, each triangle is assumed to be a right triangle. In other similar triangle problems, if the triangles are not right triangles, you must be careful that the corresponding angles in the two triangles are the same.

Les Ailes
Parce que vous
n'êtes pas un oiseau

### 3 Solving Distance Problems Involving Rational Expressions

Some distance problems are solved using equations with rational expressions. We will need the formula Distance = Rate × Time, $D = RT$, which we can write in the form $T = \dfrac{D}{R}$. In the United States, distances are usually measured in miles. In European countries, distances are usually measured in kilometers.

**Example 5** A French commuter airline flies from Paris to Avignon. Plane A flies at a speed that is 50 kilometers per hour faster than plane B. Plane A flies 500 kilometers in the amount of time that plane B flies 400 kilometers. Find the speed of each plane.

#### Solution

1. **Understand the problem.** Let $s =$ the speed of plane B in kilometers per hour. Then $s + 50 =$ the speed of plane A in kilometers per hour. Make a simple table for $D$, $R$, and $T$.

| | $D$ | $R$ | $T = \dfrac{D}{R}$ |
|---|---|---|---|
| **Plane A** | 500 | $s + 50$ | ? |
| **Plane B** | 400 | $s$ | ? |

Since $T = \dfrac{D}{R}$, for each plane we divide the expression for $D$ by the expression for $R$ and write it in the table in the column for time.

| | $D$ | $R$ | $T = \dfrac{D}{R}$ |
|---|---|---|---|
| **Plane A** | 500 | $s + 50$ | $\dfrac{500}{s + 50}$ |
| **Plane B** | 400 | $s$ | $\dfrac{400}{s}$ |

2. **Write an equation and solve.** Each plane flies the same amount of time. That is, the time for plane A equals the time for plane B.

$$\frac{500}{s + 50} = \frac{400}{s}$$

You can solve this equation using the methods in Section 6.5 or you may cross-multiply. Here we will cross-multiply.

$$\begin{aligned}
500s &= (s + 50)(400) && \text{Cross-multiply.} \\
500s &= 400s + 20{,}000 && \text{Remove parentheses.} \\
100s &= 20{,}000 && \text{Subtract } 400s \text{ from each side.} \\
s &= 200 && \text{Divide each side by 100.}
\end{aligned}$$

Plane B travels 200 kilometers per hour.

Since $s + 50 = 200 + 50 = 250$, plane A travels 250 kilometers per hour. ☐

**Student Practice 5** Two European freight trains traveled toward Paris for the same amount of time. Train A traveled 180 kilometers, while train B traveled 150 kilometers. Train A traveled 10 kilometers per hour faster than train B. What was the speed of each train?

**4** Solving Work Problems

Some applied problems involve the length of time needed to do a job. These problems are often referred to as work problems.

**Example 6** Reynaldo can sort a huge stack of mail on an old sorting machine in 9 hours. His brother Carlos can sort the same amount of mail using a newer sorting machine in 8 hours. How long would it take them to do the job working together? Express your answer in hours and minutes. Round to the nearest minute.

**Solution**

1. **Understand the problem.** Let's do a little reasoning.

   If Reynaldo can do the job in 9 hours, then in *1 hour* he could do $\frac{1}{9}$ of the job.

   If Carlos can do the job in 8 hours, then in *1 hour* he could do $\frac{1}{8}$ of the job.

   Let $x$ = the number of hours it takes Reynaldo and Carlos to do the job together. In *1 hour* together they could do $\frac{1}{x}$ of the job.

2. **Write an equation and solve.** The amount of work Reynaldo can do in 1 hour plus the amount of work Carlos can do in 1 hour must be equal to the amount of work they could do together in 1 hour.

$$\boxed{\begin{array}{c}\text{Amount of work} \\ \text{done by Reynaldo}\end{array}} + \boxed{\begin{array}{c}\text{Amount of work} \\ \text{done by Carlos}\end{array}} = \boxed{\begin{array}{c}\text{Amount of work} \\ \text{done together}\end{array}}$$

$$\frac{1}{9} \quad + \quad \frac{1}{8} \quad = \quad \frac{1}{x}$$

Let us solve for $x$. We observe that the LCD is $72x$.

$$72x\left(\frac{1}{9}\right) + 72x\left(\frac{1}{8}\right) = 72x\left(\frac{1}{x}\right) \qquad \text{Multiply each term by the LCD.}$$

$$8x + 9x = 72 \qquad \text{Simplify.}$$

$$17x = 72 \qquad \text{Combine like terms.}$$

$$x = \frac{72}{17} \qquad \text{Divide each side by 17.}$$

$$x = 4\frac{4}{17}$$

To change $\frac{4}{17}$ of an hour to minutes, we multiply.

$$\frac{4}{17}\ \text{hour} \times \frac{60\ \text{minutes}}{1\ \text{hour}} = \frac{240}{17}\ \text{minutes, which is approximately 14.118 minutes}$$

To the nearest minute this is 14 minutes. Thus doing the job together will take 4 hours and 14 minutes. ☐

 **Student Practice 6** John Tobey and Dave Wells obtained night custodian jobs at a local factory while going to college part-time. Using the buffer machine, John can buff all the floors in the building in 6 hours. Dave takes a little longer and can do all the floors in the building in 7 hours. Their supervisor bought another buffer machine. How long will it take John and Dave to do all the floors in the building working together, each with his own machine? Express your answer in hours and minutes. Round to the nearest minute.

---

### Calculator

**Reciprocals**

You can find $\frac{1}{x}$ for any value of $x$ on a scientific calculator by using the key labeled $\boxed{x^{-1}}$ or the key labeled $\boxed{1/x}$. For example, to find $\frac{1}{9}$, we use $9\ \boxed{x^{-1}}$ or $9\ \boxed{1/x}$. The display will read 0.11111111.

Therefore we can solve Example 6 as follows:

$$9\ \boxed{x^{-1}}\ \boxed{+}\ 8\ \boxed{x^{-1}}\ \boxed{=}$$

The display will read 0.2361111. Thus we have obtained the equation $0.2361111 = \frac{1}{x}$.

Now, this is equivalent to

$$x = \frac{1}{0.2361111}.$$

(Do you see why?)

Thus we enter $0.2361111\ \boxed{x^{-1}}$, and the display reads 4.2352943. If we round to the nearest hundredth, we have $x \approx 4.24$ hours, which is approximately equal to our answer of $4\frac{4}{17}$ hours.

## 6.6 Exercises  MyMathLab®

*Solve.*

1. $\dfrac{5}{11} = \dfrac{8}{x}$

2. $\dfrac{7}{14} = \dfrac{x}{9}$

3. $\dfrac{x}{17} = \dfrac{12}{5}$

4. $\dfrac{18}{x} = \dfrac{7}{3}$

5. $\dfrac{9.1}{8.4} = \dfrac{x}{6}$

6. $\dfrac{3}{x} = \dfrac{12.5}{3.2}$

7. $\dfrac{7}{x} = \dfrac{40}{130}$

8. $\dfrac{x}{12} = \dfrac{7}{3}$

**Applications**  *Use a proportion to answer exercises 9–14.*

9. **Exchange Rates** Robyn spent two months traveling in New Zealand. The day she arrived, the exchange rate was 1.3 New Zealand dollars per U.S. dollar.

    (a) If she exchanged $500 U.S. dollars when she arrived, how many New Zealand dollars did she receive?

    (b) Three days later the exchange rate of the New Zealand dollar was 1.15 New Zealand dollars per U.S. dollar. How much less money would she have received had she waited three days to exchange her money?

10. **Exchange Rates** Sean spent a semester studying in Germany. On the day he arrived in Berlin, the exchange rate for the euro was 0.77 euro per U.S. dollar. Sean converted $350 to euros that day.

    (a) How many euros did Sean receive for his $350?

    (b) On his way home Sean decided to spend a week in London. He had €200 (200 euros) that he wanted to change into British pounds. If the exchange rate was 0.83 British pounds per euro, how many pounds did he receive?

11. **Speed Units** Alfonse and Melinda are taking a drive in Mexico. They know that a speed of 100 kilometers per hour is approximately equal to 62 miles per hour. They are now driving on a Mexican road that has a speed limit of 90 kilometers per hour. How many miles per hour is the speed limit? Round to the nearest mile per hour.

12. **Baggage Weight** Dick and Anne took a trip to France. Their suitcases were weighed at the airport and the weight recorded was 39 kilograms. If 50 kilograms is equivalent to 110 pounds, how many pounds did their suitcases weigh? Round to the nearest pound.

13. **Map Scale** On a map the distance between two mountains is $3\frac{1}{2}$ inches. The actual distance between the mountains is 136 miles. Russ is camped at a location that on the map is $\frac{3}{4}$ inch from the base of the mountain. How many miles is he from the base of the mountain? Round to the nearest mile.

14. **Map Scale** John, Stephanie, Stella, Nathaniel, and Josiah are taking a trip from Denver to Pueblo. The scale on the AAA map of Colorado is approximately $\frac{3}{4}$ inch to 15 miles. If the distance from Denver to Pueblo measures 5.5 inches on the map, how far apart are the two cities?

*Geometry*  *Triangles A and B are similar. Use them to answer exercises 15–18. Leave your answers as fractions.*

Triangle A

Triangle B

▲ 15. If $x = 20$ in., $y = 29$ in., and $m = 13$ in., find the length of side $n$.

▲ 16. If $p = 12$ in., $m = 16$ in., and $z = 20$ in., find the length of side $x$.

▲ 17. If $x = 175$ meters, $n = 40$ meters, and $m = 35$ meters, find the length of side $y$.

▲ 18. If $z = 18$ cm, $y = 25$ cm, and $n = 9$ cm, find the length of side $p$.

*Geometry Just as we have discussed similar triangles, other geometric shapes can be similar. Similar geometric shapes will have sides that are proportional. Quadrilaterals abcd and ghjk are similar. Use them to answer exercises 19–22. Leave your answers as fractions.*

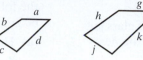

▲ **19.** If $a = 5$ ft, $d = 8$ ft, and $g = 7$ ft, find the length of side $k$.

▲ **20.** If $j = 12$ in., $k = 14$ in., and $c = 9$ in., find the length of side $d$.

▲ **21.** If $b = 20$ m, $h = 24$ m, and $d = 32$ m, find the length of side $k$.

▲ **22.** If $a = 20$ cm, $d = 24$ cm, and $k = 30$ cm, find the length of side $g$.

*Use a proportion to solve.*

▲ **23.** *Geometry* A rectangle whose width-to-length ratio is approximately 5 to 8 is called a **golden rectangle** and is said to be pleasing to the eye. Using this ratio, what should the length of a rectangular picture be if its width is to be 30 inches?

▲ **24.** *Shadows* Samantha is 5.5 feet tall and notices that she casts a shadow of 9 feet. At the same time, the new sculpture at the local park casts a shadow of 20 feet. How tall is the sculpture? Round your answer to the nearest foot.

▲ **25.** *Floral Displays* Floral designers often create arrangements where the flower height to container height ratio is 5 to 3. The FIU Art Museum wishes to create a floral display for the opening of a new show. They know they want to use an antique Chinese vase from their collection that is 13 inches high. How tall will the entire flower arrangement be if they use this standard ratio? (Round your answer to the nearest inch.)

▲ **26.** *Securing Wires* A wire line helps to secure a radio transmission tower. The wire measures 23 meters from the tower to the ground anchor pin. The wire is secured 14 meters up on the tower. If a second wire is secured 130 meters up on the tower and is extended from the tower at the same angle as the first wire, how long would the second wire need to be to reach an anchor pin on the ground? Round to the nearest meter.

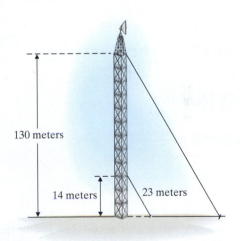

**27.** *Acceleration* Ben Hale is driving his new Toyota Camry on Interstate 90 at 45 miles per hour. He accelerates at the rate of 3 miles per hour every 2 seconds. How fast will he be traveling after accelerating for 11 seconds?

**28.** *Braking* Tim Newitt is driving a U-Haul truck to Chicago. He is driving at 55 miles per hour and has to hit the brakes because of heavy traffic. His truck slows at the rate of 2 miles per hour for every 3 seconds. How fast will he be traveling 10 seconds after he hits the brakes? Round to the nearest tenth.

**29.** *Flight Speeds* A Montreal commuter airliner travels 40 kilometers per hour faster than the television news helicopter over the city. The commuter airliner travels 1250 kilometers during the same time that the television news helicopter travels only 1050 kilometers. How fast does the commuter airliner fly? How fast does the television news helicopter fly?

**30.** *Driving Speeds* Jenny drove to Dallas while Mary drove to Houston in the same amount of time. Jenny drove 225 miles, while Mary drove 175 miles. Jenny traveled 12 miles per hour faster than Mary on her trip. What was the average speed in miles per hour for each woman?

31. *Fluff Containers* Marshmallow fluff comes in only two sizes, a $7\frac{1}{2}$-oz glass jar and a 16-oz plastic tub. At the local Stop and Shop, the 16-oz tub costs $2.19 and the $7\frac{1}{2}$-oz jar costs $1.29.

 (a) How much does marshmallow fluff in the glass jar cost per ounce? (Round your answer to the nearest cent.)

 (b) How much does marshmallow fluff in the plastic tub cost per ounce? (Round your answer to the nearest cent.)

 (c) If the marshmallow fluff company decided to add a third size, a 40-oz bucket, how much would the price be if it was at the same unit price as the 16-oz plastic tub? (*Hint*: Set up a proportion. Do *not* use your answer from part (b). Round your answer to the nearest cent.)

32. *Green Tea* Won Ling is a Chinese tea importer in Boston's Chinatown. He charges $12.25 for four sample packs of his famous green tea. The packs are in the following sizes: 25 grams, 40 grams, 50 grams, and 60 grams.

 (a) How much is Won Ling charging per gram for his green tea?

 (b) How much would you pay for a 60-gram pack if he were willing to sell that one by itself?

 (c) How much would you pay for an 800-gram package of green tea if it cost the same amount per gram?

33. *Raking Leaves* When all the leaves have fallen at Fred and Suzie's house in Concord, New Hampshire, Suzie can rake the entire yard in 6 hours. When Fred does it alone, it takes him 8 hours. How long would it take them to rake the yard together? Round to the nearest minute.

34. *Meal Preparation* To celebrate Diwali, a major Indian festival, Deepak and Alpa host a party each year for all their friends. Deepak can decorate and prepare all the food in 6 hours. Alpa takes 5 hours to decorate and prepare the food. How long would it take if they decorated and made the food together? Round your answer to the nearest minute.

## Cumulative Review

35. [4.2.2] Write in scientific notation. 0.000892465

36. [4.2.2] Write in decimal notation. $6.83 \times 10^9$

37. [4.2.1] Write with positive exponents. $\dfrac{x^{-3}y^{-2}}{z^4 w^{-8}}$

38. [4.2.1] Evaluate. $\left(\dfrac{2}{3}\right)^{-3}$

### Quick Quiz 6.6

1. Solve for $x$. $\dfrac{16.5}{2.1} = \dfrac{x}{7}$

2. While hiking in the White Mountains, Phil, Melissa, Noah, and Olivia saw a tall tree that cast a shadow 34 feet long. They observed at the same time that a 6-foot-tall person cast a shadow that was 8.5 feet long. How tall is the tree?

3. Last week Jeff Slater noted that 164 of the 205 flights to Chicago's O'Hare Airport flying out of Logan Airport departed on time. During the next week 215 flights left Logan Airport for Chicago's O'Hare Airport. If the same ratio holds, how many of those flights would he expect to depart on time?

4. **Concept Check** Mike found that his car used 18 gallons of gas to travel 396 miles. He needs to take a trip of 450 miles and wants to know how many gallons of gas it will take. He set up the equation $\frac{18}{x} = \frac{450}{396}$. Explain what error he made and how he should correctly solve the problem.

### Background: Operations Analyst

Brian is a member of the operations management team of Fitness Equipment Engineering. As a member of the team, he's responsible for identifying the cost and time associated with manufacturing new equipment.

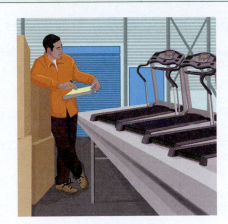

### Facts

- Recently, a new treadmill with state-of-the-art features has been approved for manufacture. Fixed daily costs are $22,000, and the cost of producing each treadmill is $350.
- The production facility has two assembly lines, line A and a newer robotic line, line B. Line A can produce $n$ treadmills in 18 hours, whereas line B can produce the same number $n$ treadmills in 14 hours.
- A workday is 8 hours.

### Tasks

1. Brian wants to create a model that will identify the average treadmill cost ($C_{ave}$) for producing $n$ number of treadmills daily. What form should this model take?

2. Production of the new treadmill is to occur using the production facility's assembly line A. Brian is proposing to his supervisor that two assembly lines be used instead of just one, and the supervisor wants to know how much time it will take to produce $n$ treadmills if the two lines are used simultaneously. Rounded to the nearest hour, how long will it take for the two lines to produce this specific number of treadmills, $n$?

3. Brian now wants to determine the average cost per hour for producing 50 treadmills and 100 treadmills, given that a workday is 8 hours. What is the cost per hour for producing each of these two amounts?

# Chapter 6 Organizer

| Topic and Procedure | Examples | You Try It |
|---|---|---|
| **Simplifying rational expressions, p. 348**<br><br>1. Factor the numerator and denominator.<br>2. Divide out any factor common to both the numerator and denominator. | $$\frac{36x^2 - 16y^2}{18x^2 + 24xy + 8y^2} = \frac{4(3x + 2y)(3x - 2y)}{2(3x + 2y)(3x + 2y)}$$ $$= \frac{2(3x - 2y)}{3x + 2y}$$ | **1.** Simplify. $\dfrac{6x^2 - 12x - 90}{3x^2 - 27}$ |
| **Multiplying rational expressions, p. 354**<br><br>1. Factor all numerators and denominators.<br>2. Simplify the resulting rational expression as described above. | $$\frac{x^2 - y^2}{x^2 + 2xy + y^2} \cdot \frac{x^2 + 4xy + 3y^2}{x^2 - 4xy + 3y^2}$$ $$= \frac{(x + y)(x - y)}{(x + y)(x + y)} \cdot \frac{(x + y)(x + 3y)}{(x - y)(x - 3y)}$$ $$= \frac{x + 3y}{x - 3y}$$ | **2.** Multiply. $\dfrac{x^2 - 4xy - 5y^2}{2x^2 - 9xy - 5y^2} \cdot \dfrac{4x^2 - y^2}{4x^2 - 4xy + y^2}$ |
| **Dividing rational expressions, p. 355**<br><br>1. Invert the second fraction and rewrite the problem as a product.<br>2. Multiply the rational expressions. | $$\frac{14x^2 + 17x - 6}{x^2 - 25} \div \frac{4x^2 - 8x - 21}{x^2 + 10x + 25}$$ $$= \frac{(2x + 3)(7x - 2)}{(x + 5)(x - 5)} \cdot \frac{(x + 5)(x + 5)}{(2x - 7)(2x + 3)}$$ $$= \frac{(7x - 2)(x + 5)}{(x - 5)(2x - 7)}$$ | **3.** Divide. $\dfrac{2x^2 + 3x - 20}{8x + 8} \div \dfrac{x^2 - 16}{4x^2 - 12x - 16}$ |
| **Adding rational expressions, p. 359**<br><br>1. If the denominators differ, factor them and determine the least common denominator (LCD).<br>2. Use multiplication to change each fraction into an equivalent one with the LCD as the denominator.<br>3. Add the numerators; put the answer over the LCD.<br>4. Simplify as needed. | $$\frac{x - 1}{x^2 - 4} + \frac{x - 1}{3x + 6} = \frac{x - 1}{(x + 2)(x - 2)} + \frac{x - 1}{3(x + 2)}$$ $$\text{LCD} = 3(x + 2)(x - 2)$$ $$\frac{x - 1}{(x + 2)(x - 2)} + \frac{x - 1}{3(x + 2)}$$ $$= \frac{x - 1}{(x + 2)(x - 2)} \cdot \frac{3}{3} + \frac{x - 1}{3(x + 2)} \cdot \frac{x - 2}{x - 2}$$ $$= \frac{3x - 3 + x^2 - 3x + 2}{3(x + 2)(x - 2)}$$ $$= \frac{x^2 - 1}{3(x + 2)(x - 2)}$$ $$= \frac{(x + 1)(x - 1)}{3(x + 2)(x - 2)}$$ | **4.** Add. $\dfrac{x + 2}{2x + 6} + \dfrac{x}{x^2 - 9}$ |
| **Subtracting rational expressions, p. 359**<br><br>Move the subtraction sign to the numerator of the second fraction. Add. Simplify if possible.<br>$$\frac{a}{b} - \frac{c}{b} = \frac{a}{b} + \frac{-c}{b}$$ | $$\frac{5x}{x - 2} - \frac{3x + 4}{x - 2} = \frac{5x}{x - 2} + \frac{-(3x + 4)}{x - 2}$$ $$= \frac{5x - 3x - 4}{x - 2}$$ $$= \frac{2x - 4}{x - 2}$$ $$= \frac{2(x - 2)}{(x - 2)} = 2$$ | **5.** Subtract. $\dfrac{9x}{x + 3} - \dfrac{3x - 18}{x + 3}$ |
| **Simplifying complex rational expressions, p. 368**<br><br>1. Add the two fractions in the numerator.<br>2. Add the two fractions in the denominator. | $$\frac{\dfrac{x}{x^2 - 4} + \dfrac{1}{x + 2}}{\dfrac{3}{x + 2} - \dfrac{4}{x - 2}}$$ $$= \frac{\dfrac{x}{(x + 2)(x - 2)} + \dfrac{1}{x + 2} \cdot \dfrac{x - 2}{x - 2}}{\dfrac{3}{x + 2} \cdot \dfrac{x - 2}{x - 2} - \dfrac{4}{x - 2} \cdot \dfrac{x + 2}{x + 2}}$$ | **6.** Simplify. $\dfrac{\dfrac{x}{x - 3} + \dfrac{2}{x + 3}}{\dfrac{1}{x - 3} + \dfrac{3}{x^2 - 9}}$ |

| Topic and Procedure | Examples | ▷ You Try It |
|---|---|---|
| 3. Divide the fraction in the numerator by the fraction in the denominator. This is done by inverting the fraction in the denominator and multiplying by the numerator.<br>4. Simplify. | $$= \frac{\dfrac{x + x - 2}{(x + 2)(x - 2)}}{\dfrac{3x - 6 - 4x - 8}{(x + 2)(x - 2)}}$$ $$= \frac{2x - 2}{(x + 2)(x - 2)} \div \frac{-x - 14}{(x + 2)(x - 2)}$$ $$= \frac{2(x - 1)}{(x + 2)(x - 2)} \cdot \frac{(x + 2)(x - 2)}{-x - 14}$$ $$= \frac{2(x - 1)}{-x - 14} \text{ or } -\frac{2(x - 1)}{x + 14} \text{ or } \frac{-2(x - 1)}{x + 14}$$ | |
| **Solving equations involving rational expressions, p. 374**<br><br>1. Determine the LCD of all denominators.<br>2. Note what values will make the LCD equal to 0. These are excluded from your solutions.<br>3. Multiply each side by the LCD, distributing as needed.<br>4. Solve the resulting polynomial equation.<br>5. Check. Be sure to exclude those values found in step 2. | $$\frac{3}{x - 2} = \frac{4}{x + 2}$$ LCD $= (x - 2)(x + 2)$. $$(x - 2)(x + 2)\frac{3}{x - 2} = \frac{4}{x + 2}(x - 2)(x + 2)$$ $$3(x + 2) = 4(x - 2)$$ $$3x + 6 = 4x - 8$$ $$-x = -14$$ $$x = 14$$ Check: $\dfrac{3}{14 - 2} \overset{?}{=} \dfrac{4}{14 + 2}$ $$\frac{3}{12} \overset{?}{=} \frac{4}{16}$$ $$\frac{1}{4} = \frac{1}{4} \checkmark$$ | **7.** Solve. $\dfrac{5x}{x^2 - 16} = \dfrac{5}{x + 4}$ |
| **Solving applied problems with proportions, p. 379**<br><br>1. Organize the data.<br>2. Write a proportion equating the respective parts. Let $x$ represent the value that is not known.<br>3. Solve the proportion. | Renee can make five cherry pies with 3 cups of flour. How many cups of flour does she need to make eight cherry pies? $$\frac{5 \text{ cherry pies}}{3 \text{ cups flour}} = \frac{8 \text{ cherry pies}}{x \text{ cups flour}}$$ $$\frac{5}{3} = \frac{8}{x}$$ $$5x = 24$$ $$x = \frac{24}{5}$$ $$x = 4\frac{4}{5}$$ $4\frac{4}{5}$ cups of flour are needed for eight cherry pies. | **8.** On the blueprint of Rob and Amy's new family room addition, the room measures 2 inches wide by 3 inches long. The actual width of the family room will be 10.5 feet. How long will the actual room be? |

# Chapter 6 Review Problems

## Section 6.1

*Simplify.*

**1.** $\dfrac{bx}{bx - by}$

**2.** $\dfrac{4x - 4y}{5y - 5x}$

**3.** $\dfrac{x^3 - 4x^2}{x^3 - x^2 - 12x}$

**4.** $\dfrac{2x^2 + 7x - 15}{25 - x^2}$

**5.** $\dfrac{2x^2 - 2xy - 24y^2}{2x^2 + 5xy - 3y^2}$

**6.** $\dfrac{4 - y^2}{3y^2 + 5y - 2}$

7. $\dfrac{5x^3 - 10x^2}{25x^4 + 5x^3 - 30x^2}$

8. $\dfrac{16x^2 - 4y^2}{4x - 2y}$

## Section 6.2

*Multiply or divide.*

9. $\dfrac{2x^2 + 6x}{3x^2 - 27} \cdot \dfrac{x^2 + 3x - 18}{4x^2 - 4x}$

10. $\dfrac{y^2 + 8y + 16}{5y^2 + 20y} \div \dfrac{y^2 + 7y + 12}{2y^2 + 5y - 3}$

11. $\dfrac{6y^2 + 13y - 5}{9y^2 + 3y} \div \dfrac{4y^2 + 20y + 25}{12y^2}$

12. $\dfrac{3xy^2 + 12y^2}{2x^2 - 11x + 5} \div \dfrac{2xy + 8y}{8x^2 + 2x - 3}$

13. $\dfrac{x^2 - 5xy - 24y^2}{2x^2 - 2xy - 24y^2} \cdot \dfrac{4x^2 + 4xy - 24y^2}{x^2 - 10xy + 16y^2}$

14. $\dfrac{2x^2 + 10x + 2}{8x - 8} \cdot \dfrac{3x - 3}{4x^2 + 20x + 4}$

## Section 6.3

*Add or subtract.*

15. $\dfrac{6}{y + 2} + \dfrac{2}{3y}$

16. $3 + \dfrac{2}{x + 1} + \dfrac{1}{x}$

17. $\dfrac{7}{x + 2} + \dfrac{3}{x - 4}$

18. $\dfrac{2}{x^2 - 9} + \dfrac{x}{x + 3}$

19. $\dfrac{x}{y} + \dfrac{3}{2y} + \dfrac{1}{y + 2}$

20. $\dfrac{4}{a} + \dfrac{2}{b} + \dfrac{3}{a + b}$

21. $\dfrac{3x + 1}{3x} - \dfrac{1}{x}$

22. $\dfrac{x + 4}{x + 2} - \dfrac{1}{2x}$

23. $\dfrac{27}{x^2 - 81} + \dfrac{3}{2(x + 9)}$

24. $\dfrac{1}{x^2 + 7x + 10} - \dfrac{x}{x + 5}$

## Section 6.4

*Simplify.*

25. $\dfrac{\dfrac{4}{3y} - \dfrac{2}{y}}{\dfrac{1}{2y} + \dfrac{1}{y}}$

26. $\dfrac{\dfrac{5}{x} + \dfrac{1}{2x}}{\dfrac{x}{4} + x}$

27. $\dfrac{w - \dfrac{4}{w}}{1 + \dfrac{2}{w}}$

28. $\dfrac{1 - \dfrac{w}{w - 1}}{1 + \dfrac{w}{1 - w}}$

29. $\dfrac{1 + \dfrac{1}{y^2 - 1}}{\dfrac{1}{y + 1} - \dfrac{1}{y - 1}}$

30. $\dfrac{\dfrac{1}{y} + \dfrac{1}{x + y}}{1 + \dfrac{2}{x + y}}$

**31.** $\dfrac{\dfrac{1}{a+b} - \dfrac{1}{a}}{b}$

**32.** $\dfrac{\dfrac{2}{a+b} - \dfrac{3}{b}}{\dfrac{1}{a+b}}$

## Section 6.5

*Solve for the variable. If there is no solution, say so.*

**33.** $\dfrac{8a-1}{6a+8} = \dfrac{3}{4}$

**34.** $\dfrac{8}{a-3} = \dfrac{12}{a+3}$

**35.** $\dfrac{2x-1}{x} - \dfrac{1}{2} = -2$

**36.** $\dfrac{5}{4} - \dfrac{1}{2x} = \dfrac{1}{x} + 2$

**37.** $\dfrac{7}{8x} - \dfrac{3}{4} = \dfrac{1}{4x} + \dfrac{1}{2}$

**38.** $\dfrac{3}{y-3} = \dfrac{3}{2} + \dfrac{y}{y-3}$

**39.** $\dfrac{3x}{x^2-4} - \dfrac{2}{x+2} = -\dfrac{4}{x-2}$

**40.** $\dfrac{3y-1}{3y} - \dfrac{6}{5y} = \dfrac{1}{y} - \dfrac{4}{15}$

**41.** $\dfrac{y+18}{y^2-16} = \dfrac{y}{y+4} - \dfrac{y}{y-4}$

**42.** $\dfrac{4}{x^2-1} = \dfrac{2}{x-1} + \dfrac{2}{x+1}$

**43.** $\dfrac{3y+1}{y^2-y} - \dfrac{3}{y-1} = \dfrac{4}{y}$

**44.** $\dfrac{3}{y-2} + \dfrac{4}{3y+2} = \dfrac{1}{2-y}$

## Section 6.6

*Solve.*

**45.** $\dfrac{x}{4} = \dfrac{7}{10}$

**46.** $\dfrac{8}{5} = \dfrac{2}{x}$

**47.** $\dfrac{33}{10} = \dfrac{x}{8}$

**48.** $\dfrac{16}{x} = \dfrac{24}{9}$

**49.** $\dfrac{13.5}{0.6} = \dfrac{360}{x}$

**50.** $\dfrac{2\frac{1}{2}}{3\frac{1}{4}} = \dfrac{7}{x}$

*Use a proportion to answer each question.*

**51. Paint Needs** A 5-gallon can of paint will cover 240 square feet. How many gallons of paint will be needed to cover 400 square feet? Round to the nearest tenth of a gallon.

**52. Recipe Ratios** Aunt Lexie uses 3 pounds of sugar to make 100 cookies. How many cookies can she make with 5 pounds of sugar? Round to the nearest whole cookie.

**53. Map Scale** On a map of Texas, the distance between El Paso and Dallas is 4 inches. The actual distance between these cities is 640 miles. Houston and Dallas are 1.5 inches apart on the same map. How many miles apart are Houston and Dallas?

**54. Travel Speeds** A train travels 180 miles in the same time that a car travels 120 miles. The speed of the train is 20 miles per hour faster than the speed of the car. Find the speed of the train and the speed of the car.

▲ 55. *Shadows* Mary takes a walk across a canyon in New Mexico. She stands 5.75 feet tall and her shadow is 3 feet long. At the same time, the shadow from the peak of the canyon wall casts a shadow that is 95 feet long. How tall is the peak of the canyon? Round to the nearest foot.

▲ 56. *Shadows* A flagpole that is 8 feet tall casts a shadow of 3 feet. At the same time of day, a tall office building in the city casts a shadow of 450 feet. How tall is the office building?

57. *Window Cleaning* As part of their spring cleaning routine, Tina and Mathias wash all of their windows, inside and out. Tina can do this job in 4 hours. When Mathias washes the windows, it takes him 6 hours. How long would it take them if they worked together?

58. *Plowing Fields* Sally runs the family farm in Boone, Iowa. She can plow the fields of the farm in 20 hours. Her daughter Brenda can plow the fields of the farm in 30 hours. If they have two identical tractors, how long would it take Brenda and Sally to plow the fields of the farm if they worked together?

## Mixed Practice

*Perform the operation indicated. Simplify.*

59. $\dfrac{a^2 + 2a - 8}{6a^2 - 3a^3}$

60. $\dfrac{4a^3 + 20a^2}{2a^2 + 13a + 15}$

61. $\dfrac{x^2 - y^2}{x^2 + 4xy + 3y^2} \cdot \dfrac{x^2 + xy - 6y^2}{x^2 + xy - 2y^2}$

62. $\dfrac{x}{x + 3} + \dfrac{9x + 18}{x^2 + 3x}$

63. $\dfrac{x - 30}{x^2 - 5x} + \dfrac{x}{x - 5}$

64. $\dfrac{a + b}{ax + ay} - \dfrac{a + b}{bx + by}$

65. $\dfrac{\dfrac{5}{3x} + \dfrac{2}{9x}}{\dfrac{3}{x} + \dfrac{8}{3x}}$

66. $\dfrac{\dfrac{4}{5y} - \dfrac{8}{y}}{y + \dfrac{y}{5}}$

67. $\dfrac{x - 3y}{x + 2y} \div \left( \dfrac{2}{y} - \dfrac{12}{x + 3y} \right)$

68. $\dfrac{7}{x + 2} = \dfrac{4}{x - 4}$

69. $\dfrac{2x - 1}{3x - 8} = \dfrac{5}{8}$

70. $2 + \dfrac{4}{b - 1} = \dfrac{4}{b^2 - b}$

# How Am I Doing? Chapter 6 Test

**MATH COACH**     MyMathLab®     You Tube

After you take this test read through the Math Coach on pages 394–395. Math Coach videos are available via MyMathLab and YouTube. Step-by-step test solutions on the Chapter Test Prep Videos are also available via MyMathLab and YouTube. (Search "TobeyBeginningAlg" and click on "Channels.")

*Perform the operation indicated. Simplify.*

**1.** $\dfrac{2ac + 2ad}{3a^2 c + 3a^2 d}$

**2.** $\dfrac{8x^2 - 2x^2 y^2}{y^2 + 4y + 4}$

**3.** $\dfrac{x^2 + 2x}{2x - 1} \cdot \dfrac{10x^2 - 5x}{12x^3 + 24x^2}$

**4.** $\dfrac{x + 2y}{12y^2} \cdot \dfrac{4y}{x^2 + xy - 2y^2}$

**MC 5.** $\dfrac{2a^2 - 3a - 2}{a^2 + 5a + 6} \div \dfrac{a^2 - 5a + 6}{a^2 - 9}$

**6.** $\dfrac{1}{a^2 - a - 2} + \dfrac{3}{a - 2}$

**7.** $\dfrac{x - y}{xy} - \dfrac{a - y}{ay}$

**MC 8.** $\dfrac{3x}{x^2 - 3x - 18} - \dfrac{x - 4}{x - 6}$

**9.** $\dfrac{\dfrac{x}{3y} - \dfrac{1}{2}}{\dfrac{4}{3y} - \dfrac{2}{x}}$

**MC 10.** $\dfrac{\dfrac{6}{b} - 4}{\dfrac{5}{bx} - \dfrac{10}{3x}}$

**11.** $\dfrac{2x^2 + 3xy - 9y^2}{4x^2 + 13xy + 3y^2}$

**12.** $\dfrac{1}{x + 4} - \dfrac{2}{x^2 + 6x + 8}$

*In questions 13–18, solve for x. Check your answers. If there is no solution, say so.*

**13.** $\dfrac{4}{3x} - \dfrac{5}{2x} = 5 - \dfrac{1}{6x}$

**MC 14.** $\dfrac{x - 3}{x - 2} = \dfrac{2x^2 - 15}{x^2 + x - 6} - \dfrac{x + 1}{x + 3}$

**15.** $3 - \dfrac{7}{x + 3} = \dfrac{x - 4}{x + 3}$

**16.** $\dfrac{3}{3x - 5} = \dfrac{7}{5x + 4}$

**17.** $\dfrac{9}{x} = \dfrac{13}{5}$

**18.** $\dfrac{9.3}{2.5} = \dfrac{x}{10}$

**19.** A random check of America West air flights last month showed that 113 of the 150 flights checked arrived on time. If the inspectors check 200 flights next month, how many can be expected to be on time? (Round your answer to the nearest whole number.)

**20.** In northern Michigan the Gunderson family heats their home with firewood. They used $100 worth of wood in 25 days. Mr. Gunderson estimates that he needs to burn wood at that rate for about 92 days during the winter. If that is so, how much will the 92-day supply of wood cost?

▲ **21.** A hiking club is trying to construct a rope bridge across a canyon. A 6-foot construction pole held upright casts a 7-foot shadow. At the same time of day, a tree at the edge of the canyon casts a shadow that exactly covers the distance that is needed for the rope bridge. The tree is exactly 87 feet tall. How long should the rope bridge be? Round to the nearest foot.

1. _____

2. _____

3. _____

4. _____

5. _____

6. _____

7. _____

8. _____

9. _____

10. _____

11. _____

12. _____

13. _____

14. _____

15. _____

16. _____

17. _____

18. _____

19. _____

20. _____

21. _____

**Total Correct:** _____

393

# MATH COACH

*Mastering the skills you need to do well on the test.*

The following problems are from the Chapter 6 Test. Here are some helpful hints to keep you from making common errors on test problems.

**Chapter 6 Test, Problem 5** Simplify. $\dfrac{2a^2 - 3a - 2}{a^2 + 5a + 6} \div \dfrac{a^2 - 5a + 6}{a^2 - 9}$

> **HELPFUL HINT** The operation of division is performed by inverting the *second fraction* and multiplying it by the first fraction. This step should be done first, before you begin factoring.

Did you keep the first fraction the same as it is written and then multiply it by $\dfrac{a^2 - 9}{a^2 - 5a + 6}$?

Yes _____ No _____

If you answered No, stop and make this correction to your work.

Were you able to factor each expression so that you obtained $\dfrac{(2a + 1)(a - 2)}{(a + 2)(a + 3)} \cdot \dfrac{(a + 3)(a - 3)}{(a - 2)(a - 3)}$?

Yes _____ No _____

If you answered No, go back and check each factoring step to see if you can get the same result. Complete the problem by dividing out any common factors.

If you answered Problem 5 incorrectly, go back and rework the problem using these suggestions.

---

**Chapter 6 Test, Problem 8** Simplify. $\dfrac{3x}{x^2 - 3x - 18} - \dfrac{x - 4}{x - 6}$

> **HELPFUL HINT** First factor the trinomial in the denominator so that you can determine the LCD of these two fractions. Then multiply by what is needed in the second fraction so that it becomes an equivalent fraction with the LCD as the denominator. When subtracting, it is a good idea to place brackets around the second numerator to avoid sign errors.

Did you factor the first denominator into $(x - 6)(x + 3)$?

Yes _____ No _____

Did you then determine that the LCD is $(x - 6)(x + 3)$?

Yes _____ No _____

If you answered No to these questions, stop and review how to factor the trinomial in the first denominator and how to find the LCD when working with polynomials as denominators.

Did you multiply the numerator and denominator of the second fraction by $(x + 3)$ and place brackets around the this product to obtain

$\dfrac{3x}{(x - 6)(x + 3)} - \dfrac{[(x - 4)(x + 3)]}{(x - 6)(x + 3)}$?

Yes _____ No _____

Did you obtain $3x - x^2 - x - 12$ in the numerator?

Yes _____ No _____

If you answered Yes, you forgot to distribute the negative sign. Rework the problem and make this correction.

As your final step, remember to combine like terms and then factor the new numerator before dividing out common factors.

Now go back and rework the problem using these suggestions.

Need more help? Watch the **MATH COACH** videos in MyMathLab® or on YouTube.

394

**Chapter 6 Test, Problem 10**   Simplify. $\dfrac{\dfrac{6}{b} - 4}{\dfrac{5}{bx} - \dfrac{10}{3x}}$

> **HELPFUL HINT** There are two ways to simplify this expression:
> 1. combine the numerators and the denominators separately, or
> 2. multiply the numerator and denominator of the complex fraction by the LCD of all the denominators.
>
> Consider both methods and choose the one that seems easiest to you. We will show the steps of the first method for this particular problem.

Did you multiply 4 by $\dfrac{b}{b}$ and then subtract $\dfrac{6}{b} - \dfrac{4b}{b}$?

Yes _____ No _____

If you answered No, remember that 4 can be written as $\dfrac{4}{1}$, and to find the LCD, you must multiply numerator and denominator by the variable $b$.

In the denominator of the complex fraction, did you obtain $3bx$ as the LCD and multiply the first fraction by $\dfrac{3}{3}$ and the second fraction by $\dfrac{b}{b}$ before subtracting the two fractions?

Yes _____ No _____

If you answered No, stop and carefully change these two fractions into equivalent fractions with $3bx$ as the denominator. Then subtract the numerators and keep the common denominator.

Were you able to rewrite the problem as follows: $\dfrac{6 - 4b}{b} \div \dfrac{15 - 10b}{3bx}$?

Yes _____ No _____

If you answered No, examine your steps carefully. Once you have one fraction in the numerator and one fraction in the denominator, you can rewrite the division as multiplication by inverting the fraction in the denominator.

If you answered Problem 10 incorrectly, go back and rework the problem using these suggestions.

---

**Chapter 6 Test, Problem 14**   Solve for $x$. $\dfrac{x - 3}{x - 2} = \dfrac{2x^2 - 15}{x^2 + x - 6} - \dfrac{x + 1}{x + 3}$

> **HELPFUL HINT** First factor any denominators that need to be factored so that you can determine the LCD of all the denominators. Verify that you are solving an equation, then multiply *each term* of the equation by the LCD. Solve the resulting equation. Check your solution.

Did you factor $x^2 + x - 6$ to get $(x + 3)(x - 2)$ and identify the LCD as $(x + 3)(x - 2)$?

Yes _____ No _____

If you answered No, go back and complete these steps again.

Did you notice that we are solving an equation and that we can multiply each term of the equation by the LCD and then multiply the binomials to obtain the equation $x^2 - 9 = (2x^2 - 15) - (x^2 - x - 2)$?

Yes _____ No _____

If you answered No, look at the step $-[(x + 1)(x - 2)]$.

After you multiplied the binomials, did you distribute the negative sign and change the sign of all terms inside the grouping symbols?

Yes _____ No _____

If you answered No, review how to multiply binomials and subtract polynomial expressions. Then combine like terms and solve the equation for $x$.

Now go back and rework the problem using these suggestions.

Need more help? Look for section examples marked with $^{M}_{C}$ to review.

395

# Cumulative Test for Chapters 0–6

*This test provides a comprehensive review of the key objectives for Chapters 0–6.*

1. _____

2. _____

3. _____

4. _____

5. _____

6. _____

7. _____

8. _____

9. _____

10. _____

11. _____

12. _____

13. _____

14. _____

15. _____

16. _____

17. _____

18. _____

19. _____

1. Add. $-\dfrac{5}{3} + \dfrac{1}{2} + \dfrac{5}{6}$

2. Divide. $\left(-4\dfrac{1}{2}\right) \div \left(5\dfrac{1}{4}\right)$

3. Evaluate $3a^2 + ab - 4b^2$ for $a = 5$ and $b = -1$.

4. Henry is thinking about buying a new car for $22,500. His state presently has a sales tax of 6.5%. In a few weeks the state will raise the sales tax to 7%. How much will he save in sales tax if he purchases the car before the sales tax rate is raised?

5. Solve. $5(x - 3) - 2(4 - 2x) = 7(x - 1) - (x - 2)$

6. Solve for $h$. $A = \pi r^2 h$

7. Solve and graph on a number line. $4(2 - x) < 3$

   $\xleftarrow{\hspace{1cm}} \underset{0}{+} \quad \underset{}{+} \quad \underset{1}{+} \quad \underset{}{+} \quad \underset{2}{+} \quad \underset{}{+} \xrightarrow{\hspace{0.5cm}} x$

8. When triple a number and 11 are added, the result is 56. What is the number?

9. The length of a rectangular patio is 3 feet less than double the width. The perimeter of the patio is 42 feet. What are the dimensions of the patio?

10. After working at a consulting form for a year, Louisa received a 4% raise. Her salary after the raise is $43,680. What was Louisa's salary before the raise?

11. Factor. $16x^4 - b^4$

12. Factor. $8a^3 - 38a^2b - 10ab^2$

13. Simplify. Write the answer with positive exponents. $\dfrac{-16a^3b^{-1}}{4a^{-2}b^5}$

14. Write in scientific notation. $0.00056$

15. Multiply. $(3x - 5)^2$

*Perform the indicated operations.*

16. $\dfrac{x^2 - 4}{x^2 - 25} \cdot \dfrac{3x^2 - 14x - 5}{3x^2 + 6x}$

17. $\dfrac{5}{2x + 4} + \dfrac{3}{x - 3}$

18. Solve for $x$. $\dfrac{x - 3}{x} = \dfrac{x + 2}{x + 3}$

19. Simplify. $\dfrac{\dfrac{3}{a} + \dfrac{2}{b}}{\dfrac{5}{a^2} - \dfrac{2}{b^2}}$

# CHAPTER 7

# Graphing and Functions

## CAREER OPPORTUNITIES

## Economist

Economists play an important role throughout society by studying the production, supply, demand, and exchange of goods and services. Identifying trends and their possible impact, these professionals help shape not only business decision making but also public policy. Many economists are employed in businesses and academic institutions. Though their job titles differ, they all share a strong background in mathematics, and they use the math presented in this chapter to assist their research, analysis, and forecasting.

To investigate how the mathematics in this chapter can help with this field, see the Career Exploration Problems on page **452**.

## 7.1 The Rectangular Coordinate System

**Student Learning Objectives**

**After studying this section, you will be able to:**

**1** Plot a point, given the coordinates. ▶

**2** Determine the coordinates of a plotted point. ▶

**3** Find ordered pairs for a given linear equation. ▶

### 1 Plotting a Point, Given the Coordinates

Often we can better understand an idea if we see a picture. This is the case with many mathematical concepts, including those relating to algebra. We can illustrate algebraic relationships with drawings called **graphs.** Before we can draw a graph, however, we need a frame of reference.

In Chapter 1 we showed that any real number can be represented on a number line. Look at the following number line. The arrow indicates the positive direction.

To form a **rectangular coordinate system,** we draw a second number line vertically. We construct it so that the 0 point on each number line is exactly at the same place. We refer to this location as the **origin.** The horizontal number line is often called the **x-axis.** The vertical number line is often called the **y-axis.** Arrows show the positive direction for each axis.

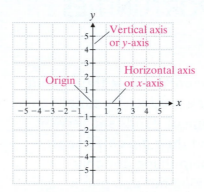

We can represent a point in this rectangular coordinate system by using an **ordered pair** of numbers. For example, $(5, 2)$ is an ordered pair that represents a point in the rectangular coordinate system. The numbers in an ordered pair are often referred to as the **coordinates** of the point. The first number is called the **x-coordinate** and it represents the distance from the origin measured along the horizontal or $x$-axis. If the $x$-coordinate is positive, we count the proper number of squares to the right (that is, in the positive direction). If the $x$-coordinate is negative, we count to the left. The second number in the pair is called the **y-coordinate** and it represents the distance from the origin measured along the $y$-axis. If the $y$-coordinate is positive, we count the proper number of squares upward (that is, in the positive direction). If the $y$-coordinate is negative, we count downward.

$$(5, 2)$$
$$x\text{-coordinate} \quad \quad y\text{-coordinate}$$

Suppose the directory for the map on the left indicated that you would find a certain street in the region C2. To find the street you would first scan across the horizontal scale until you found section C; from there you would scan up the map until you hit section 2 along the vertical scale. As we will see in the next example, plotting a point in the rectangular coordinate system is much like finding a street on a map with grids.

**Example 1** Plot the point $(5, 2)$ on a rectangular coordinate system. Label this point as $A$.

**Solution** Since the $x$-coordinate is 5, we first count 5 units to the right on the $x$-axis. Then, because the $y$-coordinate is 2, we count 2 units up from the point where we stopped on the $x$-axis. This locates the point corresponding to $(5, 2)$. We mark this point with a dot and label it $A$.

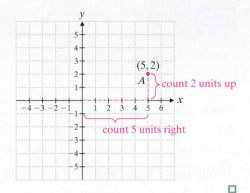

The first number indicates the x-direction — The second number indicates the y-direction

$(5, 2)$

**Student Practice 1** Plot the point $(3, 4)$ on the preceding rectangular coordinate system. Label this point as $B$.

It is important to remember that the first number in an ordered pair is the $x$-coordinate and the second number is the $y$-coordinate. The ordered pairs $(5, 2)$ and $(2, 5)$ represent different points.

**Example 2** Use the rectangular coordinate system to plot each point. Label the points $F$, $G$, and $H$, respectively.

**(a)** $(-5, -3)$      **(b)** $(2, -6)$      **(c)** $(-6, 2)$

**Solution**

**(a)** $(-5, -3)$ Notice that the $x$-coordinate, $-5$, is negative. On the coordinate grid, negative $x$-values appear to the left of the origin. Thus, we will begin by counting 5 squares to the left, starting at the origin. Since the $y$-coordinate, $-3$, is negative, we will count 3 units down from the point where we stopped on the $x$-axis.

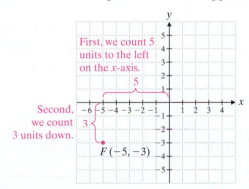

**(b)** $(2, -6)$ The $x$-coordinate is positive. Begin by counting 2 squares to the right of the origin. Then count down because the $y$-coordinate is negative.

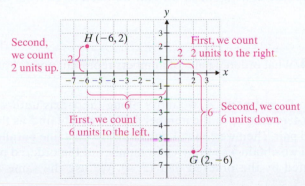

**(c)** $(-6, 2)$ The $x$-coordinate is negative. Begin by counting 6 squares to the left of the origin. Then count up because the $y$-coordinate is positive.

**Student Practice 2**

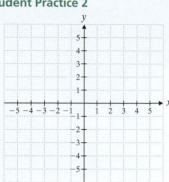

**Student Practice 2** Use the rectangular coordinate system in the margin to plot each point. Label the points $I, J$, and $K$, respectively.

(a) $(-2, -4)$ **(b)** $(-4, 5)$ **(c)** $(4, -2)$

**Example 3** Plot the following points.

$F: (0, 5)$ $G: \left(3, \frac{3}{2}\right)$ $H: (-6, 4)$ $I: (-3, -4)$
$J: (-4, 0)$ $K: (2, -3)$ $L: (6.5, -7.2)$

**Solution** These points are plotted in the figure.

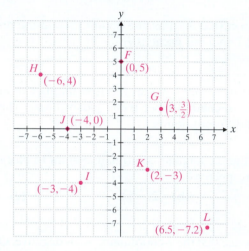

**Student Practice 3**

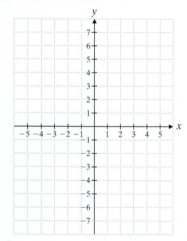

*Note:* When you are plotting decimal values like $(6.5, -7.2)$, plot the point halfway between 6 and 7 in the $x$-direction (for the 6.5) and at your best approximation of $-7.2$ in the $y$-direction. ▫

**Student Practice 3** Plot the following points. Label each point with both the letter and the ordered pair. Use the coordinate system provided in the margin.

$$A: (3, 7); B: (0, -6); C: (3, -4.5); D: \left(-\frac{7}{2}, 2\right)$$

## 2 Determining the Coordinates of a Plotted Point ▶

Sometimes we need to find the coordinates of a point that has been plotted. First, we count the units we need on the $x$-axis to get as close as possible to the point. Next we count the units up or down that we need to go from the $x$-axis to reach the point.

**Example 4** What ordered pairs of numbers represent point $A$ and point $B$ on the graph?

**Solution** To find point $A$, we move along the $x$-axis until we get as close as possible to $A$, ending up at the number 5. Thus we obtain 5 as the first number of the ordered pair. Then we count 4 units upward on a line parallel to the $y$-axis to reach $A$. So we obtain 4 as the second number of the ordered pair. Thus point $A$ is represented by the ordered pair $(5, 4)$. We use the same approach to find point $B: (-5, -3)$.

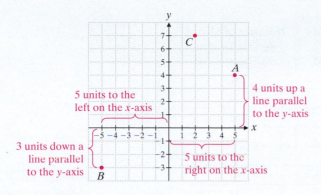

**Student Practice 4**  What ordered pair of numbers represents point $C$ on the graph in Example 4?

In examining data from real-world situations, we often find that plotting data points shows useful trends. In such cases, it is often necessary to use a different scale, one that displays only positive values.

**Example 5**  The number of motor vehicle accidents in millions is recorded in the following table for the years 2000 to 2012.

(a) Plot points that represent this data on the given coordinate system.

(b) What trends are apparent from the plotted data?

| Number of Years Since 2000 | Number of Motor Vehicle Accidents (in Millions) |
|:---:|:---:|
| 0 | 13 |
| 2 | 18 |
| 4 | 11 |
| 6 | 10 |
| 8 | 10 |
| 10 | 11 |
| 12 | 10 |

*Source: 2012 Traffic Safety Facts*

## Solution

(a)

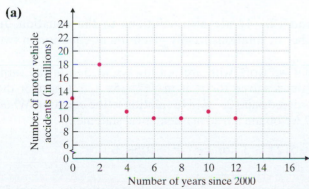

(b) From 2000 to 2002, there was a *significant* increase in the number of accidents. From 2002 to 2004, there was a *significant* decrease in the number of accidents. From 2004 to 2012, the number of accidents was relatively stable.

**Student Practice 5**  The number of motor vehicle deaths in thousands is recorded in the following table for the years 1980 to 2010.

(a) Plot points that represent this data on the given coordinate system.

(b) What trends are apparent from the plotted data?

*Continued on next page*

| Number of Years Since 1980 | Number of Motor Vehicle Deaths (in Thousands) |
|---|---|
| 0 | 51 |
| 5 | 44 |
| 10 | 45 |
| 15 | 42 |
| 20 | 42 |
| 25 | 44 |
| 30 | 33 |

*Source: 2012 Statistical Abstract; 2012 Traffic Safety Facts*

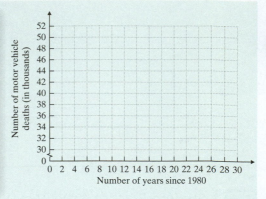

### 3 Finding Ordered Pairs for a Given Linear Equation

Equations such as $6x + 2y = 4$ and $5x + y = 3$ are called linear equations in two variables.

> A **linear equation in two variables** is an equation that can be written in the form $Ax + By = C$ where $A$, $B$, and $C$ are real numbers but $A$ and $B$ are not *both* zero.

Replacement values for $x$ and $y$ that make *true mathematical statements* of the equation are called *truth values;* and an ordered pair of these truth values is called a **solution.**

Consider the equation $6x + 2y = 4$. The ordered pair $(0, 2)$ is a solution to the equation because when we replace $x$ by 0 and $y$ by 2 in the equation, we obtain a true statement.

$$6(0) + 2(2) = 4 \quad \text{or} \quad 0 + 4 = 4$$

Likewise $(1, -1), (-2, 8)$ and $(2, -4)$ are also solutions to the equation. In fact, there is an infinite number of solutions to any given linear equation in two variables.

The linear equations that we work with are not always written in the form $Ax + By = C$, but are sometimes solved for $y$, as in $y = -5x + 3$. Consider the equation $y = -5x + 3$. The ordered pair $(2, -7)$ is a solution to the equation. When we replace $x$ by 2 and $y$ by $-7$ we obtain a true mathematical statement:

$$(-7) = -5(2) + 3 \quad \text{or} \quad -7 = -10 + 3.$$

**Example 6** Are the following ordered pairs solutions to the equation $3x + 2y = 5$?

(a) $(-1, 4)$ 　　　　　　(b) $(2, 2)$

**Solution** We replace values for $x$ and $y$ in the equation to see if we obtain a true statement.

(a) $\qquad 3x + 2y = 5$

$\qquad 3(-1) + 2(4) \stackrel{?}{=} 5$ 　　　Replace $x$ with $-1$ and $y$ with 4.

$\qquad\qquad -3 + 8 \stackrel{?}{=} 5$

$\qquad\qquad\qquad 5 = 5$ ✓ 　True statement

The ordered pair $(-1, 4)$ is a solution to $3x + 2y = 5$ because when we replace $x$ by $-1$ and $y$ by 4, we obtain a true statement.

**(b)** 
$$3x + 2y = 5$$
$$3(2) + 2(2) \stackrel{?}{=} 5 \quad \text{Replace } x \text{ with 2 and } y \text{ with 2.}$$
$$6 + 4 \stackrel{?}{=} 5$$
$$10 = 5 \quad \text{False statement}$$

The ordered pair $(2, 2)$ is *not* a solution to $3x + 2y = 5$ because when we replace $x$ by 2 and $y$ by 2, we obtain a *false* statement. ☐

**Student Practice 6** Are the following ordered pairs solutions to the equation $3x + 2y = 5$?

**(a)** $(3, -1)$ **(b)** $\left(2, -\dfrac{1}{2}\right)$

If one value of an ordered-pair solution to a linear equation is known, the other can be quickly obtained. To do so, we replace the proper variable in the equation by the known value. Then, using the methods learned in Chapter 2, we solve the resulting equation for the other variable.

**Example 7** Find the missing coordinate to complete the following ordered-pair solutions to the equation $2x + 3y = 15$.

**(a)** $(0, ?)$ **(b)** $(?, 1)$

**Solution**

**(a)** For the ordered pair $(0, ?)$, we know that $x = 0$. Replace $x$ by 0 in the equation and solve for $y$.

$$2x + 3y = 15$$
$$2(0) + 3y = 15 \quad \text{Replace } x \text{ with 0.}$$
$$0 + 3y = 15 \quad \text{Simplify.}$$
$$y = 5 \quad \text{Divide both sides by 3.}$$

Thus we have the ordered pair $(0, 5)$.

**(b)** For the ordered pair $(?, 1)$, we *do not know* the value of $x$. However, we do know that $y = 1$. So we start by replacing the variable $y$ by 1. We will end up with an equation with one variable, $x$. We can then solve for $x$.

$$2x + 3y = 15$$
$$2x + 3(1) = 15 \quad \text{Replace } y \text{ with 1.}$$
$$2x + 3 = 15 \quad \text{Simplify.}$$
$$2x = 12 \quad \text{Isolate the variable term.}$$
$$x = 6 \quad \text{Solve for } x.$$

Thus we have the ordered pair $(6, 1)$. ☐

**Student Practice 7** Find the missing coordinate to complete the following ordered-pair solutions to the equation $3x - 4y = 12$.

**(a)** $(0, ?)$ **(b)** $(?, 3)$ **(c)** $(?, -6)$

**Verbal and Writing Skills, Exercises 1–6**

*Unless otherwise indicated, assume each grid line represents one unit.*

1. What is the $x$-coordinate of the origin?

2. What is the $y$-coordinate of the origin?

3. Explain why (5, 1) is referred to as an *ordered* pair of numbers.

4. Explain how you would locate the point (4, 3) on graph paper.

5. Explain why the ordered pairs (2, 7) and (7, 2) do not represent the same point on a graph.

6. The equation $x + y = 10$ has how many ordered-pair solutions?

7. Plot the following points.
   $J: (-4, 3.5)$   $K: (6, 0)$   $L: (5, -6)$
   $M: (0, -4)$   $N: (3, 4)$   $P: (-6, 5)$

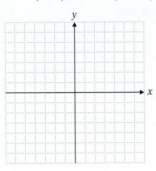

8. Plot the following points.
   $R: (-3, 0)$   $S: (3.5, 4)$   $T: (-2, -2.5)$
   $V: (0, 5)$   $W: (3, 0)$   $X: (2, -4)$

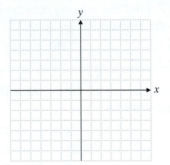

*Consider the points plotted on the graph at right.*

9. Give the coordinates for points $R$, $S$, $X$, and $Y$.

10. Give the coordinates for points $T$, $V$, $W$, and $Z$.

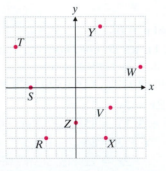

*In exercises 11 and 12, six points are plotted in each figure. List all the ordered pairs needed to represent the points.*

11.

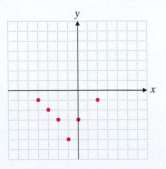

12.

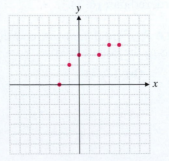

*Using Road Maps* *The map below shows a portion of New York, Connecticut, and Massachusetts. Like many maps used in driving or flying, it has horizontal and vertical grid markers for ease of use. For example, Newburgh, New York, is located in grid B3. Use the grid labels to indicate the locations of the following cities.*

**13.** Lynbrook, New York

**14.** Hampton Bays, New York

**15.** Athol, Massachusetts

**16.** Pittsfield, Massachusetts

**17.** Hartford, Connecticut

**18.** Waterbury, Connecticut

**19.** *DVD Movie Shipments* According to a large movie DVD manufacturer, the number of DVDs shipped by the manufacturer decreased significantly from 2008 to 2015. The number of DVDs shipped during these years is recorded in the following table and is measured in thousands. For example, 803 in the second column means 803 thousand, or 803,000.

  **(a)** Plot points that represent this data on the given rectangular coordinate system.

  **(b)** What trends are apparent from the plotted data?

| Number of Years Since 2008 | Number of DVDs Shipped (in Thousands) |
|:---:|:---:|
| 0 | 803 |
| 1 | 746 |
| 2 | 767 |
| 3 | 705 |
| 4 | 615 |
| 5 | 511 |
| 6 | 368 |
| 7 | 293 |

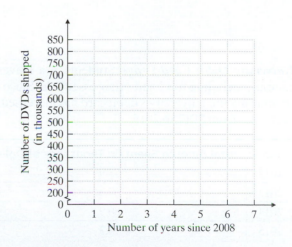

**20.** *Customers Visiting a Restaurant Chain* The number of customers visiting a large restaurant chain per year in the United States for selected years starting in 1990 is recorded in the following table. The number of customers is measured in millions. For example, 15 in the second column means 15 million, or 15,000,000 customers.

  **(a)** Plot points that represent the data on the given rectangular coordinate system.

  **(b)** What trends are apparent from the plotted data?

| Number of Years Since 1990 | Customers Visiting Restaurant Chain per year (in Millions of customers) |
|---|---|
| 0 | 15 |
| 5 | 20 |
| 10 | 21 |
| 15 | 22 |
| 20 | 22 |
| 25 | 23 |

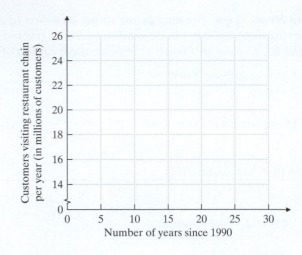

**21.** *Buying Books Online* A large university's bookstore decided to offer students the option of buying their books online. A review of online sales at the end of the fourth year indicated that online sales have increased at a significant rate. The following chart records the amount spent each year for Year 1 through Year 4.

| Year | Total Amount Spent (in Millions of Dollars) |
|---|---|
| 1 | 3.2 |
| 2 | 3.6 |
| 3 | 3.9 |
| 4 | 4.2 |

**(a)** Plot points that represent the data on the given rectangular coordinate system.

**(b)** Based on your graph, estimate the amount students will spend buying books online in year 5.

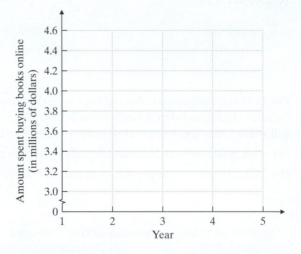

**22.** *University Debit Card Usage* Students at a university can use a prepaid debit card to make all their purchases on campus, including food and other essentials. Since the university began this program in 2012, the amount of money spent using the university's debit card has increased at a significant rate. The following chart records the amount spent each year from 2012 to 2015.

| Year | Total Amount Spent (in Thousands of Dollars) |
|---|---|
| 2012 | 53.9 |
| 2013 | 67.3 |
| 2014 | 80.9 |
| 2015 | 95.3 |

**(a)** Plot points that represent the data on the given rectangular coordinate system.

**(b)** Based on your graph, estimate the amount students spend using their debit cards in 2016.

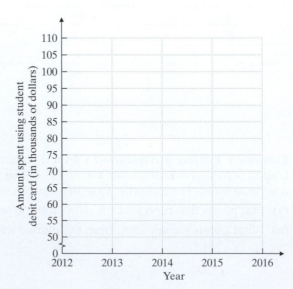

*Are the following ordered pairs solutions to the equation $2x - y = 6$?*

**23.** $(1, 0)$      **24.** $(6, 2)$      **25.** $(2, -2)$      **26.** $(0, -6)$

## To Think About, Exercises 27 and 28

**27.** After an exam, Jon and Syed discussed their answers to the following exam question. Which of the following is an ordered-paired solution to the equation $2x - 2y = 4$, $(3, 5)$ or $(5, 3)$? Jon said the answer was $(5, 3)$, and Syed stated that $(3, 5)$ is the correct answer. Who is right? Why?

**28.** On an exam, students were asked to write an ordered-pair solution for the equation $x + y = 18$. Damien, Jennifer, and Tanya compared their answers after the exam and discovered they each had different answers. Is it possible for each of them to be correct? Explain.

*Find the missing coordinate to complete the ordered-pair solution to the given linear equation.*

**29.** $y = 4x + 7$
    **(a)** $(0, \ )$
    **(b)** $(2, \ )$

**30.** $y = 6x + 5$
    **(a)** $(0, \ )$
    **(b)** $(3, \ )$

**31.** $y + 6x = 5$
    **(a)** $(-1, \ )$
    **(b)** $(3, \ )$

**32.** $y + 4x = 9$
    **(a)** $(-2, \ )$
    **(b)** $(5, \ )$

**33.** $3x - 4y = 11$
    **(a)** $(-3, \ )$
    **(b)** $( \ , 1)$

**34.** $5x - 2y = 9$
    **(a)** $(7, \ )$
    **(b)** $( \ , -7)$

**35.** $3x + 2y = -6$
    **(a)** $(-2, \ )$
    **(b)** $( \ , 3)$

**36.** $5x + 4y = 10$
    **(a)** $(-2, \ )$
    **(b)** $( \ , 0)$

**37.** $y - 1 = \dfrac{2}{7}x$
    **(a)** $(7, \ )$
    **(b)** $\left( \ , \dfrac{5}{7} \right)$

**38.** $x - 2 = \dfrac{5}{6}y$
    **(a)** $(7, \ )$
    **(b)** $\left( \ , \dfrac{6}{5} \right)$

**39.** $3x + \dfrac{1}{2}y = 7$
    **(a)** $( \ , 2)$
    **(b)** $\left( \dfrac{3}{2}, \ \right)$

**40.** $3x + \dfrac{1}{4}y = 11$
    **(a)** $( \ , 8)$
    **(b)** $\left( \dfrac{13}{4}, \ \right)$

## Cumulative Review

▲ **41.** **[1.8.2]** *Circular Swimming Pool* The circular pool at the hotel where Bob and Linda stayed in Orlando, Florida, has a radius of 19 yards. What is the area of the pool? (Use $\pi \approx 3.14$.)

**42.** **[3.2.1]** A number is doubled and then decreased by three. The result is twenty-one. What is the original number?

**43.** **[5.6.1]** Factor completely. $8x^2 - 18$

**44.** **[5.6.1]** Factor completely. $3x^2 + 9x - 54$

## Quick Quiz 7.1

1. Plot and label the following points.

   $A$: $(3, -4)$

   $B$: $(-6, -2)$

   $C$: $(0, 5)$

   $D$: $(-3, 6)$

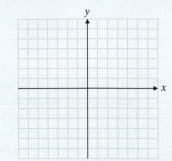

*Find the missing coordinate to complete the ordered-pair solution to the given linear equation.*

2. $y = -5x - 7$

   **(a)** $(-2, \ )$

   **(b)** $(3, \ )$

   **(c)** $(0, \ )$

3. $4x - 3y = -12$

   **(a)** $(3, \ )$

   **(b)** $( \ , -8)$

   **(c)** $( \ , 10)$

4. **Concept Check** Explain how you would find the missing coordinate to complete the ordered-pair solution to the equation $2.5x + 3y = 12$ if the ordered pair was of the form $( \ , -6)$.

# 7.2 Graphing Linear Equations ▶

## 1 Graphing a Linear Equation by Plotting Three Ordered Pairs ▶

We have seen that a solution to a linear equation in two variables is an ordered pair. The graph of an ordered pair is a point. Thus we can graph an equation by graphing the points corresponding to its ordered-pair solutions.

A linear equation in two variables has an infinite number of ordered-pair solutions. We can see that this is true by noting that we can substitute any number for $x$ in the equation and solve it to obtain a $y$-value. For example, if we substitute $x = 0, 1, 2, 3, \ldots$ into the equation $y = -x + 3$ and solve for $y$, we obtain the ordered-pair solutions $(0, 3), (1, 2), (2, 1), (3, 0), \ldots$. (If desired, substitute these values into the equation to convince yourself.) If we plot these points on a rectangular coordinate system, we notice that they fall on a straight line, as illustrated in the margin.

It turns out that all of the points corresponding to the ordered-pair solutions of $y = -x + 3$ lie on this line, and the line extends forever in both directions. A similar statement can be made about any linear equation in two variables.

> The graph of any linear equation in two variables is a straight line.

From geometry, we know that two points determine a line. Thus to graph a linear equation in two variables, we need to graph only two ordered-pair solutions of the equation and then draw the line that passes through them. Having said this, we recommend that you use three points to graph a line. Two points will determine where the line is. The third point verifies that you have drawn the line correctly. For ease in plotting, it is better if the ordered pairs contain integers.

> **TO GRAPH A LINEAR EQUATION**
> 1. Find three ordered pairs that are solutions to the equation.
> 2. Plot the points.
> 3. Draw a line through the points.

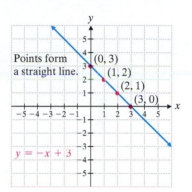

**Example 1** Find three ordered pairs that satisfy $y = -2x + 4$. Then graph the resulting straight line.

**Solution** Since we can choose any value for $x$, we choose numbers that are convenient. To organize the results, we will make a table of values. We will let $x = 0$, $x = 1$, and $x = 3$. We write these numbers under $x$ in our table of values. For each of these $x$-values, we find the corresponding $y$-value in the equation $y = -2x + 4$.

$$y = -2x + 4 \qquad y = -2x + 4 \qquad y = -2x + 4$$
$$y = -2(0) + 4 \qquad y = -2(1) + 4 \qquad y = -2(3) + 4$$
$$y = 0 + 4 \qquad y = -2 + 4 \qquad y = -6 + 4$$
$$y = 4 \qquad y = 2 \qquad y = -2$$

We record these results by placing each $y$-value in the table next to its corresponding $x$-value. Keep in mind that these values represent ordered pairs, each of which is a solution to the equation. If we plot these ordered pairs and connect the three points, we

| Table of Values | |
|:---:|:---:|
| **x** | **y** |
| 0 | 4 |
| 1 | 2 |
| 3 | −2 |

▶ **Student Practice 1** Find three ordered pairs that satisfy $y = -3x - 1$. Then graph the resulting straight line. Use the given coordinate system.

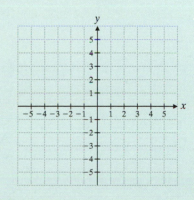

*Continued on next page*

get a straight line that is the graph of the equation $y = -2x + 4$. The graph of the equation is shown in the figure below.

Find the point $(2, 0)$ on the graph. Is it on the line? Check to verify that it is a solution to $y = -2x + 4$. Find another solution by *locating* another point on the line.

**Table of Values**

| x | y |
|---|---|
| 0 | 4 |
| 1 | 2 |
| 3 | −2 |

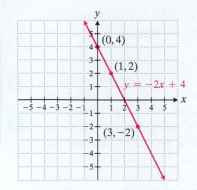

### Sidelight: Determining Values for Ordered Pairs

Why can we choose any value for either $x$ or $y$ when we are finding ordered pairs that are solutions to an equation?

The numbers for $x$ and $y$ that are solutions to the equation come in pairs. When we try to find a pair that fits, we have to start with some number. Usually, it is an $x$-value that is small and easy to work with. Then we must find the value for $y$ so that the pair of numbers $(x, y)$ is a solution to the given equation.

### Example 2 Graph $5x - 4y + 2 = 2$.

**Solution** First, we simplify the equation $5x - 4y + 2 = 2$ by subtracting 2 from each side.

$$5x - 4y + 2 - 2 = 2 - 2$$
$$5x - 4y = 0$$

Since we are free to choose any value of $x$, $x = 0$ is a natural choice. Calculate the value of $y$ when $x = 0$ and place the results in the table of values.

**Table of Values**

| x | y |
|---|---|
| 0 | 0 |
|   |   |
|   |   |

$$5(0) - 4y = 0$$
$$-4y = 0 \quad \text{Remember: Any number times 0 is 0.}$$
$$y = 0 \quad \text{Since } -4y = 0, y \text{ must equal 0.}$$

Now let's see what happens when $x = 1$.

$$5(1) - 4y = 0$$
$$5 - 4y = 0$$
$$-4y = -5$$
$$y = \frac{-5}{-4} \quad \text{or} \quad \frac{5}{4} \quad \text{This is not an easy number to graph.}$$

A better choice for a replacement of $x$ is a number that is divisible by 4. Let's see why. Let $x = 4$ and let $x = -4$.

$$5(4) - 4y = 0 \qquad\qquad 5(-4) - 4y = 0$$
$$20 - 4y = 0 \qquad\qquad -20 - 4y = 0$$
$$-4y = -20 \qquad\qquad -4y = 20$$
$$y = \frac{-20}{-4} \quad \text{or} \quad 5 \qquad\qquad y = \frac{20}{-4} \quad \text{or} \quad -5$$

Now we can put these numbers into our table of values and graph the line.

| Table of Values | |
|---|---|
| **x** | **y** |
| 0 | 0 |
| 4 | 5 |
| −4 | −5 |

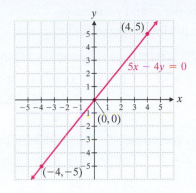

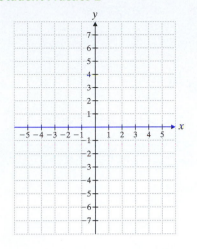

**Student Practice 2** Graph $7x + 3 = -2y + 3$ on the coordinate system in the margin.

**To Think About:** *An Alternative Approach*  In Example 2, when we picked the value of 1 for $x$, we found that the corresponding value for $y$ was a fraction. To avoid fractions, we can solve the equation for the variable $y$ *first*, then choose the values for $x$.

$$5x - 4y = 0 \qquad \text{We must isolate } y.$$

$$-4y = -5x \qquad \text{Subtract } 5x \text{ from each side.}$$

$$\frac{-4y}{-4} = \frac{-5x}{-4} \qquad \text{Divide each side by } -4.$$

$$y = \frac{5}{4}x$$

Now let $x = -4$, $x = 0$, and $x = 4$, and find the corresponding values of $y$. Explain why you would choose multiples of 4 as replacements for $x$ in this equation. Graph the equation and compare it to the graph in Example 2.

In the previous two examples we began by picking values for $x$. We could just as easily have chosen values for $y$.

## 2 Graphing a Straight Line by Plotting Its Intercepts

What values should we pick for $x$ and $y$? Which points should we use for plotting? For many straight lines it is easiest to pick the two *intercepts*. Some lines have only one intercept. We will discuss these separately.

> The **x-intercept** of a line is the point where the line crosses the $x$-axis; it has the form $(a, 0)$. The **y-intercept** of a line is the point where the line crosses the $y$-axis; it has the form $(0, b)$.

### INTERCEPT METHOD OF GRAPHING

To graph an equation using intercepts, we:

1. Find the $x$-intercept by letting $y = 0$ and solving for $x$.

2. Find the $y$-intercept by letting $x = 0$ and solving for $y$.

3. Find one additional ordered pair so that we have three points with which to plot the line.

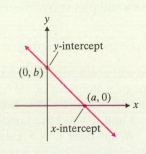

**Example 3** Complete **(a)** and **(b)** for the equation $5y - 3x = 15$.

**(a)** State the $x$- and $y$-intercepts.

**(b)** Use the intercept method to graph.

**Solution**   Substitute values in the equation $5y - 3x = 15$.

**(a)** Let $y = 0$.     $5(0) - 3x = 15$    Replace $y$ by 0.

$-3x = 15$    Simplify.

$x = -5$    Divide both sides by $-3$.

| x | y |
|---|---|
| $-5$ | 0 |

$x$-intercept

The ordered pair $(-5, 0)$ is the $x$-intercept.

Let $x = 0$.     $5y - 3(0) = 15$    Replace $x$ by 0.

$5y = 15$    Simplify.

$y = 3$    Divide both sides by 5.

| x | y |
|---|---|
| 0 | 3 |

$y$-intercept

The ordered pair $(0, 3)$ is the $y$-intercept.

**(b)** We find another ordered pair to have a third point and then graph.

Let $y = 6$.     $5(6) - 3x = 15$              Replace $y$ by 6.

$30 - 3x = 15$              Simplify.

$-3x = -15$              Subtract 30 from both sides.

$$x = \frac{-15}{-3} \quad \text{or} \quad 5$$

The ordered pair is $(5, 6)$.
Our table of values is

| x | y |
|---|---|
| $-5$ | 0 |
| 0 | 3 |
| 5 | 6 |

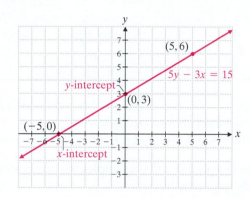

**Student Practice 3**

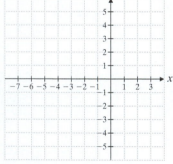

**CAUTION:** The three points on the graph must form a straight line. If the three points do not form a straight line, you made a calculation error.      ☐

 **Student Practice 3**   Use the intercept method to graph $2y - x = 6$. Use the given coordinate system.

**To Think About:** *Lines That Go Through the Origin* Can you draw all straight lines by the intercept method? Not really. Some straight lines may go through the origin and have only one intercept. If a line goes through the origin, it will have an equation of the form $Ax + By = 0$, where $A \neq 0$ or $B \neq 0$ or both. Refer to Example 2. When we simplified the equation, we obtained $5x - 4y = 0$. Notice that the graph goes through the origin, and thus there is only one intercept. In such cases you should plot two additional points besides the origin.

**3** Graphing Horizontal and Vertical Lines   ▶

You will notice that the $x$-axis is a horizontal line. It is the line $y = 0$, since for any value of $x$, the value of $y$ is 0. Try a few points. The points $(1, 0)$, $(3, 0)$, and $(-2, 0)$ all lie on the $x$-axis. Any horizontal line will be parallel to the $x$-axis. Lines such as $y = 5$ and $y = -2$ are horizontal lines.

What does $y = 5$ mean? It means that for any value of $x$, $y$ is 5. Likewise $y = -2$ means that for any value of $x$, $y = -2$.

How can we recognize the equation of a line that is horizontal, that is, parallel to the $x$-axis?

> If the graph of an equation is a straight line that is parallel to the $x$-axis (that is, a horizontal line), the equation will be of the form $y = b$, where $b$ is some real number.

**Example 4** Graph $y = -3$.

**Solution**   A solution to $y = -3$ is any ordered pair that has $y$-coordinate $-3$. The $x$-coordinate can be any number as long as $y$ is $-3$.

$(0, -3)$, $(-3, -3)$, and $(4, -3)$ are solutions to $y = -3$ since all the **$y$-values are $-3$**.

Since the $y$-coordinate of every point on the line is $-3$, it is easy to see that the horizontal line will be 3 units below the $x$-axis.

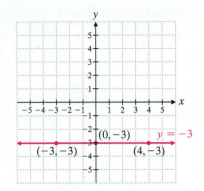

*Note:* You could write the equation $y = -3$ as $0x + y = -3$. Then it is clear that for any value of $x$, you will always obtain $y = -3$. Try it and see.   □

**Student Practice 4**   Graph $y = 2$ on the given coordinate system.

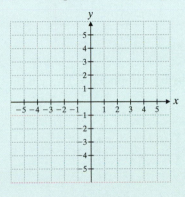

Notice that the $y$-axis is a vertical line. This is the line $x = 0$, since for any $y$, $x$ is 0. Try a few points. The points $(0, 2)$, $(0, -3)$, and $(0, \frac{1}{2})$ all lie on the $y$-axis. Any vertical line will be parallel to the $y$-axis. Lines such as $x = 2$ and $x = -3$ are vertical lines.

Think of what $x = 2$ means. It means that for any value of $y$, $x$ is 2. The graph of $x = 2$ is a vertical line two units to the right of the $y$-axis.

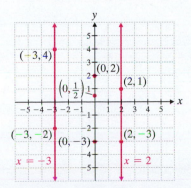

How can we recognize the equation of a line that is vertical, that is, parallel to the $y$-axis?

If the graph of an equation is a straight line that is parallel to the $y$-axis (that is, a vertical line), the equation will be of the form $x = a$, where $a$ is some real number.

**Example 5** Graph $2x + 1 = 11$.

**Solution** Notice that there is only one variable, $x$, in the equation. This is an indication that we can simplify the equation to the form $x = a$.

$$2x + 1 = 11 \quad \text{$x$ is the only variable in the equation.}$$
$$2x = 10 \quad \text{Solve for $x$.}$$
$$x = 5$$

Since the $x$-coordinate of every point on this line is 5, we can see that the vertical line will be 5 units to the right of the $y$-axis.

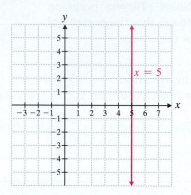

**Student Practice 5** Graph $3x + 1 = -8$ on the following coordinate system.

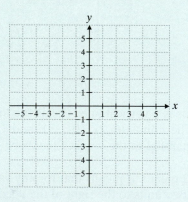

## Verbal and Writing Skills, Exercises 1–4

*Unless otherwise indicated, assume each grid line represents one unit.*

1. Is the point $(-2, 5)$ a solution to the equation $2x + 5y = 0$? Why or why not?

2. The graph of a linear equation in two variables is a _____ _____.

3. The $x$-intercept of a line is the point where the line crosses the _____.

4. The graph of the equation $y = b$ is a _____ line.

*Complete the ordered pairs so that each is a solution of the given linear equation. Then graph the equation by plotting each solution and connecting the points by a straight line.*

5. $y = x - 4$
   $(0, \ )$
   $(2, \ )$
   $(4, \ )$

6. $y = x + 4$
   $(-1, \ )$
   $(0, \ )$
   $(1, \ )$

7. $y = -2x + 1$
   $(0, \ )$
   $(-2, \ )$
   $(1, \ )$

8. $y = -3x - 4$
   $(-2, \ )$
   $(-1, \ )$
   $(0, \ )$

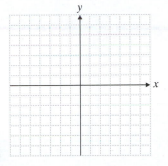
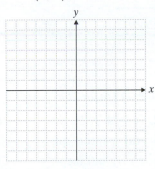
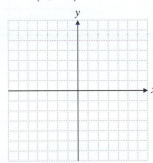

9. $y = 3x - 1$
   $(0, \ )$
   $(2, \ )$
   $(-1, \ )$

10. $y = -2x + 3$
    $(0, \ )$
    $(2, \ )$
    $(4, \ )$

11. $y = 2x - 5$
    $(0, \ )$
    $(2, \ )$
    $(4, \ )$

12. $y = 3x + 2$
    $(-1, \ )$
    $(0, \ )$
    $(1, \ )$

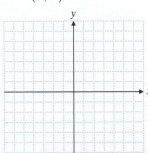

*Graph each equation by plotting three points and connecting them. Use a table of values to organize the ordered pairs.*

13. $y = -x + 3$

14. $y = -3x + 2$

15. $3x - 2y = 0$

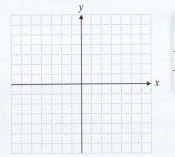
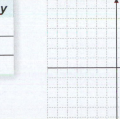

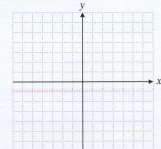

**16.** $2y - 5x = 0$

**17.** $y = -\dfrac{3}{4}x + 3$

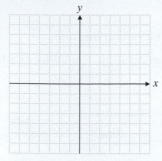

**18.** $y = \dfrac{2}{3}x + 2$

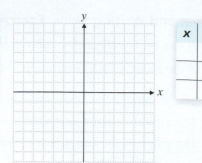

**19.** $4x + 6 + 3y = 18$

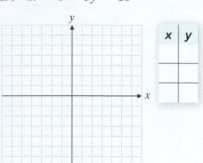

**20.** $-2x + 3 + 4y = 15$

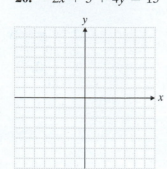

*For exercises 21–24: (**a**) state the x- and y-intercepts; (**b**) use the intercept method to graph.*

**21.** $y = 6 - 2x$

(**a**) $x$-intercept: _____

$y$-intercept: _____

(**b**)

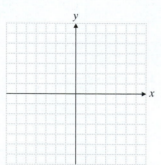

**22.** $y = 4 - 2x$

(**a**) $x$-intercept: _____

$y$-intercept: _____

(**b**)

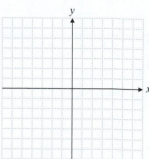

**23.** $x + 3 = 6y$

(**a**) $x$-intercept: _____

$y$-intercept: _____

(**b**)

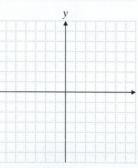

**24.** $x - 6 = 2y$

(**a**) $x$-intercept: _____

$y$-intercept: _____

(**b**)

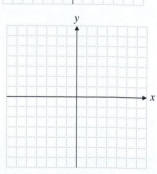

*Graph the equation. Be sure to simplify the equation before graphing it.*

**25.** $x = 4$

**26.** $y = -4$

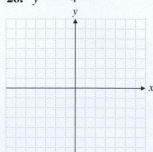

**27.** $y - 2 = 3y$

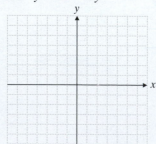

**28.** $3x - 4 = -13$

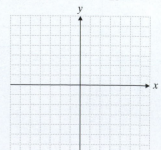

**Mixed Practice** *Graph.*

**29.** $2x + 5y - 2 = -12$

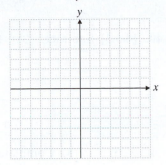

**30.** $3x - 4y - 5 = -17$

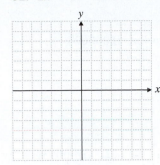

**31.** $2x + 9 = 5x$

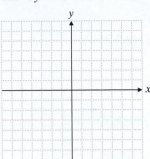

**32.** $3y + 1 = 7$

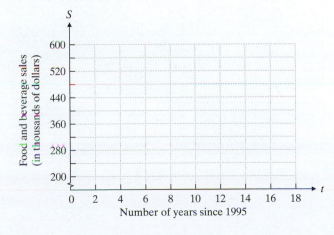

## Applications

**33.** *Cross-Country Skiing* The number of calories burned by an average person while cross-country skiing is given by the equation $C = 8m$, where $m$ is the number of minutes. (*Source:* National Center for Health Statistics.) Graph the equation for $m = 0, 15, 30, 45, 60,$ and 75.

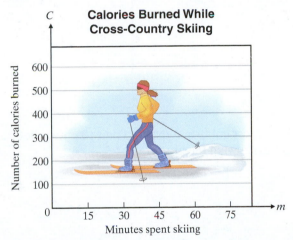

**34.** *Calories Burned While Jogging* The number of calories burned by an average person while jogging is given by the equation $C = \frac{28}{3}m$, where $m$ is the number of minutes. (*Source:* National Center for Health Statistics.) Graph the equation for $m = 0, 15, 30, 45, 60,$ and 75.

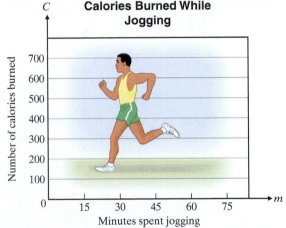

**35.** *Foreign Students in the United States* The number of foreign students enrolled in college in the United States can be approximated by the equation $S = 11t + 395$, where $t$ stands for the number of years since 1990, and $S$ is the number of foreign students (in thousands). (*Source:* www.opendoors.iienetwork.org.) Graph the equation for $t = 0, 4, 8, 16$.

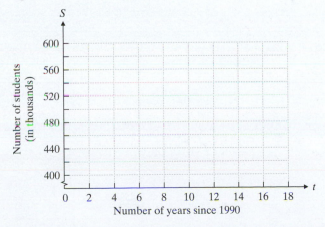

**36.** *Food and Beverage Sales* The amount of money spent at a popular amusement park on food and beverages can be approximated by the equation $S = 16.8t + 223$, where $S$ is the sales in thousands of dollars and $t$ is the number of years since 1995. Graph the equation for $t = 0, 6, 10, 14, 18$.

## Cumulative Review

**37.** **[2.3.3]** Solve. $2(x + 3) + 5x = 3x - 2$

**38.** **[4.2.1]** Simplify. $\left(\dfrac{3x^2}{2y}\right)^{-2}$

**39.** **[4.2.2]** Write in scientific notation. $0.000078$

**40.** **[4.3.3]** Subtract. $(3x^2 + 5x) - (x^2 - x + 4)$

## Quick Quiz 7.2

**1.** Graph $4y + 1 = x + 9$.

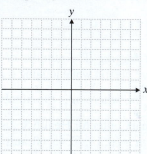

**2.** Graph $3y = 2y + 4$.

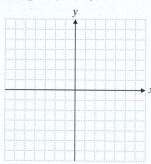

**3.** Complete (**a**) and (**b**) for the equation $y = -2x + 4$.

   (**a**) State the $x$- and $y$-intercepts.

   (**b**) Use the intercept method to graph.

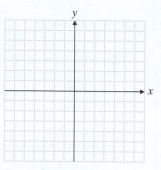

**4.** **Concept Check** In graphing the equation $3y - 7x = 0$, what is the most important ordered pair to obtain before drawing a graph of the line? Why is that ordered pair so essential to drawing the graph?

## 7.3 The Slope of a Line ▶

### 1 Finding the Slope of a Line Given Two Points on the Line ▶

We often use the word *slope* to describe the incline (the steepness) of a hill. A carpenter or a builder will refer to the *pitch* or *slope* of a roof. The slope is the change in the vertical distance (the rise) compared to the change in the horizontal distance (the run) as you go from one point to another point along the roof. If the change in the vertical distance is greater than the change in the horizontal distance, the slope will be steep. If the change in the horizontal distance is greater than the change in the vertical distance, the slope will be gentle.

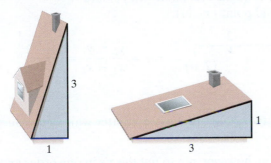

**Student Learning Objectives**

After studying this section, you will be able to:

1 Find the slope of a line given two points on the line. ▶

2 Find the slope and $y$-intercept of a line given its equation. ▶

3 Write the equation of a line given the slope and $y$-intercept. ▶

4 Graph a line using the slope and $y$-intercept. ▶

5 Find the slopes of parallel and perpendicular lines. ▶

In a coordinate plane, the **slope** of a straight line is defined by the change in $y$ divided by the change in $x$.

$$\text{slope} = \frac{\text{change in } y}{\text{change in } x}$$

Consider the line drawn through points $A$ and $B$ in the figure. If we measure the change from point $A$ to point $B$ in the $x$-direction and the $y$-direction, we will have an idea of the steepness (or the slope) of the line. From point $A$ to point $B$, the change in $y$-values is from 2 to 4, a *change of* 2. From point $A$ to point $B$, the change in $x$-values is from 1 to 5, a *change of* 4. Thus

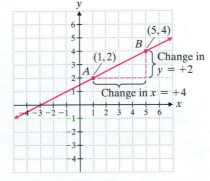

$$\text{slope} = \frac{\text{change in } y}{\text{change in } x} = \frac{2}{4} = \frac{1}{2}.$$

Informally, we can describe this move as the rise over the run: $\text{slope} = \dfrac{\text{rise}}{\text{run}}$. We now state a more formal (and more frequently used) definition.

**DEFINITION OF SLOPE OF A LINE**

The **slope** of any *nonvertical* straight line that contains the points with coordinates $(x_1, y_1)$ and $(x_2, y_2)$ is defined by the difference ratio

$$\text{slope} = m = \frac{y_2 - y_1}{x_2 - x_1} \qquad \text{where } x_2 \neq x_1.$$

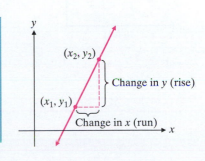

The use of subscripted terms such as $x_1, x_2$, and so on, is just a way of indicating that the first $x$-value is $x_1$ and the second $x$-value is $x_2$. Thus $(x_1, y_1)$ are the coordinates of the first point and $(x_2, y_2)$ are the coordinates of the second point. The letter $m$ is commonly used for the slope.

**Example 1** Find the slope of the line that passes through $(2, 0)$ and $(4, 2)$.

**Solution** Let $(2, 0)$ be the first point $(x_1, y_1)$ and $(4, 2)$ be the second point $(x_2, y_2)$.

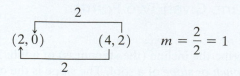

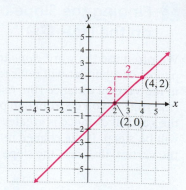

$$\text{slope} = m = \frac{y_2 - y_1}{x_2 - x_1} = \frac{2 - 0}{4 - 2} = \frac{2}{2} = 1.$$

Note that the slope of the line will be the same if we let $(4, 2)$ be the first point $(x_1, y_1)$ and $(2, 0)$ be the second point $(x_2, y_2)$.

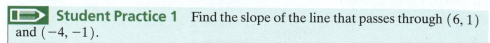

$$m = \frac{y_2 - y_1}{x_2 - x_1} = \frac{0 - 2}{2 - 4} = \frac{-2}{-2} = 1$$

Thus, given two points, it does not matter which you call $(x_1, y_1)$ and which you call $(x_2, y_2)$. ∎

**CAUTION:** Be careful, however, not to put the $x$'s in one order and the $y$'s in another order when finding the slope from two points on a line.

**Student Practice 1** Find the slope of the line that passes through $(6, 1)$ and $(-4, -1)$.

It is a good idea to have some concept of what different slopes mean. In downhill skiing, a very gentle slope used for teaching beginning skiers might drop 1 foot vertically for each 10 feet horizontally. The slope would be $\frac{1}{10}$. The speed of a skier on a hill with such a gentle slope would be only about 6 miles per hour.

A triple diamond slope for experts might drop 11 feet vertically for each 10 feet horizontally. The slope would be $\frac{11}{10}$. The speed of a skier on such an expert trail would be in the range of 60 miles per hour.

It is important to see how positive and negative slopes affect the graphs of lines.

Slope $= \frac{1}{10}$
1 foot
10 feet

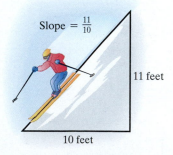
Slope $= \frac{11}{10}$
11 feet
10 feet

POSITIVE SLOPE                    NEGATIVE SLOPE

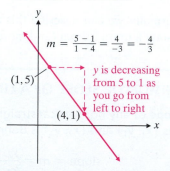

$m = \frac{5 - 2}{4 - 1} = \frac{3}{3} = 1$  (positive slope graph: $(4,5)$, $(1,2)$; $y$ is increasing from 2 to 5 as you go from left to right)

$m = \frac{5 - 1}{1 - 4} = \frac{4}{-3} = -\frac{4}{3}$  (negative slope graph: $(1,5)$, $(4,1)$; $y$ is decreasing from 5 to 1 as you go from left to right)

**1.** If the $y$-values increase as you go from left to right, the slope of the line is positive.

**2.** If the $y$-values decrease as you go from left to right, the slope of the line is negative.

**Example 2** Find the slope of the line that passes through $(-3, 2)$ and $(2, -4)$.

**Solution**  Let $(-3, 2)$ be $(x_1, y_1)$ and $(2, -4)$ be $(x_2, y_2)$.

$$m = \frac{y_2 - y_1}{x_2 - x_1} = \frac{-4 - 2}{2 - (-3)} = \frac{-4 - 2}{2 + 3} = \frac{-6}{5} = -\frac{6}{5}$$

The slope of this line is negative. We would expect this, since the $y$-value decreased from 2 to $-4$ as the $x$-value increased. What does the graph of this line look like? Plot the points and draw the line to verify. $\quad\square$

**Student Practice 2**  Find the slope of the line that passes through $(2, 0)$ and $(-1, 1)$.

**To Think About:** *Using the Slope to Describe a Line* Describe the line in Student Practice 2 by looking at its slope. Then verify by drawing the graph.

**Example 3** Find the slope of the line that passes through the given points.

**(a)** $(0, 2)$ and $(5, 2)$        **(b)** $(-4, 0)$ and $(-4, -4)$

**Solution**

**(a)** Take a moment to look at the $y$-values. What do you notice? What does this tell you about the line? Now calculate the slope.

$(0, 2)$ and $(5, 2)$      $m = \dfrac{2 - 2}{5 - 0} = \dfrac{0}{5} = 0$

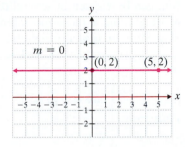

Since any two points on a horizontal line will have the same $y$-value, the slope of a horizontal line is 0.

**(b)** Take a moment to look at the $x$-values. What do you notice? What does this tell you about the line? Now calculate the slope.

$(-4, 0)$ and $(-4, -4)$    $m = \dfrac{-4 - 0}{-4 - (-4)} = \dfrac{-4}{0}$

Recall that division by 0 is undefined. The slope of a vertical line is undefined. We say that a vertical line has **no slope.**

Notice in our definition of slope that $x_2 \neq x_1$. Thus it is not appropriate to use the formula for slope for the points in part **(b)**. We did so to illustrate what would happen if $x_2 = x_1$. We get an impossible situation, $\dfrac{y_2 - y_1}{0}$. Now you can see why we include the restriction $x_2 \neq x_1$ in our definition. $\quad\square$

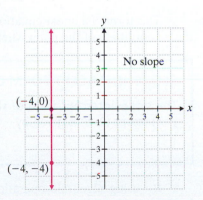

**Student Practice 3**  Find the slope of the line that passes through the given points.

**(a)** $(-5, 6)$ and $(-5, 3)$        **(b)** $(-7, -11)$ and $(3, -11)$

### SLOPE OF A STRAIGHT LINE

1. Positive slope
   Line goes upward to the right

   *Lines with positive slopes go upward as* you go from left to right.

2. Negative slope
   Line goes downward to the right

   *Lines with negative slopes go downward as* you go from left to right.

3. Zero slope
   Horizontal line

   Horizontal lines have a slope of 0.

4. Undefined slope
   Vertical line

   A vertical line is said to have undefined slope. The slope of a vertical line is not defined. In other words, a vertical line has no slope.

## 2 Finding the Slope and $y$-Intercept of a Line Given Its Equation

Recall that the equation of a line is a linear equation in two variables. This equation can be written in several different ways. A very useful form of the equation of a straight line is the slope–intercept form. This form can be derived in the following way. Suppose that a straight line with slope $m$ crosses the $y$-axis at a point $(0, b)$. Consider any other point on the line and label the point $(x, y)$. Then we have the following.

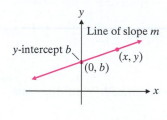

$$\frac{y_2 - y_1}{x_2 - x_1} = m \qquad \text{Definition of slope.}$$

$$\frac{y - b}{x - 0} = m \qquad \text{Substitute } (0, b) \text{ for } (x_1, y_1) \text{ and } (x, y) \text{ for } (x_2, y_2).$$

$$\frac{y - b}{x} = m \qquad \text{Simplify.}$$

$$y - b = mx \qquad \text{Multiply both sides by } x.$$

$$y = mx + b \qquad \text{Add } b \text{ to both sides.}$$

This form of a linear equation immediately reveals the slope of the line, $m$, and the $y$-coordinate of the point where the line intercepts (crosses) the $y$-axis, $b$.

### SLOPE–INTERCEPT FORM OF A LINE

The slope–intercept form of the equation of the line that has slope $m$ and $y$-intercept $(0, b)$ is given by

$$y = mx + b.$$

By using algebraic operations, we can write any linear equation in slope–intercept form and use this form to identify the slope and the $y$-intercept of the line.

**Example 4** What is the slope and the $y$-intercept of the line $5x + 3y = 2$?

**Solution** We want to solve for $y$ and get the equation in the form $y = mx + b$. We need to isolate the $y$-variable.

$$5x + 3y = 2$$

$$3y = -5x + 2 \qquad \text{Subtract } 5x \text{ from both sides.}$$

$$y = \frac{-5x + 2}{3} \qquad \text{Divide both sides by 3.}$$

$$y = -\frac{5}{3}x + \frac{2}{3}$$ Using the property $\frac{a+b}{c} = \frac{a}{c} + \frac{b}{c}$, write the right-hand side as two fractions.

$$m = -\frac{5}{3} \text{ and } b = \frac{2}{3}$$ The *slope* is $-\frac{5}{3}$.    The *y-intercept* is $\left(0, \frac{2}{3}\right)$.

*Note:* We write the $y$-intercept as an ordered pair of the form $(0, b)$. ☐

**Student Practice 4** What is the slope and the $y$-intercept of the line $4x - 2y = -5$?

### 3 Writing the Equation of a Line Given the Slope and y-Intercept ▶

If we know the slope of a line and the $y$-intercept, we can write the equation of the line, $y = mx + b$.

**Example 5** Find an equation of the line with slope $\frac{2}{5}$ and $y$-intercept $(0, -3)$.

**(a)** Write the equation in slope–intercept form, $y = mx + b$.

**(b)** Write the equation in the form $Ax + By = C$.

**Solution**

**(a)** We are given that $m = \frac{2}{5}$ and $b = -3$. Thus we have the following.

$$y = mx + b$$

$$y = \frac{2}{5}x + (-3)$$

$$y = \frac{2}{5}x - 3$$

**(b)** To write the equation in the form $Ax + By = C$, let us first clear the equation of fractions so that $A$, $B$, and $C$ are integers. Then we move the $x$-term to the left side.

$$5y = 5\left(\frac{2x}{5}\right) - 5(3)$$ Multiply each term by 5.

$$5y = 2x - 15$$ Simplify.

$$-2x + 5y = -15$$ Subtract $2x$ from each side.

$$2x - 5y = 15$$ Multiply each term by $-1$. The form $Ax + By = C$ is usually written with $A$ as a positive integer. ☐

**Student Practice 5** Find an equation of the line with slope $-\frac{3}{7}$ and $y$-intercept $\left(0, \frac{2}{7}\right)$.

**(a)** Write the equation in slope–intercept form.

**(b)** Write the equation in the form $Ax + By = C$.

### 4 Graphing a Line Using the Slope and y-Intercept ▶

If we know the slope of a line and the $y$-intercept, we can draw the graph of the line.

**Example 6** Graph the line with slope $m = \frac{2}{3}$ and $y$-intercept $(0, -3)$. Use the given coordinate system.

**Solution** Recall that the $y$-intercept is the point where the line crosses the $y$-axis. We need a point to start the graph with. So we plot the point $(0, -3)$ on the $y$-axis.

*Continued on next page*

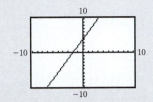

**Graphing Calculator**

**Graphing Lines**

You can graph a line given in the form $y = mx + b$ using a graphing calculator. For example, to graph $y = 2x + 4$, enter the right-hand side of the equation in the Y = editor of your calculator and graph. Choose an appropriate window to show all the intercepts. The following window is $-10$ to 10 by $-10$ to 10.

Display:

Try graphing other equations given in slope–intercept form.

Recall that slope $= \dfrac{\text{rise}}{\text{run}}$. Since the slope of this line is $\frac{2}{3}$, we will go up (rise) 2 units and go over (run) to the right 3 units from the point $(0, -3)$. Look at the figure to the right. This is the point $(3, -1)$. Plot the point. Draw a line that connects the two points $(0, -3)$ and $(3, -1)$.

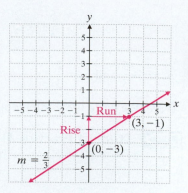

This is the graph of the line with slope $\frac{2}{3}$ and $y$-intercept $(0, -3)$. ☐

**Student Practice 6** Graph the line with slope $= \frac{3}{4}$ and $y$-intercept $(0, -1)$. Use the coordinate system in the margin.

**Student Practice 6**

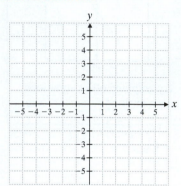

Let's summarize the process used in Example 6. If we know both the slope $m$ and the $y$-intercept $b$, we can graph the equation without performing any calculations as follows.

**Step 1.** Plot the $y$-intercept point on the graph and then begin at this point.

**Step 2.** Using the slope, $\dfrac{\text{rise}}{\text{run}}$, we *rise* in the $y$-direction. We move up if the number is positive and down if the number is negative. We *run* in the $x$-direction. We move right if the number is positive and left if the number is negative.

**Step 3.** Plot this point and draw a line through the two points.

**Example 7** Graph the equation $y = -\frac{1}{2}x + 4$. Use the following coordinate system.

**Solution**

**Step 1.** Since $b = 4$, we plot the $y$-intercept $(0, 4)$.

**Step 2.** Using $m = -\frac{1}{2}$ or $\frac{-1}{2}$, we begin at $(0, 4)$ and go down 1 unit and to the right 2 units.

**Step 3.** This is the point $(2, 3)$. Plot this point and draw a line that connects the points $(0, 4)$ and $(2, 3)$.

This is the graph of the equation $y = -\frac{1}{2}x + 4$.

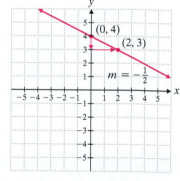

*Note:* The slope $-\frac{1}{2} = \frac{-1}{2} = \frac{1}{-2}$. Therefore, we get the same line if we write the slope as $\frac{1}{-2}$. Try it. ☐

**Student Practice 7**

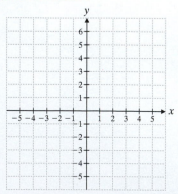

**Student Practice 7** Graph the equation $y = -\frac{2}{3}x + 5$. Use the coordinate system in the margin.

*Graphing Using Different Methods* In this section and in Section 7.2 we learned the following three ways to graph a linear equation:

1. Plot any 3 points.
2. Find the intercepts.
3. Find the slope and $y$-intercept.

The following Graphing Organizer summarizes each method and notes the advantages and disadvantages of each.

## Graphing Organizer

| Method 1: Find Any Three Ordered Pairs | Method 2: Find $x$- and $y$-Intercepts and One Additional Ordered Pair | Method 3: Write the Equation in Slope–Intercept Form, $y = mx + b$ |
|---|---|---|

**Method 1: Find Any Three Ordered Pairs**

**Graph $2y - x = 2$ by plotting three points.**

**Choose any value for either $x$ or $y$.**

| $x$ | $y$ |
|---|---|
| 1 | |
| 2 | |
| −4 | |

**Substitute to find the other value.**

$$2y - 1 = 2 \qquad 2y - 2 = 2 \qquad 2y - (-4) = 2$$
$$2y = 3 \qquad\quad 2y = 4 \qquad\quad 2y + 4 = 2$$
$$y = \frac{3}{2} \qquad\quad y = 2 \qquad\quad 2y = -2$$
$$y = -1$$

| $x$ | $y$ |
|---|---|
| 1 | $\frac{3}{2}$ |
| 2 | 2 |
| −4 | −1 |

**Plot points and draw line through points.**

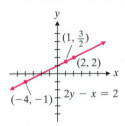

**Advantage**
1. Method is easy to remember: just choose any point for either $x$ or $y$ then solve for the other value.

**Disadvantages**
1. It may be difficult to avoid fractions.
2. Whether you start with a value for $x$ or $y$, solving for the other value can be difficult.

**You Try It**
**Graph $-3x + 2 = y$ Find and label three points.**

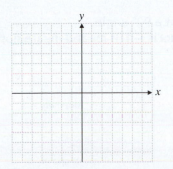

---

**Method 2: Find $x$- and $y$-Intercepts and One Additional Ordered Pair**

**Graph $2y - x = 2$. State the $x$- and $y$-intercepts.**

**Choose $x = 0$ and $y = 0$, and any other value for either $x$ or $y$.**

| $x$ | $y$ |
|---|---|
| 0 | |
| | 0 |
| −1 | |

*Notice we are still choosing 3 values, but to get the intercepts we must pick $x = 0$ and $y = 0$.*

**Substitute to find the other value.**

$$2y - 0 = 2 \qquad 2(0) - x = 2 \qquad 2y - (-1) = 2$$
$$2y = 2 \qquad\qquad -x = 2 \qquad\quad 2y + 1 = 2$$
$$y = 1 \qquad\qquad x = -2 \qquad\quad 2y = 1$$
$$y = \frac{1}{2}$$

| $x$ | $y$ | |
|---|---|---|
| 0 | 1 | $y$-intercept: $(0, 1)$ |
| −2 | 0 | $x$-intercept: $(-2, 0)$ |
| −1 | $\frac{1}{2}$ | |

**Plot points and draw line through points.**

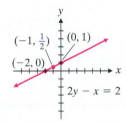

**Advantage**
1. When we substitute 0 into an equation it is easy to calculate the value of the other variable.

**Disadvantage**
1. It may be difficult to avoid fractions.

**You Try It**
**Graph $-3x + 2 = y$. Label the $x$- and $y$-intercepts.**

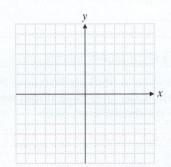

---

**Method 3: Write the Equation in Slope–Intercept Form, $y = mx + b$**

**Graph $2y - x = 2$. State $m$ and $b$. Solve for $y$.**

$$2y = x + 2$$
$$y = \frac{1}{2}x + 1$$

*Once we solve for y we can choose any method to graph.*

**Option 1. Use the slope–intercept method, $y = \frac{1}{2}x + 1$.**

Identify $b$: $b = 1$; $y$-intercept: $(0, 1)$.

Identify the slope, $m$: $m = \frac{1}{2}$

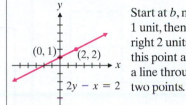

Start at $b$, move up 1 unit, then move right 2 units. Plot this point and draw a line through the two points.

**Option 2. We can also choose to find intercepts or any three ordered pairs when equation is in the form $y = mx + b$.**

$$y = \frac{1}{2}x + 1$$

**Substitute $x = 2, 0, -2$ and find $y$.**

| $x$ | $y$ | |
|---|---|---|
| 2 | 2 | Plot these points on the coordinate planes in columns 1 and 2. You should get the same line. Why? |
| 0 | 1 | |
| −2 | 0 | |

**Advantages**
1. When using the slope–intercept method, we can avoid doing any calculations.
2. When we find ordered pairs, the slope–intercept form makes it easy to see how to avoid fractions.

$y = \frac{1}{2}x + 1$ — Choosing a multiple of 2 for $x$ will avoid getting a fraction for $y$.

**Disadvantage**
1. It may be difficult to solve for $y$.

**You Try It**
**Graph $-3x + 2 = y$. State $m$ and $b$.**

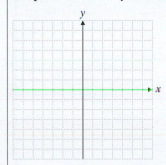

<table>
<tr><td>

**Graphing Calculator**

**Graphing Lines**

You can graph a line given in the form $y = mx + b$ using a graphing calculator. For example, to graph $y = 2x + 4$, enter the right-hand side of the equation in the $Y =$ editor of your calculator and graph. Choose an appropriate window to show all the intercepts. The following window is $-10$ to $10$ by $-10$ to $10$. Display:

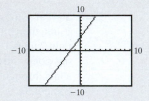

Try graphing other equations given in slope–intercept form.

</td></tr>
</table>

**5** Finding the Slopes of Parallel and Perpendicular Lines

**Parallel lines** are two straight lines that never touch. Look at the parallel lines in the figure. Notice that the slope of line $a$ is $-3$ and the slope of line $b$ is also $-3$. Why do you think the slopes must be equal? What would happen if the slope of line $b$ were $-1$? Graph it and see.

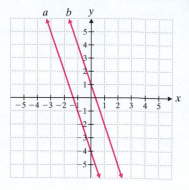

**PARALLEL LINES**

Parallel lines are two straight lines that never touch.
**Parallel lines have the same slope but different $y$-intercepts.**

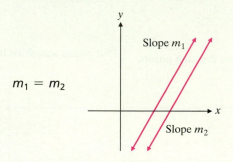

$$m_1 = m_2$$

**Perpendicular lines** are two lines that meet at a 90° angle. Look at the perpendicular lines in the figure at the left. The slope of line $c$ is $-3$. The slope of line $d$ is $\frac{1}{3}$. Notice that

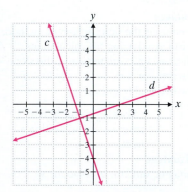

$$(-3)\left(\frac{1}{3}\right) = \left(-\frac{3}{1}\right)\left(\frac{1}{3}\right) = -1.$$

You may wish to draw several pairs of perpendicular lines to determine whether the product of their slopes is always $-1$.

**PERPENDICULAR LINES**

Perpendicular lines are two lines that meet at a 90° angle.
**Perpendicular lines have slopes whose product is $-1$.** If $m_1$ and $m_2$ are slopes of perpendicular lines, then

$$m_1 m_2 = -1 \quad \text{or} \quad m_1 = -\frac{1}{m_2}.$$

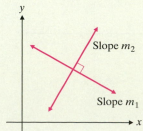

**Example 8** Line $h$ has a slope of $-\frac{2}{3}$.

(a) If line $f$ is parallel to line $h$, what is its slope?

(b) If line $g$ is perpendicular to line $h$, what is its slope?

**Solution**

(a) Parallel lines have the same slope. Line $f$ has a slope of $-\frac{2}{3}$.

(b) Perpendicular lines have slopes whose product is $-1$.

$$m_1 m_2 = -1$$

$$-\frac{2}{3}m_2 = -1 \qquad \text{Substitute } -\frac{2}{3} \text{ for } m_1.$$

$$\left(-\frac{3}{2}\right)\left(-\frac{2}{3}\right)m_2 = -1\left(-\frac{3}{2}\right) \qquad \text{Multiply both sides by } -\frac{3}{2}.$$

$$m_2 = \frac{3}{2}$$

Thus line $g$ has a slope of $\frac{3}{2}$.

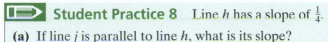

**Student Practice 8** Line $h$ has a slope of $\frac{1}{4}$.

(a) If line $j$ is parallel to line $h$, what is its slope?

(b) If line $k$ is perpendicular to line $h$, what is its slope?

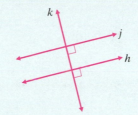

**Example 9** The equation of line $l$ is $y = -2x + 3$.

(a) What is the slope of a line that is parallel to line $l$?

(b) What is the slope of a line that is perpendicular to line $l$?

**Solution**

(a) Looking at the equation, we can see that the slope of line $l$ is $-2$. The slope of a line that is parallel to line $l$ is $-2$.

(b) Perpendicular lines have slopes whose product is $-1$.

$$m_1 m_2 = -1$$

$$(-2)m_2 = -1 \qquad \text{Substitute } -2 \text{ for } m_1.$$

$$m_2 = \frac{1}{2} \qquad \text{Because } (-2)\left(\frac{1}{2}\right) = -1.$$

The slope of a line that is perpendicular to line $l$ is $\frac{1}{2}$.

**Student Practice 9** The equation of line $n$ is $y = \frac{2}{3}x - 4$.

(a) What is the slope of a line that is parallel to line $n$?

(b) What is the slope of a line that is perpendicular to line $n$?

---

**Graphing Calculator**

**Graphing Parallel Lines**

If two equations are in the form $y = mx + b$, then it will be obvious that they are parallel because the slope will be the same. On a graphing calculator, graph both of these equations:

$$y = -2x + 6$$
$$y = -2x - 4$$

Use a window of $-10$ to $10$ for both $x$ and $y$.

Display:

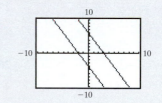

## 7.3 Exercises   MyMathLab®

**Verbal and Writing Skills, Exercises 1 and 2**

**1.** Can you find the slope of the line passing through $(5, -12)$ and $(5, -6)$? Why or why not?

**2.** Can you find the slope of the line passing through $(6, -2)$ and $(-8, -2)$? Why or why not?

*Find the slope of the straight line that passes through the given pair of points.*

**3.** $(2, 3)$ and $(5, 9)$     **4.** $(1, 4)$ and $(9, 12)$     **5.** $(5, 10)$ and $(6, 5)$     **6.** $(5, 3)$ and $(2, 15)$

**7.** $(-2, 1)$ and $(3, 4)$     **8.** $(7, 4)$ and $(-2, 8)$     **9.** $(-6, -5)$ and $(2, -7)$     **10.** $(-8, -3)$ and $(4, -9)$

**11.** $(-3, 0)$ and $(0, -4)$     **12.** $(0, 5)$ and $(5, 3)$     **13.** $(5, -1)$ and $(-7, -1)$     **14.** $(-1, 5)$ and $(-1, 7)$

**15.** $\left(\dfrac{3}{4}, -4\right)$ and $(2, -8)$     **16.** $\left(\dfrac{5}{3}, -2\right)$ and $(3, 6)$

*Find the slope and the y-intercept.*

**17.** $y = 8x + 9$     **18.** $y = 2x + 10$     **19.** $3x + y - 4 = 0$     **20.** $8x + y + 7 = 0$

**21.** $y = -\dfrac{8}{7}x + \dfrac{3}{4}$     **22.** $y = \dfrac{5}{3}x - \dfrac{4}{5}$     **23.** $y = -6x$     **24.** $y = 4x$

**25.** $y = -2$     **26.** $y = 7$     **27.** $7x - 3y = 4$     **28.** $9x - 4y = 18$

*Write the equation of the line in slope–intercept form.*

**29.** $m = \dfrac{3}{5}$, y-intercept $(0, 3)$     **30.** $m = \dfrac{2}{3}$, y-intercept $(0, 5)$     **31.** $m = 4$, y-intercept $(0, -5)$

**32.** $m = 2$, y-intercept $(0, -1)$     **33.** $m = -1$, y-intercept $(0, 0)$     **34.** $m = 1$, y-intercept $(0, 0)$

**35.** $m = -\dfrac{5}{4}$, y-intercept $\left(0, -\dfrac{3}{4}\right)$     **36.** $m = -4$, y-intercept $\left(0, \dfrac{1}{2}\right)$

*Graph the line $y = mx + b$ for the given values.*

**37.** $m = \dfrac{3}{4}, b = -4$     **38.** $m = \dfrac{1}{3}, b = -2$     **39.** $m = -\dfrac{5}{3}, b = 2$     **40.** $m = -\dfrac{3}{2}, b = 4$

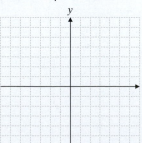
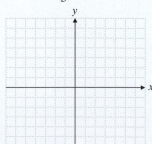
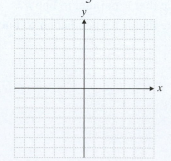
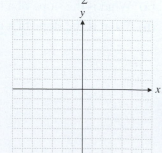

*In exercises 41–46, graph the line. You may use any method: $y = mx + b$, intercepts, or finding any three points.*

**41.** $y = \dfrac{2}{3}x + 2$

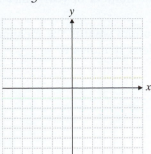

**42.** $y = \dfrac{3}{4}x + 1$

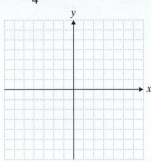

**43.** $y + 2x = 3$

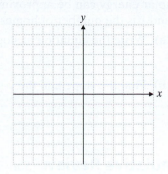

**44.** $y + 4x = 5$

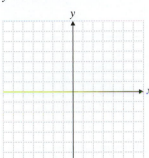

**45.** $y = 2x$

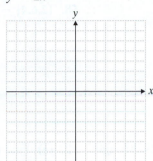

**46.** $y = 3x$

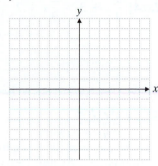

**47.** A line has a slope of $\dfrac{5}{6}$.

    **(a)** What is the slope of a line parallel to it?

    **(b)** What is the slope of a line perpendicular to it?

**48.** A line has a slope of $\dfrac{11}{4}$.

    **(a)** What is the slope of a line parallel to it?

    **(b)** What is the slope of a line perpendicular to it?

**49.** A line has a slope of $-8$.

    **(a)** What is the slope of a line parallel to it?

    **(b)** What is the slope of a line perpendicular to it?

**50.** A line has a slope of $-\dfrac{5}{8}$.

    **(a)** What is the slope of a line parallel to it?

    **(b)** What is the slope of a line perpendicular to it?

**51.** The equation of a line is $y = \dfrac{2}{3}x + 6$.

    **(a)** What is the slope of a line parallel to it?

    **(b)** What is the slope of a line perpendicular to it?

**52.** The equation of a line is $y = -\dfrac{3}{4}x - 2$.

    **(a)** What is the slope of a line parallel to it?

    **(b)** What is the slope of a line perpendicular to it?

## To Think About

**53.** Do the points $(3, -4)$, $(18, 6)$, and $(9, 0)$ all lie on the same line? If so, what is the equation of the line?

**54.** Do the points $(2, 1)$, $(-3, -2)$, and $(7, 4)$ lie on the same line? If so, what is the equation of the line?

**55.** *Cell Phone Accessory Sales* A distribution center projected that one of the fastest-growing products in the United States between 2010 and 2020 will be cell phone accessories. During this 10-year period, the number of cell phone accessories can be approximated by the equation $y = 5(7x + 125)$, where $x$ is the number of years since 2010 and $y$ is the number of accessories sold in thousands.

    **(a)** Write the equation in slope–intercept form.

    **(b)** Find the slope and the $y$-intercept.

    **(c)** In this specific equation, what is the meaning of the slope? What does it indicate?

**56.** *Solar Energy* Essex Solar Company executives projected that one of the fastest-growing sources of home energy in the United States between 2014 and 2024 will be solar energy. During this 10-year period, the number of homes using solar energy can be approximated by the equation $y = \frac{1}{10}(31x + 620)$, where $x$ is the number of years since 2014 and $y$ is the number of homes in thousands.

**(a)** Write the equation in slope–intercept form.

**(b)** Find the slope and the $y$-intercept.

**(c)** In this specific equation, what is the meaning of the slope? What does it indicate?

## Cumulative Review   *Solve for x and graph the solution.*

**57.** **[2.6.4]** $\frac{1}{4}x + 3 > \frac{2}{3}x + 2$

**58.** **[2.6.4]** $\frac{1}{2}(x + 2) \leq \frac{1}{3}x + 5$

---

## Quick Quiz 7.3

**1.** Find the slope of the straight line that passes through the points $(-2, 5)$ and $(-6, 3)$.

**2. (a)** Find the slope and $y$-intercept of $6x + 2y - 4 = 0$.
**(b)** Graph.

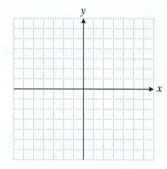

**3.** Write an equation of the line in slope–intercept form that passes through $(0, -5)$ and has a slope of $-\frac{5}{7}$.

**4. Concept Check** Consider the formula for slope: $m = \frac{y_2 - y_1}{x_2 - x_1}$. Explain why we substitute the $y$ coordinates in the numerator.

# Use Math To Save Money

**Did You Know?**  Saving only 5% of your income can go a long way toward replacing your income in retirement.

## Retirement Planning

### Understanding the Problem:

It is never too early to begin thinking about your retirement. Louis began investing money in his retirement account at the age of 25 with the goal of retiring at the age of 65. He has a job where he gets paid $42,000 per year. His company has a retirement plan where he can contribute a percentage of his salary and have it invested in a retirement account that has an average annual return of 8%.

### Making a Plan:

Louis is interested in what his monthly income will be at retirement if he saves 5% of his earnings for the next 40 years. To calculate his projected income at retirement, Louis needs to determine what the projected balance of his retirement account will be. He uses the retirement formula

$$FV = PMT \times \frac{(1 + I)^N - 1}{I}$$

with the variables defined as follows:

- $FV$ is the future value of the retirement account.
- $PMT$ is the monthly payment contributed to the retirement account.
- $I$ is the interest rate in decimal notation divided by 12, which is the amount of interest being earned each month.
- $N$ is the number of payments made into the account, which is 12 times the number of years the account is contributed to.

**Step 1:** Louis calculates what his monthly payment ($PMT$) will be if he saves 5% of his earnings.

**Task 1:** How much does Louis earn per month?

**Task 2:** What is 5% of his monthly income? This value will be the PMT.

**Step 2:** Using $I = 0.08 / 12 \approx 0.0067$ and $N = 12 \times 40 = 480$, Louis substitutes these values into the retirement formula to solve for $FV$.

**Task 3:** Substitute and solve for FV. Round your answer to the nearest dollar.

### Finding a Solution:

Louis now knows how much he can expect to have in his retirement account when he retires. He has been told that to make sure his money lasts throughout his retirement, he should withdraw 4% of his account balance the first year and

then increase that amount to keep up with inflation every year thereafter.

**Step 3:** Louis calculates his monthly income based on a withdrawal of 4% the first year. His goal is to contribute just enough to his retirement fund so that his retirement income is equal to his current working income.

**Task 4:** How much can Louis withdraw the first year?

**Task 5:** What will Louis's monthly income be the first year?

**Task 6:** Calculate the difference between Louis's current monthly income and his projected monthly income in retirement.

**Task 7:** *Does Louis need to change the percentage he is contributing to his retirement plan to reach his goal? If so, should he increase or decrease the amount?*

### Applying the Situation to Your Life:

These are simplified calculations for projecting income in retirement. Other factors like inflation, which are more complicated to calculate, need to be taken into consideration. Inflation will reduce the value of money over time so that in 40 years, $42,000 may not be enough money to support the same lifestyle. You should also expect your yearly income to increase over time, so that the amount you are saving will increase. In addition there will be other sources of income in retirement like Social Security and pensions. The above calculations will give you a good idea of how much you will have in retirement, based on your current earnings and the expected number of years until retirement.

431

## 7.4 Writing the Equation of a Line ▶

**Student Learning Objectives**

After studying this section, you will be able to:

1　Write an equation of a line given a point and the slope. ▶

2　Write an equation of a line given two points. ▶

3　Write an equation of a line given a graph of the line. ▶

### 1 Writing an Equation of a Line Given a Point and the Slope ▶

If we know the slope of a line and the $y$-intercept, we can write the equation of the line in slope–intercept form. Sometimes we are given the slope and a point on the line. We use the information to find the $y$-intercept. Then we can write the equation of the line. It may be helpful to summarize our approach.

> **TO FIND AN EQUATION OF A LINE GIVEN A POINT AND THE SLOPE**
>
> 1. Substitute the given values of $x$, $y$, and $m$ into the equation $y = mx + b$.
> 2. Solve for $b$.
> 3. Use the values of $b$ and $m$ to write the equation in the form $y = mx + b$.

**Example 1** Find an equation of the line that passes through $(-3, 6)$ with slope $-\frac{2}{3}$.

**Solution**　We are given the values $m = -\frac{2}{3}$, $x = -3$, and $y = 6$, and we must find $b$.

$$y = mx + b$$

$$6 = \left(-\frac{2}{3}\right)(-3) + b \quad \text{Substitute known values.}$$

$$6 = 2 + b \quad \text{Solve for } b.$$

$$4 = b \quad \text{Now use the values for } b \text{ and } m \text{ to write the equation in slope–intercept form.}$$

An equation of the line is $y = -\frac{2}{3}x + 4$.　□

▶ **Student Practice 1**　Find an equation of the line that passes through $(-8, 12)$ with slope $-\frac{3}{4}$.

### 2 Writing an Equation of a Line Given Two Points ▶

We can use our procedure when we are *given two points* if we first find the slope $m$ and then find $b$. Recall from Section 7.3 that if we are given two points on a line, we can find the slope $m$ using the formula $m = \dfrac{y_2 - y_1}{x_2 - x_1}$. In the next example we will use this information to find the values of $m$ and $b$ so that we can write an equation in the form $y = mx + b$.

**Mc Example 2**　Find an equation of the line that passes through $(2, 5)$ and $(6, 3)$.

**Solution**　We must find both $m$ and $b$. We first find $m$ using the formula for slope. Then we proceed as in Example 1 to find $b$.

$$m = \frac{y_2 - y_1}{x_2 - x_1}$$

$$m = \frac{3 - 5}{6 - 2} \quad \text{Substitute } (x_1, y_1) = (2, 5) \text{ and } (x_2, y_2) = (6, 3) \text{ into the formula.}$$

$$= \frac{-2}{4} = -\frac{1}{2}$$

▶ **Student Practice 2**　Find an equation of the line that passes through $(3, 5)$ and $(-1, 1)$.

Now that we know $m = -\dfrac{1}{2}$, we can find $b$. Choose either point, say $(2, 5)$, to substitute into $y = mx + b$ as in Example 1.

$$5 = -\frac{1}{2}(2) + b \quad \textcolor{red}{\text{Substitute known values into } y = mx + b.}$$
$$5 = -1 + b$$
$$6 = b$$

Use the values for $b$ and $m$ to write the equation.

An equation of the line is $y = -\dfrac{1}{2}x + 6$.

*Note:* We could have substituted the slope and the other point, $(6, 3)$, into the slope–intercept form and arrived at the same answer. Try it. ☐

As we saw in Examples 1 and 2, once we find $m$ and $b$ we can write the equation of a line. Now, if we are given the graph of a line, can we write the equation of this line? That is, from the graph can we find $m$ and $b$? We will look at this in Example 3.

## 3  Writing an Equation of a Line Given a Graph of the Line ▶

**Example 3** What is the equation of the line in the figure at the right?

**Solution** First, look for the $y$-intercept. The line crosses the $y$-axis at $(0, 4)$. Thus $b = 4$.

Second, find the slope.

$$m = \frac{\text{change in } y}{\text{change in } x} = \frac{\text{rise}}{\text{run}}$$

Look for another point on the line. We choose $(5, -2)$. Count the number of vertical units from 4 to $-2$ (rise). Count the number of horizontal units from 0 to 5 (run).

$$m = \frac{-6}{5}$$

Now, using $m = -\dfrac{6}{5}$ and $b = 4$, we can write an equation of the line.

$$y = mx + b$$
$$y = -\frac{6}{5}x + 4 \qquad ☐$$

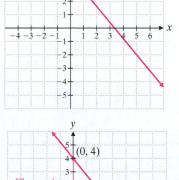

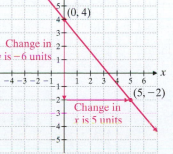

**Student Practice 3**

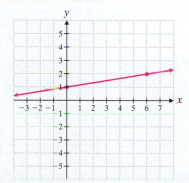

▶ **Student Practice 3** What is the equation of the line in the figure at the right?

*Find an equation of the line that has the given slope and passes through the given point.*

**1.** $m = 3, (-2, 0)$      **2.** $m = 5, (4, -1)$      **3.** $m = -2, (3, 5)$      **4.** $m = -4, (1, 1)$

**5.** $m = -3, \left(\frac{1}{2}, 2\right)$      **6.** $m = -4, \left(1, \frac{1}{5}\right)$      **7.** $m = \frac{1}{4}, (4, 5)$      **8.** $m = \frac{2}{3}, (3, -2)$

*Write an equation of the line passing through the given points.*

**9.** $(3, -12)$ and $(-4, 2)$    **10.** $(-1, 8)$ and $(0, 5)$    **11.** $(2, -6)$ and $(-1, 6)$    **12.** $(-1, 1)$ and $(5, 7)$

**13.** $(3, 5)$ and $(-1, -15)$    **14.** $(-1, -19)$ and $(2, 2)$    **15.** $\left(1, \frac{5}{6}\right)$ and $\left(3, \frac{3}{2}\right)$    **16.** $(2, 0)$ and $\left(\frac{3}{2}, \frac{1}{2}\right)$

## Mixed Practice, Exercises 17–20

**17.** Find an equation of the line with a slope of $-3$ that passes through the point $(-1, 3)$.

**18.** Find an equation of the line with a slope of $-\frac{1}{2}$ that passes through the point $(6, -2)$.

**19.** Find an equation of the line that passes through $(2, -3)$ and $(-1, 6)$.

**20.** Find an equation of the line that passes through $(1, -8)$ and $(2, -14)$.

*Write an equation of each line.*

**21.** 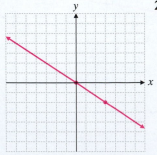    **22.** 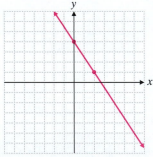    **23.** 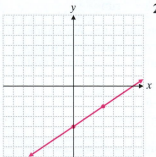    **24.**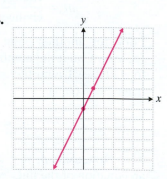

**25.** 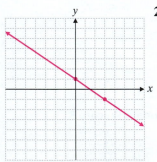    **26.** 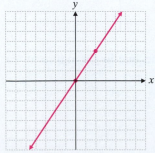    **27.** 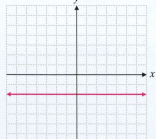    **28.**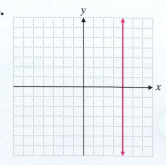

**To Think About, Exercises 29–36** *Find an equation of the line that fits each description.*

**29.** Passes through $(7, -5)$ and has zero slope

**30.** Passes through $(9, 3)$ and has undefined slope

**31.** Passes through $(4, -6)$ and is perpendicular to the $x$-axis

**32.** Passes through $(-7, 11)$ and is perpendicular to the $y$-axis

**33.** Passes through $(0, 5)$ and is parallel to $y = \frac{1}{3}x + 4$

**34.** Passes through $(0, 5)$ and is perpendicular to $y = \frac{1}{3}x + 4$

**35.** Passes through $(2, 3)$ and is perpendicular to $y = 2x - 9$

**36.** Passes through $(2, 9)$ and is parallel to $y = 5x - 3$

**37. *Population Growth*** The growth of the population of a large city during the period from 1990 to 2018 can be approximated by an equation of the form $y = mx + b$, where $x$ is the number of years since 1990 and $y$ is the population measured in thousands. Find the equation if two ordered pairs that satisfy it are $(0, 227)$ and $(10, 251)$.

**38. *Home Equity Loans*** The amount of debt outstanding on home equity loans at a bank during the period from 2000 to 2015 can be approximated by an equation of the form $y = mx + b$, where $x$ is the number of years since 2000 and $y$ is the debt measured in millions of dollars. Find the equation if two ordered pairs that satisfy it are $(1, 280)$ and $(6, 500)$.

## Cumulative Review

**39.** [6.5.1] $\dfrac{3}{t} - \dfrac{2}{t-1} = \dfrac{4}{t}$

**40.** [6.2.1] Simplify. $\dfrac{4x^2 - 25}{4x^2 - 20x + 25} \cdot \dfrac{x^2 - 9}{2x^2 + x - 15}$

**41.** [0.6.1] *Basketball Sneakers* A pair of basketball sneakers sells for \$80. The next week the sneakers go on sale for 15% off. The third week there is a coupon in the newspaper offering a 10% discount off the second week's price. How much would you have paid for the sneakers during the third week if you had used the coupon?

**42.** [3.4.1] *Cell Phone Costs* Dave and Jane Wells have a cell phone. The plan they subscribe to costs \$50 per month. It includes 200 free minutes of calling time. Each minute after the 200 is charged at the rate of \$0.21 per minute. Last month their cell phone bill was \$68.90. How many total minutes did they use their cell phone during the month?

## Quick Quiz 7.4

**1.** Write an equation in slope–intercept form of the line that passes through the point $(3, -5)$ and has a slope of $\frac{2}{3}$.

**2.** Write an equation in slope–intercept form of the line that passes through the points $(-2, 7)$ and $(-4, -5)$.

**3.** Write an equation of the line that passes through $(4, 5)$ and $(4, -2)$. What is the slope of this line?

**4. Concept Check** How would you find an equation of the line that passes through $(-2, -3)$ and has zero slope?

## 7.5 Graphing Linear Inequalities

**Student Learning Objective**

After studying this section, you will be able to:

1 Graph linear inequalities in two variables.

In Section 2.6 we discussed inequalities in one variable. Look at the inequality $x < -2$ ($x$ is less than $-2$). Some of the solutions to the inequality are $-3$, $-5$, and $-5\frac{1}{2}$. In fact all numbers to the left of $-2$ on a number line are solutions. The graph of the inequality is given in the following figure. Notice that the open circle at $-2$ indicates that $-2$ is *not* a solution.

$$x < -2$$

Now we will extend our discussion to consider linear inequalities in two variables.

### 1 Graphing Linear Inequalities in Two Variables

Consider the inequality $y \geq x$. The solution of the inequality is the set of all possible ordered pairs that when substituted into the inequality will yield a true statement. Which ordered pairs will make the statement $y \geq x$ true? Let's try some.

| $(0, 6)$ | $(-2, 1)$ | $(1, -2)$ | $(3, 5)$ | $(4, 4)$ |
|---|---|---|---|---|
| $6 \geq 0$, true | $1 \geq -2$, true | $-2 \geq 1$, false | $5 \geq 3$, true | $4 \geq 4$, true |

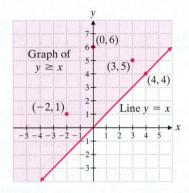

$(0, 6)$, $(-2, 1)$, $(3, 5)$, and $(4, 4)$ are solutions to the inequality $y \geq x$. In fact, every point at which the $y$-coordinate is greater than or equal to the $x$-coordinate is a solution to the inequality. This is shown by the solid line and the shaded region in the graph at the left.

Is there an easier way to graph a linear inequality in two variables? It turns out that we can graph such an inequality by first graphing the associated linear equation and then testing one point that is not on that line. That is, we can change the inequality symbol to an equals sign and graph the equation. If the inequality symbol is $\geq$ or $\leq$, we use a solid line to indicate that the points on the line are included in the solution of the inequality. If the inequality symbol is $>$ or $<$, we use a dashed line to indicate that the points on the line are not included in the solution of the inequality. Then we test one point that is not on the line. If the point is a solution to the inequality, we shade the region on the side of the line that includes the point. If the point is not a solution, we shade the region on the other side of the line.

$\mathbb{M}_{\mathbb{C}}$ **Example 1** Graph $5x + 3y > 15$.

**Solution** We begin by graphing the line $5x + 3y = 15$. You may use any method discussed previously to graph the line. Since there is no equals sign in the inequality, we will draw a *dashed line* to indicate that the line is *not* part of the solution set.

Look for a test point. The easiest point to test is $(0, 0)$. Substitute $(0, 0)$ for $(x, y)$ in the inequality.

$$5x + 3y > 15$$
$$5(0) + 3(0) > 15$$
$$0 > 15 \quad \text{false}$$

$(0, 0)$ is *not* a solution. Shade the region on the side of the line that does *not* include $(0, 0)$.

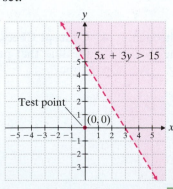

**Student Practice 1**

Graph $x - y \geq -10$.

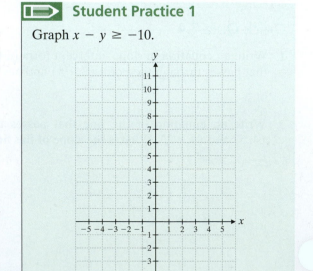

**GRAPHING LINEAR INEQUALITIES**

1. Replace the inequality symbol by an equality symbol. Graph the line.
   (a) The line will be solid if the inequality is $\geq$ or $\leq$.
   (b) The line will be dashed if the inequality is $>$ or $<$.

2. Test the point $(0, 0)$ in the inequality if $(0, 0)$ does not lie on the graphed line in step 1.
   (a) If the inequality is true, shade the region on the side of the line that includes $(0, 0)$.
   (b) If the inequality is false, shade the region on the side of the line that does not include $(0, 0)$.

3. If the point $(0, 0)$ is a point on the line, choose another test point and proceed accordingly.

**Example 2** Graph $2y \leq -3x$.

**Solution**

**Step 1** Graph $2y = -3x$. Since $\leq$ is used, the line will be a solid line.

**Step 2** We see that the line passes through $(0, 0)$.

**Step 3** Choose another test point. We will choose $(-3, -3)$.

$$2y \leq -3x$$
$$2(-3) \leq -3(-3)$$
$$-6 \leq 9 \quad \text{true}$$

Shade the region that includes $(-3, -3)$, that is, the region below the line. □

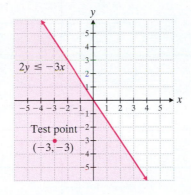

**Student Practice 2**

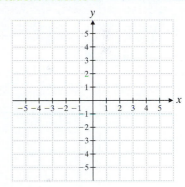

> **Student Practice 2** Graph $2y > x$ on the coordinate system in the margin.

If we are graphing the inequality $x < -2$ on a coordinate plane, the solution will be a region. Notice that this is very different from the solution of $x < -2$ on a number line discussed earlier.

**Example 3** Graph $x < -2$.

**Solution**

**Step 1** Graph $x = -2$. Since $<$ is used, the line will be dashed.

**Step 2** Test $(0, 0)$ in the inequality.
$$x < -2$$
$$0 < -2 \quad \text{false}$$

Shade the region that does not include $(0, 0)$, that is, the region to the left of the line $x = -2$. Observe that every point in the shaded region has an $x$-value that is less than $-2$. □

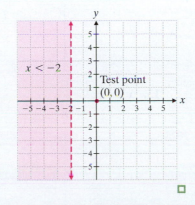

**Student Practice 3**

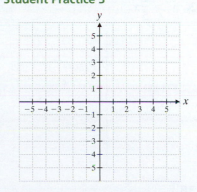

> **Student Practice 3** Graph $y \geq -3$ on the coordinate system in the margin.

## 7.5 Exercises  MyMathLab®

### Verbal and Writing Skills, Exercises 1 and 2

**1.** Does it matter what point you use as your test point? Justify your response.

**2.** Explain when to use a solid line or a dashed line when graphing a linear inequality in two variables.

*Graph the region described by the inequality.*

**3.** $y \geq 4x$

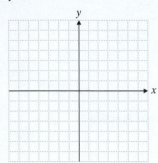

**4.** $y \leq -2x$

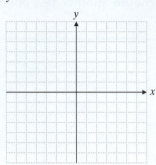

**5.** $2x - 3y < 6$

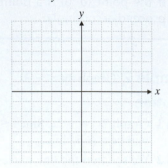

**6.** $3x + 2y < -6$

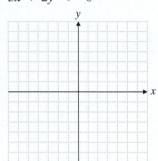

**7.** $2x - y \geq 3$

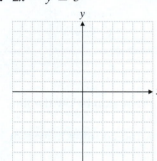

**8.** $3x - y \geq 4$

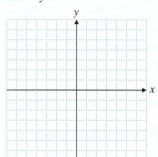

**9.** $y < 2x - 4$

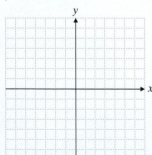

**10.** $y > 1 - 3x$

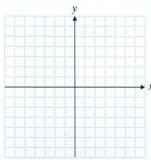

**11.** $y < -\dfrac{1}{2}x$

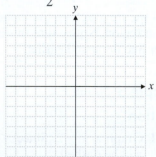

**12.** $y > \dfrac{1}{5}x$

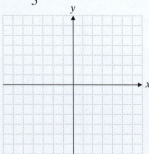

**13.** $x \geq 2$

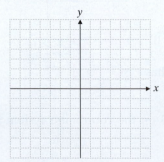

**14.** $y \leq -2$

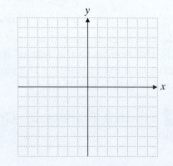

**15.** $2x - 3y + 6 \geq 0$

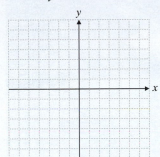

**16.** $3x + 4y - 8 \leq 0$

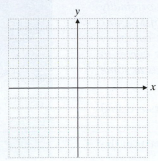

**17.** $x > -2y$

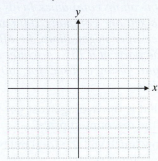

**18.** $x < -3y$

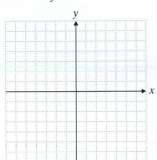

**19.** $2x > 3 - y$

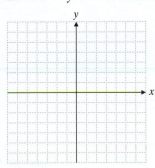

**20.** $x > 4 - y$

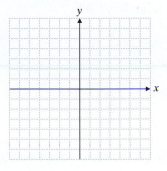

**21.** $2x \geq -3y$

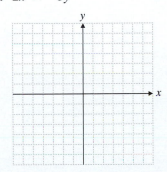

**22.** $3x \leq -2y$

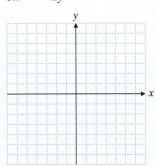

## Cumulative Review   *Perform the operations in the proper order.*

**23.** [1.5.1] $6(2) + 10 \div (-2)$

**24.** [1.5.1] $3(-3) + 2(12 - 15)^2 \div 9$

*Evaluate for $x = -2$ and $y = 3$.*

**25.** [1.8.1] $2x^2 + 3xy - 2y^2$

**26.** [1.8.1] $x^3 - 5x^2 + 3y - 1$

***Satellite Parts* [0.5.3]** *Brian sells high-tech parts to satellite communications companies. In his negotiations he originally offers to sell one company 200 parts for a total of $22,400. However, after negotiations, he offers to sell that company the same parts at a 15% discount if the company agrees to sign a purchasing contract for 200 additional parts at some future date.*

**27.** What is the average cost per part if the parts are sold at the discounted price?

**28.** How much will the total bill be for the 200 parts at the discounted price?

## Quick Quiz 7.5

**1.** When you graph the inequality $y > -3x + 1$, should you use a solid line for a boundary or a dashed line for a boundary? Why?

**2.** Graph the region described by $3y \leq -7x$.

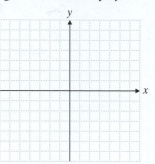

**3.** Graph the region described by $-5x + 2y > -3$.

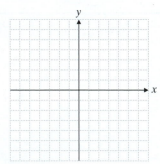

**4.** **Concept Check** Explain how you would determine if you should shade the region above the line or below the line if you were to graph the inequality $y > -3x + 4$ using $(0, 0)$ as a test point.

# 7.6  Functions

## 1  Understanding the Meanings of a Relation and a Function

Thus far you have studied linear equations in two variables. You have seen that such an equation can be represented by a table of values, by the algebraic equation itself, or by a graph.

| $x$ | 0 | 1 | 3 |
|---|---|---|---|
| $y$ | 4 | 1 | −5 |

$y = -3x + 4$

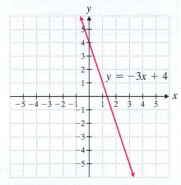

The solutions to the linear equation are all the ordered pairs that satisfy the equation (make the equation true). They are all the points that lie on the graph of the line. These ordered pairs can be represented in a table of values. Notice the relationship between the ordered pairs. We can choose any value for $x$. But once we have chosen a value for $x$, the value of $y$ is determined. For the preceding equation $y = -3x + 4$, if $x$ is 0, then $y$ must be 4. We say that $x$ is the **independent variable** and that $y$ is the **dependent variable**.

Mathematicians call such a pairing of two values a *relation*.

---

### DEFINITION OF A RELATION

A **relation** is any set of ordered pairs.

---

All the *first* coordinates in all of the ordered pairs of the relation make up the **domain** of the relation. All the *second* coordinates in all of the ordered pairs make up the **range** of the relation. Notice that the definition of a relation is very broad. Some relations cannot be described by an equation. These relations may simply be a set of discrete ordered pairs.

**Example 1** State the domain and range of the relation.

$$\{(5, 7), (9, 11), (10, 7), (12, 14)\}$$

**Solution**  The domain consists of all the first coordinates in the ordered pairs.

Domain

$$\{(5, 7), (9, 11), (10, 7), (12, 14)\}$$

Range

The range consists of all the second coordinates in the ordered pairs. We usually list the values of a domain or range from smallest to largest.

The domain is $\{5, 9, 10, 12\}$.

The range is $\{7, 11, 14\}$.    We list 7 only once.  □

**Student Practice 1**  State the domain and range of the relation.

$$\{(-3, -5), (3, 5), (0, -5), (20, 5)\}$$

Some relations have the special property that no two different ordered pairs have the same first coordinate. Such relations are called **functions**. The relation $y = -3x + 4$ is a function. If we substitute a value for $x$, we get just one value for $y$. Thus no two ordered pairs will have the same $x$-coordinate and different $y$-coordinates.

**DEFINITION OF A FUNCTION**

A **function** is a relation in which no two different ordered pairs have the same first coordinate.

**Example 2** Determine whether the relation is a function.

(a) $\{(3, 9), (4, 16), (5, 9), (6, 36)\}$      (b) $\{(7, 8), (9, 10), (12, 13), (7, 14)\}$

**Solution**

(a) Look at the ordered pairs. No two ordered pairs have the same first coordinate. Thus this set of ordered pairs defines a function. Note that the ordered pairs $(3, 9)$ and $(5, 9)$ have the same second coordinate, but the relation is still a function. It is the first coordinates that cannot be the same.

(b) Look at the ordered pairs. Two different ordered pairs, $(7, 8)$ and $(7, 14)$, have the same first coordinate. Thus this relation is *not* a function.    □

**Student Practice 2** Determine whether the relation is a function.

(a) $\{(-5, -6), (9, 30), (-3, -3), (8, 30)\}$

(b) $\{(60, 30), (40, 20), (20, 10), (60, 120)\}$

A functional relationship is often what we find when we analyze two sets of data. Look at the following table of values, which compares Celsius temperature with Fahrenheit temperature. Is there a relationship between degrees Fahrenheit and degrees Celsius? Is the relation a function?

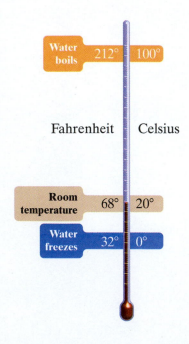

| **Temperature** | | | | |
|---|---|---|---|---|
| **°F** | 23 | 32 | 41 | 50 |
| **°C** | −5 | 0 | 5 | 10 |

Since every Fahrenheit temperature produces a unique Celsius temperature, we would expect this to be a function. We can verify our assumption by looking at the formula $C = \frac{5}{9}(F - 32)$ and its graph. The formula is a linear equation, and its graph is a line with slope $\frac{5}{9}$ and $y$-intercept at about $-17.8$. The relation is a function. In the equation given here, notice that the *dependent variable* is $C$, since the value of $C$ *depends* on the value of $F$. We say that $F$ is the *independent variable*. The *domain* can be described as the set of possible values of the independent variable. The *range* is the set of corresponding values of the dependent variable. Scientists believe that the coldest temperature possible is approximately $-273°C$. In Fahrenheit, they call this temperature **absolute zero**. Thus,

Domain = {all possible Fahrenheit temperatures from absolute zero to infinity}

Range = {all corresponding Celsius temperatures from $-273°C$ to infinity}.

**Example 3** Each of the following tables contains some data pertaining to a relation. Determine whether the relation suggested by the table is a function. If it is a function, identify the domain and range.

**(a) Circle**

| Radius | 1 | 2 | 3 | 4 | 5 |
|---|---|---|---|---|---|
| Area | 3.14 | 12.56 | 28.26 | 50.24 | 78.5 |

**(b) $4000 Loan at 8% for a Minimum of One Year**

| Time (yr) | 1 | 2 | 3 | 4 | 5 |
|---|---|---|---|---|---|
| Interest | $320 | $665.60 | $1038.85 | $1441.96 | $1877.31 |

**Solution**

**(a)** Looking at the table, we see that no two different ordered pairs have the same first coordinate. The area of a circle is a function of the length of the radius.

Next we need to identify the independent variable to determine the domain. Sometimes it is easier to identify the dependent variable. Here we notice that the area of the circle depends on the length of the radius. Thus radius is the independent variable. Since a negative length does not make sense, the radius cannot be a negative number. Although only integer radius values are listed in the table, the radius of a circle can be any nonnegative real number.

$$\text{Domain} = \{\text{all nonnegative real numbers}\}$$
$$\text{Range} = \{\text{all nonnegative real numbers}\}$$

**(b)** No two different ordered pairs have the same first coordinate. Interest is a function of time.

Since the amount of interest paid on a loan depends on the number of years (term of the loan), interest is the dependent variable and time is the independent variable. Negative numbers do not apply in this situation. Although the table includes only integer values for the time, the length of a loan in years can be any real number that is greater than or equal to 1.

$$\text{Domain} = \{\text{all real numbers greater than or equal to 1}\}$$
$$\text{Range} = \{\text{all positive real numbers greater than or equal to \$320}\} \quad \square$$

▭▶ **Student Practice 3** Determine whether the relation suggested by the table is a function. If it is a function, identify the domain and the range.

**(a) 28 Mpg at $4.16 per Gallon**

| Distance | 0 | 28 | 42 | 56 | 70 |
|---|---|---|---|---|---|
| Cost | $0 | $4.16 | $6.24 | $8.32 | $10.40 |

**(b) Store's Inventory of Shirts**

| Number of Shirts | 5 | 10 | 5 | 2 | 8 |
|---|---|---|---|---|---|
| Price of Shirt | $20 | $25 | $30 | $45 | $50 |

**To Think About:** *Is It a Function?* Look at the following bus schedule. Determine whether the relation is a function. Which is the independent variable? Explain your choice.

**Bus Schedule**

| Bus Stop | Main St. | 8th Ave. | 42nd St. | Sunset Blvd. | Cedar Lane |
|---|---|---|---|---|---|
| Time | 7:00 | 7:10 | 7:15 | 7:30 | 7:39 |

## 2 Graphing Simple Nonlinear Equations

Thus far in this chapter we have graphed linear equations in two variables. We now turn to graphing a few nonlinear equations. We will need to plot more than three points to get a good idea of what the graph of a nonlinear equation will look like.

**Example 4** Graph $y = x^2$.

**Solution** Begin by constructing a table of values. We select values for $x$ and then determine by the equation the corresponding values of $y$. We will include negative values for $x$ as well as positive values. We then plot the ordered pairs and connect the points with a smooth curve.

**Student Practice 4**

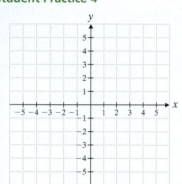

| $x$ | $y = x^2$ | $y$ |
|---|---|---|
| $-2$ | $y = (-2)^2 = 4$ | $4$ |
| $-1$ | $y = (-1)^2 = 1$ | $1$ |
| $0$ | $y = (0)^2 = 0$ | $0$ |
| $1$ | $y = (1)^2 = 1$ | $1$ |
| $2$ | $y = (2)^2 = 4$ | $4$ |

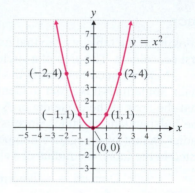

This type of curve is called a *parabola*.

**Student Practice 4** Graph $y = x^2 - 2$ on the coordinate system in the margin.

Some equations are solved for $x$. Usually, in those cases we pick values of $y$ and then obtain the corresponding values of $x$ from the equation.

**Example 5** Graph $x = y^2 + 2$.

**Solution** Since the equation is solved for $x$, we start by picking a value of $y$. We will find the value of $x$ from the equation in each case. We will select $y = -2$ first. Then we substitute it into the equation to obtain $x$. For convenience in graphing, we will repeat the $y$ column at the end so that it is easy to write the ordered pairs $(x, y)$.

**Student Practice 5**

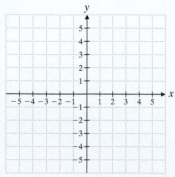

| $y$ | $x = y^2 + 2$ | $x$ | $y$ |
|---|---|---|---|
| $-2$ | $x = (-2)^2 + 2 = 4 + 2 = 6$ | $6$ | $-2$ |
| $-1$ | $x = (-1)^2 + 2 = 1 + 2 = 3$ | $3$ | $-1$ |
| $0$ | $x = (0)^2 + 2 = 0 + 2 = 2$ | $2$ | $0$ |
| $1$ | $x = (1)^2 + 2 = 1 + 2 = 3$ | $3$ | $1$ |
| $2$ | $x = (2)^2 + 2 = 4 + 2 = 6$ | $6$ | $2$ |

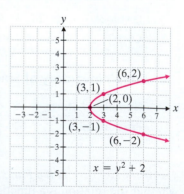

**Student Practice 5** Graph $x = y^2 - 1$ on the coordinate system in the margin.

If the equation involves fractions with variables in the denominator, we must use extra caution. Remember that you may never divide by zero.

**Example 6** Graph $y = \dfrac{4}{x}$.

**Solution** It is important to note that $x$ cannot be zero because division by zero is not defined. $y = \frac{4}{0}$ is not allowed! Observe that when we draw the graph we get two separate branches that do not touch.

| $x$ | $y = \dfrac{4}{x}$ | $y$ |
|---|---|---|
| $-4$ | $y = \dfrac{4}{-4} = -1$ | $-1$ |
| $-2$ | $y = \dfrac{4}{-2} = -2$ | $-2$ |
| $-1$ | $y = \dfrac{4}{-1} = -4$ | $-4$ |
| $0$ | We cannot divide by zero. | There is no value. |
| $1$ | $y = \dfrac{4}{1} = 4$ | $4$ |
| $2$ | $y = \dfrac{4}{2} = 2$ | $2$ |
| $4$ | $y = \dfrac{4}{4} = 1$ | $1$ |

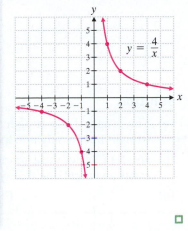

**Student Practice 6**

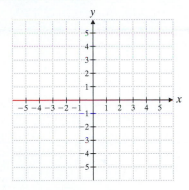

**Student Practice 6** Graph $y = \dfrac{6}{x}$ on the coordinate system in the margin.

**3** **Determining Whether a Graph Represents a Function**

Can we tell whether a graph represents a function? Recall that a function cannot have two different ordered pairs with the same first coordinate. That is, each value of $x$ must have a separate, unique value of $y$. Look at the graph below of the function $y = x^2$. Each $x$-value has a unique $y$-value. Now look at the graph of $x = y^2 + 2$ below. At $x = 6$ there are two $y$-values, 2 and $-2$. In fact, for every $x$-value greater than 2, there are two $y$-values. $x = y^2 + 2$ is not a function.

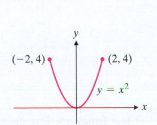

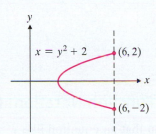

Observe that we can draw a vertical line through $(6, 2)$ and $(6, -2)$. Any graph that is not a function will have at least one region in which a vertical line will cross the graph more than once.

**Graphing Calculator**

**Graphing Nonlinear Equations**

You can graph nonlinear equations solved for $y$ using a graphing calculator. For example, graph $y = x^2 - 2$ on a graphing calculator using an appropriate window. Display:

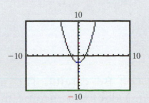

Try graphing $y = \dfrac{5}{x}$.

> **VERTICAL LINE TEST**
>
> If a vertical line can intersect the graph of a relation more than once, the relation is not a function. If no such line can be drawn, then the relation is a function.

**Example 7** Determine whether each of the following is the graph of a function.

(a) 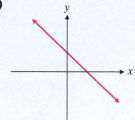 (b) 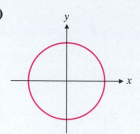 (c)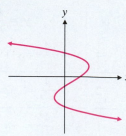

**Solution**

(a) The graph of the straight line is a function. Any vertical line will cross this straight line in only one location.

(b) and (c) Each of these graphs is not the graph of a function. In each case there exists a vertical line that will cross the curve in more than one place.

(a) 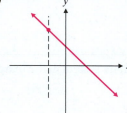 (b) 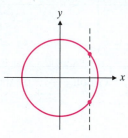 (c)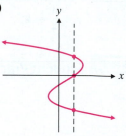

**Student Practice 7** Determine whether each of the following is the graph of a function.

(a) 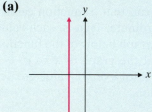 (b) 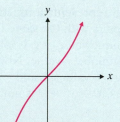 (c)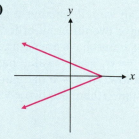

CAUTION: Be careful. The notation $f(x)$ does not mean $f$ multiplied by $x$.

## 4  Using Function Notation

We have seen that an equation like $y = 2x + 7$ is a function. For each value of $x$, the equation assigns a unique value to $y$. We could say, "$y$ is a function of $x$." If we name the function $f$, this statement can be symbolized by using the **function notation** $y = f(x)$. Many times we avoid using the $y$-variable completely and write the function as $f(x) = 2x + 7$.

**Example 8** If $f(x) = 3x^2 - 4x + 5$, find each of the following.

(a) $f(-2)$        (b) $f(4)$        (c) $f(0)$

**Solution**

(a) $f(-2) = 3(-2)^2 - 4(-2) + 5 = 3(4) - 4(-2) + 5$
$= 12 + 8 + 5 = 25$

(b) $f(4) = 3(4)^2 - 4(4) + 5 = 3(16) - 4(4) + 5$
$= 48 - 16 + 5 = 37$

(c) $f(0) = 3(0)^2 - 4(0) + 5 = 3(0) - 4(0) + 5 = 0 - 0 + 5 = 5$ □

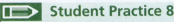 **Student Practice 8**

If $f(x) = -2x^2 + 3x - 8$, find each of the following.

(a) $f(2)$

(b) $f(-3)$

(c) $f(0)$

When evaluating a function, it is helpful to place parentheses around the value that is being substituted for $x$. Taking the time to do this will minimize sign errors in your work.

Some functions are useful in medicine, anthropology, and forensic science. For example, the approximate length of a man's femur (thigh bone) is given by the function $f(x) = 0.53x - 17.03$, where $x$ is the height of the man in inches. If a man is 68 inches tall, $f(68) = 0.53(68) - 17.03 = 19.01$. A man 68 inches tall would have a femur length of approximately 19.01 inches.

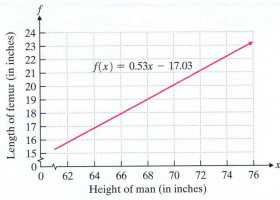

*Source*: www.nsbri.org

**Sidelight: Linear Versus Nonlinear**
When graphing, keep the following facts in mind.

The graph of a linear equation will be a straight line.

The graph of a nonlinear equation will *not* be a straight line.

The graphs in Examples 4, 5, and 6 are *not* straight lines because the equations are *not* linear equations. The graph shown above *is* a straight line because its equation *is* a linear equation.

## 7.6 Exercises MyMathLab®

### Verbal and Writing Skills, Exercises 1–6

1. What are the three ways you can describe a function?

2. What is the difference between a function and a relation?

3. The domain of a function is the set of _____ _____ of the _____ variable.

4. The range of a function is the set of _____ _____ of the _____ variable.

5. How can you tell whether a graph is the graph of a function?

6. Without drawing a graph, how could you tell if the equation $x = y^2$ is a function or not?

(a) *Find the domain and range of the relation.* (b) *Determine whether the relation is a function.*

7. $\left\{ \left( \frac{3}{7}, 4 \right), \left( 3, \frac{3}{7} \right), \left( -3, \frac{3}{7} \right), \left( \frac{3}{7}, -1 \right) \right\}$

8. $\left\{ \left( \frac{2}{3}, -4 \right), (-4, 5), \left( \frac{2}{3}, 2 \right), (5, -4) \right\}$

9. $\{ (6, 2.5), (3, 1.5), (0, 0.5) \}$

10. $\{ (6, 1), (7, 8), (5, 1) \}$

11. $\{ (12, 1), (14, 3), (1, 12), (9, 12) \}$

12. $\{ (7.2, 8), (7.3, 6), (9, 5.8), (4, 6) \}$

13. $\{ (3, 75), (5, 95), (3, 85), (7, 100) \}$

14. $\{ (85, 3), (95, 11), (110, 15), (110, 20) \}$

*Graph the equation.*

15. $y = x^2 + 3$

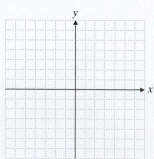

16. $y = x^2 - 1$

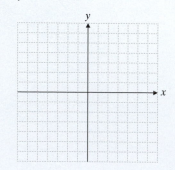

17. $y = 2x^2$

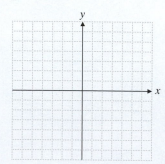

**18.** $y = \dfrac{1}{2}x^2$

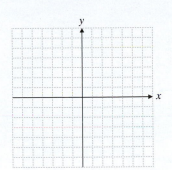

**19.** $x = -2y^2$

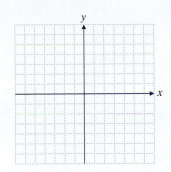

**20.** $x = \dfrac{1}{2}y^2$

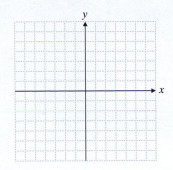

**21.** $x = y^2 - 4$

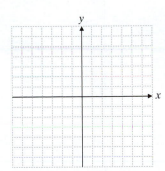

**22.** $x = 2y^2$

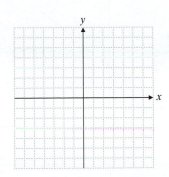

**23.** $y = \dfrac{2}{x}$

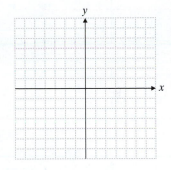

**24.** $y = -\dfrac{2}{x}$

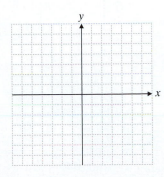

**25.** $y = \dfrac{4}{x^2}$

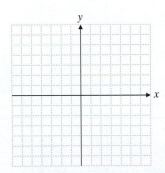

**26.** $y = -\dfrac{6}{x^2}$

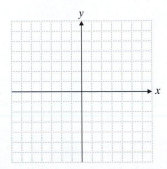

**27.** $x = (y + 1)^2$

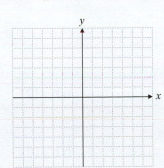

**28.** $y = (x - 3)^2$

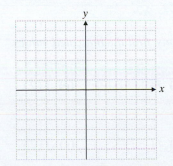

**29.** $y = \dfrac{4}{x - 2}$

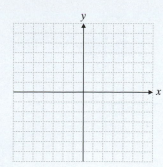

**30.** $x = \dfrac{2}{y + 1}$

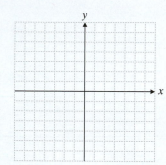

*Determine whether each relation is a function.*

**31.**

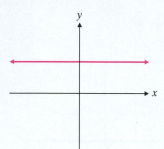

**32.**

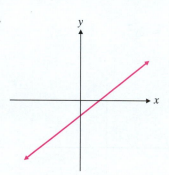

**33.**

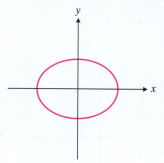

**34.**

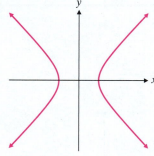

**35.**

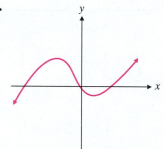

**36.**

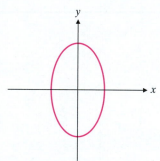

**37.**

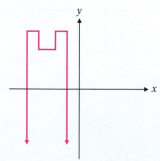

**38.**

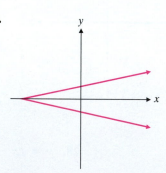

*Given the following functions, find the indicated values.*

**39.** $f(x) = 2 - 3x$
    **(a)** $f(-8)$      **(b)** $f(0)$      **(c)** $f(2)$

**40.** $f(x) = 5 - 2x$
    **(a)** $f(-3)$      **(b)** $f(1)$      **(c)** $f(2)$

**41.** $f(x) = 2x^2 - x + 3$
    **(a)** $f(0)$      **(b)** $f(-3)$      **(c)** $f(2)$

**42.** $f(x) = -x^2 - 2x + 3$
    **(a)** $f(-1)$      **(b)** $f(2)$      **(c)** $f(0)$

## Applications

**43.** *Pet Ownership* A market research company predicted that pet ownership will increase in the United States during the period 2014 to 2024. The approximate number of pet owners measured in thousands could be predicted by the function $f(x) = 0.02x^2 + 0.08x + 31.6$, where $x$ is the number of years since 2014. Predict $f(0), f(4)$, and $f(10)$. Graph the function. What pattern do you observe?

**44.** *Kentucky Population* During a population growth period in Kentucky, from 1995 to 2005, the approximate population of the state measured in thousands could be predicted by the function $f(x) = -0.64x^2 + 30.2x + 3860$, where $x$ is the number of years since 1995. (*Source:* www.census.gov.) Find $f(0), f(5)$, and $f(10)$. Use the function to predict the population in 2015 and 2017. Graph the function. What pattern do you observe?

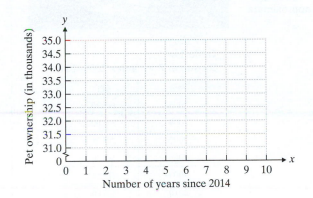

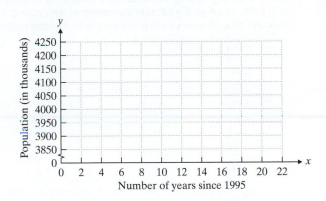

## Cumulative Review   *Simplify.*

**45.** **[1.6.1]** $-4x(2x^2 - 3x + 8)$

**46.** **[1.6.1]** $5a(ab + 6b - 2a)$

**47.** **[1.7.2]** $-7x + 10y - 12x - 8y - 2$

**48.** **[1.7.2]** $3x^2y - 6xy^2 + 7xy + 6x^2y$

---

## Quick Quiz 7.6

**1.** Is this relation a function? $\{(5,7), (7,5), (5,5)\}$ Why?

**2.** For $f(x) = 3x^2 - 4x + 2$:

    **(a)** Find $f(-3)$.     **(b)** Find $f(4)$.

**3.** For $g(x) = \dfrac{7}{x-3}$:

    **(a)** Find $g(2)$.     **(b)** Find $g(-5)$.

**4.** **Concept Check** In the relation $\{(3,4), (5,6), (3,8), (2,9)\}$, why is there a different number of elements in the domain than in the range?

## Background: Economist

The U.S. Bureau of Labor Statistics collects and analyzes data related to economic and social issues important to the American public's interests. Salina, an economist for the bureau, is working on a project involving the consumer price index (CPI), a value indicating the change in the price of a defined set of goods and services over time. She's examining the CPI and its trends over the past decade to see if it's possible to predict future CPI values.

## Facts

Salina has assembled data about the consumer price index over the past ten years. See Table 1.

## Tasks

1. Salina wants to display the data she's assembled in graphical form to get a general sense of when the CPI was increasing or decreasing. Record the approximate location of each year's data point on the line graph outline below.

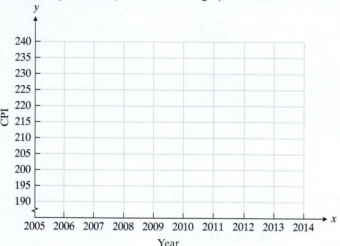

*Source:* U.S. Bureau of Labor Statistics

**Table 1**
**Consumer Price Index 2005–2014**

| Year | CPI |
|------|--------|
| 2005 | 195.30 |
| 2006 | 201.60 |
| 2007 | 207.34 |
| 2008 | 215.30 |
| 2009 | 214.54 |
| 2010 | 218.06 |
| 2011 | 224.94 |
| 2012 | 229.59 |
| 2013 | 232.96 |
| 2014 | 236.74 |

*Source:* U.S. Bureau of Labor Statistics

2. Salina now computes the average rate of change, given by slope, for three time periods: 2005–2008, 2008–2009, and 2010–2014. (Use the precise data points from the table, and round the final values to the nearest hundredth when necessary.) What is the average rate of change for each of these three periods?

3. Now, Salina wants to write the equation of the line that connects the first and last data points to describe the relation of years to CPI. What is this equation?

4. Salina uses the equation she found to approximate the CPI for two future years, if trends continue. How much will the set of goods and services purchased for $236.74 in 2014 cost in 2016? How much will it cost in 2020?

# Chapter 7 Organizer

| Topic and Procedure | Examples | 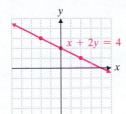 You Try It |
|---|---|---|

**Graphing straight lines by plotting three ordered pairs, p. 409**

An equation of the form

$$Ax + By = C$$

has a graph that is a straight line.

To graph such an equation, plot any three points; two give the line and the third checks it.

Graph $x + 2y = 4$.

| x | y |
|---|---|
| 0 | 2 |
| 2 | 1 |
| -2 | 3 |

**1.** Graph $2x - y = 6$.

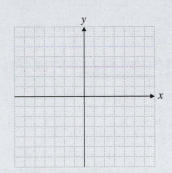

---

**Graphing straight lines by plotting intercepts, p. 411**

1. Let $x = 0$ to find the $y$-intercept.
2. Let $y = 0$ to find the $x$-intercept.
3. Find another ordered pair to have a third point.

For the equation $3x + 2y = 6$:

(a) State the intercepts.

(b) Graph using the intercept method.

| x | y |
|---|---|
| 0 | 3 |
| 2 | 0 |
| 4 | -3 |

$y$-intercept: $(0, 3)$
$x$-intercept: $(2, 0)$

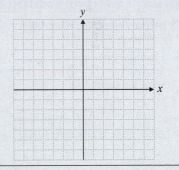

**2.** For the equation $-2x + 4y = -8$:

(a) State the intercepts.

(b) Graph using the intercept method.

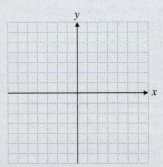

---

**Graphing horizontal and vertical lines, p. 412**

The graph of $x = a$, where $a$ is a real number, is a vertical line.

The graph of $y = b$, where $b$ is a real number, is a horizontal line.

Graph each line.

(a) $x = 3$

(b) $y = -4$

(a)

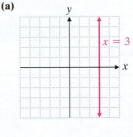

(b)

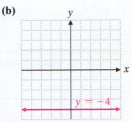

**3.** Graph each line.

(a) $x = -2$

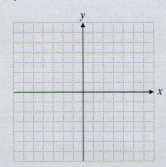

(b) $y = 5$

---

453

| Topic and Procedure | Examples | 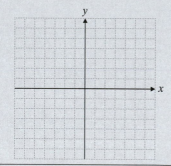 You Try It |
|---|---|---|
| **Finding the slope given two points, p. 419**<br><br>Nonvertical lines passing through distinct points $(x_1, y_1)$ and $(x_2, y_2)$ have slope<br>$$m = \frac{y_2 - y_1}{x_2 - x_1}.$$<br>The slope of a horizontal line is 0. The slope of a vertical line is undefined. | What is the slope of the line through $(2, 8)$ and $(5, 1)$?<br>$$m = \frac{1 - 8}{5 - 2} = -\frac{7}{3}$$ | **4.** What is the slope of the line through $(-2, 5)$ and $(0, 1)$? |
| **Finding the slope and y-intercept of a line given the equation, p. 422**<br><br>1. Rewrite the equation in the form $y = mx + b$.<br>2. The slope is $m$.<br>3. The $y$-intercept is $(0, b)$. | Find the slope and $y$-intercept.<br>$$3x - 4y = 8$$<br>$$-4y = -3x + 8$$<br>$$y = \frac{3}{4}x - 2$$<br>The slope is $\frac{3}{4}$.<br>The $y$-intercept is $(0, -2)$. | **5.** Find the slope and $y$-intercept.<br>$5x + 3y = 9$ |
| **Finding an equation of a line given the slope and y-intercept, p. 423**<br><br>The slope–intercept form of the equation of a line is<br>$$y = mx + b.$$<br>The slope is $m$ and the $y$-intercept is $(0, b)$. | Find an equation of the line with $y$-intercept $(0, 7)$ and with slope $m = 3$.<br>$$y = 3x + 7$$ | **6.** Find an equation of the line with $y$-intercept $(0, -3)$ and with slope $m = 2$. |
| **Graphing a line using slope and y-intercept, p. 423**<br><br>1. Plot the $y$-intercept.<br>2. Starting from $(0, b)$, plot a second point using the slope.<br>$$\text{slope} = \frac{\text{rise}}{\text{run}}$$<br>3. Draw a line that connects the two points. | Graph $y = -4x + 1$.<br>First plot the $y$-intercept at $(0, 1)$.<br>Slope $= -4$ or $\frac{-4}{1}$<br>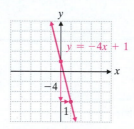 | **7.** Graph $y = 3x - 4$.<br>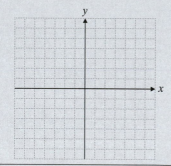 |
| **Finding the slopes of parallel and perpendicular lines, p. 426**<br><br>Parallel lines have the same slope. Perpendicular lines have slopes whose product is $-1$. | Line $q$ has a slope of 2.<br>The slope of a line parallel to $q$ is 2.<br>The slope of a line perpendicular to $q$ is $-\frac{1}{2}$. | **8.** Line $q$ has a slope of 3.<br>**(a)** What is the slope of a line parallel to line $q$?<br>**(b)** What is the slope of a line perpendicular to line $q$? |
| **Finding an equation of the line through a point with a given slope, p. 432**<br><br>1. Write the slope–intercept form of the equation of a line: $y = mx + b$.<br>2. Find $m$ (if not given).<br>3. Substitute the known values into the equation $y = mx + b$.<br>4. Solve for $b$.<br>5. Use the values of $m$ and $b$ to write the slope–intercept form of the equation. | Find an equation of the line through $(3, 2)$ with slope $m = \frac{4}{5}$.<br>$$y = mx + b \qquad 2 = \frac{4}{5}(3) + b$$<br>$$2 = \frac{12}{5} + b$$<br>$$-\frac{2}{5} = b$$<br>An equation is $y = \frac{4}{5}x - \frac{2}{5}$. | **9.** Find an equation of the line through $(1, 3)$ with slope $m = -\frac{1}{2}$. |
| **Finding an equation of the line through two points, pp. 432–433**<br><br>1. Find the slope.<br>2. Use the procedure for finding the equation of a line when given a point and the slope. | Find an equation of the line through $(3, 2)$ and $(13, 10)$.<br>$$m = \frac{y_2 - y_1}{x_2 - x_1} = \frac{10 - 2}{13 - 3} = \frac{8}{10} = \frac{4}{5}$$<br>We are given the point $(3, 2)$.<br>$$y = mx + b \qquad 2 = \frac{4}{5}(3) + b$$<br>$$2 = \frac{12}{5} + b$$<br>$$-\frac{2}{5} = b$$<br>The equation is $y = \frac{4}{5}x - \frac{2}{5}$. | **10.** Find an equation of the line through $(3, 2)$ and $(-1, 0)$. |

| Topic and Procedure | Examples | ▶ You Try It |
|---|---|---|
| **Graphing linear inequalities, p. 436**<br><br>1. Graph as if it were an equation. If the inequality symbol is $>$ or $<$, use a dashed line. If the inequality symbol is $\geq$ or $\leq$, use a solid line.<br>2. Look for a test point. The easiest test point is $(0,0)$, unless the line passes through $(0,0)$. In that case, choose another test point.<br>3. Substitute the coordinates of the test point into the inequality.<br>4. If it is a true statement, shade the region on the side of the line containing the test point. If it is a false statement, shade the region on the side of the line that does *not* contain the test point. | Graph $y \geq 3x + 2$.<br>Graph the line $y = 3x + 2$. Use a solid line.<br><br>Test $(0,0)$. $\quad 0 \geq 3(0) + 2$<br>$\qquad\qquad\qquad 0 \geq 2 \quad$ false<br><br>Shade the region that does not contain $(0,0)$.<br><br>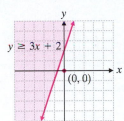 | **11.** Graph $y \leq 2x - 4$.<br>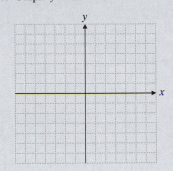 |
| **Determining whether a relation is a function, p. 442**<br><br>A function is a relation in which no two different ordered pairs have the same first coordinate. | Is this relation a function?<br>$$\{(5,7),(3,8),(5,10)\}$$<br>It is *not* a function since $(5,7)$ and $(5,10)$ are two different ordered pairs with the same $x$-coordinate, 5. | **12.** Is this relation a function?<br>$$\{(2,-1),(3,0),(-1,4),(0,4)\}$$ |
| **Graphing nonlinear equations, p. 444**<br><br>Make a table of values to find several ordered pairs that lie on the graph. Plot the ordered pairs and connect the points with a smooth curve. | Graph $y = x^2 - 5$.<br><br>| $x$ | $y$ |<br>|---|---|<br>| $-2$ | $-1$ |<br>| $-1$ | $-4$ |<br>| $0$ | $-5$ |<br>| $1$ | $-4$ |<br>| $2$ | $-1$ |<br>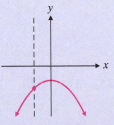 | **13.** Graph $y = (x + 2)^2$.<br>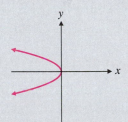 |
| **Determining whether a graph represents a function, p. 445**<br><br>If a vertical line can intersect the graph of a relation more than once, the relation is not a function. If no such line exists, the relation is a function. | Does this graph represent a function?<br>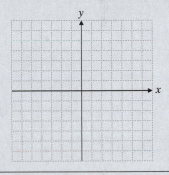<br>Yes. Any vertical line will intersect it at most once. | **14.** Does this graph represent a function?<br>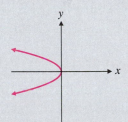 |
| **Evaluating a function, p. 446**<br><br>To evaluate a function, substitute the given value of $x$ into the expression. | If $f(x) = -3x + 8$, find $f(3)$ and $f(-4)$.<br>**(a)** $f(3) = -3(3) + 8 = -9 + 1 = -1$<br>**(b)** $f(-4) = -3(-4) + 8 = 12 + 8 = 20$ | **15.** If $f(x) = x^2 - x$, find<br>**(a)** $f(5)$ and<br>**(b)** $f(-2)$. |

# Chapter 7 Review Problems

## Section 7.1

**1.** Plot and label the following points.

$A: (2, -3)$
$B: (-1, 0)$
$C: (3, 2)$
$D: (-2, -3)$

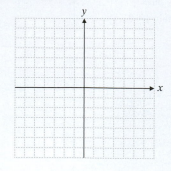

**2.** Give the coordinates of each point.

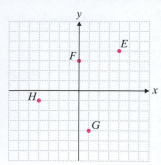

*Complete the ordered pairs so that each is a solution to the given equation.*

**3.** $y = 7 - 3x$
  **(a)** $(0, \ )$   **(b)** $(\ , 10)$

**4.** $2x + 5y = 12$
  **(a)** $(1, \ )$   **(b)** $(\ , 4)$

**5.** $x = 6$
  **(a)** $(\ , -1)$   **(b)** $(\ , 3)$

## Section 7.2

**6.** Graph $5y + x = -15$.

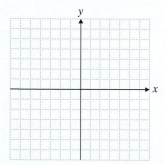

**7.** Graph $2y + 4x = -8 + 2y$.

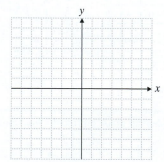

**8.** Graph $3y = 2x + 6$ and label the intercepts.

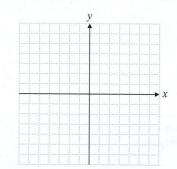

## Section 7.3

**9.** Find the slope of the line passing through $(5, -3)$ and $\left(2, -\frac{1}{2}\right)$.

**10.** The equation of a line is $y = \frac{3}{5}x - 2$. What is the slope of a line perpendicular to that line?

**11.** Find the slope and $y$-intercept of the line $9x - 11y + 15 = 0$.

**12.** Write an equation of the line with slope $-\frac{1}{2}$ and $y$-intercept $(0, 3)$.

**13.** Graph $y = -\frac{1}{2}x + 3$.

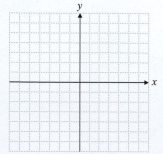

**14.** Graph $2x - 3y = -12$.

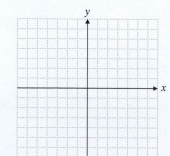

**15.** Graph $y = -2x$.

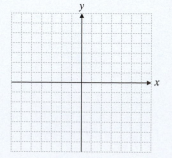

## Section 7.4

**16.** Write an equation of the line passing through $(3, -4)$ having a slope of $-6$.

**17.** Write an equation of the line passing through $(-1, 4)$ having a slope of $-\frac{1}{3}$.

**18.** Write an equation of the line passing through $(2, 5)$ having a slope of 1.

**19.** Write an equation of the line passing through $(3, 7)$ and $(-6, 7)$.

*Write an equation of the graph.*

**20.**

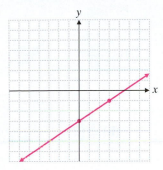

**21.**

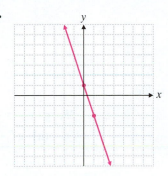

**22.**

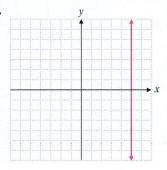

## Section 7.5

**23.** Graph $y < \frac{1}{3}x + 2$.

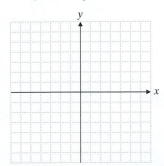

**24.** Graph $3y + 2x \geq 12$.

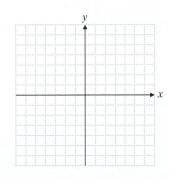

**25.** Graph $x \leq 2$.

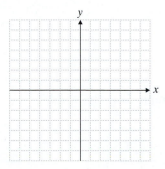

## Section 7.6

*Determine the domain and range of the relation. Determine whether the relation is a function.*

**26.** $\{(5, -6), (-6, 5), (-5, 5), (-6, -6)\}$

**27.** $\{(2, -3)(5, -3)(6, 4)(-2, 4)\}$

*In exercises 28–30, determine whether the graphs represent a function.*

**28.**

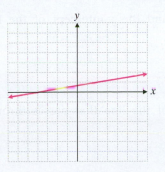

**29.**

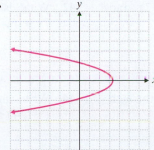

**30.**

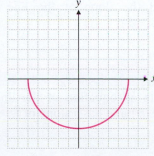

**31.** Graph $y = x^2 - 5$.

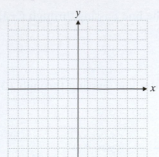

**32.** Graph $x = y^2 + 3$.

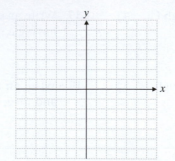

**33.** Graph $y = (x - 3)^2$.

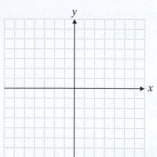

*Given the following functions, find the indicated values.*

**34.** $f(x) = 7 - 6x$

   **(a)** $f(0)$            **(b)** $f(-4)$

**35.** $g(x) = -2x^2 + 3x + 4$

   **(a)** $g(-1)$          **(b)** $g(3)$

**36.** $f(x) = \dfrac{2}{x + 4}$

   **(a)** $f(-2)$          **(b)** $f(6)$

**37.** $f(x) = x^2 - 2x + \dfrac{3}{x}$

   **(a)** $f(-1)$          **(b)** $f(3)$

## Mixed Practice

**38.** Graph $5x + 3y = -15$.

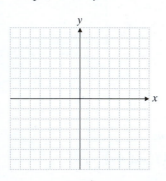

**39.** Graph $y = \frac{3}{4}x - 3$.

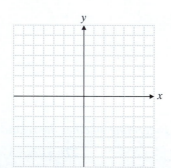

**40.** Graph $y < -2x + 1$.

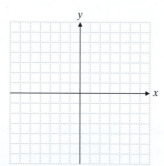

**41.** Find the slope of the line through $(2, -7)$ and $(-3, -5)$.

**42.** What is the slope and the $y$-intercept of the line $7x + 6y - 10 = 0$?

**43.** Write an equation of the line that passes through $(3, -5)$ and has a slope of $\frac{2}{3}$.

**44.** Write an equation of the line that passes through $(-1, 4)$ and $(2, 1)$.

## Applications

***Monthly Electric Bill*** *Russ and Norma Camp found that their monthly electric bill could be calculated by the equation* $y = 30 + 0.09x$. *In this equation y represents the amount of the monthly bill in dollars and x represents the number of kilowatt-hours used during the month.*

**45.** What would be their monthly bill if they used 2000 kilowatt-hours of electricity?

**46.** What would be their monthly bill if they used 1600 kilowatt-hours of electricity?

**47.** Write the equation in the form $y = mx + b$, and determine the numerical value of the $y$-intercept. What is the significance of this $y$-intercept? What does it tell us?

**48.** If the equation is placed in the form $y = mx + b$, what is the numerical value of the slope? What is the significance of this slope? What does it tell us?

**49.** If Russ and Norma have a monthly bill of $147, how many kilowatt-hours of electricity did they use?

**50.** If Russ and Norma have a monthly bill of $246, how many kilowatt-hours of electricity did they use?

***Job Losses in Manufacturing*** *Economists in the Labor Department are concerned about the continued job loss in the manufacturing industry. The number of people employed in manufacturing jobs in the United States can be predicted by the equation* $y = -269x + 17{,}020$, *where x is the number of years since 1994 and y is the number of employees in* **thousands** *in the manufacturing industry. Use this data to answer the following questions. (Source: www.bls.gov)*

**51.** How many people were employed in manufacturing in 1994? In 2008? In 2014?

**52.** Use your answers for 51 to draw a graph of the equation $y = -269x + 17{,}020$.

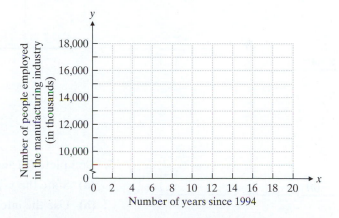

**53.** What is the slope of this equation? What is its significance?

**54.** What is the $y$-intercept of this equation? What is its significance?

**55.** Use the equation to predict in what year the number of manufacturing jobs will be 11,102,000.

**56.** Use the equation to predict in what year the number of manufacturing jobs will be 10,295,000.

459

# How Am I Doing? Chapter 7 Test

MATH COACH          MyMathLab®          You Tube

After you take this test read through the Math Coach on pages 462–463. Math Coach videos are available via MyMathLab and YouTube. Step-by-step test solutions on the Chapter Test Prep Videos are also available via MyMathLab and YouTube. (Search "TobeyBeginningAlg" and click on "Channels.")

**1.** Plot and label the following points.

$B: (6, 1)$     $C: (-4, -3)$
$D: (-3, 0)$     $E: (5, -2)$

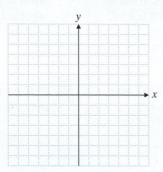

**2.** Graph the line $6x - 3 = 5x - 2y$.

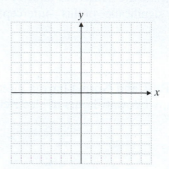

**3.** Graph $-3x + 9 = 6x$.

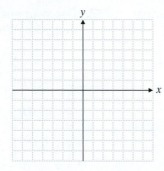

ᴹᴄ **4.** Graph $y = \dfrac{2}{3}x - 4$.

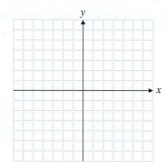

**5.** Complete **(a)** and **(b)** for the equation $4x + 2y = -8$.

   **(a)** State the $x$- and $y$-intercepts.

   **(b)** Use the intercept method to graph.

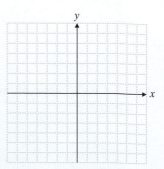

**6.** Find the slope of the line that passes through $(8, 6)$ and $(-3, -5)$.

**7.** What is the slope and the $y$-intercept of the line $3x + 2y - 5 = 0$?

1. _____

2. _____

3. _____

4. _____

5. (a) _____

  (b) _____

6. _____

7. _____

**8.** Find an equation of the line with slope $\frac{3}{4}$ and $y$-intercept $(0, -6)$.

**9. (a)** Write an equation for the line $f$ that passes through $(4, -2)$ and has a slope of $\frac{1}{2}$.

  **(b)** What is the slope of a line perpendicular to line $f$?

$\mathbb{MC}$ **10.** Find the equation for the line passing through $(5, -4)$ and $(-3, 8)$.

**11.** Graph the region described by $4y \le 3x$.

$\mathbb{MC}$ **12.** Graph the region described by $-3x - 2y > 10$.

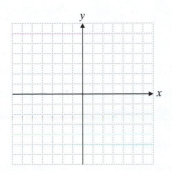
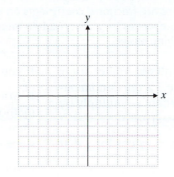

**13.** Is this relation a function? $\{(2, -8), (3, -7), (2, 5)\}$ Why?

**14.** Look at the relation graphed below. Is this relation a function? Why?

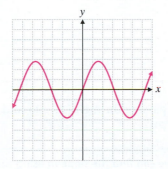

**15.** Graph $y = 2x^2 - 3$.

| x | y |
|----|----|
| $-2$ | |
| $-1$ | |
| $0$ | |
| $1$ | |
| $2$ | |

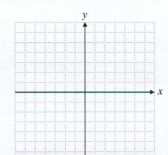

**16.** For $f(x) = -x^2 - 2x - 3$:
  **(a)** Find $f(0)$.
  **(b)** Find $f(-2)$.

**8.** _____ □

**9. (a)** _____ □

**(b)** _____ □

**10.** _____ □

**11.** _____ □

**12.** _____ □

**13.** _____ □

**14.** _____ □

**15.** _____ □

**16. (a)** _____ □

**(b)** _____ □

**Total Correct:** □

461

# MATH COACH

*Mastering the skills you need to do well on the test.*

The following problems are from the Chapter 7 Test. Here are some helpful hints to keep you from making common errors on test problems.

**Chapter 7 Test, Problem 4**   Graph $y = \frac{2}{3}x - 4$.

**HELPFUL HINT** Find three ordered pairs that are solutions to the equation. Plot those 3 points. Then draw a line through the points. When the equation is solved for $y$ and there are fractional coefficients on $x$, it is sometimes a good idea to choose values for $x$ that result in integer values for $y$. This will make graphing easier.

Choosing values for $x$ that result in integer values for $y$ means choosing 0 or multiples of 3 since 3 is the denominator of the fractional coefficient on $x$. When we multiply by these values, the result becomes an integer.

Did you choose values for $x$ that result in values for $y$ that are not fractions?

Yes _____ No _____

If you answered No, try using $x = 0$, $x = 3$, and $x = 6$. Solve the equation for $y$ in each case to find the $y$-coordinate. Remember that it will make the graphing process easier if you choose values for $x$ that will clear the fraction from the equation.

Did you plot the three points and connect them with a line?

Yes _____ No _____

If you answered No, go back and complete this step.

If you feel more comfortable using the slope–intercept method to solve this problem, simply identify the $y$-intercept from the equation in $y = mx + b$ form, and then use the slope $m$ to find two other points.

If you answered Problem 4 incorrectly, go back and rework the problem using these suggestions.

---

**Chapter 7 Test, Problem 10**   Find the equation for the line passing through $(5, -4)$ and $(-3, 8)$.

**HELPFUL HINT** When given two points $(x_1, y_1)$ and $(x_2, y_2)$, we can find the slope $m$ using the slope formula $m = \frac{y_2 - y_1}{x_2 - x_1}$. Then you can substitute $m$ into the equation $y = mx + b$ along with the coordinates of one of the points to find $b$, the $y$-intercept.

When you substituted the points into the slope formula to find $m$, did you obtain either $m = \frac{8 - (-4)}{-3 - 5}$ or $m = \frac{-4 - 8}{5 - (-3)}$?

Yes _____ No _____

If you answered No, check your work to make sure you substituted the points correctly. Be careful to avoid any sign errors.

Need more help? Watch the **MATH COACH** videos in MyMathLab® or on YouTube™.

462

Did you use $m = -\dfrac{3}{2}$ and either of the points given when you substituted the values into the equation $y = mx + b$ to find the value of $b$?

Yes _____ No _____

If you answered No, stop and make a careful substitution for $m = -\dfrac{3}{2}$ and either $x = 5$ and $y = -4$ or $x = -3$ and $y = 8$. See if you can solve the resulting equation for $b$.

Now go back and rework the problem using these suggestions.

**Chapter 7 Test, Problem 12**    Graph the region described by $-3x - 2y > 10$.

> **HELPFUL HINT** First graph the equation $-3x - 2y = 10$. Determine if the line should be solid or dashed. Then pick a test point to see if it satisfies the inequality $-3x - 2y > 10$. If the test point satisfies the inequality, shade the side of the line on which the point lies. If the test point does not satisfy the inequality, shade the opposite side of the line.

Examine your work. Does the line $-3x - 2y = 10$ pass through the point $(0, -5)$?

Yes _____ No _____

If you answered No, substitute $x = 0$ into the equation and solve for $y$. Check the calculations for each of the points you plotted to find the graph of this equation.

Did you draw a solid line?

Yes _____ No _____

If you answered Yes, look at the inequality symbol. Remember that we only use a solid line with the symbols $\leq$ and $\geq$. A dashed line is used for $<$ and $>$.

Did you shade the area above the dashed line?

Yes _____ No _____

If you answered Yes, stop now and use $(0, 0)$ as a test point and substitute it into the inequality $-3x - 2y > 10$. Then use the Helpful Hint to determine which side to shade.

If you answered Problem 12 incorrectly, go back and rework the problem using these suggestions.

**Chapter 7 Test, Problem 16(a) and 16(b)**

For $f(x) = -x^2 - 2x - 3$:    **(a)** Find $f(0)$.    **(b)** Find $f(-2)$.

> **HELPFUL HINT** Replace $x$ with the number indicated. It is a good idea to place parentheses around the value to avoid any sign errors. Then use the order of operations to evaluate the function in each case.

Did you replace $x$ with $0$ and write

$$f(0) = -(0)^2 - 2(0) - 3?$$

Yes _____ No _____

If you answered No, take time to go over your steps one more time, remembering that $0$ times any number is $0$.

Did you replace $x$ with $-2$ and write

$$f(-2) = -(-2)^2 - 2(-2) - 3?$$

Yes _____ No _____

If you answered No, go over your steps again, remembering to place parentheses around $-2$. Note that $(-2)^2 = 4$ and therefore $-(-2)^2 = -4$.

Now go back and rework the problem again using these suggestions.

Need more help? Look for section examples marked with $^{M}_{C}$ to review.

CHAPTER 8

# Systems of Equations

## CAREER OPPORTUNITIES

## Marketing Manager

Marketing and sales managers must be aware of both the popularity of what they're selling and its availability—also known as supply and demand. Effectively pricing products by balancing consumers' desire for the product with what they're willing to pay for it is a key responsibility of marketing and sales managers. These professionals have to be sensitive to trends and understand systems of supply and demand equations. Using the math in this chapter, they're positioned to make business decisions that lead to increased profitability.

To investigate how the mathematics in this chapter can help with this field, see the Career Exploration Problems on page 500.

## 8.1 Solving a System of Equations in Two Variables by Graphing

In Chapter 7 we examined linear equations and inequalities in two variables. In this chapter we will examine systems of linear equations. Two or more equations in several variables that are considered simultaneously are called a **system of equations.** Recall that the graph of a linear equation is a straight line. If we have the graphs of two linear equations on one coordinate system, how might they relate to one another?

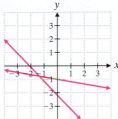 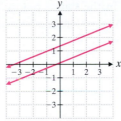 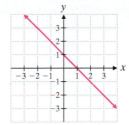

The lines may intersect.     The lines may be parallel.     The lines may coincide.

The solutions to a system of linear equations are those points that both lines have in common. Note that parallel lines have no points in common. Thus the second system has no solution. Intersecting lines intersect at one point. Thus the first system has one solution. Lines that coincide have an infinite number of points in common. Thus the third system has an infinite number of solutions. You can determine the solution to a system of equations by graphing.

### 1 Using Graphing to Solve a System of Linear Equations with a Unique Solution

To solve a system of equations by graphing, graph both equations on the same coordinate system. If the lines intersect, the system has a unique solution. The ordered pair of coordinates of the point of intersection is the solution to the system of equations.

**Example 1** Solve by graphing.

$$-2x + 3y = 6$$
$$2x + 3y = 18$$

**Solution** Graph both equations on the same coordinate system and determine the point of intersection.

We will graph the system by first obtaining a table of values for each equation.

First we solve the first equation for $y$.

$$-2x + 3y = 6$$
$$3y = 2x + 6$$
$$y = \frac{2}{3}x + 2$$

Now that we have the slope–intercept form of the equation, we can easily see what values to substitute for $x$ so that the $y$-values are not fractions.

| Let $x = 0$. | $y = \frac{2}{3}(0) + 2 = 0 + 2 = 2$ |
| Let $x = -3$. | $y = \frac{2}{3}(-3) + 2 = -2 + 2 = 0$ |
| Let $x = 6$. | $y = \frac{2}{3}(6) + 2 = 4 + 2 = 6$ |

| x | y |
|---|---|
| 0 | 2 |
| -3 | 0 |
| 6 | 6 |

This allows us to use the three sets of ordered pairs to graph $-2x + 3y = 6$.

*Continued on next page*

*Continued on next page*

### Student Learning Objectives

After studying this section, you will be able to:

1 Use graphing to solve a system of linear equations with a unique solution.

2 Graph a system of linear equations to determine the type of solution.

3 Solve applied problems by graphing.

### Graphing Calculator

**Graphing a System of Linear Equations**

To solve the system
$$3x + 4y = -28$$
$$x - 4y = 12,$$
we first need to solve each equation for $y$. This will give us the equivalent system
$$y = -0.75x - 7$$
$$y = 0.25x - 3.$$
We graph this system with a scale of $-10$ to $10$ for both $x$ and $y$.
Display:

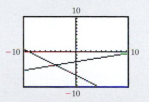

If we use the Zoom function, we will locate the intersection at $x = -4, y = -4$.

Next we solve the second equation for $y$. $\quad 2x + 3y = 18$
$$3y = -2x + 18$$
$$y = -\frac{2}{3}x + 6$$

| x | y |
|---|---|
| 0 | 6 |
| −3 | 8 |
| 6 | 2 |

Let $x = 0$. $\qquad y = -\frac{2}{3}(0) + 6 = 0 + 6 = 6$

Let $x = -3$. $\qquad y = -\frac{2}{3}(-3) + 6 = 2 + 6 = 8$

Let $x = 6$. $\qquad y = -\frac{2}{3}(6) + 6 = -4 + 6 = 2$

This allows us to use these three ordered pairs to graph $2x + 3y = 18$. The graph of the system is shown in the figure below. The lines intersect at the point for which $x = 3$ and $y = 4$. Thus the unique solution to the system of equations is $(3, 4)$. Always check your answer.

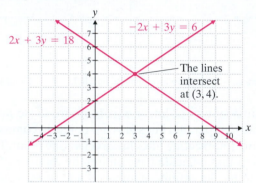

**Student Practice 1**

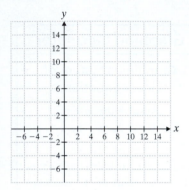

*Check.* $\qquad -2x + 3y = 6 \qquad\qquad 2x + 3y = 18$

$$-2(3) + 3(4) \stackrel{?}{=} 6 \qquad\qquad 2(3) + 3(4) \stackrel{?}{=} 18$$

$$6 = 6 \ \checkmark \qquad\qquad\qquad 18 = 18 \ \checkmark$$

A linear system of equations that has one solution is said to be **consistent.** ◻

**Student Practice 1** Solve by graphing. $\qquad x + y = 12$
$$-x + y = \ 4$$

**2** Graphing a System of Linear Equations to Determine the Type of Solution

Two lines in a plane may intersect, may never intersect (be parallel lines), or may be the same line. The corresponding system of equations will have one solution, no solution, or an infinite number of solutions, respectively. Thus far we have focused on systems of linear equations that have a unique solution. We now draw your attention to the other two cases.

**Example 2** Solve by graphing.

$$3x - y = \ \ 1$$
$$3x - y = -7$$

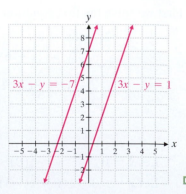

**Student Practice 2**

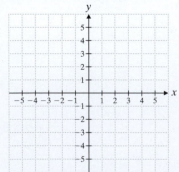

**Solution** We graph both equations on the same coordinate system. The graph is at the right. Notice that the lines are parallel. They do not intersect. Hence there is *no solution* to this system of equations. A system of linear equations that has no solution is called **inconsistent.** ◻

**Student Practice 2** Solve by graphing.

$$4x + 2y = 8$$
$$-6x - 3y = 6$$

**Example 3** Solve by graphing.

$$x + y = 4$$
$$3x + 3y = 12$$

**Solution**  We graph both equations on the same coordinate system at the right. Notice that the equations represent the same line (coincide). The coordinates of every point on the line will satisfy both equations. That is, every point on the line represents a solution to $x + y = 4$ and to $3x + 3y = 12$. Thus there is *an infinite number of solutions* to this system. Such equations are said to be **dependent.**

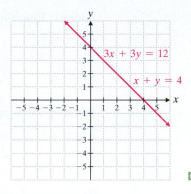

**Student Practice 3**  Solve by graphing.

$$3x - 9y = 18$$
$$-4x + 12y = -24$$

**Student Practice 3**

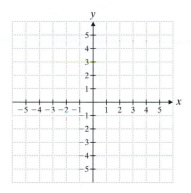

**To Think About:**  *Determining the Type of Solution Without Graphing*  Without graphing you can tell that the system in Example 2 represents parallel lines because both lines have the *same slope*. Verify this fact by writing each equation in slope–intercept form. What can you say about the *y*-intercepts for each of these lines? Now look at the system in Example 3 and describe how, without graphing, you could determine that this system is dependent (lines coincide) rather than inconsistent (lines don't intersect).

## 3  Solving Applied Problems by Graphing

**Example 4**  Walter and Barbara need some plumbing repairs done at their house. They called two plumbing companies for estimates of the work that needs to be done. Roberts Plumbing and Heating charges $40 for a house call and then $35 per hour for labor. Instant Plumbing Repairs charges $70 for a house call and then $25 per hour for labor.

**(a)** Create a cost equation for each company, where $y$ is the total cost of plumbing repairs and $x$ is the number of hours of labor. Write a system of equations.

**(b)** Graph the two equations using the values $x = 0, 3,$ and $6$.

**(c)** Determine from your graph how many hours of plumbing repairs would be required for the two companies to charge the same.

**(d)** Determine from your graph which company charges less if the estimated amount of time to complete the plumbing repairs is 4 hours.

**Solution**

**(a)** For each company we will obtain a cost equation.

| Total cost of plumbing | = | Cost of house call | + | cost per hour | × | number of labor hours | |
|---|---|---|---|---|---|---|---|
| $y$ | = | 40 | + | 35 | × | $x$ | Roberts Plumbing and Heating charges $40 for a house call and $35 per hour. |
| $y$ | = | 70 | + | 25 | × | $x$ | Instant Plumbing charges $70 for a house call and $25 per hour. |

This yields the following system of equations.

$$y = 40 + 35x$$
$$y = 70 + 25x$$

*Continued on next page*

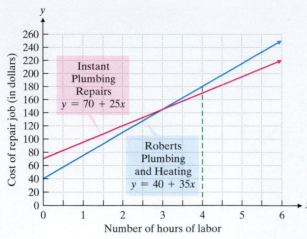

**(b)** We will graph the system by first obtaining a table of values for each equation using $x = 0, 3$, and 6.

**Roberts Plumbing and Heating** $y = 40 + 35x$

Let $x = 0$. $\quad y = 40 + 35(0) = 40$

Let $x = 3$. $\quad y = 40 + 35(3) = 145$

Let $x = 6$. $\quad y = 40 + 35(6) = 250$

| x | y |
|---|---|
| 0 | 40 |
| 3 | 145 |
| 6 | 250 |

**Instant Plumbing Repairs** $y = 70 + 25x$

Let $x = 0$. $\quad y = 70 + 25(0) = 70$

Let $x = 3$. $\quad y = 70 + 25(3) = 145$

Let $x = 6$. $\quad y = 70 + 25(6) = 220$

| x | y |
|---|---|
| 0 | 70 |
| 3 | 145 |
| 6 | 220 |

The graph of the system is to the left.

**(c)** We see that the graphs of the two lines intersect at (3, 145). Thus the two companies will charge the same if 3 hours of plumbing repairs are required.

**(d)** We draw a dashed line at $x = 4$. We see that the blue line, representing Roberts Plumbing and Heating, is higher than the red line, representing Instant Plumbing Repairs after 3 hours. Thus the cost would be less if Walter and Barbara use Instant Plumbing Repairs for 4 hours of work. □

**Student Practice 4** Ken and Joan Thompson need some electrical work done on their new house. They obtained estimates from two companies. Bill Tupper's Electrical Service charges $100 for a house call and $30 per hour for labor. Wire for Hire charges $50 for a house call and $40 per hour for labor.

**(a)** Create a cost equation for each company where $y$ is the total cost of the electrical work and $x$ is the number of hours of labor. Write a system of equations.

**(b)** Graph the two equations using the values $x = 0, 4$, and 8.

**(c)** Determine from your graph how many hours of electrical repairs would be required for the two companies to charge the same.

**(d)** Determine from your graph which company charges less if the estimated amount of time to complete the electrical repairs to the house is 6 hours.

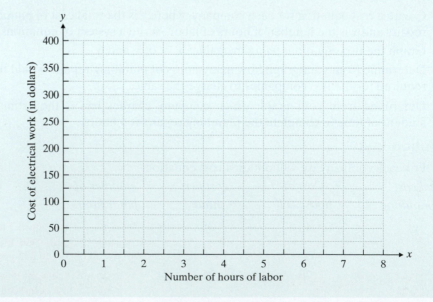

**Verbal and Writing Skills, Exercises 1–6**

1. In the system $y = 2x - 5$ and $y = 2x + 6$, the lines have the same slope but different $y$-intercepts. What does this tell you about the graph of this system?

2. In the system $y = -3x + 4$ and $y = -3x + 4$, the lines have the same slope and the same $y$-intercepts. What does this tell you about the graph of the system?

3. Before you graph a system of equations, if you notice that the lines have different slopes, you can conclude that they _____ and that the system has _____ solution.

4. *Engineering Assistant* Noah is working this semester as an engineering assistant. He has been given a situation to analyze. He notices that two straight lines have the same slope. Can he conclude that the lines are parallel? Why?

5. *Engineering Drafting* Lexi has a job this semester doing drafting tasks for an engineering firm. She is considering two straight lines that have different slopes, but they have the same $y$-intercept. What can she conclude about the graphs of these lines? What can she conclude about the solution of the system?

6. *Gymnastics Presentation Designer* Olivia is a gymnast. She is working this semester designing the layout for several college gymnastics competitions. In order to mark the boundaries for the competition area, she is using the two equations $y = 22x - 17$ and $y = 22x + 150$. When she solves this system of equations, what result will she obtain? Why?

*Solve by graphing. If there isn't a unique solution to the system, state the reason.*

7. $x - y = 3$
$x + y = 5$

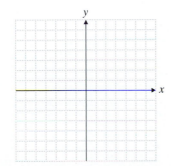

8. $x - y = 6$
$-x - y = 4$

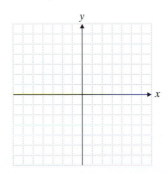

9. $y = -3x$
$y = 2x + 5$

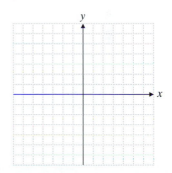

10. $x + 4y = 6$
$-x + 2y = 12$

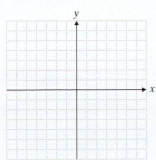

11. $4x + y = 5$
$3x - 2y = 12$

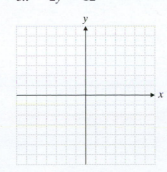

12. $x + 2y = 10$
$3x - 2y = 6$

**13.** $-2x + y - 3 = 0$
$\phantom{-2}4x + y + 3 = 0$

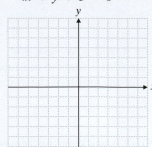

**14.** $\phantom{2}x - 2y - 10 = 0$
$\phantom{2}2x + 3y - \phantom{1}6 = 0$

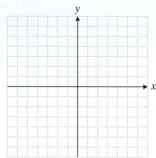

**15.** $3x - 2y = -18$
$2x + 3y = \phantom{-1}14$

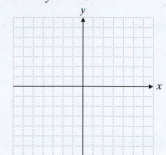

**Mixed Practice** *Solve by graphing. If there isn't a unique solution to the system, state the reason.*

**16.** $\phantom{-}3x + 2y = -10$
$-2x + 3y = \phantom{-}24$

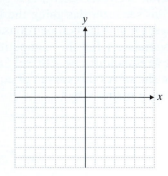

**17.** $y = \phantom{-}\dfrac{3}{4}x + 7$
$y = -\dfrac{1}{2}x + 2$

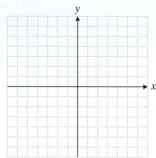

**18.** $y = \dfrac{5}{7}x - 2$
$y = \dfrac{1}{3}x + \dfrac{2}{3}$

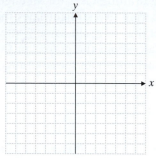

**19.** $\phantom{-}3x - 2y = -4$
$-9x + 6y = -9$

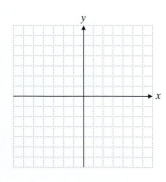

**20.** $\phantom{-}4x - 6y = \phantom{-}8$
$-2x + 3y = -4$

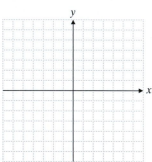

**21.** $y - 2x - 6 = 0$
$\dfrac{1}{2}y - 3 = x$

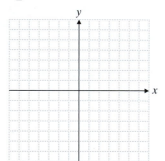

**22.** $2y + x - 6 = 0$
$y + \dfrac{1}{2}x \phantom{ww} = 4$

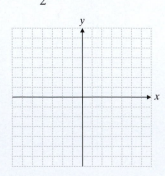

**23.** $y = \dfrac{1}{2}x - 2$
$y = \dfrac{2}{3}x - 1$

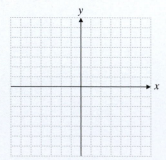

**24.** $2y + 6x - 8 = 0$
$y = \dfrac{3}{2}x + 4$

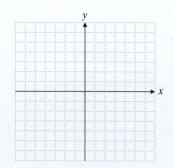

## Applications

**25.** ***Cell Phone Charges*** Jason's cell phone carrier offers two options for cell phone plans. One of the differences in the options refers to the charges for exceeding the allotment of minutes each month. The 10-10 Plan charges 7 cents per minute for each minute Jason exceeds his allotment. The Tier-2 Plan charges 3 cents per minute for each minute over his allotment, plus 36 cents per call made after exceeding his allotment.

**(a)** Create a cost equation for each plan, where $y$ is the total cost for each phone call made after exceeding the allotment and $x$ is the number of minutes the phone call lasts. Write a system of equations.

**(b)** Graph the two equations using the values $x = 1, 4, 8$, and $12$. Fill in a table for those values of $x$ and $y$.

**(c)** Determine from the graph the length of a phone call, in minutes, when both plans would cost Jason the same amount of money for a call made after exceeding his monthly allotment.

**(d)** Determine from your graph which plan would cost Jason less per call if he typically makes phone calls of 10 minutes or more after exceeding his monthly allotment.

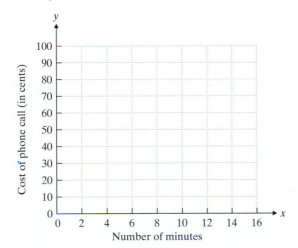

**26.** ***Landscaping Costs*** Fred and Amy want to have shrubs planted in front of their new house. They have obtained estimates from two landscaping companies. Camp Property Care charges $100 for an initial consultation and $60 an hour for labor. Manchester Landscape Designs charges $200 for an initial consultation and $40 an hour for labor.

**(a)** Create a cost equation for each company where $y$ is the total cost of the landscaping work and $x$ is the number of hours of labor. Write a system of equations.

**(b)** Graph the two equations using the values $x = 0, 5$, and $10$. Fill in a table for those values of $x$ and $y$.

**(c)** Determine from your graph how many hours of labor would be required for the cost of landscaping from each of the two companies to be the same.

**(d)** Determine from your graph which company would charge less if 7 hours of landscaping are required.

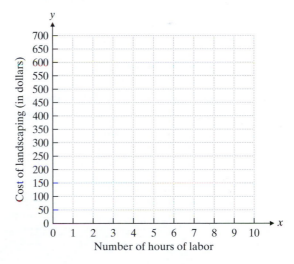

*If you have a graphing calculator, solve each system by graphing the equations and estimating the point of intersection to the nearest hundredth. You will need to adjust the window for the appropriate x- and y-values in order to see the region where the straight lines intersect.*

**27.** $34x + 26y = 1224.36$
$52x - 17y = 3221.28$

**28.** $65x + 49y = -5252.07$
$19x - 58y = 3092.01$

**29.** $y = 67x + 1998$
$y = 13x + 756$

**30.** $y = -14x + 588$
$y = 55x + 1623$

## Cumulative Review

**31. [2.4.1]** Solve. $\dfrac{2}{3}x - \dfrac{1}{15} = \dfrac{3}{5}$

**32. [2.4.1]** Solve. $\dfrac{2}{5}x + 2 = 6$

**33. [2.5.1]** Solve for $x$. $3x - 2y = 6$

**34. [2.5.1]** Solve for $y$. $4x + 2y = 16$

---

## Quick Quiz 8.1

**1.** Solve the following system by graphing.
$$x - 3y = 6$$
$$4x + 3y = 9$$

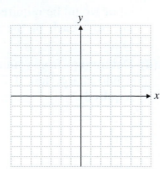

**2.** Find the slope and the $y$-intercept of each of the following lines.
$$y = \dfrac{3}{5}x + 2$$
$$5y - 3x = 35$$
What does this tell you if you want to find the solution to the system of equations by the graphing method?

**3.** What will happen when you try to find the solution of the following system of equations by graphing?
$$y = 3x - 2$$
$$-6x + 2y = -4$$

**4. Concept Check** You are attempting to find the solution to this system of equations by graphing.
$$-3x + 4y = -16$$
$$0 = 6x - 8y + 16$$
What will you discover?

# 8.2 Solving a System of Equations in Two Variables by the Substitution Method ▶

## 1 Solving a System of Two Linear Equations with Integer Coefficients by the Substitution Method ▶

Since the solution to a system of two linear equations may be a point where the two lines intersect, we would expect the solution to be an ordered pair. Since the point must lie on both lines, the ordered pair $(x, y)$ must satisfy both equations. For example, the solution to the system $3x + 2y = 6$ and $x + y = 3$ is $(0, 3)$. Let's check.

$$3x + 2y = 6 \qquad\qquad x + y = 3$$
$$3(0) + 2(3) \overset{?}{=} 6 \qquad\qquad 0 + 3 \overset{?}{=} 3$$
$$6 = 6 \checkmark \qquad\qquad 3 = 3 \checkmark$$

Given a system of linear equations, we can develop algebraic methods for finding a solution. These methods reduce the system to one equation in one variable, which we already know how to solve. Once we have solved for this variable, we can substitute this known value into any one of the original equations and solve for the second variable.

The first method we will discuss is the **substitution method.**

> **PROCEDURE FOR SOLVING A SYSTEM OF EQUATIONS BY THE SUBSTITUTION METHOD**
>
> 1. Solve one of the two equations for one variable. If possible, solve for a variable with a coefficient of 1 or −1.
> 2. Substitute the expression from step 1 into the *other* equation.
> 3. You now have one equation with one variable. Solve this equation to find the value for that one variable.
> 4. Substitute this value for the variable into one of the original or equivalent equations to obtain a value for the other variable.
> 5. Check the solution in both original equations to verify your results.

Student Learning Objectives

After studying this section, you will be able to:

1 Solve a system of two linear equations with integer coefficients by the substitution method. ▶

2 Solve a system of two linear equations with fractional coefficients by the substitution method. ▶

Mᴄ **Example 1** Find the solution. Check your answer.

$$x - 2y = \phantom{-}7 \quad \textbf{(1)}$$
$$-5x + 4y = -5 \quad \textbf{(2)}$$

### Solution

**Step 1** Solve one equation for one variable.

$x - 2y = 7$     Equation **(1)** is the easiest in which to isolate a variable.

$x = 7 + 2y$     Add $2y$ to both sides to solve for $x$.

**Step 2** Now substitute this expression into the other equation.

$-5x + 4y = -5$     Write equation **(2)**.

$-5(7 + 2y) + 4y = -5$     Substitute the value $7 + 2y$ for $x$ in this equation.

Now we have one equation in one variable, $y$.

**Step 3** Solve this equation.

$-35 - 10y + 4y = -5$     Remove the parentheses.

$-35 - 6y = -5$     Combine like terms.

$-6y = 30$     Add 35 to both sides.

$y = -5$     Divide both sides by −6.

⟶ **Student Practice 1** Find the solution. Be sure to check your answer.

$$5x + 3y = 19$$
$$2x - \phantom{3}y = 12$$

*Continued on next page*

**Step 4** We will now use the value $y = -5$ in one of the equivalent equations to find the value for $x$.

$x = 7 + 2y$     The easiest equation to use is the one we found in step 1.

$x = 7 + 2(-5)$     Replace $y$ by $-5$.

$x = 7 + (-10)$     Simplify.

$x = -3$     Solve for $x$.

The solution is $(-3, -5)$.

**Step 5** *Check.* To be sure that we have the correct solution, we will need to check that the values obtained for $x$ and $y$ can be substituted into *both original* equations to obtain true mathematical statements.

$$x - 2y = 7 \quad \textbf{(1)} \qquad\qquad -5x + 4y = -5 \quad \textbf{(2)}$$
$$-3 - 2(-5) \stackrel{?}{=} 7 \qquad\qquad -5(-3) + 4(-5) \stackrel{?}{=} -5$$
$$-3 + 10 \stackrel{?}{=} 7 \qquad\qquad 15 - 20 \stackrel{?}{=} -5$$
$$7 = 7 \ \checkmark \qquad\qquad -5 = -5 \ \checkmark \quad \square$$

### 2   Solving a System of Two Linear Equations with Fractional Coefficients by the Substitution Method ▶

If a system of equations contains fractions, clear the equations of fractions *before* performing any other steps.

$^{M}_{C}$ **Example 2** Find the solution. Check your answer.

$$\frac{3}{2}x + y = \frac{5}{2} \quad \textbf{(1)}$$
$$-y + 2x = -1 \quad \textbf{(2)}$$

**Solution**   We want to clear the first equation of fractions. We observe that the LCD of the fractions is 2.

$$2\left(\frac{3}{2}x\right) + 2(y) = 2\left(\frac{5}{2}\right) \qquad \text{Multiply each term of equation \textbf{(1)} by the LCD.}$$

$$3x + 2y = 5 \qquad \text{This is equivalent to } \frac{3}{2}x + y = \frac{5}{2}; \text{ and we label it \textbf{(3)}.}$$

The new system is as follows.

$$3x + 2y = 5 \quad \textbf{(3)}$$
$$-y + 2x = -1 \quad \textbf{(2)}$$

Now follow the five-step procedure.

**Step 1** Solve for one variable.

$-y + 2x = -1$     Equation **(2)** is the easiest one in which to isolate a variable.

$-y = -1 - 2x$     Add $-2x$ to both sides.

$y = 1 + 2x$     Multiply each term by $-1$.

**Step 2** Substitute the resulting expression into equation **(3)**.

$$3x + 2(1 + 2x) = 5 \quad \textbf{(3)}$$

---

▷ **Student Practice 2**

Find the solution. Be sure to check your answer.

$$\frac{1}{3}x - \frac{1}{2}y = 1$$
$$x + 4y = -8$$

**Step 3** Solve this equation for the variable.

$$3x + 2 + 4x = 5 \quad \text{Remove parentheses.}$$
$$7x + 2 = 5 \quad \text{Simplify.}$$
$$7x = 3 \quad \text{Add } -2 \text{ to each side.}$$
$$x = \frac{3}{7} \quad \text{Divide both sides by 7.}$$

**Step 4** Find the value of the second variable, $y$.

$$y = 1 + 2x \qquad \text{We will use the equation } y = 1 + 2x.$$
$$y = 1 + 2\left(\frac{3}{7}\right) \quad \text{Replace } x \text{ with } \frac{3}{7}.$$
$$y = 1 + \frac{6}{7} = \frac{13}{7} \quad 1 = \frac{7}{7}; \ \frac{7}{7} + \frac{6}{7} = \frac{13}{7}$$

The solution to the system is $\left(\frac{3}{7}, \frac{13}{7}\right)$.

**Step 5** *Check.* Be sure to check the solutions in each *original* equation.

$$\frac{3}{2}x + y = \frac{5}{2} \quad \textbf{(1)} \qquad\qquad -y + 2x = -1 \quad \textbf{(2)}$$

$$\frac{3}{2}\left(\frac{3}{7}\right) + \frac{13}{7} \overset{?}{=} \frac{5}{2} \qquad\qquad -\frac{13}{7} + 2\left(\frac{3}{7}\right) \overset{?}{=} -1$$

$$\frac{9}{14} + \frac{13}{7} \overset{?}{=} \frac{5}{2} \qquad\qquad -\frac{13}{7} + \frac{6}{7} \overset{?}{=} -\frac{7}{7}$$

$$\frac{9}{14} + \frac{26}{14} \overset{?}{=} \frac{35}{14} \qquad\qquad -\frac{7}{7} = -\frac{7}{7} \ \checkmark$$

$$\frac{35}{14} = \frac{35}{14} \ \checkmark \qquad\qquad\qquad \square$$

## Graphing Calculator

### Finding an Approximate Solution to a System of Equations

We can solve systems of equations graphically by using a graphing calculator. For example, to solve the system of equations in Example 2, first rewrite each equation in slope–intercept form.

$$y = -\frac{3}{2}x + \frac{5}{2}$$
$$y = \ \ 2x + 1$$

Then graph $y_1 = -\frac{3}{2}x + \frac{5}{2}$ and $y_2 = 2x + 1$ on the same screen. Display:

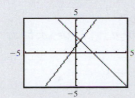

Next you can use the Trace and Zoom features to find the intersection of the two lines. Some graphing calculators have a command to find and calculate the intersection. Display:

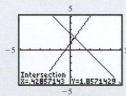

Rounded to four decimal places, the solution is (0.4286, 1.8571). Observe that

$$\frac{3}{7} = 0.4285714\ldots$$

and $\frac{13}{7} = 1.8571428\ldots,$

so the answer agrees with the answer found in Example 2.

## 8.2 Exercises  MyMathLab®

*Find the solution to each system of equations by the substitution method. Check your answers. Write your solution in the form $(x, y)$, $(p, q)$, or $(s, t)$.*

1. $5x + 2y = 22$
$x = 3y + 1$

2. $x = 7 - 2y$
$-x + 3y = 3$

3. $2x + y = 4$
$2x - y = 0$

4. $3x - 5y = -9$
$x - 2y = -3$

5. $5x + 2y = 5$
$3x + y = 4$

6. $2x + y = 4$
$3x + 2y = 5$

7. $4x - 3y = -9$
$x - y = -3$

8. $3x - y = 3$
$x + 2y = -6$

9. $p + 2q - 4 = 0$
$7p - q - 3 = 0$

10. $7s + 2t - 10 = 0$
$5s - t + 5 = 0$

11. $3x - y - 9 = 0$
$8x + 5y - 1 = 0$

12. $8x + 2y - 7 = 0$
$-2x - y + 2 = 0$

13. $\frac{5}{3}x + \frac{1}{3}y = -3$
$-2x + 3y = 24$

14. $-x + y = -4$
$\frac{3}{7}x + \frac{2}{3}y = 5$

15. $4x + 5y = 2$
$\frac{1}{5}x + y = -\frac{7}{5}$

16. $x + 2y = -6$
$\frac{2}{3}x + \frac{1}{3}y = -3$

17. $\frac{4}{7}x + \frac{2}{7}y = 2$
$3x + y = 13$

18. $x - 4y = 11$
$\frac{2}{3}x + \frac{1}{2}y = 1$

**Mixed Practice** *Find the solution to each system of equations. Write your answer in the form $(x, y)$ or $(a, b)$.*

19. $2a - 3b = 0$
$3a + b = 22$

20. $a - 4b = 21$
$5a + b = 0$

21. $3x - y = 3$
$x + 3y = 11$

22. $4x - y = 9$
$3x - 5y = 11$

23. $\frac{3}{2}x + \frac{y}{2} = -\frac{1}{2}$
$-2x - y = -1$

24. $\frac{x}{3} - \frac{2}{3}y = -\frac{4}{3}$
$5x - 4y = 4$

25. $\dfrac{2}{5}x + \dfrac{3}{5}y = \dfrac{21}{5}$
$y + 10 = 3(x + 2)$

26. $\dfrac{2}{7}x + \dfrac{3}{7}y = -\dfrac{13}{7}$
$y + 21 = 2(x + 3)$

27. $5(x - 1) = 4(y + 1) - 9$
$3(x + 1) + 5(y + 2) = 50$

28. $2(y + 1) + 11 = -3(x - 1)$
$4(x + 1) \qquad = -3(4 + y)$

## Applications

29. ***Construction Costs*** Tim Martinez is involved in a large construction project in the city. He needs to rent a heavy construction crane for several weeks. He is considering renting from one of two companies. Boston Construction will rent a crane for an initial delivery charge of $1500 and a rental fee of $900 per week. North End Contractors will rent a crane for an initial delivery charge of $500 and a rental fee of $1000 per week.

    (a) Create a cost equation for each company where $y$ is the total cost of renting a crane and $x$ is the number of weeks the crane is rented. Write a system of equations.

    (b) Solve the system of equations by the substitution method to find out how many weeks would be required for the two companies to charge the same. What would the cost be for each company?

    (c) Tim remembers that last year he considered renting from the same two companies. For the number of weeks he needed the crane, the cost of renting from one company was $4000 more than the other. How many weeks was the rental last year? Which company was less expensive for that period last year?

30. ***Snow Removal Costs*** Fred Driscoll has just moved from the sunny Southwest to the wintry Northeast, outside Buffalo, New York. His farmhouse has a very long driveway and Fred has little experience with snow. He decides to hire a snow removal company for the winter. Lake Erie Plowing requires a $50 nonrefundable reservation fee at the start of the winter and charges $90 every time they come out to plow. Adirondack Plowing requires a reservation fee of $100 but only charges $80 each time they come out to plow.

    (a) Create a cost equation for each plowing company where $y$ is the total cost of plowing and $x$ is the number of times Fred needs to be plowed out. Write a system of equations.

    (b) Solve the system of equations by the substitution method to find out how many times Fred would have to be plowed out for the cost of the plowing companies to be the same. How many times would this be? What would be the cost for each company?

    (c) Fred calculated that if he chose one company over the other, based on the average number of times that his neighbors claimed he would need to be plowed, he would save himself $80. How many times do his neighbors expect Fred to need plowing? Which company should he hire to save that money?

## To Think About

31. How many equations do you think you would need to solve a system with three unknowns? With seven unknowns? Explain your reasoning.

32. The point where the graphs of two linear equations intersect is the _____ to the system of linear equations.

33. The solution to a system of two equations must satisfy _____ equations.

34. How many solutions will a system of equations have if the graphs of the lines intersect? Justify your answer.

35. How many solution(s) will a system of equations have if the graphs of the equations are parallel lines? Why?

## Cumulative Review

**36. [7.3.2] (a)** Find the slope and $y$-intercept of $7x + 4y = 12$.

**[7.3.4] (b)** Graph.

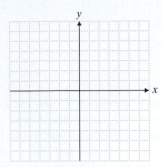

**37. [7.5.1]** Graph. $6x + 3y \geq 9$

**38. [3.4.1]** *Gas Mileage* Joe purchased a 2015 Corvette Z06 as soon as the car was available. It has an 18.5-gallon gasoline tank and gets 22 miles per gallon on the highway. The average price for gasoline was $2.75 per gallon in July 2015. Joe travels 60 miles roundtrip on the highway each day to work and back home. How much does gasoline cost per day for Joe to drive to work?

**39. [0.5.4]** *Holiday Spending* In 2014, the average consumer spent about $125.00 on Halloween candy, decorations, and costumes. The average amount spent on costumes was $47.50. What percent of the total was spent on decorations and candy? If the average 18–24-year-old spent 20% more on Halloween expenses than the average consumer, how much did the average 18–24-year-old consumer spend on Halloween? (*Source*: Forbes.com)

Quick Quiz 8.2 *Find the solution to each system of equations by the substitution method. Write your solution in the form $(x, y)$.*

**1.** $10x + 3y = 8$
$2x + y = 2$

**2.** $-3x - y = -9$
$-x + 2y = -10$

**3.** $-14x - 4y + 20 = 0$
$5x - y + 5 = 0$

**4. Concept Check** Explain how you would solve the following system by the substitution method.
$$3x - y = -4$$
$$-2x + 4y = 36$$

## STEPS TO SUCCESS Your Math Coach

*Why is the Math Coach important?* To learn mathematics you must master each skill as you proceed through the course. Completing the Math Coach after you finish each How Am I Doing? Chapter Test will help you master each building block for the skills needed to succeed in mathematics.

*How do I use the Math Coach?* Take the Chapter Test without any assistance. Then place a ✓ beside each problem you answered incorrectly. Next, read the Math Coach to help you identify and fix your errors. If you missed any other problems, find an example similar to that problem and try to correct it

yourself. Otherwise, contact your instructor or a tutor to see how to complete it correctly.

**Making it personal:** Try to correct all the problems you answered incorrectly on your in-class exam. Make it a priority to get help with the ones you cannot rework correctly. Make a list of the types of errors you often make. Describe how you can avoid them in the future. ▼

## 8.3 Solving a System of Equations in Two Variables by the Addition Method ▶

### 1 Using the Addition Method to Solve a System of Two Linear Equations with Integer Coefficients ▶

The substitution method is useful for solving a system of equations when the coefficient of one variable is 1. In the following system, the variable $x$ in the first equation has a coefficient of 1, and we can easily solve for $x$.

$$x - 2y = 7 \Rightarrow x = 7 + 2y$$
$$-5x + 4y = -5$$

This makes the substitution method a natural choice for solving this system. But we may not always have 1 as a coefficient of a variable. We need a method of solving systems of equations that will work for integer coefficients, fractional coefficients, and decimal coefficients. One such method is the **addition method.**

**Example 1** Solve by addition.

$$5x + 2y = 7 \quad \textbf{(1)}$$
$$3x - y = 13 \quad \textbf{(2)}$$

**Solution** We would like the coefficients of the $y$-terms to be opposites. One term should be $+2y$ and the other $-2y$. Thus, we multiply each term of equation **(2)** by 2.

$$2(3x) - 2(y) = 2(13) \qquad \text{Multiply each term of equation \textbf{(2)} by 2.}$$
$$5x + 2y = 7 \quad \textbf{(1)}$$
$$6x - 2y = 26 \quad \textbf{(3)} \qquad \text{We label this equation \textbf{(3)}.}$$
$$11x \quad = 33 \qquad \text{Add the two equations. This will eliminate the } y\text{-variable.}$$
$$x = 3 \qquad \text{Divide both sides by 11.}$$

We substitute $x = 3$ into one of the original equations to find $y$.

$$5(x) + 2y = 7 \qquad \text{Arbitrarily, we pick equation \textbf{(1)}.}$$
$$5(3) + 2y = 7 \qquad \text{Substitute for } x.$$
$$15 + 2y = 7 \qquad \text{Remove parentheses.}$$
$$2y = -8 \qquad \text{Subtract 15 from both sides.}$$
$$y = -4 \qquad \text{Divide both sides by 2.}$$

The solution is $(3, -4)$.

*Check.* Replace $x$ by 3 and $y$ by $-4$ in *both* original equations.

$$5x + 2y = 7 \quad \textbf{(1)} \qquad\qquad 3x - y = 13 \quad \textbf{(2)}$$
$$5(3) + 2(-4) \overset{?}{=} 7 \qquad\qquad 3(3) - (-4) \overset{?}{=} 13$$
$$15 - 8 \overset{?}{=} 7 \qquad\qquad 9 + 4 \overset{?}{=} 13$$
$$7 = 7 \ \checkmark \qquad\qquad 13 = 13 \ \checkmark \qquad\qquad \square$$

▷ **Student Practice 1** Solve by addition.

$$3x + y = 7$$
$$5x - 2y = 8$$

Notice that when we added the two equations together, one variable was eliminated. The addition method is therefore often called the **elimination method.**

For convenience, we will make a list of the steps we use to solve a system of equations by the addition method.

---

**Student Learning Objectives**

After studying this section, you will be able to:

1 Use the addition method to solve a system of two linear equations with integer coefficients. ▶

2 Use the addition method to solve a system of two linear equations with fractional coefficients. ▶

3 Use the addition method to solve a system of two linear equations with decimal coefficients. ▶

> **PROCEDURE FOR SOLVING A SYSTEM OF EQUATIONS BY THE ADDITION METHOD**
>
> 1. Multiply each term of one or both equations by some nonzero integer so that the coefficients of one of the variables are opposites.
> 2. Add the equations of this new system so that one variable is eliminated.
> 3. Solve the resulting equation for the remaining variable.
> 4. Substitute the value found in step 3 into one of the original or equivalent equations to find the value of the other variable.
> 5. Check your solution in both of the original equations.

Care should be used when the solution of a system contains fractions.

**Example 2** Solve by addition.

$$3x + 4y = 17 \quad \textbf{(1)}$$
$$2x + 7y = 19 \quad \textbf{(2)}$$

**Solution** Suppose we want the $x$-terms to have opposite coefficients. One equation could have $+6x$ and the other $-6x$. This would happen if we multiply equation **(1)** by $+2$ and equation **(2)** by $-3$.

$$(2)(3x) + (2)(4y) = (2)(17) \qquad \text{Multiply each term of equation \textbf{(1)} by 2.}$$
$$(-3)(2x) + (-3)(7y) = (-3)(19) \qquad \text{Multiply each term of equation \textbf{(2)} by } -3.$$

Often these multiplication steps can be done mentally.

We now have an equivalent system of equations labeled **(3)** and **(4)**.

$$6x + 8y = 34 \quad \textbf{(3)}$$
$$\underline{-6x - 21y = -57} \quad \textbf{(4)} \qquad \text{The coefficients of the } x\text{-terms are opposites.}$$
$$-13y = -23 \qquad \text{Add the two equations to eliminate the variable } x.$$
$$y = \frac{-23}{-13} = \frac{23}{13} \qquad \text{Divide both sides by } -13.$$

Substitute $y = \dfrac{23}{13}$ into one of the original or equivalent equations to find $x$.

$$3x + 4\left(\frac{23}{13}\right) = 17 \quad \textbf{(1)} \qquad \text{We pick equation \textbf{(1)}.}$$

$$3x + \frac{92}{13} = 17 \qquad \text{Solve for } x.$$

$$13(3x) + 13\left(\frac{92}{13}\right) = 13(17) \qquad \text{Multiply both sides by 13 to clear the fraction.}$$

$$39x + 92 = 221 \qquad \text{Remove the parentheses.}$$
$$39x = 129 \qquad \text{Subtract 92 from both sides.}$$
$$x = \frac{43}{13} \qquad \text{Divide both sides by 39 and reduce. The solution is } \left(\frac{43}{13}, \frac{23}{13}\right).$$

**CAUTION:** A common error occurs when students forget that we want to obtain two terms with *opposite signs* that when added will equal zero. If all the coefficients in the equations are positive, such as in Example 2, it will be necessary to multiply *one* of the equations by a negative number.

*Alternative solution:* You could also have eliminated the $y$-variable. For example, if you multiply equation **(1)** by 7 and equation **(2)** by $-4$, you obtain the equivalent system shown here.

$$21x + 28y = 119$$
$$-8x - 28y = -76$$

You can add these equations to eliminate the $y$-variable. Since the numbers involved in this approach are somewhat larger, it is probably wiser to eliminate the $x$-variable in this example. $\qquad \square$

**Student Practice 2**　Solve by addition.

$$4x + 5y = 17$$
$$3x + 7y = 12$$

## 2 Using the Addition Method to Solve a System of Two Linear Equations with Fractional Coefficients ▶

If the system of equations has fractional coefficients, you should first clear each equation of the fractions. To do so, you will need to multiply each term in the equation by the LCD of the fractions.

**Example 3** Solve.

$$x - \frac{5}{2}y = \frac{5}{2} \quad \textbf{(1)}$$
$$\frac{4}{3}x + y = \frac{23}{3} \quad \textbf{(2)}$$

**Solution**

$$2\,(x) - 2\left(\frac{5}{2}y\right) = 2\left(\frac{5}{2}\right) \qquad \text{Multiply each term of equation \textbf{(1)} by 2.}$$
$$2x - 5y = 5$$

$$3\left(\frac{4}{3}x\right) + 3\,(y) = 3\left(\frac{23}{3}\right) \qquad \text{Multiply each term of equation \textbf{(2)} by 3.}$$
$$4x + 3y = 23$$

We now have an equivalent system of equations that does not contain fractions.

$$2x - 5y = 5 \quad \textbf{(3)}$$
$$4x + 3y = 23 \quad \textbf{(4)}$$

Let us eliminate the $x$-variable. We want the coefficients of $x$ to be opposites.

$$(-2)(2x) - (-2)5y = (-2)(5) \qquad \text{Multiply each term of equation \textbf{(3)} by } -2.$$

$$
\begin{array}{ll}
-4x + 10y = -10 \quad \textbf{(5)} & \text{We now have an equivalent system of equations.} \\
\underline{4x + 3y = 23} \quad \textbf{(4)} & \text{The coefficients of the } x\text{-terms are opposites.} \\
13y = 13 & \text{Add the two equations to eliminate the variable } x. \\
y = 1 & \text{Divide both sides by 13.}
\end{array}
$$

Since the original equations contain fractions, it will simplify calculations to use one of the equivalent equations to find the value of $x$.

$$
\begin{array}{ll}
4x + 3(1) = 23 & \text{Substitute } y = 1 \text{ into equation \textbf{(4)}.} \\
4x + 3 = 23 & \text{Remove the parentheses.} \\
4x = 20 & \text{Add } -3 \text{ to both sides.} \\
x = 5 & \text{Divide both sides by 4.}
\end{array}
$$

The solution is $(5, 1)$.

*Check.* Check the solution in both of the *original* equations.　□

 **Student Practice 3**　Solve.

$$\frac{2}{3}x - \frac{3}{4}y = 3$$
$$-2x + y = 6$$

### 3 Using the Addition Method to Solve a System of Two Linear Equations with Decimal Coefficients ▶

Some linear equations will have decimal coefficients. It will be easier to work with the equations if we change the decimal coefficients to integer coefficients. To do so, we will multiply each term of the equation by a power of 10.

$\mathbb{M}_{\mathbb{C}}$ **Example 4** Solve.

$$0.12x + 0.05y = -0.02 \quad (1)$$
$$0.08x - 0.03y = -0.14 \quad (2)$$

**Solution** Since the decimals are hundredths, we will multiply each term of both equations by 100.

$$100(0.12x) + 100(0.05y) = 100(-0.02)$$
$$100(0.08x) - 100(0.03y) = 100(-0.14)$$
$$12x + 5y = -2 \quad (3) \qquad \text{We now have an equivalent system of equations that has integer coefficients.}$$
$$8x - 3y = -14 \quad (4)$$

We will eliminate the variable $y$. We want the coefficients of the $y$-terms to be opposites.

$$36x + 15y = -6 \qquad \text{Multiply equation (3) by 3 and equation (4) by 5.}$$
$$\underline{40x - 15y = -70}$$
$$76x \qquad = -76 \qquad \text{Add the equations.}$$
$$x = -1 \qquad \text{Solve for } x.$$
$$12(-1) + 5y = -2 \qquad \text{Replace } x \text{ by } -1 \text{ in equation (3), and solve for } y.$$
$$-12 + 5y = -2 \qquad \text{Remove the parentheses.}$$
$$5y = -2 + 12 \quad \text{Add 12 to both sides.}$$
$$5y = 10 \qquad \text{Simplify.}$$
$$y = 2 \qquad \text{Divide both sides by 5.}$$

The solution is $(-1, 2)$.

*Check.* We substitute $x = -1$ and $y = 2$ into the original equations. Most students probably would rather not use equation (1) and equation (2)! However, *we must check the solutions in the original equations* if we want to be sure of our answers. It is possible to make an error going from equations (1) and (2) to equations (3) and (4). The solution could satisfy equations (3) and (4), but might not satisfy equations (1) and (2).

$$0.12x + 0.05y = -0.02 \quad (1) \qquad\qquad 0.08x - 0.03y = -0.14 \quad (2)$$
$$0.12(-1) + 0.05(2) \overset{?}{=} -0.02 \qquad 0.08(-1) - 0.03(2) \overset{?}{=} -0.14$$
$$-0.12 + 0.10 \overset{?}{=} -0.02 \qquad\qquad -0.08 - 0.06 \overset{?}{=} -0.14$$
$$-0.02 = -0.02 \ \checkmark \qquad\qquad\qquad -0.14 = -0.14 \ \checkmark \ \square$$

▷ **Student Practice 4** Solve.

$$0.2x + 0.3y = -0.1$$
$$0.5x - 0.1y = -1.1$$

**CAUTION:** A common error when solving a system is to find the value for $x$ and then stop. You have not solved a system in $x$ and $y$ until you find both the $x$- and the $y$-values that make both equations true.

## 8.3 Exercises MyMathLab®

### Verbal and Writing Skills, Exercises 1–4

Look at the following systems of equations. Decide which variable to eliminate in each system, and explain how you would eliminate that variable.

**1.** $3x + 2y = 5$
$5x - y = 3$

**2.** $2x - 9y = 1$
$2x + 3y = 2$

**3.** $4x - 3y = 10$
$5x + 4y = 0$

**4.** $7x + 6y = 13$
$-2x + 5y = 3$

Find the solution by the addition method. Check your answers.

**5.** $-x + y = -5$
$-3x - y = -3$

**6.** $x + y = 6$
$3x - y = 10$

**7.** $2x + 3y = 1$
$x - 2y = 4$

**8.** $3x + y = 6$
$5x - 3y = -4$

**9.** $6x - y = -7$
$6x + 2y = 5$

**10.** $5x + y = -1$
$4x + 3y = 8$

**11.** $5x - 15y = 9$
$-x + 10y = 1$

**12.** $-4x + 5y = -16$
$8x + y = -1$

**13.** $2x + 5y = 2$
$3x + y = 3$

**14.** $5x - 3y = 14$
$2x - y = 6$

**15.** $8x + 6y = -2$
$10x - 9y = -8$

**16.** $4x + 9y = 0$
$8x - 5y = -23$

**17.** $2x + 3y = -8$
$5x + 4y = -34$

**18.** $5x - 2y = 6$
$6x + 7y = 26$

**19.** $4x + 3y = 1$
$2x - 5y = -19$

**20.** $-5x + 4y = -13$
$11x + 6y = -1$

**21.** $12x - 6y = -2$
$-9x - 7y = -10$

**22.** $2x - 4y = -22$
$-6x + 3y = 3$

### Verbal and Writing Skills, Exercises 23–26

Before you solve a system, you often need to simplify the equation(s). Explain how you would change the fractional coefficients to integer coefficients in each system, but **do not solve** the system.

**23.** $\frac{1}{4}x - \frac{3}{4}y = 2$
$2x + 3y = 1$

**24.** $10x + 5y = 18$
$\frac{4}{9}x - \frac{2}{3}y = 4$

**25.** $\frac{1}{2}x + \frac{2}{3}y = \frac{1}{3}$

$\frac{3}{4}x - \frac{4}{5}y = 2$

**26.** $\frac{1}{4}x + \frac{1}{3}y = 15$

$\frac{1}{5}x + \frac{3}{10}y = 13$

*Find the solution. Check your answers.*

**27.** $x + \frac{5}{4}y = \frac{9}{4}$

$\frac{2}{5}x - y = \frac{3}{5}$

**28.** $\frac{2}{3}x + y = 2$

$x + \frac{1}{2}y = 7$

**29.** $\frac{x}{6} + \frac{y}{2} = -\frac{1}{2}$

$x - 9y = 21$

**30.** $\frac{3}{2}x - \frac{y}{8} = -1$

$16x + 3y = -28$

**31.** $\frac{5}{6}x + y = -\frac{1}{3}$

$-8x + 9y = 28$

**32.** $\frac{2}{9}x - \frac{1}{3}y = 1$

$4x + 9y = -2$

**33.** $\frac{2}{3}x + \frac{3}{5}y = -\frac{1}{5}$

$\frac{1}{4}x + \frac{1}{3}y = \frac{1}{4}$

**34.** $\frac{x}{5} + \frac{1}{2}y = \frac{9}{10}$

$\frac{x}{3} + \frac{3}{4}y = \frac{5}{4}$

## Verbal and Writing Skills, Exercises 35–37

*Explain how you would change each system to an equivalent system with integer coefficients. Write the equivalent system.*

**35.** $0.5x - 0.3y = 0.1$

$5x + 3y = 6$

**36.** $0.08x + y = 0.05$

$2x - 0.1y = 3$

**37.** $4x + 0.5y = 9$

$0.2x - 0.05y = 1$

*Solve for x and y using the addition method.*

**38.** $0.2x - 0.3y = 0.4$

$0.3x - 0.4y = 0.9$

**39.** $0.2x + 0.3y = 0.4$

$0.5x + 0.4y = 0.3$

**40.** $0.02x - 0.04y = 0.26$

$0.07x - 0.09y = 0.66$

**41.** $0.04x - 0.03y = 0.05$

$0.05x + 0.08y = -0.76$

**42.** $0.4x - 5y = -1.2$

$-0.03x + 0.5y = 0.14$

**43.** $-0.6x - 0.08y = -4$

$3x + 2y = 4$

**Mixed Practice** *Solve for x and y using the addition method.*

**44.** $3x + y = -2$
$6x - 5y = 17$

**45.** $3x - y = -8$
$-9x + 2y = 18$

**46.** $\frac{1}{2}x + 2y = 15$
$3x + \frac{1}{5}y = 31$

**47.** $\frac{1}{3}x - \frac{1}{4}y = 0$
$\frac{1}{6}x - \frac{1}{2}y = -6$

**48.** $0.3x + 0.1y = 0.7$
$0.04x + 0.03y = -0.14$

**49.** $0.05x - 0.02y = 0.16$
$0.7x + 0.3y = -2.4$

**To Think About** *Find the solution.*

**50.** $4(x - 2y) = 5 - (y - 3x)$
$-5(x + 1) = y - 6x$

**51.** $3(3x + 2) = y + 5x$
$4 = -2(x - y)$

## Cumulative Review

**52.** **[0.5.3]** *Air Traffic* At any given moment of the day, between 5000 and 7000 commercial aircraft are flying in U.S. airspace. If 89% of the air traffic is flying over the contiguous states, how many commercial airplanes are flying over Alaska and Hawaii?

**53.** **[3.4.1]** *Used Car Purchase* A used car is priced at $5800, if you pay cash. An installment plan requires $1000 down, plus $230 a month for 24 months. How much more would you pay for the car under the installment plan?

*Solve for the variable.*

**54.** **[2.4.1]** $\frac{1}{3}(4 - 2x) = \frac{1}{2}x$

**55.** **[2.3.3]** $2(y - 3) - (2y + 4) = -6y$

**Quick Quiz 8.3** *Find the solution to the system by the addition method. Express your answer in the form $(x, y)$.*

**1.** $3x + 5y = -1$
$-5x + 4y = -23$

**2.** $-7x + 3y = -31$
$4x + 6y = 10$

**3.** $\frac{1}{3}x + y = \frac{8}{3}$
$\frac{4}{5}x - \frac{2}{5}y = \frac{18}{5}$

**4.** **Concept Check** Explain how you would obtain an equivalent system that does not have decimals if you wanted to solve the following system.
$0.5x + 0.4y = -3.4$
$0.02x + 0.03y = -0.08$

# Use Math to Save Money

**Did You Know?** The government will pay interest on certain student loans while you are in college.

## Choosing a Student Loan

### Understanding the Problem:

Alicia needs a $10,000 college loan. A credit union offers a 20-year private loan with a 4.65% fixed rate. A subsidized federal loan provides a 20-year loan at a 6% fixed rate with the government paying interest during school (4.5 years). Which option is the better deal?

### Making a Plan:

**Step 1:** Notice that the two interest rates are different for each loan. Also, the total number of payments is different because the government will pay for Alicia's interest for 4.5 years out of the 20-year life of the subsidized loan.

**Task 1:** What is the total amount that Alicia must pay back for each loan?

**Task 2:** What is surprising about the results of Task 1?

**Step 2:** The private loan requires that Alicia make more payments overall, and she must start making payments immediately. The subsidized loan allows for fewer payments, and she can wait 6 months after graduation to begin making payments.

**Task 3:** What is the monthly payment for each loan?

**Task 4:** Which loan offers the lowest monthly payment?

### Finding the Solution:

**Step 3:** Let's compare the loans.

**Task 5:** What are the values for the missing places in this table?

**Task 6:** Which option provides the best deal? Explain your reasoning.

### Applying the Solution to Your Life:

When choosing between the two loans, students must consider the total amount that Alicia must repay, whether the government will pay interest while she is in school, and when she starts making her monthly payments. To apply for a subsidized loan, students must fill out a Free Application for Federal Student Aid (FAFSA) form. Approval for subsidized loans is based on financial need, but you do not need a high credit score. On the other hand, a private loan from a bank or credit union will require a good credit score. Credit unions sometimes offer lower-cost private loans than banks. Make sure to check into all of your options before making a decision.

| Type of Loan | Total Amount | Number of Monthly Payments | Monthly Payment Amount | When Payments Begin |
|---|---|---|---|---|
| Private loan | | | | Immediately |
| Subsidized loan | | | | 6 months after graduation |

486

# 8.4 Review of Methods for Solving Systems of Equations ▶

## 1 Choosing an Appropriate Method to Solve a System of Equations Algebraically ▶

At this point we will review the algebraic methods for solving systems of linear equations and discuss the advantages and disadvantages of each method.

| Method | Advantage | Disadvantage |
|---|---|---|
| Substitution | Works well if one or more variables have a coefficient of 1 or $-1$. | Often becomes difficult to use if no variable has a coefficient of 1 or $-1$. |
| Addition | Works well if equations have fractional or decimal coefficients. Works well if no variable has a coefficient of 1 or $-1$. | None |

**Student Learning Objectives**

**After studying this section, you will be able to:**

1 Choose an appropriate method to solve a system of equations algebraically. ▶

2 Use algebraic methods to identify inconsistent and dependent systems. ▶

**Example 1** Select a method and solve each system of equations.

**(a)** $x + y = 3080$
$2x + 3y = 8740$

**(b)** $5x - 1 = 2(y + 9)$
$-3x = 7(5 - y)$

### Solution

**(a)** Since there are $x$- and $y$-values that have coefficients of 1, we will select the substitution method.

$$y = 3080 - x \qquad \text{Solve the first equation for } y.$$
$$2x + 3(3080 - x) = 8740 \qquad \text{Substitute the expression into the second equation.}$$
$$2x + 9240 - 3x = 8740 \qquad \text{Remove parentheses.}$$
$$-1x = -500 \qquad \text{Simplify.}$$
$$x = 500 \qquad \text{Divide each side by } -1.$$

$$y = 3080 - x$$
$$= 3080 - 500 \qquad \text{Substitute 500 for } x.$$
$$y = 2580 \qquad \text{Simplify.}$$

The solution is $(500, 2580)$.

**(b)** For each of the equations, we first remove parentheses and then simplify so that all $x$ terms and all $y$ terms are on the left and all the numbers are on the right side of the equals sign.

$$5x - 1 = 2(y + 9) \qquad\qquad -3x = 7(5 - y)$$
$$5x - 1 = 2y + 18 \qquad\qquad -3x = 35 - 7y$$
$$5x = 2y + 19 \qquad\qquad -3x + 7y = 35$$
$$5x - 2y = 19$$

Because none of the $x$- or $y$-variables has a coefficient of 1 or $-1$, we select the addition method. We choose to eliminate the $y$-variable. Thus, we would like the coefficients of $y$ to be $-14$ and 14.

$$7(5x) - 7(2y) = 7(19) \qquad \text{Multiply each term of the first equation by 7.}$$
$$2(-3x) + 2(7y) = 2(35) \qquad \text{Multiply each term of the second equation by 2.}$$
$$35x - 14y = 133 \qquad \text{We now have an equivalent system of equations.}$$
$$\underline{-6x + 14y = \phantom{0}70}$$
$$29x \phantom{-14y} = 203 \qquad \text{Add the two equations.}$$
$$x = 7 \qquad \text{Divide each side by 29.}$$

*Continued on next page*

Substitute $x = 7$ into one of the original or equivalent equations.

$$5x - 2y = 19$$
$$5(7) - 2y = 19$$
$$35 - 2y = 19 \qquad \text{Solve for } y.$$
$$-2y = -16$$
$$y = 8$$

The solution is $(7, 8)$. ◻

**Student Practice 1**    Select a method and solve each system of equations.

(a) $3x + 5y = 1485$
$\quad\ \ x + 2y = \ \ 564$

(b) $7x - 3 = -6(y - 7)$
$\quad\quad\ -5y = -2(3x + 1)$

## 2 Using Algebraic Methods to Identify Inconsistent and Dependent Systems ▶

Recall that an inconsistent system of linear equations is a system of parallel lines. Since parallel lines never intersect, the system has no solution. Can we determine this algebraically?

**Example 2** Solve algebraically.

$$3x - y = -1$$
$$3x - y = -7$$

**Solution**    Clearly, the addition method would be very convenient in this case.

$$3x - y = -1 \qquad \text{Keep the first equation unchanged.}$$
$$\underline{-3x + y = +7} \qquad \text{Multiply each term in the second equation by } -1.$$
$$0 = \ \ 6 \qquad \text{Add the two equations.}$$

Notice we have $0 = 6$, which we know is not true. The statement $0 = 6$ is inconsistent with known mathematical facts. No possible $x$- and $y$-values can make this equation true. Thus there is **no solution to this system of equations.**

This example shows us that, if we obtain a mathematical statement that is not true (inconsistent with known facts), we can identify the entire system as **inconsistent.** If graphed, these lines would be parallel. ◻

**Student Practice 2**    Solve algebraically.

$$4x + 2y = 2$$
$$-6x - 3y = 6$$

What happens if we try to solve a dependent system of equations algebraically? Recall that a dependent system consists of two lines that coincide (are the same line).

Mc **Example 3** Solve algebraically.

$$x + \ \ y = \ \ 4$$
$$3x + 3y = 12$$

**Solution**    Let us use the substitution method.

$$y = 4 - x \qquad \text{Solve the first equation for } y.$$
$$3x + 3(4 - x) = 12 \qquad \text{Substitute } 4 - x \text{ for } y \text{ in the second equation.}$$
$$3x + 12 - 3x = 12 \qquad \text{Remove parentheses.}$$
$$12 = 12$$

**Student Practice 3**

Solve algebraically.
$$3x - \ \ 9y = \ \ 18$$
$$-4x + 12y = -24$$

Notice we have $12 = 12$, which is always true.

Thus when we solved this system, we obtained an equation that is true for any value of $x$. An equation that is always true is called an **identity.** This means all the solutions of one equation of the system are also solutions of the other equation.

Thus the lines coincide, and there is an **infinite number of solutions to this system of equations.** □

Example 3 shows us that, if we obtain a mathematical statement that is always true (an identity), we can identify the equations as **dependent.** There is an unlimited number of solutions to a system that has dependent equations.

**To Think About:** *Two Linear Equations with Two Variables* Now is a good time to look back over what we have learned. When you graph a system of two linear equations, what possible kinds of graphs will you obtain?

What will happen when you try to solve a system of two linear equations using algebraic methods? How many solutions are possible in each case? The following chart may help you to organize your answers to these questions.

| Graph | Number of Solutions | Algebraic Interpretation |
|---|---|---|
| Two lines intersect at one point $(6, -3)$ | **One unique solution** | You obtain one value for $x$ and one value for $y$. For example, $$x = 6, \quad y = -3.$$ |
| Parallel lines | **No solution** | You obtain an equation that is inconsistent with known facts. For example, $$0 = 6.$$ The system of equations is inconsistent. |
| Lines coincide | **Infinite number of solutions** | You obtain an equation that is always true. For example, $$8 = 8.$$ The equations are dependent. |

Think about each of the three possibilities in this chart. You need to understand each one. As you do each of exercises 5–28, you will need to identify which of the three possibilities is involved.

## 8.4 Exercises  MyMathLab®

### Verbal and Writing Skills, Exercises 1–4

1. If there is no solution to a system of linear equations, the graphs of the equations are _____. Solving the system algebraically, you will obtain an equation that is _____ with known facts.

2. If an algebraic attempt at solving a system of linear equations results in an identity, the system is said to be _____. There are an _____ number of solutions, and the graphs of the lines _____.

3. If there is exactly one solution, the graphs of the equations _____. This system is said to be _____ and _____.

4. A student solves a system of equations and obtains the equation $-36x = 0$. He is unsure of what to do next. What should he do next? What should he conclude about the system of equations?

*If possible, solve by an algebraic method (substitution or addition method, without the use of graphing). Otherwise, state that the problem has no solution or an infinite number of solutions.*

5. $-5x + 3y = -26$
   $4x - 2y = 20$

6. $2x + 5y = -14$
   $-3x + 4y = -25$

7. $3x - 4y = 2$
   $-5x + 6y = -12$

8. $6x - 5y = 36$
   $-7x - 2y = 5$

9. $2x - 4y = 5$
   $-4x + 8y = 9$

10. $5x + 4y = 7$
    $10x + 8y = 13$

11. $-5x + 2y = 2$
    $15x - 6y = -6$

12. $15x - 10y = 20$
    $-3x + 2y = -4$

13. $5x - 3y = 13$
    $7x + 2y = 43$

14. $-4x + 5y = 10$
    $6x - 7y = 8$

15. $3x - 2y = 70$
    $0.6x + 0.5y = 50$

16. $5x - 3y = 10$
    $0.9x + 0.4y = 30$

17. $0.2x - 0.3y = 0.1$
    $-0.5x + 0.8y = 0$

18. $0.05x + 0.04y = 0.12$
    $0.03x - 0.08y = 0.28$

19. $\frac{4}{3}x + \frac{1}{2}y = 1$
    $\frac{1}{3}x - y = -\frac{1}{2}$

20. $\frac{3}{5}x - \frac{3}{4}y = \frac{1}{5}$
    $2x + \frac{3}{2}y = 2$

21. $\frac{2}{3}x + \frac{1}{6}y = 2$
    $\frac{1}{4}x - \frac{1}{2}y = -\frac{3}{4}$

22. $\frac{1}{2}x + \frac{2}{5}y = \frac{9}{10}$
    $\frac{2}{3}x + \frac{2}{15}y = \frac{1}{2}$

**Mixed Practice** *Solve by any algebraic method. If there is not one solution to a system of equations, state the reason.*

23. $6x + 5y = 3$
    $5x + 3y = 6$

24. $2x = 3(y - 4) + 24$
    $y + 8 = 2(x - 3) + 6$

25. $\frac{3}{4}x + y = 4$
    $x - \frac{7}{5}y = \frac{13}{5}$

26. $0.3x + 0.2y = 0.2$
    $0.5x - 0.3y = 3.5$

**27.** $x - 2y + 2 = 0$
$3(x - 2y + 1) = 15$

**28.** $3(x - 2y + 2) = -3$
$-\dfrac{1}{3}x + \dfrac{2}{3}y = 1$

**To Think About** *Remove parentheses and solve the system for $(a, b)$.*

**29.** $2(a + 3) = b + 1$
$3(a - b) = a + 1$

**30.** $3(a + b) - 2 = b + 2$
$a - 2(b - 1) = 3a - 4$

**Cumulative Review** *Solve for x.*

**31.** **[2.4.1]** $8x + 15\left(\dfrac{2}{3}x - \dfrac{2}{5}\right) = -10$

**32.** **[2.4.1]** $0.2(x + 0.3) + x = 0.3(x + 0.5)$

*Last year, federal income taxes for single filers in some situations could be estimated using the following chart.*

| If taxable income is over... | But not over... | Estimated income tax is... |
|---|---|---|
| $0 | $8350 | 10% of amount over $0 |
| $8350 | $33,950 | $835.00 + 15% of amount over $8350 |
| $33,950 | $82,250 | $4675.00 + 25% of amount over $33,950 |
| $82,250 | $171,550 | $16,750.00 + 28% of amount over $82,250 |

*Source: www.irs.gov*

**33.** **[0.5.3]** Last year, Margot earned $27,600. Find the estimated federal income tax she had to pay.

**34.** **[0.5.3]** Last year, Dean earned $35,900. During the year, a total of $5100 was deducted from his paychecks for federal income tax. Was this more or less than what he should have paid? Estimate the amount he still owed or his federal income tax refund.

**Quick Quiz 8.4** *Solve if possible. Otherwise, state that the problem has no solution or has an infinite number of solutions.*

**1.** $12 - x = 3(x + y)$
$3x + 14 = 4x + 5(x + y)$

**2.** $-4x + 6y = 2$
$12y = 8x + 12$

**3.** $\dfrac{1}{2}x - \dfrac{1}{4}y = 1$
$\dfrac{4}{3}x + 4y = 12$

**4. Concept Check** Explain how you would remove the fractions in order to solve the following system.
$\dfrac{3}{10}x + \dfrac{2}{5}y = \dfrac{1}{2}$
$\dfrac{1}{8}x - \dfrac{3}{16}y = -\dfrac{1}{2}$

## 8.5 Solving Word Problems Using Systems of Equations

**Student Learning Objective**

After studying this section, you will be able to:

1 Use a system of equations to solve word problems.

### 1 Using a System of Equations to Solve Word Problems

Word problems can be solved in a variety of ways. In Chapter 3 and throughout the text, you have used one equation with one unknown to describe situations found in real-life applications. Some of these same problems can be solved using two equations in two unknowns. The method you use will depend on what you find works best for you. Become familiar with both methods before you decide.

**Example 1** A worker in a large post office is trying to verify the rate at which two electronic card-sorting machines operate. Yesterday the first machine sorted for 3 minutes and the second machine sorted for 4 minutes. The total workload both machines processed during that time period was 10,300 cards. Two days ago the first machine sorted for 2 minutes and the second machine for 3 minutes. The total workload both machines processed during that time period was 7400 cards. Can you determine the number of cards per minute sorted by each machine?

**Solution**

1. *Understand the problem.* The number of cards processed by two different machines on two different days provides the basis for two linear equations with two unknowns.

   The two equations will represent what occurred on two different days:

   **(1)** yesterday and          **(2)** two days ago.

   The number of cards sorted by each machine is what we need to find.

   Let $x =$ the number of cards per minute sorted by the *first machine* and
   $y =$ the number of cards per minute sorted by the *second machine*.

2. *Write the equations.* Yesterday the first machine sorted for 3 minutes and the second machine sorted for 4 minutes, processing a total of 10,300 cards.

   | *first machine* | | *second machine* | | *total no. of cards* |
   |---|---|---|---|---|
   | 3(no. of cards per minute) | | 4(no. of cards per minute) | | |
   | $3x$ | $+$ | $4y$ | $=$ | $10{,}300$    **(1)** |

   Two days ago the first machine sorted for 2 minutes and the second machine sorted for 3 minutes, processing a total of 7400 cards.

   | *first machine* | | *second machine* | | *total no. of cards* |
   |---|---|---|---|---|
   | 2(no. of cards per minute) | | 3(no. of cards per minute) | | |
   | $2x$ | $+$ | $3y$ | $=$ | $7400$    **(2)** |

3. *Solve and state the answer.* We will use the addition method to solve this system.

$$3x + 4y = 10{,}300 \quad \textbf{(1)}$$
$$2x + 3y = 7400 \quad \textbf{(2)}$$

$$6x + 8y = 20{,}600 \qquad \text{Multiply equation \textbf{(1)} by 2.}$$
$$\underline{-6x - 9y = -22{,}200} \qquad \text{Multiply equation \textbf{(2)} by } -3.$$
$$-1y = -1600 \qquad \text{Add the two equations to eliminate } x \text{ and solve for } y.$$

$$y = 1600$$
$$2x + 3(1600) = 7400 \qquad \text{Substitute the value of } y \text{ into equation \textbf{(2)}.}$$
$$2x + 4800 = 7400 \qquad \text{Simplify and solve for } x.$$
$$2x = 2600$$
$$x = 1300$$

Thus, the first machine sorts 1300 cards per minute and the second machine sorts 1600 cards per minute.

**4. Check.** We can verify each statement in the problem.

Yesterday: Were 10,300 cards sorted?

$$3(1300) + 4(1600) \overset{?}{=} 10,300$$

$$3900 + 6400 \overset{?}{=} 10,300$$

$$10,300 = 10,300 \ \checkmark$$

Two days ago: Were 7400 cards sorted?

$$2(1300) + 3(1600) \overset{?}{=} 7400$$

$$2600 + 4800 \overset{?}{=} 7400$$

$$7400 = 7400 \ \checkmark \qquad \qquad \square$$

**Student Practice 1** After a recent disaster, the Red Cross brought in a pump to evacuate water from a basement. After 3 hours a larger pump was brought in and the two ran for an additional 5 hours, removing 49,000 gallons of water. The next day, after running both pumps for 3 hours, the large one was taken to another site. The small pump finished the job after another 2 hours. If on the second day 30,000 gallons were pumped, how much water does each pump remove per hour?

**Example 2** Fred is considering working part-time for a company that sells custom garage doors. Fred found out that all starting part-time representatives receive the same annual base salary and a standard commission of a certain percentage of the sales they make during the first year. He has been told that one representative sold $50,000 worth of garage doors her first year and that she earned $14,000. He was able to find out that another representative sold $80,000 worth of garage doors and that he earned $17,600. Determine the base salary and the commission rate of a beginning part-time sales representative.

**Solution** What is unknown? *The base salary (b) and the commission rate (r).* How do we proceed? *We must represent two equations, one for each sales representative, in the form*

$$\boxed{\text{base salary}} + \boxed{\text{commission}} = \boxed{\text{total earnings}} \quad \textit{Note:} \text{commission} = \text{value of sales} \times \text{commission rate}$$

We set up our two equations and let $b = $ *base salary* and $r = $ *commission rate.*

*base salary + commission = total earnings*

$$b \ + \ 50,000r \ = 14,000 \quad \textbf{(1)} \quad \text{Earnings for the first sales representative}$$

$$b \ + \ 80,000r \ = 17,600 \quad \textbf{(2)} \quad \text{Earnings for the second sales representative}$$

We will use the addition method to solve these equations.

$$-b - 50,000r = -14,000 \quad \text{Multiply equation } \textbf{(1)} \text{ by } -1.$$

$$\underline{b + 80,000r = \ \ 17,600} \quad \text{Leave equation } \textbf{(2)} \text{ unchanged.}$$

$$30,000r = \ \ 3600 \quad \text{Add the two equations and solve for } r.$$

$$r = \frac{3600}{30,000}$$

$$r = 0.12 \quad \text{The commission rate is 12\%.}$$

$$b + (50,000)(0.12) = 14,000 \quad \text{Substitute this value into equation } \textbf{(1)}.$$

$$b + 6000 = 14,000 \quad \text{Simplify and solve for } b.$$

$$b = 8000 \quad \text{The base salary is \$8000 per year.}$$

Thus the base salary is $8000 per year, and the commission rate is 12%. $\quad \square$

*Continued on next page*

**Student Practice 2**    Rick has two cars, one using premium gas and the other using regular gas. Last week Rick bought 7 gallons of unleaded premium and 8 gallons of unleaded regular gasoline and paid $47.15. This week he purchased 8 gallons of unleaded premium and 4 gallons of unleaded regular and paid $38.20. He forgot to record how much each type of fuel cost per gallon. Can you determine these values?

**Example 3** A lab technician is required to prepare 200 liters of a solution. The prepared solution must contain 42% fungicide. The technician wishes to combine a solution that contains 30% fungicide with a solution that contains 50% fungicide. How much of each solution should he use?

**Solution**    *Understand the problem.*  The technician needs to make one solution from two different solutions. We need to find the amount (number of liters) of each solution that will give us 200 liters of a 42% solution. The two unknowns are easy to identify.

Let $x$ = the number of liters of the 30% solution needed and
$y$ = the number of liters of the 50% solution needed.

Now, what are the two equations? We know that the 200 liters of the final solution will be made by combining $x$ and $y$.

$$x + y = 200 \quad \textbf{(1)}$$

The second piece of information concerns the percent of fungicide in each solution (30% of $x$ liters, 50% of $y$ liters, and 42% of 200 liters). See the picture at the left.

$$0.3x + 0.5y = 0.42(200) \quad \textbf{(2)}$$

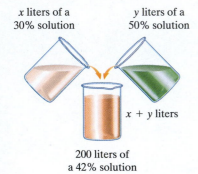

$x$ liters of a
30% solution

$y$ liters of a
50% solution

$x + y$ liters

200 liters of
a 42% solution

Thus our system of equations is

$$x + \quad y = 200 \quad \textbf{(1)}$$
$$0.3x + 0.5y = \quad 84. \quad \textbf{(2)}$$

We will solve this system by the addition method.

$$
\begin{array}{ll}
1.0x + 1.0y = \quad\ \ 200 & \text{Rewrite equation \textbf{(1)} in an equivalent form.}\\
-0.6x - 1.0y = -168 & \text{Multiply equation \textbf{(2)} by } -2.\\
\hline
0.4x \qquad\quad = 32 & \text{Solve for } x.\\
\dfrac{0.4x}{0.4} = \dfrac{32}{0.4} & \\
x = 80 & \\
80 + y = 200 & \text{Substitute the value } x = 80 \text{ into one of the}\\
y = 120 & \text{original equations and solve for } y.
\end{array}
$$

The technician should use 80 liters of the 30% fungicide and 120 liters of the 50% fungicide to obtain the required solution.

*Check.*  Do the amounts total 200 liters?

$$80 + 120 = 200 \ \checkmark$$

Do the amounts yield a 42% strength mixture?

$$0.30x + 0.50y = 0.42(200)$$
$$0.30(80) + 0.50(120) \overset{?}{=} 84$$
$$24 + 60 = 84 \ \checkmark$$

**Student Practice 3**    Dwayne Laboratories needs to ship 4000 liters of $H_2SO_4$ (sulfuric acid). Dwayne keeps in stock solutions that are 20% $H_2SO_4$ and 80% $H_2SO_4$. If the strength of the solution shipped is to be 65%, how should Dwayne's stock be mixed?

**Example 4** Mike recently rode his boat on Lazy River. He took a 48-mile trip up the river traveling against the current in exactly 3 hours. He refueled and made the return trip in exactly 2 hours. What was the speed of his boat in still water and the speed of the current in the river?

**Solution** *Understand the problem.* You may want to draw a picture to see how these speeds relate to one another.

Let $b$ = the speed of the boat in still water in miles/hour and

$c$ = the speed of the river current in miles/hour.

*Against the current*
When we travel against the current, the current is slowing us down. Since the current's speed opposes the boat's speed in still water, we must subtract: $b - c$.

*With the current*
When we travel with the current, the current is helping us travel forward. So the current's speed is added to the boat's speed in still water, and we add: $b + c$.

*Write a system of two equations.* To help us write our equations, we organize our information in a chart and use the formula distance = rate × time or $D = RT$.

Distance was 48 miles in 3 hours — Current
Actual speed of the boat in still water: $b$ mph
Actual traveling speed: $b - c$ mph
Speed of the current: $c$ mph

Distance was 48 miles in 2 hours — Current
Actual speed of the boat in still water: $b$ mph
Actual traveling speed: $b + c$ mph
Speed of the current: $c$ mph

Using the rows of the chart, we obtain a system of equations.

|  | D | = R | · T |
|---|---|---|---|
| Traveling against current | 48 | $(b - c)$ | 3 |
| Traveling with current | 48 | $(b + c)$ | 2 |

$48 = (b - c) \cdot 3$

$48 = (b + c) \cdot 2$

When we remove the parentheses, we obtain the following system.

$$48 = 3b - 3c \quad \textbf{(1)}$$
$$48 = 2b + 2c \quad \textbf{(2)}$$

This system is most easily solved by the addition method.

$$96 = 6b - 6c \quad \text{Multiply equation \textbf{(1)} by 2.}$$
$$\underline{144 = 6b + 6c} \quad \text{Multiply equation \textbf{(2)} by 3.}$$
$$240 = 12b$$
$$20 = b$$

$48 = 2b + 2c \qquad \textbf{(2)}$ Substitute the value of $b = 20$ into one of the original
$48 = 2(20) + 2c$ equations and solve for $c$. We'll use equation **(2)**.
$48 = 40 + 2c$
$8 = 2c$
$4 = c$

Thus the speed of Mike's boat in still water was 20 miles/hour and the speed of the current in Lazy River was 4 miles/hour.

*Check.* Can you verify these answers? ▫

**Student Practice 4** Mr. Caminetti traveled downstream in his boat a distance of 72 miles in a time of 3 hours. His return trip up the river against the current took 4 hours. Find the speed of the boat in still water. Find the speed of the current in the river.

**Applications** *Solve using two equations with two variables.*

**1. *Airline Travel*** Martha took the research department from her company on a round trip to Chicago to meet with a potential client. Including Martha, a total of 15 people took the trip. She was able to purchase coach tickets for $350 and first-class tickets for $1250. She used her total budget for airfare for the trip, which was $11,550. How many first-class tickets did she buy? How many coach tickets did she buy?

**2. *Baseball*** Jeff had to purchase tickets for 14 people in his office to go to a Red Sox game. He paid $1074 to buy the 14 tickets. Infield grandstand seats cost $78. Right field roof seats cost $75. How many of each kind of ticket did he purchase?

**3. *Small Business Management*** Gregg manages a small cleaning business. He has 16 employees. The cleaners get paid $12 per hour. The cleaning supervisors get paid $19 per hour. His budget has set aside a payroll expense of $220 per hour if all of his employees are working. How many cleaners and how many supervisors does he employ?

**4. *Owner of a Local Moving Company*** Margaret owns a local moving company that services primarily the local university population. She has some pickup trucks and some moving vans. She has a total of 11 trucks. She charges $250 per day for rental of her pickup trucks. She charges $380 per day for rental of her moving vans. Yesterday, all 11 of her trucks were rented. She collected a total of $3790 yesterday for the rental of the trucks. How many pickup trucks and how many moving vans does she have?

**5. *Income of a Nurse*** Barbara worked as a registered nurse in Paris last summer. The first week she was there she worked a total of 23 hours at a city hospital. She was paid $17 an hour for day shift and $24 an hour for night shift work. She earned $461 for her first week of work. How many hours did she spend on the day shift? How many hours did she spend on the night shift?

**6. *Income of a Substitute Teacher*** Ken worked as a substitute teacher last month. He substituted a total of 20 days. The first part of the month he worked in Tucson, Arizona, where he was paid $85 a day. Then he moved to Oregon, where he was paid $145 a day to substitute. During the month he earned $2420. How many days did he substitute in Tucson? How many days did he substitute in Oregon?

▲ **7. *Home Office Construction*** The basement of your house is only partially finished. Currently, there is a room that you use for storage. You want to convert this space to a home office, but you want more space than is available in the room now. You decide to knock down two walls and enlarge the room. The existing room has a perimeter of 38 feet. If you extend the width of the room by 5 feet and make the new length 6 feet less than double the original length, the new perimeter will be 56 feet. What are the dimensions of the existing storage room? What are the dimensions of the new room?

existing room

enlarged room

▲ 8. *Playground Dimensions* KidTime Daycare has plans to expand its playground. Presently it takes 1050 feet of fencing to enclose its rectangular playground. The expansion plans call for the daycare to triple the width of the playground and to double the length. Then the center will require 2550 feet of fencing. What is the length and width of the current playground?

9. *Automobile Radiators* During her beginning mechanics course, Rose learned that her leaking radiator holds 16 quarts of water. She temporarily patched the leak when the radiator was partially full. At that time it was 50% antifreeze. After she filled the remaining space in the radiator with an 80% antifreeze solution and waited a little while, she found that it was 65% antifreeze. How many quarts of solution were in the radiator just before Rose added antifreeze? How many quarts of 80% antifreeze solution did she add?

10. *Manufacturing Chocolate* A candy company wants to produce 20 kilograms of a special 45% fat content chocolate to conform with customer dietary demands. To obtain this, a 50% fat content Hawaiian chocolate is combined with a 30% fat content domestic chocolate in a special melting process. How many kilograms of the 50% fat content chocolate and how many kilograms of the 30% fat content chocolate are used to make the required 20 kilograms?

11. *Job Earnings* Anita's friend Chloe is encouraging her to apply for a position at Patio & BBQ Depot, where Chloe's roommate Josie is currently employed as a sales representative and earning a very good income. Josie has stated that each sales representative is paid the same annual base salary plus a standard commission of a certain percentage of the sales they make during the year. The first year Josie earned $41,000 and sold $55,000 worth of products, which was the average for a first-year salesperson. Josie was told that in the second year, sales double on average, resulting in annual earnings of $52,000. If Anita becomes a sales representative at Patio & BBQ Depot, what is the base pay and commission rate she should expect for that position?

12. *Job Earnings* Scott Hammond is considering applying for a sales position with a medical supply company. He saw a brochure that stated that all sales representatives are paid the same base annual salary plus a certain percentage of the sales they make each year. The brochure stated that in the first year of employment, the average sales are $16,000 and the total yearly income is $39,000. The brochure emphasized that if sales representatives network to build up clientele, the second year's sales usually increase to 4 times that of the first year, and yearly earnings increase to $51,000. Determine the base salary and the commission rate for the sales position.

13. *Airline Travel* An airplane traveled between two cities that are 3000 kilometers apart. The trip against the wind took 6 hours. The return trip with the benefit of the wind was 5 hours. What was the wind speed in kilometers per hour? What was the speed of the plane in still air?

14. *Airline Travel* Fred Blum often travels on business. His airplane flies between Boston and Cleveland. During a roundtrip flight, the plane flew a distance of 630 miles each way. The trip with a tailwind (the wind traveling the same direction as the plane) took 3 hours. The return trip traveling against the wind took 3.5 hours. Find the speed of the wind. Find the speed of the airplane in still air.

15. ***Hiking Club*** The hiking club wants to sell 50 pounds of an energy snack food that will cost them only $1.80 per pound. The mixture will be made up of nuts that cost $2.00 per pound and raisins that cost $1.50 per pound. How much of each should they buy to obtain 50 pounds of the desired mixture?

16. ***Flower Costs*** How many bunches of local daisies that cost $2.50 per bunch should be mixed with imported daisies that cost $3.50 per bunch to obtain 100 bunches of daisies that will cost $2.90 per bunch?

**Mixed Practice** *Solve the following applied problems using a system of equations.*

17. ***Airplane Flight*** Walter and Mary Jones flew from West Palm Beach to Denver to visit their daughter in college. The flight of the plane was a distance of 2000 miles. It took 5 hours to fly west against the jet stream. Coming back home with the jet stream the trip only took 4 hours. What was the speed of the jet stream? What was the speed of the plane in still air?

18. ***Canoe Trip*** Melissa and Phil lead an Outdoor Adventures Club that sponsored a canoe trip. The participants paddled 4 miles upstream from the canoe rental headquarters to a popular swimming and picnicking spot. They ate lunch, rested, and headed back down to where they'd started. The trip upstream took 2 hours and the trip back took only 1 hour. How fast were they paddling on average? How fast was the current?

19. ***Small Business Owner*** Joseph Lanza, owner of Lanza Electronics and Home Appliances, is preparing to open a second store in Los Angeles, California. Joseph pays the manager of his original Orange County store a weekly base pay of $700 plus 25% of the store's total weekly sales. The marketing data indicates that with a good sales force in the new L. A. store, the total weekly sales will be the same as the Orange County store. To motivate sales at the new store, Joseph will pay the manager a smaller base pay of $400 and a higher commission rate of 35%. He predicts that both managers will have the same total sales and earn the same weekly pay with this salary plan. Determine the amount of weekly sales and the manager's weekly salary at each store with this plan.

20. ***Sales Positions*** While attending college, Victor works as a part-time sales representative for J & L Tools and is paid a monthly base pay of $500 plus 15% of the total monthly sales. His friend works for Tool Supply Outlet and recommended that Victor apply for a part-time sales position with the Outlet company, which pays a lower base pay of $250 and a higher commission rate of 20%. He explained to Victor that although the first year, his monthly sales amount and income would remain the same as at his current position, the higher commission paid at the Outlet would allow him to earn more as he became more experienced in future years. Determine the total amount of sales and earnings each month Victor should expect for the first year at the Tool Supply Outlet.

21. ***Manager of a Car Wash*** Carlos manages a car wash while he goes to college. He has a contract with the local police department to keep the vehicles clean. He charges one price to wash a police cruiser. He charges a slightly higher price to wash a police SUV. Last week he washed 8 police cruisers and 5 police SUVs and charged the department $93. This week he washed 9 police cruisers and 7 police SUVs and charged the department $117. How much does he charge the department to clean one police cruiser? How much does he charge the department to clean one police SUV?

22. ***Helicopter Flight Scheduler*** Katrina is the scheduler for a sightseeing helicopter service on the island of Kauai. Yesterday she scheduled 15 helicopter trips for mountain view sightseeing and 7 helicopter trips for coastal view sightseeing. Today, she scheduled 9 helicopter trips for mountain view sightseeing and 5 helicopter trips for coastal view sightseeing. Yesterday the company earned $20,600 for all the helicopter trips. Today the company earned $13,000 for all the helicopter trips. How much money does the company charge for one mountain view sightseeing trip? How much money does the company charge for one coastal view sightseeing trip?

## Cumulative Review

**23. [7.3.1]** Find the slope of the line passing through $(3,-4)$ and $(-1,-2)$.

**24. [7.3.2]** What is the slope and the $y$-intercept of the line defined by the equation $3x + 4y = -8$?

**25. [7.4.2]** Find an equation of the line passing through $(2, 6)$ and $(-2,1)$. Write the equation in slope–intercept form.

**26. [7.3.3]** Find an equation of the line with slope of $-2$ and $y$-intercept $(0, 10)$.

## Quick Quiz 8.5

**1.** A college football game at Hampton University is played to a sell-out crowd of 18,500 people. Student tickets are $9 while general admission tickets are $15. A total of $217,500 was collected from ticket sales. How many people purchased student tickets? How many people purchased general admission tickets?

**2.** Last year the computer lab purchased 8 Dell PCs and 12 Apple iMacs. The cost of these computers was $20,800. This year the computer company said it would sell computers to the lab at the same cost per computer. This year the lab purchased 12 Dell PCs and 7 Apple iMacs for $18,000. How much did each Dell PC cost? How much did each Apple iMac cost?

**3.** The Salzmans chartered a jet from Denver to Richmond, Virginia. The flight plan required that the plane travel exactly 1500 miles. The trip was aided by the jet stream and took only 3 hours. The return trip of 1500 miles was traveling against the jet stream. That trip took 5 hours. What is the speed of the plane in still air? What is the speed of the jet stream?

**4. Concept Check** A new company is making a drink that is 45% pure fruit juice. They make a test batch of 500 gallons of the new drink. They are using some juice that is 50% pure fruit juice and some juice that is 30% pure fruit juice. They want to find out how many gallons of each of these two types they will need. Explain how you would set up a system of two equations using the variables $x$ and $y$ to solve this problem.

## Background: Marketing Manager

Dave has responsibility for identifying the price for a smartphone at which the supply will equal the demand. As a sales market manager for a large telecommunications company, he knows that as the price of a product increases, the demand for that product will lessen. However, he also knows that if a product can be sold at a higher price, suppliers are receptive to manufacturing a larger supply of the product. Given the importance of having supply equal demand, finding the appropriate price for the smartphone can mean the difference between a profit and a loss.

## Facts

- Market research data shows that the monthly demand equation for the smartphone is

$$3x + 12p - 2400 = 0,$$

where $x$ is the number of smartphones demanded monthly (in hundreds) and $p$ is their price. The monthly supply equation for this same smartphone is

$$3x - 20p + 1000 = 0,$$

where $x$ is the number to be sold monthly (in hundreds) at the price $p$ (in dollars).
- Fixed monthly expenses are \$512,820, and the cost of producing each smartphone is \$92.39. Expenses are therefore expressed as $\$92.39x + \$512,820$.

Revenue is expressed as $\$106.25x$.

## Tasks

1. Dave needs to determine the optimal price per phone and quantity of phones to make available for sale. What are those two amounts?

2. Dave now wants to determine the break-even point; that is, how many units must be sold in order for his company's revenues to match its expenses? He again develops the answer to this question using a system of equations. He already knows that revenue is expressed as $\$106.25x$, and expenses are expressed as $\$92.39x + \$512,820$. How many units must be sold to meet the break-even point? Round to the nearest thousand.

# Chapter 8 Organizer

| Topic and Procedure | Examples |  You Try It |
|---|---|---|
| ***Solving a system of equations by graphing, pp. 465–466***<br><br>Graph both equations on the same coordinate axis. One of the following will be true.<br><br>1. The lines intersect. The point of intersection is the solution to the system. Such a system is consistent. | **1.**<br>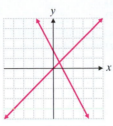 | **1.** Graph each system of equations. How many solutions does each system have?<br>**(a)** $y = 2x + 5$<br>    $y = 2x - 3$<br>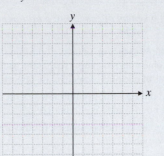 |
| 2. The lines are parallel. There is no point of intersection. The system has no solution. Such a system is inconsistent. | **2.**<br>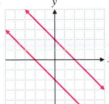 | **(b)**    $x - y = 4$<br>  $-2x + 2y = -8$<br>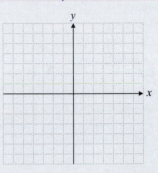 |
| 3. The lines coincide. Every point on the line represents a solution. There is an infinite number of solutions. Such a system has dependent equations. | **3.**<br>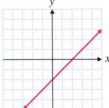 | **(c)** $y = -x - 1$<br>    $y = \phantom{-}x - 3$<br>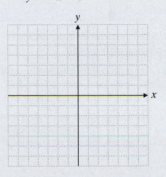 |
| ***Solving a system of equations: substitution method, pp. 473–474***<br><br>1. Solve for one variable in terms of the other variable in one of the two equations.<br>2. Substitute the expression you obtain for this variable into the other equation.<br>3. Solve this equation to find the value of the variable.<br>4. Substitute this value for the variable into one of the original or equivalent equations to obtain a value for the other variable.<br>5. Check the solution in both original equations. | Solve.   $2x + 3y = -6$<br>          $3x - y = 13$<br><br>Solve the second equation for $y$.<br>$$-y = -3x + 13$$<br>$$y = \phantom{-}3x - 13$$<br>Substitute this expression into the first equation.<br>$$2x + 3(3x - 13) = -6$$<br>$$2x + 9x - 39 = -6$$<br>$$11x - 39 = -6$$<br>$$11x = 33$$<br>$$x = 3$$<br>Substitute $x = 3$ into $y = 3x - 13$ to find $y$.<br>$$y = 3(3) - 13 = 9 - 13 = -4$$<br>$$y = -4$$ | **2.** Solve by the substitution method.<br>$$x - 3y = \phantom{-}5$$<br>$$-2x + 7y = -9$$ |

501

| Topic and Procedure | Examples | ⟹ You Try It |
|---|---|---|
| | *Check.*<br><br>$2(3) + 3(-4) \stackrel{?}{=} -6 \qquad 3(3) - (-4) \stackrel{?}{=} 13$<br>$6 - 12 \stackrel{?}{=} -6 \qquad\qquad 9 - (-4) \stackrel{?}{=} 13$<br>$-6 = -6 \checkmark \qquad\qquad 9 + 4 \stackrel{?}{=} 13$<br>$\qquad\qquad\qquad\qquad\qquad\qquad 13 = 13 \checkmark$<br><br>The solution is $(3, -4)$. | |
| *Solving a system of equations: addition method, pp. 479–480*<br><br>1. Multiply each term of one or both equations by some nonzero integer so that the coefficients of one of the variables are opposites.<br><br>2. Add the equations of this new system so that one variable is eliminated.<br><br>3. Solve the resulting equation for the remaining variable.<br><br>4. Substitute the value found into one of the original or equivalent equations to find the value of the other variable.<br><br>5. Check your solution in both of the original equations. | Solve.  $3x - 2y = 0$  **(1)**<br>$\qquad\quad -4x + 3y = 1$  **(2)**<br><br>Eliminate the $x$'s: a common multiple of 3 and 4 is 12. Multiply equation **(1)** by 4 and equation **(2)** by 3; then add.<br><br>$\qquad\qquad 12x - 8y = 0$<br>$\qquad\qquad \underline{-12x + 9y = 3}$<br>$\qquad\qquad\qquad\qquad y = 3$<br><br>Substitute $y = 3$ into either equation and solve for $x$. Let us pick $3x - 2y = 0$.<br><br>$3x - 2(3) = 0 \qquad 3x - 6 = 0 \qquad x = 2$<br>*Check.*<br>$3(2) - 2(3) \stackrel{?}{=} 0 \qquad -4(2) + 3(3) \stackrel{?}{=} 1$<br>$6 - 6 \stackrel{?}{=} 0 \qquad\qquad -8 + 9 \stackrel{?}{=} 1$<br>$0 = 0 \checkmark \qquad\qquad\qquad 1 = 1 \checkmark$<br>The solution is $(2, 3)$. | **3.** Solve by the addition method.<br><br>$\qquad 2x - 5y = \phantom{-}8$<br>$\qquad 4x + \phantom{5}y = -6$ |
| *Choosing a method to solve a system of equations, p. 487*<br><br>1. Substitution works well if at least one variable has a coefficient of 1 or $-1$.<br><br>2. Addition works well for integer, fractional, or decimal coefficients. | For each system of equations, decide if the addition or substitution method would be easier to use. Then solve the system.<br><br>$x + \phantom{2}y = 8$<br>$2x - 3y = 9$  Use substitution.<br><br>Solution $\left(\dfrac{33}{5}, \dfrac{7}{5}\right)$<br><br>$\dfrac{3}{5}x - \dfrac{1}{4}y = 17$<br><br>$\dfrac{1}{5}x + \dfrac{3}{4}y = -11$  Use addition.<br><br>Solution $(20, -20)$ | **4.** For each system of equations, decide if the addition or substitution method would be easier to use. Then solve the system.<br><br>**(a)** $-3x + \phantom{2}y = -4$<br>$\qquad\phantom{-}5x - 2y = \phantom{-}8$<br><br>**(b)** $\dfrac{1}{2}x + \dfrac{1}{3}y = \phantom{-1}0$<br><br>$\qquad \dfrac{5}{6}x - \dfrac{2}{3}y = -11$ |
| *Identifying inconsistent and dependent systems algebraically, p. 488*<br><br>1. If you obtain an equation that is inconsistent with known facts, such as $0 = 2$, the system is inconsistent. There is no solution.<br><br>2. If you obtain an equation that is always true, such as $0 = 0$, the equations are dependent. There is an infinite number of solutions. | Solve.  $x + 2y = \phantom{1}1$<br>$\qquad\quad 3x + 6y = 12$<br><br>Multiply the first equation by $-3$.<br><br>$\qquad\qquad -3x - 6y = -3$<br>$\qquad\qquad \underline{\phantom{-}3x + 6y = 12}$<br>$\qquad\qquad\qquad\qquad 0 = 9$<br><br>0 is not equal to 9, so there is **no solution**. The system is inconsistent.<br><br>Solve.  $2x + \phantom{3}y = 1$<br>$\qquad\quad 6x + 3y = 3$<br><br>Multiply the first equation by $-3$.<br><br>$\qquad\qquad -6x - 3y = -3$<br>$\qquad\qquad \underline{\phantom{-}6x + 3y = \phantom{-}3}$<br>$\qquad\qquad\qquad\qquad 0 = 0$<br><br>Since 0 is always equal to 0, the equations are dependent. There is an **infinite number of solutions**. | **5.** Solve each system.<br><br>**(a)** $\qquad x - 2y = \phantom{-}4$<br>$\qquad -2x + 4y = -8$<br><br>**(b)** $-9x - 3y = \phantom{-}5$<br>$\qquad\phantom{-}3x + \phantom{3}y = -1$ |

| Topic and Procedure | Examples |  You Try It |
|---|---|---|
| **Solving word problems using systems of equations, p. 492**<br><br>1. Understand the problem. Choose a variable to represent each unknown quantity.<br><br>2. Write a system of equations in two variables.<br><br>3. Solve the system of equations and state the answer.<br><br>4. Check the answer. | Apples sell for $0.35 per pound. Oranges sell for $0.40 per pound. Nancy bought 12 pounds of apples and oranges for $4.45. How many pounds of each fruit did she buy?<br><br>Let $x$ = number of pounds of apples and<br>$\quad y$ = number of pounds of oranges.<br><br>12 pounds in all were purchased. $x + y = 12$<br>The purchase cost is $4.45.<br>$0.35x + 0.40y = 4.45$ Multiply the second equation by 100 to obtain the following system.<br><br>$$x + \quad y = \ 12$$<br>$$35x + 40y = 445$$<br><br>Multiply the first equation by $-35$ and add the equations.<br><br>$$\begin{array}{r} -35x - 35y = -420 \\ 35x + 40y = \quad 445 \\ \hline 5y = \quad 25 \\ y = 5 \end{array}$$<br><br>Substitute $y = 5$ into $x + y = 12$.<br>$$x + 5 = 12$$<br>$$x = 7$$<br><br>Nancy purchased 7 pounds of apples and 5 pounds of oranges.<br><br>*Check.* Are there 12 pounds of apples and oranges?<br>$$7 + 5 = 12 \ \checkmark$$<br><br>Would this purchase cost $4.45?<br>$$0.35(7) + 0.40(5) \overset{?}{=} 4.45$$<br>$$2.45 + 2.00 \overset{?}{=} 4.45$$<br>$$4.45 = 4.45 \ \checkmark$$ | **6.** Green peppers sell for $0.65 per pound. Onions sell for $0.75 per pound. Manuel bought a total of 8 pounds of green peppers and onions for $5.70. How many pounds of each vegetable did he buy? |

# Chapter 8 Review Problems

## Section 8.1

*Solve by graphing the lines and finding the point of intersection.*

**1.** $2x + 3y = 0$
  $-x + 3y = 9$

**2.** $-3x + y = -2$
  $-2x - y = -8$

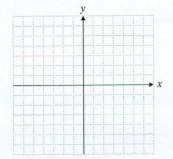

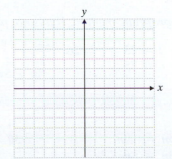

503

**3.** $2x - y = 6$
$6x + 3y = 6$

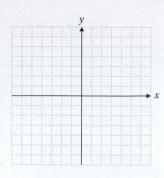

**4.** $2x - y = 1$
$3x + y = -6$

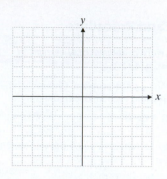

## Section 8.2

*Solve by the substitution method. Check your solution.*

**5.** $x + y = 6$
$-2x + y = -3$

**6.** $x + 3y = 18$
$2x + y = 11$

**7.** $3x - 2y = 3$
$x - \dfrac{1}{3}y = 8$

**8.** $0.5x + y = 16$
$4x - 2y = 8$

## Section 8.3

*Solve by the addition method. Check your solution.*

**9.** $6x - 2y = 10$
$2x + 3y = 7$

**10.** $4x - 5y = -4$
$-3x + 2y = 3$

**11.** $5x - 4y = 2$
$-3x + 10y = -5$

**12.** $9x + 2y = -5$
$12x - y = 8$

## Section 8.4

*Solve by any appropriate method. If it is not possible to obtain one solution, state the reason. Write the solution in the form $(x, y)$.*

**13.** $6x - y = 33$
$6x + 7y = 9$

**14.** $2x + 5y = 29$
$-3x + 10y = -26$

**15.** $7x + 3y = 2$
$-8x - 7y = 2$

**16.** $4x + 5y = 4$
$-5x - 8y = 2$

**17.** $7x - 2y = 1$
$-5x + 3y = -7$

**18.** $2x - 4y = 6$
$-3x + 6y = 7$

**19.** $4x - 7y = 8$
$5x + 9y = 81$

**20.** $2x - 9y = 0$
$3x + 5 = 6y$

**21.** $2x + 10y = 1$
$-4x - 20y = -2$

**22.** $1 + x - y = y + 4$
$4(x - y) = 3 - x$

**23.** $3x + y = 9$
$x - 2y = 10$

**24.** $2(x + 3) = y + 4$
$4x - 2y = -4$

## Mixed Practice

*Solve by any appropriate method. If it is not possible to obtain the solution, state the reason.*

**25.** $5x + 4y + 3 = 23$
$8x - 3y - 4 = 75$

**26.** $\dfrac{2x}{3} - \dfrac{3y}{4} = \dfrac{7}{12}$
$8x + 5y = 9$

**27.** $\dfrac{1}{5}a + \dfrac{1}{2}b = 6$
$\dfrac{3}{5}a - \dfrac{1}{2}b = 2$

**28.** $\dfrac{2}{3}a + \dfrac{3}{5}b = -17$
$\dfrac{1}{2}a - \dfrac{1}{3}b = -1$

**29.** $4.8 + 0.6m = 0.9n$
$0.2m - 0.3n = 1.6$

**30.** $8.4 - 0.8m = 0.4n$
$0.2m + 0.1n = 2.1$

**31.** $6s - 4t = 5$
$4s - 5t = -2$

32. $10s + 3t = 4$
$4s - 2t = -5$

33. $3(x + 2) = -2 - (x + 3y)$
$3(x + y) = 3 - 2(y - 1)$

34. $13 - x = 3(x + y) + 1$
$14 + 2x = 5(x + y) + 3x$

35. $0.2b = 1.4 - 0.3a$
$0.1b + 0.6 = 0.5a$

36. $0.3a = 1.1 - 0.2b$
$0.3b = 0.4a - 0.9$

37. $\dfrac{b}{5} = \dfrac{2}{5} - \dfrac{a - 3}{2}$
$4(a - b) = 3b - 2(a - 2)$

## Section 8.5

*Solve.*

38. **Museum of Modern Art** Admission to the Museum of Modern Art in New York City is $25 for adults and $18 for students. One Sunday, a total of 186 adults and students visited the museum. The receipts for that day totaled $4230. How many adults visited the museum that day? How many students?

39. **Fruit Packaging** A farmer near Napa has 252 pounds of apples to take to the farmer's market. He packages them into 2-lb and 5-lb bags to sell. If the number of 5-lb bags he used was double the number of 2-lb bags, how many of each type of bags did he prepare to bring to market?

40. **Airplane Travel** A plane travels 1500 miles in 5 hours with the benefit of a tailwind. On the return trip it requires 6 hours to fly against the wind. Can you find the speed of the plane in still air and the wind speed?

41. **Chemical Solutions** A chemist has 20% and 30% acid solutions and needs 40 liters of a 25% acid solution. How much of each solution should he mix to get the 25% solution?

42. **Scout Tree Sale** A boy scout troop is selling Christmas trees to raise money for a trip. On the first day they sold 79 trees. The balsam firs sold for $23 and the Norwegian pines were $28. The receipts for the day totaled $1942. How many trees of each type were sold?

43. **Sanding Roads in Maine** Carl manages a highway department garage in Maine. A mixture of salt and sand is stored in the garage for use during the winter. Carl needs 24 tons of a salt/sand mixture that is 25% salt. He will combine shipments of 15% salt and shipments of 30% salt to achieve the desired 24 tons. How much of each type should he use?

44. **Water Ski Boat** A water ski boat traveled 23 kilometers/hour going with the current. It went back in the opposite direction against the current and traveled only 15 kilometers/hour. What was the speed of the boat? What was the speed of the current?

45. **Manufacturing Production** Two printers were used for a 5-hour job of printing 15,000 labels. The next day, a second run of the same labels took 7 hours because one printer broke down after 2 hours. How many labels can each printer process per hour?

## To Think About

46. Lexi and Olivia are competing in the West Chicago cross country race. Lexi starts at the starting line, running at a pace of $n$ miles per hour. Olivia starts the race 3 miles into the course running at a pace of $m$ miles per hour. They have a maximum of 4 hours to catch up to each other. What are the two possible paces that Lexi and Olivia can run at to intersect in the maximum of four hours?

# How Am I Doing? Chapter 8 Test

MATH COACH　　　MyMathLab®　　　You Tube™

After you take this test read through the Math Coach on pages 507–508. Math Coach videos are available via MyMathLab and YouTube. Step-by-step test solutions on the Chapter Test Prep Videos are also available via MyMathLab and YouTube. (Search "TobeyBeginningAlg" and click on "Channels.")

*Solve by the method specified.*

MC **1.** Substitution method
$$3x - y = -5$$
$$-2x + 5y = -14$$

**2.** Addition method
$$3x + 4y = 7$$
$$2x + 3y = 6$$

**3.** Graphing method
$$2x - y = 4$$
$$4x + y = 2$$

**4.** Any method
$$x + 3y = 12$$
$$2x - 4y = 4$$

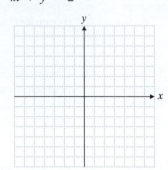

*Solve by any method. If there is not one solution to a system, state why.*

**5.** $2x - y = 5$
$-x + 3y = 5$

**6.** $2x + 3y = 13$
$3x - 5y = 10$

MC **7.** $\dfrac{2}{3}x - \dfrac{1}{5}y = 2$
$\dfrac{4}{3}x + 4y = 4$

**8.** $3x - 6y = 5$
$-\dfrac{1}{2}x + y = \dfrac{7}{2}$

MC **9.** $5x - 2 = y$
$10x = 4 + 2y$

MC **10.** $0.3x + 0.2y = 0$
$1.0x + 0.5y = -0.5$

**11.** $2(x + y) = 2(1 - y)$
$5(-x + y) = 2(23 - y)$

**12.** $3(x - y) = 12 + 2y$
$8(y + 1) = 6x - 7$

*Solve.*

**13.** Twice one number plus three times a second is one. Twice the second plus three times the first is nine. Find the numbers.

**14.** Both $8 and $12 tickets were sold for a basketball game. In all, 30,500 people paid for admission, resulting in a "gate" of $308,000. How many of each type of ticket were sold?

**15.** Five shirts and three pairs of slacks cost $172. Three of the same shirts and four pairs of the same slacks cost $156. How much does each shirt cost? How much is a pair of slacks?

**16.** You are producing a commemorative booklet for the 25th anniversary of your church. Ace Printers charges a $200 typesetting fee and $0.25 per booklet. Wong Printers charges a $250 typesetting fee but only $0.20 per booklet. How many booklets must you order for the cost to be the same? What is the cost?

**17.** A jet plane flew 2000 kilometers against the wind in a time of 5 hours. It refueled and flew back the same distance with the wind in 4 hours. Find the wind speed in kilometers per hour. Find the speed of the jet in still air in kilometers per hour.

1. _____ ☐
2. _____ ☐
3. _____ ☐
4. _____ ☐
5. _____ ☐
6. _____ ☐
7. _____ ☐
8. _____ ☐
9. _____ ☐
10. _____ ☐
11. _____ ☐
12. _____ ☐
13. _____ ☐
14. _____ ☐
15. _____ ☐
16. _____ ☐
17. _____ ☐

**Total Correct:** ☐

# MATH COACH

*Mastering the skills you need to do well on the test.*

The following problems are from the Chapter 8 Test. Here are some helpful hints to keep you from making common errors on test problems.

**Chapter 8 Test, Problem 1** Solve by the substitution method.

$$3x - y = -5$$
$$-2x + 5y = -14$$

> **HELPFUL HINT** If one equation contains $-y$, it is easier to solve for $y$ by adding $+y$ to each side of that equation. Use this result to substitute for $y$ in the second equation.

Did you add $y$ to each side of the first equation and then add 5 to each side to obtain $3x + 5 = y$?

Yes _____ No _____

If you answered No, consider why solving for $y$ in the first equation is the most logical first step when using the substitution method to solve this system.

Did you substitute $(3x + 5)$ for $y$ in the second equation to obtain $-2x + 5(3x + 5) = -14$ and then simplify and solve for $x$?

Yes _____ No _____

If you answered No, stop and perform these steps. Remember to substitute your final value of $x$ into one of the original equations to find $y$.

If you answered Problem 1 incorrectly, go back and rework the problem using these suggestions.

---

**Chapter 8 Test, Problem 7** Solve by any method.

$$\frac{2}{3}x - \frac{1}{5}y = 2$$
$$\frac{4}{3}x + 4y = 4$$

> **HELPFUL HINT** Find the LCD of the fractions in the top equation. Multiply each term of that equation by this LCD. Find the LCD of the fractions in the bottom equation. Multiply each term of that equation by this LCD. Now use these two new equations to solve the system.

Did you find that 15 is the LCD of the top equation? Did you multiply all three terms of the top equation by 15 to obtain $10x - 3y = 30$?

Yes _____ No _____

If you answered No to these questions, consider why the LCD is 15 and then carefully *multiply each term* of the top equation by 15 to find the correct result.

Did you find that 3 is the LCD of the bottom equation? Did you multiply all three terms of the bottom equation by 3 to obtain $4x + 12y = 12$?

Yes _____ No _____

If you answered No to these questions, consider why the LCD is 3 and then carefully *multiply each term* of the bottom equation by 3 to find the correct result. Then use these two new equations to solve the system.

Now go back and rework the problem again using these suggestions.

Need more help? Watch the **MATH COACH** videos in MyMathLab® or on You Tube .

507

## Chapter 8 Test, Problem 9

Solve by any method. If there is not one solution to a system, state why.

$$5x - 2 = y$$
$$10x = 4 + 2y$$

> **HELPFUL HINT** When you try to solve a system and get a false statement such as $3 = 0$, then the system has no solution and is inconsistent. When you try to solve a system and get a statement that is always true such as $4 = 4$, then the system has an infinite number of solutions and is dependent.

Since the top equation is already solved for $y$, it makes sense to use the substitution method.

Did you substitute $(5x - 2)$ for $y$ in the bottom equation to obtain $10x = 4 + 2(5x - 2)$?

Yes _____ No _____

If you answered No, go back and make this substitution. Did you simplify this equation to obtain $0 = 0$?

Yes _____ No _____

If you used the addition method, you should still obtain $0 = 0$. Did you determine that this system has an infinite number of solutions?

Yes _____ No _____

If you answered No to these questions, examine your work for any errors and review the definition of inconsistent and dependent systems. In your final description, state whether the system is inconsistent or dependent and also state whether the system has no solution or an infinite number of solutions.

If you answered Problem 9 incorrectly, go back and rework the problem using these suggestions.

---

## Chapter 8 Test, Problem 10  Solve by any method.

$$0.3x + 0.2y = 0$$
$$1.0x + 0.5y = -0.5$$

> **HELPFUL HINT** Since the decimals are tenths, multiply each term of each equation by 10 to find an equivalent system of equations without decimal coefficients. Then solve the system.

Did you multiply both equations by 10 to obtain the system

$$3x + 2y = 0$$
$$10x + 5y = -5?$$

Yes _____ No _____

If you answered No, carefully go through each equation and move the decimal point one place to the right for each number. This represents multiplying by 10. See if you get the above result.

With the new system, a good approach is to multiply the top equation by 5 and the bottom equation by $-2$. If you follow these steps, do you get the system

$$15x + 10y = 0$$
$$-20x - 10y = 10?$$

Yes _____ No _____

If you answered No, stop and carefully multiply every term of the top equation by 5. Remember that $5 \times 0 = 0$.

Next, multiply every term of the bottom equation by $-2$ and complete the problem. Watch out for $+$ and $-$ signs. Remember to substitute your final value for $x$ into either of the original equations to find $y$.

Now go back and rework the problem using these suggestions.

Need more help? Look for section examples marked with MC to review.

CHAPTER 9

# Radicals

## CAREER OPPORTUNITIES
## Graphic Designer

Graphic design professionals create designs that are appealing to look at while communicating an intended message. Their designs, when successfully produced, provoke the formation of relationships between viewers and the brand being conveyed. Whether the context is educational, commercial, or social, executing effective design work involves problem solving and math. These professionals use math to create compositions that are functional, appealing, informative, and even inspirational.

To investigate how the mathematics in this chapter can help with this field, see the Career Exploration Problems on page **551**.

## 9.1 Square Roots ▶

### 1 Evaluating the Square Root of a Perfect Square ▶

How long is the side of a square whose area is 4? Recall the formula for the area of a square.

area of a square $= s^2$    Our question then becomes, what number times itself is 4?

$$s^2 = 4$$
$$s = 2 \quad \text{because } (2)(2) = 4$$
$$s = -2 \quad \text{because } (-2)(-2) = 4 \text{ (Note: Even though } -2 \cdot -2 = 4,$$
$$\text{the side of a square cannot be a negative number)}$$

We say that 2 is a **square root** of 4 because $(2)(2) = 4$. We can also say that $-2$ is a square root of 4 because $(-2)(-2) = 4$. Note that 4 is a **perfect square.** A square root of a perfect square is an integer.

The symbol $\sqrt{\phantom{x}}$ is used in mathematics for the **principal square root** of a number, which is the nonnegative square root. The symbol itself $\sqrt{\phantom{x}}$ is called the **radical sign.** If we want to find the negative square root of a number, we use the symbol $-\sqrt{\phantom{x}}$.

---

**DEFINITION OF PRINCIPAL SQUARE ROOT**

For all nonnegative numbers $N$, the principal square root of $N$ (written $\sqrt{N}$) is defined to be the nonnegative number $a$ if and only if $a^2 = N$.

---

Notice in the definition that we did not use the words "for all *positive* numbers $N$" because we also want to include the square root of 0. $\sqrt{0} = 0$ since $0^2 = 0$.

Let us now examine the use of this definition with a few examples.

**Example 1** Find.    **(a)** $\sqrt{144}$    **(b)** $-\sqrt{9}$

**Solution**

**(a)** $\sqrt{144} = 12$ since $(12)^2 = (12)(12) = 144$ and $12 \geq 0$.

**(b)** The symbol $-\sqrt{9}$ is read "the opposite of the square root of 9."
Because $\sqrt{9} = 3$, we have $-\sqrt{9} = -3$. □

▶ **Student Practice 1** Find.    **(a)** $\sqrt{64}$    **(b)** $-\sqrt{121}$

The number beneath the radical sign is called the **radicand.** The radicand will not always be an integer.

**Example 2** Find.

**(a)** $\sqrt{\dfrac{1}{4}}$    **(b)** $-\sqrt{\dfrac{4}{9}}$

**Solution**

**(a)** $\sqrt{\dfrac{1}{4}} = \dfrac{1}{2}$ since $\left(\dfrac{1}{2}\right)^2 = \left(\dfrac{1}{2}\right)\left(\dfrac{1}{2}\right) = \dfrac{1}{4}$.    **(b)** $-\sqrt{\dfrac{4}{9}} = -\dfrac{2}{3}$ since $\sqrt{\dfrac{4}{9}} = \dfrac{2}{3}$. □

▶ **Student Practice 2** Find.

**(a)** $\sqrt{\dfrac{9}{25}}$    **(b)** $-\sqrt{\dfrac{121}{169}}$

**Example 3** Find.   **(a)** $\sqrt{0.09}$     **(b)** $\sqrt{1600}$     **(c)** $\sqrt{225}$

**Solution**

**(a)** $\sqrt{0.09} = 0.3$ since $(0.3)(0.3) = 0.09$.

Notice that $\sqrt{0.09}$ is *not* 0.03 because $(0.03)(0.03) = 0.0009$. Remember to count the decimal places when you multiply decimals. When finding the square root of a decimal, you should multiply to check your answer.

**(b)** $\sqrt{1600} = 40$ since $(40)(40) = 1600$.     **(c)** $\sqrt{225} = 15$     ◻

▭➡ **Student Practice 3**   Find.

**(a)** $-\sqrt{0.0036}$     **(b)** $\sqrt{2500}$     **(c)** $\sqrt{196}$

### 2 Approximating the Square Root of a Number That Is Not a Perfect Square ▶

Not all numbers are perfect squares. How can we find the square root of 2? That is, what number times itself equals 2? $\sqrt{2}$ is an irrational number. It cannot be written as $\frac{p}{q}$, where $p$ and $q$ are integers and $q \neq 0$. $\sqrt{2}$ is a nonterminating, nonrepeating decimal, and the best we can do is approximate its value. To do so, we can use the square root key on a calculator.

To use a calculator, enter 2 and push the $\boxed{\sqrt{x}}$ or $\boxed{\sqrt{\phantom{x}}}$ key. Or you might need to first push the $\boxed{\sqrt{x}}$ or $\boxed{\sqrt{\phantom{x}}}$ key and then the 2.

$$\boxed{2}\ \boxed{\sqrt{x}}\ \text{Display: } \boxed{1.4142136}$$

$$\text{OR } \sqrt{x}\ \boxed{2}\ \text{Display: } \boxed{1.4142136}$$

Remember, this is only an approximation. Most calculators are limited to a ten- or twelve-digit display.

**Example 4** Use a calculator to approximate. Round to the nearest thousandth.

**(a)** $\sqrt{28}$     **(b)** $\sqrt{191}$

**Solution**

**(a)** Using a calculator, $\boxed{\sqrt{x}}$ 28 gives 5.291502622. Thus $\sqrt{28} \approx 5.292$.

**(b)** Using a calculator, $\boxed{\sqrt{x}}$ 191 gives 13.82027496. Thus $\sqrt{191} \approx 13.820$.     ◻

▭➡ **Student Practice 4**   Approximate. Round to the nearest thousandth.

**(a)** $\sqrt{13}$     **(b)** $\sqrt{35}$     **(c)** $\sqrt{127}$

On a professional softball diamond, the distance from home plate to second base is exactly $\sqrt{7200}$ feet. Approximate this value using a calculator.

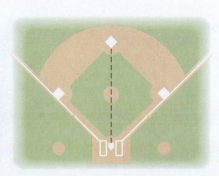

## 9.1 Exercises   MyMathLab®

**Verbal and Writing Skills, Exercises 1–4**

**1.** Define a principal square root in your own words.

**2.** Is $\sqrt{576} = 24$? Why or why not?

**3.** Is $\sqrt{0.9} = 0.3$? Why or why not?

**4.** How would you find $\sqrt{90}$?

*Find the two square roots of each number.*

**5.** 9

**6.** 4

**7.** 49

**8.** 81

*Find the square root. Do not use a calculator.*

**9.** $\sqrt{16}$

**10.** $\sqrt{100}$

**11.** $\sqrt{144}$

**12.** $\sqrt{25}$

**13.** $-\sqrt{36}$

**14.** $-\sqrt{25}$

**15.** $\sqrt{0.81}$

**16.** $\sqrt{0.49}$

**17.** $\sqrt{\dfrac{36}{121}}$

**18.** $\sqrt{\dfrac{16}{81}}$

**19.** $\sqrt{\dfrac{49}{169}}$

**20.** $\sqrt{\dfrac{9}{196}}$

**21.** $\sqrt{625}$

**22.** $\sqrt{1225}$

**23.** $-\sqrt{10{,}000}$

**24.** $-\sqrt{90{,}000}$

**Mixed Practice** *Find the square root. Do not use a calculator.*

**25.** $\sqrt{169}$

**26.** $\sqrt{225}$

**27.** $-\sqrt{\dfrac{1}{64}}$

**28.** $-\sqrt{\dfrac{4}{25}}$

**29.** $\sqrt{\dfrac{9}{16}}$

**30.** $-\sqrt{\dfrac{16}{49}}$

**31.** $\sqrt{0.0036}$

**32.** $\sqrt{0.0009}$

**33.** $\sqrt{19{,}600}$

**34.** $\sqrt{28{,}900}$

**35.** $-\sqrt{289}$

**36.** $-\sqrt{361}$

*Use a calculator to approximate to the nearest thousandth.*

**37.** $\sqrt{27}$

**38.** $\sqrt{35}$

**39.** $\sqrt{62}$

**40.** $\sqrt{51}$

**41.** $-\sqrt{183}$

**42.** $-\sqrt{123}$

**43.** $-\sqrt{195}$

**44.** $-\sqrt{214}$

### Applications

▲ **45.** *Tree Care* A tree is 12 feet tall. A wire that stretches from a point that is 5 feet from the base of the tree to the top of the tree is $\sqrt{169}$ feet long. Exactly how many feet long is the wire?

▲ **46.** *Extension Ladder* A garage is 15 feet tall. A ladder that stretches from a point that is 8 feet from the base of the garage to the top of the garage is $\sqrt{289}$ feet long. Exactly how many feet long is the ladder?

*Time for an Object to Fall* *If we ignore wind resistance, the time in seconds it takes for an object to fall s feet is given by* $t = \frac{1}{4}\sqrt{s}$. *Thus an object will fall 100 feet in* $t = \frac{1}{4}\sqrt{100} = \frac{1}{4}(10) = 2.5$ *seconds.*

47. How long would it take an object dropped off a building to fall 900 feet?

48. How long would it take an object dropped out of a tower to fall 2500 feet?

**To Think About: Finding Higher Roots** *In addition to finding the square root of a number, we can also find the cube root of a number. To find the cube root of N (which is written as $\sqrt[3]{N}$), we find a number that, when cubed, will equal N. For example, $\sqrt[3]{8} = 2$ since $2^3 = 8$. We can also find the cube root of a negative number—for example, $\sqrt[3]{-8} = -2$ since $(-2)^3 = -8$. The cube root of a positive number is positive. The cube root of a negative number is negative. Therefore, each real number has only one real cube root.*

*The concept of roots can be extended to fourth and fifth roots. For example, $\sqrt[4]{16} = 2$ since $2^4 = 16$. Also, $\sqrt[5]{32} = 2$ since $2^5 = 32$.*

*Find each higher root.*

49. $\sqrt[3]{27}$

50. $\sqrt[3]{125}$

51. $\sqrt[3]{-64}$

52. $\sqrt[4]{256}$

53. $\sqrt[4]{81}$

54. $\sqrt[4]{625}$

55. $\sqrt[5]{243}$

56. $\sqrt[5]{1024}$

57. Is there a real number that equals $\sqrt[4]{-16}$ Why?

58. Is there a real number that equals $\sqrt[4]{-81}$ Why?

## Cumulative Review

59. **[8.4.1]** Solve. $3x + 2y = 8$
$7x - 3y = 11$

60. **[8.4.2]** Solve if possible. $2x - 3y = 1$
$-8x + 12y = 4$

61. **[8.5.1]** *Snowboarding* A snowboard racer bought three new snowboards and two pairs of racing goggles last month for $850. This month, she bought four snowboards and three pairs of goggles for $1150. How much did each snowboard cost? How much did each pair of goggles cost?

62. **[8.5.1]** *Flying with a Wind* With a tailwind, the Qantas flight from Los Angeles to Sydney, Australia, covered 7280 miles in 14 hours, nonstop. The return flight of 7140 miles took 17 hours because of a headwind. If the airspeed of the plane remained constant and if the return flight encountered a constant headwind equal to the earlier tailwind, what was the plane's airspeed in still air?

**Quick Quiz 9.1** *Find the square root. Do not use a calculator.*

1. $\sqrt{121}$

2. $\sqrt{\dfrac{25}{64}}$

3. $\sqrt{0.09}$

4. **Concept Check** Explain how you would go about determining $\sqrt{32{,}400}$ without using a calculator.

# 9.2 Simplifying Radical Expressions ▶

## 1 Simplifying a Radical Expression with a Radicand That Is a Perfect Square ▶

We know that $\sqrt{25} = \sqrt{5^2} = 5$, $\sqrt{9} = \sqrt{3^2} = 3$, and $\sqrt{36} = \sqrt{6^2} = 6$. If we write the radicand as a square of a nonnegative number, the square root is this nonnegative base. That is, $\sqrt{x^2} = x$ when $x$ is nonnegative. We can use this idea to simplify radicals.

**Example 1** Find.

(a) $\sqrt{9^4}$ 　　　　(b) $\sqrt{126^2}$ 　　　　(c) $\sqrt{17^6}$

**Solution**

(a) Using the law of exponents, we rewrite $9^4$ as a square.

$$\sqrt{9^4} = \sqrt{(9^2)^2} \qquad \text{Raise a power to a power: } (9^2)^2 = 9^4.$$
$$= 9^2 \qquad\qquad \text{To check, multiply: } (9^2)(9^2) = 9^4.$$

(b) $\sqrt{126^2} = 126$

(c) $\sqrt{17^6} = \sqrt{(17^3)^2} \qquad \text{Raise a power to a power: } (17^3)^2 = 17^6.$
$\qquad = 17^3 \qquad\qquad \text{To check, multiply: } (17^3)(17^3) = 17^6.$ ☐

▶ **Student Practice 1** Find.

(a) $\sqrt{6^2}$ 　　　　(b) $\sqrt{13^4}$ 　　　　(c) $\sqrt{18^{12}}$

This same concept can be used with variable expressions. We will restrict the value of each variable to be nonnegative if the variable is under the radical sign. Thus *all variable radicands in this chapter* are assumed to *represent positive numbers or zero*.

**Example 2** Find.

(a) $\sqrt{x^6}$ 　　　　　　(b) $\sqrt{y^{24}}$

**Solution**

(a) $\sqrt{x^6} = \sqrt{(x^3)^2} \qquad \text{By the law of exponents, } (x^3)^2 = x^6.$
$\qquad = x^3$

(b) $\sqrt{y^{24}} = \sqrt{(y^{12})^2} = y^{12}$ ☐

▶ **Student Practice 2** Find.

(a) $\sqrt{y^{18}}$ 　　　　(b) $\sqrt{x^{30}}$

In each of these examples we found the square root of a perfect square, and thus the radical sign disappeared. When you find the square root of a perfect square, do *not* leave a radical sign in your answer.

Often the coefficient of a variable radical will not be written in exponent form. You will want to be able to recognize the squares of numbers from 1 to 15, 20, and 25. If you can, you will be able to find many square roots faster mentally than by using a calculator.

Before you begin, you need to know the multiplication rule for square roots.

**MULTIPLICATION RULE FOR SQUARE ROOTS**

For all nonnegative numbers $a$ and $b$,

$$\sqrt{a} \cdot \sqrt{b} = \sqrt{ab} \quad \text{and} \quad \sqrt{ab} = \sqrt{a} \cdot \sqrt{b}.$$

**Example 3** Find.

(a) $\sqrt{225x^2}$      (b) $\sqrt{x^6 y^{14}}$      (c) $\sqrt{169x^8 y^{10}}$

**Solution**

(a) $\sqrt{225x^2} = 15x$   Using the multiplication rule for square roots, $\sqrt{225x^2} = \sqrt{225}\sqrt{x^2} = 15x$.

(b) $\sqrt{x^6 y^{14}} = x^3 y^7$      (c) $\sqrt{169x^8 y^{10}} = 13x^4 y^5$

To check each answer, square it. The result should be the expression under the radical sign. ◻

**Student Practice 3** Find.

(a) $\sqrt{625y^4}$      (b) $\sqrt{x^{16} y^{22}}$      (c) $\sqrt{121x^{12} y^6}$

## 2 Simplifying a Radical Expression with a Radicand That Is Not a Perfect Square ▶

Most of the time when we encounter a square root, the radicand is not a perfect square. Thus, when we simplify the square root algebraically, we must retain the radical sign. Still, being able to simplify such radical expressions is a useful skill because it allows us to combine them, as we shall see in Section 9.3.

**Example 4** Simplify.

(a) $\sqrt{20}$      (b) $\sqrt{50}$      (c) $\sqrt{48}$

**Solution**   To begin, when you look at a radicand, look for factors that are perfect squares.

(a) $\sqrt{20} = \sqrt{4 \cdot 5}$   Note that 4 is a perfect square, and we can write 20 as $4 \cdot 5$.

$\qquad = \sqrt{4}\sqrt{5}$   Recall that $\sqrt{4} = 2$.

$\qquad = 2\sqrt{5}$

(b) $\sqrt{50} = \sqrt{25 \cdot 2}$      (c) $\sqrt{48} = \sqrt{16 \cdot 3}$

$\qquad = \sqrt{25}\sqrt{2}$           $= \sqrt{16}\sqrt{3}$

$\qquad = 5\sqrt{2}$              $= 4\sqrt{3}$   ◻

**Student Practice 4** Simplify.

(a) $\sqrt{98}$      (b) $\sqrt{12}$      (c) $\sqrt{75}$

The same procedure can be used with square root radical expressions containing variables. The key is to think of squares. That is, think, "How can I rewrite the expression so that it contains a perfect square factor?"

**Example 5** Simplify.

(a) $\sqrt{x^3}$      (b) $\sqrt{x^7 y^9}$

**Solution**

(a) Recall that the law of exponents tells us to add the exponents when multiplying two expressions. Because we want one of these expressions to be a perfect square, we think of the exponent 3 as a sum of an even number and $1: 3 = 2 + 1$.

$\sqrt{x^3} = \sqrt{x^2}\sqrt{x}$   Because $(x^2)(x) = x^3$ by the law of exponents.

$\qquad = x\sqrt{x}$

*Continued on next page*

In general, to simplify a square root that has a variable with an exponent in the radicand, first write the square root as a product of the form $\sqrt{x^n}\sqrt{x}$, where $n$ is the largest possible even exponent.

**(b)** $\sqrt{x^7y^9} = \sqrt{x^6}\sqrt{x}\sqrt{y^8}\sqrt{y}$

$\qquad = x^3\sqrt{x}\, y^4\sqrt{y}$

$\qquad = x^3y^4\sqrt{xy}$

In the final simplified form, we place the variable factors with exponents first and the radical factor second.  □

**Student Practice 5**   Simplify.

**(a)** $\sqrt{x^{11}}$  **(b)** $\sqrt{x^5y^3}$

In summary, to simplify a square root radical, you factor each part of the expression under the radical sign and simplify the perfect squares.

**Example 6**  Simplify.

**(a)** $\sqrt{12y^5}$  **(b)** $\sqrt{18x^3y^7w^{10}}$

**Solution**

**(a)** $\sqrt{12y^5} = \sqrt{4\cdot 3\cdot y^4\cdot y}$

$\qquad = 2y^2\sqrt{3y}$  Remember, $y^4 = (y^2)^2$ and $\sqrt{(y^2)^2} = y^2$.

**(b)** $\sqrt{18x^3y^7w^{10}} = \sqrt{9\cdot 2\cdot x^2\cdot x\cdot y^6\cdot y\cdot w^{10}}$

$\qquad = 3xy^3w^5\sqrt{2xy}$  □

**Student Practice 6**   Simplify.

**(a)** $\sqrt{48x^{11}}$  **(b)** $\sqrt{121x^6y^7z^8}$

**Sidelight:  A Topic for Further Thought**

Let's look at the radicand carefully. Does the radicand need to be a nonnegative number? What if we write $\sqrt{-4}$? Is there a number that you can square to get $-4$?

Obviously, there is *no real number* that you can square to get $-4$. We know that $(2)^2 = 4$ and $(-2)^2 = 4$. Any real number that is squared will be nonnegative. We therefore conclude that $\sqrt{-4}$ does not represent a real number. Because we want to work with real numbers, our definition of $\sqrt{n}$ requires that $n$ be nonnegative. Thus $\sqrt{-4}$ is *not a real number*.

You may encounter a more sophisticated number system called **complex numbers** in a higher-level mathematics course. In the complex number system, negative numbers have square roots.

## 9.2 Exercises  MyMathLab®

### Verbal and Writing Skills, Exercises 1–4

1. Is $\sqrt{6^2} = (\sqrt{6})^2$?

2. Why is it *not correct* to say
$$\sqrt{30} = \sqrt{(-6)(-5)} = \sqrt{-6}\sqrt{-5}?$$

3. Is $\sqrt{(-9)^2} = -\sqrt{9^2}$?

4. Is $\sqrt{2^4} = (\sqrt{2})^4$?

*Simplify. Leave the answer in exponent form.*

5. $\sqrt{8^2}$

6. $\sqrt{2^2}$

7. $\sqrt{18^4}$

8. $\sqrt{11^6}$

9. $\sqrt{10^8}$

10. $\sqrt{5^{12}}$

11. $\sqrt{33^8}$

12. $\sqrt{51^{16}}$

13. $\sqrt{5^{140}}$

14. $\sqrt{6^{200}}$

*Simplify. Assume that all variables represent positive numbers.*

15. $\sqrt{x^{12}}$

16. $\sqrt{x^{30}}$

17. $\sqrt{t^{18}}$

18. $\sqrt{a^{18}}$

19. $\sqrt{y^{26}}$

20. $\sqrt{w^{14}}$

21. $\sqrt{36x^8}$

22. $\sqrt{64x^6}$

23. $\sqrt{144x^2}$

24. $\sqrt{196y^{10}}$

25. $\sqrt{x^6y^4}$

26. $\sqrt{x^2y^{18}}$

27. $\sqrt{16x^2y^{20}}$

28. $\sqrt{64a^8b^4}$

29. $\sqrt{100a^{10}b^6}$

30. $\sqrt{81x^6y^{10}}$

31. $\sqrt{24}$

32. $\sqrt{27}$

33. $\sqrt{45}$

34. $\sqrt{44}$

35. $\sqrt{18}$

36. $\sqrt{32}$

37. $\sqrt{72}$

38. $\sqrt{60}$

39. $\sqrt{90}$

40. $\sqrt{75}$

41. $\sqrt{128}$

42. $\sqrt{96}$

43. $\sqrt{8x^3}$

44. $\sqrt{12y^5}$

45. $\sqrt{27w^5}$

46. $\sqrt{25x^5}$

47. $\sqrt{32z^9}$

48. $\sqrt{24y^4}$

49. $\sqrt{28x^5y^7}$

50. $\sqrt{20x^4y^5}$

### Mixed Practice *Simplify. Assume all variables represent positive numbers.*

51. $\sqrt{48y^3w}$

52. $\sqrt{12x^2y^3}$

53. $\sqrt{75x^2y^3}$

54. $\sqrt{50x^7y^6}$

55. $\sqrt{64x^{10}}$

56. $\sqrt{49x^6}$

57. $\sqrt{y^7}$

58. $\sqrt{w^9}$

**59.** $\sqrt{x^4 y^{14}}$

**60.** $\sqrt{x^{16} y^2}$

**61.** $\sqrt{135 x^5 y^7}$

**62.** $\sqrt{180 a^5 b c^2}$

**63.** $\sqrt{72 a^2 b^4 c^5}$

**64.** $\sqrt{80 a^3 b c^5}$

**65.** $\sqrt{169 a^3 b^8 c^5}$

**66.** $\sqrt{225 a^7 b^5 c^{10}}$

**To Think About** *Simplify. Assume that all variables represent positive numbers.*

**67.** $\sqrt{(x+5)^2}$

**68.** $\sqrt{x^2 + 10x + 25}$

**69.** $\sqrt{16x^2 + 8x + 1}$

**70.** $\sqrt{16x^2 + 24x + 9}$

**71.** $\sqrt{x^2 y^2 + 14xy^2 + 49y^2}$

**72.** $\sqrt{16x^2 + 16x + 4}$

## Cumulative Review

**73.** **[8.2.1]** Solve. $3x - 2y = 14$
$$y = x - 5$$

**74.** **[2.4.1]** Solve. $\dfrac{1}{2}(x - 4) = \dfrac{1}{3}x - 5$

**75.** **[5.3.2]** Factor. $2x^2 + 12xy + 18y^2$

**76.** **[5.4.1]** Factor. $2x^2 - x - 21$

**77.** **[6.2.2]** Divide. $\dfrac{x^2 + 6x + 9}{2x^2 y - 18y} \div \dfrac{6xy + 18y}{3x^2 y - 27y}$

**78.** **[6.3.3]** Add. $\dfrac{6}{2x - 4} + \dfrac{1}{12x + 4}$

Quick Quiz 9.2 *Simplify. Assume that all variables represent positive numbers.*

**1.** $\sqrt{120}$

**2.** $\sqrt{81 x^3 y^4}$

**3.** $\sqrt{135 x^8 y^5}$

**4.** **Concept Check** Explain how you would simplify the following expression.
$$\sqrt{y^6 z^7}$$

# 9.3 Adding and Subtracting Radical Expressions ▶

## 1 Adding and Subtracting Radical Expressions with Like Radicals ▶

Recall that when you simplify an algebraic expression, you combine like terms. $8b + 5b = 13b$ because both terms contain the same variable, $b$, and you can add the coefficients. This same idea is used when combining radicals. To add or subtract square root radicals, the numbers or the algebraic expressions under the radical signs must be the same. That is, the radicals must be **like radicals.**

**Student Learning Objectives**

**After studying this section, you will be able to:**

1 Add and subtract radical expressions with like radicals. ▶

2 Add and subtract radical expressions that must first be simplified. ▶

**Example 1** Combine.

(a) $5\sqrt{2} - 8\sqrt{2}$

(b) $7\sqrt{a} + 3\sqrt{a} - 5\sqrt{a}$

**Solution** First check to see if the radicands are the same. If they are, you can combine the radicals.

(a) $5\sqrt{2} - 8\sqrt{2} = (5 - 8)\sqrt{2} = -3\sqrt{2}$

(b) $7\sqrt{a} + 3\sqrt{a} - 5\sqrt{a} = (7 + 3 - 5)\sqrt{a} = 5\sqrt{a}$ ▫

**Student Practice 1** Combine.

(a) $7\sqrt{11} + 4\sqrt{11}$

(b) $4\sqrt{t} - 7\sqrt{t} + 6\sqrt{t} - 2\sqrt{t}$

Be careful. Not all the terms in an expression will have like radicals.

**Example 2** Combine. $5\sqrt{2a} + 3\sqrt{a} - 7\sqrt{2} + 3\sqrt{2a}$

**Solution** The only terms that have the same radicand are $5\sqrt{2a}$ and $3\sqrt{2a}$. These terms may be combined. All other terms stay the same.

$$5\sqrt{2a} + 3\sqrt{a} - 7\sqrt{2} + 3\sqrt{2a} = 8\sqrt{2a} + 3\sqrt{a} - 7\sqrt{2}$$ ▫

**Student Practice 2** Combine. $3\sqrt{x} - 2\sqrt{xy} - 5\sqrt{y} + 7\sqrt{xy}$

## 2 Adding and Subtracting Radical Expressions That Must First Be Simplified ▶

Sometimes it is necessary to simplify one or more radicals before their terms can be combined. Be sure to combine only like radicals.

**Example 3** Combine.

(a) $2\sqrt{3} + \sqrt{12}$

(b) $\sqrt{12} - \sqrt{27} + \sqrt{50}$

**Solution**

(a) $2\sqrt{3} + \sqrt{12} = 2\sqrt{3} + \sqrt{4 \cdot 3}$    Look for perfect square factors.

$\qquad\qquad\quad = 2\sqrt{3} + 2\sqrt{3}$    Simplify and combine like radicals.

$\qquad\qquad\quad = 4\sqrt{3}$

(b) $\sqrt{12} - \sqrt{27} + \sqrt{50} = \sqrt{4 \cdot 3} - \sqrt{9 \cdot 3} + \sqrt{25 \cdot 2}$

$\qquad\qquad\qquad\qquad = 2\sqrt{3} - 3\sqrt{3} + 5\sqrt{2}$

$\qquad\qquad\qquad\qquad = -\sqrt{3} + 5\sqrt{2}$ or $5\sqrt{2} - \sqrt{3}$    Only $2\sqrt{3}$ and $-3\sqrt{3}$ can be combined. ▫

**Student Practice 3** Combine.

(a) $\sqrt{50} - \sqrt{18} + \sqrt{98}$

(b) $\sqrt{12} + \sqrt{18} - \sqrt{50} + \sqrt{27}$

**Mc Example 4** Combine. $\sqrt{2a} + \sqrt{8a} + \sqrt{27a}$

**Solution**

$$\sqrt{2a} + \sqrt{8a} + \sqrt{27a}$$
$$= \sqrt{2a} + \sqrt{4 \cdot 2a} + \sqrt{9 \cdot 3a} \quad \text{Look for perfect square factors.}$$
$$= \sqrt{2a} + 2\sqrt{2a} + 3\sqrt{3a} \quad \text{Be careful. } \sqrt{2a} \text{ and } \sqrt{3a} \text{ are \textbf{not} like radicals. The expressions under the radical signs must be the same.}$$
$$= 3\sqrt{2a} + 3\sqrt{3a}$$

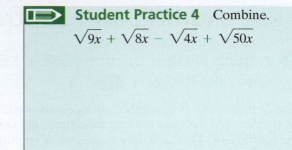

**Student Practice 4** Combine.

$$\sqrt{9x} + \sqrt{8x} - \sqrt{4x} + \sqrt{50x}$$

Special care should be taken if the original radical has a numerical coefficient. If the radical can be simplified, the two resulting numerical coefficients should be multiplied.

**Example 5** Combine. $3\sqrt{20} + 4\sqrt{45} - 2\sqrt{80}$

**Solution**

$$3\sqrt{20} + 4\sqrt{45} - 2\sqrt{80} = 3 \cdot \sqrt{4} \cdot \sqrt{5} + 4 \cdot \sqrt{9} \cdot \sqrt{5} - 2 \cdot \sqrt{16} \cdot \sqrt{5}$$
$$= 3 \cdot 2 \cdot \sqrt{5} + 4 \cdot 3 \cdot \sqrt{5} - 2 \cdot 4 \cdot \sqrt{5}$$
$$= 6\sqrt{5} + 12\sqrt{5} - 8\sqrt{5}$$
$$= 10\sqrt{5}$$

 **Student Practice 5** Combine.

$$\sqrt{27} - 4\sqrt{3} + 2\sqrt{75}$$

**Example 6** Combine. $3a\sqrt{8a} + 2\sqrt{50a^3}$

**Solution**

$$3a\sqrt{8a} + 2\sqrt{50a^3} = 3a\sqrt{4}\sqrt{2a} + 2\sqrt{25a^2}\sqrt{2a}$$
$$= 3a \cdot 2 \cdot \sqrt{2a} + 2 \cdot 5a \cdot \sqrt{2a}$$
$$= 6a\sqrt{2a} + 10a\sqrt{2a}$$
$$= 16a\sqrt{2a}$$

(*Note:* If you are unsure of the last step, show the use of the distributive property in performing the addition.)

$$6a\sqrt{2a} + 10a\sqrt{2a} = (6 + 10)a\sqrt{2a}$$
$$= 16a\sqrt{2a}$$

 **Student Practice 6** Combine.

$$2a\sqrt{12a} + 3\sqrt{27a^3}$$

## 9.3 Exercises MyMathLab®

### Verbal and Writing Skills, Exercises 1 and 2

1. List the steps involved in simplifying an expression such as $\sqrt{75x} + \sqrt{48x} + \sqrt{16x^2}$.

2. Mario said that, since $\sqrt{12}$ and $\sqrt{3}$ are not like radicals, $\sqrt{12} + \sqrt{3}$ cannot be simplified. What do you think?

*Combine if possible. Do not use a calculator.*

3. $3\sqrt{5} - \sqrt{5} + 4\sqrt{5}$

4. $-\sqrt{3} + 5\sqrt{3} + 6\sqrt{3}$

5. $\sqrt{2} + 8\sqrt{3} - 5\sqrt{3} + 4\sqrt{2}$

6. $\sqrt{5} + 2\sqrt{6} - 3\sqrt{5} - \sqrt{6}$

7. $\sqrt{7} + \sqrt{28}$

8. $\sqrt{6} - \sqrt{96}$

9. $\sqrt{80} + 4\sqrt{20}$

10. $\sqrt{80} + 2\sqrt{20}$

11. $3\sqrt{8} - 5\sqrt{2}$

12. $2\sqrt{27} - 4\sqrt{3}$

13. $-\sqrt{2} + \sqrt{18} + \sqrt{98}$

14. $-\sqrt{27} + 7\sqrt{3} - \sqrt{108}$

15. $\sqrt{25} + \sqrt{72} + 3\sqrt{12}$

16. $4\sqrt{12} + 2\sqrt{16} - \sqrt{44}$

17. $3\sqrt{20} - \sqrt{50} + \sqrt{18}$

18. $\sqrt{20} + \sqrt{27} - \sqrt{4}$

19. $8\sqrt{2x} + \sqrt{50x}$

20. $6\sqrt{5y} - \sqrt{20y}$

21. $1.2\sqrt{3x} - 0.5\sqrt{12x}$

22. $-1.5\sqrt{2a} + 0.2\sqrt{8a}$

23. $\sqrt{20y} + 2\sqrt{45y} - \sqrt{5y}$

24. $\sqrt{72w} - 3\sqrt{2w} - \sqrt{50w}$

### Mixed Practice *Combine if possible. Do not use a calculator.*

25. $\sqrt{50} - 3\sqrt{8}$

26. $\sqrt{75} - 5\sqrt{27}$

27. $3\sqrt{28x} - 5x\sqrt{63x}$

28. $-3b\sqrt{75b} + 2\sqrt{48b}$

29. $5\sqrt{8x^3} - 3x\sqrt{50x}$

30. $-3\sqrt{72y^3} + y\sqrt{18y}$

521

**31.** $5x\sqrt{48} - 2x\sqrt{75}$

**32.** $3x\sqrt{20} + x\sqrt{125}$

**33.** $2\sqrt{6y^3} - 2y\sqrt{54}$

**34.** $2x\sqrt{12x^3} - 6\sqrt{5x^2}$

**35.** $2x\sqrt{80x} + 10\sqrt{45x^3}$

**36.** $-2\sqrt{108x^3} + 3x\sqrt{12x}$

## Applications

▲ **37.** *Mars Mission* A planned unmanned mission to Mars may possibly require a small vehicle to make scientific measurements as it travels along the perimeter of a region like the one shown below. If the region is rectangular and the dimensions are as labeled, find the exact distance to be traveled by the vehicle in one trip around the perimeter.

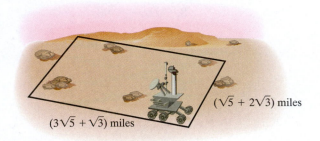

$(\sqrt{5} + 2\sqrt{3})$ miles

$(3\sqrt{5} + \sqrt{3})$ miles

▲ **38.** *Mars Mission* Suppose that on the second day of the mission to Mars, the vehicle must travel along the perimeter of a right triangle with dimensions as labeled below. Find the exact distance to be traveled by the vehicle in one trip around the perimeter.

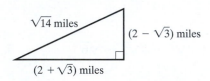

$\sqrt{14}$ miles

$(2 - \sqrt{3})$ miles

$(2 + \sqrt{3})$ miles

## Cumulative Review

**39.** **[2.6.4]** Solve and graph on a number line. $7 - 3x \le 11$

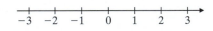

**40.** **[8.5.1]** *Toy Store Purchases* Melanie went holiday shopping for her nieces and nephews. She bought 12 games for $121.88. Some of the games were $8.99 and some were $12.49. How many of each did Melanie buy?

*Simplify.*

**41.** **[4.2.1]** $\left(\dfrac{2x^2y}{3xy^3}\right)^{-2}$

**42.** **[4.1.3]** $(-3a^2b^0x^4)^3$

---

**Quick Quiz 9.3** *Combine if possible. Do not use a calculator.*

**1.** $\sqrt{98} + 3\sqrt{8}$

**2.** $5\sqrt{3} + 2\sqrt{75} - 3\sqrt{48}$

**3.** $\sqrt{9a} + \sqrt{8a} + \sqrt{49a} + \sqrt{32a}$

**4.** **Concept Check** Explain how you would combine $3\sqrt{8x^3} + 5x\sqrt{98x}$.

## 9.4 Multiplying Radical Expressions ▶

### 1 Multiplying Monomial Radical Expressions ▶

Recall the basic rule for multiplication of square root radicals,

$$\sqrt{a}\sqrt{b} = \sqrt{ab}.$$

We will use this concept to multiply square root radical expressions. Note that the direction "multiply" means to express the product in simplest form.

**Example 1** Multiply. $\sqrt{7}\sqrt{14x}$

**Solution**

$$\sqrt{7}\sqrt{14x} = \sqrt{98x}$$

We do *not* stop here, because the radical $\sqrt{98x}$ can be simplified.

$$\sqrt{98x} = \sqrt{49 \cdot 2x} = \sqrt{49}\sqrt{2x} = 7\sqrt{2x} \qquad \square$$

▷ **Student Practice 1** Multiply. $\sqrt{3a}\sqrt{6a}$

**Example 2** Multiply.

(a) $(2\sqrt{3})(5\sqrt{7})$ 

(b) $(2a\sqrt{3a})(3\sqrt{6a})$

**Solution**

(a) $(2\sqrt{3})(5\sqrt{7}) = 10\sqrt{21}$    Multiply the coefficients: $(2)(5) = 10$.
Multiply the radicals: $\sqrt{3}\sqrt{7} = \sqrt{21}$.

(b) $(2a\sqrt{3a})(3\sqrt{6a}) = 6a\sqrt{18a^2}$    Multiply the coefficients and multiply the radicals.

$$= 6a\sqrt{9}\sqrt{2}\sqrt{a^2} \quad \text{Simplify } \sqrt{18a^2}.$$

$$= 18a^2\sqrt{2} \quad \text{Multiply } (6a)(3)(a). \qquad \square$$

▷ **Student Practice 2** Multiply.

(a) $(2\sqrt{3})(5\sqrt{5})$ 

(b) $(4\sqrt{3x})(2x\sqrt{6x})$

**Example 3** Find the area of a farm field that measures $\sqrt{9400}$ feet long and $\sqrt{4800}$ feet wide. Express your answer as a simplified radical.

**Solution** We multiply.

$$(\sqrt{9400})(\sqrt{4800}) = (10\sqrt{94})(40\sqrt{3})$$

$$= 400\sqrt{282} \text{ square feet } (\text{ft}^2) \qquad \square$$

▷ **Student Practice 3** Find the area of a computer chip that measures $\sqrt{180}$ millimeters long and $\sqrt{150}$ millimeters wide.

**Student Learning Objectives**

After studying this section, you will be able to:

**1** Multiply monomial radical expressions. ▶

**2** Multiply a monomial radical expression by a polynomial. ▶

**3** Multiply two polynomial radical expressions. ▶

### 2 Multiplying a Monomial Radical Expression by a Polynomial ▶

Recall that a binomial consists of two terms. $\sqrt{2} + \sqrt{5}$ is a binomial. We use the distributive property when multiplying a binomial by another factor.

**Example 4** Multiply and simplify. $\sqrt{5}\left(\sqrt{2} + 3\sqrt{10}\right)$

**Solution**

$$\sqrt{5}\left(\sqrt{2} + 3\sqrt{10}\right)$$
$$= \sqrt{5}\sqrt{2} + 3\sqrt{5}\sqrt{10}$$
$$= \sqrt{10} + 15\sqrt{2}$$

In similar fashion, we can multiply a trinomial by another factor. ◻

▶ **Student Practice 4** Multiply and simplify.

$$\sqrt{6}\left(\sqrt{3} + 2\sqrt{8}\right)$$

We discover that $\sqrt{5} \cdot \sqrt{5} = 5$ and $\sqrt{3} \cdot \sqrt{3} = 3$. These are specific cases of the multiplication rule for square root radicals.

> For any nonnegative real number $a$,
> $$\sqrt{a} \cdot \sqrt{a} = a.$$

**Example 5** Multiply and simplify. $\sqrt{a}\left(3\sqrt{a} - 2\sqrt{5}\right)$

**Solution**

$$\sqrt{a}\left(3\sqrt{a} - 2\sqrt{5}\right)$$
$$= 3\sqrt{a}\sqrt{a} - 2\sqrt{5}\sqrt{a} \quad \text{Use the distributive property.}$$
$$= 3a - 2\sqrt{5a}$$

◻

▶ **Student Practice 5** Multiply and simplify.

$$2\sqrt{x}\left(4\sqrt{x} - x\sqrt{2}\right)$$

Be sure to simplify all radicals after multiplying.

### 3 Multiplying Two Polynomial Radical Expressions ▶

Recall the FOIL method used to multiply two binomials. The same method can be used for radical expressions.

*Algebraic Expressions*

$$(2x + y)(x - 2y) = 2x^2 - 4xy + xy - 2y^2$$
$$= 2x^2 - 3xy - 2y^2$$

*Radical Expressions*

$$(2\sqrt{3} + \sqrt{5})(\sqrt{3} - 2\sqrt{5}) = 2\sqrt{9} - 4\sqrt{15} + \sqrt{15} - 2\sqrt{25}$$
$$= 2(3) - 3\sqrt{15} - 2(5)$$
$$= 6 - 3\sqrt{15} - 10$$
$$= -4 - 3\sqrt{15}$$

Let's look at this procedure more closely.

**Example 6** Multiply. $(\sqrt{2} + 5)(\sqrt{2} - 3)$

**Solution**

$$(\sqrt{2} + 5)(\sqrt{2} - 3)$$

$= \sqrt{4} - 3\sqrt{2} + 5\sqrt{2} - 15$   Multiply to obtain the four products.

$= 2 + 2\sqrt{2} - 15$   Simplify $\sqrt{4}$; combine the middle terms.

$= -13 + 2\sqrt{2}$

**Student Practice 6** Multiply.
$$(\sqrt{2} + \sqrt{3})(2\sqrt{2} - \sqrt{3})$$

Be especially watchful if there is a number outside the radical sign. You must multiply the numerical coefficients very carefully.

 **Example 7** Multiply. $(\sqrt{2} - 3\sqrt{6})(\sqrt{2} + \sqrt{6})$

**Solution**

$$(\sqrt{2} - 3\sqrt{6})(\sqrt{2} + \sqrt{6})$$

$= \sqrt{4} + \sqrt{12} - 3\sqrt{12} - 3\sqrt{36}$   Multiply to obtain the four products.

$= 2 - 2\sqrt{12} - 18$   Simplify; combine the middle terms.

$= -16 - 4\sqrt{3}$   Simplify $-2\sqrt{12}$.

**Student Practice 7** Multiply.
$$(\sqrt{6} + \sqrt{3})(\sqrt{6} + 2\sqrt{3})$$

If an expression with radicals is squared, write the expression as the product of two binomials and multiply.

**Example 8** Multiply. $(2\sqrt{3} - \sqrt{6})^2$

**Solution**
$$(2\sqrt{3} - \sqrt{6})^2$$
$$= (2\sqrt{3} - \sqrt{6})(2\sqrt{3} - \sqrt{6})$$
$$= 4\sqrt{9} - 2\sqrt{18} - 2\sqrt{18} + \sqrt{36}$$
$$= 12 - 4\sqrt{18} + 6$$
$$= 18 - 12\sqrt{2}$$

**Student Practice 8** Multiply.
$$(3\sqrt{5} - \sqrt{10})^2$$

## 9.4 Exercises   MyMathLab®

*Multiply. Be sure to simplify any radicals in your answer. Do not use a calculator.*

1. $\sqrt{3}\sqrt{10}$

2. $\sqrt{5}\sqrt{7}$

3. $\sqrt{2}\sqrt{10}$

4. $\sqrt{3}\sqrt{15}$

5. $\sqrt{18}\sqrt{3}$

6. $\sqrt{28}\sqrt{2}$

7. $(4\sqrt{5})(3\sqrt{2})$

8. $(3\sqrt{7})(2\sqrt{10})$

9. $\sqrt{5}\sqrt{10a}$

10. $\sqrt{5b}\sqrt{20}$

11. $(3\sqrt{10x})(2\sqrt{5x})$

12. $(4\sqrt{6x})(5\sqrt{3x})$

13. $(-4\sqrt{8a})(-2\sqrt{3a})$

14. $(4\sqrt{12a})(2\sqrt{6a})$

15. $(-3\sqrt{ab})(2\sqrt{b})$

16. $(-2\sqrt{x})(5\sqrt{xy})$

17. $\sqrt{5}(\sqrt{2} + \sqrt{3})$

18. $\sqrt{2}(\sqrt{6} + \sqrt{7})$

19. $\sqrt{3}(2\sqrt{6} + 5\sqrt{15})$

20. $\sqrt{6}(5\sqrt{12} - 4\sqrt{3})$

21. $2\sqrt{x}(\sqrt{x} - 8\sqrt{5})$

22. $-3\sqrt{b}(2\sqrt{a} + 3\sqrt{b})$

23. $\sqrt{6}(\sqrt{2} - 3\sqrt{6} + 2\sqrt{10})$

24. $\sqrt{10}(\sqrt{5} - 3\sqrt{10} + 5\sqrt{2})$

25. $(2\sqrt{3} + \sqrt{6})(\sqrt{3} - 2\sqrt{6})$

26. $(3\sqrt{5} + \sqrt{3})(\sqrt{5} + \sqrt{3})$

27. $(5 + 3\sqrt{2})(3 + \sqrt{2})$

28. $(4 - 3\sqrt{2})(5 + \sqrt{2})$

29. $(2\sqrt{7} - 3\sqrt{3})(\sqrt{7} + \sqrt{3})$

30. $(5\sqrt{6} - \sqrt{2})(\sqrt{6} + 3\sqrt{2})$

31. $(\sqrt{3} + 2\sqrt{6})(2\sqrt{3} - \sqrt{6})$

32. $(\sqrt{2} + 3\sqrt{10})(2\sqrt{2} - \sqrt{10})$

33. $(3\sqrt{7} - \sqrt{8})(\sqrt{8} + 2\sqrt{7})$

34. $(\sqrt{12} - \sqrt{5})(\sqrt{5} + 2\sqrt{12})$

35. $(2\sqrt{6} - 1)^2$

36. $(3\sqrt{2} + 2)^2$

37. $(\sqrt{3} + 5\sqrt{2})^2$

38. $(\sqrt{5} - 2\sqrt{3})^2$

39. $(3\sqrt{5} - \sqrt{3})^2$

40. $(2\sqrt{3} + 5\sqrt{5})^2$

**Mixed Practice**  *Multiply. Be sure to simplify any radicals in your answer.*

41. $(3\sqrt{5})(2\sqrt{8})$

42. $(3\sqrt{10})(\sqrt{6})$

43. $\sqrt{7}(\sqrt{2} + 3\sqrt{14} - \sqrt{6})$

44. $\sqrt{6}(\sqrt{6} + 5\sqrt{10} - \sqrt{3})$

45. $(5\sqrt{2} - 3\sqrt{7})(3\sqrt{2} + \sqrt{7})$

46. $(4\sqrt{6} + 5\sqrt{3})(\sqrt{6} - 4\sqrt{3})$

526

**47.** $5\sqrt{x}\left(-2\sqrt{y} + 4\sqrt{x} - \sqrt{xy}\right)$

**48.** $4\sqrt{a}\left(3\sqrt{2} - \sqrt{a} - 5\sqrt{b}\right)$

*Multiply the following binomials. Be sure to simplify your answer. Do not use a calculator.*

**49.** $\left(3x\sqrt{y} + \sqrt{5}\right)\left(3x\sqrt{y} - \sqrt{5}\right)$

**50.** $\left(5a\sqrt{2} - 3\sqrt{5}\right)\left(5a\sqrt{2} + 3\sqrt{5}\right)$

**Applications** *Construction Worker Damian is a construction worker and is building a wheelchair ramp with the dimensions as listed on the diagram. The gray shaded triangle is a right triangle.*

▲ **51.** Find the area of the shaded portion of the wheelchair ramp.

▲ **52.** Find the perimeter of the shaded portion of the wheelchair ramp. Leave your answer as a radical in simplified form.

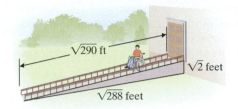

$\sqrt{290}$ ft

$\sqrt{2}$ feet

$\sqrt{288}$ feet

## Cumulative Review

**53.** **[5.5.1]** Factor. $64a^2 - 25b^2$

**54.** **[5.5.2]** Factor. $4x^2 - 20x + 25$

**55.** **[1.8.2]** *Speed in Knots* Ships often measure their speed in knots (or nautical miles per hour). The regular mile (statute mile) that we are more familiar with is 5280 feet. A nautical mile is about 6076 feet. Use the formula of the form $m = 1.15k$, where $m$ is the number of statute miles, $k$ is the number of nautical miles, to find out how fast a Coast Guard cutter is going in miles per hour if it is traveling at 35 knots. Round your answer to the nearest tenth.

**56.** **[0.5.3]** *Car Insurance* Phil and Melissa LaBelle found that the collision and theft portion of their car insurance could be reduced by 15% if they purchased an auto security device. They purchased and had one installed for $350. The yearly collision and theft portion of their car insurance is $280. How many years will it take for the security device to pay for itself?

**Quick Quiz 9.4** *Multiply. Be sure to simplify any radicals in your answer.*

**1.** $\left(2\sqrt{x}\right)\left(4\sqrt{y}\right)\left(3\sqrt{xy}\right)$

**2.** $\sqrt{3}\left(\sqrt{8} + 2\sqrt{3} - 4\sqrt{5}\right)$

**3.** $\left(4\sqrt{2} - 5\sqrt{3}\right)\left(2\sqrt{2} + 3\sqrt{3}\right)$

**4.** **Concept Check** Explain how to simplify $\left(3\sqrt{3} - 2\right)^2$.

# Use Math to Save Money

**Did You Know?** You can save money by purchasing store brand products.

## Food and Rice Prices

### Understanding the Problem:

Many large stores and chains have their own brands of products. Store brand products often cost much less than the equivalent name brand products. For example, a one-pound bag of rice from a national name brand can cost $2.50. A one-pound bag of the same rice from a store brand might cost only $1.00.

### Making a Plan:

Lucy and her family enjoy rice as part of their dinner four times a week. Lucy has a family of four that consumes approximately 2/3 cup of rice with each meal. Lucy wants to calculate the cost of rice for her family and see if she can save money.

**Step 1:** Lucy needs to know how much rice her family consumes in a week.

**Task 1:** How many cups of rice does Lucy's family eat in a week?

**Task 2:** If each cup of raw rice weighs approximately 21 ounces, what is the weight in ounces of the rice Lucy's family eats in a week?

**Task 3:** There are 16 ounces in a pound. Find the number of pounds of raw rice Lucy's family eats each week.

**Step 2:** Lucy notices that her supermarket has its own brand of rice that is much less expensive than the brand she normally buys. She usually buys name brand rice for $2.66 per pound. The store brand is only 88 cents per pound.

**Task 4:** Find out how much Lucy spends per week on the name brand rice.

**Task 5:** Find out how much Lucy could save per week by buying the same amount of store brand rice.

**Task 6:** Find the percent savings that Lucy would get by buying the store brand rice instead of the name brand.

### Finding a Solution:

**Step 3:** Lucy realizes that she can have about the same percent of savings each week on all the food she buys by choosing the store brand products over the name brand.

**Task 7:** If Lucy typically spends $162 per week on groceries by buying the name brand products, how much could she save in a year by purchasing all store brand products?

**Task 8:** Lucy finds that her family still prefers some of the name brand products. She continues to buy some of the name brand products and some of the store brands. She finds that her weekly grocery bill is reduced to around $130 per week. If this continues, how much will she save in the course of a year?

### Applying the Situation to Your Life:

You should calculate how much you spend on groceries every week. Then see how you can reduce that by purchasing the store brand. You may not change to the store brand for all your shopping, but for every case where you do, you can save money.

## 9.5 Dividing Radical Expressions ▶

### 1 Using the Quotient Rule for Square Roots to Simplify a Fraction Involving Radicals ▶

Just as there is a multiplication rule for square roots, there is a quotient rule for square roots. The rules are similar.

> **QUOTIENT RULE FOR SQUARE ROOTS**
>
> For all positive numbers $a$ and $b$,
> $$\frac{\sqrt{a}}{\sqrt{b}} = \sqrt{\frac{a}{b}} \quad \text{and} \quad \sqrt{\frac{a}{b}} = \frac{\sqrt{a}}{\sqrt{b}}.$$

This can be used to divide square root radicals or to simplify square root radical expressions involving division. We will be using both parts of the quotient rule.

**Example 1** Simplify.

**(a)** $\dfrac{\sqrt{75}}{\sqrt{3}}$ 　　　　　　**(b)** $\sqrt{\dfrac{25}{36}}$

**Solution**

**(a)** We notice that 3 is a factor of 75. We use the quotient rule to rewrite the expression.

$$\frac{\sqrt{75}}{\sqrt{3}} = \sqrt{\frac{75}{3}}$$
$$= \sqrt{25} \quad \text{Divide.}$$
$$= 5 \quad\quad\;\; \text{Simplify.}$$

**(b)** Since both 25 and 36 are perfect squares, we will rewrite this as the quotient of square roots.

$$\sqrt{\frac{25}{36}} = \frac{\sqrt{25}}{\sqrt{36}}$$
$$= \frac{5}{6} \qquad\qquad\qquad\;\; \square$$

▶ **Student Practice 1** Simplify.

**(a)** $\dfrac{\sqrt{98}}{\sqrt{2}}$ 　　　　　　**(b)** $\sqrt{\dfrac{81}{100}}$

**Example 2** Simplify. $\sqrt{\dfrac{20}{x^6}}$

**Solution**

$$\sqrt{\frac{20}{x^6}} = \frac{\sqrt{20}}{\sqrt{x^6}} = \frac{\sqrt{4}\sqrt{5}}{x^3} = \frac{2\sqrt{5}}{x^3} \quad \text{Don't forget to simplify } \sqrt{20} \text{ as } 2\sqrt{5}. \quad \square$$

▶ **Student Practice 2** Simplify.
$$\sqrt{\frac{50}{a^4}}$$

### 2 Rationalizing the Denominator of a Fraction with a Square Root in the Denominator ▶

Sometimes, when calculating with fractions that contain radicals, it is advantageous to have an integer in the denominator. If a fraction has a radical in the denominator, we **rationalize the denominator.** That is, we multiply to change the fraction to an equivalent one that has an integer in the denominator. Remember, we do not want to change the value of the fraction. Thus we will multiply the numerator and the denominator by the same number.

**Example 3** Simplify. $\dfrac{3}{\sqrt{2}}$

**Solution** Think, "What times 2 will make a perfect square?"

$$\frac{3}{\sqrt{2}} = \frac{3}{\sqrt{2}} \times 1 = \frac{3}{\sqrt{2}} \times \frac{\sqrt{2}}{\sqrt{2}} = \frac{3\sqrt{2}}{\sqrt{4}} = \frac{3\sqrt{2}}{2}$$

**▶ Student Practice 3** Simplify.

$$\frac{9}{\sqrt{7}}$$

(Note that a fraction is not simplified unless the denominator is rationalized.)

When rationalizing a denominator containing a square root radical, we will want to use the smallest possible radical that will yield the square root of a perfect square. Very often we will not use the radical that is in the denominator.

**Example 4** Simplify.

(a) $\dfrac{\sqrt{7}}{\sqrt{8}}$ 　　　　　　　　(b) $\dfrac{3x}{\sqrt{x^3}}$

**Solution**

(a) Think, "What times 8 will make a perfect square?"

$$\frac{\sqrt{7}}{\sqrt{8}} = \frac{\sqrt{7}}{\sqrt{8}} \times \frac{\sqrt{2}}{\sqrt{2}} = \frac{\sqrt{14}}{\sqrt{16}} = \frac{\sqrt{14}}{4}$$

(b) Think, "What times $x^3$ will give an even exponent?"

$$\frac{3x}{\sqrt{x^3}} = \frac{3x}{\sqrt{x^3}} \times \frac{\sqrt{x}}{\sqrt{x}} = \frac{3x\sqrt{x}}{\sqrt{x^4}} = \frac{3x\sqrt{x}}{x^2} = \frac{3\sqrt{x}}{x}$$

We could have simplified $\sqrt{x^3}$ before we rationalized the denominator.

$$\frac{3x}{\sqrt{x^3}} = \frac{3x}{x\sqrt{x}} \times \frac{\sqrt{x}}{\sqrt{x}} = \frac{3x\sqrt{x}}{x\sqrt{x^2}} = \frac{3\sqrt{x}}{x}$$

**▶ Student Practice 4** Simplify.

(a) $\dfrac{\sqrt{2}}{\sqrt{12}}$ 　　　　　　　　(b) $\dfrac{6a}{\sqrt{a^7}}$

**Example 5** Simplify. $\dfrac{\sqrt{2}}{\sqrt{27x}}$

**Solution**   Since it is not apparent what we should multiply 27 by to obtain a perfect square, we will begin by simplifying the denominator.

$$\frac{\sqrt{2}}{\sqrt{27x}} = \frac{\sqrt{2}}{3\sqrt{3x}}$$

Now it is easy to see that we multiply by $\dfrac{\sqrt{3x}}{\sqrt{3x}}$ to rationalize the denominator.

$$= \frac{\sqrt{2}}{3\sqrt{3x}} \times \frac{\sqrt{3x}}{\sqrt{3x}}$$

$$= \frac{\sqrt{6x}}{3\sqrt{9x^2}}$$

$$= \frac{\sqrt{6x}}{9x}$$

**Student Practice 5**   Simplify.

$$\frac{\sqrt{5}}{\sqrt{8x}}$$

### 3 Rationalizing the Denominator of a Fraction with a Binomial Denominator Containing at Least One Square Root ▶

Sometimes the denominator of a fraction is a binomial with a square root radical term. $\dfrac{1}{5 - 3\sqrt{2}}$ is such a fraction. How can we eliminate the radical term in the denominator? Recall that when we multiply $(a + b)(a - b)$, we obtain the square of $a$ minus the square of $b$: $a^2 - b^2$. For example, $(x + 3)(x - 3) = x^2 - 9$. The resulting terms are perfect squares. We can use this idea to eliminate the radical in $5 - 3\sqrt{2}$.

$$(5 - 3\sqrt{2})(5 + 3\sqrt{2}) = 5^2 + 15\sqrt{2} - 15\sqrt{2} - (3\sqrt{2})^2 = 5^2 - (3\sqrt{2})^2 = 25 - 18 = 7$$

Expressions like $(5 - 3\sqrt{2})$ and $(5 + 3\sqrt{2})$ are called **conjugates.**

**Example 6** Simplify.

**(a)** $\dfrac{2}{\sqrt{3} - 4}$

**(b)** $\dfrac{\sqrt{x}}{\sqrt{5} + \sqrt{3}}$

**Solution**

**(a)** The conjugate of $\sqrt{3} - 4$ is $\sqrt{3} + 4$.

$$\frac{2}{\sqrt{3} - 4} \cdot \frac{\sqrt{3} + 4}{\sqrt{3} + 4} = \frac{2\sqrt{3} + 8}{(\sqrt{3})^2 + 4\sqrt{3} - 4\sqrt{3} - 4^2}$$

$$= \frac{2\sqrt{3} + 8}{3 - 16}$$

$$= \frac{2\sqrt{3} + 8}{-13}$$

$$= -\frac{2\sqrt{3} + 8}{13}$$

*Continued on next page*

(b) $\dfrac{\sqrt{x}}{\sqrt{5} + \sqrt{3}}$

The conjugate of $\sqrt{5} + \sqrt{3}$ is $\sqrt{5} - \sqrt{3}$.

$$\frac{\sqrt{x}}{\sqrt{5} + \sqrt{3}} \cdot \frac{\sqrt{5} - \sqrt{3}}{\sqrt{5} - \sqrt{3}} = \frac{\sqrt{5x} - \sqrt{3x}}{(\sqrt{5})^2 - \sqrt{15} + \sqrt{15} - (\sqrt{3})^2}$$

$$= \frac{\sqrt{5x} - \sqrt{3x}}{5 - 3}$$

$$= \frac{\sqrt{5x} - \sqrt{3x}}{2}$$

Be careful not to combine $\sqrt{5x}$ and $\sqrt{3x}$ in the numerator. They are not like radicals. □

**Student Practice 6** Simplify.

(a) $\dfrac{4}{\sqrt{3} + \sqrt{5}}$

(b) $\dfrac{\sqrt{a}}{\sqrt{10} - 3}$

Note that in Example 7 the numerator is a binomial.

 **Example 7** Rationalize the denominator. $\dfrac{\sqrt{3} + \sqrt{2}}{\sqrt{3} - \sqrt{2}}$

**Solution**

The conjugate of $\sqrt{3} - \sqrt{2}$ is $\sqrt{3} + \sqrt{2}$.

$$\frac{\sqrt{3} + \sqrt{2}}{\sqrt{3} - \sqrt{2}} \cdot \frac{\sqrt{3} + \sqrt{2}}{\sqrt{3} + \sqrt{2}}$$

$$= \frac{\sqrt{9} + \sqrt{6} + \sqrt{6} + \sqrt{4}}{(\sqrt{3})^2 - (\sqrt{2})^2} \qquad \text{Multiply.}$$

$$= \frac{3 + 2\sqrt{6} + 2}{3 - 2} \qquad \text{Simplify and combine like terms.}$$

$$= \frac{5 + 2\sqrt{6}}{1}$$

$$= 5 + 2\sqrt{6} \qquad □$$

**Student Practice 7** Rationalize the denominator.

$$\frac{\sqrt{3} + \sqrt{5}}{\sqrt{3} - \sqrt{5}}$$

Note that in Examples 6 and 7 we used a raised dot · to indicate multiplication instead of the ×. You need to be familiar with both forms of multiplication notation.

## 9.5 Exercises  MyMathLab®

*Simplify. Be sure to rationalize all denominators. Do not use a calculator.*

1. $\dfrac{\sqrt{12}}{\sqrt{3}}$

2. $\dfrac{\sqrt{3}}{\sqrt{27}}$

3. $\dfrac{\sqrt{7}}{\sqrt{63}}$

4. $\dfrac{\sqrt{52}}{\sqrt{13}}$

5. $\dfrac{\sqrt{72}}{\sqrt{2}}$

6. $\dfrac{\sqrt{3}}{\sqrt{48}}$

7. $\dfrac{\sqrt{6}}{\sqrt{x^4}}$

8. $\dfrac{\sqrt{12}}{\sqrt{x^2}}$

9. $\dfrac{\sqrt{18}}{\sqrt{a^4}}$

10. $\dfrac{\sqrt{24}}{\sqrt{b^6}}$

11. $\dfrac{3}{\sqrt{7}}$

12. $\dfrac{4}{\sqrt{6}}$

13. $\dfrac{8}{\sqrt{15}}$

14. $\dfrac{9}{\sqrt{14}}$

15. $\dfrac{x\sqrt{x}}{\sqrt{2}}$

16. $\dfrac{\sqrt{3y}}{\sqrt{15}}$

17. $\dfrac{\sqrt{8}}{\sqrt{x}}$

18. $\dfrac{\sqrt{12}}{\sqrt{a}}$

19. $\dfrac{6}{\sqrt{28}}$

20. $\dfrac{3}{\sqrt{27}}$

21. $\dfrac{6}{\sqrt{a}}$

22. $\dfrac{8}{\sqrt{b}}$

23. $\dfrac{x}{\sqrt{2x^5}}$

24. $\dfrac{y}{\sqrt{5x^3}}$

25. $\dfrac{\sqrt{18}}{\sqrt{2x^3}}$

26. $\dfrac{\sqrt{64}}{\sqrt{4x^5}}$

27. $\sqrt{\dfrac{3}{5}}$

28. $\sqrt{\dfrac{1}{10}}$

29. $\sqrt{\dfrac{10}{21}}$

30. $\sqrt{\dfrac{15}{17}}$

31. $\dfrac{9}{\sqrt{32x}}$

32. $\dfrac{3}{\sqrt{50x}}$

*Rationalize the denominator. Simplify your answer. Do not use a calculator.*

33. $\dfrac{4}{\sqrt{3}-1}$

34. $\dfrac{2}{\sqrt{6}+1}$

35. $\dfrac{8}{\sqrt{10}+\sqrt{2}}$

36. $\dfrac{3}{\sqrt{5}-\sqrt{2}}$

37. $\dfrac{\sqrt{2}}{\sqrt{2}-\sqrt{7}}$

38. $\dfrac{\sqrt{3}}{\sqrt{3}+\sqrt{5}}$

39. $\dfrac{\sqrt{7}}{\sqrt{8}+\sqrt{7}}$

40. $\dfrac{\sqrt{5}}{\sqrt{5}+\sqrt{6}}$

41. $\dfrac{3x}{2\sqrt{2}-\sqrt{5}}$

42. $\dfrac{4x}{2\sqrt{7}+2\sqrt{6}}$

43. $\dfrac{\sqrt{x}}{\sqrt{6}+\sqrt{2}}$

44. $\dfrac{\sqrt{x}}{\sqrt{11}+\sqrt{5}}$

45. $\dfrac{\sqrt{10}-\sqrt{3}}{\sqrt{10}+\sqrt{3}}$

46. $\dfrac{\sqrt{13}+\sqrt{2}}{\sqrt{13}-\sqrt{2}}$

47. $\dfrac{4\sqrt{7}+3}{\sqrt{5}-\sqrt{2}}$

48. $\dfrac{4\sqrt{3}+2\sqrt{5}}{\sqrt{5}-\sqrt{3}}$

49. $\dfrac{4\sqrt{3}+2}{\sqrt{8}-\sqrt{6}}$

50. $\dfrac{3\sqrt{5}+4}{\sqrt{15}-\sqrt{3}}$

51. $\dfrac{x-25}{\sqrt{x}+5}$

52. $\dfrac{a-16}{\sqrt{a}-4}$

## Applications

▲ 53. *Geometry* A pyramid has a volume $V$ and a height $h$. The length of one side of the square base is $s = \sqrt{\dfrac{3V}{h}}$.

(a) Simplify this expression by rationalizing the denominator.

(b) Find the length of the side of a pyramid with a volume of 36 cubic inches and a height of 12 inches.

▲ 54. *Geometry* A cylinder has a volume $V$ and a height $h$. The radius of the cylinder is $r = \sqrt{\dfrac{V}{\pi h}}$.

(a) Simplify this expression by rationalizing the denominator.

(b) Find the radius of a cylindrical storage silo of volume 6250 cubic feet and height $10\pi$ feet.

▲ 55. *Relay Tower* A reflector sheet for a microwave relay tower is shaped like a rectangle. The area of the rectangle is 3 square meters. Ideally the length should be exactly $\left(\sqrt{5} + 2\right)$ meters long. What should be the width of the rectangle? Express the answer exactly using radical expressions.

▲ 56. *Mechanical Engineer* Rama is a mechanical engineer and needs the approximate dimensions of the rectangle in exercise 55. Approximate the length and the width of that rectangle to the nearest hundredth.

## Cumulative Review

57. **[2.3.3]** Solve for $x$. Round your answer to the nearest hundredth. $-2(x - 5) = 8x - 3(x + 1) + 18$

58. **[1.8.1]** Evaluate $a^2 - b \div a + a^3 - b$ for $a = -1$ and $b = 4$.

---

**Quick Quiz 9.5** *Simplify. Be sure to rationalize all denominators.*

1. $\dfrac{3}{\sqrt{10}}$

2. $\dfrac{5x}{3\sqrt{2} + \sqrt{5}}$

3. $\dfrac{\sqrt{6} + \sqrt{5}}{\sqrt{6} - \sqrt{5}}$

4. **Concept Check** Explain how you would simplify $\dfrac{x - 4}{\sqrt{x} + 2}$.

# 9.6 The Pythagorean Theorem and Radical Equations

## 1 Using the Pythagorean Theorem to Solve Applied Problems

**Student Learning Objectives**

After studying this section, you will be able to:

**1** Use the Pythagorean Theorem to solve applied problems.

**2** Solve radical equations.

In ancient Greece, mathematicians studied a number of properties of right triangles. They proved that the square of the length of the longest side of a right triangle is equal to the sum of the squares of the lengths of the other two sides. This property is called the Pythagorean Theorem, in honor of the Greek mathematician Pythagoras (ca. 590 B.C.). The shorter sides, $a$ and $b$, are referred to as the **legs** of the right triangle. The longest side, $c$, is called the **hypotenuse** of the right triangle. We state the theorem as follows.

> **PYTHAGOREAN THEOREM**
>
> In any right triangle, if $c$ is the length of the hypotenuse and $a$ and $b$ are the lengths of the two legs, then
> $$c^2 = a^2 + b^2.$$

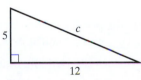

If we know any two sides of a right triangle, we can find the third side using this theorem.

Notice that the hypotenuse is always opposite the right angle. The smallest side is opposite the smallest angle.

Recall that in Section 9.1 we showed that if $s^2 = 4$, then $s = 2$ or $-2$. In general, if $x^2 = a$, then $x = \pm\sqrt{a}$. This is sometimes called "taking the square root of each side of an equation." The abbreviation $\pm\sqrt{a}$ means $+\sqrt{a}$ or $-\sqrt{a}$. It is read "plus or minus $\sqrt{a}$."

▲ **Example 1** The ramp to Tony Pitkin's barn rises 5 feet over a horizontal distance of 12 feet. How long is the ramp? That is, find the length of the hypotenuse of a right triangle whose legs are 5 feet and 12 feet.

**Solution**

1. **Understand the problem.** Draw and label a diagram.

2. **Write an equation.** Write the Pythagorean Theorem. Substitute the known values into the equation.
$$c^2 = a^2 + b^2$$
$$c^2 = 5^2 + 12^2$$

3. **Solve and state the answer.**
$$c^2 = 25 + 144$$
$$c^2 = 169$$
$$c = \pm\sqrt{169} \quad \text{Take the square root of each side of the equation.}$$
$$= \pm 13$$

The hypotenuse is 13 feet. We do not use $-13$ because length is not negative. Thus the length of the ramp is 13 feet.

4. **Check.** Substitute the values for $a$, $b$, and $c$ into the Pythagorean Theorem and evaluate. Does $13^2 = 5^2 + 12^2$? Yes. The solution checks. ✓ ☐

▲ **Student Practice 1** Find the length of the hypotenuse of a right triangle with legs of 9 centimeters and 12 centimeters.

Sometimes you will need to find one of the legs of a right triangle given the hypotenuse and the other leg.

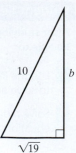

▲ **Example 2** Find the unknown leg of a right triangle that has a hypotenuse of 10 yards and one leg of $\sqrt{19}$ yards.

**Solution** Draw a diagram. The diagram is in the margin on the left.

$$c^2 = a^2 + b^2 \qquad \text{Write the Pythagorean Theorem.}$$
$$10^2 = (\sqrt{19})^2 + b^2 \qquad \text{Substitute the known values into the equation.}$$
$$100 = 19 + b^2 \qquad \text{Square the numbers.}$$
$$81 = b^2 \qquad \text{Subtract 19 from both sides.}$$
$$\pm 9 = b \qquad \text{Take the square root of each side of the equation.}$$

The leg is 9 yards long. ◻

▲  **Student Practice 2** The hypotenuse of a right triangle is $\sqrt{17}$ meters. One leg is 1 meter long. Find the length of the other leg.

Sometimes the answer to a problem will be an irrational number such as $\sqrt{33}$. $\sqrt{33}$ is an exact answer. It is best to leave the answer in radical form unless you are asked for an approximate answer. To find the approximate value, use a calculator.

▲ **Example 3** A 25-foot ladder is placed against a building. The foot of the ladder is 8 feet from the wall. At approximately what height does the top of the ladder touch the building? Round to the nearest tenth.

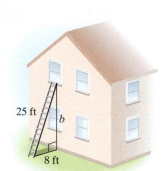

**Solution**

$$c^2 = a^2 + b^2$$
$$25^2 = 8^2 + b^2$$
$$625 = 64 + b^2$$
$$561 = b^2$$
$$\pm\sqrt{561} = b$$

We want only the positive value for the distance, so $b = \sqrt{561}$. Using a calculator, we have 561 $\boxed{\sqrt{}}$ gives 23.68543856. Rounding, we obtain $b \approx 23.7$.

The ladder touches the building at a height of approximately 23.7 feet. ◻

**Student Practice 3**

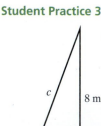

▲  **Student Practice 3** A support line is placed 3 meters away from the base of an 8-meter pole. If the support line is attached to the top of the pole and pulled tight (assume that it is a straight line), how long is the support line from the ground to the pole? Round to the nearest tenth.

## 2 Solving Radical Equations ▶

A **radical equation** is an equation with a variable in one or more of the radicands. A square root radical equation can be simplified by squaring each side of the equation. The apparent solution to a radical equation *must* be checked by substitution into the *original* equation.

For some square root radical equations, you will need to *isolate the radical before* you square each side of the equation.

**Example 4** Solve and check. $1 + \sqrt{5x - 4} = 5$

**Solution** $\quad 1 + \sqrt{5x - 4} = 5 \qquad$ We want to isolate the radical first.

$\qquad\qquad \sqrt{5x - 4} = 4 \qquad$ Subtract 1 from each side to isolate the radical.

$\qquad\qquad (\sqrt{5x - 4})^2 = (4)^2 \qquad$ Square each side.

$\qquad\qquad 5x - 4 = 16 \qquad$ Simplify and solve for $x$.

$\qquad\qquad 5x = 20$

$\qquad\qquad x = 4$

*Check.* $\qquad 1 + \sqrt{5(4) - 4} \overset{?}{=} 5$

$\qquad\qquad 1 + \sqrt{20 - 4} \overset{?}{=} 5$

$\qquad\qquad 1 + \sqrt{16} \overset{?}{=} 5$

$\qquad\qquad 1 + 4 = 5 \quad \checkmark \quad$ Thus 4 is the solution. $\qquad\qquad \square$

**Student Practice 4** Solve and check. $\sqrt{3x - 2} - 7 = 0$

Some problems will have an apparent solution, but that solution does not always check. An apparent solution that does not satisfy the original equation is called an **extraneous root.**

**Example 5** Solve and check. $\sqrt{x + 3} = -7$

**Solution** $\qquad\qquad \sqrt{x + 3} = -7$

$\qquad\qquad (\sqrt{x + 3})^2 = (-7)^2 \qquad$ Square each side.

$\qquad\qquad x + 3 = 49 \qquad$ Simplify and solve for $x$.

$\qquad\qquad x = 46$

*Check.* $\qquad \sqrt{46 + 3} \overset{?}{=} -7$

$\qquad\qquad \sqrt{49} \overset{?}{=} -7$

$\qquad\qquad 7 \neq -7! \qquad$ Does not check! The apparent solution is an extraneous root.

There is no solution to Example 5. $\qquad\qquad \square$

**Student Practice 5** Solve and check. $\sqrt{5x + 4} + 2 = 0$

Wait a minute! Is the original equation in Example 5 possible? No. $\sqrt{x + 3} = -7$ is an impossible statement. We defined the $\sqrt{\phantom{x}}$ symbol to mean the positive square root of the radicand. Thus, it cannot be equal to a negative number. If we had noticed this in the first step, we could have written down "no solution" immediately.

When you square both sides of an equation, you may obtain a quadratic equation. In such cases, you may obtain two apparent solutions, and you must check both of them.

**Example 6** Solve and check. $\sqrt{3x + 1} = x + 1$.

**Solution** $\qquad \sqrt{3x + 1} = x + 1$

$\qquad\qquad (\sqrt{3x + 1})^2 = (x + 1)^2 \qquad$ Square each side.

$\qquad\qquad 3x + 1 = x^2 + 2x + 1 \qquad$ Simplify and set the equation equal to 0.

$\qquad\qquad 0 = x^2 - x$

$\qquad\qquad 0 = x(x - 1) \qquad$ Factor the quadratic equation.

$\qquad x = 0 \quad \text{or} \quad x - 1 = 0 \qquad$ Set each factor equal to 0 and solve.

$\qquad\qquad x = 1$

*Continued on next page*

*Check.* If $x = 0$: $\quad \sqrt{3(0) + 1} \stackrel{?}{=} 0 + 1 \qquad$ If $x = 1$: $\quad \sqrt{3(1) + 1} \stackrel{?}{=} 1 + 1$

$$\sqrt{0 + 1} \stackrel{?}{=} 0 + 1 \qquad\qquad \sqrt{3 + 1} \stackrel{?}{=} 2$$

$$\sqrt{1} \stackrel{?}{=} 1 \qquad\qquad\qquad \sqrt{4} \stackrel{?}{=} 2$$

$$1 = 1 \checkmark \qquad\qquad\qquad\qquad 2 = 2 \checkmark$$

Thus 0 and 1 are both solutions.

It is possible in these cases for both apparent solutions to check, for both to be extraneous, or for one to check and the other to be extraneous. ☐

**▶ Student Practice 6** Solve and check.

$$-2 + \sqrt{6x - 1} = 3x - 2$$

**Ⓜ️ Example 7** Solve and check. $2 + \sqrt{2x - 1} = x$

**Solution**

$$2 + \sqrt{2x - 1} = x$$

$$\sqrt{2x - 1} = x - 2 \qquad \text{First we isolate the radical.}$$

$$(\sqrt{2x - 1})^2 = (x - 2)^2 \qquad \text{Square each side.}$$

$$2x - 1 = x^2 - 4x + 4$$

$$0 = x^2 - 6x + 5 \qquad \text{Simplify and set the equation to 0.}$$

$$0 = (x - 5)(x - 1) \qquad \text{Solve for } x.$$

$$x - 5 = 0 \quad \text{or} \quad x - 1 = 0$$

$$x = 5 \qquad\qquad x = 1$$

*Check.* If $x = 5$: $\quad 2 + \sqrt{2(5) - 1} \stackrel{?}{=} 5 \quad$ If $x = 1$: $\quad 2 + \sqrt{2(1) - 1} \stackrel{?}{=} 1$

$$\sqrt{10 - 1} \stackrel{?}{=} 5 - 2 \qquad\qquad \sqrt{2 - 1} \stackrel{?}{=} 1 - 2$$

$$\sqrt{9} \stackrel{?}{=} 3 \qquad\qquad\qquad \sqrt{1} \stackrel{?}{=} -1$$

$$3 = 3 \qquad\qquad\qquad\qquad 1 \neq -1$$

It does not check. In this case 1 is an extraneous root.

Thus only 5 is a solution to this equation. ☐

**▶ Student Practice 7**

Solve and check.

$$2 - x + \sqrt{x + 4} = 0$$

Example 8 is somewhat different in format. Take a minute to see how it compares to Examples 6 and 7.

**Example 8** Solve and check. $\sqrt{3x + 3} = \sqrt{5x - 1}$

**Solution** Here there are two radicals. Each radical is already isolated.

$$\sqrt{3x + 3} = \sqrt{5x - 1}$$

$$(\sqrt{3x + 3})^2 = (\sqrt{5x - 1})^2 \qquad \text{Square each side.}$$

$$3x + 3 = 5x - 1 \qquad \text{Simplify and solve for } x.$$

$$3 = 2x - 1$$

$$4 = 2x$$

$$2 = x$$

*Check.* $\quad \sqrt{3(2) + 3} \stackrel{?}{=} \sqrt{5(2) - 1}$

$$\sqrt{6 + 3} \stackrel{?}{=} \sqrt{10 - 1}$$

$$\sqrt{9} = \sqrt{9} \quad \checkmark \quad \text{Thus 2 is the solution.} \qquad ☐$$

**▶ Student Practice 8** Solve and check. $\sqrt{2x + 1} = \sqrt{x + 5}$

## 9.6 Exercises  MyMathLab®

Use the Pythagorean Theorem to find the length of the third side of each right triangle. Leave any irrational answers in radical form.

▲ 1.
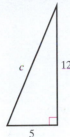

▲ 2.
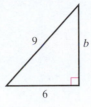

▲ 3.
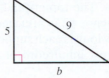

▲ 4.
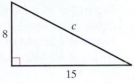

Exercises 5–14 refer to a right triangle with legs a and b and hypotenuse c. Find the exact length of the unknown side.

▲ 5. $a = 7, b = 7$  Find $c$.

▲ 6. $a = 10, b = 6$  Find $c$.

▲ 7. $a = \sqrt{11}, b = 8$  Find $c$.

▲ 8. $a = \sqrt{21}, b = 5$  Find $c$.

▲ 9. $c = 15, b = 13$  Find $a$.

▲ 10. $c = 20, b = 15$  Find $a$.

▲ 11. $c = \sqrt{82}, a = 5$  Find $b$.

▲ 12. $a = \sqrt{6}, c = 6$  Find $b$.

▲ 13. $c = 13.8, b = 9.42$
Find $a$ to the nearest hundredth.

▲ 14. $a = 4.5, b = 6.34$
Find $c$ to the nearest hundredth.

**Applications** Draw a diagram and use the Pythagorean Theorem to solve. Round to the nearest tenth.

▲ 15. **Tenting** A guy rope is attached to the top of a tent pole. The tent pole is 8 feet tall. If the guy rope is pegged into the ground 5 feet from the tent, how long is the guy rope?

▲ 16. **Softball** A softball diamond is a square. Each side of the square is 20 yards long. How far is it from home plate to second base? *Hint:* Draw the diagonal.

▲ 17. **Flying a Kite** A kite is flying on a string that is 100 feet long and is fastened to the ground at the other end. Assume that the string is a straight line. How high is the kite if it is flying above a point that is 20 feet away from where it is fastened to the ground?

▲ 18. **Boat Mooring** A boat must be moored 25 feet away from a dock. The boat will be 10 feet below the level of the dock at low tide. What is the minimum length of rope needed to go from the boat to the dock at low tide?

*Radio Tower Constructor* *Jay is the lead radio tower constructor for his company. In order to estimate the cost of erecting a radio tower, he needs to know how many feet of support cable are needed.*

▲ **19.** Ground anchor #2 is exactly 54 feet from the base of the tower (dashed line in diagram). The bottom ground support cable from level 6 to ground anchor #2 is attached to the tower at a height of 130 feet. The top ground support cable from level 6 to ground anchor #2 is attached to the tower at a height of 135 feet. How long is each support cable to level 6? (Round to the nearest hundredth of a foot.)

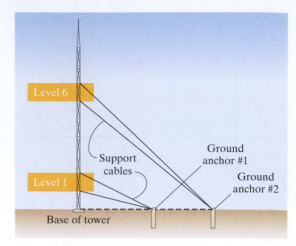

▲ **20.** Ground anchor #1 is exactly 28 feet from the base of the tower (dashed line in diagram). The bottom ground support cable from level 1 to ground anchor #1 is attached to the tower at a height of 42 feet. The top ground support cable from level 1 to ground anchor #1 is attached to the tower at a height of 47 feet. How long is each support cable to level 1? (Round to the nearest hundredth of a foot.)

*Solve for the variable. Check your solutions.*

**21.** $\sqrt{x + 2} = 3$

**22.** $\sqrt{x - 8} = 4$

**23.** $\sqrt{2x + 7} = 5$

**24.** $\sqrt{2x - 5} = 7$

**25.** $\sqrt{2x + 2} = \sqrt{3x - 5}$

**26.** $\sqrt{4x + 1} = \sqrt{x + 16}$

**27.** $\sqrt{2x} - 5 = 4$

**28.** $\sqrt{13x - 2} = 5$

**29.** $\sqrt{3x + 10} = x$

**30.** $\sqrt{6x - 5} = x$

**31.** $\sqrt{5y + 1} = y + 1$

**32.** $\sqrt{3y + 25} = y + 5$

**33.** $\sqrt{x + 3} = 3x - 1$

**34.** $\sqrt{2x + 3} = 2x - 9$

**35.** $\sqrt{5y + 4} - y = 2$

**36.** $\sqrt{5y - 9} + 1 = y$

**37.** $\sqrt{6y + 1} - 3y = y$

**38.** $\sqrt{16 - 4x} + 5 = x$

**To Think About** *Solve for x.*

**39.** $\sqrt{12x + 1} = 2\sqrt{6x} - 1$

**40.** $\sqrt{x - 2} + 3 = \sqrt{4x + 1}$

**41.** Explain why can we state that our answer to $\sqrt{x} = -2$ is "no solution" without doing any calculations.

## Cumulative Review

**42.** **[6.5.1]** Solve. $\dfrac{5x}{x - 4} = 5 + \dfrac{4x}{x - 4}$

**43.** **[4.5.2]** Multiply mentally. $(7x - 2)^2$

**44.** **[7.6.4]** $f(x) = 2x^2 - 3x + 6$
Find $f(-2)$.

**45.** **[7.6.3]** Is $x^2 + y^2 = 9$ a relation or a function?

---

## Quick Quiz 9.6

**1.** In a right triangle with legs $a$ and $b$ and a hypotenuse of $c$, find the exact length of hypotenuse $c$ if $a = \sqrt{3}$ and $b = 5$.

*Solve. Verify your solutions.*

**2.** $8 - \sqrt{5x - 6} = 0$

**3.** $\sqrt{6x + 1} = x - 1$

**4.** **Concept Check** A student attempted to solve the equation $2 + \sqrt{1 - 8x} = x$ and found values of $x = -3$ and $x = -1$. Explain how you would check these values to see if they are solutions to the original equation.

---

## 👣 STEPS TO SUCCESS How Do I Prepare for the Final Exam?

To do well on the final exam, you should begin to prepare many weeks before the exam. Cramming for any test, especially the final one, often causes anxiety and fatigue and impairs your performance. **Schedule at least five study sessions now.** During any of these sessions if you come across a problem or topic you do not understand, even after seeking assistance, place an * beside it, then move on to another topic. When you finish reviewing all topics for the final exam, return to this topic and try again.

**Making it personal:** Plan your study sessions now.

Date of your final exam: _____
Find one or more study partners. Names of study partners: _____

Session 1   Date of session: _____
**Review all your exams.** Rework the problems you answered incorrectly.

Session 2   Date of session: _____
**Complete the Cumulative Test for Chapters 0–6.**

Session 3   Date of session: _____
**Get help.** Check with a tutor, study partners, or your instructor for those problems you cannot rework correctly.

Session 4   Date of session: _____
**Start the Practice Final Exam.** Complete all chapters on the exam that you have already covered in class. **Get help** with those problems you do not understand.

Session 5   Date of session: _____
**Finish the Practice Final Exam.** Do this immediately after your instructor covers the last topic for the semester. Then **revisit the * topics**, and get help with those you still do not understand.

**Get a good night's sleep the night before the final exam!**

# 9.7 Word Problems Involving Radicals: Direct and Inverse Variation ▶

**Student Learning Objectives**

After studying this section, you will be able to:

1 Solve problems involving direct variation. ▶

2 Solve problems involving inverse variation. ▶

3 Identify the graphs of variation equations. ▶

## 1 Solving Problems Involving Direct Variation ▶

Often in daily life there is a relationship between two measurable quantities. For example, we turn up the thermostat and the heating bill increases. We say that the heating bill varies directly as the temperature on the thermostat. If $y$ **varies directly as** $x$, then $y = kx$, where $k$ is a constant. The constant is often called the **constant of variation.** Consider the following example.

**Example 1** Cliff works part-time in a local supermarket while going to college. His salary varies directly as the number of hours worked. Last week he earned $50.75 for working 7 hours. This week he earned $79.75. How many hours did he work?

**Solution**  Let  $S =$ his salary,

$h =$ the number of hours he worked, and

$k =$ the constant of variation.

Since his salary varies directly as the number of hours he worked, we write

$$S = k \cdot h.$$

We can find the constant $k$ by substituting the known values of $S = 50.75$ and $h = 7$.

$$50.75 = k \cdot 7 = 7k$$

$$\frac{50.75}{7} = \frac{7k}{7} \qquad \text{Solve for } k.$$

$$7.25 = k \qquad \text{The constant of variation is 7.25.}$$

$$S = 7.25h \qquad \text{Replace } k \text{ in the variation equation by 7.25.}$$

How many hours did he work to earn $79.75?

$$S = 7.25h \quad \text{The direct variation equation with } k = 7.25.$$

$$79.75 = 7.25h \quad \text{Substitute 79.75 for } S \text{ and solve for } h.$$

$$\frac{79.75}{7.25} = \frac{7.25h}{7.25}$$

$$11 = h$$

Cliff worked 11 hours this week. ◻

▶ **Student Practice 1**  The *change* in temperature measured on the Celsius scale varies directly as the change measured on the Fahrenheit scale. The change in temperature from freezing water to boiling water is 100° on the Celsius scale and 180° on the Fahrenheit scale. If the Fahrenheit temperature drops 20°, what will be the change in temperature on the Celsius scale?

---

**SOLVING A DIRECT VARIATION PROBLEM**

1. Write the direct variation equation.
2. Solve for the constant $k$ by substituting in known values.
3. Replace $k$ in the direct variation equation by the value obtained in step 2.
4. Solve for the desired value.

**Example 2** If $y$ varies directly as $x$, and $y = 20$ when $x = 3$, find the value of $y$ when $x = 21$.

**Solution**

**Step 1**   $y = kx$

**Step 2**   To find $k$, we substitute $y = 20$ and $x = 3$.

$$20 = k(3)$$
$$20 = 3k$$
$$\frac{20}{3} = k$$

**Step 3**   We now write the variation equation with $k$ replaced by $\frac{20}{3}$.

$$y = \frac{20}{3}x$$

**Step 4**   We replace $x$ by 21 and find $y$.

$$y = \left(\frac{20}{\cancel{3}_1}\right)(\cancel{21}^7) = (20)(7) = 140$$

Thus $y = 140$ when $x = 21$.                                                   □

▶ **Student Practice 2**   If $y$ varies directly as $x$, and $y = 18$ when $x = 5$, find the value of $y$ when $x = \frac{20}{23}$.

A number of real-life situations can be described by direct variation equations. The following table shows some of the more common forms of the direct variation equation. In each case, $k =$ the constant of variation.

**SAMPLE DIRECT VARIATION SITUATIONS**

| Verbal Description | Variation Equation |
|---|---|
| $y$ varies directly as $x$ | $y = kx$ |
| $b$ varies directly as the square of $c$ | $b = kc^2$ |
| $l$ varies directly as the cube of $m$ | $l = km^3$ |
| $V$ varies directly as the square root of $h$ | $V = k\sqrt{h}$ |

**Example 3** In a certain class of racing cars, the maximum speed varies directly as the square root of the horsepower of the engine. If a car with 225 horsepower can achieve a maximum speed of 120 mph, what speed could it achieve with 256 horsepower?

**Solution**   Let   $V =$ the maximum speed,

$h -$ the horsepower of the engine, and

$k =$ the constant of variation.

**Step 1**   Since the maximum speed ($V$) varies directly as the square root of the horsepower of the engine,

$$V = k\sqrt{h}.$$

**Step 2**   $120 = k\sqrt{225}$   Substitute known values of $V$ and $h$.

$120 = k \cdot 15$   Solve for $k$.

$8 = k$

**Step 3**   Now we can write the direct variation equation with the known value for $k$.

$$V = 8\sqrt{h}$$

*Continued on next page*

**Step 4** $V = 8\sqrt{256}$ Substitute the value of $h = 256$.

$= (8)(16)$ Solve for $V$.

$= 128$

Thus a car with 256 horsepower could achieve a maximum speed of 128 mph. □

**Student Practice 3** A certain type of car has a stopping distance that varies directly as the square of its speed. If this car is traveling at a speed of 20 mph on an ice-covered road, it can stop in 60 feet. If the car is traveling at 40 mph on an icy road, what will its stopping distance be?

## 2 Solving Problems Involving Inverse Variation

If one variable is a constant multiple of the reciprocal of another, the two variables are said to **vary inversely.** If $y$ varies inversely as $x$, we express this by the equation $y = \dfrac{k}{x}$, where $k$ is the constant of variation. Inverse variation problems can be solved by a four-step procedure similar to that used for direct variation problems.

---

**SOLVING AN INVERSE VARIATION PROBLEM**

1. Write the inverse variation equation.
2. Solve for the constant $k$ by substituting in known values.
3. Replace $k$ in the inverse variation equation by the value obtained in step 2.
4. Solve for the desired value.

---

**Example 4** If $y$ varies inversely as $x$, and $y = 12$ when $x = 7$, find the value of $y$ when $x = \dfrac{2}{3}$.

**Solution**

**Step 1** $y = \dfrac{k}{x}$

**Step 2** $12 = \dfrac{k}{7}$ Substitute known values of $x$ and $y$ to find $k$.

$84 = k$

**Step 3** $y = \dfrac{84}{x}$

**Step 4** To find $y$ when $x = \dfrac{2}{3}$, we substitute.

$$y = \frac{84}{\dfrac{2}{3}}$$

$$= \frac{84}{1} \div \frac{2}{3}$$

$$= \frac{\overset{42}{\cancel{84}}}{1} \cdot \frac{3}{\underset{1}{\cancel{2}}}$$

$$= 42 \cdot 3 = 126$$

Thus $y = 126$ when $x = \dfrac{2}{3}$. □

**Student Practice 4** If $y$ varies inversely as $x$, and $y = 8$ when $x = 15$, find the value of $y$ when $x = \dfrac{3}{5}$.

**Example 5** A car manufacturer is thinking of reducing the size of the wheel used in a subcompact car. The number of times a car wheel must turn to cover a given distance varies inversely as the radius of the wheel. (Notice that this says that the smaller the wheel, the more times it must turn to cover a given distance.) A wheel with a radius of 0.35 meter must turn 400 times to cover a specified distance on a test track. How many times would it have to turn if the radius were reduced to 0.30 meter (see the sketch)?

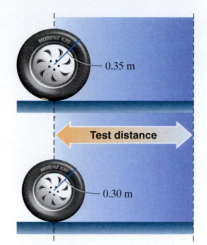

**Solution**   Let   $n = $ the number of times the car wheel turns,

   $r = $ the radius of the wheel, and

   $k = $ the constant of variation.

**Step 1**   Since the number of turns varies inversely as the radius, we can write the following.

$$n = \frac{k}{r}$$   Write the variation equation.

**Step 2**   $400 = \dfrac{k}{0.35}$   Substitute known values of $n$ and $r$ to find $k$.

$140 = k$

**Step 3**   $n = \dfrac{140}{r}$   Use the variation equation where $k$ is known.

How many times must the wheel turn if the radius is 0.30 meter?

**Step 4**   $n = \dfrac{140}{0.30}$   Substitute 0.30 for $r$.

$n = 466\dfrac{2}{3}$

The wheel would have to turn $466\dfrac{2}{3}$ times to cover the same distance if the radius were only 0.30 meter.   ☐

**Student Practice 5**   Over the last three years, the market research division of a calculator company found that the volume of sales of scientific calculators varies inversely as the price of the calculator. One year 120,000 calculators were sold at $30 each. How many calculators were sold the next year when the price was $24 for each calculator?

The following table contains various forms of the inverse variation equation. In each case, $k = $ the constant of variation.

**SAMPLE INVERSE VARIATION SITUATIONS**

| *Verbal Description* | *Variation Equation* |
|---|---|
| $y$ varies inversely as $x$ | $y = \dfrac{k}{x}$ |
| $b$ varies inversely as the square of $c$ | $b = \dfrac{k}{c^2}$ |
| $l$ varies inversely as the cube of $m$ | $l = \dfrac{k}{m^3}$ |
| $d$ varies inversely as the square root of $t$ | $d = \dfrac{k}{\sqrt{t}}$ |

**Example 6** The illumination of a light source varies inversely as the square of the distance from the source. The illumination measures 25 candlepower when a certain light is 4 meters away. Find the illumination when the light is 8 meters away (see figure).

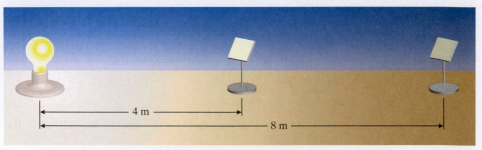

**Solution** Let $I$ = the measurement of illumination,

$d$ = the distance from the light source, and

$k$ = the constant of variation.

**Step 1** Since the illumination varies inversely as the square of the distance,

$$I = \frac{k}{d^2}.$$

**Step 2** We evaluate the constant by substituting the given values.

$$25 = \frac{k}{4^2} \quad \text{Substitute } I = 25 \text{ and } d = 4.$$

$$25 = \frac{k}{16} \quad \text{Simplify and solve for } k.$$

$$400 = k$$

**Step 3** We may now write the variation equation with the constant evaluated.

$$I = \frac{400}{d^2}$$

**Step 4** $I = \dfrac{400}{8^2}$     Substitute a distance of 8 meters.

$$= \frac{400}{64} \quad \text{Square 8.}$$

$$= \frac{25}{4} = 6.25$$

The illumination is 6.25 candlepower when the light source is 8 meters away.

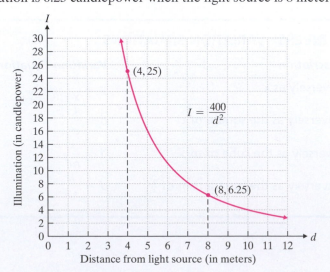

**Student Practice 6**   If the amount of power in an electrical circuit is held constant, the resistance in the circuit varies inversely as the square of the amount of current. If the amount of current is 0.01 ampere, the resistance is 800 ohms. What is the resistance if the amount of current is 0.02 ampere?

## 3  Identifying the Graphs of Variation Equations

The graphs of direct variation equations are shown in the following chart. Notice that as $x$ increases, $y$ also increases.

| Variation Statement | Equation | Graph |
|---|---|---|
| 1. $y$ varies directly as $x$ | $y = kx$ | 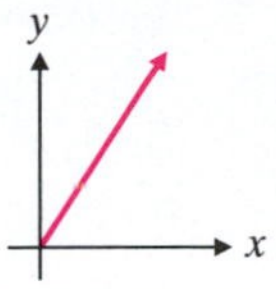 |
| 2. $y$ varies directly as $x^2$ | $y = kx^2$ | 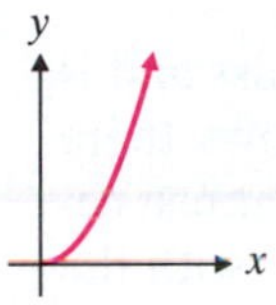 |
| 3. $y$ varies directly as $x^3$ | $y = kx^3$ | 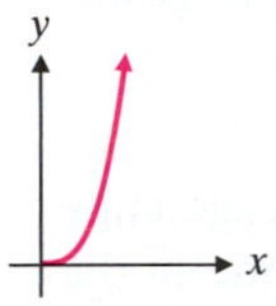 |
| 4. $y$ varies directly as the square root of $x$ | $y = k\sqrt{x}$ | 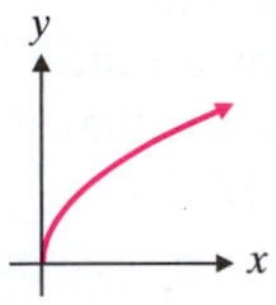 |

The graphs of inverse variation equations are shown in the following chart. Notice that as $x$ increases, $y$ decreases.

| Variation Statement | Equation | Graph |
|---|---|---|
| 5. $y$ varies inversely as $x$ | $y = \dfrac{k}{x}$ | 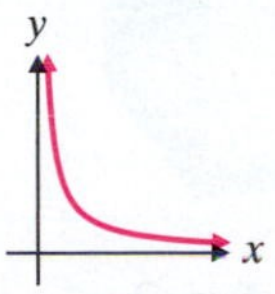 |
| 6. $y$ varies inversely as $x^2$ | $y = \dfrac{k}{x^2}$ | 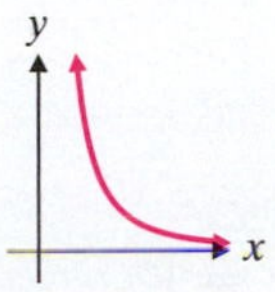 |

**1.** If $y$ varies directly as $x$, and $y = 9$ when $x = 2$, find $y$ when $x = 16$.

**2.** If $y$ varies directly as $x$, and $y = 5$ when $x = 3$, find $y$ when $x = 18$.

**3.** If $y$ varies directly as the cube of $x$, and $y = 12$ when $x = 2$, find $y$ when $x = 7$.

**4.** If $y$ varies directly as the square root of $x$, and $y = 9$ when $x = 16$, find $y$ when $x = 49$.

**5.** If $y$ varies directly as the square of $x$, and $y = 900$ when $x = 25$, find $y$ when $x = 30$.

**6.** If $y$ varies directly as the square of $x$, and $y = 3200$ when $x = 8$, find $y$ when $x = 6$.

## Applications

**7.** *Nutritionist* Sylvia is a registered nutritionist and is reviewing a patient's food intake. She knows there are 120 calories in 30 grams of a certain breakfast cereal. The number of calories varies directly with the weight of the cereal. How many calories will 75 grams of cereal have?

**8.** *Stretching a Spring* Hooke's Law states that the force needed to stretch a spring varies directly with the amount the spring is stretched. If a force of 80 pounds is needed to stretch a spring 10 inches, how far will a force of 20 pounds stretch the same spring?

▲**9.** *Storage Cube* The time it takes to fill a storage cube with sand varies directly as the cube of the side of the box. A storage cube that is 2.0 meters on each side (inside dimensions) can be filled in 7 minutes by a sand loader. How long will it take to fill a storage cube that is 4.0 meters on each side (inside dimensions)?

**10.** *Weight on the Moon* The weight of an object on the surface of the moon varies directly as the weight of the object on the surface of Earth. An astronaut with his protective suit weighs 175 lb on Earth's surface; on the moon and wearing the same suit, he will weigh 29 lb. If his moon rover vehicle weighs 2000 pounds on Earth, how much will it weigh on the moon? Round to the nearest tenth.

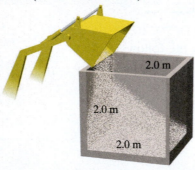

**11.** If $y$ varies inversely as $x$, and $y = 18$ when $x = 5$, find $y$ when $x = 8$.

**12.** If $y$ varies inversely as $x$, and $y = 21$ when $x = 4$, find $y$ when $x = 18$.

**13.** If $y$ varies inversely as $x$, and $y = \frac{1}{3}$ when $x = 12$, find $y$ when $x = 4$.

**14.** If $y$ varies inversely as $x$, and $y = \frac{1}{10}$ when $x = 30$, find $y$ when $x = 1$.

**15.** If $y$ varies inversely as the square of $x$, and $y = 30$ when $x = 2$, find $y$ when $x = 9$.

**16.** If $y$ varies inversely as the cube of $x$, and $y = \frac{1}{16}$ when $x = 4$, find $y$ when $x = \frac{1}{2}$.

## Applications

**17. *Ice Cube Melting*** The amount of time in minutes that it takes for an ice cube to melt varies inversely as the temperature of the water that the ice cube is placed in. When an ice cube is placed in 60°F water, it takes 2.3 minutes to melt. How long would it take for an ice cube to melt if it were placed in 40°F water?

**18. *Textile Mills*** The textile mills in Lowell, Massachusetts, were operated with water power. The Merrimack River and its canals powered huge turbines that generated the electricity for the mills. In each mill room, a single driveshaft supplied power to all machines by means of belts that ran to pulleys on the machines. The driveshaft turned at a constant speed, but the pulleys were different diameters, which provided for the different speeds of each machine. The speed of each machine varied inversely as the diameter of its pulley. If a machine with a 60-centimeter diameter pulley turned at 1000 rpm, what was the speed of a machine with an 80-centimeter diameter pulley?

**19. *Variation in Weight on Earth*** The weight of an object near Earth's surface varies inversely as the square of its distance from the center of Earth. A Mini Cooper weighs about 2700 pounds on Earth's surface. This is approximately 4000 miles from the center of Earth. How much will the object weigh 8000 miles from the center of Earth?

**20. *Radio Waves*** The wavelength, $w$, of a radio wave varies inversely as the frequency, $f$. A radio wave with a frequency of 500 kilohertz has a length of 720 meters. What is the length of a radio wave with a frequency of 2000 kilohertz?

**21. *Distance Vision in a Boat*** A person on the mast of a sailing ship 81 meters above the water can see approximately 12.3 kilometers out to sea. In general, the distance, $d$, you can see varies directly as the square root of your height, $h$, above the water. In this case, the equation is $d = \frac{15}{11}\sqrt{h}$, where $d$ is measured in kilometers and $h$ is measured in meters.

**(a)** Find the distances you can see when $h = 121$ m, 36 m, and 4 m. Round to the nearest tenth.

**(b)** Graph the equation using values of $h$ of 121 m, 81 m, 36 m, and 4 m.

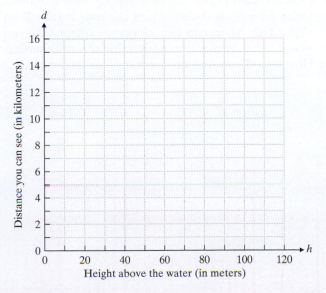

**22. *Computer Files*** Suppose a computer has a hard drive that can hold 2000 files if each file is 100,000 bytes long. The number of files, $f$, the hard drive can hold varies inversely as the size, $s$, of a file. In this case, the equation is $f = \dfrac{200,000,000}{s}$, where $f$ is the number of files and $s$ is the size of a file in bytes.

**(a)** Find the number of files that can be held if $s = 400,000$ bytes, $s = 200,000$ bytes, and $s = 50,000$ bytes.

**(b)** Graph the equation using the values $s = 400,000$ bytes, $s = 200,000$ bytes, $s = 100,000$ bytes, and $s = 50,000$ bytes.

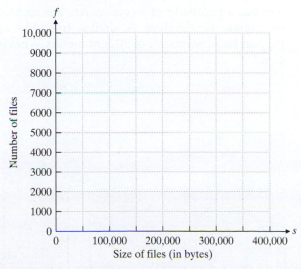

**To Think About** *Write an equation to describe the variation and solve.*

23. *Planetary Motion* Kepler's Third Law of Planetary Motion states that the square of the time (in Earth years) it takes a planet to complete one revolution of the sun varies directly with the cube of the distance between the planet and the sun. Astronomers use 1 astronomical unit (AU) as a measure of the distance from Earth to the sun.

   (a) Using this astronomical unit, find the constant $k$ for the direct variation equation that represents Kepler's Third Law.

   (b) Pluto is 39.5 times as far from the sun as Earth is. Find the approximate length of the Plutonian year in terms of Earth years. Round to the nearest hundredth.

24. *Sail Designer* Anne sews custom sails for sailboats. She knows that an important element of sailboat design is the calculation of the force of wind on the sail. The force varies directly with the square of the wind speed and directly with the area of the sail. For a certain class of racing boat, the force on 100 square feet of sail is 280 pounds when the wind speed is 6 mph. What would be the force on the sail if it were rigged for a storm (the amount of sail exposed to the wind is reduced by lashing part of the sail to the boom) so that only 60 square feet of sail were exposed and the wind speed were 40 mph? Round to the nearest whole lb.

## Cumulative Review

25. [6.5.1] Solve for $a$. $\dfrac{80,000}{320} = \dfrac{120,000}{320 + a}$

26. [7.2.1] Graph the equation $S = kh$, where $k = 4.80$. Use $h$ for the horizontal axis and $S$ for the vertical. Assume that $h$ and $S$ are nonnegative.

## Quick Quiz 9.7

1. The intensity of illumination from a light source varies inversely with the square of the distance from the light source. When a photoelectric cell is placed 2 inches from a light source, the intensity is 45 lumens. Then it is moved so that it receives only 5 lumens. How far is it from the light source?

2. The number of calories in a breakfast bar varies directly with the weight of the bar. There are 150 calories in a bar weighing 40 grams. How many calories are in a bar weighing 75 grams?

3. The areas of similar triangle-shaped supports vary directly as the squares of their perimeters. A support that has a perimeter of 8 centimeters has an area of 6.4 square centimeters. A similar triangle-shaped support has a perimeter of 20 centimeters. What is its area?

4. **Concept Check** The distance a body falls from rest varies directly as the square of the time it falls. If an object falls 64 feet in two seconds, explain how you would write an equation to describe this relationship. How would you find the constant $k$ in this equation?

## Background: Graphic Designer

Seabreeze Design has been commissioned to develop signage for a newly built 55-year-old-plus residential community. Each sign must include the community name along with its logo against the backdrop of an oceanside vista. Because of what she's learned, Erin has decided to use what's called the golden ratio in her design work for the rectangular-shaped signs she plans to produce.

The golden ratio is a value that functions as a guide for the shape, size, and proportion of building edifices, paintings, and even signs. This particular ratio, a rectangle's length $l$ to its width $w$, is considered by many to render results that are the most pleasing to the eye, and this is why so many design professionals use it in their work.

### Facts

The golden ratio is expressed as

$$\frac{l}{w} = \frac{2}{\sqrt{5} - 1}.$$

Three rectangular-shaped signs are to be constructed, each meeting the following requirements:

- One sign has a width measuring $2\frac{1}{2}$ feet.
- A second sign must have a length of no more than 80 inches in order for it to meet town zoning requirements, and its total area must be no more than 25 square feet.
- The scenic background of the third sign has a diagonal measuring 8 feet in length.

### Tasks

1. Erin begins her work by first rationalizing the denominator of the golden ratio and then calculating its value. What is this measurement, rounded to the nearest thousandth?

2. Erin now must determine the length of a rectangular sign to be constructed with a width measuring $2\frac{1}{2}$ feet. What is this length, rounded to the nearest hundredth of a foot?

3. Erin decides a suitable length for the second sign to be 75 inches. If her design meets the requirement of being a golden rectangle, will it meet the town's zoning restriction for total area?

4. Once again, Erin wants to construct the third rectangular sign using the golden ratio. Keeping in mind the diagonal length of 8 feet, what should the length and width of this rectangle approximately be? (Round the final value to the nearest tenth.)

# Chapter 9 Organizer

| Topic and Procedure | Examples | ⟹ You Try It |
|---|---|---|
| **Square roots, p. 510**<br>1. The square roots of perfect squares up to $\sqrt{225} = 15$ (or so) should be memorized. Approximate values of others can be found by using a calculator.<br>2. Negative numbers do not have real number square roots. | Evaluate, if possible.<br>(a) $\sqrt{196} = 14$  (b) $\sqrt{169} = 13$<br>(c) $\sqrt{200} \approx 14.142$<br>(d) Approximate $\sqrt{13} \approx 3.606$<br>(e) $\sqrt{-5}$  This is not a real number. | 1. Evaluate, if possible.<br>(a) $\sqrt{144}$  (b) $\sqrt{225}$<br>(c) $\sqrt{18}$<br>(d) $\sqrt{-10}$<br>(e) $\sqrt{150}$ |
| **Simplifying radicals, p. 514**<br>Assuming that $a$ and $b$ are nonnegative,<br>$$\sqrt{ab} = \sqrt{a}\sqrt{b}$$<br>Square root radicals are simplified by taking the square roots of all factors that are perfect squares and leaving all other factors under the radical sign. | Simplify.<br>(a) $\sqrt{48} = \sqrt{16}\sqrt{3} = 4\sqrt{3}$<br>(b) $\sqrt{12x^2y^3z} = \sqrt{2^2x^2y^2}\sqrt{3yz}$<br>$\quad = 2xy\sqrt{3yz}$<br>(c) $\sqrt{a^{11}} = \sqrt{a^{10}\cdot a} = \sqrt{a^{10}}\sqrt{a} = a^5\sqrt{a}$ | 2. Simplify.<br>(a) $\sqrt{75}$<br>(b) $\sqrt{8a^3b^2c^4}$<br>(c) $\sqrt{x^{15}}$ |
| **Adding and subtracting radicals, p. 519**<br>Simplify all radicals and then add or subtract like radicals. | Simplify. $\sqrt{12} + \sqrt{18} + \sqrt{27} + \sqrt{50}$<br>$= \sqrt{4}\sqrt{3} + \sqrt{9}\sqrt{2} + \sqrt{9}\sqrt{3} + \sqrt{25}\sqrt{2}$<br>$= 2\sqrt{3} + 3\sqrt{2} + 3\sqrt{3} + 5\sqrt{2}$<br>$= 5\sqrt{3} + 8\sqrt{2}$ | 3. Simplify. $\sqrt{50} + \sqrt{45} + \sqrt{32} - \sqrt{80}$ |
| **Multiplying radicals, p. 523**<br>Square root radicals are multiplied like polynomials. Assuming that $a$ and $b$ are nonnegative, apply the rule $\sqrt{a}\sqrt{b} = \sqrt{ab}$ and simplify all results. | Multiply. $(2\sqrt{6} + \sqrt{3})(\sqrt{6} - 3\sqrt{3})$<br>$= 2\sqrt{36} - 6\sqrt{18} + \sqrt{18} - 3\sqrt{9}$<br>$= 2(6) - 5\sqrt{18} - 3(3)$<br>$= 12 - 5\sqrt{9}\sqrt{2} - 9$<br>$= 12 - 5(3)\sqrt{2} - 9$<br>$= 3 - 15\sqrt{2}$ | 4. Multiply. $(3\sqrt{5} - \sqrt{2})(\sqrt{5} - 2\sqrt{2})$ |
| **Dividing radicals, p. 529**<br>For positive numbers $a$ and $b$,<br>$$\frac{\sqrt{a}}{\sqrt{b}} = \sqrt{\frac{a}{b}} \quad \text{and} \quad \sqrt{\frac{a}{b}} = \frac{\sqrt{a}}{\sqrt{b}}.$$ | Divide.<br>(a) $\dfrac{\sqrt{80}}{\sqrt{5}} = \sqrt{\dfrac{80}{5}} = \sqrt{16} = 4$<br>(b) $\sqrt{\dfrac{144}{25}} = \dfrac{\sqrt{144}}{\sqrt{25}} = \dfrac{12}{5}$ | 5. Divide.<br>(a) $\dfrac{\sqrt{200}}{\sqrt{8}}$  (b) $\sqrt{\dfrac{196}{9}}$ |
| **Conjugates, p. 531**<br>Note that the product of the sum and difference of two square root radicals is equal to the difference of their squares and thus contains no radicals. The sum and difference are called conjugates. | $(3\sqrt{6} + \sqrt{10})$ and $(3\sqrt{6} - \sqrt{10})$ are conjugates.<br>Multiply $(3\sqrt{6} + \sqrt{10})(3\sqrt{6} - \sqrt{10})$.<br>$= 9\sqrt{6}^2 - \sqrt{10}^2$<br>$= 9\cdot 6 - 10 = 54 - 10 = 44$ | 6. (a) What is the conjugate of $\sqrt{3} - 2\sqrt{6}$?<br>(b) Multiply $\sqrt{3} - 2\sqrt{6}$ and its conjugate. |
| **Rationalizing denominators, pp. 530–531**<br>To simplify a fraction with a square root radical in the denominator, the radical (in the denominator) must be removed.<br>1. Multiply the denominator by the smallest radical that will yield the square root of a perfect square. Multiply the numerator by the same quantity.<br>2. Multiply the denominator by its conjugate if it is a binomial. Multiply the numerator by the same quantity. | Rationalize the denominator.<br>(a) $\dfrac{\sqrt{10}}{\sqrt{6}} = \dfrac{\sqrt{10}\cdot\sqrt{6}}{\sqrt{6}\cdot\sqrt{6}} = \dfrac{2\sqrt{15}}{6} = \dfrac{\sqrt{15}}{3}$<br>(b) $\dfrac{\sqrt{3}}{\sqrt{3} - \sqrt{2}} = \dfrac{\sqrt{3}(\sqrt{3} + \sqrt{2})}{(\sqrt{3} - \sqrt{2})(\sqrt{3} + \sqrt{2})}$<br>$\qquad = \dfrac{\sqrt{9} + \sqrt{6}}{3 - 2} = 3 + \sqrt{6}$ | 7. Rationalize the denominator.<br>(a) $\dfrac{\sqrt{15}}{\sqrt{10}}$<br>(b) $\dfrac{\sqrt{5}}{\sqrt{2} + \sqrt{3}}$ |
| **Radicals with fractions, p. 530**<br>Note that $\sqrt{\dfrac{5}{3}}$ is the same as $\dfrac{\sqrt{5}}{\sqrt{3}}$ and so must be rationalized. | Simplify. $\sqrt{\dfrac{5}{3}} = \dfrac{\sqrt{5}}{\sqrt{3}}\cdot\dfrac{\sqrt{3}}{\sqrt{3}} = \dfrac{\sqrt{15}}{3}$ | 8. (a) Rewrite $\sqrt{\dfrac{6}{5}}$ as a fraction with a radical in both the numerator and denominator.<br>(b) Rationalize your result from (a). |

552

| Topic and Procedure | Examples |  You Try It |
|---|---|---|
| **Pythagorean Theorem, p. 535**<br>In any right triangle with hypotenuse of length $c$ and legs of lengths $a$ and $b$,<br>$c^2 = a^2 + b^2$. | Find $c$ to the nearest tenth if $a = 7$ and $b = 9$.<br>$c^2 = 7^2 + 9^2 = 49 + 81 = 130$<br>$c = \sqrt{130} \approx 11.4$ | **9.** Find $b$ to the nearest tenth if $a = 6$ and $c = 11$. |
| **Radical equations, pp. 536–538**<br>To solve an equation containing a square root radical:<br>1. Isolate the radical by itself on one side of the equation.<br>2. Square both sides.<br>3. Solve the resulting equation.<br>4. Check all apparent solutions. Extraneous roots may have been introduced in step 2. | Solve. $\sqrt{3x + 2} - 4 = 8$<br>$\sqrt{3x + 2} = 12$<br>$(\sqrt{3x + 2})^2 = 12^2$<br>$3x + 2 = 144$<br>$3x = 142$<br>$x = \dfrac{142}{3}$<br>Check. $\sqrt{3\left(\dfrac{142}{3}\right) + 2} \stackrel{?}{=} 12$<br>$\sqrt{144} = 12 \ \checkmark$<br>So $x = \dfrac{142}{3}$. | **10.** Solve. $12 = \sqrt{4x - 5} + 7$ |
| **Variation: direct, p. 542**<br>If $y$ varies directly with $x$, there is a constant of variation $k$ such that $y = kx$. Once $k$ is determined, the value of $y$ or $x$ can easily be computed. | $y$ varies directly with $x$. When $x = 2$, $y = 7$.<br>$y = kx$<br>$7 = k(2)$    Substituting<br>$k = \dfrac{7}{2}$    Solving<br>$y = \dfrac{7}{2}x$<br>What is $y$ when $x = 8$?<br>$y = \dfrac{7}{2}x = \dfrac{7}{2} \cdot 8 = 28$ | **11.** $w$ varies directly with $t$. When $t = 4$, $w = 6$. What is $w$ when $t = 10$? |
| **Variation: inverse, p. 544**<br>If $y$ varies inversely with $x$, there is a constant of variation $k$ such that<br>$y = \dfrac{k}{x}$. | $y$ varies inversely with $x$. When $x$ is 5, $y$ is 12.<br>$y = \dfrac{k}{x}$<br>$12 = \dfrac{k}{5}$    Substituting<br>$k = 60$    Solving<br>$y = \dfrac{60}{x}$    Substituting<br>What is $y$ when $x$ is 15?<br>$y = \dfrac{60}{15} = 4$ | **12.** $m$ varies inversely with $g$. When $g = 3$, $m = 8$. What is $m$ when $g$ is 32? |

# Chapter 9 Review Problems

## Section 9.1

*Simplify or evaluate, if possible.*

**1.** $\sqrt{64}$      **2.** $-\sqrt{121}$      **3.** $-\sqrt{144}$      **4.** $\sqrt{289}$      **5.** $\sqrt{0.04}$

**6.** $\sqrt{0.49}$      **7.** $\sqrt{\dfrac{1}{100}}$      **8.** $\sqrt{\dfrac{64}{81}}$

*Use a calculator to approximate. Round to the nearest thousandth, if necessary.*

**9.** $\sqrt{105}$ **10.** $\sqrt{198}$ **11.** $\sqrt{77}$ **12.** $\sqrt{88}$

## Section 9.2

*Simplify.*

**13.** $\sqrt{28}$ **14.** $\sqrt{125}$ **15.** $\sqrt{40}$

**16.** $\sqrt{80}$ **17.** $\sqrt{y^{10}}$ **18.** $\sqrt{x^5 y^6}$

**19.** $\sqrt{16x^3 y^5}$ **20.** $\sqrt{98x^4 y^6}$ **21.** $\sqrt{12x^5}$

**22.** $\sqrt{72x^9}$ **23.** $\sqrt{120a^3 b^4 c^5}$ **24.** $\sqrt{121a^6 b^4 c}$

**25.** $\sqrt{56x^7 y^9}$ **26.** $\sqrt{99x^{13} y^7}$

## Section 9.3

*Simplify.*

**27.** $\sqrt{5} - \sqrt{20} + \sqrt{80}$ **28.** $5\sqrt{6} - \sqrt{24} + 2\sqrt{54}$

**29.** $x\sqrt{3} + 3x\sqrt{3} + \sqrt{27x^2}$ **30.** $a\sqrt{2} + \sqrt{12a^2} + a\sqrt{98}$

**31.** $5\sqrt{5} - 6\sqrt{20} + 2\sqrt{10}$ **32.** $2\sqrt{40} - 2\sqrt{90} + 3\sqrt{28}$

## Section 9.4

*Simplify.*

**33.** $\left(2\sqrt{x}\right)\left(3\sqrt{x^3}\right)$ **34.** $\left(-5\sqrt{a}\right)\left(2\sqrt{ab}\right)$

**35.** $\left(-4x\sqrt{x}\right)\left(2x\sqrt{x}\right)$ **36.** $\sqrt{3}\left(\sqrt{48} - 8\sqrt{3}\right)$

**37.** $\sqrt{5}\left(\sqrt{6} - 2\sqrt{5} + \sqrt{10}\right)$ **38.** $\left(\sqrt{11} + 2\right)\left(2\sqrt{11} - 1\right)$

**39.** $\left(\sqrt{10} + 3\right)\left(3\sqrt{10} - 1\right)$

**40.** $\left(2 + 3\sqrt{6}\right)\left(4 - 2\sqrt{3}\right)$

**41.** $\left(5 - \sqrt{2}\right)\left(3 - \sqrt{12}\right)$

**42.** $\left(2\sqrt{3} + 3\sqrt{6}\right)^2$

**43.** $\left(5\sqrt{2} - 2\sqrt{6}\right)^2$

## Section 9.5

*Rationalize the denominator.*

**44.** $\dfrac{1}{\sqrt{3x}}$

**45.** $\dfrac{2y}{\sqrt{5}}$

**46.** $\sqrt{\dfrac{5}{6}}$

**47.** $\sqrt{\dfrac{6}{11}}$

**48.** $\dfrac{\sqrt{x^3}}{\sqrt{5x}}$

**49.** $\dfrac{3}{\sqrt{5} + \sqrt{2}}$

**50.** $\dfrac{2}{\sqrt{6} - \sqrt{3}}$

**51.** $\dfrac{1 - \sqrt{5}}{2 + \sqrt{5}}$

**52.** $\dfrac{1 - \sqrt{3}}{3 + \sqrt{3}}$

## Section 9.6

*In exercises 53–58, use the Pythagorean Theorem to find the length of the unknown side of a right triangle with legs a and b and hypotenuse c.*

▲ **53.** $a = 5, b = 8$

▲ **54.** $c = \sqrt{11}, b = 3$

▲ **55.** $c = 5, a = 3.5$

▲ **56.** ***Flagpole*** A flagpole is 24 meters tall. A man stands 18 meters from the base of the pole. How far is it from the feet of the man to the top of the pole?

▲ **57.** ***Distances Between Cities*** Kingman is a city in Arizona that is 143 miles west of Flagstaff. Fountain Hills is 120 miles south of Flagstaff. How far is it from Kingman to Fountain Hills? Round your answer to the nearest tenth of a mile.

▲ **58.** ***Softball Diamond*** The four bases in a fast-pitch soft-ball field form a square. The length of the diagonal (the distance from home plate to second base) is $\sqrt{7200}$ feet. How long is each side of the square?

*Solve. Be sure to verify your answers.*

**59.** $\sqrt{5x - 1} = 8$

**60.** $\sqrt{1 - 5x} = \sqrt{9 - x}$

**61.** $\sqrt{-5 + 2x} = \sqrt{1 + x}$

**62.** $\sqrt{10x + 9} = -1 + 2x$

**63.** $\sqrt{2x - 5} = 10 - x$

**64.** $6 - \sqrt{5x - 1} = x + 1$

## Section 9.7

**65.** If $y$ varies directly with the square of $x$, and $y = 27$ when $x = 3$, find the value of $y$ when $x = 4$.

**66.** If $y$ varies inversely with the square root of $x$, and $y = 2$ when $x = 25$, find the value of $y$ when $x = 100$.

**67.** If $y$ varies inversely with the cube of $x$, and $y = 4$ when $x = 2$, find the value of $y$ when $x = 4$.

**68.** *Illumination from a Light Source* The intensity of illumination from a light source varies inversely as the square of the distance from the light source. When an object is 4 meters from a light source, the intensity is 45 lumens. What is the intensity at a distance of 12 meters?

**69.** *Skid Marks of a Car* When an automobile driver slams on the brakes, the length of skid marks on the road varies directly with the square of the speed of the car. At 30 mph a certain car had skid marks 40 feet long. How long will the skid marks be if the car travels at 55 mph? Round to the nearest foot.

**70.** *Horsepower of a Boat* The horsepower that is needed to drive a racing boat through water varies directly with the cube of the speed of the boat. What will happen to the horsepower requirement if someone wants to double the maximum speed of a given boat?

**Mixed Practice** *Find the square root.*

**71.** $\sqrt{\dfrac{36}{121}}$

**72.** $\sqrt{0.0004}$

*Simplify.*

**73.** $\sqrt{98x^6y^3}$

*Combine.*

**74.** $3\sqrt{27} - 2\sqrt{75} + \sqrt{48}$

*Multiply and simplify your answer.*

**75.** $\left(\sqrt{3} - 2\sqrt{5}\right)\left(2\sqrt{5} + 3\sqrt{2}\right)$

**76.** $\left(4\sqrt{3} + 3\right)^2$

*Rationalize the denominator.*

**77.** $\dfrac{5}{\sqrt{12}}$

**78.** $\dfrac{\sqrt{3} + \sqrt{6}}{2\sqrt{3} + \sqrt{2}}$

*Solve.*

**79.** $\sqrt{2x - 3} = 9$

**80.** $\sqrt{10x + 5} = 2x + 1$

# How Am I Doing? Chapter 9 Test

MATH COACH    MyMathLab®    YouTube™

After you take this test read through the Math Coach on pages 558–559. Math Coach videos are available via MyMathLab and YouTube. Step-by-step test solutions on the Chapter Test Prep Videos are also available via MyMathLab and YouTube. (Search "TobeyBeginningAlg" and click on "Channels.")

*Evaluate.*

1. $\sqrt{121}$

2. $\sqrt{\dfrac{9}{100}}$

*Simplify.*

3. $\sqrt{48x^2y^7}$

4. $\sqrt{100x^3yz^4}$

*Combine and simplify.*

5. $8\sqrt{3} + 5\sqrt{27} - 5\sqrt{48}$

Mc 6. $\sqrt{4a} + \sqrt{8a} + \sqrt{36a} + \sqrt{18a}$

*Multiply.*

7. $(2\sqrt{a})(3\sqrt{b})(2\sqrt{ab})$

8. $\sqrt{5}(\sqrt{10} + 2\sqrt{3} - 3\sqrt{5})$

9. $(2\sqrt{3} + 5)^2$

Mc 10. $(4\sqrt{2} - \sqrt{5})(3\sqrt{2} + \sqrt{5})$

*In questions 11–14, simplify.*

11. $\sqrt{\dfrac{x}{5}}$

12. $\dfrac{3}{\sqrt{12}}$

Mc 13. $\dfrac{\sqrt{3} + 4}{5 + \sqrt{3}}$

14. $\dfrac{3a}{\sqrt{5} + \sqrt{2}}$

15. Use a calculator to approximate $\sqrt{156}$. Round your answer to the nearest hundredth.

*Find the missing side by using the Pythagorean Theorem.*

▲ 16.

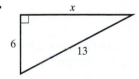

▲ 17.

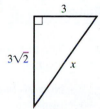

*Solve. Verify your solutions.*

18. $6 - \sqrt{2x + 1} = 0$

Mc 19. $x = 5 + \sqrt{x + 7}$

*Solve.*

20. The intensity of illumination from a light source varies inversely with the square of the distance from the light source. When a photoelectric cell is placed 8 inches from a light source, the intensity is 12 lumens. Then it is moved so that it receives only 3 lumens. How far is it from the light source?

21. Commission on the sale of office supplies varies directly with the amount of the sale. If a salesman earns $23.40 on sales of $780, what does he earn on sales of $2859?

22. The area of an equilateral triangle varies directly as the square of its perimeter. If the perimeter is 12 centimeters, the area is 6.93 square centimeters. What is the area of an equilateral triangle whose perimeter is 21 centimeters? Round to the nearest tenth.

1. _____ ☐
2. _____ ☐
3. _____ ☐
4. _____ ☐
5. _____ ☐
6. _____ ☐
7. _____ ☐
8. _____ ☐
9. _____ ☐
10. _____ ☐
11. _____ ☐
12. _____ ☐
13. _____ ☐
14. _____ ☐
15. _____ ☐
16. _____ ☐
17. _____ ☐
18. _____ ☐
19. _____ ☐
20. _____ ☐
21. _____ ☐
22. _____ ☐

**Total Correct:** ☐

# MATH COACH

*Mastering the skills you need to do well on the test.*

The following problems are from the Chapter 9 Test. Here are some helpful hints to keep you from making common errors on test problems.

**Chapter 9 Test, Problem 6** Combine and simplify. $\sqrt{4a} + \sqrt{8a} + \sqrt{36a} + \sqrt{18a}$

> **HELPFUL HINT** You must simplify each radical first. Then you can only combine like terms for those terms where the expression inside the radical sign is *exactly* the same.

Did you identify that 4 and 36 are perfect squares and then simplify the *first* radical to $2\sqrt{a}$ and the *third* radical to $6\sqrt{a}$?

Yes _____ No _____

If you answered No, notice that the two radicals can be rewritten as $\sqrt{4} \cdot \sqrt{a}$ and $\sqrt{36} \cdot \sqrt{a}$. You can then take the square root of 4 and the square root of 36. Go back and complete this step again.

Did you look for perfect squares and then rewrite the *second* radical as $\sqrt{4 \cdot 2 \cdot a}$ and the *fourth* radical as $\sqrt{9 \cdot 2 \cdot a}$?

Yes _____ No _____

If you answered No, remember that we want to factor the radicand into products of perfect squares whenever possible. Now the square roots of 4 and 9 can be taken to simplify these radicals further.

The last step is to combine any radicals where the expression inside the radical is exactly the same.

If you answered Problem 6 incorrectly, go back and rework the problem using these suggestions.

---

**Chapter 9 Test, Problem 10** Multiply. $(4\sqrt{2} - \sqrt{5})(3\sqrt{2} + \sqrt{5})$

> **HELPFUL HINT** Remember that you can use the FOIL method to multiply any two binomials. This also applies to binomial radical expressions. Be careful to separate numbers outside the radical and numbers inside the radical when completing the multiplication.

When you multiplied the *first two terms* and simplified that product, did you get 24?

Yes _____ No _____

If you answered No, remember we must separately multiply the numbers outside the radicals and then multiply the numbers inside the radical signs. Be sure to simplify radical expressions such as $\sqrt{4}$.

When you multiplied the *outer terms*, did you get $4\sqrt{10}$?

Yes _____ No _____

If you answered No, notice that $4\sqrt{2} \cdot \sqrt{5} = 4 \cdot 1\sqrt{2 \cdot 5} = 4\sqrt{10}$.

The next steps are to multiply the *inner terms*, multiply the *last two terms*, and then combine any like terms.

Now go back and rework the problem using these suggestions.

Need more help? Watch the **MATH COACH** videos in MyMathLab® or on You Tube.

558

**Chapter 9 Test, Problem 13**  Simplify. $\dfrac{\sqrt{3} + 4}{5 + \sqrt{3}}$

> **HELPFUL HINT** When the denominator of a fraction contains a radical expression, we multiply both the numerator and the denominator of the fraction by the conjugate of the denominator.

Did you identify the conjugate of the denominator as $5 - \sqrt{3}$?

Yes _____ No _____

Did you multiply the numerator and denominator by $5 - \sqrt{3}$ to obtain the expression $\dfrac{(\sqrt{3} + 4)(5 - \sqrt{3})}{(5 + \sqrt{3})(5 - \sqrt{3})}$?

Yes _____ No _____

If you answered No to these questions, go back and review the definition of conjugate and try these two steps again.

After completing the multiplication, did you get the unsimplified expression $\dfrac{5\sqrt{3} - \sqrt{9} + 20 - 4\sqrt{3}}{25 - 5\sqrt{3} + 5\sqrt{3} - \sqrt{9}}$?

Yes _____ No _____

If you answered No, slowly go through the step of multiplying the two binomials in the numerator. Then carefully multiply the two binomials in the denominator.

After completing these steps, you can look for square roots to evaluate. Then combine any like terms separately in both the numerator and denominator. Your final result will still be a fraction.

If you answered Problem 13 incorrectly, go back and rework the problem using these suggestions.

---

**Chapter 9 Test, Problem 19**  Solve. Verify your solutions. $x = 5 + \sqrt{x + 7}$

> **HELPFUL HINT** Always isolate the radical on one side of the equation before squaring each side of the equation. Be sure to check your results for extraenous roots.

Did you first isolate the radical to get $x - 5 = \sqrt{x + 7}$?

Yes _____ No _____

When you squared each side of this equation, did you multiply and then simplify to obtain the equation $x^2 - 10x + 25 = x + 7$?

Yes _____ No _____

If you answered No to these questions, notice that $(x - 5)^2 = (x - 5)(x - 5)$ on the left side. Also remember that $(\sqrt{x + 7})^2 = \sqrt{x + 7} \cdot \sqrt{x + 7} = x + 7$.

Stop now and complete these calculations.

Did you transform the equation to $x^2 - 11x + 18 = 0$?

Yes _____ No _____

If you answered No, try to get all the terms on the left and only 0 on the right. Then factor the resulting quadratic equation.

When you check both possible answers in the original equation, did both answers work?

If you answered Yes, carefully replace $x$ by 2 in the equation. Remember that when you obtain $2 = 5 + 3$ you can see immediately that this is a invalid equation. The value $x = 2$ does not check. It is an extraneous root.

Yes _____ No _____

Now go back and rework the problem using these suggestions.

Need more help? Look for section examples marked with $\overset{M}{C}$ to review.

# Quadratic Equations

## CAREER OPPORTUNITIES

### Civil Engineer

A region's transportation system, particularly its roads and bridges, is vital to the well-being of that area. Used by commuters traveling to and from work, truckers transporting goods and materials, and vacationers sightseeing and traveling to a favorite destination, the capacity and condition of roads and bridges directly affect the quality of life in that region. Civil engineers are the professionals who design, create, and maintain these systems.

To investigate how the mathematics in this chapter can help with this field, see the Career Exploration Problems on page **597**.

# 10.1 Introduction to Quadratic Equations

## 1 Writing a Quadratic Equation in Standard Form

In Section 5.7 we introduced quadratic equations. A **quadratic equation** is a polynomial equation of degree two. For example,

$$3x^2 + 5x - 7 = 0, \quad \frac{1}{2}x^2 - 5x = 0, \quad 2x^2 = 5x + 9,$$

and $3x^2 = 8$ are all quadratic equations.

> The **standard form of a quadratic equation** is $ax^2 + bx + c = 0$, where $a$, $b$, and $c$ are real numbers, and $a \neq 0$.

You will need to be able to recognize the real numbers that represent $a$, $b$, and $c$ in specific equations. It will be easier to do so if these equations are placed in standard form. Note that it is usually easier to work with quadratic equations if $a$ is positive, so in cases where $a$ is negative, we recommend that you multiply the entire equation by $-1$. This is a suggestion and not a requirement.

**Example 1** Place each quadratic equation in standard form and identify the real numbers $a$, $b$, and $c$.

**(a)** $5x^2 - 6x + 3 = 0$     This equation is in standard form.

**Solution**   $ax^2 + bx + c = 0$    Match each term to the standard form.

$a = 5, \quad b = -6, \quad c = 3$

**(b)** $2x^2 + 5x = 4$     The right-hand side is not zero. It is not in standard form.

**Solution**   $2x^2 + 5x - 4 = 0$    Add $-4$ to each side of the equation.

$ax^2 + bx + c = 0$    Match each term to the standard form.

$a = 2, \quad b = 5, \quad c = -4$

**(c)** $-2x^2 + 15x + 4 = 0$     It is easier to work with quadratic equations if the first term is not negative.

**Solution**   $2x^2 - 15x - 4 = 0$    Multiply each term on both sides of the equation by $-1$.

$ax^2 + bx + c = 0$

$a = 2, \quad b = -15, \quad c = -4$

**(d)** $7x^2 - 9x = 0$     This equation is in standard form.

**Solution**   $ax^2 + bx + c = 0$    Note that the constant term is missing,

$a = 7, \quad b = -9, \quad c = 0$    so we know $c = 0$.     ☐

> **Student Practice 1** Place each quadratic equation in standard form and identify the real numbers $a$, $b$, and $c$. If $a < 0$, you can multiply each term of the equation by $-1$ to obtain an equivalent equation. This is not required, but most students find it helpful to have the first term be a positive number.
>
> **(a)** $2x^2 + 12x - 9 = 0$       **(b)** $7x^2 = 6x - 8$
>
> **(c)** $-x^2 - 6x + 3 = 0$       **(d)** $10x^2 - 12x = 0$

**Student Learning Objectives**

After studying this section, you will be able to:

1 Write a quadratic equation in standard form.

2 Solve quadratic equations of the form $ax^2 + bx = 0$ by factoring.

3 Solve quadratic equations of the form $ax^2 + bx + c = 0$ by factoring.

4 Solve applied problems involving quadratic equations.

## 2 Solving Quadratic Equations of the Form $ax^2 + bx = 0$ by Factoring ▶

Notice that the terms in a quadratic equation of the form $ax^2 + bx = 0$ both have $x$ as a factor. To solve such equations, begin by factoring out the $x$ and any common numerical factor. Thus you may be able to factor out an $x$ or an expression such as $5x$. Remember that you want to remove the *greatest common factor*. So if $5x$ is a common factor of each term, be sure to factor it out. Then use the zero factor property discussed in Section 5.7. This property states that if $a \cdot b = 0$, then either $a = 0$ or $b = 0$. Once each factor is set equal to 0, solve for $x$.

**Example 2** Solve. $7x^2 + 9x - 2 = -8x - 2$

**Solution**

$$7x^2 + 9x - 2 = -8x - 2 \quad \text{The equation is not in standard form.}$$
$$7x^2 + 9x - 2 + 8x + 2 = 0 \quad \text{Add } 8x + 2 \text{ to each side.}$$
$$7x^2 + 17x = 0 \quad \text{Combine like terms.}$$
$$x(7x + 17) = 0 \quad \text{Factor.}$$
$$x = 0 \quad 7x + 17 = 0 \quad \text{Set each factor equal to zero.}$$
$$7x = -17$$
$$x = 0 \text{ and } x = -\frac{17}{7} \quad \text{The solutions are 0 and } -\frac{17}{7}$$

One of the solutions to the preceding equation is 0. This will always be true of equations of this form. One root will be zero. The other root will be a nonzero real number. □

**▶ Student Practice 2** Solve. $2x^2 - 7x - 6 = 4x - 6$

## 3 Solving Quadratic Equations of the Form $ax^2 + bx + c = 0$ by Factoring ▶

If an equation you are trying to solve contains fractions, clear the fractions by multiplying each term by the least common denominator of all the fractions in the equation. Sometimes an equation does not look like a quadratic equation. However, it takes the quadratic form once the fractions have been cleared. A note of caution: Always check a possible solution in the original equation. Any apparent solution that would make the denominator of any fraction in the original equation 0 is not a valid solution.

**Example 3** Solve and check. $8x - 6 + \dfrac{1}{x} = 0$

**Solution**

$$8x - 6 + \frac{1}{x} = 0 \quad \text{The equation has a fractional term.}$$
$$x(8x) - (x)(6) + x\left(\frac{1}{x}\right) = x(0) \quad \text{Multiply each term by the LCD, which is } x.$$
$$8x^2 - 6x + 1 = 0 \quad \text{Simplify. The equation is now in standard form.}$$
$$(4x - 1)(2x - 1) = 0 \quad \text{Factor.}$$
$$4x - 1 = 0 \quad 2x - 1 = 0 \quad \text{Set each factor equal to 0.}$$
$$4x = 1 \quad 2x = 1 \quad \text{Solve each equation for } x.$$
$$x = \frac{1}{4}, \quad x = \frac{1}{2}$$

*Check.*  Checking fractional roots is more difficult, but you should be able to do it if you work carefully.

$$\text{If } x = \frac{1}{4}: \quad 8\left(\frac{1}{4}\right) - 6 + \frac{1}{\frac{1}{4}} = 2 - 6 + 4 = 0 \quad \left(\text{since } 1 \div \frac{1}{4} = 1 \cdot \frac{4}{1} = 4\right)$$

$$0 = 0 \checkmark$$

$$\text{If } x = \frac{1}{2}: \quad 8\left(\frac{1}{2}\right) - 6 + \frac{1}{\frac{1}{2}} = 4 - 6 + 2 = 0 \quad \left(\text{since } 1 \div \frac{1}{2} = 1 \cdot \frac{2}{1} = 2\right)$$

$$0 = 0 \checkmark$$

Both roots check, so $\frac{1}{2}$ and $\frac{1}{4}$ are the two roots that satisfy the original equation

$$8x - 6 + \frac{1}{x} = 0.$$

It is also correct to write the answers in decimal form as 0.5 and 0.25. This will speed up the checking process, especially if you use a scientific calculator when performing the check.  □

 **Student Practice 3**  Solve and check.

$$3x + 10 - \frac{8}{x} = 0$$

If a quadratic equation is given to us in standard form, we examine it to see if it can be factored. If it is possible to factor the quadratic equation, then we use the zero factor property to find the solutions.

**Example 4**  Solve and check. $6x^2 + 7x - 10 = 0$

**Solution**

$$(6x - 5)(x + 2) = 0 \qquad \text{Factor the quadratic equation.}$$

$$6x - 5 = 0 \qquad x + 2 = 0 \quad \text{Set each factor equal to zero and solve for } x.$$

$$6x = 5 \qquad\qquad x = -2$$

$$x = \frac{5}{6}$$

Thus the solutions to the equation are $\frac{5}{6}$ and $-2$.

Check each solution in the original equation.

$$6\left(\frac{5}{6}\right)^2 + 7\left(\frac{5}{6}\right) - 10 \overset{?}{=} 0 \qquad\qquad 6(-2)^2 + 7(-2) - 10 \overset{?}{=} 0$$

$$6\left(\frac{25}{36}\right) + 7\left(\frac{5}{6}\right) - 10 \overset{?}{=} 0 \qquad\qquad 6(4) + 7(-2) - 10 \overset{?}{=} 0$$

$$\frac{25}{6} + \frac{35}{6} - 10 \overset{?}{=} 0 \qquad\qquad 24 + (-14) - 10 \overset{?}{=} 0$$

$$\frac{60}{6} - 10 \overset{?}{=} 0 \qquad\qquad 10 - 10 = 0 \checkmark$$

$$10 - 10 = 0 \checkmark \qquad\qquad\qquad □$$

**Student Practice 4**  Solve and check. $2x^2 + 9x - 18 = 0$

It is important to place a quadratic equation in standard form before factoring. This will sometimes involve removing parentheses and combining like terms.

Ⓜ️ **Example 5** Solve. $x + (x - 6)(x - 2) = 2(x - 1)$

**Solution**

$$x + x^2 - 2x - 6x + 12 = 2x - 2 \quad \text{Remove parentheses.}$$
$$x^2 - 7x + 12 = 2x - 2 \quad \text{Combine like terms.}$$
$$x^2 - 9x + 14 = 0 \quad \text{Add } -2x + 2 \text{ to each side.}$$
$$(x - 7)(x - 2) = 0 \quad \text{Factor.}$$
$$x = 7, \quad x = 2 \quad \text{Use the zero factor property.}$$

The solutions to the equation are 7 and 2. The check is left to the student. □

**Student Practice 5** Solve.

$$-4x + (x - 5)(x + 1) = 5(1 - x)$$

**Example 6** Solve. $\dfrac{8}{x + 1} - \dfrac{x}{x - 1} = \dfrac{2}{x^2 - 1}$

**Solution** The LCD $= (x + 1)(x - 1)$. We multiply each term by the LCD.

$$(x + 1)(x - 1)\left(\frac{8}{x + 1}\right) - (x + 1)(x - 1)\left(\frac{x}{x - 1}\right) = (x + 1)(x - 1)\left[\frac{2}{(x + 1)(x - 1)}\right]$$

$$8(x - 1) - x(x + 1) = 2 \quad \text{Simplify.}$$
$$8x - 8 - x^2 - x = 2 \quad \text{Remove parentheses.}$$
$$-x^2 + 7x - 8 = 2 \quad \text{Combine like terms.}$$
$$-x^2 + 7x - 8 - 2 = 0 \quad \text{Subtract 2 from each side.}$$
$$-x^2 + 7x - 10 = 0 \quad \text{Simplify.}$$
$$x^2 - 7x + 10 = 0 \quad \text{Multiply each term by } -1.$$
$$(x - 5)(x - 2) = 0 \quad \text{Factor and solve for } x.$$
$$x = 5, \quad x = 2 \quad \text{Use the zero factor property.}$$

*Check.* The check is left to the student. □

**Student Practice 6** Solve.

$$\frac{10x + 18}{x^2 + x - 2} = \frac{3x}{x + 2} + \frac{2}{x - 1}$$

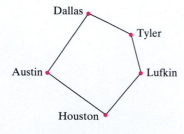

**4** **Solving Applied Problems Involving Quadratic Equations** ▶

A manager of a truck delivery service is determining the possible routes he may have to send trucks on during a given day. His delivery service must have the capability to send trucks between any two of the following Texas cities: Austin, Dallas, Tyler, Lufkin, and Houston. He made a rough map of the cities.

What is the maximum number of routes he may have to send trucks on in a given day, if there is a route between every two cities? One possible way to figure this out is to draw a straight line between every two cities. Let us assume that no three cities lie on a straight line. Therefore, the route between any two cities is distinct from all other routes. The figure to the left shows that we can draw exactly 10 lines, so the answer is that there are 10 truck routes. What if the manager has to service 18 cities? How many truck routes will there be then?

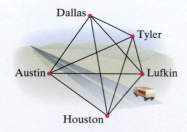

As the number of cities increases, it becomes quite difficult (and time consuming) to use a drawing to determine the number of routes. In a more advanced math course, it can be proved that the maximum number of truck routes ($t$) can be found by using the number of cities ($n$) in the following equation.

$$t = \frac{n^2 - n}{2}$$

Thus, to find the number of truck routes for 18 cities, we merely substitute 18 for $n$.

$$t = \frac{(18)^2 - (18)}{2}$$
$$= \frac{324 - 18}{2}$$
$$= \frac{306}{2}$$
$$= 153$$

Thus 153 truck routes are needed to service 18 cities if no three of those cities lie on a straight line.

**Example 7** A truck delivery company can handle a maximum of 36 truck routes in one day, and every two cities have a distinct truck route between them. How many separate cities can the truck company service?

**Solution**

$$t = \frac{n^2 - n}{2}$$

$$36 = \frac{n^2 - n}{2}$$     Substitute 36 for $t$, the number of truck routes.

$$72 = n^2 - n$$     Multiply both sides of the equation by 2.

$$0 = n^2 - n - 72$$     Subtract 72 from each side to obtain the standard form of a quadratic equation.

$$0 = (n - 9)(n + 8)$$     Factor.

$n - 9 = 0$     $n + 8 = 0$     Set each factor equal to zero.

$n = 9,$     $n = -8$     Solve for $n$.

We reject $-8$ as a meaningless solution for this problem. We cannot have a negative number of cities. Thus our answer is that the company can service 9 cities.

*Check.* If the truck company services $n = 9$ cities, does that really result in a maximum number of $t = 36$ truck routes?

$$t = \frac{n^2 - n}{2}$$     The equation in which $t =$ the number of truck routes and $n =$ the number of cities to be connected.

$$36 \stackrel{?}{=} \frac{(9)^2 - (9)}{2}$$     Substitute $t = 36$ and $n = 9$.

$$36 \stackrel{?}{=} \frac{81 - 9}{2}$$

$$36 \stackrel{?}{=} \frac{72}{2}$$

$$36 = 36 \checkmark$$

**Student Practice 7** How many cities can the company service with 28 truck routes?

## 10.1 Exercises  MyMathLab®

*Write in standard form. Determine the values of a, b, and c.*

**1.** $x^2 + 8x + 7 = 0$

**2.** $x^2 + 6x - 5 = 0$

**3.** $8x^2 - 11x = 0$

**4.** $6x^2 + 16x = 0$

**5.** $x^2 + 15x - 7 = 12x + 8$

**6.** $7x^2 - 11x + 9 = 2x - 5$

*Using the factoring method, solve for the roots of each quadratic equation. Be sure to place your equation in standard form before factoring.*

**7.** $20x^2 - 4x = 0$

**8.** $12x^2 - 9x = 0$

**9.** $4x^2 - x = -9x$

**10.** $5x^2 + x = 21x$

**11.** $11x^2 - 13x = 8x - 3x^2$

**12.** $34x^2 + 15x = 4x + x^2$

**13.** $x^2 - 3x - 28 = 0$

**14.** $x^2 + 4x - 21 = 0$

**15.** $2x^2 + 11x - 6 = 0$

**16.** $2x^2 - 9x - 5 = 0$

**17.** $3x^2 - 16x + 5 = 0$

**18.** $5x^2 - 21x + 4 = 0$

**19.** $x^2 = 5x + 14$

**20.** $x^2 = 4x + 32$

**21.** $12x^2 + 17x + 6 = 0$

**22.** $14x^2 + 29x + 12 = 0$

**23.** $a^2 + 9a + 39 = 7 - 3a$

**24.** $m^2 + 15m + 33 = 5 - m$

**25.** $6y^2 = -7y + 3$

**26.** $8y^2 = -2y + 3$

**27.** $25x^2 - 60x + 36 = 0$

### Mixed Practice, Exercises 28–38

**28.** $9x^2 - 24x + 16 = 0$

**29.** $(x - 5)(x + 2) = -10$

**30.** $(x - 6)(x + 1) = 8$

566

**31.** $3x^2 + 8x - 10 = -6x - 10$

**32.** $5x^2 - 7x + 12 = -8x + 12$

**33.** $y(y - 7) = 2(5 - 2y)$

**34.** $x(x - 4) = 3(x - 2)$

**35.** $(x - 6)(x - 4) = 3$

**36.** $(y + 6)(y - 1) = 8$

**37.** $2x(x + 3) = (3x + 1)(x + 1)$

**38.** $2(2x + 1)(x + 1) = x(7x + 1)$

*Find the LCD and multiply each term by the LCD. Then solve. Check your solutions.*

**39.** $\dfrac{x}{3} + \dfrac{2}{3} = \dfrac{5}{x}$

**40.** $\dfrac{x}{4} - \dfrac{7}{x} = -\dfrac{3}{4}$

**41.** $\dfrac{x}{7} - \dfrac{2}{x} = -\dfrac{5}{7}$

**42.** $\dfrac{x}{5} - \dfrac{7}{x} = \dfrac{2}{5}$

**43.** $\dfrac{5}{x + 2} = \dfrac{2x - 1}{5}$

**44.** $\dfrac{12}{x - 2} = \dfrac{2x - 3}{3}$

**45.** $\dfrac{4}{x} + \dfrac{3}{x + 5} = 2$

**46.** $\dfrac{7}{x} + \dfrac{6}{x - 1} = 2$

**47.** $\dfrac{24}{x^2 - 4} = 1 + \dfrac{2x - 6}{x - 2}$

**48.** $\dfrac{5x + 7}{x^2 - 9} = 1 + \dfrac{2x - 8}{x - 3}$

## To Think About

**49.** Why can a quadratic equation in standard form with $c = 0$ always be solved?

**50.** Martha solved $(x + 3)(x - 2) = 14$ as follows.

$$x + 3 = 14 \qquad x - 2 = 14$$
$$x = 11 \qquad x = 16$$

Josette said this had to be wrong because these values do not check. Explain what is wrong with Martha's method.

**51.** The equation $ax^2 - 7x + c = 0$ has the roots $-\dfrac{3}{2}$ and 6. Find the values of $a$ and $c$.

**52.** The equation $2x^2 + bx - 15 = 0$ has the roots $-3$ and $\dfrac{5}{2}$. Find the value of $b$.

**Applications** *Business Revenue The revenue for producing the label for the new Starblazer video game is given by the formula $R = -2.5n^2 + 80n - 280$, where $R$ is the revenue (in dollars) and $n$ is the number of labels produced (in hundreds).*

**53.** If the revenue for producing labels is $320, what are the two values of the two different numbers of labels that may be produced?

**54.** If the revenue for producing labels is $157.50, what are the two values of the different numbers of labels that may be produced?

**55.** What is the average of the two answers you obtained in problem 53? Use the revenue equation to find the revenue for producing that number of labels. What is significant about that answer?

**56.** What is the average of the two answers you obtained in problem 54? Use the revenue equation to find the revenue for producing that number of labels. What is significant about that answer?

**57.** Find the revenue for producing 1500 and 1700 labels. Compare the two revenues with the results from 53 and 55. What can you conclude?

**58.** Find the revenue for producing 1400 and 1800 labels. Compare the two revenues with the results from 54 and 56. What can you conclude?

## Cumulative Review

**59.** **[6.4.1]** Simplify. $\dfrac{\dfrac{1}{x} + \dfrac{3}{x-2}}{\dfrac{4}{x^2 - 4}}$

**60.** **[6.3.3]** Add the fractions. $\dfrac{2x}{3x-5} + \dfrac{2}{3x^2 - 11x + 10}$

---

**Quick Quiz 10.1** *Using the factoring method, solve for the roots of each quadratic equation.*

**1.** $6x^2 - x = 15$

**2.** $x^2 = 2(7x - 24)$

**3.** $x^2 = 2 - \dfrac{7}{3}x$

**4.** **Concept Check** In the following problem, explain how you would place the quadratic equation in standard form.
$$2 = \frac{5}{x+1} + \frac{3}{x-1}$$

## 10.2 Using the Square Root Property and Completing the Square to Find Solutions ▶

### 1 Solving Quadratic Equations Using the Square Root Property ▶

Recall that the quadratic equation $x^2 = a$ has two possible solutions, $x = \sqrt{a}$ or $x = -\sqrt{a}$, where $a$ is a nonnegative real number. That is, $x = \pm\sqrt{a}$. This basic idea is called the **square root property.**

> **SQUARE ROOT PROPERTY**
>
> If $x^2 = a$, then $x = \sqrt{a}$ or $x = -\sqrt{a}$, for all nonnegative real numbers $a$.

**Example 1** Solve.

(a) $x^2 = 49$      (b) $x^2 = 20$      (c) $5x^2 = 125$

**Solution**

(a) $x^2 = 49$

$x = \pm\sqrt{49}$

$x = \pm 7$

(b) $x^2 = 20$

$x = \pm\sqrt{20}$

$x = \pm 2\sqrt{5}$

(c) $5x^2 = 125$

$x^2 = 25$      Divide both sides by 5 before taking the square root.

$x = \pm\sqrt{25}$

$x = \pm 5$       ☐

**Student Practice 1** Solve.

(a) $x^2 = 1$      (b) $x^2 - 50 = 0$      (c) $3x^2 = 81$

**Example 2** Solve. $3x^2 + 5x = 18 + 5x + x^2$

**Solution** Simplify the equation by placing all the variable terms on the left and the constants on the right.

$3x^2 + 5x = 18 + 5x + x^2$

$2x^2 = 18$    Add $-5x - x^2$ to both sides

$x^2 = 9$     Divide both sides by 2 before taking the square root.

$x = \pm 3$       ☐

**Student Practice 2** Solve. $4x^2 - 5 = 319$

     The square root property can also be used if the term that is squared is a binomial.

     Sometimes when we use the square root property, we obtain an irrational number. This will occur if the number on the right side of the equation is not a perfect square.

**Example 3** Solve. $(3x + 1)^2 = 8$

**Solution** $(3x + 1)^2 = 8$

$$3x + 1 = \pm\sqrt{8} \quad \text{Take the square root of both sides.}$$

$$3x + 1 = \pm 2\sqrt{2} \quad \text{Simplify as much as possible.}$$

Now we must solve the two equations expressed by the plus or minus statement.

$$3x + 1 = +2\sqrt{2} \qquad\qquad 3x + 1 = -2\sqrt{2}$$

$$3x = -1 + 2\sqrt{2} \qquad\qquad 3x = -1 - 2\sqrt{2}$$

$$x = \frac{-1 + 2\sqrt{2}}{3}, \qquad\qquad x = \frac{-1 - 2\sqrt{2}}{3}$$

The roots of this quadratic equation are irrational numbers. They are

$$\frac{-1 + 2\sqrt{2}}{3} \quad \text{and} \quad \frac{-1 - 2\sqrt{2}}{3} \quad \text{or} \quad \frac{-1 \pm 2\sqrt{2}}{3}.$$

We cannot simplify these roots further, so we leave them in this form. ☐

 **Student Practice 3** Solve.

$$(2x - 3)^2 = 12$$

## 2 Solving Quadratic Equations by Completing the Square ▶

If a quadratic equation is not in a form where we can use the square root property, we can rewrite the equation by a method called **completing the square** to make it so. That is, we rewrite the equation with a perfect square on the left side so that we have an equation of the form $(ax + b)^2 = c$.

**Example 4** Solve. $x^2 + 12x = 4$

**Solution** Write out the equation $x^2 + 12x = 4$. Think of a way to determine what number to add to each side.

$$x^2 + 12x = 4$$

$$(x + 6)(x + 6) = 4 + 36 \quad \text{We need a +36 to complete the square.}$$

$$x^2 + 12x + 36 = 4 + 36 \quad \text{Add +36 to both sides of the equation.}$$

$$(x + 6)^2 = 40 \quad \text{Write the left side as a perfect square.}$$

$$x + 6 = \pm\sqrt{40} \quad \text{Now we solve for } x.$$

$$x + 6 = +2\sqrt{10}, \qquad x + 6 = -2\sqrt{10}$$

$$x = -6 + 2\sqrt{10}, \qquad x = -6 - 2\sqrt{10}$$

The two roots are $-6 + 2\sqrt{10}$ and $-6 - 2\sqrt{10}$ or the roots can be written as $-6 \pm 2\sqrt{10}$. ☐

**Student Practice 4** Solve. $x^2 + 10x = 3$

Completing the square is a little more difficult when the coefficient of $x$ is an odd number. Let's see what happens.

**Example 5** Solve. $x^2 - 3x - 1 = 0$

**Solution**   Write the equation so that all the $x$-terms are on the left and the constants are on the right.

$x^2 - 3x \qquad = 1$       The missing number when added to itself must be $-3$.

$(x \qquad)(x \qquad)$       Since $-\dfrac{3}{2} + \left(-\dfrac{3}{2}\right) = -\dfrac{6}{2} = -3$, the missing number is $-\dfrac{3}{2}$.

$\left(x - \dfrac{3}{2}\right)\left(x - \dfrac{3}{2}\right)$       Complete the pattern.

$x^2 - 3x + \dfrac{9}{4} = 1 + \dfrac{9}{4}$   Complete the square by adding $\left(-\dfrac{3}{2}\right)\left(-\dfrac{3}{2}\right) = \dfrac{9}{4}$ to each side.

$\left(x - \dfrac{3}{2}\right)^2 = \dfrac{13}{4}$       Now we solve for $x$.

$$x - \dfrac{3}{2} = \pm\sqrt{\dfrac{13}{4}}$$

$$x - \dfrac{3}{2} = \pm\dfrac{\sqrt{13}}{2}$$

$x - \dfrac{3}{2} = +\dfrac{\sqrt{13}}{2},$ \qquad\qquad $x - \dfrac{3}{2} = -\dfrac{\sqrt{13}}{2}$

$x = \dfrac{3}{2} + \dfrac{\sqrt{13}}{2},$ \qquad\qquad $x = \dfrac{3}{2} - \dfrac{\sqrt{13}}{2}$

$x = \dfrac{3 + \sqrt{13}}{2},$ \qquad\qquad $x = \dfrac{3 - \sqrt{13}}{2}$

The two roots are $\dfrac{3 \pm \sqrt{13}}{2}$.                                     □

**Student Practice 5**   Solve.

$$x^2 - 5x = 7$$

If $a$, the coefficient of the squared variable, is not 1, we divide all terms of the equation by $a$ so that the coefficient of the squared variable will be 1.

Let us summarize for future reference the steps we perform in order to solve a quadratic equation by completing the square.

**COMPLETING THE SQUARE**

1. Put the equation in the form $ax^2 + bx = c$. If $a \neq 1$, divide each term by $a$.

2. Square $\dfrac{b}{2}$ and add the result to both sides of the equation.

3. Factor the left side (a perfect-square trinomial).

4. Use the square root property.

5. Solve the equations.

6. Check the solutions in the original equation.

**Example 6** Solve. $4x^2 + 4x - 3 = 0$

**Solution**

**Step 1** $\qquad 4x^2 + 4x = 3 \qquad$ Place the constant on the right. This puts the equation in the form $ax^2 + bx = c$.

$\qquad\qquad x^2 + x = \dfrac{3}{4} \qquad$ Divide all terms by 4.

**Step 2** $\qquad x^2 + 1x = \dfrac{3}{4} \qquad$ Take one-half the coefficient of $x$ and square it. $\left(\dfrac{1}{2}\right)^2 = \dfrac{1}{4}$.

$\qquad x^2 + x + \dfrac{1}{4} = \dfrac{3}{4} + \dfrac{1}{4} \qquad$ Add $\dfrac{1}{4}$ to each side.

**Step 3** $\qquad \left(x + \dfrac{1}{2}\right)^2 = 1 \qquad$ Factor the left side.

**Step 4** $\qquad x + \dfrac{1}{2} = \pm\sqrt{1} \qquad$ Use the square root property.

**Step 5** $\qquad x + \dfrac{1}{2} = \pm 1$

$$x + \dfrac{1}{2} = 1 \qquad\qquad x + \dfrac{1}{2} = -1$$

$$x = \dfrac{2}{2} - \dfrac{1}{2} \qquad\qquad x = -\dfrac{2}{2} - \dfrac{1}{2}$$

$$x = \dfrac{1}{2} \qquad\qquad x = -\dfrac{3}{2}$$

Thus the two roots are $\dfrac{1}{2}$ and $-\dfrac{3}{2}$.

**Step 6** $\qquad 4\left(\dfrac{1}{2}\right)^2 + 4\left(\dfrac{1}{2}\right) - 3 \overset{?}{=} 0 \qquad$ Check.

$$4\left(\dfrac{1}{4}\right) + 4\left(\dfrac{1}{2}\right) - 3 \overset{?}{=} 0$$

$$1 + 2 - 3 = 0 \quad \checkmark$$

$$4\left(-\dfrac{3}{2}\right)^2 + 4\left(-\dfrac{3}{2}\right) - 3 \overset{?}{=} 0$$

$$4\left(\dfrac{9}{4}\right) + 4\left(-\dfrac{3}{2}\right) - 3 \overset{?}{=} 0$$

$$9 - 6 - 3 = 0 \quad \checkmark$$

Both values check. $\qquad\qquad\qquad\qquad\qquad\qquad$ ☐

**Student Practice 6** Solve.

$$8x^2 + 2x - 1 = 0$$

This method of completing the square will enable us to solve any quadratic equation that has real number roots. It is usually faster, however, to factor the quadratic equation if the polynomial is factorable.

## 10.2 Exercises   MyMathLab®

*Solve using the square root property.*

1. $x^2 = 49$

2. $x^2 = 121$

3. $x^2 = 72$

4. $x^2 = 120$

5. $x^2 - 28 = 0$

6. $x^2 - 32 = 0$

7. $5x^2 = 45$

8. $3x^2 = 48$

9. $6x^2 = 120$

10. $2x^2 = 160$

11. $3x^2 - 375 = 0$

12. $4x^2 - 252 = 0$

13. $5x^2 + 13 = 73$

14. $6x^2 + 5 = 53$

15. $13x^2 + 17 = 82$

16. $6x^2 - 13 = 23$

17. $(x - 7)^2 = 16$

18. $(x - 6)^2 = 100$

19. $(x + 4)^2 = 6$

20. $(x - 5)^2 = 36$

21. $(2x + 5)^2 = 2$

22. $(3x - 4)^2 = 6$

23. $(3x - 1)^2 = 7$

24. $(2x + 3)^2 = 10$

25. $(5x + 1)^2 = 18$

26. $(8x + 5)^2 = 20$

27. $(4x - 5)^2 = 54$

28. $(5x - 3)^2 = 50$

*Solve by completing the square.*

29. $x^2 - 6x = 11$

30. $x^2 - 2x = 4$

31. $x^2 + 6x + 7 = 0$

32. $x^2 - 12x - 4 = 0$

33. $x^2 - 12x - 5 = 0$

34. $x^2 + 16x + 30 = 0$

35. $x^2 - 7x = 0$

36. $x^2 = 9x$

37. $5x^2 = 25x$

**38.** $2x^2 - 3x = 20$

**39.** $2x^2 - 7x = 9$

**40.** $2x^2 + 7x = 4$

## Cumulative Review

**41.** **[8.3.1]** Solve. $3a - 5b = 8$
$\qquad\qquad\qquad 5a - 7b = 8$

**42.** **[6.5.1]** Solve. $\dfrac{x}{x+3} - \dfrac{2}{x} = \dfrac{x-2}{x^2+3x}$

**43.** **[3.2.2]** *Hunting Knife* While hiking in the Grand Canyon, Tom Slicklen had to use his hunting knife to clear a fallen branch from the path. Find the pressure on the blade of his knife if the knife is 8 in. long and sharpened to an edge 0.01 in. wide and he applies a force of 16 pounds. Use the formula $P = \frac{F}{A}$, where $P$ is the pressure on the blade of the knife, $F$ is the force in pounds placed on the knife, and $A$ is the area exposed to an object by the sharpened edge of the knife.

**44.** **[0.6.1]** *Tire Inspector* Darlene inspected a sample of 100 new steel-belted radial tires. 13 of the tires had defects in workmanship, while 9 of the tires had defects in materials. Of those defective tires, 6 had defects in both workmanship and materials. How many of the 100 tires had any kind of defect? (Assume that the only defects found were those in materials or workmanship.)

**Quick Quiz 10.2** *Solve using the square root property.*

**1.** $4x^2 + 13 = 37$

**2.** $(3x - 2)^2 = 40$

**3.** Solve by completing the square. $x^2 + 8x - 10 = 0$

**4.** **Concept Check** Explain the first three steps of how you would solve the following problem by completing the square.

$$5x^2 - 10x + 2 = 0$$

## 10.3 Using the Quadratic Formula to Find Solutions ▶

### 1 Solving a Quadratic Equation Using the Quadratic Formula ▶

To find solutions of a quadratic equation, you can factor or you can complete the square when the trinomial is not factorable. If we use the method of completing the square on the general equation $ax^2 + bx + c = 0$, we can derive a general formula for the solution of any quadratic equation. This is called the **quadratic formula.**

**QUADRATIC FORMULA**

The roots of any quadratic equation of the form $ax^2 + bx + c = 0$, where $a$, $b$, and $c$ are real numbers and $a \neq 0$, are

$$x = \frac{-b \pm \sqrt{b^2 - 4ac}}{2a}.$$

**Student Learning Objectives**

After studying this section, you will be able to:

1 Solve a quadratic equation using the quadratic formula. ▶

2 Find a decimal approximation for the real roots of a quadratic equation. ▶

3 Determine whether a quadratic equation has no real solutions. ▶

All you will need to know are the values of $a$, $b$, and $c$ to solve a quadratic equation using the quadratic formula.

**Example 1**  Solve using the quadratic formula. $3x^2 + 10x + 7 = 0$

**Solution**  In our given equation, $3x^2 + 10x + 7 = 0$, we have $a = 3$ (the coefficient of $x^2$), $b = 10$ (the coefficient of $x$), and $c = 7$ (the constant term).

$$x = \frac{-b \pm \sqrt{b^2 - 4ac}}{2a} = \frac{-10 \pm \sqrt{(10)^2 - 4(3)(7)}}{2(3)}$$

Write the quadratic formula and substitute values for $a$, $b$, and $c$.

$$= \frac{-10 \pm \sqrt{100 - 84}}{6}$$

Simplify.

$$= \frac{-10 \pm \sqrt{16}}{6} = \frac{-10 \pm 4}{6}$$

$$x = \frac{-10 + 4}{6} = \frac{-6}{6} = -1$$

Using the positive sign.

$$x = \frac{-10 - 4}{6} = \frac{-14}{6} = -\frac{7}{3}$$

Using the negative sign.

Thus the two solutions are $-1$ and $-\frac{7}{3}$.

[*Note:* Here we obtain rational roots. We would obtain the same answer by factoring $3x^2 + 10x + 7 = 0$ as $(3x + 7)(x + 1) = 0$ and setting each factor $= 0$.] ☐

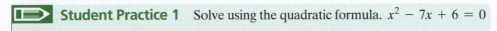 **Student Practice 1**  Solve using the quadratic formula. $x^2 - 7x + 6 = 0$

Often the roots will be irrational numbers. Sometimes these roots can be simplified. You should always leave your answer in simplest form.

**Example 2**  Solve. $2x^2 - 4x - 9 = 0$

**Solution**  $a = 2$,  $b = -4$,  $c = -9$

$$x = \frac{-b \pm \sqrt{b^2 - 4ac}}{2a} = \frac{-(-4) \pm \sqrt{(-4)^2 - 4(2)(-9)}}{2(2)}$$

$$= \frac{4 \pm \sqrt{16 + 72}}{4} = \frac{4 \pm \sqrt{88}}{4} \quad \text{Do not stop here!}$$

*Continued on next page*

Notice that we can simplify $\sqrt{88}$.

$$x = \frac{4 \pm 2\sqrt{22}}{4} = \frac{2(2 \pm \sqrt{22})}{4} = \frac{\overset{1}{2}(2 \pm \sqrt{22})}{\underset{2}{4}} = \frac{2 \pm \sqrt{22}}{2}$$

Be careful. Here we were able to divide the numerator and denominator by 2 because 2 was a factor of every term. ◻

**Student Practice 2** Solve. $3x^2 - 8x + 3 = 0$

A quadratic equation *must* be written in the standard form $ax^2 + bx + c = 0$ *before* the quadratic formula can be used. Several algebraic steps may be needed to accomplish this objective. Also, since it is much easier to use the formula if $a$, $b$, and $c$ are integers, we will avoid using fractional values.

$\mathbb{M}\mathbb{C}$ **Example 3** Solve. $x^2 = 5 - \frac{3}{4}x$

**Solution** First we obtain an equivalent equation that does not have fractions.

$$x^2 = 5 - \frac{3}{4}x$$

$$4(x^2) = 4(5) - 4\left(\frac{3}{4}x\right) \quad \text{Multiply each term by the LCD, 4.}$$

$$4x^2 = 20 - 3x \quad \text{Simplify.}$$

$$4x^2 + 3x - 20 = 0 \quad \text{Add } 3x - 20 \text{ to each side.}$$

$$a = 4, \quad b = 3, \quad c = -20$$

$$x = \frac{-3 \pm \sqrt{(3)^2 - 4(4)(-20)}}{2(4)} \quad \begin{array}{l}\text{Substitute } a = 4, b = 3, \text{ and}\\ c = -20 \text{ into the quadratic formula.}\end{array}$$

$$x = \frac{-3 \pm \sqrt{9 + 320}}{8}$$

$$x = \frac{-3 \pm \sqrt{329}}{8} \qquad ◻$$

**Student Practice 3** Solve.

$$x^2 = 7 - \frac{3}{5}x$$

## 2 Finding a Decimal Approximation for the Real Roots of a Quadratic Equation ▶

Whenever the directive for solving an equation states that you are to "approximate" the solution, you can simplify the answer using a calculator. When you approximate a solution, replace "=" with "≈" in the solution.

**Example 4** Find the roots of $3x^2 - 5x = 7$. Approximate to the nearest thousandth.

**Solution** $3x^2 - 5x = 7$

$$3x^2 - 5x - 7 = 0 \quad \text{Place the equation in standard form.}$$

$$a = 3, \quad b = -5, \quad c = -7$$

$$x = \frac{-(-5) \pm \sqrt{(-5)^2 - 4(3)(-7)}}{2(3)} = \frac{5 \pm \sqrt{25 + 84}}{6}$$

$$= \frac{5 \pm \sqrt{109}}{6}$$

Using most scientific calculators, we can find $\dfrac{5 + \sqrt{109}}{6}$ with these keystrokes.

$$5 \boxed{+} 109 \boxed{\sqrt{}} \boxed{=} \boxed{\div} 6 \boxed{=} \; 2.5733844$$

Rounding to the nearest thousandth, we have $x \approx 2.573$.

To find $\dfrac{5 - \sqrt{109}}{6}$, we use the following keystrokes.

$$5 \boxed{-} 109 \boxed{\sqrt{}} \boxed{=} \boxed{\div} 6 \boxed{=} \; -0.9067178$$

Rounding to the nearest thousandth, we have $x \approx -0.907$.

If you do not have a calculator with a square root key, look up $\sqrt{109}$ in a square root table online. $\sqrt{109} \approx 10.440$. Using this result, we have

$$x \approx \frac{5 + 10.440}{6} = \frac{15.440}{6} \approx 2.573 \quad \text{(rounded to the nearest thousandth)}$$

and

$$x \approx \frac{5 - 10.440}{6} = \frac{-5.440}{6} \approx -0.907 \quad \text{(rounded to the nearest thousandth)}.$$

If you have a graphing calculator, there is an alternate method for finding a decimal approximation of the roots of a quadratic equation. In Section 10.4, you will cover graphing quadratic equations. You will discover that finding the roots of a quadratic equation of the form $ax^2 + bx + c = 0$ is equivalent to finding the values of $x$ on the graph of $y = ax^2 + bx + c$, where $y = 0$. These may be approximated quite quickly with a graphing calculator. $\square$

▐➡ **Student Practice 4**   Find the roots of $2x^2 = 13x + 5$. Approximate to the nearest thousandth.

**3** **Determining Whether a Quadratic Equation Has No Real Solutions**

ᴹ꜀ **Example 5**   Solve. $2x^2 + 5 = -3x$

**Solution**   The equation is not in standard form.

$2x^2 + 5 = -3x$    We must add $3x$ to both sides.

$2x^2 + 3x + 5 = 0$    We can now find $a, b,$ and $c$.

$a = 2, \quad b = 3, \quad c = 5$    Substitute these values into the quadratic formula.

$$x = \frac{-3 \pm \sqrt{(3)^2 - 4(2)(5)}}{2(2)} = \frac{-3 \pm \sqrt{9 - 40}}{4} = \frac{-3 \pm \sqrt{-31}}{4}$$

There is no real number that is $\sqrt{-31}$. Since we are using only real numbers in this book, we say there is no solution to the problem. (*Note:* In more advanced math courses, these types of numbers, called complex numbers, will be studied.) $\square$

▐➡ **Student Practice 5**   Solve.

$$5x^2 + 2x = -3$$

We can tell whether the roots of any given quadratic equation are real numbers. Look at the quadratic formula.

$$\frac{-b \pm \sqrt{b^2 - 4ac}}{2a}$$

The expression under the radical sign is called the **discriminant.**

$$b^2 - 4ac \text{ is the discriminant.}$$

If the discriminant is a negative number, the roots are not real numbers, and there is no real number solution (no real roots) to the equation.

If the discriminant is a positive number, the roots are real numbers, and there are two real number solutions (two real roots) to the equation. In addition, if $b^2 - 4ac$ is a perfect square, the two real number solutions are rational numbers. If $b^2 - 4ac$ is not a perfect square, the two real number solutions are irrational. If the discriminant is equal to 0, there is one real number solution (one root) to the equation.

**Example 6** Determine whether $3x^2 = 5x - 4$ has real number solution(s).

**Solution** First, we place the equation in standard form. Then we need only check the discriminant.

$$3x^2 = 5x - 4$$
$$3x^2 - 5x + 4 = 0$$

$a = 3, \quad b = -5, \quad c = 4$    Substitute the values for $a$, $b$, and $c$ into the discriminant and evaluate.

$$b^2 - 4ac = (-5)^2 - 4(3)(4) = 25 - 48 = -23$$

The discriminant is negative. Thus $3x^2 = 5x - 4$ has no real number solution(s). ☐

**Student Practice 6** Determine whether each equation has real number solution(s).

(a) $5x^2 = 3x + 2$          (b) $2x^2 + 5 = 4x$

***How Do I Know Which Method to Use?*** This three-step plan can help you decide which method to use when you do not have specific directions to use a particular method.

1. Remove all parentheses. Clear any fractions by multiplying each term by the LCD. Combine like terms. Put the equation in the form $ax^2 + bx + c = 0$.
2. If you can quickly see a way to factor the expression, use the factoring method. For most students this is a faster way to do the problem.
3. If you are unable to factor the problem after a few minutes, then use the quadratic formula. The quadratic formula takes a little longer but it has the advantage that it works in all cases. The factoring method only works if the expression can be factored.

When you do exercises 35–46, use this three-step plan to assist you.

## 10.3 Exercises  MyMathLab®

**Verbal and Writing Skills, Exercises 1–4**

1. In the equation $3x^2 + 4x - 7 = 0$, what can we learn from the discriminant?

2. In the equation $7x^2 - 6x + 2 = 0$, what can we learn from the discriminant?

3. Is the equation $4x^2 = -5x + 6$ in standard form? If not, place it in standard form and find $a$, $b$, and $c$.

4. Is the equation $8x = -9x^2 + 2$ in standard form? If not, place it in standard form and find $a$, $b$, and $c$.

*Solve using the quadratic formula. If there are no real roots, say so.*

5. $x^2 + 4x + 1 = 0$

6. $x^2 - 2x - 4 = 0$

7. $x^2 - 3x - 8 = 0$

8. $x^2 - x - 8 = 0$

9. $4x^2 + 7x - 2 = 0$

10. $2x^2 + 12x - 5 = 0$

11. $2x^2 = 3x + 20$

12. $1 = 5x - 2x^2$

13. $6x^2 - 3x = 1$

14. $9x - 2 = 3x^2$

15. $x + \dfrac{3}{2} = 3x^2$

16. $\dfrac{2}{3}x^2 + x = \dfrac{1}{2}$

17. $\dfrac{x}{2} + \dfrac{5}{x} = \dfrac{7}{2}$

18. $\dfrac{1}{4} + \dfrac{5}{x} = \dfrac{x}{4}$

19. $5x^2 + 6x + 2 = 0$

20. $3x^2 - 4x + 2 = 0$

21. $4y^2 = 10y - 3$

22. $8y^2 = 8y - 1$

579

**23.** $\dfrac{d^2}{2} + \dfrac{5d}{6} - 2 = 0$

**24.** $\dfrac{k^2}{4} - \dfrac{5k}{6} = \dfrac{2}{3}$

**25.** $3x^2 - 2x + 5 = 0$

**26.** $4x^2 + 3x + 2 = 0$

**27.** $4x^2 - 12x + 9 = 0$

**28.** $9x^2 - 24x + 16 = 0$

*Use the quadratic formula to find the roots. Find a decimal approximation to the nearest thousandth.*

**29.** $x^2 + 5x - 2 = 0$

**30.** $x^2 + 8x - 8 = 0$

**31.** $2x^2 - 7x - 5 = 0$

**32.** $3x^2 + 4x - 6 = 0$

**33.** $5x^2 + 10x + 1 = 0$

**34.** $2x^2 + 9x + 2 = 0$

**Mixed Practice** *Solve for the variable. Choose the method you feel will work best.*

**35.** $6x^2 - 13x + 6 = 0$

**36.** $6x^2 - x - 2 = 0$

**37.** $3(x^2 + 1) = 10x$

**38.** $3 + x(x + 2) = 18$

**39.** $(t + 5)(t - 3) = 7$

**40.** $(b + 4)(b - 7) = 4$

**41.** $y^2 - \dfrac{2}{5}y = 2$

**42.** $y^2 - 3 = \dfrac{8}{3}y$

**43.** $3x^2 - 13 = 0$

**44.** $4x^2 + 7x = 0$

**45.** $x(x - 2) = 7$

**46.** $(4x - 1)^2 = 24$

**To Think About**

▲ **47.** *Swimming Pool Design* John and Chris Maney have designed a new in-ground swimming pool for their backyard. The pool is rectangular and measures 30 feet by 20 feet. They have also designed a tile border to go around the pool. In the diagram below, this border is $x$ feet wide. The tile border covers 216 square feet in area. What is the width $x$ of this tile border? (*Hint:* Find an expression for the area of the large rectangle and subtract an expression for the area of the small rectangle.)

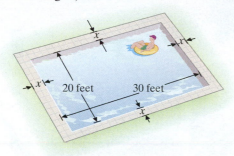

▲ **48.** *Lynn Campus Entrance* The Alumni Association of North Shore Community College has designed a new garden for the front entrance of the Lynn campus. It is rectangular and measures 40 feet by 30 feet. The association has designed a concrete walkway to go around this rectangular garden. In the diagram below, this concrete walkway is $x$ feet wide. The entire concrete walkway covers 456 square feet in area. What is the width $x$ of this concrete walkway? (*Hint:* Find an expression for the area of the large rectangle and subtract an expression for the area of the small rectangle.)

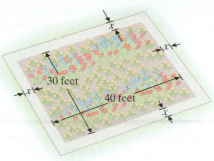

## Cumulative Review

**49.** **[4.6.2]** Divide. $(x^3 + 8x^2 + 17x + 6) \div (x + 3)$

**50.** **[7.4.2]** Find the equation of the line passing through $(-8, 2)$ and $(5, -3)$.

**51.** **[7.2.2]** Graph and label the $x$- and $y$-intercepts.
$2x + y = 4$

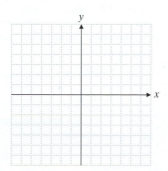

---

**Quick Quiz 10.3** *Solve by using the quadratic formula.*

**1.** $9x^2 + 6x - 1 = 0$

**2.** $4x^2 - 4x = 19$

**3.** $2 + x = 3x(x + 1)$

**4.** **Concept Check** Explain how you would determine if the quadratic equation $5x^2 - 8x + 9 = 0$ has real solutions or has no real solutions.

# Use Math to Save Money

**Did You Know?** You can save money by turning down your thermostat this winter.

## Adjust the Thermostat

### Understanding the Problem:

Mark heats his home with oil. He is interested in how much he needs to budget for the heating season and how changing the settings on his thermostat will affect that amount.

### Making a Plan:

Mark needs to calculate the cost of his current oil usage and potential savings.

**Step 1:** During the winter of 2009 Mark paid $2.95 per gallon for home heating oil. The price has since increased to $3.55 per gallon during the winter of 2015.

**Task 1:** The oil company will deliver only 100 gallons of heating oil or more. Determine the cost of 100 gallons in December 2009 and November 2015.

**Task 2:** Mark knows he will need a 100-gallon heating oil delivery every month during the winter (November and December 2015 and January, February, March 2016). What can Mark expect to pay in home heating oil costs for these five months of winter?

### Making a Decision:

**Step 2:** Mark knows that for every 1° change in his thermostat setting, he can see a 2% change in his utility bills.

**Task 3:** Mark typically keeps his thermostat at 72°. How much will he save on his heating costs if he turns the thermostat down to 68° for the entire winter?

**Task 4:** How low would Mark have to set his thermostat to save one month's worth of heating costs ($355)?

### Applying the Situation to Your Life:

If you lower your thermostat a few degrees, you too can realize savings on your heating bill. If you are not comfortable with the temperature set lower, you can still turn the heat down when you are away at work or school, as well as when you are sleeping. A programmable thermostat that will adjust the temperature depending on the time of day would make it easy to do that.

**Task 5:** At what temperature do you keep your thermostat?

**Task 6:** Calculate your savings if you lower your thermostat during the winter.

# 10.4 Graphing Quadratic Equations ▶

## 1 Graphing Equations of the Form $y = ax^2 + bx + c$ Using Ordered Pairs ▶

Recall that the graph of a linear equation is always a straight line. What is the graph of a quadratic equation? What does the graph look like if one of the variable terms is squared? We will use point plotting to look for a pattern. We begin with a table of values.

**Example 1** Graph.  (a) $y = x^2$    (b) $y = -x^2$

**Solution**

(a) $y = x^2$

| x | y |
|----|----|
| $-3$ | 9 |
| $-2$ | 4 |
| $-1$ | 1 |
| 0 | 0 |
| 1 | 1 |
| 2 | 4 |
| 3 | 9 |

(b) $y = -x^2$

| x | y |
|----|----|
| $-3$ | $-9$ |
| $-2$ | $-4$ |
| $-1$ | $-1$ |
| 0 | 0 |
| 1 | $-1$ |
| 2 | $-4$ |
| 3 | $-9$ |

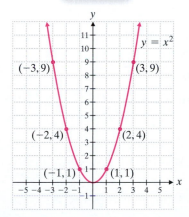

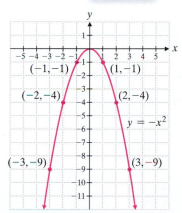

**Student Learning Objectives**

**After studying this section, you will be able to:**

**1** Graph equations of the form $y = ax^2 + bx + c$ using ordered pairs. ▶

**2** Graph equations of the form $y = ax^2 + bx + c$ using the vertex formula. ▶

▷ **Student Practice 1**   Graph.   (a) $y = 2x^2$    (b) $y = -2x^2$

(a) $y = 2x^2$

| x | y |
|----|----|
| $-2$ | |
| $-1$ | |
| 0 | |
| 1 | |
| 2 | |

(b) $y = -2x^2$

| x | y |
|----|----|
| $-2$ | |
| $-1$ | |
| 0 | |
| 1 | |
| 2 | |

**Student Practice 1**

(a)

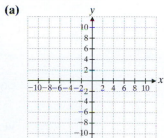

(b)

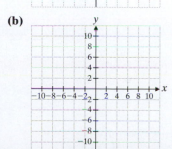

**To Think About:** *When the Coefficient of $x^2$ Is Negative*   What happens to the graph when the coefficient of $x^2$ is negative? What happens when the coefficient of $x^2$ is an integer greater than 1 or less than $-1$? Graph $y = 3x^2$ and $y = -3x^2$. What happens when $-1 < a < 1$? Graph $y = \frac{1}{3}x^2$ and $y = -\frac{1}{3}x^2$.

The curves that you have been graphing are called **parabolas.** What happens to the graph of a parabola as the quadratic equation that describes it changes?

In the preceding example, the equations we worked with were relatively simple. They provided us with an understanding of the general shape of the graph of a quadratic equation. We have seen how changing the equation affects the graph of that equation. Let's look at an equation that is slightly more difficult to graph.

The **vertex** is the lowest point on a parabola that opens upward, or the highest point on a parabola that opens downward.

**Example 2** Graph $y = x^2 - 2x$. Identify the coordinates of the vertex.

**Solution**

| x | y |
|----|----|
| -2 | 8 |
| -1 | 3 |
| 0 | 0 |
| 1 | -1 |
| 2 | 0 |
| 3 | 3 |
| 4 | 8 |

Vertex

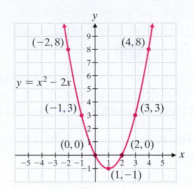

The vertex is at $(1, -1)$.

**Student Practice 2**

**(a)**

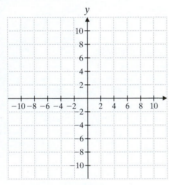

**(b)**

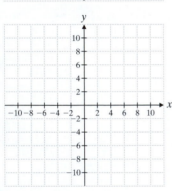

Notice that the $x$-intercepts are at $(0, 0)$ and $(2, 0)$. The $x$-coordinates are the solutions to the equation $x^2 - 2x = 0$.

$$x^2 - 2x = 0$$
$$x(x - 2) = 0$$
$$x = 0, \quad x = 2$$

**Student Practice 2** Graph. For each table of values, use $x$-values from $-4$ to $2$. Identify the coordinates of the vertex.

**(a)** $y = x^2 + 2x$

**(b)** $y = -x^2 - 2x$

**(a)**

| x | y |
|----|----|
| -4 | |
| -3 | |
| -2 | |
| -1 | |
| 0 | |
| 1 | |
| 2 | |

**(b)**

| x | y |
|----|----|
| -4 | |
| -3 | |
| -2 | |
| -1 | |
| 0 | |
| 1 | |
| 2 | |

**2 Graphing Equations of the Form $y = ax^2 + bx + c$ Using the Vertex Formula**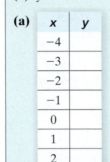

Finding the vertex is the key to graphing a parabola. Let's take a look at some equations and identify the vertex of each parabola.

Equation: $y = x^2$ $\quad y = x^2 - 3$ $\quad y = (x - 1)^2$ $\qquad y = (x + 2)^2$ $\qquad y = x^2 - 2x$

$\qquad\qquad\qquad\qquad\qquad\qquad\qquad\qquad y = x^2 - 2x + 1 \quad y = x^2 + 4x + 4$

Vertex: $\quad (0, 0)$ $\qquad (0, -3)$ $\qquad\quad (1, 0)$ $\qquad\qquad\quad (-2, 0)$ $\qquad\qquad (1, -1)$

Notice that the $x$-coordinate of the vertex of the parabola for each of the first two equations is 0. You may also notice that these two equations do not have middle terms. That is, for $y = ax^2 + bx + c, b = 0$. This is not true of the last three equations. As you may have already guessed, there is a relationship between the quadratic equation and the $x$-coordinate of the vertex of the parabola it describes.

> **VERTEX OF A PARABOLA**
>
> The $x$-coordinate of the vertex of the parabola described by $y = ax^2 + bx + c$ is
> $$x = \frac{-b}{2a}.$$

Once we have the $x$-coordinate, we can find the $y$-coordinate using the equation. We can plot the point and draw a general sketch of the curve. The sign of the $x^2$-term will tell us whether the graph opens upward or downward. That is, if $a$ is positive the parabola opens upward, and if $a$ is negative it opens downward.

**Example 3** $y = -x^2 - 2x + 3$. Determine the vertex and the $x$-intercepts. Then sketch the graph.

**Solution** We will first determine the coordinates of the vertex. We begin by finding the $x$-coordinate.

$$x = \frac{-b}{2a} = \frac{-(-2)}{2(-1)} = \frac{2}{-2} = -1$$

$y = -(-1)^2 - 2(-1) + 3$   Determine $y$ when $x$ is $-1$.

$\quad = 4$   The point $(-1, 4)$ is the vertex of the parabola.

Now we will determine the $x$-intercepts, those points where the graph crosses the $x$-axis. This occurs when $y = 0$. These are the solutions to the equation $-x^2 - 2x + 3 = 0$. Note that this equation is factorable.

$-x^2 - 2x + 3 = 0$   Find the solutions to the equation.

$0 = x^2 + 2x - 3$   It is easier to factor if the first term is positive.

$0 = (x - 1)(x + 3)$   Factor.

$x - 1 = 0 \quad x + 3 = 0$   Solve for $x$.

$x = 1, \quad\quad x = -3$   The $x$-intercepts are at $(1, 0)$ and at $(-3, 0)$.

We now have enough information to sketch the graph. Since $a = -1$, the parabola opens downward and the vertex $(-1, 4)$ is the highest point. The graph crosses the $x$-axis at $(-3, 0)$ and at $(1, 0)$.

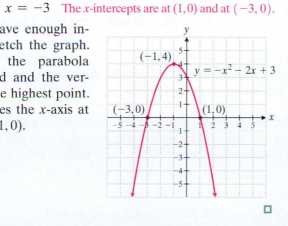

**Student Practice 3** $y = x^2 - 4x - 5$.
Determine the vertex and the $x$-intercepts. Then sketch the graph.

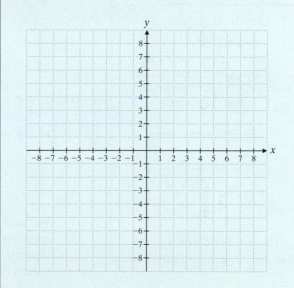

**Example 4** A person throws a ball into the air and its distance from the ground in feet is given by the equation $h = -16t^2 + 64t + 5$, where $t$ is measured in seconds.

(a) Graph the equation $h = -16t^2 + 64t + 5$.

(b) What physical significance does the vertex of the graph have?

(c) After how many seconds will the ball hit the ground?

*Continued on next page*

**Solution**

**(a)** Since the equation is a quadratic equation, the graph is a parabola. $a$ is negative. Thus the parabola opens downward and the vertex is the highest point. We begin by finding the coordinates of the vertex. Our equation, $h = -16t^2 + 64t + 5$, has ordered pairs $(t, h)$ instead of $(x, y)$.

The $t$-coordinate of the vertex is

$$t = \frac{-b}{2a} = \frac{-64}{2(-16)} = \frac{-64}{-32} = 2.$$

Substitute $t = 2$ into the equation to find the $h$-coordinate of the vertex.

$$h = -16(2)^2 + 64(2) + 5 = -64 + 128 + 5 = 69$$

The vertex $(t, h)$ is $(2, 69)$.

To sketch this graph, it would be helpful to know the $h$-intercept, that is, the point where $t = 0$.

$$h = -16t^2 + 64t + 5$$
$$h = -16(0)^2 + 64(0) + 5 = 5$$

The $h$-intercept is at $(0, 5)$.

The $t$-intercept is the point where $h = 0$. We will use the quadratic formula to find $t$ when $h$ is 0, that is, to find the roots of the quadratic equation $-16t^2 + 64t + 5 = 0$.

$$t = \frac{-64 \pm \sqrt{(64)^2 - 4(-16)(5)}}{2(-16)}$$

$$t = -\frac{64 \pm \sqrt{4096 + 320}}{-32} = -\frac{64 \pm \sqrt{4416}}{-32} = \frac{8 \pm \sqrt{69}}{4}$$

Using a calculator, we have the following.

$$t = \frac{8 + 8.307}{4} \approx 4.077, \qquad t = \frac{8 - 8.307}{4} \approx -0.077$$

$$t \approx 4.1 \quad (\text{to the nearest tenth})$$

We will not use negative $t$-values. We are studying the height of the ball from the time it is thrown ($t = 0$ seconds) to the time that it hits the ground (approximately $t = 4.1$ seconds). It is not useful to find points with negative values of $t$ because they have no situation.

**(b)** The vertex represents the greatest height of the ball. The ball is 69 feet above the ground 2 seconds after it is thrown. See the graph in the margin.

**(c)** The ball will hit the ground approximately 4.1 seconds after it is thrown. This is the point at which the graph intersects the $t$-axis, the positive root of the equation. See the graph in the margin. □

**⟹ Student Practice 4** A ball is thrown upward with a speed of 32 feet per second from a distance 10 feet above the ground. The height of the ball in feet is given by the equation $h = -16t^2 + 32t + 10$, where $t$ is measured in seconds.

**(a)** Graph the equation $h = -16t^2 + 32t + 10$.

**(b)** What is the greatest height of the ball?

**(c)** After how many seconds will the ball hit the ground?

**To Think About:** *What If the x-intercept Is the Vertex?* In some equations you will find that the $x$-intercept is the vertex of the parabola. In those cases, you may find it helpful to plot a few additional points to assist you with your graph. Remember this idea when you do exercises 27 and 28 in the following exercise set.

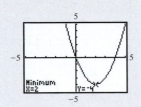

**Graphing Calculator**

**Quadratic Equations**

Graph $y = x^2 - 4x$.
Use the Trace and Zoom features. Use the feature that calculates the minimum point to find the vertex.

Thus the vertex is at $(2, -4)$.

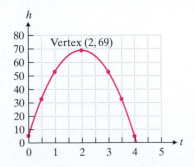

Vertex $(2, 69)$

**Student Practice 4**

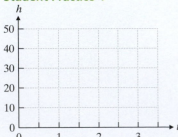

## 10.4 Exercises  MyMathLab®

**Verbal and Writing Skills, Exercises 1–4**

1. Nathaniel and Josiah wanted to know why the graph of $y = x^2 + 5$ does not have any $x$-intercepts. Can you explain why?

2. Stella said that a parabola can have zero, one, or two vertices. Is she correct?

3. When you have a parabola in the form $y = ax^2 + bx + c$ and you substitute $b = 0$ into the equation $x = \dfrac{-b}{2a}$, what result takes place?

4. If $a < 0$ in the parabola in the form $y = ax^2 + bx + c$, what property is observed?

*Graph each equation.*

5. $y = x^2 + 2$

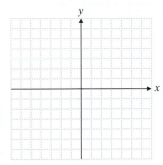

6. $y = x^2 + 4$

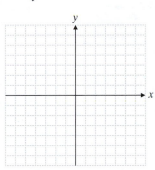

7. $y = -\frac{1}{3}x^2$

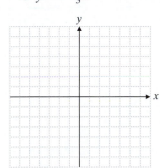

8. $y = -\frac{1}{4}x^2$

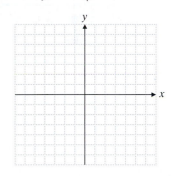

9. $y = 3x^2 - 1$

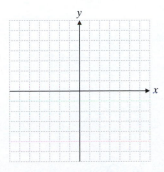

10. $y = 3x^2 + 2$

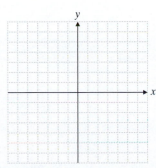

11. $y = (x - 2)^2$

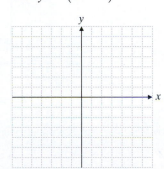

12. $y = (x + 1)^2$

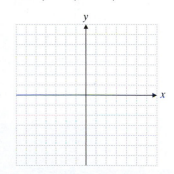

13. $y = -\frac{1}{2}x^2 + 4$

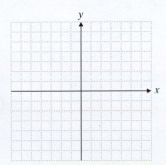

14. $y = -\frac{1}{3}x^2 + 6$

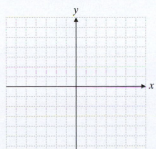

15. $y = \frac{1}{2}(x - 3)^2$

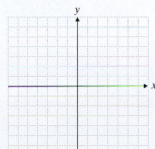

16. $y = \frac{1}{2}(x + 2)^2$

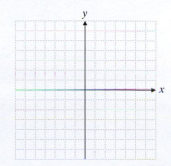

*Graph each equation. Identify the coordinates of the vertex.*

**17.** $y = x^2 + 4x$

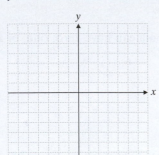

**18.** $y = -x^2 + 4x$

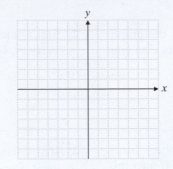

**19.** $y = -x^2 - 4x$

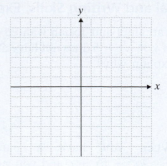

**20.** $y = x^2 - 4x$

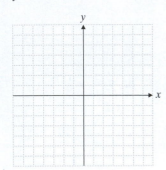

**21.** $y = -2x^2 + 8x$

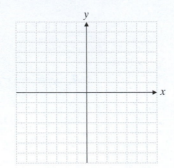

**22.** $y = x^2 + 6x$

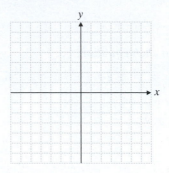

*Determine the vertex and the x-intercepts. Then sketch the graph.*

**23.** $y = x^2 + 2x - 3$

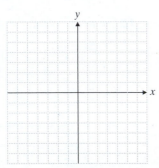

**24.** $y = x^2 - 4x + 3$

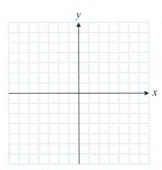

**25.** $y = -x^2 - 6x - 5$

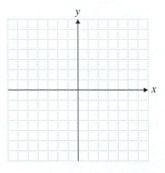

**26.** $y = -x^2 + 8x - 15$

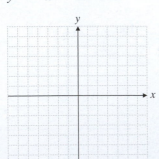

**27.** $y - 9 = x^2 + 6x$

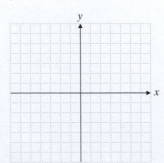

**28.** $y + 1 = 2x - x^2$

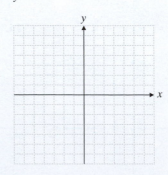

## Applications

**29. *Miniature Rockets*** A miniature rocket is launched so that its height $h$ in meters after $t$ seconds is given by the equation $h = -4.9t^2 + 39.2t + 4$.

(a) Draw a graph of the equation.

(b) How high is the rocket after 2 seconds?

(c) What is the maximum height the rocket will attain?

(d) How long after the launch will the rocket hit the ground? Round to the nearest tenth.

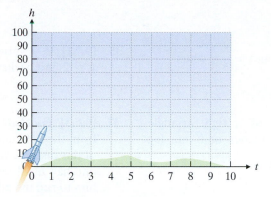

**30. *Miniature Rockets*** A miniature rocket is launched so that its height $h$ in meters after $t$ seconds is given by the equation $h = -4.9t^2 + 19.6t + 10$.

(a) Draw a graph of the equation.

(b) How high is the rocket after 3 seconds?

(c) What is the maximum height the rocket will attain?

(d) How long after the launch will the rocket hit the ground? Round to the nearest hundredth.

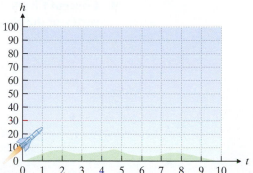

**31. *Mosquito Population*** The number of mosquitoes in some areas varies with the amount of rain. The number, $N$, of mosquitoes measured in millions in Hamilton, Massachusetts, in May can be predicted by the equation $N = 9x - x^2$ where $x$ is the number of inches of rain during that month. (*Source*: Massachusetts Environmental Protection Agency)

(a) Graph the equation.

(b) How many inches of rain produce the maximum number of mosquitoes in May?

(c) What is the maximum number of mosquitoes?

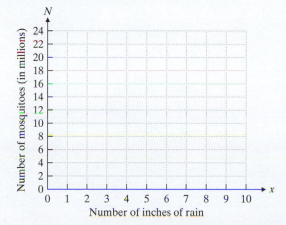

*Use your graphing calculator to find the vertex, the y-intercept, and the x-intercepts of the parabolas described by the following equations. Round to the nearest thousandth when necessary.*

**32.** $y = -164x^2 - 263x + 1235$

**33.** $y = -355x^2 - 182x + 1635$

## Cumulative Review

**34.** **[9.7.2]** If $y$ varies inversely as $x$, and $y = \frac{1}{3}$ when $x = 21$, find $y$ when $x = 7$.

**35.** **[9.7.1]** If $y$ varies directly as $x^2$, and $y = 12$ when $x = 2$ find $y$ when $x = 5$.

**36.** **[9.6.1]** Find the unknown leg of a right triangle that has hypotenuse of $\sqrt{130}$ inches and one leg 9 inches.

**37.** **[6.6.3]** *Football* A football player picks up a ball from the field and runs 60 yards. He runs the first half of this distance at 20 feet per second. He runs the second half at 24 feet per second. How long does it take him to run the 60 yards?

**Quick Quiz 10.4** *Consider the equation* $y = x^2 - 6x + 5$.

**1.** What is the vertex of this curve?

**2.** What are the $x$-intercepts of this curve?

**3.** Graph the equation.

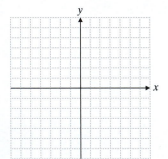

**4.** **Concept Check** Explain how you would find the vertex of the equation $y = 4x^2 + 16x - 2$.

## 10.5 Formulas and Applied Problems ▶

### 1 Solving Applied Problems Using Quadratic Equations ▶

**Student Learning Objective**

After studying this section, you will be able to:

1 Solve applied problems using quadratic equations. ▶

We can use what we have learned about quadratic equations to solve word problems. Some word problems will involve geometry.

Recall that the Pythagorean Theorem states that in a right triangle, $a^2 + b^2 = c^2$, where $a$ and $b$ are the lengths of the legs and $c$ is the length of the hypotenuse. We will use this theorem to solve word problems involving right triangles.

▲ **Example 1** The hypotenuse of a right triangle is 25 meters in length. One leg is 17 meters longer than the other. Find the length of each leg.

**Solution**

1. **Understand the problem.** Draw a picture.

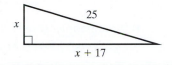

2. **Write an equation.** The problem involves a right triangle. Use the Pythagorean Theorem.

$$a^2 + b^2 = c^2$$
$$x^2 + (x + 17)^2 = 25^2$$

3. **Solve and state the answer.**

$$x^2 + (x + 17)^2 = 25^2$$
$$x^2 + x^2 + 34x + 289 = 625$$
$$2x^2 + 34x - 336 = 0$$
$$x^2 + 17x - 168 = 0$$
$$(x + 24)(x - 7) = 0$$
$$x + 24 = 0 \qquad x - 7 = 0$$
$$x = -24 \qquad x = 7$$

(Note that $x = -24$ is not a valid solution for this particular word problem.) One leg is 7 meters in length. The other leg is $x + 17 = 7 + 17 = 24$ meters in length.

4. **Check.** Check the conditions in the original problem.

Do the two legs differ by 17 meters?

$$24 - 7 \overset{?}{=} 17 \qquad 17 = 17 \ \checkmark$$

Do the sides of the triangle satisfy the Pythagorean Theorem?

$$7^2 + 24^2 \overset{?}{=} 25^2$$
$$49 + 576 \overset{?}{=} 625$$
$$625 = 625 \ \checkmark$$

▲ ✏ **Student Practice 1** The hypotenuse of a right triangle is 30 meters in length. One leg is 6 meters shorter than the other leg. Find the length of each leg.

Sometimes it may help to use two variables in the initial part of the problem. Then one variable can be eliminated by substitution.

**Example 2** The ski club rented a bus to travel to Mount Snow. The members agreed to share the cost of $180 equally. On the day of the ski trip, three members were sick with the flu and could not go. This increased the share of each person going on the trip by $10. How many people originally planned to attend?

**Solution**

1. *Understand the problem.*

$$\text{Let } s = \text{the number of students in the ski club}$$
$$c = \text{the cost for each student in the original group}$$

2. *Write an equation(s).*

$$\text{number of students} \times \text{cost per student} = \text{total cost}$$
$$s \cdot c = 180 \qquad \textbf{(1)}$$

If three people were sick, the number of students dropped by three, but the cost for each increased by $10. The total is still $180. Therefore, we have the following.

$$(s - 3)(c + 10) = 180 \qquad \textbf{(2)}$$

3. *Solve and state the answer.*

$$c = \frac{180}{s} \qquad \text{Solve equation \textbf{(1)} for } c.$$

$$(s - 3)\left(\frac{180}{s} + 10\right) = 180 \qquad \text{Substitute } \frac{180}{s} \text{ for } c \text{ in equation \textbf{(2)}.}$$

$$(s - 3)\left(\frac{180 + 10s}{s}\right) = 180 \qquad \text{Add } \frac{180}{s} + 10. \text{ Use the LCD, which is } s.$$

$$(s - 3)(180 + 10s) = 180s \qquad \text{Multiply both sides of the equation by } s.$$

$$10s^2 + 150s - 540 = 180s \qquad \text{Multiply the binomials.}$$

$$10s^2 - 30s - 540 = 0$$

$$s^2 - 3s - 54 = 0$$

$$(s - 9)(s + 6) = 0$$

$$s = 9, \qquad s = -6$$

The number of students originally going on the ski trip was 9.

4. *Check.* Would the cost increase by $10 if the number of students dropped from nine to six? Nine people in the club would mean that each would pay $20 $(180 \div 9 = 20)$.

   If the number of students was reduced by three, there were six people who took the trip. Their cost was $30 each $(180 \div 6 = 30)$. The increase is $10.

$$20 + 10 \overset{?}{=} 30$$
$$30 = 30 \quad \checkmark \qquad \qquad \square$$

▶ **Student Practice 2** Several friends from the University of New Mexico decided to charter a sport fishing boat for the afternoon while on a vacation at the coast. They agreed to share the rental cost of $240 equally. At the last minute, two of them could not go on the trip. This raised the cost for each person in the group by $4. How many people originally planned to charter the boat?

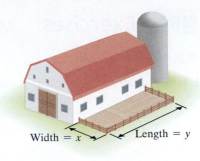

▲ **Example 3** Minette is fencing a garden that borders the back of a large barn. She has 120 feet of fencing. She would like a rectangular garden that measures 1350 square feet in area. She wants to use the back of the barn, so she needs to use fencing on three sides only. What dimensions should she use for her garden?

**Solution**

1. *Understand the problem.* Draw a picture.

   Let $x = $ the width of the garden in feet and

   $y = $ the length of the garden in feet.

Width = $x$     Length = $y$

2. *Write an equation(s).* Look at the drawing. The 120 feet of fencing will be needed for the width twice (two sides) and the length once (one side).

$$120 = 2x + y \quad \textbf{(1)}$$

The area formula is $A = (\text{width})(\text{length})$.

$$1350 = xy \quad \textbf{(2)}$$

3. *Solve and state the answer.*

$$y = 120 - 2x \qquad \text{Solve for } y \text{ in equation \textbf{(1)}.}$$
$$1350 = x(120 - 2x) \qquad \text{Substitute for } y \text{ in equation \textbf{(2)}.}$$
$$1350 = 120x - 2x^2 \qquad \text{Solve for } x.$$
$$2x^2 - 120x + 1350 = 0$$
$$x^2 - 60x + 675 = 0$$
$$(x - 15)(x - 45) = 0$$
$$x = 15, \qquad x = 45$$

To find the dimensions of the garden we can substitute $x = 15$ and $x = 45$ into either equation **(1)** or **(2)** to find $y$. We will use equation **(1)**.

**First Solution:** If the width is 15 feet, the length is 90 feet.

**Check.**
$$15 + 15 + 90 = 120 \quad \checkmark$$
$$(15)(90) = 1350 \quad \checkmark$$

**Second Solution:** If the width is 45 feet, the length is 30 feet.

**Check.**
$$45 + 45 + 30 = 120 \quad \checkmark$$
$$(45)(30) = 1350 \quad \checkmark$$

First solution

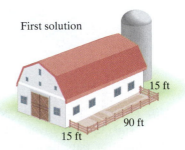

15 ft, 90 ft, 15 ft

Thus we see that Minette has two choices that satisfy the given requirements (see the figures). She would probably choose the shape that is the most practical for her. (If her barn is shorter than 90 feet in length, she could not use the first solution!)

*Remark:* Is there another way to do this problem? Yes. Use only the variable $x$ to represent the length and width. If you let $x = $ the width and $120 - 2x = $ the length, you can immediately write equation **(3)**, which is

Second solution

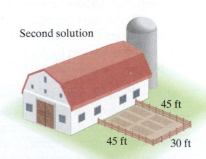

45 ft, 45 ft, 30 ft

$$1350 = x(120 - 2x). \qquad \textbf{(3)}$$

However, many students find that this is difficult to do, and they would prefer to use two variables. Work these problems in the way that is easiest for you. ☐

▲ **Student Practice 3** An Arizona rancher has a small fenced-in area against a canyon wall. Therefore, he only needs three sides of fencing to make a rectangular holding area for cattle. He has 160 feet of fencing available. He wants the rectangular area to measure 3150 square feet. What dimensions should he use for the rectangular holding area?

## 10.5 Exercises   MyMathLab®

**Applications** *Solve.*

▲ 1. ***Geometry*** A triangle is used in a sign. The base of a triangle is 5 centimeters shorter than the altitude. The area of the triangle is 88 square centimeters. Find the lengths of the base and altitude.

▲ 2. ***Geometry*** A garden park is in the shape of a right triangle. One leg of the right triangle is 7 feet longer than the other. The hypotenuse is 17 feet in length. Find the length of each leg.

▲ 3. ***Garden Expansion*** Michael and Diane wish to expand the size of their front garden, which is square. They measure the front yard and decide they can increase the width by 4 feet and the length by 9 feet. If the area of the newly expanded garden is 266 square feet, what were the dimensions of the old garden?

▲ 4. ***Addition to a House*** Tawan and Keisha have a home with a very small rectangular room on the first floor that they use for a living room. This room is 2 feet longer than it is wide. They have decided to put on an addition to the house that will give them a larger rectangular room to meet the need for an adequate living room for their family. The length of the living room will increase by 10 feet. The width of the living room will increase by 8 feet. The area of the new, expanded living room will be 252 square feet. What are the dimensions of the present living room before the addition is built?

5. ***Gift Expense*** To help out Nina and Tom with their new twins, the employees at Tom's office decided to give the new parents a gift certificate worth $420 to furnish the babies' room. The employees planned to share the cost equally. At the last minute, seven people from the executive staff chipped in, lowering the cost per person by $5. How many people originally planned to give the gift?

6. ***Septic Systems*** Everyone in the School Street Townhouse Association agreed to equally share the cost of a new $6000 septic system. However, before the system was installed, five additional units were built and the five new families were asked to contribute their share. This lowered the contribution of each family by $200. How many families originally agreed to share the cost of the new septic system?

7. ***Music Cost*** A fraternity was sponsoring a spring fling, for which the $400 DJ cost was to be split evenly among those members attending. If 20 more members than were expected came, reducing the cost per person by $1, find the number of fraternity members that attended.

8. ***Beach House Rental*** The Cornell University *Wild Roses* Ultimate team took a spring break trip to Savannah, Georgia, where they all agreed to chip in and rent a beach house for the week. The cost of the rental for the week was $1500, and each girl was assessed an equal amount. Just before they left New York, three more girls decided to go. The captain said that each girl would now pay $25 less. How many girls originally planned to go on the trip?

▲ 9. *Security Services* A security service is enclosing a rectangular area on one side of Fenway Park in Boston. Three sides of fencing are used, since the fourth side of the area is formed by a building. The enclosed area measures 1250 square feet. Exactly 100 feet of fencing is used to fence in three sides of this rectangle. What are the possible dimensions that could have been used to construct this area?

▲ 10. *Landscaping* Christopher Smith is a landscaper who needs to fence in a garden, to protect it from the rabbits who love to feast on the lettuce. He plans on planting the garden next to the barn, so that he only needs to fence in three sides of the garden. If Christopher has 80 feet of fencing available, and he wants the garden to have an area of 750 square feet, what are the possible dimensions for the garden?

11. *Jet Test Pilot* A pilot is testing a new experimental craft. The cruising speed is classified information. In a recent test over the desert, the jet traveled 2400 miles. The pilot revealed that if he had increased his speed 200 mph, the total trip would have taken 1 hour less. Determine the cruising speed of the jet.

▲ 12. *Building Contractor* A gymnasium floor is being covered by square shock-absorbing tiles. The old gym floor required 864 square tiles. The new tiles are 2 inches larger in both length and width than the old tiles. The contractor says the new flooring will require only 600 tiles. What is the length of a side of one of the new shock-absorbing tiles? (*Hint*: Since the area is the same in each case, write an expression for area with old tiles and one with new tiles. Set the two expressions equal to each other.)

*Use the following information to solve exercises 13–18.*

*Business Expansion* Two college graduates opened a chain of Quick Print stores in southern California. The chain rapidly expanded, and many new stores were opened. The number of Quick Print Centers in operation during any year from 2008 to 2015 is given by the equation $y = -2.5x^2 + 22.5x + 50$, where $x$ is the number of years since 2008.

13. How many Quick Print Centers were in operation in 2008?

14. How many Quick Print Centers were in operation in 2014?

15. How many more Quick Print Centers were in operation in 2012 than in 2010?

16. How many fewer Quick Print Centers were in operation in 2015 than in 2013?

17. During what two years were there 100 Quick Print Centers?

18. During what two years were there 85 Quick Print Centers?

*Cannon Shell Path* Big Bertha, *an infamous World War I cannon, fired shells over an amazingly long distance. The path of a shell from this monstrous cannon can be modeled by the formula* $y = -0.018x^2 + 1.336x$, *where $x$ is the horizontal distance the cannon shell traveled in miles and $y$ is the height in miles the shell would reach on its flight.*

19. How far could *Big Bertha* send a shell? Round to the nearest whole number.

20. (a) How far into its journey would the shell achieve maximum height? Round to the nearest whole number.

(b) What was the maximum height in miles? Round to the nearest whole number.

## Cumulative Review

**21.** [2.6.4] Graph on a number line the solution to $3x + 11 \geq 9x - 4$.

**22.** [9.5.3] Rationalize the denominator. $\dfrac{\sqrt{3}}{\sqrt{5} + 2}$

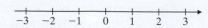

▲ **23.** [3.5.3] *Recycling Bin* A recycling bin that has a length of 3 feet, a width of 2 feet, and a height of 2 feet has three separate sections inside for sorting. Section 1 is for paper, section 2 is for glass and plastic, and section 3 is for metal. Sections 1 and 2 have identical measurements, but section 3 has $\frac{2}{3}$ the volume of section 1 or 2. If all three sections have the same width and height, how long is section 3?

### Quick Quiz 10.5

**1.** The hypotenuse of a right triangle is 13 yards long. One leg is 7 yards longer than the other leg. Find the length of each leg.

*When an object is thrown upward, its height (S) in meters is given approximately by the quadratic equation*
$$S = -5t^2 + vt + h$$

where $v$ = the initial upward velocity in meters per second,

$t$ = the time of flight in seconds, and

$h$ = the height above level ground from which the object is thrown.

*Suppose a ball is thrown upward with a velocity of 25 meters per second at a height of 30 meters above the ground.*

**2.** How long after it is thrown will the ball hit the ground?

**3.** What is the maximum height of the ball?

**4.** **Concept Check** The hiking club needs to raise $720 through dues paid by each member. However, 4 people dropped out of the club so the dues went up by $6 for each member. Explain how you could set up an equation to find out how many members were in the club prior to the dropout.

## Background: Civil Engineer

Michelle is a civil engineer working as the project manager supervising the repair and maintenance of a large suspension toll bridge. Her inspection of the bridge's cable system reveals the need to make a number of improvements, two of which she hopes to quickly complete so she can devote her attention to the creation of a much-needed parking lot near the bridge's entrance.

## Facts

- Cable joining two towers of the suspension bridge is represented by the equation

$$y = 0.013x^2 - 1.872x + 80,$$

where $y$ is the height in feet of the cable from the roadway and $x$ is the distance in feet from the left tower.
- The rectangular-shaped parking lot is located before the on-ramp to the bridge. The entire length of the lot's entrance will be edged by security barriers, and the perimeter of the remaining three sides totals 220 feet.

## Tasks

1. Michelle has decided that safety signals must be placed on the roadway where the height of the cable is at its lowest point from the roadway. Where in relation to the left tower is this, and how far from the roadway is the cable? (When necessary, round final answers to the nearest whole number.)

2. Michelle's inspection revealed that vertical bracing should be installed where the cable is 30 feet from the roadway. Where in relation to the left tower is the cable 30 feet from the roadway? (Round the final answers to the nearest whole number.)

3. Michelle needs to begin work constructing the rectangular-shaped parking lot. What dimensions of this rectangular parking lot will maximize its area?

# Chapter 10 Organizer

| Topic and Procedure | Examples | ▶ You Try It |
|---|---|---|
| **Solving quadratic equations by factoring, p. 562**<br><br>1. Clear the equation of fractions, if necessary.<br>2. Write as $ax^2 + bx + c = 0$.<br>3. Factor.<br>4. Set each factor equal to 0.<br>5. Solve the resulting equations. | Solve. $x + \dfrac{1}{2} = \dfrac{3}{2x}$<br><br>Multiply each term by LCD $= 2x$.<br><br>$$2x(x) + 2x\left(\dfrac{1}{2}\right) = 2x\left(\dfrac{3}{2x}\right)$$<br>$$2x^2 + x = 3$$<br>$$2x^2 + x - 3 = 0$$<br>$$(2x + 3)(x - 1) = 0$$<br>$$2x + 3 = 0 \qquad x - 1 = 0$$<br>$$2x = -3 \qquad x = 1$$<br>$$x = -\dfrac{3}{2}$$ | **1.** Solve. $x + \dfrac{1}{2x} = \dfrac{9}{4}$ |
| **Solving quadratic equations: taking the square root of each side of the equation, p. 569**<br><br>Begin solving quadratic equations of the form $ax^2 - c = 0$ by solving for $x^2$. Then use the property that if $x^2 = a$, $x = \pm\sqrt{a}$. This amounts to taking the square root of each side of the equation. | Solve by using the square root property.<br>$2x^2 + 1 = 99$<br><br>$$2x^2 = 99 - 1$$<br>$$2x^2 = 98$$<br>$$\dfrac{2x^2}{2} = \dfrac{98}{2}$$<br>$$x^2 = 49$$<br>$$x = \pm\sqrt{49} = \pm 7$$ | **2.** Solve by using the square root property. $3x^2 - 4 = 104$ |
| **Solving quadratic equations: completing the square, p. 570**<br><br>1. Put the equation in the form $ax^2 + bx = c$. If $a \neq 1$, divide each term by $a$.<br>2. Square $\dfrac{b}{2}$ and add the result to both sides of the equation.<br>3. Factor the left side (a perfect-square trinomial).<br>4. Use the square root property.<br>5. Solve the equations.<br>6. Check the solutions in the original equation. | Solve by completing the square.<br>$5x^2 - 3x - 7 = 0$<br><br>$$5x^2 - 3x = 7$$<br>$$x^2 - \dfrac{3}{5}x = \dfrac{7}{5}$$<br>$$x^2 - \dfrac{3}{5}x + \dfrac{9}{100} = \dfrac{7}{5} + \dfrac{9}{100}$$<br>$$\left(x - \dfrac{3}{10}\right)^2 = \dfrac{149}{100}$$<br>$$x - \dfrac{3}{10} = \pm\dfrac{\sqrt{149}}{10}$$<br>$$x = \dfrac{3}{10} \pm \dfrac{\sqrt{149}}{10}$$<br>$$x = \dfrac{3 \pm \sqrt{149}}{10}$$ | **3.** Solve by completing the square.<br>$2x^2 + x - 6 = 0$ |
| **Solving quadratic equations: quadratic formula, p. 575**<br><br>1. Put the equation in standard form: $ax^2 + bx + c = 0$.<br>2. Carefully determine the values of $a, b$, and $c$.<br>3. Substitute these into the formula<br>$$x = \dfrac{-b \pm \sqrt{b^2 - 4ac}}{2a}.$$<br>4. Simplify. | Solve using the quadratic formula.<br>$3x^2 - 4x - 8 = 0$<br>$a = 3, \; b = -4, \; c = -8$<br>$$x = \dfrac{-(-4) \pm \sqrt{(-4)^2 - 4(3)(-8)}}{2(3)}$$<br>$$= \dfrac{4 \pm \sqrt{16 + 96}}{6}$$<br>$$= \dfrac{4 \pm \sqrt{112}}{6} = \dfrac{4 \pm \sqrt{16 \cdot 7}}{6}$$<br>$$= \dfrac{4 \pm 4\sqrt{7}}{6} = \dfrac{2(2 \pm 2\sqrt{7})}{2 \cdot 3} = \dfrac{2 \pm 2\sqrt{7}}{3}$$ | **4.** Solve using the quadratic formula.<br>$5x^2 + 2x - 4 = 0$ |
| **The discriminant, pp. 577–578**<br><br>The *discriminant* in the quadratic formula is $b^2 - 4ac$. If it is negative, the quadratic equation has no real solutions. If the discriminant is positive, there are two real solutions. If the discriminant is equal to 0, there is one real solution. | What kinds of solutions does this equation have?<br>$3x^2 + 2x + 1 = 0$<br><br>$$b^2 - 4ac = 2^2 - 4 \cdot 3 \cdot 1$$<br>$$= 4 - 12 = -8$$<br><br>This equation has no real solutions. | **5.** What kinds of solutions does this equation have? $x^2 - 2x + 1 = 0$ |

598

| Topic and Procedure | Examples | ➡ You Try It |
|---|---|---|
| **Properties of parabolas, pp. 584–585** <br><br> 1. The graph of $y = ax^2 + bx + c$ is a parabola. It opens up if $a > 0$ and down if $a < 0$. <br> 2. The vertex of a parabola is its lowest (if $a > 0$) or highest (if $a < 0$) point. <br><br> Its $x$-coordinate is $x = \dfrac{-b}{2a}$. | The vertex of $y = 3x^2 + 4x - 11$ has $x$-coordinate <br><br> $$x = \frac{-b}{2a} = \frac{-4}{6} = -\frac{2}{3}.$$ <br><br> The parabola opens upward. | **6.** The graph of $y = -x^2 + 6x - 5$ is a parabola. <br><br> **(a)** Determine the vertex. <br> **(b)** Does the parabola open upward or downward? |
| **Graphing parabolas, p. 585** <br><br> Graph the vertex of a parabola and several other points to get a good idea of its graph. Usually, we find the $x$-intercepts if any exist. The following procedure is helpful. <br><br> 1. Determine if $a < 0$ or $a > 0$. <br> 2. Find the vertex. <br> 3. Find the $x$-intercepts. <br> 4. Plot one or two extra points if necessary. <br> 5. Draw the graph. | Graph. $y = x^2 - 2x - 8$ <br> $a = 1, b = -2, c = -8; a > 0$, so the parabola opens upward. <br> The $x$-coordinate of the vertex is <br> $$x = \frac{-b}{2a} = \frac{-(-2)}{2} = 1.$$ <br> If $x = 1, y = (1)^2 - 2(1) - 8 = -9.$ <br> The vertex is at $(1, -9)$. <br> To find the $x$-intercepts, let $y = 0$. <br> $$0 = x^2 - 2x - 8$$ <br> $$0 = (x - 4)(x + 2)$$ <br> $$x = 4, \quad x = -2$$ <br> The $x$-intercepts are $(4, 0)$ and $(-2, 0)$. <br> 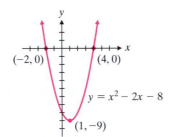 | **7.** Graph. $y = x^2 - 3x - 10$ <br> 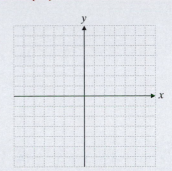 |
| **Formulas and applied problems, p. 591** <br><br> Some word problems require the use of a quadratic equation. You may want to follow this four-step procedure for problem solving. <br><br> 1. Understand the problem. <br> 2. Write an equation. <br> 3. Solve and state the answer. <br> 4. Check. | The area of a rectangle is 48 square inches. The length is three times the width. Find the dimensions of the rectangle. <br><br> 1. Draw a picture. <br> 2. $(3w)w = 48$ <br> 3. $3w^2 = 48$ <br>     $w^2 = 16$ <br>     $w = 4$ inches, $l = 12$ inches <br> 4. $(4)(12) = 48$ | **8.** The length of a rectangle is one more than twice the width. The area is 55 square meters. Find the dimensions of the rectangle. |

# Chapter 10 Review Problems

## Section 10.1

*Write in standard form. Determine the values of a, b, and c. Do not solve.*

**1.** $9x^2 + 3x = -5x^2 + 16$ 　　　　　　　　　　 **2.** $3x(4x + 1) = -x(5 - x)$

**3.** $\dfrac{3}{x^2} - \dfrac{6}{x} - 2 = 0$

**4.** $\dfrac{x}{x+2} - \dfrac{3}{x-2} = 5$

*Place in standard form. Solve by factoring.*

**5.** $x^2 + 26x + 25 = 0$

**6.** $x^2 + 6x - 55 = 0$

**7.** $(y + 4)(y + 1) = 18$

**8.** $9x^2 - 24x + 16 = 0$

**9.** $6x^2 = 13x - 5$

**10.** $x^2 + \dfrac{1}{6}x - 2 = 0$

**11.** $\dfrac{1}{2}x^2 = \dfrac{3}{4}x - \dfrac{1}{4}$

**12.** $1 + \dfrac{13}{12x} - \dfrac{1}{3x^2} = 0$

**13.** $x^2 + 9x + 5 = x + 5$

**14.** $x^2 - 2x + 7 = 7 + x$

**15.** $2 - \dfrac{5}{x+1} = \dfrac{3}{x-1}$

**16.** $5 + \dfrac{24}{2-x} = \dfrac{24}{2+x}$

## Section 10.2

*Solve by taking the square root of each side.*

**17.** $x^2 - 8 = 41$

**18.** $x^2 + 11 = 50$

**19.** $3x^2 + 6 = 60$

**20.** $2x^2 - 5 = 43$

**21.** $(x - 4)^2 = 7$

**22.** $(x - 2)^2 = 3$

**23.** $(3x + 2)^2 = 28$

**24.** $(2x - 1)^2 = 32$

*Solve by completing the square.*

**25.** $x^2 + 8x + 7 = 0$

**26.** $x^2 + 14x + 33 = 0$

**27.** $-5x^2 + 30x - 35 = 0$

**28.** $3x^2 + 6x - 6 = 0$

**29.** $2x^2 + 10x - 3 = 0$

## Section 10.3

*Solve using the quadratic formula.*

**30.** $x^2 + 4x - 6 = 0$

**31.** $x^2 + 4x - 8 = 0$

**32.** $2x^2 - 7x + 4 = 0$

**33.** $2x^2 + 5x - 6 = 0$

**34.** $3x^2 - 8x - 4 = 0$     **35.** $4x^2 - 2x + 11 = 0$     **36.** $5x^2 - x = 2$     **37.** $3x^2 + 3x = 1$

## Mixed Practice

*Solve by any method. If there is no real number solution, say so.*

**38.** $2x^2 - 9x + 10 = 0$     **39.** $4x^2 - 4x - 3 = 0$     **40.** $25x^2 + 10x + 1 = 0$     **41.** $2x^2 - 11x + 12 = 0$

**42.** $3x^2 - 6x + 2 = 0$     **43.** $5x^2 - 7x = 8$     **44.** $4x^2 + 4x = x^2 + 5$     **45.** $5x^2 + 7x + 1 = 0$

**46.** $x^2 = 9x + 3$     **47.** $12x^2 + 3x + 10 = 0$     **48.** $2x^2 - 1 = 35$     **49.** $\dfrac{(y + 2)^2}{5} + 2y = -9$

**50.** $3x^2 + 1 = 6 - 8x$     **51.** $8y(y + 1) = 7y + 9$     **52.** $\dfrac{y^2 + 5}{2y} = \dfrac{2y - 1}{3}$     **53.** $\dfrac{5x^2}{2} = x - \dfrac{7x^2}{2}$

## Section 10.4

*Graph each quadratic equation. Label the vertex.*

**54.** $y = 2x^2$                          **55.** $y = x^2 + 4$                        **56.** $y = x^2 - 3$

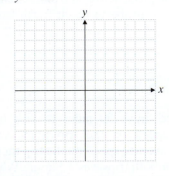     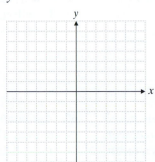    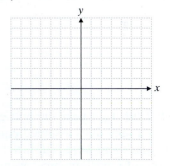

**57.** $y = -\dfrac{1}{2}x^2$                **58.** $y = x^2 - 3x - 4$            **59.** $y = \dfrac{1}{2}x^2 - 2$

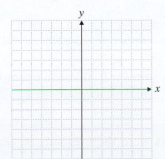     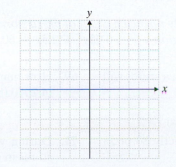    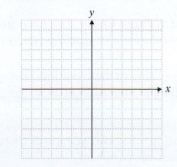

601

*Determine the vertex and the x-intercepts. Then sketch the graph.*

**60.** $y = -x^2 + 4x + 5$

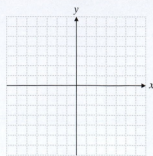

**61.** $y = -x^2 - 6x - 8$

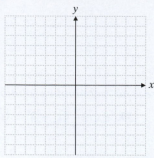

## Section 10.5

*Solve.*

▲ **62.** ***Dog Pen Fencing*** Ken is building a dog pen against his house. He has 15 feet of fencing and wants to pen in an area of 28 square feet. What are the possible dimensions of the pen?

▲ **63.** ***Geometry*** The hypotenuse of a right triangle is 20 centimeters. One leg of the right triangle is 4 centimeters shorter than the other. What is the length of each leg of the right triangle?

▲ **64.** ***TV Screen Dimensions*** A television designer believes that the dimensions (length and width) of a TV screen are as important to sales as the size (measured by the diagonal) of the TV. If the diagonal of a newly designed TV is 30 inches and the length is 6 inches more than the width, what are the dimensions of the TV?

▲ **65.** ***Space Probe*** Suppose an advanced alien civilization sends a space ship to orbit around our planet. They send a probe to Earth to do research. The equation for the height of the probe is $h = -225t^2 + 202,500$, where $h$ is the distance in feet from the surface of Earth $t$ seconds after the probe leaves the mother ship. How long will it take the probe to reach Earth from the mother ship?

**66.** ***Golf Club*** Last year the golf club on campus raised $720 through dues assigned equally to each member. This year there were four fewer members. As a result, the dues for each member went up by $6. How many members were in the golf club last year?

**67.** ***Automobile Trip*** Alice drove 90 miles to visit her cousin. Her average speed on the trip home was 15 mph faster than her average speed on the trip there. Her total travel time for both trips was 3.5 hours. What was her average rate of speed on each trip?

*Use the following information to solve exercises 68–71.*

***Lacorazza Pizza Chain*** *The Lacorazza family opened a chain of pizza stores in 1980. The number of pizza stores in operation during any year is given by the equation $y = -0.05x^2 + 2.5x + 60$, where x is the number of years since 1980.*

**68.** According to the equation, how many pizza stores will still be in operation in 2030?

**69.** How many fewer pizza stores will still be in operation in 2040 than in 2030?

**70.** During what two years will there be exactly 80 pizza stores in operation?

**71.** During what two years were there exactly 90 pizza stores in operation?

**602**

# How Am I Doing? Chapter 10 Test

**MATH COACH**     **MyMathLab®**     **You Tube**

After you take this test read through the Math Coach on pages 604–605. Math Coach videos are available via MyMathLab and YouTube. Step-by-step test solutions on the Chapter Test Prep Videos are also available via MyMathLab and YouTube. (Search "TobeyBeginningAlg" and click on "Channels.")

*Solve by any desired method. If there is no real number solution, say so.*

**MC 1.** $5x^2 + 7x = 4$

**2.** $3x^2 + 13x = 10$

**MC 3.** $2x^2 = 2x - 5$

**4.** $4x^2 - 19x = 4x - 15$

**5.** $12x^2 + 11x = 5$

**6.** $18x^2 + 32 = 48x$

**7.** $2x^2 - 11x + 3 = 5x + 3$

**8.** $5x^2 + 7 = 52$

**MC 9.** $2x(x - 6) = 6 - x$

**10.** $x^2 - x = \dfrac{3}{4}$

*Graph each quadratic equation. Locate the vertex.*

**11.** $y = 3x^2 - 6x$

**MC 12.** $y = -x^2 + 8x - 12$

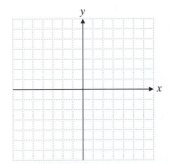

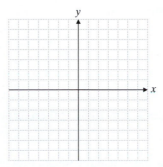

*Solve.*

▲ **13.** The hypotenuse of a right triangle is 15 meters in length. One leg is 3 meters longer than the other leg. Find the length of each leg.

**14.** When an object is thrown upward, its height ($S$) in meters is given (approximately) by the quadratic equation

$$S = -5t^2 + vt + h,$$

where $v =$ the initial upward velocity in meters per second,

$t =$ the time of flight in seconds, and

$h =$ the height above level ground from which the object is thrown.

Suppose that a ball is thrown upward with a velocity of 33 meters/second at a height of 14 meters above the ground. How long after it is thrown will it strike the ground?

1. _____ ☐

2. _____ ☐

3. _____ ☐

4. _____ ☐

5. _____ ☐

6. _____ ☐

7. _____ ☐

8. _____ ☐

9. _____ ☐

10. _____ ☐

11. _____ ☐

12. _____ ☐

13. _____ ☐

14. _____ ☐

**Total Correct:** ___

# MATH COACH

*Mastering the skills you need to do well on the test.*

Students often make the same types of errors when they do the Chapter 10 Test. Here are some helpful hints to keep you from making those common errors on test problems.

**Chapter 10 Test, Problem 1**   Solve by any desired method. If there is no real number solution, say so. $5x^2 + 7x = 4$

> **HELPFUL HINT** We often use the quadratic formula when other methods of factoring, square roots, or completing the square are not practical. Make sure that you rewrite the equation in $ax^2 + bx + c = 0$ form before completing any other steps.

Did you rewrite the equation as $5x^2 + 7x - 4 = 0$?

Yes _____ No _____

Can you see that this equation cannot be factored?

Yes _____ No _____

If you answered No to these questions, review how to write a quadratic equation in standard form. Try to break the equation into possible factors to discover that the equation cannot be factored.

Did you identify $a = 5, b = 7$, and $c = -4$?

Yes _____ No _____

Did you substitute these values into the quadratic formula

to obtain $x = \dfrac{-7 \pm \sqrt{(7)^2 - 4(5)(-4)}}{2(5)}$ ?

Yes _____ No _____

If you answered No to these questions, go back and substitute the correct values into the quadratic formula. Be careful with signs. Simplify the expression further to find the possible solutions.

If you answered Problem 1 incorrectly, go back and rework the problem using these suggestions.

---

**Chapter 10 Test, Problem 3**   Solve by any desired method. If there is no real number solution, say so.
$2x^2 = 2x - 5$

> **HELPFUL HINT** If you use the quadratic formula and obtain the square root of a negative number, then you know that there is no real number solution to the quadratic equation.

Did you rewrite the equation as $2x^2 - 2x + 5 = 0$?

Yes _____ No _____

Did you identify $a = 2, b = -2$, and $c = 5$?

Yes _____ No _____

If you answered No to these questions, stop and review how to write a quadratic equation in standard form and how to identify the values of $a, b$, and $c$. Be careful with signs.

When you substituted the values for $a, b$, and $c$ in the quadratic

formula, did you get $x = \dfrac{-(-2) \pm \sqrt{(-2)^2 - 4(2)(5)}}{2(2)}$?

Yes _____ No _____

If you answered No, stop, carefully make the required substitutions, and then simplify. You should get a negative value for the radicand.

Now go back and rework the problem using these suggestions.

Need more help? Watch the **MATH COACH** videos in MyMathLab® or on You Tube™.

**Chapter 10 Test, Problem 9** Solve by any desired method. If there is no real number solution, say so.
$2x(x - 6) = 6 - x$

> **HELPFUL HINT** Be sure to remove the parentheses and combine like terms first. Then write the equation in $ax^2 + bx + c = 0$ form before choosing your method of solution.

When you removed the parentheses, did you get $2x^2 - 12x = 6 - x$?

Yes _____ No _____

Next, did you combine like terms and write the equation as $2x^2 - 11x - 6 = 0$ ?

Yes _____ No _____

If you answered No to these questions, go back and complete these steps again. Be careful to avoid sign errors.

Did you determine that you could factor $2x^2 - 11x - 6$ ?

Yes _____ No _____

Did you choose the first term in one set of parentheses as $2x$ and the first term in the other set of parentheses as $x$?

Yes _____ No _____

If you answered No to these questions, try to factor the expression on the left side of the equation again.

Remember to set each set of parentheses equal to 0 and then solve for $x$.

If you answered Problem 9 incorrectly, go back and rework this problem using these suggestions.

---

**Chapter 10 Test, Problem 12** Graph the quadratic equation. Locate the vertex. $y = -x^2 + 8x - 12$

> **HELPFUL HINT** Make sure that the quadratic equation is written in $y = ax^2 + bx + c$ form. Use the vertex formula to find the vertex. Next, find the intercepts by setting $y = 0$ and $x = 0$. Notice that when $a < 0$, the graph shows a parabola opening downward.

Did you identify that the equation is written in $y = ax^2 + bx + c$ form and that $a = -1$, $b = 8$, and $c = -12$ ?

Yes _____ No _____

Did you determine that the $x$-coordinate of the vertex is 4?

Yes _____ No _____

If you answered No to these questions, remember that the vertex formula for the $x$-coordinate of the vertex point is $x = \dfrac{-b}{2a}$. You must then substitute this value for $x$ in the original equation to find the $y$-coordinate for the vertex point. Graph this point on the coordinate plane.

Did you let $y = 0$ to find the $x$-intercepts and let $x = 0$ to find the $y$-intercepts?

Yes _____ No _____

If you answered No, remember to follow these steps and solve for the other variable in each case to find the intercept points. Graph these points on the same coordinate plane with the vertex.

Once you know that the parabola opens downward and you have located the vertex and intercept points, you can graph the quadratic equation. For this problem, your vertex point should be the highest point on the graph.

Now go back and rework this problem using these suggestions.

Need more help? Look for section examples marked with $^{M}\!C$ to review.

# Practice Final Examination

1. _____

2. _____

3. _____

4. _____

5. _____

6. _____

7. _____

8. _____

9. _____

10. _____

11. _____

12. _____

13. _____

14. _____

15. _____

16. _____

17. _____

*The following questions cover the content of Chapters 0–10. Follow the directions for each question and simplify your answers.*

1. Simplify.
$$-2x + 3y\{7 - 2[x - (4x + y)]\}$$

2. Evaluate if $x = -2$ and $y = 3$.
$$2x^2 - 3xy - 4y$$

3. Simplify. $(-3x^2y)(-6x^3y^4)$

4. Combine like terms.
$$5x^2y - 6xy + 8xy - 3x^2y - 10xy$$

5. Solve for $x$. $2(x + 4) - 6 = 4x - 3$

6. Solve for $x$.
$$\frac{1}{2}(x + 4) - \frac{2}{3}(x - 7) = 4x$$

7. Solve for $x$ and graph the resulting inequality on a number line.
$$5x + 3 - (4x - 2) \le 6x - 8$$

*Multiply.*

8. $(2x + 1)(x - 2)$

9. $(2x + 3)^2$

10. $(x + 3y)(x - 3y)$

11. $(2x + y)(x^2 - 3xy + 2y^2)$

12. Factor completely. $4x^2 - 18x - 10$

13. Factor completely. $3x^3 - 9x^2 - 30x$

14. Combine. $\dfrac{2}{x - 3} - \dfrac{3}{x^2 - x - 6} + \dfrac{4}{x + 2}$

15. Simplify. $\dfrac{\dfrac{3}{x} + \dfrac{5}{2x}}{1 + \dfrac{2}{x + 2}}$

16. Solve for $x$. $\dfrac{2}{x + 2} = \dfrac{4}{x - 2} + \dfrac{3x}{x^2 - 4}$

17. Find the slope of the line and then graph the line. $5x - 2y - 3 = 0$

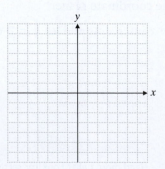

**18.** Find the equation of the line that has a slope of $-\dfrac{3}{4}$ and passes through the point $(-2, 5)$.

▲ **19.** Two quarter-circles are attached to a rectangle. The dimensions are indicated in the figure. Find the area of the region. Use 3.14 as an approximation for $\pi$.

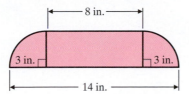

**20.** Solve for $x$ and $y$.

$2x + 7y = 4$

$-3x - 5y = 5$

**21.** Solve for $a$ and $b$.

$$a - \frac{3}{4}b = \frac{1}{4}$$

$$\frac{3}{2}a + \frac{1}{2}b = -\frac{9}{2}$$

**22.** Simplify and combine.

$\sqrt{45x^3} + 2x\sqrt{20x} - 6\sqrt{5x^3}$

**23.** Multiply and simplify.

$\sqrt{6}\left(3\sqrt{2} - 2\sqrt{6} + 4\sqrt{3}\right)$

**24.** Simplify by rationalizing the denominator. $\dfrac{\sqrt{3} + \sqrt{7}}{\sqrt{5} - \sqrt{7}}$

**25.** Solve for $x$. $12x^2 - 5x - 2 = 0$

**26.** Solve for $y$. $2y^2 = 6y - 1$

**27.** Solve for $x$. $4x^2 + 3 = 19$

**28.** Graph. $y = x^2 + 6x + 8$

▲ **29.** In the right triangle shown, sides $b$ and $c$ are given. Find the length of side $a$.

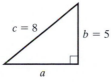

**30.** A number is tripled and then increased by 6. The result is 21. What is the original number?

▲ **31.** A rectangular region has a perimeter of 38 meters. The length of the rectangle is 2 meters shorter than double the width. What are the dimensions of the rectangle?

**32.** A benefit concert was held on campus to raise scholarship money. A total of 360 tickets were sold. Admission prices were $5 for reserved seats and $3 for general admission. Total receipts were $1480. How many reserved-seat tickets were sold? How many general admission tickets were sold?

▲ **33.** The area of a triangle is 68 square meters. The altitude of the triangle is one meter longer than double the base of the triangle. Find the altitude and base of the triangle.

18. _____

19. _____

20. _____

21. _____

22. _____

23. _____

24. _____

25. _____

26. _____

27. _____

28. _____

29. _____

30. _____

31. _____

32. _____

33. _____

# Appendix A Table of Square Roots

| $x$ | $\sqrt{x}$ | $x$ | $\sqrt{x}$ | $x$ | $\sqrt{x}$ | $x$ | $\sqrt{x}$ | $x$ | $\sqrt{x}$ |
|---|---|---|---|---|---|---|---|---|---|
| 1 | 1.000 | 41 | 6.403 | 81 | 9.000 | 121 | 11.000 | 161 | 12.689 |
| 2 | 1.414 | 42 | 6.481 | 82 | 9.055 | 122 | 11.045 | 162 | 12.728 |
| 3 | 1.732 | 43 | 6.557 | 83 | 9.110 | 123 | 11.091 | 163 | 12.767 |
| 4 | 2.000 | 44 | 6.633 | 84 | 9.165 | 124 | 11.136 | 164 | 12.806 |
| 5 | 2.236 | 45 | 6.708 | 85 | 9.220 | 125 | 11.180 | 165 | 12.845 |
| 6 | 2.449 | 46 | 6.782 | 86 | 9.274 | 126 | 11.225 | 166 | 12.884 |
| 7 | 2.646 | 47 | 6.856 | 87 | 9.327 | 127 | 11.269 | 167 | 12.923 |
| 8 | 2.828 | 48 | 6.928 | 88 | 9.381 | 128 | 11.314 | 168 | 12.961 |
| 9 | 3.000 | 49 | 7.000 | 89 | 9.434 | 129 | 11.358 | 169 | 13.000 |
| 10 | 3.162 | 50 | 7.071 | 90 | 9.487 | 130 | 11.402 | 170 | 13.038 |
| 11 | 3.317 | 51 | 7.141 | 91 | 9.539 | 131 | 11.446 | 171 | 13.077 |
| 12 | 3.464 | 52 | 7.211 | 92 | 9.592 | 132 | 11.489 | 172 | 13.115 |
| 13 | 3.606 | 53 | 7.280 | 93 | 9.644 | 133 | 11.533 | 173 | 13.153 |
| 14 | 3.742 | 54 | 7.348 | 94 | 9.695 | 134 | 11.576 | 174 | 13.191 |
| 15 | 3.873 | 55 | 7.416 | 95 | 9.747 | 135 | 11.619 | 175 | 13.229 |
| 16 | 4.000 | 56 | 7.483 | 96 | 9.798 | 136 | 11.662 | 176 | 13.266 |
| 17 | 4.123 | 57 | 7.550 | 97 | 9.849 | 137 | 11.705 | 177 | 13.304 |
| 18 | 4.243 | 58 | 7.616 | 98 | 9.899 | 138 | 11.747 | 178 | 13.342 |
| 19 | 4.359 | 59 | 7.681 | 99 | 9.950 | 139 | 11.790 | 179 | 13.379 |
| 20 | 4.472 | 60 | 7.746 | 100 | 10.000 | 140 | 11.832 | 180 | 13.416 |
| 21 | 4.583 | 61 | 7.810 | 101 | 10.050 | 141 | 11.874 | 181 | 13.454 |
| 22 | 4.690 | 62 | 7.874 | 102 | 10.100 | 142 | 11.916 | 182 | 13.491 |
| 23 | 4.796 | 63 | 7.937 | 103 | 10.149 | 143 | 11.958 | 183 | 13.528 |
| 24 | 4.899 | 64 | 8.000 | 104 | 10.198 | 144 | 12.000 | 184 | 13.565 |
| 25 | 5.000 | 65 | 8.062 | 105 | 10.247 | 145 | 12.042 | 185 | 13.601 |
| 26 | 5.099 | 66 | 8.124 | 106 | 10.296 | 146 | 12.083 | 186 | 13.638 |
| 27 | 5.196 | 67 | 8.185 | 107 | 10.344 | 147 | 12.124 | 187 | 13.675 |
| 28 | 5.292 | 68 | 8.246 | 108 | 10.392 | 148 | 12.166 | 188 | 13.711 |
| 29 | 5.385 | 69 | 8.307 | 109 | 10.440 | 149 | 12.207 | 189 | 13.748 |
| 30 | 5.477 | 70 | 8.367 | 110 | 10.488 | 150 | 12.247 | 190 | 13.784 |
| 31 | 5.568 | 71 | 8.426 | 111 | 10.536 | 151 | 12.288 | 191 | 13.820 |
| 32 | 5.657 | 72 | 8.485 | 112 | 10.583 | 152 | 12.329 | 192 | 13.856 |
| 33 | 5.745 | 73 | 8.544 | 113 | 10.630 | 153 | 12.369 | 193 | 13.892 |
| 34 | 5.831 | 74 | 8.602 | 114 | 10.677 | 154 | 12.410 | 194 | 13.928 |
| 35 | 5.916 | 75 | 8.660 | 115 | 10.724 | 155 | 12.450 | 195 | 13.964 |
| 36 | 6.000 | 76 | 8.718 | 116 | 10.770 | 156 | 12.490 | 196 | 14.000 |
| 37 | 6.083 | 77 | 8.775 | 117 | 10.817 | 157 | 12.530 | 197 | 14.036 |
| 38 | 6.164 | 78 | 8.832 | 118 | 10.863 | 158 | 12.570 | 198 | 14.071 |
| 39 | 6.245 | 79 | 8.888 | 119 | 10.909 | 159 | 12.610 | 199 | 14.107 |
| 40 | 6.325 | 80 | 8.944 | 120 | 10.954 | 160 | 12.649 | 200 | 14.142 |

Unless the value of $\sqrt{x}$ ends in 000, all values are rounded to the nearest thousandth.

# Appendix B Metric Measurement and Conversion of Units

## Student Learning Objectives

**After studying this section, you will be able to:**

1 Convert from one metric unit of measurement to another.

2 Convert between metric and U.S. units of measure.

3 Convert from one type of unit of measure to another.

## 1 Converting from One Metric Unit of Measurement to Another

Metric measurements are becoming more common in the United States. The metric system is used in many parts of the world and in the sciences. The basic unit of length in the metric system is the meter. For smaller measurements, centimeters are commonly used. There are 100 centimeters in 1 meter. For larger measurements, kilometers are commonly used. There are 1000 meters in 1 kilometer. The following tables give the metric units of measurement for length.

**METRIC LENGTH**

| | | |
|---|---|---|
| *1 kilometer (km) | = | 1000 meters |
| 1 hectometer (hm) | = | 100 meters |
| 1 dekameter (dam) | = | 10 meters |
| *1 meter (m) | = | 1 meter |
| 1 decimeter (dm) | = | 0.1 meter |
| *1 centimeter (cm) | = | 0.01 meter |
| *1 millimeter (mm) | = | 0.001 meter |

The four most common units of metric measurement are indicated by an asterisk.

**METRIC WEIGHT**

| | | |
|---|---|---|
| 1 kilogram (kg) | = | 1000 grams |
| 1 gram (g) | = | 1 gram |
| 1 milligram (mg) | = | 0.001 gram |

**METRIC VOLUME**

| | | |
|---|---|---|
| 1 kiloliter (kL) | = | 1000 liters |
| 1 liter (L) | = | 1 liter |
| 1 milliliter (mL) | = | 0.001 liter |

One way to convert from one metric unit to another is to multiply by 1. We know mathematically that multiplying by 1 will yield an equivalent expression since $(a)(1) = a$. In our calculations when solving word problems, we treat dimension symbols much as we treat variables in algebra.

**Example 1** Fred drove a distance of 5.4 kilometers. How many meters is that?

**Solution** $5.4 \text{ km} \cdot 1 = 5.4 \text{ km} \cdot \dfrac{1000 \text{ m}}{1 \text{ km}} = \dfrac{(5.4)(1000)}{1} \cdot \dfrac{\text{km}}{\text{km}} \cdot \text{m}$

$$= 5400 \text{ m}$$

The distance is 5400 meters.

**Student Practice 1** June's room is 3.4 meters wide. How many centimeters is that?

**Example 2** A chemist measured 67 milliliters of a solution. How many liters is that?

**Solution** $67 \text{ mL} \cdot \dfrac{0.001 \text{ L}}{1 \text{ mL}} = 0.067 \text{ L}$

The chemist measured 0.067 liter of the solution. □

**Student Practice 2** The container has 125 liters of water. How many kiloliters is that?

## 2 Converting Between Metric and U.S. Units of Measure ▶

The most common relationships between the U.S. and metric systems are listed in the following table. Most of these values are approximate.

**METRIC CONVERSION RATIOS**

LENGTH:
- 1 inch = 2.54 centimeters
- 39.37 inches = 1 meter
- 1 mile = 1.61 kilometers
- 0.62 mile = 1 kilometer

WEIGHT:
- 1 pound = 454 grams
- 2.20 pounds = 1 kilogram
- 1 ounce = 28.35 grams
- 0.0353 ounce = 1 gram

LIQUID CAPACITY:
- 1 quart = 946 milliliters
- 1.06 quarts = 1 liter
- 1 gallon = 3.785 liters

**Example 3** A box weighs 190 grams. How many ounces is that? Round to the nearest hundredth.

**Solution** $190 \text{ g} \cdot \dfrac{0.0353 \text{ oz}}{1 \text{ g}} = 6.707 \text{ oz}$

$\approx 6.71 \text{ oz}$   (rounded to the nearest hundredth)

The box weighs approximately 6.71 ounces. □

**Student Practice 3** A bag of groceries weighs 5.72 pounds. How many kilograms is that?

**Example 4** Juanita drives 23 kilometers to work each day. How many miles is the trip? Round to the nearest mile.

**Solution** $23 \text{ km} \cdot \dfrac{0.62 \text{ mi}}{1 \text{ km}} = 14.26 \text{ mi} \approx 14 \text{ mi}$ (rounded to the nearest mile)

Juanita drives approximately 14 miles to work each day. □

**Student Practice 4** Carlos installed an electrical connection that is 8.00 centimeters long. How many inches long is the connection? Round to the nearest hundredth.

**Example 5** Anita purchased 42.0 gallons of gasoline for her car last month. How many liters did she purchase? Round to the nearest liter.

**Solution** $42.0 \text{ gal} \cdot \dfrac{3.785 \text{ L}}{1 \text{ gal}} = 158.97 \text{ L} \approx 159 \text{ L}$ (rounded to the nearest liter)

Anita purchased approximately 159 liters of gasoline last month. □

**Student Practice 5** Warren purchased a 3-liter bottle of Coca-Cola. How many quarts of Coca-Cola is that?

### 3 Converting from One Type of Unit of Measure to Another

Sometimes you will need to convert from one type of unit of measure to another. For example, you may need to convert days to minutes or miles per hour to feet per second. Recall the U.S. units of measure.

| Length | Time |
|---|---|
| 12 inches = 1 foot | 60 seconds = 1 minute |
| 3 feet = 1 yard | 60 minutes = 1 hour |
| 5280 feet = 1 mile | 24 hours = 1 day |
| 1760 yards = 1 mile | 7 days = 1 week |
| **Weight** | **Volume** |
| 16 ounces = 1 pound | 2 cups = 1 pint |
| 2000 pounds = 1 ton | 2 pints = 1 quart |
| | 4 quarts = 1 gallon |

**Example 6** A car was traveling at 50.0 miles per hour. How many feet per second was the car traveling? Round to the nearest tenth of a foot per second.

**Solution** $\dfrac{50 \text{ mi}}{\text{hr}} \cdot \dfrac{5280 \text{ ft}}{1 \text{ mi}} \cdot \dfrac{1 \text{ hr}}{60 \text{ min}} \cdot \dfrac{1 \text{ min}}{60 \text{ sec}} = \dfrac{(50)(5280) \text{ ft}}{(60)(60) \text{ sec}} = \dfrac{73.333 \ldots \text{ ft}}{\text{sec}}$

The car was traveling at 73.3 feet per second, rounded to the nearest tenth of a foot per second. □

**Student Practice 6** A speeding car was traveling at 70.0 miles per hour. How many feet per second was the car traveling? Round to the nearest tenth of a foot per second.

## Appendix B Exercises MyMathLab®

1. How many meters are in 34 km?
2. How many meters are in 128 km?
3. How many centimeters are in 57 m?
4. How many centimeters are in 46 m?
5. How many millimeters are in 25 cm?
6. How many millimeters are in 63 cm?
7. How many meters are in 563 mm?
8. How many meters are in 831 mm?
9. How many milligrams are in 29.4 g?
10. How many milligrams are in 75.2 g?
11. How many kilograms are in 98.4 g?
12. How many kilograms are in 62.7 g?
13. How many milliliters are in 7 L?
14. How many milliliters are in 12 L?
15. How many kiloliters are in 4 mL?
16. How many kiloliters are in 3 mL?

*Use the table of metric conversion ratios to find each of the following. Round all answers to the nearest tenth.*

17. How many inches are in 4.2 cm?
18. How many inches are in 3.8 cm?

19. How many kilometers are in 14 mi?
20. How many kilometers are in 13 mi?

21. How many meters are in 110 in.?
22. How many meters are in 150 in.?

23. How many centimeters are in 7 in.?
24. How many centimeters are in 9 in.?

25. A box weighing 2.4 lb would weigh how many grams?
26. A box weighing 1.6 lb would weigh how many grams?

27. A man weighs 78 kg. How many pounds does he weigh?
28. A woman weighs 52 kg. How many pounds does she weigh?

29. Ferrante purchased 3 qt of milk. How many liters is that?
30. Wong Tin purchased 1 gal of milk. How many liters is that?

*Answer the following questions. Round all answers to the nearest hundredth when necessary.*

31. How many inches are in 3050 miles?
32. How many inches are in 4500 miles?

33. A truck traveled at a speed of 40 feet per second. How many miles per hour is that?
34. A car traveled at a speed of 55 feet per second. How many miles per hour is that?

35. How many years are in 3,500,000 seconds? (Use the approximate value that 365 days = 1 year.)
36. How many years are in 2,800,000 seconds? (Use the approximate value that 365 days = 1 year.)

# Appendix C Interpreting Data from Tables, Charts, and Graphs

## Student Learning Objectives

After studying this section, you will be able to interpret data from:

1 Tables
2 Charts
3 Pictographs
4 Bar Graphs
5 Line Graphs
6 Pie Graphs and Circle Graphs

## 1 Tables

A table is a device used to organize information into categories. Using it, you can readily find details about each category.

### Example 1

#### Table of Nutritive Values of Certain Popular "Fast Foods"

| Type of Sandwich | Calories | Protein (g) | Fat (g) | Cholesterol (g) | Sodium (mg) |
|---|---|---|---|---|---|
| Burger King Whopper | 630 | 27 | 38 | 90 | 880 |
| McDonald's Big Mac | 500 | 25 | 26 | 100 | 890 |
| Wendy's Bacon Cheeseburger | 440 | 22 | 25 | 65 | 870 |
| Burger King BK Broiler (chicken) | 280 | 20 | 10 | 50 | 770 |
| McDonald's McChicken | 415 | 19 | 20 | 50 | 830 |
| Wendy's Grilled Chicken | 290 | 24 | 7 | 60 | 670 |

*Source:* U.S. Government Agencies and Food Manufacturers

**(a)** Which food item has the least amount of fat per serving?

**(b)** Which beef item has the least amount of calories per serving?

**(c)** How much more protein does a Burger King Whopper have than a McDonald's Big Mac?

### Solution

**(a)** The least amount of fat, 7 grams, is in the Wendy's Grilled Chicken Sandwich.

**(b)** The least amount of calories, 440, for a beef sandwich is in the Wendy's Bacon Cheeseburger.

**(c)** The Burger King Whopper has 2 grams of protein more than the McDonald's Big Mac.

### Student Practice 1

**(a)** Which sandwich has the lowest level of sodium?

**(b)** Which sandwich has the highest level of cholesterol?

## 2 Charts

A chart is a device used to organize information in which not every category is the same. Example 2 illustrates a chart containing different types of data.

**Example 2** The following chart shows how people in Topsfield indicated they spend their free time.

**Survey of Use of Leisure Time**

| Category | Activity | Hours spent per week |
|---|---|---|
| Single men | Gym | 6 |
| | Outdoor sports | 4 |
| | Dating | 7 |
| | Watching pro sports | 12 |
| | Reading & TV | 3 |
| Single women | Gym | 4 |
| | Outdoor sports | 2 |
| | Dating | 7 |
| | Time with friends | 10 |
| | Reading & TV | 9 |
| Couples | Time with family | 21 |
| | Time as a couple | 8 |
| | Time with friends | 4 |
| | Reading & TV | 9 |
| Children | Watching TV | 28 |
| | Playing outside | 8 |
| | Reading | 1 |

Use the chart to answer the following questions about people in Topsfield.

**(a)** What is the average amount of time a couple spends with family during the week?

**(b)** How much more time do children spend watching TV than playing outside?

**(c)** What activity do single men spend most of their time doing?

### Solution

**(a)** The average amount of time a couple spends with family is 21 hours per week.

**(b)** Children spend 20 more hours per week watching TV than playing outside.

**(c)** Single men spend more time per week watching pro sports (12 hr) than on any other activity.  ☐

### ▶ Student Practice 2

**(a)** What two categories do single women spend the most time doing?

**(b)** What do couples spend most of their time doing?

**(c)** What is the most significant numerical difference in terms of the number of hours per week spent by single women versus single men?

### 3 Pictographs

A pictograph uses a visually appropriate symbol to represent an amount of items. A pictograph is used in Example 3.

**Example 3** Consider the following pictograph.

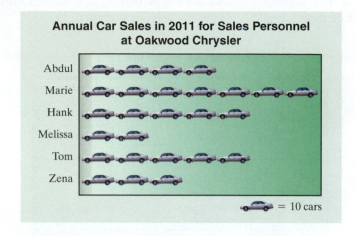

**(a)** How many cars did Melissa sell in 2011?

**(b)** Who sold the greatest number of cars?

**(c)** How many more cars did Tom sell than Zena?

### Solution

**(a)** Melissa sold $2 \times 10 = 20$ cars.

**(b)** Marie sold the greatest number of cars.

**(c)** Tom sold $5 \times 10 = 50$ cars. Zena sold $3 \times 10 = 30$ cars. Now, $50 - 30 = 20$. Therefore, Tom sold 20 cars more than Zena. ☐

### Student Practice 3

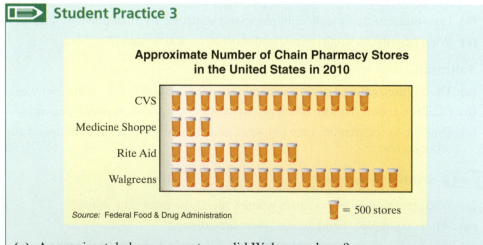

**(a)** Approximately how many stores did Walgreens have?

**(b)** Approximately how many more stores did Rite Aid have than Medicine Shoppe?

**(c)** What was the combined number of Medicine Shoppe stores and CVS stores?

## 4 Bar Graphs

A bar graph is helpful for making comparisons and noting changes or trends. A scale is provided so that the height of each bar indicates a specific number. A bar graph is displayed in Example 4. A bar graph may be represented horizontally or vertically. In either case the basic concepts of interpreting a bar graph are the same.

**Example 4** The approximate population of California by year is given in the following bar graph.

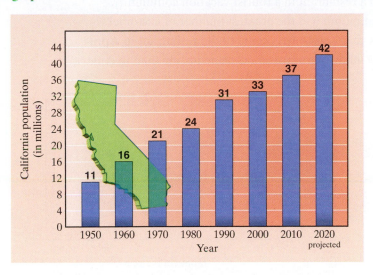

(a) What is the projected population of California in 2020?

(b) How much greater was the population of California in 1980 than in 1970?

### Solution

(a) The projected population of California in 2020 is about 42 million.

(b) In 1980 it was 24 million. In 1970 it was 21 million. The population of California was approximately 3 million more people in 1980 than in 1970. □

**Student Practice 4** The following bar graph depicts the number of accidents for U.S. air carriers on scheduled flight service.

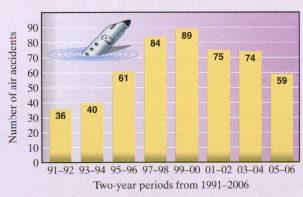

Source: National Transportation Safety Board

(a) What two-year period had the greatest number of accidents?

(b) What was the increase in the number of accidents from the 1995–1996 period to the 1997–1998 period?

(c) What was the decrease in the number of accidents from the 1999–2000 period to the 2001–2002 period?

## 5 Line Graphs

A line graph is often used to display data when significant changes or trends are present. In a line graph, only a few points are actually plotted from measured values. The points are then connected by straight lines in order to show a trend. The intervening values between points may not exactly lie on the line. A line graph is displayed in Example 5.

**Example 5** The following line graph shows the number of customers per month coming to a restaurant in a tourist vacation community.

(a) What month had the greatest number of customers?

(b) How many customers came to the restaurant in April?

(c) How many more customers came in July than in August?

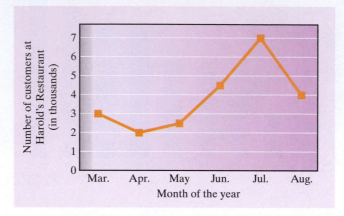

### Solution

(a) More customers came during the month of July.

(b) Approximately 2000 customers came in April.

(c) In July there were 7000 customers, while in August there were 4000 customers. Thus, there were 3000 more customers in July than in August.

**Student Practice 5** The quality of the air is measured by the Air Quality Index (AQI). Air is tested daily in metropolitan areas and is reported as one of the following categories: good, moderate, unhealthy for sensitive groups, or unhealthy. The following line graph indicates the number of days that the air quality was considered unhealthy in the city of Baltimore during a 12-year period.

(a) What was the number of days in which the air was considered unhealthy in Baltimore in 2001?

(b) In what year did Baltimore have the most days in which the air was considered unhealthy?

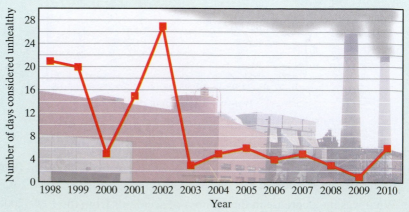

*Source:* U.S. Environmental Protection Agency

## 6 Pie Graphs and Circle Graphs

A pie graph or a circle graph indicates how a whole quantity is divided into parts. These graphs help you to visualize the size of the relative proportions of parts. Each piece of the pie or circle is called a sector. Example 6 uses a pie graph.

**Example 6** Together, the Great Lakes form the largest body of fresh water in the world. The total area of these five lakes is about 290,000 square miles, almost all of which is suitable for boating. The percentage of this total area taken up by each of the Great Lakes is shown in the pie graph.

(a) What percentage of the area is taken up by Lake Michigan?

(b) What lake takes up the largest percentage of area?

(c) How many square miles are taken up by Lake Huron and Lake Michigan together?

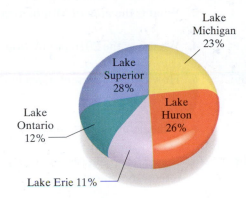

## Solution

(a) Lake Michigan takes up 23% of the area.

(b) Lake Superior takes up the largest percentage.

(c) If we add $26 + 23$, we get 49. Thus Lake Huron and Lake Michigan together take up 49% of the total area.

$49\%$ of $290,000 = (0.49)(290,000) = 142,100$ square miles.  □

**Student Practice 6**   Seattle receives on average about 37 inches of rain per year. However, the amount of rainfall per month varies significantly. The percent of rainfall that occurs during each quarter of the year is shown by the circle graph.

(a) What percent of the rain in Seattle falls between April and June?

(b) Forty percent of the rainfall occurs in what three-month period?

(c) How many inches of rain fall in Seattle from January to March?

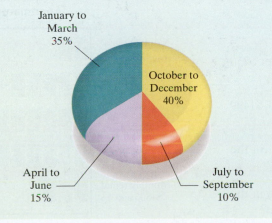

*In each case, study carefully the appropriate visual display, then answer the questions. Consider the following table in answering exercises 1–10.*

### Table of Facts of the Rocky Mountain States

| State | Area in Square Miles | Date Admitted to the Union | 2010 Population | Number of Representatives in U.S. Congress | Popular Name |
|---|---|---|---|---|---|
| Colorado | 104,247 | 1876 | 5,044,930 | 7 | Centennial State |
| Idaho | 83,557 | 1890 | 1,573,499 | 2 | Gem State |
| Montana | 147,138 | 1889 | 994,416 | 1 | Treasure State |
| Nevada | 110,540 | 1864 | 2,709,432 | 3 | Silver State |
| Utah | 84,916 | 1896 | 2,770,765 | 3 | Beehive State |
| Wyoming | 97,914 | 1890 | 568,300 | 1 | Equality State |

1. What is the area of Utah in square miles?

2. What is the area of Montana in square miles?

3. What was the 2010 population of Colorado?

4. What was the 2010 population of Nevada?

5. How many representatives in the U.S. Congress come from Idaho?

6. How many representatives in the U.S. Congress come from Wyoming?

7. What is the popular name for Montana?

8. What is the popular name for Utah?

9. Which of these six states was the first one to be admitted to the Union?

10. In what year did two of these six states both get admitted to the Union?

*Use this pictograph to answer exercises 11–16.*

11. How many homes were built in Tarrant County in the year 2011?

12. How many homes were built in Essex County in the year 2011?

13. In what county were the most homes built?

14. How many more homes were built in Tarrant County than Waverly County?

15. How many homes were built in Essex County and Northface County combined?

16. How many homes were built in DuPage County and Waverly County combined?

*Use this pictograph to answer exercises 17–20.*

17. How many apartment units were rented for under $250 per month?

18. How many apartment units were rented for $800–$1499 per month?

19. How many more apartment units were available in the $500–$799 range than in the $1500 and up range?

20. How many more apartment units were available in the $800–$1499 range than in the $1500 and up range?

**Number of New Homes Built in 2011 in Each of Five Counties**

**Approximate Number of Apartments in U.S. in 2007**

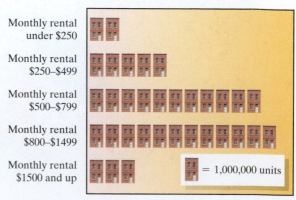

*Source:* U.S. Bureau of the Census

*Use this bar graph to answer exercises 21–26.*

**21.** What was the population in Texas in 1950?

**22.** What is the projected population in Texas in 2020?

**23.** Between what two decades on this graph was the increase in population the greatest?

**24.** Between what two decades was the increase in population the smallest?

**25.** How many more people lived in Texas in 1970 than in 1950?

**26.** How many more people lived in Texas in 1980 than in 1960?

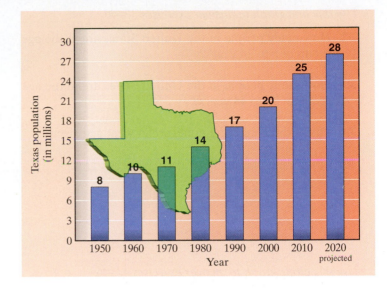

*Use this bar graph to answer exercises 27–30.*

**27.** According to the bar graph, how many people watched a sports event at least once in the last 12 months?

**28.** According to the bar graph, how many people were involved in gardening at least once in the last 12 months?

**29.** What two activities were the most common?

**30.** What two activities were the least common?

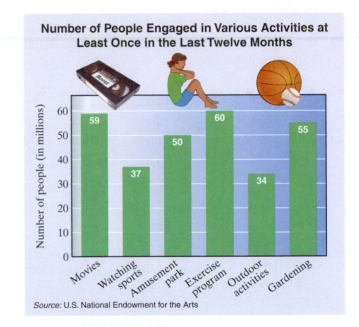

**Number of People Engaged in Various Activities at Least Once in the Last Twelve Months**

*Source:* U.S. National Endowment for the Arts

*Use this line graph to answer exercises 31–36.*

**31.** What was the profit in 2008?

**32.** What was the profit in 2007?

**33.** How much greater was the profit in 2011 than 2010?

**34.** In what year did the smallest profit occur?

**35.** Between what two years did the profit decrease the most?

**36.** Between what two years did the profit increase the most?

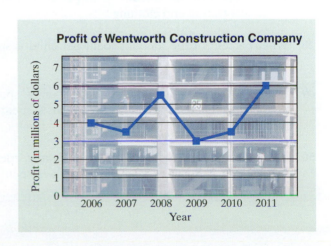

**Profit of Wentworth Construction Company**

*Use this line graph to answer exercises 37–40.*

**37.** How many people in the United States are in the age group of 5 to 24 years?

**38.** How many people in the United States are in the age group of 25 to 44 years?

**39.** Thirty-three million people are in what age group?

**40.** Fifty-one million people are in what age group?

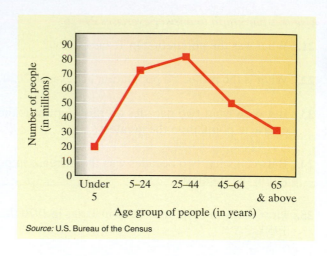

*Source:* U.S. Bureau of the Census

*Use this circle graph to answer exercises 41–46.*

**41.** What percent of the world's population is either Hindu or Buddhist?

**42.** What percent of the world's population is either Muslim or nonreligious?

**43.** What percent of the world's population is *not* Muslim?

**44.** What percent of the world's population is *not* Christian?

**45.** If there were approximately 6.5 billion people in the world in 2006, how many of them would we expect to be Hindu?

**46.** If there were approximately 6.9 billion people in the world in 2010, how many of them would we expect to be Muslim?

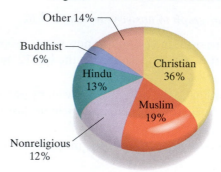

**Religious Faith Distribution in the World**

*Source:* United Nations Statistical Division

*Use this circle graph to answer exercises 47–50.*

**47.** What percent of the amount spent per child is spent on transportation and clothing?

**48.** What percent of the amount spent per child is spent on food, health care, and housing?

**49.** If the average two-child American family spends $12,350 per child each year, how much is spent on food per child?

**50.** If the average two-child American family spends $12,350 per child each year, how much is spent on child care and education per child?

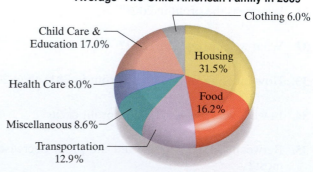

**Distribution of Spending Per Child by "Average" Two-Child American Family in 2009**

*Source:* United Nations Statistical Division

# Appendix D The Point–Slope Form of a Line

## 1 Using the Point–Slope Form of the Equation of a Line

Recall from Section 7.1 that we defined a line as an equation that can be written in the form $Ax + By = C$. This form is called the **standard form of the equation of a line** and although this form tells us that the graph is a straight line, it reveals little about the line. A more useful form of the equation was introduced in Section 7.3 called the **slope–intercept form**, $y = mx + b$. This form immediately reveals the slope and $y$-intercept of a line, and also allows us to easily write the equation of a line when we know its slope and $y$-intercept. But, what happens if we know the slope of a line and a point on the line that is not the $y$-intercept? Can we write the equation of the line? By the definition of slope, we have the following:

$$m = \frac{y - y_1}{x - x_1}$$

$$m(x - x_1) = y - y_1$$

That is, $y - y_1 = m(x - x_1)$.

This is the point–slope form of the equation of a line.

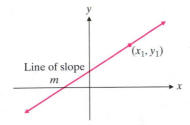

> **Student Learning Objectives**
>
> After studying this section, you will be able to:
>
> **1** Use the point–slope form of the equation of a line.
>
> **2** Write the equation of the line passing through a given point that is parallel or perpendicular to a given line.

---

**POINT–SLOPE FORM**

The **point–slope form** of the equation of a line is $y - y_1 = m(x - x_1)$, where $m$ is the slope and $(x_1, y_1)$ are the coordinates of a known point on the line.

---

### *Write the Equation of a Line Given Its Slope and One Point on the Line*

**Example 1** Find an equation of the line that has slope $-\frac{3}{4}$ and passes through the point $(-6, 1)$. Express your answer in standard form.

**Solution** Since we don't know the $y$-intercept, we can't use the slope–intercept form easily. Therefore, we use the point–slope form.

$$y - y_1 = m(x - x_1)$$

$$y - 1 = -\frac{3}{4}[x - (-6)] \qquad \text{Substitute the given values.}$$

$$y - 1 = -\frac{3}{4}x - \frac{9}{2} \qquad \text{Simplify. (Do you see how we did this?)}$$

$$4y - 4(1) = 4\left(-\frac{3}{4}x\right) - 4\left(\frac{9}{2}\right) \qquad \text{Multiply each term by the LCD 4.}$$

$$4y - 4 = -3x - 18 \qquad \text{Simplify.}$$

$$3x + 4y = -18 + 4 \qquad \text{Add } 3x + 4 \text{ to each side.}$$

$$3x + 4y = -14 \qquad \text{Add like terms.}$$

The equation in standard form is $3x + 4y = -14$. ☐

*Continued on next page*

> **Student Practice 1**  Find an equation of the line that passes through $(5, -2)$ and has a slope of $\frac{3}{4}$. Express your answer in standard form.

*Write the Equation of a Line Given Two Points on the Line*  We can use the point–slope form to find the equation of a line if we are given two points. Carefully study the following example. Be sure you understand each step. You will encounter this type of problem frequently.

**Graphing Calculator**

**Using Linear Regression to Find an Equation**

Many graphing calculators, such as the TI-84 Plus, will find the equation of a line in slope–intercept form if you enter the points as a collection of data and use the Regression feature. We would enter the data from Example 2 as follows:

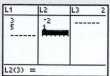

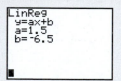

The output of the calculator uses the notation $y = ax + b$ instead of $y = mx + b$.

```
LinReg
y=ax+b
a=1.5
b=-6.5
```

Thus, our answer to Example 2 using the graphing calculator would be $y = 1.5x - 6.5$.

**Example 2**  Find an equation of the line that passes through $(3, -2)$ and $(5, 1)$. Express your answer in slope–intercept form.

**Solution**  First we find the slope.

$$m = \frac{y_2 - y_1}{x_2 - x_1} = \frac{1 - (-2)}{5 - 3} = \frac{1 + 2}{2} = \frac{3}{2}$$

Now we substitute the value of the slope and the coordinates of either point into the point–slope equation. Let's use $(5, 1)$.

$$y - y_1 = m(x - x_1)$$

$$y - 1 = \frac{3}{2}(x - 5) \qquad \text{Substitute } m = \frac{3}{2} \text{ and } (x_1, y_1) = (5, 1).$$

$$y - 1 = \frac{3}{2}x - \frac{15}{2} \qquad \text{Remove parentheses.}$$

$$y = \frac{3}{2}x - \frac{15}{2} + 1 \qquad \text{Add 1 to each side of the equation.}$$

$$y = \frac{3}{2}x - \frac{15}{2} + \frac{2}{2}$$

$$y = \frac{3}{2}x - \frac{13}{2} \qquad \text{Add the two fractions and simplfy.} \qquad \square$$

> **Student Practice 2**  Find an equation of the line that passes through $(-4, 1)$ and $(-2, -3)$. Express your answer in slope–intercept form.

Before we go further, we want to point out that these various forms of the equation of a straight line are just that—*forms* for convenience. We are *not* using different equations each time, nor should you simply try to memorize the different variations without understanding when to use them. They can easily be derived from the definition of slope, as we have seen. And remember, you can *always* use the definition of slope to find the equation of a line. You may find it helpful to review Examples 1 and 2 for a few minutes before going ahead to Example 3. It is important to see how each example is different.

### 2 Writing the Equation of a Parallel or Perpendicular Line

Let us now look at parallel and perpendicular lines. If we are given the equation of a line and a point not on the line, we can find the equation of a second line that passes through the given point and is parallel or perpendicular to the first line. We can do this because we know that the slopes of parallel lines are equal and that the slopes of perpendicular lines are negative reciprocals of each other.

We begin by finding the slope of the given line. Then we use the point–slope form to find the equation of the second line. Study each step of the following example carefully.

**Example 3** Find an equation of the line passing through the point $(-2, -4)$ and parallel to the line $2x + 5y = 8$. Express the answer in standard form.

**Solution** First we need to find the slope of the line $2x + 5y = 8$. We do this by writing the equation in slope–intercept form.

$$5y = -2x + 8$$

$$y = -\frac{2}{5}x + \frac{8}{5}$$

The slope of the given line is $-\frac{2}{5}$. Since parallel lines have the same slope, the slope of the unknown line is also $-\frac{2}{5}$. Now we substitute $m = -\frac{2}{5}$ and the coordinates of the point $(-2, -4)$ into the point–slope form of the equation of a line.

$$y - y_1 = m(x - x_1)$$

$$y - (-4) = -\frac{2}{5}[x - (-2)] \qquad \text{Substitute.}$$

$$y + 4 = -\frac{2}{5}(x + 2) \qquad \text{Simplify.}$$

$$y + 4 = -\frac{2}{5}x - \frac{4}{5} \qquad \text{Remove parentheses.}$$

$$5y + 5(4) = 5\left(-\frac{2}{5}x\right) - 5\left(\frac{4}{5}\right) \qquad \text{Multiply each term by the LCD 5.}$$

$$5y + 20 = -2x - 4 \qquad \text{Simplify.}$$

$$2x + 5y = -4 - 20 \qquad \text{Add } 2x - 20 \text{ to each side.}$$

$$2x + 5y = -24 \qquad \text{Simplify.}$$

$2x + 5y = -24$ is an equation of the line passing through the point $(-2, -4)$ and parallel to the line $2x + 5y = 8$. ☐

**Student Practice 3** Find an equation of the line passing through $(4, -5)$ and parallel to the line $5x - 3y = 10$. Express the answer in standard form.

An extra step is needed if the desired line is to be perpendicular to the given line. Carefully note the approach in Example 4.

**Example 4** Find an equation of the line that passes through the point $(2, -3)$ and is perpendicular to the line $3x - y = -12$. Express the answer in standard form.

**Solution** To find the slope of the line $3x - y = -12$, we rewrite it in slope–intercept form.

$$-y = -3x - 12$$
$$y = 3x + 12$$

This line has a slope of 3. Therefore, the slope of a line perpendicular to this line is the negative reciprocal $-\frac{1}{3}$.

Now substitute the slope $m = -\frac{1}{3}$ and the coordinates of the point $(2, -3)$ into the point–slope form of the equation.

$$y - y_1 = m(x - x_1)$$

$$y - (-3) = -\frac{1}{3}(x - 2) \qquad \text{Substitute.}$$

$$y + 3 = -\frac{1}{3}(x - 2) \qquad \text{Simplify.}$$

$$y + 3 = -\frac{1}{3}x + \frac{2}{3} \qquad \text{Remove parentheses.}$$

$$3y + 3(3) = 3\left(-\frac{1}{3}x\right) + 3\left(\frac{2}{3}\right) \qquad \text{Multiply each term by the LCD 3.}$$

$$3y + 9 = -x + 2 \qquad \text{Simplify.}$$

$$x + 3y = 2 - 9 \qquad \text{Add } x - 9 \text{ to each side.}$$

$$x + 3y = -7 \qquad \text{Simplify.}$$

$x + 3y = -7$ is an equation of the line that passes through the point $(2, -3)$ and is perpendicular to the line $3x - y = -12$. □

**Student Practice 4** Find an equation of the line that passes through $(-4, 3)$ and is perpendicular to the line $6x + 3y = 7$. Express the answer in standard form.

## Appendix D  Exercises  MyMathLab®

*Find an equation of the line that passes through the given point and has the given slope. Express your answer in slope–intercept form.*

1. $(6, 4), m = -\dfrac{2}{3}$

2. $(4, 6), m = -\dfrac{1}{2}$

3. $(-7, -2), m = 5$

4. $(8, 0), m = -3$

5. $(6, 0), m = -\dfrac{1}{5}$

6. $(0, -1), m = -\dfrac{5}{3}$

*Find an equation of the line passing through the pair of points. Write the equation in slope–intercept form.*

7. $(-4, -1)$ and $(3, 4)$

8. $(7, -2)$ and $(-1, -3)$

9. $\left(\dfrac{1}{2}, -3\right)$ and $\left(\dfrac{7}{2}, -5\right)$

10. $\left(\dfrac{7}{6}, 1\right)$ and $\left(-\dfrac{1}{3}, 0\right)$

11. $(12, -3)$ and $(7, -3)$

12. $(4, 8)$ and $(-3, 8)$

*Find an equation of the line satisfying the conditions given. Express your answer in standard form.*

13. Parallel to $5x - y = 4$ and passing through $(-2, 0)$

14. Parallel to $3x - y = -5$ and passing through $(-1, 0)$

15. Parallel to $x = 3y - 8$ and passing through $(5, -1)$

16. Parallel to $2y + x = 7$ and passing through $(-5, -4)$

17. Perpendicular to $2y = -3x$ and passing through $(6, -1)$

18. Perpendicular to $y = 5x$ and passing through $(4, -2)$

19. Perpendicular to $x + 7y = -12$ and passing through $(-4, -1)$

20. Perpendicular to $x - 4y = 2$ and passing through $(3, -1)$

## To Think About

*Without graphing determine whether the following pairs of lines are (a) parallel, (b) perpendicular, or (c) neither parallel nor perpendicular.*

21. $-3x + 5y = 40$
    $5y + 3x = 17$

22. $5x - 6y = 19$
    $6x + 5y = -30$

23. $y = -\dfrac{3}{4}x - 2$
    $6x + 8y = -5$

24. $y = \dfrac{2}{3}x + 6$
    $-2x - 3y = -12$

25. $y = \dfrac{5}{6}x - \dfrac{1}{3}$
    $6x + 5y = -12$

26. $y = \dfrac{3}{7}x - \dfrac{1}{14}$
    $14y + 6x = 3$

# Solutions to Student Practice Problems

## Chapter 0    0.1 Student Practice

1. (a) $\dfrac{10}{16} = \dfrac{2 \times 5}{2 \times 2 \times 2 \times 2} = \dfrac{5}{2 \times 2 \times 2} = \dfrac{5}{8}$

   (b) $\dfrac{24}{36} = \dfrac{2 \times 2 \times 2 \times 3}{2 \times 2 \times 3 \times 3} = \dfrac{2}{3}$

   (c) $\dfrac{36}{42} = \dfrac{2 \times 2 \times 3 \times 3}{2 \times 3 \times 7} = \dfrac{2 \times 3}{7} = \dfrac{6}{7}$

2. (a) $\dfrac{4}{12} = \dfrac{2 \times 2 \times 1}{2 \times 2 \times 3} = \dfrac{1}{3}$

   (b) $\dfrac{25}{125} = \dfrac{5 \times 5 \times 1}{5 \times 5 \times 5} = \dfrac{1}{5}$

   (c) $\dfrac{73}{146} = \dfrac{73 \times 1}{73 \times 2} = \dfrac{1}{2}$

3. (a) $\dfrac{18}{6} = \dfrac{2 \times 3 \times 3}{2 \times 3 \times 1} = 3$

   (b) $\dfrac{146}{73} = \dfrac{73 \times 2}{73 \times 1} = 2$

   (c) $\dfrac{28}{7} = \dfrac{2 \times 2 \times 7}{7} = 2 \times 2 = 4$

4. 56 out of $154 = \dfrac{56}{154} = \dfrac{2 \times 7 \times 2 \times 2}{2 \times 7 \times 11} = \dfrac{4}{11}$

5. (a) $\dfrac{12}{7} = 12 \div 7 \quad 7\overline{)12}$ remainder... 
   $$\dfrac{12}{7} = 12 \div 7 \quad 7\overline{\smash{)}12} \;\; \dfrac{7}{5} \text{ Remainder}$$
   $\dfrac{12}{7} = 1\dfrac{5}{7}$

   (b) $\dfrac{20}{5} = 20 \div 5 \quad 5\overline{\smash{)}20} \;\; \dfrac{20}{0} \text{ Remainder}$
   $\dfrac{20}{5} = 4$

6. (a) $3\dfrac{2}{5} = \dfrac{(3 \times 5) + 2}{5} = \dfrac{15 + 2}{5} = \dfrac{17}{5}$

   (b) $1\dfrac{3}{7} = \dfrac{(1 \times 7) + 3}{7} = \dfrac{7 + 3}{7} = \dfrac{10}{7}$

   (c) $2\dfrac{6}{11} = \dfrac{(2 \times 11) + 6}{11} = \dfrac{22 + 6}{11} = \dfrac{28}{11}$

   (d) $4\dfrac{2}{3} = \dfrac{(4 \times 3) + 2}{3} = \dfrac{12 + 2}{3} = \dfrac{14}{3}$

7. (a) $\dfrac{3}{8} = \dfrac{?}{24}; \quad 8 \times 3 = 24$
   $\dfrac{3 \times 3}{8 \times 3} = \dfrac{9}{24}$

   (b) $\dfrac{5}{6} = \dfrac{?}{30}; \quad 6 \times 5 = 30$
   $\dfrac{5 \times 5}{6 \times 5} = \dfrac{25}{30}$

   (c) $\dfrac{2}{7} = \dfrac{?}{56}; \quad 7 \times 8 = 56$
   $\dfrac{2 \times 8}{7 \times 8} = \dfrac{16}{56}$

## 0.2 Student Practice

1. (a) $\dfrac{3}{6} + \dfrac{2}{6} = \dfrac{3 + 2}{6} = \dfrac{5}{6}$

   (b) $\dfrac{3}{11} + \dfrac{8}{11} = \dfrac{3 + 8}{11} = \dfrac{11}{11} = 1$

   (c) $\dfrac{1}{8} + \dfrac{2}{8} + \dfrac{1}{8} = \dfrac{1 + 2 + 1}{8} = \dfrac{4}{8} = \dfrac{1}{2}$

   (d) $\dfrac{5}{9} + \dfrac{8}{9} = \dfrac{5 + 8}{9} = \dfrac{13}{9}$ or $1\dfrac{4}{9}$

2. (a) $\dfrac{11}{13} - \dfrac{6}{13} = \dfrac{11 - 6}{13} = \dfrac{5}{13}$

   (b) $\dfrac{8}{9} - \dfrac{2}{9} = \dfrac{8 - 2}{9} = \dfrac{6}{9} = \dfrac{2}{3}$

3. The LCD is 56, since 56 is exactly divisible by 8 and 7. There is no smaller number that is exactly divisible by 8 and 7.

4. Find the LCD using prime factors.
   $\dfrac{8}{35}$ and $\dfrac{6}{15}$
   $35 = 7 \cdot 5$
   $15 = \;\;\;\; 5 \cdot 3$
   $LCD = 7 \cdot 5 \cdot 3 = 105$

5. Find the LCD of $\dfrac{5}{12}$ and $\dfrac{7}{30}$.
   $12 = 3 \cdot 2 \cdot 2$
   $30 = 3 \cdot \;\; 2 \cdot 5$
   $LCD = 3 \cdot 2 \cdot 2 \cdot 5 = 60$

6. Find the LCD of $\dfrac{2}{27}, \dfrac{1}{18},$ and $\dfrac{5}{12}$.
   $27 = 3 \cdot 3 \cdot 3$
   $18 = \;\; 3 \cdot 3 \cdot 2$
   $12 = \;\;\;\;\;\; 3 \cdot 2 \cdot 2$
   $LCD = 3 \cdot 3 \cdot 3 \cdot 2 \cdot 2 = 108$

7. From Student Practice 3, the LCD is 56.
   $\dfrac{5}{7} = \dfrac{5 \times 8}{7 \times 8} = \dfrac{40}{56} \qquad \dfrac{1}{8} = \dfrac{1 \times 7}{8 \times 7} = \dfrac{7}{56}$
   $\dfrac{5}{7} + \dfrac{1}{8} = \dfrac{40}{56} + \dfrac{7}{56} = \dfrac{47}{56}$
   $\dfrac{47}{56}$ of the farm fields were planted in corn or soybeans.

8. We can see by inspection that both 5 and 25 divide exactly into 50. Thus 50 is the LCD.
   $\dfrac{4}{5} = \dfrac{4 \times 10}{5 \times 10} = \dfrac{40}{50} \qquad \dfrac{6}{25} = \dfrac{6 \times 2}{25 \times 2} = \dfrac{12}{50}$
   $\dfrac{4}{5} + \dfrac{6}{25} + \dfrac{1}{50} = \dfrac{40}{50} + \dfrac{12}{50} + \dfrac{1}{50} = \dfrac{40 + 12 + 1}{50} = \dfrac{53}{50}$ or $1\dfrac{3}{50}$

9. Add $\dfrac{1}{49} + \dfrac{3}{14}$
   First find the LCD.
   $49 = 7 \cdot 7$
   $14 = \;\;\;\; 7 \cdot 2$
   $LCD = 7 \cdot 7 \cdot 2 = 98$
   Then change to equivalent fractions and add.
   $\dfrac{1}{49} = \dfrac{1 \times 2}{49 \times 2} = \dfrac{2}{98} \qquad \dfrac{3}{14} = \dfrac{3 \times 7}{14 \times 7} = \dfrac{21}{98}$
   $\dfrac{1}{49} + \dfrac{3}{14} = \dfrac{2}{98} + \dfrac{21}{98} = \dfrac{23}{98}$

10. $\dfrac{1}{12} - \dfrac{1}{30}$
    First find the LCD.
    $12 = 2 \cdot 2 \cdot 3$
    $30 = \;\; 2 \cdot 3 \cdot 5$
    $LCD = 2 \cdot 2 \cdot 3 \cdot 5 = 60$
    Then change to equivalent fractions and subtract.
    $\dfrac{1}{12} = \dfrac{1 \times 5}{12 \times 5} = \dfrac{5}{60} \qquad \dfrac{1}{30} = \dfrac{1 \times 2}{30 \times 2} = \dfrac{2}{60}$
    $\dfrac{1}{12} - \dfrac{1}{30} = \dfrac{5}{60} - \dfrac{2}{60} = \dfrac{3}{60} = \dfrac{1}{20}$

11. $\dfrac{2}{3} + \dfrac{3}{4} - \dfrac{3}{8}$

First find the LCD.

$$3 = 3$$
$$4 = \quad 2\cdot2$$
$$8 = \quad 2\cdot2\cdot2$$

$$LCD = 3\cdot2\cdot2\cdot2 = 24$$

$$\dfrac{2}{3} = \dfrac{2\times8}{3\times8} = \dfrac{16}{24} \qquad \dfrac{3}{4} = \dfrac{3\times6}{4\times6} = \dfrac{18}{24} \qquad \dfrac{3}{8} = \dfrac{3\times3}{8\times3} = \dfrac{9}{24}$$

Now combine the fractions.

$$\dfrac{2}{3} + \dfrac{3}{4} - \dfrac{3}{8} = \dfrac{16}{24} + \dfrac{18}{24} - \dfrac{9}{24} = \dfrac{25}{24} \text{ or } 1\dfrac{1}{24}$$

12. **(a)** The LCD of 3 and 5 is 15.

$$1\dfrac{2}{3} = \dfrac{5}{3} = \dfrac{5\times5}{3\times5} = \dfrac{25}{15} \qquad 2\dfrac{4}{5} = \dfrac{14}{5} = \dfrac{14\times3}{5\times3} = \dfrac{42}{15}$$

$$1\dfrac{2}{3} + 2\dfrac{4}{5} = \dfrac{25}{15} + \dfrac{42}{15} = \dfrac{67}{15} = 4\dfrac{7}{15}$$

**(b)** The LCD of 4 and 3 is 12.

$$5\dfrac{1}{4} = \dfrac{21}{4} = \dfrac{21\times3}{4\times3} = \dfrac{63}{12} \qquad 2\dfrac{2}{3} = \dfrac{8}{3} = \dfrac{8\times4}{3\times4} = \dfrac{32}{12}$$

$$5\dfrac{1}{4} - 2\dfrac{2}{3} = \dfrac{63}{12} - \dfrac{32}{12} = \dfrac{31}{12} = 2\dfrac{7}{12}$$

13.

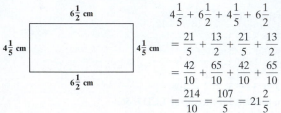

$$4\dfrac{1}{5} + 6\dfrac{1}{2} + 4\dfrac{1}{5} + 6\dfrac{1}{2}$$
$$= \dfrac{21}{5} + \dfrac{13}{2} + \dfrac{21}{5} + \dfrac{13}{2}$$
$$= \dfrac{42}{10} + \dfrac{65}{10} + \dfrac{42}{10} + \dfrac{65}{10}$$
$$= \dfrac{214}{10} = \dfrac{107}{5} = 21\dfrac{2}{5}$$

The perimeter is $21\dfrac{2}{5}$ cm.

## 0.3 Student Practice

1. **(a)** $\dfrac{2}{7} \times \dfrac{5}{11} = \dfrac{2\cdot5}{7\cdot11} = \dfrac{10}{77}$

   **(b)** $\dfrac{1}{5} \times \dfrac{7}{10} = \dfrac{1\cdot7}{5\cdot10} = \dfrac{7}{50}$

   **(c)** $\dfrac{9}{5} \times \dfrac{1}{4} = \dfrac{9\cdot1}{5\cdot4} = \dfrac{9}{20}$

   **(d)** $\dfrac{8}{9} \times \dfrac{3}{10} = \dfrac{8\cdot3}{9\cdot10} = \dfrac{24}{90} = \dfrac{4}{15}$

2. **(a)** $\dfrac{3}{5} \times \dfrac{4}{3} = \dfrac{3\cdot4}{5\cdot3} = \dfrac{4\cdot3}{5\cdot3} = \dfrac{4}{5}$

   **(b)** $\dfrac{9}{10} \times \dfrac{5}{12} = \dfrac{3\cdot3}{2\cdot5} \times \dfrac{5}{2\cdot2\cdot3} = \dfrac{3}{8}$

3. **(a)** $4 \times \dfrac{2}{7} = \dfrac{4}{1} \times \dfrac{2}{7} = \dfrac{4\cdot2}{1\cdot7} = \dfrac{8}{7} \text{ or } 1\dfrac{1}{7}$

   **(b)** $12 \times \dfrac{3}{4} = \dfrac{12}{1} \times \dfrac{3}{4} = \dfrac{4\cdot3}{1} \times \dfrac{3}{4} = \dfrac{9}{1} = 9$

4. Multiply. $5\dfrac{3}{5}$ times $3\dfrac{3}{4}$

$$5\dfrac{3}{5} \times 3\dfrac{3}{4} = \dfrac{28}{5} \times \dfrac{15}{4} = \dfrac{4\cdot7}{5} \times \dfrac{3\cdot5}{4} = \dfrac{21}{1} = 21$$

The area of the field is 21 square miles.

5. $3\dfrac{1}{2} \times \dfrac{1}{14} \times 4 = \dfrac{7}{2} \times \dfrac{1}{14} \times \dfrac{4}{1} = \dfrac{7}{2} \times \dfrac{1}{2\cdot7} \times \dfrac{2\cdot2}{1} = 1$

6. **(a)** $\dfrac{2}{5} \div \dfrac{1}{3} = \dfrac{2}{5} \times \dfrac{3}{1} = \dfrac{6}{5} \text{ or } 1\dfrac{1}{5}$

   **(b)** $\dfrac{12}{13} \div \dfrac{4}{3} = \dfrac{12}{13} \times \dfrac{3}{4} = \dfrac{4\cdot3}{13} \times \dfrac{3}{4} = \dfrac{9}{13}$

7. **(a)** $\dfrac{3}{7} \div 6 = \dfrac{3}{7} \div \dfrac{6}{1} = \dfrac{3}{7} \times \dfrac{1}{6} = \dfrac{3}{7} \times \dfrac{1}{2\cdot3} = \dfrac{1}{14}$

   **(b)** $8 \div \dfrac{2}{3} = \dfrac{8}{1} \div \dfrac{2}{3} = \dfrac{8}{1} \times \dfrac{3}{2} = \dfrac{2\cdot4}{1} \times \dfrac{3}{2} = \dfrac{12}{1} = 12$

8. **(a)** $\dfrac{\frac{3}{11}}{\frac{5}{7}} = \dfrac{3}{11} \div \dfrac{5}{7} = \dfrac{3}{11} \times \dfrac{7}{5} = \dfrac{21}{55}$

   **(b)** $\dfrac{\frac{12}{5}}{\frac{8}{15}} = \dfrac{12}{5} \div \dfrac{8}{15} = \dfrac{12}{5} \times \dfrac{15}{8} = \dfrac{3\cdot4}{5} \times \dfrac{3\cdot5}{2\cdot4} = \dfrac{9}{2} \text{ or } 4\dfrac{1}{2}$

9. **(a)** $1\dfrac{2}{5} \div 2\dfrac{1}{3} = \dfrac{7}{5} \div \dfrac{7}{3} = \dfrac{7}{5} \times \dfrac{3}{7} = \dfrac{3}{5}$

   **(b)** $4\dfrac{2}{3} \div 7 = \dfrac{14}{3} \div \dfrac{7}{1} = \dfrac{14}{3} \times \dfrac{1}{7} = \dfrac{2\cdot7}{3} \times \dfrac{1}{7} = \dfrac{2}{3}$

   **(c)** $\dfrac{1\frac{1}{5}}{1\frac{2}{7}} = 1\dfrac{1}{5} \div 1\dfrac{2}{7} = \dfrac{6}{5} \div \dfrac{9}{7} = \dfrac{6}{5} \times \dfrac{7}{9} = \dfrac{2\cdot3}{5} \times \dfrac{7}{3\cdot3} = \dfrac{14}{15}$

10. $64 \div 5\dfrac{1}{3} = \dfrac{64}{1} \div \dfrac{16}{3} = \dfrac{64}{1} \times \dfrac{3}{16} = \dfrac{4\cdot16}{1} \times \dfrac{3}{16} = \dfrac{12}{1} = 12$

He can fill 12 jars.

11. $126 \div 5\dfrac{1}{4} = \dfrac{126}{1} \div \dfrac{21}{4} = \dfrac{126}{1} \times \dfrac{4}{21} = \dfrac{6\cdot21}{1} \times \dfrac{4}{21} = 24$

The car got 24 miles per gallon.

## 0.4 Student Practice

1. **(a)** $1; 0.9 = \dfrac{9}{10}$ = nine tenths

   **(b)** $2; 0.09 = \dfrac{9}{100}$ = nine hundredths

   **(c)** $3; 0.731 = \dfrac{731}{1000}$ = seven hundred thirty-one thousandths

   **(d)** $3; 1.371 = 1\dfrac{371}{1000}$ = one and three hundred seventy-one thousandths

   **(e)** $4; 0.0005 = \dfrac{5}{10{,}000}$ = five ten-thousandths

2. **(a)** $\dfrac{3}{8} = 0.375$

$$\begin{array}{r} 0.375 \\ 8\overline{)3.000} \\ \underline{2\ 4} \\ 60 \\ \underline{56} \\ 40 \\ \underline{40} \\ 0 \end{array}$$

   **(b)** $\dfrac{7}{200} = 0.035$

$$\begin{array}{r} 0.035 \\ 200\overline{)7.000} \\ \underline{6\ 00} \\ 1\ 000 \\ \underline{1\ 000} \\ 0 \end{array}$$

   **(c)** $\dfrac{33}{20} = 1.65$

$$\begin{array}{r} 1.65 \\ 20\overline{)33.00} \\ \underline{20} \\ 13\ 0 \\ \underline{12\ 0} \\ 1\ 00 \\ \underline{1\ 00} \\ 0 \end{array}$$

3. **(a)** $\dfrac{1}{6} = 0.1666\ldots$ or $0.1\overline{6}$

$$\begin{array}{r} 0.166 \\ 6\overline{)1.000} \\ \underline{6} \\ 40 \\ \underline{36} \\ 40 \\ \underline{36} \\ 4 \end{array}$$

   **(b)** $\dfrac{5}{11} = 0.454545\ldots$ or $0.\overline{45}$

$$\begin{array}{r} 0.4545 \\ 11\overline{)5.0000} \\ \underline{4\ 4} \\ 60 \\ \underline{55} \\ 50 \\ \underline{44} \\ 60 \\ \underline{55} \\ 5 \end{array}$$

**4. (a)** $0.8 = \dfrac{8}{10} = \dfrac{4}{5}$     **(b)** $0.88 = \dfrac{88}{100} = \dfrac{22}{25}$

**(c)** $0.45 = \dfrac{45}{100} = \dfrac{9}{20}$     **(d)** $0.148 = \dfrac{148}{1000} = \dfrac{37}{250}$

**(e)** $0.612 = \dfrac{612}{1000} = \dfrac{153}{250}$     **(f)** $0.016 = \dfrac{16}{1000} = \dfrac{2}{125}$

**5. (a)** 
$$
\begin{array}{r} 3.12 \\ +5.08 \\ \hline 8.20 \end{array}
$$
**(b)** 
$$
\begin{array}{r} 152.003 \\ -136.118 \\ \hline 15.885 \end{array}
$$
**(c)** 
$$
\begin{array}{r} 1.1 \\ 3.16 \\ +5.123 \\ \hline 9.383 \end{array}
$$
**(d)** 
$$
\begin{array}{r} 1.0052 \\ -0.1234 \\ \hline 0.8818 \end{array}
$$

**6. (a)** 
$$
\begin{array}{r} 0.0610 \\ 5.0008 \\ +1.3000 \\ \hline 6.3618 \end{array}
$$
**(b)** 
$$
\begin{array}{r} 18.000 \\ -0.126 \\ \hline 17.874 \end{array}
$$

**7.** 
$$
\begin{array}{r} 0.5 \quad \text{(one decimal place)} \\ \times\, 0.3 \quad \text{(one decimal place)} \\ \hline 0.15 \quad \text{(two decimal places)} \end{array}
$$

**8.** 
$$
\begin{array}{r} 0.12 \quad \text{(two decimal places)} \\ \times\, 0.4 \quad \text{(one decimal place)} \\ \hline 0.048 \quad \text{(three decimal places)} \end{array}
$$

**9. (a)** 
$$
\begin{array}{r} 1.23 \quad \text{(two decimal places)} \\ \times\, 0.005 \quad \text{(three decimal places)} \\ \hline 0.00615 \quad \text{(five decimal places)} \end{array}
$$
**(b)** 
$$
\begin{array}{r} 0.003 \quad \text{(three decimal places)} \\ \times\, 0.00002 \quad \text{(five decimal places)} \\ \hline 0.00000006 \quad \text{(eight decimal places)} \end{array}
$$

**10.** 
$$
\begin{array}{r} 5.26 \\ 6\overline{)31.56} \\ \underline{30} \\ 1\,5 \\ \underline{1\,2} \\ 36 \\ \underline{36} \\ 0 \end{array}
$$
Each box of paper costs $5.26.

**11.** 
$$
0.06_\wedge\overline{)\,1800.00_\wedge}
$$
$$
\begin{array}{r} 300\,00. \\ \underline{18} \\ 00000 \end{array}
$$
Thus, $1800 \div 0.06 = 30{,}000$.

**12.** 
$$
4.9_\wedge\overline{)\,0.0_\wedge1764}
$$
$$
\begin{array}{r} 0.0036 \\ \underline{147} \\ 294 \\ \underline{294} \\ 0 \end{array}
$$
Thus, $0.01764 \div 4.9 = 0.0036$.

**13. (a)** $0.0016 \times 100 = 0.16$
Move decimal point 2 places to the right.
**(b)** $2.34 \times 1000 = 2340$
Move decimal point 3 places to the right.
**(c)** $56.75 \times 10{,}000 = 567{,}500$
Move decimal point 4 places to the right.

**14. (a)** $\dfrac{5.82}{10} = 0.582$ (Move decimal point 1 place to the left.)

**(b)** $\dfrac{123.4}{1000} = 0.1234$ (Move decimal point 3 places to the left.)

**(c)** $\dfrac{0.00614}{10{,}000} = 0.000000614$ (Move decimal point 4 places to the left.)

## 0.5 Student Practice

**1. (a)** $0.92 = 92\%$
**(b)** $0.0736 = 7.36\%$
**(c)** $0.7 = 0.70 = 70\%$
**(d)** $0.0003 = 0.03\%$
**2. (a)** $3.04 = 304\%$
**(b)** $5.186 = 518.6\%$
**(c)** $2.1 = 2.10 = 210\%$
**3. (a)** $7\% = 0.07$
**(b)** $9.3\% = 0.093$
**(c)** $131\% = 1.31$
**(d)** $0.04\% = 0.0004$
**4.** Change the percents to decimals and multiply.
**(a)** 18% of $50 = 0.18 \times 50 = 9$
**(b)** 4% of $64 = 0.04 \times 64 = 2.56$
**(c)** 156% of $35 = 1.56 \times 35 = 54.6$
**5. (a)** 4.2% of $38{,}000 = 0.042 \times 38{,}000 = 1596$
His raise is $1596.
**(b)** $38{,}000 + 1596 = 39{,}596$
His new salary is $39,596.
**6.** $\dfrac{37}{148}$ reduces to $\dfrac{37 \cdot 1}{37 \cdot 4} = \dfrac{1}{4} = 0.25 = 25\%$

**7. (a)** $\dfrac{24}{48} = \dfrac{1}{2} = 0.5 = 50\%$

**(b)** $\dfrac{4}{25} = 0.16 = 16\%$

**8.** $\dfrac{430}{1256} = \dfrac{215}{628} \approx 0.342 \approx 34\%$

**9.** Round 128,621 to 100,000. Round 378 to 400.
$100{,}000 \times 400 = 40{,}000{,}000$
**10.** Round $12\frac{1}{2}$ feet to 10 feet. Round $9\frac{3}{4}$ feet to 10 feet.
First room: $10 \times 10 = 100$ square feet
Round $11\frac{1}{4}$ feet to 10 feet. Round $18\frac{1}{2}$ feet to 20 feet.
Second room: $10 \times 20 = 200$ square feet
$100 + 200 = 300$
The estimate is 300 square feet.
**11. (a)** Round 422.8 miles to 400 miles. Round 19.3 gallons to 20 gallons.
$$
20\overline{)400}
$$
(quotient 20)
Roberta's truck gets about 20 miles per gallon.

**(b)** Round 3862 miles to 4000 miles.
$$
20\overline{)4000}
$$
(quotient 200)
She will use about 200 gallons of gas for her trip.
Round $3.69\frac{9}{10}$ to $4.00.
$200 \times \$4 = \$800$
The estimated cost is $800.

## 0.6 Student Practice

**1.** Mathematics Blueprint for Problem Solving

| Gather the Facts | What Am I Solving for? | What Must I Calculate? | Key Points to Remember |
|---|---|---|---|
| Living room measures $16\frac{1}{2}$ ft by $10\frac{1}{2}$ ft. | Area of room in square feet. | Multiply $16\frac{1}{2}$ ft by $10\frac{1}{2}$ ft to get the area in square feet. | 9 sq feet = 1 sq yard |
| | Area of room in square yards. | | |
| The carpet costs $20.00 per square yard. | Cost of the carpet. | Divide the number of square feet by 9 to get the number of square yards. | |
| | | Multiply the number of square yards by $20.00. | |

$16\frac{1}{2}$ ft $\times 10\frac{1}{2}$ ft $= \frac{33}{2}$ ft $\times \frac{21}{2}$ ft $= \frac{693}{4}$ or $173\frac{1}{4}$ square feet

$173\frac{1}{4} \div 9 = \frac{693}{4} \div \frac{9}{1} = \frac{693}{4} \times \frac{1}{9} = \frac{77}{4} = 19\frac{1}{4}$

$19\frac{1}{4}$ square yards of carpet are needed.

$19\frac{1}{4} \times 20 = \frac{77}{4} \times \frac{20}{1} = 385$

It will cost a minimum of $385.
**Check:** Estimate area of room: $20 \times 10 = 200$ square feet
Estimate area in square yards: $200 \div 10 = 20$
Estimate the cost: $20 \times 20 = \$400.00$
This is close to our answer of $385.00. Our answer seems reasonable.
**To Think About:** 6.5 sq ft; 52.5 sq ft; $468

# Chapter 1  ## 1.1 Student Practice

**1.**

| | Number | Integer | Rational Number | Irrational Number | Real Number |
|---|---|---|---|---|---|
| **(a)** | $-\frac{2}{5}$ | | X | | X |
| **(b)** | $1.515151\ldots$ | | X | | X |
| **(c)** | $-8$ | X | X | | X |
| **(d)** | $\pi$ | | | X | X |

**2. (a)** Population growth of 1259 is $+1259$.
**(b)** Depreciation of $763 is $-763$.
**(c)** Wind-chill factor of minus 10 is $-10$.

**3. (a)** The additive inverse (opposite) of $\frac{2}{5}$ is $-\frac{2}{5}$.
**(b)** The additive inverse (opposite) of $-1.92$ is $+1.92$.
**(c)** The opposite of a loss of 12 yards on a football play is a gain of 12 yards on the play.

**4. (a)** $|-7.34| = 7.34$
**(b)** $\left|\frac{5}{8}\right| = \frac{5}{8}$
**(c)** $\left|\frac{0}{2}\right| = 0$

**5. (a)** $37 + 19$
$37 + 19 = 56$
$37 + 19 = +56$
**(b)** $-23 + (-35)$
$23 + 35 = 58$
$-23 + (-35) = -58$

**6.** $-\frac{3}{5} + \left(-\frac{4}{7}\right)$
$-\frac{21}{35} + \left(-\frac{20}{35}\right)$
$\frac{21}{35} + \frac{20}{35} = \frac{41}{35}$ or $1\frac{6}{35}$
$-\frac{21}{35} + \left(-\frac{20}{35}\right) = -\frac{41}{35}$ or $-1\frac{6}{35}$

**7.** $-12.7 + (-9.38)$
$12.7 + 9.38 = 22.08$
$-12.7 + (-9.38) = -22.08$

**8.** $-7 + (-11) + (-33)$
$= -18 + (-33)$
$= -51$

**9.** $-9 + 15$
$15 - 9 = 6$
$-9 + 15 = +6$ or $6$

**2. (a)** $\frac{10}{55} \approx 0.182 \approx 18\%$
**(b)** $\frac{3,660,000}{13,240,000} \approx 0.276 \approx 28\%$
**(c)** $\frac{15}{55} \approx 0.273 \approx 27\%$
**(d)** $\frac{3,720,000}{13,240,000} \approx 0.281 \approx 28\%$
**(e)** We notice that 18% of the company's sales force is located in the Northwest, and they were responsible for 28% of the sales volume. The percent of sales compared to the percent of sales force is about 150%. 27% of the company's sales force is located in the Southwest, and they were responsible for 28% of the sales volume. The percent of sales compared to the percent of sales force is approximately 100%. It would appear that the Northwest sales force is more effective.
**To Think About:** 88; 35

**10.** $-\frac{5}{12} + \frac{7}{12} + \left(-\frac{11}{12}\right)$
$= \frac{2}{12} + \left(-\frac{11}{12}\right) = -\frac{9}{12} = -\frac{3}{4}$

**11.** $-6.3 + (-8.0) + 3.5$
$= -14.3 + 3.5$
$= -10.8$

**12.**   $-6$   $+5$
      $-7$   $+5$
      $\underline{-2}$   $\underline{+3}$
      $-15$   $13$
$-6 + 5 + (-7) + (-2) + 5 + 3 = -15 + 13 = -2$

**13. (a)** $-2.9 + (-5.7) = -8.6$
**(b)** $\frac{2}{3} + \left(-\frac{1}{4}\right)$
$= \frac{8}{12} + \left(-\frac{3}{12}\right) = \frac{5}{12}$

## 1.2 Student Practice

**1.** $9 - (-3) = 9 + (+3) = 12$
**2.** $-12 - (-5) = -12 + (+5) = -7$
**3. (a)** $\frac{5}{9} - \frac{7}{9} = \frac{5}{9} + \left(-\frac{7}{9}\right) = -\frac{2}{9}$
**(b)** $-\frac{5}{21} - \left(-\frac{3}{7}\right) = -\frac{5}{21} + \frac{3}{7} = -\frac{5}{21} + \frac{9}{21} = \frac{4}{21}$
**4.** $-17.3 - (-17.3) = -17.3 + 17.3 = 0$
**5. (a)** $-21 - 9$
$= -21 + (-9)$
$= -30$
**(b)** $17 - 36$
$= 17 + (-36)$
$= -19$
**(c)** $12 - (-15)$
$= 12 + 15$
$= 27$
**(d)** $\frac{3}{5} - 2$
$= \frac{3}{5} + (-2)$
$= \frac{3}{5} + \left(-\frac{10}{5}\right) = -\frac{7}{5}$ or $-1\frac{2}{5}$

**6.** $350 - (-186) = 350 + 186 = 536$
The helicopter is 536 feet from the sunken vessel.

## 1.3 Student Practice

**1. (a)** $(-6)(-2) = 12$
**(b)** $(7)(9) = 63$
**(c)** $\left(-\frac{3}{5}\right)\left(\frac{2}{7}\right) = -\frac{6}{35}$
**(d)** $\left(\frac{5}{6}\right)(-7) = \left(\frac{5}{6}\right)\left(-\frac{7}{1}\right) = -\frac{35}{6}$ or $-5\frac{5}{6}$
**2.** $(-5)(-2)(-6) = (+10)(-6) = -60$

3. (a) positive; $-2(-3) = 6$
   (b) negative; $(-1)(-3)(-2) = +3(-2) = -6$
   (c) positive;
      $-4\left(-\frac{1}{4}\right)(-2)(-6) = +1(-2)(-6) = -2(-6) = +12$ or $12$
4. (a) $-36 \div (-2) = 18$
   (b) $49 \div 7 = 7$
   (c) $\frac{50}{-10} = -5$
   (d) $\frac{-39}{13} = -3$
5. (a) The numbers have the same sign, so the result will be positive.
      Divide the absolute values.

$$1.8_{\wedge} \overline{\smash{)}12.6_{\wedge}} \quad \frac{7.}{\underline{12\,6}_{\wedge}}$$

   $-12.6 \div (-1.8) = 7$
   (b) The numbers have different signs, so the result will be negative.
      Divide the absolute values.

$$0.9_{\wedge} \overline{\smash{)}0.4\,5} \quad \frac{0.5}{\underline{4^{\wedge}5}}$$

   $0.45 \div (-0.9) = -0.5$

6. $-\frac{5}{16} \div \left(-\frac{10}{13}\right) = \left(-\frac{5}{16}\right)\left(-\frac{13}{10}\right) = \left(-\frac{\overset{1}{\cancel{5}}}{16}\right)\left(-\frac{13}{\underset{2}{\cancel{10}}}\right) = \frac{13}{32}$

7. (a) $\dfrac{\dfrac{-12}{4}}{-\dfrac{4}{5}} = -\frac{12}{1} \div \left(-\frac{4}{5}\right) = \left(-\frac{\overset{3}{\cancel{12}}}{1}\right)\left(-\frac{5}{\underset{1}{\cancel{4}}}\right) = 15$

   (b) $\dfrac{-\dfrac{2}{9}}{\dfrac{8}{13}} = -\frac{2}{9} \div \frac{8}{13} = -\frac{\overset{1}{\cancel{2}}}{9}\left(\frac{13}{\underset{4}{\cancel{8}}}\right) = -\frac{13}{36}$

8. (a) $6(-10) = -60$
      The team lost approximately 60 yards with plays that were considered medium losses.
   (b) $7(15) = 105$
      The team gained approximately 105 yards with plays that were considered medium gains.
   (c) $-60 + 105 = 45$
      A total of 45 yards were gained during plays that were medium losses or medium gains.

## 1.4 Student Practice

1. (a) $6(6)(6)(6) = 6^4$
   (b) $-2(-2)(-2)(-2)(-2) = (-2)^5$
   (c) $108(108)(108) = 108^3$
   (d) $-11(-11)(-11)(-11)(-11) = (-11)^6$
   (e) $(w)(w)(w) = w^3$
   (f) $(z)(z)(z)(z) = z^4$
2. (a) $3^5 = (3)(3)(3)(3)(3) = 243$
   (b) $2^2 = (2)(2) = 4$
      $3^3 = (3)(3)(3) = 27$
      $2^2 + 3^3 = 4 + 27 = 31$
3. (a) $(-3)^3 = -27$
   (b) $(-2)^6 = +64$
   (c) $-2^4 = -(2^4) = -16$
   (d) $-(6^3) = -216$
4. (a) $\left(\frac{1}{3}\right)^3 = \left(\frac{1}{3}\right)\left(\frac{1}{3}\right)\left(\frac{1}{3}\right) = \frac{1}{27}$
   (b) $(0.3)^4 = (0.3)(0.3)(0.3)(0.3) = 0.0081$
   (c) $\left(\frac{3}{2}\right)^4 = \left(\frac{3}{2}\right)\left(\frac{3}{2}\right)\left(\frac{3}{2}\right)\left(\frac{3}{2}\right) = \frac{81}{16}$
   (d) $3^4 = (3)(3)(3)(3) = 81$
      $4^2 = (4)(4) = 16$
      $(3)^4(4)^2 = (81)(16) = 1296$
   (e) $4^2 - 2^4 = 16 - 16 = 0$

## 1.5 Student Practice

1. $25 \div 5 \cdot 6 + 2^3$
   $= 25 \div 5 \cdot 6 + 8$
   $= 5 \cdot 6 + 8$
   $= 30 + 8$
   $= 38$
2. $(-4)^3 - 2^6 = -64 - 64 = -128$
3. $6 - (8 - 12)^2 + 8 \div 2$
   $= 6 - (-4)^2 + 8 \div 2$
   $= 6 - (16) + 8 \div 2$
   $= 6 - 16 + 4$
   $= -10 + 4$
   $= -6$
4. $\left(-\frac{1}{7}\right)\left(-\frac{14}{5}\right) + \left(-\frac{1}{2}\right) \div \left(\frac{3}{4}\right)^2$
   $= \left(-\frac{1}{7}\right)\left(-\frac{14}{5}\right) + \left(-\frac{1}{2}\right)\left(\frac{16}{9}\right)$
   $= \frac{2}{5} + \left(-\frac{8}{9}\right)$
   $= \frac{2 \cdot 9}{5 \cdot 9} + \left(-\frac{8 \cdot 5}{9 \cdot 5}\right)$
   $= \frac{18}{45} + \left(-\frac{40}{45}\right)$
   $= -\frac{22}{45}$

## 1.6 Student Practice

1. (a) $3(x + 2y) = 3x + 3(2y) = 3x + 6y$
   (b) $-2(a + 3b) = -2(a) + (-2)(+3b) = -2a - 6b$
2. (a) $-(-3x + y) = (-1)(-3x + y)$
      $= (-1)(-3x) + (-1)(y)$
      $= 3x - y$
3. (a) $\frac{3}{5}(a^2 - 5a + 25) = \left(\frac{3}{5}\right)(1a^2) + \left(\frac{3}{5}\right)(-5a) + \left(\frac{3}{5}\right)(25)$
      $= \frac{3}{5}a^2 - 3a + 15$
   (b) $2.5(x^2 - 3.5x + 1.2)$
      $= (2.5)(1x^2) + (2.5)(-3.5x) + (2.5)(1.2)$
      $= 2.5x^2 - 8.75x + 3$
4. $-4x(x - 2y + 3) = (-4)(x)(x) + (-4)(x)(-2)(y) + (-4)(x)(3)$
   $= -4x^2 + 8xy - 12x$
5. $(3x^2 - 2x)(-4) = (3x^2)(-4) + (-2x)(-4) = -12x^2 + 8x$
6. $400(6x + 9y) = 400(6x) + 400(9y)$
   $= 2400x + 3600y$
   The area of the field in square feet is $2400x + 3600y$.

## 1.7 Student Practice

1. (a) $5a$ and $8a$ are like terms.
      $2b$ and $-4b$ are like terms.
   (b) $y^2$ and $-7y^2$ are like terms. These are the only like terms.
2. (a) $16y^3 + 9y^3 = (16 + 9)y^3 = 25y^3$
   (b) $5a + 7a + 4a = (5 + 7 + 4)a = 16a$
3. $-8y^2 - 9y^2 + 4y^2 = (-8 - 9 + 4)y^2 = -13y^2$
4. (a) $1.3x + 3a - 9.6x + 2a = -8.3x + 5a$
   (b) $5ab - 2ab^2 - 3a^2b + 6ab = 11ab - 2ab^2 - 3a^2b$
   (c) There are no like terms in the expression
      $7x^2y - 2xy^2 - 3x^2y^2 - 4xy$, so no terms can be combined.
5. $5xy - 2x^2y + 6xy^2 - xy - 3xy^2 - 7x^2y$
   $= 5xy - xy - 2x^2y - 7x^2y + 6xy^2 - 3xy^2$
   $= 4xy - 9x^2y + 3xy^2$

**6.** $\frac{1}{7}a^2 - \frac{5}{12}b + 2a^2 - \frac{1}{3}b$

$\frac{1}{7}a^2 + 2a^2 = \frac{1}{7}a^2 + \frac{2}{1}a^2 = \frac{1}{7}a^2 + \frac{2 \cdot 7}{1 \cdot 7}a^2$

$\qquad = \frac{1}{7}a^2 + \frac{14}{7}a^2 = \frac{15}{7}a^2$

$-\frac{5}{12}b - \frac{1}{3}b = -\frac{5}{12}b - \frac{1 \cdot 4}{3 \cdot 4}b = -\frac{5}{12}b - \frac{4}{12}b$

$\qquad = -\frac{9}{12}b = -\frac{3}{4}b$

Thus, our solution is $\frac{15}{7}a^2 - \frac{3}{4}b$.

**7.** $5a(2 - 3b) - 4(6a + 2ab) = 10a - 15ab - 24a - 8ab$
$\qquad\qquad\qquad\qquad\qquad\quad = -14a - 23ab$

## 1.8  Student Practice

**1.** $4 - \frac{1}{2}x = 4 - \frac{1}{2}(-8)$

$\qquad = 4 + 4$

$\qquad = 8$

**2. (a)** $4x^2 = 4(-3)^2 = 4(9) = 36$
  **(b)** $(4x)^2 = [4(-3)]^2 = (-12)^2 = 144$

**3.** $2x^2 - 3x = 2(-2)^2 - 3(-2)$

$\qquad\qquad = 2(4) - 3(-2)$

$\qquad\qquad = 8 + 6$

$\qquad\qquad = 14$

**4.** $6a + 4ab^2 - 5 = 6\left(-\frac{1}{6}\right) + 4\left(-\frac{1}{6}\right)(3)^2 - 5$

$\qquad\qquad\qquad = -1 + 4\left(-\frac{1}{6}\right)(9) - 5$

$\qquad\qquad\qquad = -1 + (-6) - 5$

$\qquad\qquad\qquad = -1 + (-6) + (-5)$

$\qquad\qquad\qquad = -12$

**5.** Area of a triangle is $A = \frac{1}{2}ab$
altitude = 3 meters (m)
base = 7 meters (m)

$A = \frac{1}{2}(3\,\text{m})(7\,\text{m})$

$\quad = \frac{1}{2}(3)(7)(\text{m})(\text{m})$

$\quad = \left(\frac{3}{2}\right)(7)(\text{m})^2$

$\quad = \frac{21}{2}(\text{m})^2$

$\quad = 10.5$ square meters

## Chapter 2   2.1 Student Practice

**1.** $\qquad x + 14 = 23$
$x + 14 + (-14) = 23 + (-14)$
$\qquad\quad x + 0 = 9$
$\qquad\qquad x = 9$
***Check.*** $\quad x + 14 = 23$
$\qquad\qquad 9 + 14 \overset{?}{=} 23$
$\qquad\qquad\qquad 23 = 23$ ✓

**2.** $\qquad 17 = x - 5$        ***Check.*** $\quad 17 = x - 5$
$\quad 17 + 5 = x - 5 + 5$        $\qquad 17 \overset{?}{=} 22 - 5$
$\qquad 22 = x + 0$        $\qquad 17 = 17$ ✓
$\qquad 22 = x$

**3.** $\quad 0.5 - 1.2 = x - 0.3$        ***Check.*** $\quad 0.5 - 1.2 = x - 0.3$
$\qquad -0.7 = x - 0.3$        $\qquad 0.5 - 1.2 \overset{?}{=} -0.4 - 0.3$
$-0.7 + 0.3 = x - 0.3 + 0.3$        $\qquad -0.7 = -0.7$ ✓
$\qquad -0.4 = x$

**4.** $\quad x + 8 = -22 + 6$
$\quad -2 + 8 \overset{?}{=} -22 + 6$
$\qquad\quad 6 \neq -16$        This is not true.

**6.** Area of a circle is
$A = \pi r^2$
$r = 3$ meters
$A \approx 3.14(3\,\text{m})^2$
$\quad \approx 3.14(9)(\text{m})^2$
$\quad \approx 28.26$ square meters

**7.** Formula

$C = \frac{5}{9}(F - 32)$

$\quad = \frac{5}{9}(68 - 32)$

$\quad = \frac{5}{9}(36)$

$\quad = 5(4)$

$\quad = 20°$ Celsius

The temperature is 20° Celsius or 20°C.

**8.** Use the formula.
$k \approx 1.61(r)$
$\quad \approx 1.61(35)$    Replace $r$ by 35.
$\quad \approx 56.35$
The truck is traveling at approximately 56.35 kilometers per hour.
It is violating the minimum speed law.

## 1.9 Student Practice

**1.** $5[4x - 3(y - 2)]$
$\quad = 5[4x - 3y + 6]$
$\quad = 20x - 15y + 30$

**2.** $-3[2a - (3b - c) + 4d]$
$\quad = -3[2a - 3b + c + 4d]$
$\quad = -6a + 9b - 3c - 12d$

**3.** $3[4x - 2(1 - x)] - [3x + (x - 2)]$
$\quad = 3[4x - 2 + 2x] - [3x + x - 2]$
$\quad = 3[6x - 2] - [4x - 2]$
$\quad = 18x - 6 - 4x + 2$
$\quad = 14x - 4$

**4.** $-2\{5x - 3[2x - (3 - 4x)]\}$
$\quad = -2\{5x - 3[2x - 3 + 4x]\}$
$\quad = -2\{5x - 3[6x - 3]\}$
$\quad = -2\{5x - 18x + 9\}$
$\quad = -2\{-13x + 9\}$
$\quad = 26x - 18$

Thus $x = -2$ is not a solution. Solve to find the solution.
$\quad x + 8 = -22 + 6$
$\quad x + 8 = -16$
$x + 8 - 8 = -16 - 8$
$\qquad x = -24$

**5.** $\qquad \frac{1}{20} - \frac{1}{2} = x + \frac{3}{5}$

$\quad \frac{1}{20} - \frac{1 \cdot 10}{2 \cdot 10} = x + \frac{3 \cdot 4}{5 \cdot 4}$

$\qquad \frac{1}{20} - \frac{10}{20} = x + \frac{12}{20}$

$\qquad\qquad -\frac{9}{20} = x + \frac{12}{20}$

$-\frac{9}{20} + \left(-\frac{12}{20}\right) = x + \frac{12}{20} + \left(-\frac{12}{20}\right)$

$\qquad\qquad -\frac{21}{20} = x$

$\qquad\qquad x = -\frac{21}{20}$ or $-1\frac{1}{20}$

***Check.***

$\frac{1}{20} - \frac{1}{2} = x + \frac{3}{5}$

$\frac{1}{20} - \frac{1}{2} \overset{?}{=} -\frac{21}{20} + \frac{3}{5}$

$\frac{1}{20} - \frac{10}{20} \overset{?}{=} -\frac{21}{20} + \frac{12}{20}$

$\qquad -\frac{9}{20} = -\frac{9}{20}$ ✓

## 2.2 Student Practice

**1.** $\frac{1}{8}x = -2$

$8\left(\frac{1}{8}x\right) = 8(-2)$

$\left(\frac{8}{1}\right)\left(\frac{1}{8}\right)x = -16$

$x = -16$

**2.** $9x = 72$

$\frac{9x}{9} = \frac{72}{9}$

$x = 8$

**3.** $6x = 50$

$\frac{6x}{6} = \frac{50}{6}$

$x = \frac{25}{3} \text{ or } 8\frac{1}{3}$

**4.** $-27x = 54$

$\frac{-27x}{-27} = \frac{54}{-27}$

$x = -2$

**5.** $-x = 36$

$-1x = 36$

$\frac{-1x}{-1} = \frac{36}{-1}$

$x = -36$

**6.** $-51 = 3x - 9x$

$-51 = -6x$

$\frac{-51}{-6} = \frac{-6x}{-6}$

$\frac{17}{2} = x$

**7.** $16.2 = 5.2x - 3.4x$

$16.2 = 1.8x$

$\frac{16.2}{1.8} = \frac{1.8x}{1.8}$

$9 = x$

## 2.3 Student Practice

**1.** $9x + 2 = 38$

$9x + 2 + (-2) = 38 + (-2)$

$9x = 36$

$\frac{9x}{9} = \frac{36}{9}$

$x = 4$

**Check.** $9(4) + 2 \overset{?}{=} 38$

$36 + 2 \overset{?}{=} 38$

$38 = 38$ ✓

**2.** $13x = 2x - 66$

$13x + (-2x) = 2x + (-2x) - 66$

$11x = -66$

$\frac{11x}{11} = \frac{-66}{11}$

$x = -6$

**3.** $3x + 2 = 5x + 2$

$3x + (-3x) + 2 = 5x + (-3x) + 2$

$2 = 2x + 2$

$2 + (-2) = 2x + 2 + (-2)$

$0 = 2x$

$\frac{0}{2} = \frac{2x}{2}$

$0 = x$

**Check.** $3(0) + 2 \overset{?}{=} 5(0) + 2$

$2 = 2$ ✓

**4.** $-z + 8 - z = 3z + 10 - 3$

$-2z + 8 = 3z + 7$

$-2z + 2z + 8 = 3z + 2z + 7$

$8 = 5z + 7$

$8 + (-7) = 5z + 7 + (-7)$

$1 = 5z$

$\frac{1}{5} = \frac{5z}{5}$

$\frac{1}{5} = z$

**5.** $4x - (x + 3) = 12 - 3(x - 2)$

$4x - x - 3 = 12 - 3x + 6$

$3x - 3 = -3x + 18$

$3x + 3x - 3 = -3x + 3x + 18$

$6x - 3 = 18$

$6x - 3 + 3 = 18 + 3$

$6x = 21$

$\frac{6x}{6} = \frac{21}{6}$

$x = \frac{7}{2}$

**Check.** $4\left(\frac{7}{2}\right) - \left(\frac{7}{2} + 3\right) \overset{?}{=} 12 - 3\left(\frac{7}{2} - 2\right)$

$14 - \frac{13}{2} \overset{?}{=} 12 - 3\left(\frac{3}{2}\right)$

$\frac{28}{2} - \frac{13}{2} \overset{?}{=} \frac{24}{2} - \frac{9}{2}$

$\frac{15}{2} = \frac{15}{2}$ ✓

**6.** $4(-2x - 3) = -5(x - 2) + 2$

$-8x - 12 = -5x + 10 + 2$

$-8x - 12 = -5x + 12$

$-8x + 8x - 12 = -5x + 8x + 12$

$-12 = 3x + 12$

$-12 - 12 = 3x + 12 - 12$

$-24 = 3x$

$\frac{-24}{3} = \frac{3x}{3}$

$-8 = x$

**7.** $0.3x - 2(x + 0.1) = 0.4(x - 3) - 1.1$

$0.3x - 2x - 0.2 = 0.4x - 1.2 - 1.1$

$-1.7x - 0.2 = 0.4x - 2.3$

$-1.7x + 1.7x - 0.2 = 0.4x + 1.7x - 2.3$

$-0.2 = 2.1x - 2.3$

$-0.2 + 2.3 = 2.1x - 2.3 + 2.3$

$2.1 = 2.1x$

$\frac{2.1}{2.1} = \frac{2.1x}{2.1}$

$1 = x$

**8.** $5(2z - 1) + 7 = 7z - 4(z + 3)$

$10z - 5 + 7 = 7z - 4z - 12$

$10z + 2 = 3z - 12$

$10z - 3z + 2 = 3z - 3z - 12$

$7z + 2 = -12$

$7z + 2 - 2 = -12 - 2$

$7z = -14$

$\frac{7z}{7} = \frac{-14}{7}$

$z = -2$

**Check.** $5[2(-2) - 1] + 7 \overset{?}{=} 7(-2) - 4[(-2) + 3]$

$5[-4 - 1] + 7 \overset{?}{=} 7(-2) - 4[1]$

$5(-5) + 7 \overset{?}{=} -14 - 4$

$-25 + 7 \overset{?}{=} -18$

$-18 = -18$ ✓

## 2.4 Student Practice

**1.** $\frac{3}{8}x - \frac{3}{2} = \frac{1}{4}x$

$8\left(\frac{3}{8}x - \frac{3}{2}\right) = 8\left(\frac{1}{4}x\right)$

$\left(\frac{8}{1}\right)\left(\frac{3}{8}\right)(x) - \left(\frac{8}{1}\right)\left(\frac{3}{2}\right) = \left(\frac{8}{1}\right)\left(\frac{1}{4}\right)(x)$

$3x - 12 = 2x$

$3x + (-3x) - 12 = 2x + (-3x)$

$-12 = -x$

$12 = x$

**2.** $\frac{5x}{4} - 1 = \frac{3x}{4} + \frac{1}{2}$

$4\left(\frac{5x}{4}\right) - 4(1) = 4\left(\frac{3x}{4}\right) + 4\left(\frac{1}{2}\right)$

$5x - 4 = 3x + 2$

$5x - 3x - 4 = 3x - 3x + 2$

$2x - 4 = 2$

$2x - 4 + 4 = 2 + 4$

$2x = 6$

$\frac{2x}{2} = \frac{6}{2}$

$x = 3$

**Check.** $\frac{5(3)}{4} - 1 \overset{?}{=} \frac{3(3)}{4} + \frac{1}{2}$

$\frac{15}{4} - 1 \overset{?}{=} \frac{9}{4} + \frac{1}{2}$

$\frac{15}{4} - \frac{4}{4} \overset{?}{=} \frac{9}{4} + \frac{2}{4}$

$\frac{11}{4} = \frac{11}{4}$ ✓

**3.**
$$\frac{x+6}{9} = \frac{x}{6} + \frac{1}{2}$$
$$\frac{x}{9} + \frac{6}{9} = \frac{x}{6} + \frac{1}{2}$$
$$18\left(\frac{x}{9}\right) + 18\left(\frac{6}{9}\right) = 18\left(\frac{x}{6}\right) + 18\left(\frac{1}{2}\right)$$
$$2x + 12 = 3x + 9$$
$$2x - 2x + 12 = 3x - 2x + 9$$
$$12 = x + 9$$
$$12 - 9 = x + 9 - 9$$
$$3 = x$$

**4.**
$$\frac{1}{2}(x + 5) = \frac{1}{5}(x - 2) + \frac{1}{2}$$
$$\frac{x}{2} + \frac{5}{2} = \frac{x}{5} - \frac{2}{5} + \frac{1}{2}$$
$$10\left(\frac{x}{2}\right) + 10\left(\frac{5}{2}\right) = 10\left(\frac{x}{5}\right) - 10\left(\frac{2}{5}\right) + 10\left(\frac{1}{2}\right)$$
$$5x + 25 = 2x - 4 + 5$$
$$5x + 25 = 2x + 1$$
$$5x - 2x + 25 = 2x - 2x + 1$$
$$3x + 25 = 1$$
$$3x + 25 - 25 = 1 - 25$$
$$3x = -24$$
$$\frac{3x}{3} = \frac{-24}{3}$$
$$x = -8$$

**Check.** $\frac{1}{2}[(-8) + 5] \stackrel{?}{=} \frac{1}{5}[(-8) - 2] + \frac{1}{2}$

$$\frac{1}{2}(-3) \stackrel{?}{=} \frac{1}{5}(-10) + \frac{1}{2}$$
$$-\frac{3}{2} \stackrel{?}{=} -2 + \frac{1}{2}$$
$$-\frac{3}{2} \stackrel{?}{=} -\frac{4}{2} + \frac{1}{2}$$
$$-\frac{3}{2} = -\frac{3}{2} \checkmark$$

**5.**
$$2.8 = 0.3(x - 2) + 2(0.1x - 0.3)$$
$$2.8 = 0.3x - 0.6 + 0.2x - 0.6$$
$$10(2.8) = 10(0.3x) - 10(0.6) + 10(0.2x) - 10(0.6)$$
$$28 = 3x - 6 + 2x - 6$$
$$28 = 5x - 12$$
$$28 + 12 = 5x - 12 + 12$$
$$40 = 5x$$
$$\frac{40}{5} = \frac{5x}{5}$$
$$8 = x$$

## 2.5 Student Practice

**1.**
$$d = rt$$
$$4565 = 8.3r$$
$$\frac{4565}{8.3} = \frac{8.3r}{8.3}$$
$$550 = r$$
The average rate of speed was 550 miles per hour.

**2.** Solve for $m$.
$$E = mc^2$$
$$\frac{E}{c^2} = \frac{mc^2}{c^2}$$
$$\frac{E}{c^2} = m$$

**3.** $8 - 2y + 3x = 0$
$$8 - 2y = -3x$$
$$-2y = -3x - 8$$
$$\frac{-2y}{-2} = \frac{-3x - 8}{-2}$$

$$y = \frac{-3x}{-2} + \frac{-8}{-2}$$
$$y = \frac{3}{2}x + 4$$

**4.** $H = \frac{3}{4}(2y + x)$
$$H = \frac{6y}{4} + \frac{3x}{4}$$
$$4(H) = 4\left(\frac{6y}{4}\right) + 4\left(\frac{3x}{4}\right)$$
$$4H = 6y + 3x$$
$$4H - 6y = 3x$$
$$\frac{4H - 6y}{3} = \frac{3x}{3}$$
$$\frac{4H - 6y}{3} = x \text{ or } \frac{4H}{3} - 2y = x$$

## 2.6 Student Practice

**1. (a)** $7 > 2$    **(b)** $-2 > -4$    **(c)** $-1 < 2$
    **(d)** $-8 < -5$    **(e)** $0 > -2$    **(f)** $5 > -3$

**2. (a)** $x$ is greater than 5

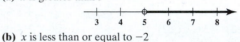

    **(b)** $x$ is less than or equal to $-2$

    **(c)** $x$ is less than 3 (or 3 is greater than $x$)

    **(d)** $x$ is greater than or equal to $-\frac{3}{2}$

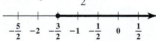

**3. (a)** Since the temperature can never exceed 180 degrees, then the temperature must always be less than or equal to 180 degrees. Thus, $t \le 180$.
    **(b)** Since the debt must be less than 15,000, we have $d < 15,000$.

**4. (a)**    $7 > 2$          **(b)** $-3 < -1$
       $-14 < -4$             $3 > 1$
    **(c)** $-10 \ge -20$        **(d)** $-15 \le -5$
       $1 \le 2$                $3 \ge 1$

**5.**    $8x - 2 < 3$
$$8x - 2 + 2 < 3 + 2$$
$$8x < 5$$
$$\frac{8x}{8} < \frac{5}{8}$$
$$x < \frac{5}{8}$$

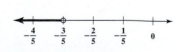

**6.**    $4 - 5x > 7$
$$4 - 4 - 5x > 7 - 4$$
$$-5x > 3$$
$$\frac{-5x}{-5} < \frac{3}{-5}$$
$$x < -\frac{3}{5}$$

**7.**    $\frac{1}{2}x + 3 < \frac{2}{3}x$
$$6\left(\frac{1}{2}x\right) + 6(3) < 6\left(\frac{2}{3}x\right)$$
$$3x + 18 < 4x$$
$$3x - 4x + 18 < 4x - 4x$$
$$-x + 18 < 0$$
$$-x + 18 - 18 < 0 - 18$$
$$-x < -18$$
$$\frac{-x}{-1} > \frac{-18}{-1}$$
$$x > 18$$

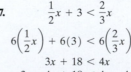

**8.**  $\frac{1}{2}(3-x) \le 2x + 5$

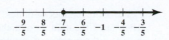

$$\frac{3}{2} - \frac{1}{2}x \le 2x + 5$$

$$2\left(\frac{3}{2}\right) - 2\left(\frac{1}{2}x\right) \le 2(2x) + 2(5)$$

$$3 - x \le 4x + 10$$

$$3 - x - 4x \le 4x - 4x + 10$$

$$3 - 5x \le 10$$

$$3 - 3 - 5x \le 10 - 3$$

$$-5x \le 7$$

$$\frac{-5x}{-5} \ge \frac{7}{-5}$$

$$x \ge -\frac{7}{5}$$

**9.**

$$2000n - 700,000 \ge 2,500,000$$

$$2000n - 700,000 + 700,000 \ge 2,500,000 + 700,000$$

$$2000n \ge 3,200,000$$

$$\frac{2000n}{2000} \ge \frac{3,200,000}{2000}$$

$$n \ge 1600$$

# Chapter 3   3.1 Student Practice

**1. (a)** $x + 4$   **(b)** $3x$   **(c)** $x - 8$   **(d)** $\frac{1}{4}x$

**2. (a)** $3x + 8$   **(b)** $3(x + 8)$   **(c)** $\frac{1}{3}(x + 4)$

**3.** Let $a = $ Ann's hours per week.
Then $a - 17 = $ Marie's hours per week.

**4.**

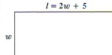

width $= w$
length $= 2w + 5$

**5.**  Number of degrees in 1st angle $= s - 16$
Number of degrees in 2nd angle $= s$
Number of degrees in 3rd angle $= 2s$

**6.** Let $x = $ the number of students in the fall.

$\frac{2}{3}x = $ the number of students in the spring.

$\frac{1}{5}x = $ the number of students in the summer.

## 3.2 Student Practice

**1.** Let $x = $ the unknown number

$$\frac{3}{4}x = -81$$

$$4\left(\frac{3}{4}x\right) = 4(-81)$$

$$3x = -324$$

$$\frac{3x}{3} = \frac{-324}{3}$$

$$x = -108$$

The number is $-108$.

**2.** Let $x = $ the unknown number.

$$3x - 2 = 49$$

$$3x - 2 + 2 = 49 + 2$$

$$3x = 51$$

$$\frac{3x}{3} = \frac{51}{3}$$

$$x = 17$$

The number is 17.

**3.** Let $x = $ the first number.
Then the second number is $3x - 12$.

$$x + (3x - 12) = 24$$

$$4x - 12 = 24$$

$$4x = 36$$

$$x = 9$$

First number is 9. Second number $= 3(9) - 12 = 15$.

**4.** Let rainfall in Canada $= x$
Then rainfall in Texas $= 3x - 14$

$$3x - 14 = 43$$

$$3x = 57$$

$$x = 19$$

19 inches of rain was recorded in Canada.

**5. (a)**    $d = rt$

$$220 = 4r$$

$$55 = r$$

Leaving the city, her average speed was 55 mph.

**(b)**    $d = rt$

$$225 = 4.5r$$

$$50 = r$$

On the return trip, her average speed was 50 mph.

**(c)** She traveled 5 mph faster on the trip leaving the city.

**6.** Let $x = $ her final exam score. Since the final counts as two tests, divide her total score by 6.

$$\frac{78 + 80 + 100 + 96 + x + x}{6} = 90$$

$$\frac{354 + 2x}{6} = 90$$

$$6\left(\frac{354 + 2x}{6}\right) = 6(90)$$

$$354 + 2x = 540$$

$$2x = 186$$

$$x = 93$$

She needs a 93 on the final exam.

## 3.3 Student Practice

**1.** Let $x = $ the length of the short piece.
Then $x + 17 = $ the length of the long piece.

$$x + (x + 17) = 89$$

$$2x + 17 = 89$$

$$2x + 17 - 17 = 89 - 17$$

$$2x = 72$$

$$\frac{2x}{2} = \frac{72}{2}$$

$$x = 36$$

$$x + 17 = 36 + 17 = 53$$

The length of the short piece is 36 feet and the length of the long piece is 53 feet.

**2.** Let $x = $ the second family's bill.
Then $x + 360 = $ the first family's bill and
$2x - 200 = $ the third family's bill.

$$x + (x + 360) + (2x - 200) = 3960$$

$$4x + 160 = 3960$$

$$4x = 3800$$

$$x = 950$$

$$x + 360 = 950 + 360 = 1310$$

$$2x - 200 = 2(950) - 200 = 1900 - 200 = 1700$$

The first family's bill was $1310, the second family's bill was $950, and the third family's bill was $1700.

3. Let $x =$ the length of the second side.
Then $x - 30 =$ the length of the first side and
$\frac{1}{2}x =$ the length of the third side.

$$x + (x - 30) + \frac{1}{2}x = 720$$

$$\frac{5}{2}x - 30 = 720$$
$$5x - 60 = 1440$$
$$5x = 1500$$
$$x = 300$$
$$x - 30 = 300 - 30 = 270$$
$$\frac{1}{2}x = \frac{1}{2}(300) = 150$$

The first side is 270 meters, the second side is 300 meters, and the third side is 150 meters.

## 3.4 Student Practice

1. Let $m =$ the number of miles.
$$3(25) + (0.20)m = 350$$
$$75 + 0.20m = 350$$
$$0.20m = 275$$
$$\frac{0.20m}{0.20} = \frac{275}{0.20}$$
$$m = 1375$$
He can travel 1375 miles for \$350.

2. Let $x =$ the cost of the chef's knives he sold.
$$0.38x = 17{,}100$$
$$\frac{0.38x}{0.38} = \frac{17{,}100}{0.38}$$
$$x = 45{,}000$$
He sold \$45,000 worth of chef's knives last year.

3. Let $x =$ the price last year.
$$x + 0.07x = 19{,}795$$
$$1.07x = 19{,}795$$
$$\frac{1.07x}{1.07} = \frac{19{,}795}{1.07}$$
$$x = 18{,}500$$
A similar model would have cost \$18,500 last year.

4. $I = prt$
$$I = (7000)(0.12)(1)$$
$$I = 840$$
The interest charge on borrowing \$7000 for one year at a simple interest rate of 12% is \$840.

5. Let $x =$ the amount invested at 9%.
Then $8000 - x$ was invested at 7%.
$$0.09x + 0.07(8000 - x) = 630$$
$$0.09x + 560 - 0.07x = 630$$
$$0.02x + 560 = 630$$
$$0.02x = 70$$
$$\frac{0.02x}{0.02} = \frac{70}{0.02}$$
$$x = 3500$$
$$8000 - x = 8000 - 3500 = 4500$$
Therefore, she invested \$3500 at 9% and \$4500 at 7%.

6. Let $x =$ the number of dimes.
Then $x + 5 =$ the number of quarters.
$$0.10x + 0.25(x + 5) = 5.10$$
$$0.10x + 0.25x + 1.25 = 5.10$$
$$0.35x + 1.25 = 5.10$$
$$0.35x = 3.85$$
$$\frac{0.35x}{0.35} = \frac{3.85}{0.35}$$
$$x = 11$$
$$x + 5 = 11 + 5 = 16$$
Therefore, she has 11 dimes and 16 quarters.

7. Let $x =$ the number of dimes.
Then $2x =$ the number of nickels and
$x + 4 =$ the number of quarters.
$$0.05(2x) + 0.10x + 0.25(x + 4) = 2.35$$
$$0.10x + 0.10x + 0.25x + 1 = 2.35$$
$$0.45x + 1 = 2.35$$
$$0.45x = 1.35$$
$$\frac{0.45x}{0.45} = \frac{1.35}{0.45}$$
$$x = 3$$
$$2x = 2(3) = 6$$
$$x + 4 = 3 + 4 = 7$$
Therefore, the boy has 3 dimes, 6 nickels, and 7 quarters.

## 3.5 Student Practice

1. $A = \frac{1}{2}ab$
$$= \frac{1}{2}(20 \text{ in.})(14 \text{ in.})$$
$$= (10)(14)(\text{in.})(\text{in.})$$
$$= 140 \text{ square inches}$$
The area of the triangle is 140 square inches.

2. $$A = lw$$
$$120 \text{ (yd)}^2 = (l)(8 \text{ yd})$$
$$\frac{120 \text{ (yd)(yd)}}{8 \text{ yd}} = \frac{8 \text{ yd}}{8 \text{ yd}}l$$
$$15 \text{ yd} = l$$
The length of the field is 15 yards.

3. $$A = \frac{1}{2}a(b_1 + b_2)$$
$$256 \text{ (ft)}^2 = \frac{1}{2}a(12 \text{ ft} + 20 \text{ ft})$$
$$256 \text{ (ft)(ft)} = \frac{1}{2}(a)(32 \text{ ft})$$
$$256 \text{ (ft)(ft)} = (16 \text{ ft})a$$
$$16 \text{ ft} = a$$
The altitude is 16 feet.

4. $C = 2\pi r$
$$C = 2(3.14)(15 \text{ m})$$
$$C = 94.2 \text{ m}$$
Rounded to the nearest meter, the circumference is approximately 94 meters.

5. $P = a + b + c$
$$P = 15 \text{ cm} + 15 \text{ cm} + 15 \text{ cm}$$
$$= 45 \text{ cm}$$
The perimeter of the triangle is 45 cm.

6.
$$132° + x + x = 180°$$
$$132° + 2x = 180°$$
$$2x = 48°$$
$$x = 24°$$
Both angles measure 24°.

7. Surface area $= 4\pi r^2$
$$= 4(3.14)(5 \text{ m})^2$$
$$= 314 \text{ m}^2$$
Rounded to the nearest square meter, the surface area is approximately 314 square meters.

8. $V = \pi r^2 h$
$$= (3.14)(3 \text{ ft})^2(4 \text{ ft})$$
$$= 113.04 \text{ ft}^3$$
Rounded to the nearest cubic foot, approximately 113 cubic feet of sand can be stored in the drum.

9. Calculate the area of the pool.
$A = lw$
$A = (12 \text{ ft})(8 \text{ ft}) = 96 \text{ ft}^2$.
Now add 6 feet to the length and 6 feet to the width of the pool and calculate the area.

$A = lw$
$A = (18 \text{ ft})(14 \text{ ft}) = 252 \text{ ft}^2.$
Now subtract the areas.
$252 \text{ ft}^2 - 96 \text{ ft}^2 = 156 \text{ ft}^2$ at $12 per square foot.
$156 \times 12 = 1872$
The cost would be $1872.

## 3.6 Student Practice

1. **(a)** height $\leq 6$    **(b)** speed $> 65$
   **(c)** area $\geq 560$    **(d)** profit margin $\geq 50$

2. Let $x =$ the number of pounds the fourth rope holds.
$$\frac{1050 + 1250 + 950 + x}{4} \geq 1100$$
$$\frac{3250 + x}{4} \geq 1100$$
$$x + 3250 \geq 4400$$
$$x \geq 1150$$
The rope must hold at least 1150 pounds.

3. Let $x =$ the value of the systems that she sells.
$$1400 + 0.02x > 2200$$
$$0.02x > 800$$
$$x > 40{,}000$$
Rita must sell more than $40,000 worth of products each month.

## Chapter 4   4.1 Student Practice

1. **(a)** $a^7 \cdot a^5 = a^{7+5} = a^{12}$    **(b)** $w^{10} \cdot w = w^{10+1} = w^{11}$
2. **(a)** $x^3 \cdot x^9 = x^{3+9} = x^{12}$    **(b)** $3^7 \cdot 3^4 = 3^{7+4} = 3^{11}$
   **(c)** $a^3 \cdot b^2 = a^3 \cdot b^2$ (cannot be simplified)
3. **(a)** $(7a^8)(a^4) = (7 \cdot 1)(a^8 \cdot a^4)$
   $= 7(a^8 \cdot a^4)$
   $= 7a^{12}$
   **(b)** $(3y^2)(-2y^3) = (3)(-2)(y^2 \cdot y^3) = -6y^5$
   **(c)** $(-4x^3)(-5x^2) = (-4)(-5)(x^3 \cdot x^2) = 20x^5$

4. $(2xy)\left(-\dfrac{1}{4}x^2y\right)(6xy^3) = (2)\left(-\dfrac{1}{4}\right)(6)(x \cdot x^2 \cdot x)(y \cdot y \cdot y^3)$
   $= -3x^4y^5$

5. **(a)** $\dfrac{10^{13}}{10^7} = 10^{13-7} = 10^6$

   **(b)** $\dfrac{x^{11}}{x} = x^{11-1} = x^{10}$

   **(c)** $\dfrac{y^{18}}{y^8} = y^{18-8} = y^{10}$

6. **(a)** $\dfrac{c^3}{c^4} = \dfrac{1}{c^{4-3}} = \dfrac{1}{c^1} = \dfrac{1}{c}$

   **(b)** $\dfrac{10^{31}}{10^{56}} = \dfrac{1}{10^{56-31}} = \dfrac{1}{10^{25}}$

   **(c)** $\dfrac{z^{15}}{z^{21}} = \dfrac{1}{z^{21-15}} = \dfrac{1}{z^6}$

7. **(a)** $\dfrac{-7x^7}{-21x^9} = \dfrac{1}{3x^{9-7}} = \dfrac{1}{3x^2}$

   **(b)** $\dfrac{15x^{11}}{-3x^4} = -5x^{11-4} = -5x^7$

   **(c)** $\dfrac{23b^8}{46b^9} = \dfrac{1}{2b^{9-8}} = \dfrac{1}{2b}$

8. **(a)** $\dfrac{r^7s^9}{s^{10}} = \dfrac{r^7}{s^{10-9}} = \dfrac{r^7}{s}$

   **(b)** $\dfrac{12x^5y^6}{-24x^3y^8} = \dfrac{x^{5-3}}{-2y^{8-6}} = -\dfrac{x^2}{2y^2}$

9. **(a)** $\dfrac{10^7}{10^7} = 1$

   **(b)** $\dfrac{12a^4}{2a^4} = 6\left(\dfrac{a^4}{a^4}\right) = 6(1) = 6$

10. **(a)** $\dfrac{-20a^3b^8c^4}{28a^3b^7c^5} = -\dfrac{5a^0b}{7c} = -\dfrac{5(1)b}{7c} = -\dfrac{5b}{7c}$

    **(b)** $\dfrac{5x^0y^6}{10x^4y^8} = \dfrac{5(1)y^6}{10x^4y^8} = \dfrac{1}{2x^4y^2}$

11. $\dfrac{(-6ab^5)(3a^2b^4)}{16a^5b^7} = \dfrac{-18a^3b^9}{16a^5b^7} = -\dfrac{9b^2}{8a^2}$

12. **(a)** $(a^4)^3 = a^{4 \cdot 3} = a^{12}$
    **(b)** $(10^5)^2 = 10^{5 \cdot 2} = 10^{10}$
    **(c)** $(-1)^{15} = -1$

13. **(a)** $(3xy)^3 = (3)^3x^3y^3 = 27x^3y^3$
    **(b)** $(yz)^{37} = y^{37}z^{37}$
    **(c)** $(-3x^3)^2 = (-3)^2(x^3)^2 = 9x^6$

14. **(a)** $\left(\dfrac{x}{5}\right)^3 = \dfrac{x^3}{5^3} = \dfrac{x^3}{125}$

    **(b)** $\dfrac{(4a)^2}{(ab)^6} = \dfrac{4^2a^2}{a^6b^6} = \dfrac{16a^2}{a^6b^6} = \dfrac{16}{a^4b^6}$

15. $\left(\dfrac{-2x^3y^0z}{4xz^2}\right)^5 = \left(\dfrac{-x^2}{2z}\right)^5 = \dfrac{(-1)^5(x^2)^5}{2^5z^5} = -\dfrac{x^{10}}{32z^5}$

## 4.2 Student Practice

1. **(a)** $x^{-12} = \dfrac{1}{x^{12}}$    **(b)** $w^{-5} = \dfrac{1}{w^5}$    **(c)** $z^{-2} = \dfrac{1}{z^2}$

2. **(a)** $4^{-3} = \dfrac{1}{4^3} = \dfrac{1}{64}$    **(b)** $2^{-4} = \dfrac{1}{2^4} = \dfrac{1}{16}$

3. **(a)** $\dfrac{3}{w^{-4}} = 3w^4$    **(b)** $\dfrac{x^{-6}y^4}{z^{-2}} = \dfrac{y^4z^2}{x^6}$    **(c)** $x^{-6}y^{-5} = \dfrac{1}{x^6y^5}$

4. **(a)** $(2x^4y^{-5})^{-2} = 2^{-2}x^{-8}y^{10} = \dfrac{y^{10}}{2^2x^8} = \dfrac{y^{10}}{4x^8}$

   **(b)** $\dfrac{y^{-3}z^{-4}}{y^2z^{-6}} = \dfrac{z^6}{y^2y^3z^4} = \dfrac{z^6}{y^5z^4} = \dfrac{z^2}{y^5}$

5. **(a)** $78{,}200 = 7.82 \times 10{,}000 = 7.82 \times 10^4$
   Notice we moved the decimal point 4 places to the left.
   **(b)** $4{,}786{,}000 = 4.786 \times 1{,}000{,}000 = 4.786 \times 10^6$
6. **(a)** $0.98 = 9.8 \times 10^{-1}$
   **(b)** $0.000092 = 9.2 \times 10^{-5}$
7. **(a)** $1.93 \times 10^6 = 1.93 \times 1{,}000{,}000 = 1{,}930{,}000$

   **(b)** $8.562 \times 10^{-5} = 8.562 \times \dfrac{1}{100{,}000} = 0.00008562$

8. $30{,}900{,}000{,}000{,}000{,}000$ meters $= 3.09 \times 10^{16}$ meters
9. **(a)** $(56{,}000)(1{,}400{,}000{,}000) = (5.6 \times 10^4)(1.4 \times 10^9)$
   $= (5.6)(1.4)(10^4)(10^9)$
   $= 7.84 \times 10^{13}$

   **(b)** $\dfrac{0.000111}{0.00000037} = \dfrac{1.11 \times 10^{-4}}{3.7 \times 10^{-7}}$
   $= \dfrac{1.11}{3.7} \times \dfrac{10^{-4}}{10^{-7}}$
   $= \dfrac{1.11}{3.7} \times \dfrac{10^7}{10^4}$
   $= 0.3 \times 10^3$
   $= 3.0 \times 10^2$

**10.** $159 \text{ parsecs} = (159 \text{ parsecs}) \dfrac{(3.09 \times 10^{13} \text{ kilometers})}{1 \text{ parsec}}$

$= 491.31 \times 10^{13} \text{ kilometers}$

$d = r \times t$

$491.31 \times 10^{13} \text{ km} = \dfrac{50{,}000 \text{ km}}{1 \text{ hr}} \times t$

$4.9131 \times 10^{15} \text{ km} = \dfrac{5 \times 10^4 \text{ km}}{1 \text{ hr}} \times t$

$\dfrac{4.9131 \times 10^{15} \text{ km} \,(1 \text{ hr})}{5.0 \times 10^4 \text{ km}} = t$

$0.98262 \times 10^{11} \text{ hr} = t$

It would take the probe about $9.83 \times 10^{10}$ hours.

## 4.3 Student Practice

**1. (a)** This polynomial is of degree 5. It has two terms, so it is a binomial.
**(b)** This polynomial is of degree 7, since the sum of the exponents is $3 + 4 = 7$. It has one term, so it is a monomial.
**(c)** This polynomial is of degree 3. It has three terms, so it is a trinomial.

**2.** $(-8x^3 + 3x^2 + 6) + (2x^3 - 7x^2 - 3)$
$= [-8x^3 + 2x^3] + [3x^2 + (-7x^2)] + [6 + (-3)]$
$= [(-8 + 2)x^3] + [(3 - 7)x^2] + [6 - 3]$
$= -6x^3 + (-4x^2) + 3$
$= -6x^3 - 4x^2 + 3$

**3.** $\left(-\dfrac{1}{3}x^2 - 6x - \dfrac{1}{12}\right) + \left(\dfrac{1}{4}x^2 + 5x - \dfrac{1}{3}\right)$

$= \left[-\dfrac{1}{3}x^2 + \dfrac{1}{4}x^2\right] + [-6x + 5x] + \left[-\dfrac{1}{12} + \left(-\dfrac{1}{3}\right)\right]$

$= \left[\left(-\dfrac{1}{3} + \dfrac{1}{4}\right)x^2\right] + [(-6 + 5)x] + \left[-\dfrac{1}{12} + \left(-\dfrac{1}{3}\right)\right]$

$= \left[\left(-\dfrac{4}{12} + \dfrac{3}{12}\right)x^2\right] + [-x] + \left[-\dfrac{1}{12} - \dfrac{4}{12}\right]$

$= -\dfrac{1}{12}x^2 - x - \dfrac{5}{12}$

**4.** $(3.5x^3 - 0.02x^2 + 1.56x - 3.5) + (-0.08x^2 - 1.98x + 4)$
$= 3.5x^3 + (-0.02 - 0.08)x^2 + (1.56 - 1.98)x + (-3.5 + 4)$
$= 3.5x^3 - 0.1x^2 - 0.42x + 0.5$

**5.** $(5x^3 - 15x^2 + 6x - 3) - (-4x^3 - 10x^2 + 5x + 13)$
$= (5x^3 - 15x^2 + 6x - 3) + (4x^3 + 10x^2 - 5x - 13)$
$= (5 + 4)x^3 + (-15 + 10)x^2 + (6 - 5)x + (-3 - 13)$
$= 9x^3 - 5x^2 + x - 16$

**6.** $(x^3 - 7x^2y + 3xy^2 - 2y^3) - (2x^3 + 4xy - 6y^3)$
$= (x^3 - 7x^2y + 3xy^2 - 2y^3) + (-2x^3 - 4xy + 6y^3)$
$= (1 - 2)x^3 - 7x^2y + 3xy^2 - 4xy + (-2 + 6)y^3$
$= -x^3 - 7x^2y + 3xy^2 - 4xy + 4y^3$

**7. (a)** 1990 is 5 years later than 1985, so $x = 5$.
$0.16(5) + 20.7 = 0.8 + 20.7 = 21.5$
We estimate that the average light truck in 1990 obtained 21.5 miles per gallon.
**(b)** 2025 is 40 years later than 1985, so $x = 40$.
$0.16(40) + 20.7 = 6.4 + 20.7 = 27.1$
We estimate that the average light truck in 2025 will obtain 27.1 miles per gallon.

## 4.4 Student Practice

**1.** $4x^3(-2x^2 + 3x) = 4x^3(-2x^2) + 4x^3(3x)$
$= 4(-2)(x^3 \cdot x^2) + (4 \cdot 3)(x^3 \cdot x)$
$= -8x^5 + 12x^4$

**2. (a)** $-3x(x^2 + 2x - 4) = -3x^3 - 6x^2 + 12x$
**(b)** $6xy(x^3 + 2x^2y - y^2) = 6x^4y + 12x^3y^2 - 6xy^3$

**3.** $(-6x^3 + 4x^2 - 2x)(-3xy) = 18x^4y - 12x^3y + 6x^2y$

**4.** $(5x - 1)(x - 2) = 5x^2 - 10x - x + 2 = 5x^2 - 11x + 2$

**5.** $(8a - 5b)(3a - b) = 24a^2 - 8ab - 15ab + 5b^2$
$= 24a^2 - 23ab + 5b^2$

**6.** $(3a + 2b)(2a - 3c) = 6a^2 - 9ac + 4ab - 6bc$

**7.** $(3x - 2y)(3x - 2y) = 9x^2 - 6xy - 6xy + 4y^2$
$= 9x^2 - 12xy + 4y^2$

**8.** $(2x^2 + 3y^2)(5x^2 + 6y^2) = 10x^4 + 12x^2y^2 + 15x^2y^2 + 18y^4$
$= 10x^4 + 27x^2y^2 + 18y^4$

**9.** $A = (\text{length})(\text{width}) = (7x + 3)(2x - 1)$
$= 14x^2 - 7x + 6x - 3$
$= 14x^2 - x - 3$
There are $(14x^2 - x - 3)$ square feet in the room.

## 4.5 Student Practice

**1.** $(6x + 7)(6x - 7) = (6x)^2 - (7)^2 = 36x^2 - 49$

**2.** $(3x - 5y)(3x + 5y) = (3x)^2 - (5y)^2 = 9x^2 - 25y^2$

**3. (a)** $(4a - 9b)^2 = (4a)^2 - 2(4a)(9b) + (9b)^2$
$= 16a^2 - 72ab + 81b^2$
**(b)** $(5x + 4)^2 = (5x)^2 + 2(5x)(4) + (4)^2 = 25x^2 + 40x + 16$

**4.**
$$\begin{array}{r} 4x^3 - 2x^2 + \phantom{0}x \\ \underline{x^2 + 3x \phantom{{}+0}- 2} \\ -8x^3 + 4x^2 - 2x \\ 12x^4 - \phantom{0}6x^3 + 3x^2 \phantom{{}- 2x} \\ \underline{4x^5 - \phantom{0}2x^4 + \phantom{0}x^3 \phantom{{}+ 7x^2 - 2x}} \\ 4x^5 + 10x^4 - 13x^3 + 7x^2 - 2x \end{array}$$

**5.** $(2x^2 + 5x + 3)(x^2 - 3x - 4)$
$= 2x^2(x^2 - 3x - 4) + 5x(x^2 - 3x - 4) + 3(x^2 - 3x - 4)$
$= 2x^4 - 6x^3 - 8x^2 + 5x^3 - 15x^2 - 20x + 3x^2 - 9x - 12$
$= 2x^4 - x^3 - 20x^2 - 29x - 12$

**6.** $(3x - 2)(2x + 3)(3x + 2) = (3x - 2)(3x + 2)(2x + 3)$
$= [(3x)^2 - 2^2](2x + 3)$
$= (9x^2 - 4)(2x + 3)$

$$\begin{array}{r} 9x^2 - \phantom{00}4 \\ \underline{2x + \phantom{0}3} \\ 27x^2 + 0x - 12 \\ \underline{18x^3 + \phantom{0}0x^2 - 8x \phantom{{}- 12}} \\ 18x^3 + 27x^2 - 8x - 12 \end{array}$$

Thus we have
$(3x - 2)(2x + 3)(3x + 2) = 18x^3 + 27x^2 - 8x - 12.$

## 4.6 Student Practice

**1.** $\dfrac{15y^4 - 27y^3 - 21y^2}{3y^2} = \dfrac{15y^4}{3y^2} - \dfrac{27y^3}{3y^2} - \dfrac{21y^2}{3y^2} = 5y^2 - 9y - 7$

**2.**
$$\begin{array}{r} x^2 + \phantom{0}6x + 7 \phantom{00000} \\ x + 4 \overline{)\,x^3 + 10x^2 + 31x + 25} \\ \underline{x^3 + \phantom{0}4x^2} \phantom{0000000000} \\ 6x^2 + 31x \phantom{0000} \\ \underline{6x^2 + 24x} \phantom{0000} \\ 7x + 25 \\ \underline{7x + 28} \\ -3 \end{array}$$

Ans: $x^2 + 6x + 7 + \dfrac{-3}{x + 4}$

**3.**
$$\begin{array}{r} 2x^2 + \phantom{0}x + 1 \phantom{00000} \\ x - 1 \overline{)\,2x^3 - \phantom{0}x^2 + 0x + 1} \\ \underline{2x^3 - 2x^2} \phantom{00000000} \\ x^2 + 0x \phantom{0000} \\ \underline{x^2 - \phantom{0}x} \phantom{0000} \\ x + 1 \\ \underline{x - 1} \\ 2 \end{array}$$

Ans: $2x^2 + x + 1 + \dfrac{2}{x - 1}$

**4.**
$$\begin{array}{r} 5x^2 + \phantom{0}x - 2 \phantom{0000} \\ 4x - 3 \overline{)\,20x^3 - 11x^2 - 11x + 6} \\ \underline{20x^3 - 15x^2} \phantom{0000000000} \\ 4x^2 - 11x \phantom{0000} \\ \underline{4x^2 - \phantom{0}3x} \phantom{0000} \\ -8x + 6 \\ \underline{-8x + 6} \\ 0 \end{array}$$

Ans: $5x^2 + x - 2$

**Check.** $(4x - 3)(5x^2 + x - 2)$
$= 20x^3 + 4x^2 - 8x - 15x^2 - 3x + 6$
$= 20x^3 - 11x^2 - 11x + 6$

# Chapter 5 5.1 Student Practice

1. (a) $21a - 7b = 7(3a - b)$ because $7(3a - b) = 21a - 7b$.
   (b) $5xy + 8x = x(5y + 8)$ because $x(5y + 8) = 5xy + 8x$.
2. 4 is the greatest numerical common factor and $a$ is a factor of each term. Thus, $4a$ is the greatest common factor.
   $12a^2 + 16ab^2 - 12a^2b = 4a(3a + 4b^2 - 3ab)$
3. (a) The largest integer common to both terms is 8.
   $16a^3 - 24b^3 = 8(2a^3 - 3b^3)$
   (b) $r^3$ is common to all the terms.
   $r^3s^2 - 4r^4s + 7r^5 = r^3(s^2 - 4rs + 7r^2)$
4. We can factor $9, a, b^2,$ and $c$ out of each term.
   $18a^3b^2c - 27ab^3c^2 - 45a^2b^2c^2 = 9ab^2c(2a^2 - 3bc - 5ac)$
5. We can factor $6, x,$ and $y^2$ out of each term.
   $30x^3y^2 - 24x^2y^2 + 6xy^2 = 6xy^2(5x^2 - 4x + 1)$
   **Check:** $6xy^2(5x^2 - 4x + 1) = 30x^3y^2 - 24x^2y^2 + 6xy^2$
6. $3(a + 5b) + x(a + 5b) = (a + 5b)(3 + x)$
7. $8y(9y^2 - 2) - (9y^2 - 2) = 8y(9y^2 - 2) - 1(9y^2 - 2)$
   $= (9y^2 - 2)(8y - 1)$
8. $\pi b^2 - \pi a^2 = \pi(b^2 - a^2)$

## 5.2 Student Practice

1. $3y(2x - 7) - 8(2x - 7) = (2x - 7)(3y - 8)$
2. $6x^2 - 15x + 4x - 10 = 3x(2x - 5) + 2(2x - 5)$
   $= (2x - 5)(3x + 2)$
3. $ax + 2a + 4bx + 8b = a(x + 2) + 4b(x + 2)$
   $= (x + 2)(a + 4b)$
4. $6a^2 + 5bc + 10ab + 3ac = 6a^2 + 10ab + 3ac + 5bc$
   $= 2a(3a + 5b) + c(3a + 5b)$
   $= (3a + 5b)(2a + c)$
5. $6xy + 14x - 15y - 35 = 2x(3y + 7) - 15y - 35$
   $= 2x(3y + 7) - 5(3y + 7)$
   $= (3y + 7)(2x - 5)$
6. $3x + 6y - 5ax - 10ay = 3(x + 2y) - 5a(x + 2y)$
   $= (x + 2y)(3 - 5a)$
7. $10ad + 27bc - 6bd - 45ac = 10ad - 6bd - 45ac + 27bc$
   $= 2d(5a - 3b) - 9c(5a - 3b)$
   $= (5a - 3b)(2d - 9c)$
   **Check:** $(5a - 3b)(2d - 9c) = 10ad - 45ac - 6bd + 27bc$
   $= 10ad + 27bc - 6bd - 45ac$

## 5.3 Student Practice

1. The two numbers that you can multiply to get 12 and add to get 8 are 6 and 2. $x^2 + 8x + 12 = (x + 6)(x + 2)$
2. The two numbers that have a product of 30 and a sum of 17 are 2 and 15. $x^2 + 17x + 30 = (x + 2)(x + 15)$
3. The two numbers that have a product of $+18$ and a sum of $-11$ must both be negative. The numbers are $-9$ and $-2$.
   $x^2 - 11x + 18 = (x - 9)(x - 2)$
4. The two numbers whose product is 24 and whose sum is $-11$ are $-8$ and $-3$. $x^2 - 11x + 24 = (x - 8)(x - 3)$ or $(x - 3)(x - 8)$
5. The two numbers whose product is $-24$ and whose sum is $-5$ are $-8$ and $+3$. $x^2 - 5x - 24 = (x - 8)(x + 3)$ or $(x + 3)(x - 8)$
6. The two numbers whose product is $-60$ and whose sum is $+17$ are $+20$ and $-3$. $y^2 + 17y - 60 = (y + 20)(y - 3)$
   **Check:** $(y + 20)(y - 3) = y^2 - 3y + 20y - 60 = y^2 + 17y - 60$
7. List the possible factors of 60.

| Factors of 60 | Difference | |
|---|---|---|
| 60 and 1 | 59 | |
| 30 and 2 | 28 | |
| 20 and 3 | 17 | |
| 15 and 4 | 11 | |
| 12 and 5 | 7 | ← Desired Value |
| 10 and 6 | 4 | |

For the coefficient of the middle term to be $-7$, we must add $-12$ and $+5$. $x^2 - 7x - 60 = (x - 12)(x + 5)$

8. $a^4 = (a^2)(a^2)$
   The two numbers whose product is $-42$ and whose sum is 1 are 7 and $-6$.
   $a^4 + a^2 - 42 = (a^2 + 7)(a^2 - 6)$
9. $3x^2 + 45x + 150 = 3(x^2 + 15x + 50)$
   $= 3(x + 5)(x + 10)$
10. $4x^2 - 8x - 140 = 4(x^2 - 2x - 35)$
    $= 4(x + 5)(x - 7)$
11. $x(x + 1) - 4(5) = x^2 + x - 20$
    $= (x + 5)(x - 4)$

## 5.4 Student Practice

1. To get a first term of $2x^2$, the coefficients of $x$ in the factors must be 2 and 1. To get a last term of 5, the constants in the factors must be 1 and 5. Possibilities:
   $(2x + 1)(x + 5) = 2x^2 + 11x + 5$
   $(2x + 5)(x + 1) = 2x^2 + 7x + 5$
   Thus $2x^2 + 7x + 5 = (2x + 5)(x + 1)$ or $(x + 1)(2x + 5)$.
2. The different factorizations of 9 are $(3)(3)$ and $(1)(9)$. The only factorization of 7 is $(1)(7)$.

| Possible Factors | Middle Term | Correct? |
|---|---|---|
| $(3x - 7)(3x - 1)$ | $-24x$ | No |
| $(9x - 7)(x - 1)$ | $-16x$ | No |
| $(9x - 1)(x - 7)$ | $-64x$ | Yes |

The correct answer is $(9x - 1)(x - 7)$ or $(x - 7)(9x - 1)$.

3. The only factorization of 3 is $(3)(1)$. The different factorizations of 14 are $(14)(1)$ and $(7)(2)$.

| Possible Factors | Middle Term | Correct Factors? |
|---|---|---|
| $(3x + 14)(x - 1)$ | $+11x$ | No |
| $(3x + 1)(x - 14)$ | $-41x$ | No |
| $(3x + 7)(x - 2)$ | $+x$ | No (wrong sign) |
| $(3x + 2)(x - 7)$ | $-19x$ | No |

To get the correct sign on the middle term, reverse the signs of the constants in the factors.
The correct answer is $(3x - 7)(x + 2)$ or $(x + 2)(3x - 7)$.

4. We list some of the options and check to see which one yields the middle term $+7x$.

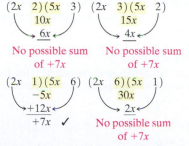

Only one of the possibilities yields $+7x$, so we can write the factors $(2x - 1)(5x + 6)$.
**Check:** $(2x - 1)(5x + 6) = 10x^2 + 7x - 6$ ✓

5. The grouping number is $3(-4) = -12$. The two numbers with a product of $-12$ and a sum of 4 are 6 and $-2$. Write $4x$ as the sum $6x + (-2x)$ or $6x - 2x$.
   $3x^2 + 4x - 4 = 3x^2 + 6x - 2x - 4$
   $= 3x(x + 2) - 2(x + 2)$
   $= (x + 2)(3x - 2)$
6. The grouping number is $9 \cdot 7 = 63$. The two numbers with a product of 63 and a sum of $-64$ are $-63$ and $-1$. Write $-64x$ as the sum $-63x + (-1x)$ or $-63x - 1x$.
   $9x^2 - 64x + 7 = 9x^2 - 63x - x + 7$
   $= 9x(x - 7) - 1(x - 7)$
   $= (x - 7)(9x - 1)$

**7.** $8x^2 + 8x - 6 = 2(4x^2 + 4x - 3)$
$\qquad = 2(2x - 1)(2x + 3)$
**8.** $24x^2 - 38x + 10 = 2(12x^2 - 19x + 5)$
$\qquad = 2(4x - 5)(3x - 1)$

## 5.5 Student Practice

**1.** $64x^2 - 1 = (8x + 1)(8x - 1)$ because $64x^2 = (8x)^2$ and $1 = (1)^2$.
**2.** $36x^2 - 49 = (6x + 7)(6x - 7)$ because $36x^2 = (6x)^2$ and
$49 = (7)^2$.
**3.** $100x^2 - 81y^2 = (10x + 9y)(10x - 9y)$
**4.** $x^8 - 1 = (x^4 + 1)(x^4 - 1)$
$\qquad = (x^4 + 1)(x^2 + 1)(x^2 - 1)$
$\qquad = (x^4 + 1)(x^2 + 1)(x + 1)(x - 1)$
**5.** The first and last terms are perfect squares: $x^2 = (x)^2$ and $25 = (5)^2$.
The middle term, $10x$, is twice the product of $x$ and $5$.
$x^2 + 10x + 25 = (x + 5)^2$
**6.** $30x = 2(5x \cdot 3)$
Also note the negative sign. $25x^2 - 30x + 9 = (5x - 3)^2$
**7. (a)** $25x^2 = (5x)^2, 36y^2 = (6y)^2$, and $60xy = 2(5x \cdot 6y)$.
$25x^2 + 60xy + 36y^2 = (5x + 6y)^2$
**(b)** $64x^6 = (8x^3)^2, 9 = (3)^2$, and $48x^3 = 2(8x^3 \cdot 3)$.
$64x^6 - 48x^3 + 9 = (8x^3 - 3)^2$
**8.** $9x^2 = (3x)^2$ and $4 = (2)^2$, but $15x \neq 2(3x \cdot 2) = 12x$.
$9x^2 + 15x + 4 = (3x + 1)(3x + 4)$
**9.** $20x^2 - 45 = 5(4x^2 - 9)$
$\qquad = 5(2x + 3)(2x - 3)$
**10.** $75x^2 - 60x + 12 = 3(25x^2 - 20x + 4)$
$\qquad = 3(5x - 2)^2$

## 5.6 Student Practice

**1. (a)** $3x^2 - 36x + 108$
$\qquad = 3(x^2 - 12x + 36)$
$\qquad = 3(x - 6)^2$
**(b)** $9x^4y^2 - 9y^2$
$\qquad = 9y^2(x^4 - 1)$
$\qquad = 9y^2(x^2 + 1)(x^2 - 1)$
$\qquad = 9y^2(x^2 + 1)(x + 1)(x - 1)$

**(c)** $12x - 9 - 4x^2 = -4x^2 + 12x - 9$
$\qquad = -(4x^2 - 12x + 9)$
$\qquad = -(2x - 3)^2$
**2.** $5x^3 - 20x + 2x^2 - 8 = 5x(x^2 - 4) + 2(x^2 - 4)$
$\qquad = (5x + 2)(x^2 - 4)$
$\qquad = (5x + 2)(x + 2)(x - 2)$

**3. (a)** $x^2 - 9x - 8$
The factorizations of $-8$ are $(-2)(4), (2)(-4), (-8)(1)$, and $(-1)(8)$.
None of these pairs will add up to be the coefficient of the middle term. Thus the polynomial cannot be factored. It is prime.
**(b)** $25x^2 + 82x + 4$
Check to see if this is a perfect-square trinomial.
$2[(5)(2)] = 2(10) = 20$
This is not the coefficient of the middle term. The grouping number is 100. No factors add to 82. It is prime.

## 5.7 Student Practice

**1.** $10x^2 - x - 2 = 0$
$(5x + 2)(2x - 1) = 0$
$5x + 2 = 0 \qquad 2x - 1 = 0$
$\quad 5x = -2 \qquad\quad 2x = 1$
$\quad\; x = -\dfrac{2}{5} \qquad\;\; x = \dfrac{1}{2}$

*Check:*
$\quad 10\left(-\dfrac{2}{5}\right)^2 - \left(-\dfrac{2}{5}\right) - 2 \overset{?}{=} 0 \qquad 10\left(\dfrac{1}{2}\right)^2 - \dfrac{1}{2} - 2 \overset{?}{=} 0$
$\quad 10\left(\dfrac{4}{25}\right) + \dfrac{2}{5} - 2 \overset{?}{=} 0 \qquad\quad 10\left(\dfrac{1}{4}\right) - \dfrac{1}{2} - 2 \overset{?}{=} 0$
$\qquad\quad \dfrac{8}{5} + \dfrac{2}{5} - 2 \overset{?}{=} 0 \qquad\qquad\quad \dfrac{5}{2} - \dfrac{1}{2} - 2 \overset{?}{=} 0$
$\qquad\quad \dfrac{10}{5} - \dfrac{10}{5} \overset{?}{=} 0 \qquad\qquad\qquad\quad \dfrac{4}{2} - 2 \overset{?}{=} 0$
$\qquad\qquad\quad 0 = 0 \qquad\qquad\qquad\qquad\quad 2 - 2 \overset{?}{=} 2$
$\qquad\qquad\qquad\qquad\qquad\qquad\qquad\qquad\quad 0 = 0$

Thus $-\frac{2}{5}$ and $\frac{1}{2}$ are both roots of the equation.
**2.** $3x^2 + 11x - 4 = 0$
$(3x - 1)(x + 4) = 0$
$3x - 1 = 0 \qquad x + 4 = 0$
$\quad 3x = 1 \qquad\qquad x = -4$
$\quad\; x = \dfrac{1}{3}$

Thus $\frac{1}{3}$ and $-4$ are both roots of the equation.
**3.** $\quad 7x^2 + 11x = 0$
$x(7x + 11) = 0$
$x = 0 \quad 7x + 11 = 0$
$\qquad\qquad 7x = -11$
$\qquad\qquad\; x = \dfrac{-11}{7}$

Thus $0$ and $-\frac{11}{7}$ are both roots of the equation.
**4.** $\qquad x^2 - 6x + 4 = -8 + x$
$\qquad x^2 - 7x + 12 = 0$
$\quad (x - 3)(x - 4) = 0$
$\quad x - 3 = 0 \qquad x - 4 = 0$
$\qquad\; x = 3 \qquad\qquad x = 4$
The roots are 3 and 4.

**5.** $\dfrac{2x^2 - 7x}{3} = 5$
$3\left(\dfrac{2x^2 - 7x}{3}\right) = 3(5)$
$2x^2 - 7x = 15$
$2x^2 - 7x - 15 = 0$
$(2x + 3)(x - 5) = 0$
$2x + 3 = 0 \qquad x - 5 = 0$
$\quad x = -\dfrac{3}{2} \qquad\;\; x = 5$

The roots are $-\frac{3}{2}$ and 5.
**6.** Let $w = $ width, then
$3w + 2 = $ length.
$(3w + 2)w = 85$
$3w^2 + 2w = 85$
$3w^2 + 2w - 85 = 0$
$(3w + 17)(w - 5) = 0$
$3w + 17 = 0 \qquad w - 5 = 0$
$\quad w = -\dfrac{17}{3} \qquad\;\; w = 5$

The only valid answer is width $= 5$ meters
length $= 3(5) + 2 = 17$ meters

**7.** Let $b$ = base.

$b - 3$ = altitude.

$$\frac{b(b - 3)}{2} = 35$$

$$\frac{b^2 - 3b}{2} = 35$$

$$b^2 - 3b = 70$$

$$b^2 - 3b - 70 = 0$$

$$(b + 7)(b - 10) = 0$$

$$b + 7 = 0$$

$$b = -7$$

This is not a valid answer.

$$b - 10 = 0$$

$$b = 10$$

Thus the base = 10 centimeters

altitude = $10 - 3 = 7$ centimeters

**8.**
$$-5t^2 + 45 = 0$$
$$-5(t^2 - 9) = 0$$
$$t^2 - 9 = 0$$
$$(t + 3)(t - 3) = 0$$
$$t + 3 = 0 \qquad t - 3 = 0$$
$$t = -3 \qquad t = 3$$

$t = -3$ is not a valid answer.

Thus it will be 3 seconds before he breaks the water's surface.

# Chapter 6   6.1 Student Practice

**1.** $\dfrac{28}{63} = \dfrac{7 \cdot 4}{7 \cdot 9} = \dfrac{4}{9}$

**2.** $\dfrac{12x - 6}{14x - 7} = \dfrac{6(2x - 1)}{7(2x - 1)} = \dfrac{6}{7}$

**3.** $\dfrac{4x^2 - 9}{2x^2 - x - 3} = \dfrac{(2x - 3)(2x + 3)}{(2x - 3)(x + 1)} = \dfrac{2x + 3}{x + 1}$

**4.** $\dfrac{x^3 - 16x}{x^3 - 2x^2 - 8x} = \dfrac{x(x^2 - 16)}{x(x^2 - 2x - 8)} = \dfrac{x(x + 4)(x - 4)}{x(x + 2)(x - 4)} = \dfrac{x + 4}{x + 2}$

**5.** $\dfrac{8x - 20}{15 - 6x} = \dfrac{4(2x - 5)}{-3(-5 + 2x)} = \dfrac{4}{-3} = -\dfrac{4}{3}$

**6.** $\dfrac{4x^2 + 3x - 10}{25 - 16x^2} = \dfrac{(4x - 5)(x + 2)}{(5 + 4x)(5 - 4x)} = \dfrac{(4x - 5)(x + 2)}{-1(-5 + 4x)(5 + 4x)}$

$\qquad = \dfrac{x + 2}{-1(5 + 4x)} = -\dfrac{x + 2}{5 + 4x}$

**7.** $\dfrac{x^2 - 8xy + 15y^2}{2x^2 - 11xy + 5y^2} = \dfrac{(x - 3y)(x - 5y)}{(x - 5y)(2x - y)} = \dfrac{x - 3y}{2x - y}$

**8.** $\dfrac{25a^2 - 16b^2}{10a^2 + 3ab - 4b^2} = \dfrac{(5a + 4b)(5a - 4b)}{(5a + 4b)(2a - b)} = \dfrac{5a - 4b}{2a - b}$

## 6.2 Student Practice

**1.** $\dfrac{6x^2 + 7x + 2}{x^2 - 7x + 10} \cdot \dfrac{x^2 + 3x - 10}{2x^2 + 11x + 5}$

$\qquad = \dfrac{(2x + 1)(3x + 2)}{(x - 5)(x - 2)} \cdot \dfrac{(x + 5)(x - 2)}{(2x + 1)(x + 5)} = \dfrac{3x + 2}{x - 5}$

**2.** $\dfrac{2y^2 - 6y - 8}{y^2 - y - 2} \cdot \dfrac{y^2 - 5y + 6}{2y^2 - 32}$

$\qquad = \dfrac{2(y^2 - 3y - 4)}{(y - 2)(y + 1)} \cdot \dfrac{(y - 3)(y - 2)}{2(y^2 - 16)}$

$\qquad = \dfrac{2(y - 4)(y + 1)}{(y - 2)(y + 1)} \cdot \dfrac{(y - 3)(y - 2)}{2(y + 4)(y - 4)} = \dfrac{y - 3}{y + 4}$

**3.** $\dfrac{x^2 + 5x + 6}{x^2 + 8x} \div \dfrac{2x^2 + 5x + 2}{2x^2 + x}$

$\qquad = \dfrac{x^2 + 5x + 6}{x^2 + 8x} \cdot \dfrac{2x^2 + x}{2x^2 + 5x + 2}$

$\qquad = \dfrac{(x + 2)(x + 3)}{x(x + 8)} \cdot \dfrac{x(2x + 1)}{(2x + 1)(x + 2)} = \dfrac{x + 3}{x + 8}$

**4.** $\dfrac{x + 3}{x - 3} \div (9 - x^2) = \dfrac{x + 3}{x - 3} \div \dfrac{9 - x^2}{1} = \dfrac{x + 3}{x - 3} \cdot \dfrac{1}{9 - x^2}$

$\qquad = \dfrac{x + 3}{x - 3} \cdot \dfrac{1}{(3 + x)(3 - x)}$

$\qquad = \dfrac{1}{(x - 3)(3 - x)}$

# 6.3 Student Practice

**1.** $\dfrac{2s + t}{2s - t} + \dfrac{s - t}{2s - t} = \dfrac{2s + t + s - t}{2s - t} = \dfrac{3s}{2s - t}$

**2.** $\dfrac{b}{(a - 2b)(a + b)} - \dfrac{2b}{(a - 2b)(a + b)}$

$\qquad = \dfrac{b - 2b}{(a - 2b)(a + b)}$

$\qquad = \dfrac{-b}{(a - 2b)(a + b)}$

**3.** $\dfrac{7}{6x + 21}, \dfrac{13}{10x + 35}$

$\qquad 6x + 21 = 3(2x + 7)$

$\qquad 10x + 35 = 5(2x + 7)$

$\qquad \text{LCD} = 3 \cdot 5 \cdot (2x + 7) = 15(2x + 7)$

**4. (a)** $\dfrac{3}{50xy^2z}, \dfrac{19}{40x^3yz}$

$50xy^2z = 2 \cdot \qquad 5 \cdot 5 \cdot x \cdot \qquad y \cdot y \cdot z$

$40x^3yz = 2 \cdot 2 \cdot 2 \cdot 5 \cdot \mid x \cdot x \cdot x \cdot y \cdot \mid z$

$\qquad \downarrow \downarrow \downarrow \downarrow \quad \downarrow \downarrow \downarrow \downarrow \quad \downarrow$

$\qquad 2 \cdot 2 \cdot 2 \cdot 5 \cdot 5 \cdot x \cdot x \cdot x \cdot y \cdot y \cdot z$

$\qquad \text{LCD} = 2^3 \cdot 5^2 \cdot x^3 \cdot y^2 \cdot z = 200x^3y^2z$

**(b)** $\dfrac{2}{x^2 + 5x + 6}, \dfrac{6}{3x^2 + 5x - 2}$

$\qquad x^2 + 5x + 6 = (x + 3)(x + 2)$

$\qquad 3x^2 + 5x - 2 = \qquad (x + 2)(3x - 1)$

$\qquad \text{LCD} = (x + 3)(x + 2)(3x - 1)$

**5.** $\text{LCD} = abc$

$\dfrac{7}{a} + \dfrac{3}{abc} = \dfrac{7}{a} \cdot \dfrac{bc}{bc} + \dfrac{3}{abc} = \dfrac{7bc}{abc} + \dfrac{3}{abc} = \dfrac{7bc + 3}{abc}$

**6.** $a^2 - 4b^2 = (a + 2b)(a - 2b)$

$\text{LCD} = (a + 2b)(a - 2b)$

$\dfrac{2a - b}{a^2 - 4b^2} + \dfrac{2}{a + 2b}$

$\qquad = \dfrac{2a - b}{(a + 2b)(a - 2b)} + \dfrac{2}{(a + 2b)} \cdot \dfrac{a - 2b}{a - 2b}$

$\qquad = \dfrac{2a - b}{(a + 2b)(a - 2b)} + \dfrac{2a - 4b}{(a + 2b)(a - 2b)}$

$\qquad = \dfrac{2a - b + 2a - 4b}{(a + 2b)(a - 2b)}$

$\qquad = \dfrac{4a - 5b}{(a + 2b)(a - 2b)}$

**7.** $\dfrac{7a}{a^2 + 2ab + b^2} + \dfrac{4}{a^2 + ab}$

$\qquad = \dfrac{7a}{(a + b)^2} + \dfrac{4}{a(a + b)} \qquad \text{LCD} = a(a + b)^2$

$\qquad = \dfrac{7a}{(a + b)^2} \cdot \dfrac{a}{a} + \dfrac{4}{a(a + b)} \cdot \dfrac{(a + b)}{(a + b)}$

$\qquad = \dfrac{7a^2}{a(a + b)^2} + \dfrac{4(a + b)}{a(a + b)^2} = \dfrac{7a^2 + 4a + 4b}{a(a + b)^2}$

**8.** $\dfrac{x+7}{3x-9} - \dfrac{x-6}{x-3} = \dfrac{x+7}{3(x-3)} - \dfrac{x-6}{x-3}$

$= \dfrac{x+7}{3(x-3)} - \dfrac{x-6}{x-3}\cdot\dfrac{3}{3} = \dfrac{x+7-3(x-6)}{3(x-3)}$

$= \dfrac{x+7-3x+18}{3(x-3)} = \dfrac{-2x+25}{3(x-3)}$

**9.** $\dfrac{x-2}{x^2-4} - \dfrac{x+1}{2x^2+4x} = \dfrac{x-2}{(x+2)(x-2)} - \dfrac{x+1}{2x(x+2)}$

$= \dfrac{x-2}{(x+2)(x-2)}\cdot\dfrac{2x}{2x} - \dfrac{x+1}{2x(x+2)}\cdot\dfrac{x-2}{x-2}$

$= \dfrac{2x(x-2)-(x+1)(x-2)}{2x(x+2)(x-2)} = \dfrac{2x^2-4x-x^2+x+2}{2x(x+2)(x-2)}$

$= \dfrac{x^2-3x+2}{2x(x+2)(x-2)} = \dfrac{(x-1)(x-2)}{2x(x+2)(x-2)} = \dfrac{x-1}{2x(x+2)}$

## 6.4 Student Practice

**1.** $\dfrac{\dfrac{1}{a}+\dfrac{1}{a^2}}{\dfrac{2}{b^2}} = \dfrac{\dfrac{1}{a}\cdot\dfrac{a}{a}+\dfrac{1}{a^2}}{\dfrac{2}{b^2}} = \dfrac{\dfrac{a+1}{a^2}}{\dfrac{2}{b^2}} = \dfrac{a+1}{a^2} \div \dfrac{2}{b^2}$

$= \dfrac{a+1}{a^2}\cdot\dfrac{b^2}{2} = \dfrac{b^2(a+1)}{2a^2}$

**2.** $\dfrac{\dfrac{1}{a}+\dfrac{1}{b}}{\dfrac{x}{2}-\dfrac{5}{y}} = \dfrac{\dfrac{1}{a}\cdot\dfrac{b}{b}+\dfrac{1}{b}\cdot\dfrac{a}{a}}{\dfrac{x}{2}\cdot\dfrac{y}{y}-\dfrac{5}{y}\cdot\dfrac{2}{2}} = \dfrac{\dfrac{b+a}{ab}}{\dfrac{xy-10}{2y}} = \dfrac{b+a}{ab}\cdot\dfrac{2y}{xy-10} = \dfrac{2y(b+a)}{ab(xy-10)}$

**3.** $\dfrac{\dfrac{x}{x^2+4x+3}+\dfrac{2}{x+1}}{x+1} = \dfrac{\dfrac{x}{(x+1)(x+3)}+\dfrac{2}{x+1}\cdot\dfrac{x+3}{x+3}}{x+1}$

$= \dfrac{\dfrac{x+2x+6}{(x+1)(x+3)}}{(x+1)} = \dfrac{3x+6}{(x+1)(x+3)}\cdot\dfrac{1}{(x+1)}$

$= \dfrac{3(x+2)}{(x+1)^2(x+3)}$

**4.** $\dfrac{\dfrac{6}{x^2-y^2}}{\dfrac{1}{x-y}+\dfrac{3}{x+y}} = \dfrac{\dfrac{6}{x^2-y^2}}{\dfrac{1}{x-y}\cdot\dfrac{x+y}{x+y}+\dfrac{3}{x+y}\cdot\dfrac{x-y}{x-y}}$

$= \dfrac{\dfrac{6}{x^2-y^2}}{\dfrac{x+y}{(x+y)(x-y)}+\dfrac{3x-3y}{(x+y)(x-y)}} = \dfrac{\dfrac{6}{(x+y)(x-y)}}{\dfrac{4x-2y}{(x+y)(x-y)}}$

$= \dfrac{6}{(x+y)(x-y)}\cdot\dfrac{(x+y)(x-y)}{2(2x-y)} = \dfrac{2\cdot3}{2(2x-y)} = \dfrac{3}{2x-y}$

**5.** The LCD of all the denominators is $3x^2y$.

$\dfrac{\dfrac{2}{3x^2}-\dfrac{3}{y}}{\dfrac{5}{xy}-4} = \dfrac{3x^2y\left(\dfrac{2}{3x^2}-\dfrac{3}{y}\right)}{3x^2y\left(\dfrac{5}{xy}-4\right)} = \dfrac{3x^2y\left(\dfrac{2}{3x^2}\right)-3x^2y\left(\dfrac{3}{y}\right)}{3x^2y\left(\dfrac{5}{xy}\right)-3x^2y(4)} = \dfrac{2y-9x^2}{15x-12x^2y}$

**6.** The LCD of all the denominators is $(x+y)(x-y)$.

$\dfrac{\dfrac{6}{x^2-y^2}}{\dfrac{7}{x-y}+\dfrac{3}{x+y}}$

$= \dfrac{(x+y)(x-y)\left(\dfrac{6}{(x+y)(x-y)}\right)}{(x+y)(x-y)\left(\dfrac{7}{x-y}\right)+(x+y)(x-y)\left(\dfrac{3}{x+y}\right)}$

$= \dfrac{6}{7(x+y)+3(x-y)}$

$= \dfrac{6}{7x+7y+3x-3y}$

$= \dfrac{6}{10x+4y}$

$= \dfrac{2\cdot3}{2(5x+2y)}$

$= \dfrac{3}{5x+2y}$

## 6.5 Student Practice

**1.** $\dfrac{3}{x}+\dfrac{4}{5} = -\dfrac{2}{x}$   LCD $= 5x$    *Check.*

$5x\left(\dfrac{3}{x}\right) + 5x\left(\dfrac{4}{5}\right) = 5x\left(-\dfrac{2}{x}\right)$

$15 + 4x = -10$

$4x = -10 - 15$

$4x = -25$

$x = -\dfrac{25}{4} \text{ or } -6\dfrac{1}{4} \text{ or } -6.25$

$\dfrac{3}{-\dfrac{25}{4}} + \dfrac{4}{5} \overset{?}{=} -\dfrac{2}{-\dfrac{25}{4}}$

$-\dfrac{12}{25} + \dfrac{4}{5} \overset{?}{=} \dfrac{8}{25}$

$-\dfrac{12}{25} + \dfrac{20}{25} \overset{?}{=} \dfrac{8}{25}$

$\dfrac{8}{25} = \dfrac{8}{25}$ ✓

**2.** LCD $= (2x+1)(x+2)$

$\dfrac{6}{2x+1} = \dfrac{2}{x+2}$

$(2x+1)(x+2)\left(\dfrac{6}{2x+1}\right) = (2x+1)(x+2)\left(\dfrac{2}{x+2}\right)$

$6(x+2) = 2(2x+1)$

$6x+12 = 4x+2$

$2x+12 = 2$

$2x = -10$

$x = -5$

*Check.*

$\dfrac{4}{2\left(-\dfrac{5}{2}\right)+1} \overset{?}{=} \dfrac{6}{2\left(-\dfrac{5}{2}\right)-1}$

$\dfrac{4}{-5+1} \overset{?}{=} \dfrac{6}{-5-1}$

$\dfrac{4}{-4} \overset{?}{=} \dfrac{6}{-6}$

$-1 = -1$ ✓

**3.** $\dfrac{x-1}{x^2-4} = \dfrac{2}{x+2}+\dfrac{4}{x-2}$

$\dfrac{x-1}{(x+2)(x-2)} = \dfrac{2}{x+2}+\dfrac{4}{x-2}$

$(x+2)(x-2)\left[\dfrac{x-1}{(x+2)(x-2)}\right]$

$= (x+2)(x-2)\left(\dfrac{2}{x+2}\right) + (x+2)(x-2)\left(\dfrac{4}{x-2}\right)$

$x-1 = 2(x-2)+4(x+2)$

$x-1 = 2x-4+4x+8$

$x-1 = 6x+4$

$-5x-1 = 4$

$-5x = 5$

$x = -1$

*Check.* $\dfrac{-1-1}{(-1)^2-4} \overset{?}{=} \dfrac{2}{-1+2}+\dfrac{4}{-1-2}$

$\dfrac{-2}{-3} \overset{?}{=} \dfrac{2}{1}+\dfrac{4}{-3}$

$\dfrac{2}{3} \overset{?}{=} \dfrac{6}{3}-\dfrac{4}{3}$

$\dfrac{2}{3} = \dfrac{2}{3}$ ✓

**4.** The LCD is $x + 1$.

$$\frac{2x}{x+1} = \frac{-2}{x+1} + 1$$

$$(x+1)\left(\frac{2x}{x+1}\right) = (x+1)\left(\frac{-2}{x+1}\right) + (x+1)(1)$$

$$2x = -2 + x + 1$$

$$2x = x - 1$$

$$x = -1 \quad \text{(but see the check)}$$

**Check.** 
$$\frac{2(-1)}{-1+1} \overset{?}{=} \frac{-2}{-1+1} + 1$$

$$\frac{-2}{0} \overset{?}{=} \frac{-2}{0} + 1$$

These expressions are not defined; therefore, there is **no solution** to this equation.

## 6.6 Student Practice

**1.** Let $x =$ the number of hours it will take to drive 315 miles.

$$\frac{8}{420} = \frac{x}{315}$$

$$8(315) = 420x$$

$$\frac{2520}{420} = x$$

$$6 = x$$

It would take Brenda 6 hours to drive 315 miles.

**2.** Let $x =$ the distance represented by $2\frac{1}{2}$ inches.

$$\frac{\frac{5}{8}}{30} = \frac{2\frac{1}{2}}{x}$$

$$\frac{5}{8}x = 30\left(2\frac{1}{2}\right)$$

$$\frac{5}{8}x = 30\left(\frac{5}{2}\right)$$

$$\frac{5}{8}x = 75$$

$$8\left(\frac{5}{8}x\right) = 8(75)$$

$$5x = 600$$

$$x = 120$$

Therefore $2\frac{1}{2}$ inches represents 120 miles.

**3.** 
$$\frac{13}{x} = \frac{16}{18}$$

$$13(18) = 16x$$

$$234 = 16x$$

$$\frac{234}{16} = x$$

Side $x$ has length 
$$\frac{117}{8} = 14\frac{5}{8} \text{ cm.}$$

**4.** 
$$\frac{6}{7} = \frac{x}{38.5}$$

$$6(38.5) = 7x$$

$$231 = 7x$$

$$x = 33$$

The height of the flagpole is 33 feet.

**5.** Let $x =$ the speed of train B. Then train A time $= \dfrac{180}{x+10}$ and train B time $= \dfrac{150}{x}$.

$$\frac{180}{x+10} = \frac{150}{x}$$

$$180x = 150(x+10)$$

$$180x = 150x + 1500$$

$$30x = 1500$$

$$x = 50$$

Train B traveled 50 kilometers per hour. Train A traveled $50 + 10 = 60$ kilometers per hour.

**6.**

| | Number of Hours | Part of the Job Done in One Hour |
|---|---|---|
| John | 6 hours | $\frac{1}{6}$ |
| Dave | 7 hours | $\frac{1}{7}$ |
| John & Dave Together | $x$ | $\frac{1}{x}$ |

$\text{LCD} = 42x$

$$\frac{1}{6} + \frac{1}{7} = \frac{1}{x}$$

$$42x\left(\frac{1}{6}\right) + 42x\left(\frac{1}{7}\right) = 42x\left(\frac{1}{x}\right)$$

$$7x + 6x = 42$$

$$13x = 42$$

$$x = 3\frac{3}{13}$$

$$\frac{3}{13} \text{ hour} \times \frac{60 \text{ min}}{1 \text{ hour}} = \frac{180}{13} \text{ min} \approx 13.846 \text{ min}$$

Thus, doing the job together will take 3 hours and 14 minutes.

## Chapter 7   7.1 Student Practice

**1.** Point $B$ is 3 units to the right on the $x$-axis and 4 units up from the point where we stopped on the $x$-axis.

**2. (a)** Begin by counting 2 squares to the left, starting at the origin. Since the $y$-coordinate is negative, count 4 units down from the point where we stopped on the $x$-axis. Label the point $I$.

**(b)** Begin by counting 4 squares to the left of the origin. Then count 5 units up because the $y$-coordinate is positive. Label the point $J$.

**(c)** Begin by counting 4 units to the right of the origin. Then count 2 units down because the $y$-coordinate is negative. Label the point $K$.

**3.** The points are plotted in the figure.

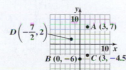

**4.** To find the point $C$ move along the $x$-axis to get as close as possible to $C$. We end up at 2. Thus the first number of the ordered pair is 2. Then count 7 units upward on a line parallel to the $y$-axis to reach $C$. So the second number of the ordered pair is 7. Thus, point $C$ is represented by $(2, 7)$.

**5. (a)**

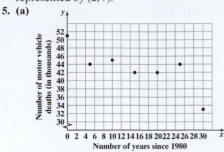

**(b)** Motor vehicle deaths were significantly high in 1980. During 1985–2005, the number of motor vehicle deaths was relatively stable. There was a significant decrease in the number of deaths in 2010.

**6. (a)** Replace $x$ with 3 and $y$ with $-1$.

$$3x + 2y = 5$$
$$3(3) + 2(-1) \stackrel{?}{=} 5$$
$$9 - 2 \stackrel{?}{=} 5$$
$$7 = 5 \quad \text{False}$$

The ordered pair $(3, -1)$ is not solution to $3x + 2y = 5$.

**(b)** Replace $x$ with 2 and $y$ with $-\dfrac{1}{2}$.

$$3x + 2y = 5$$
$$3(2) + 2\left(-\dfrac{1}{2}\right) \stackrel{?}{=} 5$$
$$6 + (-1) \stackrel{?}{=} 5$$
$$5 = 5 \quad \checkmark \text{True}$$

The ordered pair $\left(2, -\dfrac{1}{2}\right)$ is a solution to $3x + 2y = 5$.

**7. (a)** Replace $x$ by 0 in the equation.

$$3x - 4y = 12$$
$$3(0) - 4y = 12$$
$$0 - 4y = 12$$
$$y = -3$$

The ordered pair is $(0, -3)$.

**(b)** Replace the variable $y$ by 3.

$$3x - 4y = 12$$
$$3x - 4(3) = 12$$
$$3x - 12 = 12$$
$$3x = 24$$
$$x = 8$$

The ordered pair is $(8, 3)$.

**(c)** Replace the variable $y$ by $-6$.

$$3x - 4y = 12$$
$$3x - 4(-6) = 12$$
$$3x + 24 = 12$$
$$3x = -12$$
$$x = -4$$

The ordered pair is $(-4, -6)$.

## 7.2 Student Practice

**1.** Graph $y = -3x - 1$.

Let $x = 0$.
$$y = -3x - 1$$
$$y = -3(0) - 1$$
$$y = -1$$

Let $x = 1$.
$$y = -3x - 1$$
$$y = -3(1) - 1$$
$$y = -4$$

Let $x = -1$.
$$y = -3x - 1$$
$$y = -3(-1) - 1$$
$$y = 2$$

Plot the ordered pairs $(0, -1), (1, -4), (-1, 2)$

**2.**
$$7x + 3 = -2y + 3$$
$$7x + 3 - 3 = -2y + 3 - 3$$
$$7x = -2y$$
$$7x + 2y = -2y + 2y$$
$$7x + 2y = 0$$

Let $x = 0$.
$$7(0) + 2y = 0$$
$$2y = 0$$
$$y = 0$$

Let $x = -2$.
$$7(-2) + 2y = 0$$
$$-14 + 2y = 0$$
$$2y = 14$$
$$y = 7$$

Let $x = 2$.
$$7(2) + 2y = 0$$
$$14 + 2y = 0$$
$$2y = -14$$
$$y = -7$$

Graph the ordered pairs $(0, 0), (-2, 7)$, and $(2, -7)$.

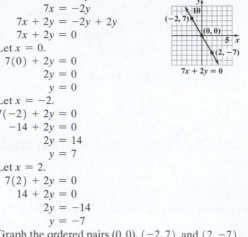

**3.** $2y - x = 6$

Find the two intercepts.

Let $y = 0$. Let $x = 0$.
$$2(0) - x = 6 \qquad 2y - 0 = 6$$
$$-x = 6 \qquad 2y = 6$$
$$x = -6 \qquad y = 3$$

The $x$-intercept is $(-6, 0)$, the $y$-intercept is $(0, 3)$.

Find a third point.

Let $y = 1$.
$$2(1) - x = 6$$
$$2 - x = 6$$
$$-x = 4$$
$$x = -4$$

Graph the ordered pairs $(-6, 0), (0, 3)$, and $(-4, 1)$.

**4.** $y = 2$

This line is parallel to the $x$-axis. It is a horizontal line 2 units above the $x$-axis.

**5.** $3x + 1 = -8$

Solve for $x$.
$$3x + 1 - 1 = -8 - 1$$
$$3x = -9$$
$$x = -3$$

This line is parallel to the $y$-axis. It is a vertical line 3 units to the left of the $y$-axis.

## 7.3 Student Practice

**1.** $m = \dfrac{y_2 - y_1}{x_2 - x_1} = \dfrac{-1 - 1}{-4 - 6} = \dfrac{-2}{-10} = \dfrac{1}{5}$

**2.** $m = \dfrac{y_2 - y_1}{x_2 - x_1} = \dfrac{1 - 0}{-1 - 2} = \dfrac{1}{-3} = -\dfrac{1}{3}$

**3. (a)** $m = \dfrac{3 - 6}{-5 - (-5)} = \dfrac{-3}{0}$

$\dfrac{-3}{0}$ is undefined. Therefore there is no slope and the line is a vertical line through $x = -5$.

**(b)** $m = \dfrac{-11 - (-11)}{3 - (-7)} = \dfrac{0}{10} = 0$

$m = 0$. The line is a horizontal line through $y = -11$.

**4.** Solve for $y$.
$$4x - 2y = -5$$
$$-2y = -4x - 5$$
$$y = \dfrac{-4x - 5}{-2}$$
$$y = 2x + \dfrac{5}{2} \qquad \text{Slope} = 2 \qquad y\text{-intercept} = \left(0, \dfrac{5}{2}\right)$$

**5. (a)** $y = mx + b$

$m = -\dfrac{3}{7}$   $y\text{-intercept} = \left(0, \dfrac{2}{7}\right), b = \dfrac{2}{7}$

$y = -\dfrac{3}{7}x + \dfrac{2}{7}$

**(b)**    $y = -\dfrac{3}{7}x + \dfrac{2}{7}$

$7(y) = 7\left(-\dfrac{3}{7}x\right) + 7\left(\dfrac{2}{7}\right)$

$7y = -3x + 2$

$3x + 7y = 2$

**6.** $y\text{-intercept} = (0, -1)$. Thus the coordinates of the $y$-intercept for this line are $(0, -1)$. Plot the point. Slope is $\dfrac{\text{rise}}{\text{run}}$. Since the slope for this line is $\dfrac{3}{4}$, we will go up (rise) 3 units and go over (run) 4 units to the right from the point $(0, -1)$. This is the point $(4, 2)$.

**7.** $y = -\dfrac{2}{3}x + 5$

The $y$-intercept is $(0, 5)$ since $b = 5$. Plot the point $(0, 5)$. The slope is $-\dfrac{2}{3} = \dfrac{-2}{3}$. Begin at $(0, 5)$, go down 2 units and to the right 3 units. This is the point $(3, 3)$. Draw the line that connects the points $(0, 5)$ and $(3, 3)$.

**8. (a)** Parallel lines have the same slope. Line $j$ has a slope of $\dfrac{1}{4}$.

**(b)** Perpendicular lines have slopes whose product is $-1$.

$m_1 m_2 = -1$

$\dfrac{1}{4}m_2 = -1$

$4\left(\dfrac{1}{4}\right)m_2 = -1(4)$

$m_2 = -4$

Thus line $k$ has a slope of $-4$.

**9. (a)** The slope of line $n$ is $\dfrac{2}{3}$. The slope of a line that is parallel to line $n$ is $\dfrac{2}{3}$.

**(b)** $m_1 m_2 = -1$

$\dfrac{2}{3}m_2 = -1$

$m_2 = -\dfrac{3}{2}$

The slope of a line that is perpendicular to line $n$ is $-\dfrac{3}{2}$.

## 7.4 Student Practice

**1.**   $y = mx + b$

$12 = -\dfrac{3}{4}(-8) + b$

$12 = 6 + b$

$6 = b$

An equation of the line is $y = -\dfrac{3}{4}x + 6$.

**2.** Find the slope.

$m = \dfrac{y_2 - y_1}{x_2 - x_1} = \dfrac{1 - 5}{-1 - 3} = \dfrac{-4}{-4} = 1$

Using either of the two points given, substitute $x$ and $y$ values into the equation $y = mx + b$.

$m = 1$   $x = 3$   and   $y = 5$.

$y = mx + b$

$5 = 1(3) + b$

$5 = 3 + b$

$2 = b$

An equation of the line is $y = x + 2$.

**3.** The $y$-intercept is $(0, 1)$. Thus $b = 1$. Look for another point on the line. We choose $(6, 2)$. Count the number of vertical units from 1 to 2 (rise). Count the number of horizontal units from 0 to 6 (run) $m = \dfrac{1}{6}$.

Now we can write an equation of the line.

$y = mx + b$

$y = \dfrac{1}{6}x + 1$

## 7.5 Student Practice

**1.** Graph $x - y \geq -10$.

Begin by graphing the line $x - y = -10$. Use any method discussed previously. Since there is an equals sign in the inequality, draw a solid line to indicate that the line is part of the solution set. The easiest test point is $(0, 0)$. Substitute $x = 0, y = 0$ in the inequality.

$x - y \geq -10$

$0 - 0 \geq -10$

$0 \geq -10$ true

Therefore, shade the side of the line that includes the point $(0, 0)$.

**2. Step 1** Graph $2y = x$. Since $>$ is used, the line should be a dashed line.

**Step 2** The line passes through $(0, 0)$.

**Step 3** Choose another test point, say $(-1, 1)$.

$2y > x$

$2(1) > -1$

$2 > -1$   true

Shade the region that includes $(-1, 1)$, that is, the region above the line.

**3. Step 1** Graph $y = -3$. Since $\geq$ is used, the line should be solid.

**Step 2** Test $(0, 0)$ in the inequality.

$y \geq -3$

$0 \geq -3$   true

Shade the region that includes $(0, 0)$, that is, the region above the line $y = -3$.

## 7.6 Student Practice

1. The domain consists of all the first coordinates of the ordered pairs. The domain is $\{-3, 0, 3, 20\}$. The range consists of all the second coordinates of the ordered pairs. The range is $\{-5, 5\}$.

2. (a) Look at the ordered pairs. No two ordered pairs have the same first coordinate. Thus this set of ordered pairs defines a function.
   (b) Look at the ordered pairs. Two different ordered pairs, $(60, 30)$ and $(60, 120)$, have the same first coordinate. Thus this relation is not a function.

3. (a) Looking at the table, we see that no two different ordered pairs have the same first coordinate. The cost of gasoline is a function of the distance traveled.
      Note that cost depends on distance. Thus distance is the independent variable. Since a negative distance does not make sense, the domain is $\{$all nonnegative real numbers$\}$.
      The range is $\{$all nonnegative real numbers$\}$.
   (b) Looking at the table, we see two ordered pairs, $(5, 20)$ and $(5, 30)$, have the same first coordinate. Thus this relation is not a function.

4. Construct a table, plot the ordered pairs, and connect the points.

| $x$ | $y = x^2 - 2$ | $y$ |
|---|---|---|
| $-2$ | $y = (-2)^2 - 2 = 2$ | $2$ |
| $-1$ | $y = (-1)^2 - 2 = -1$ | $-1$ |
| $0$ | $y = (0)^2 - 2 = -2$ | $-2$ |
| $1$ | $y = (1)^2 - 2 = -1$ | $-1$ |
| $2$ | $y = (2)^2 - 2 = 2$ | $2$ |

5. Select values of $y$ and then substitute them into the equation to obtain values of $x$.

| $y$ | $x = y^2 - 1$ | $x$ | $y$ |
|---|---|---|---|
| $-2$ | $x = (-2)^2 - 1 = 3$ | $3$ | $-2$ |
| $-1$ | $x = (-1)^2 - 1 = 0$ | $0$ | $-1$ |
| $0$ | $x = (0)^2 - 1 = -1$ | $-1$ | $0$ |
| $1$ | $x = (1)^2 - 1 = 0$ | $0$ | $1$ |
| $2$ | $x = (2)^2 - 1 = 3$ | $3$ | $2$ |

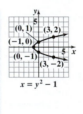

6. $y = \dfrac{6}{x}$

| $x$ | $y = \dfrac{6}{x}$ | $y$ |
|---|---|---|
| $-3$ | $y = \dfrac{6}{-3} = -2$ | $-2$ |
| $-2$ | $y = \dfrac{6}{-2} = -3$ | $-3$ |
| $-1$ | $y = \dfrac{6}{-1} = -6$ | $-6$ |
| $0$ | We cannot divide by 0. | |
| $1$ | $y = \dfrac{6}{1} = 6$ | $6$ |
| $2$ | $y = \dfrac{6}{2} = 3$ | $3$ |
| $3$ | $y = \dfrac{6}{3} = 2$ | $2$ |

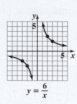

7. (a) The graph of a vertical line is not a function.
   (b) This curve is a function. Any vertical line will cross the curve in only one location.
   (c) This curve is not the graph of a function. There exist vertical lines that will cross the curve in more than one place.

8. $f(x) = -2x^2 + 3x - 8$
   (a) $f(2) = -2(2)^2 + 3(2) - 8$
       $= -2(4) + 3(2) - 8$
       $= -8 + 6 - 8$
       $= -10$
   (b) $f(-3) = -2(-3)^2 + 3(-3) - 8$
       $= -2(9) + 3(-3) - 8$
       $= -18 - 9 - 8$
       $= -35$
   (c) $f(0) = -2(0)^2 + 3(0) - 8$
       $= -2(0) + 3(0) - 8$
       $= 0 + 0 - 8$
       $= -8$

## Chapter 8  8.1 Student Practice

1. $x + y = 12$
   $-x + y = 4$
   Graph the equations on the same coordinate system.
   The lines intersect at the point $x = 4$, $y = 8$. Thus the solution to the system of equations is $(4, 8)$.

2. $4x + 2y = 8$
   $-6x - 3y = 6$
   Graph both equations on the same coordinate system.
   These lines are parallel. They do not intersect. Hence there is no solution to this system of equations.

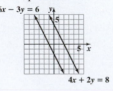

3. $3x - 9y = 18$
   $-4x + 12y = -24$
   Graph both equations on the same coordinate system.
   Notice both equations represent the same line. Thus there is an infinite number of solutions to this system.

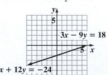

4. (a) Bill Tupper's Electrical Service charges $100 for a house call and $30 per hour. Thus we obtain the first equation, $y = 100 + 30x$.
      Wire for Hire charges $50 for a house call and $40 per hour. Thus we obtain the second equation, $y = 50 + 40x$.
   (b) **Bill Tupper's Electrical Service** $y = 100 + 30x$
      Let $x = 0$.  $y = 100 + 30(0) = 100$
      Let $x = 4$.  $y = 100 + 30(4) = 220$
      Let $x = 8$.  $y = 100 + 30(8) = 340$

| $x$ | $y$ |
|---|---|
| $0$ | $100$ |
| $4$ | $220$ |
| $8$ | $340$ |

**Wire for Hire** $y = 50 + 40x$
Let $x = 0$. $y = 50 + 40(0) = 50$
Let $x = 4$. $y = 50 + 40(4) = 210$
Let $x = 8$. $y = 50 + 40(8) = 370$

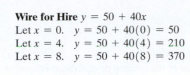

| $x$ | $y$ |
|---|---|
| 0 | 50 |
| 4 | 210 |
| 8 | 370 |

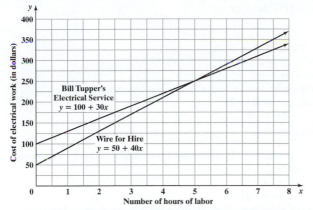

**(c)** We see that the graphs of the two lines intersect at $(5, 250)$. Thus the two companies will charge the same if 5 hours of electrical repairs are required.

**(d)** We draw a dashed line at $x = 6$. We see the line representing Wire for Hire is higher than the line representing Bill Tupper's Electrical Service after 5 hours. Thus the cost would be less if they use Bill Tupper's Electrical Service for 6 hours of work.

## 8.2 Student Practice

**1.** $5x + 3y = 19$ **(1)**
$2x - y = 12$ **(2)**

**Step 1** Solve equation **(2)** for $y$.
$2x - y = 12$ **(2)**
$-y = -2x + 12$
$y = 2x - 12$

**Step 2** Now substitute $2x - 12$ for $y$ in equation **(1)**.
$5x + 3y = 19$ **(1)**
$5x + 3(2x - 12) = 19$

**Step 3** Solve this equation.
$5x + 6x - 36 = 19$
$11x - 36 = 19$
$11x = 55$
$x = 5$

**Step 4** Now obtain the value for the second variable.
$2x - y = 12$ **(2)**
$2(5) - y = 12$
$10 - y = 12$
$-y = 2$
$y = -2$
The solution is $(5, -2)$.

**Step 5** Check.
$5x + 3y = 19$ **(1)** $\qquad$ $2x - y = 12$ **(2)**
$5(5) + 3(-2) \stackrel{?}{=} 19$ $\qquad$ $2(5) - (-2) \stackrel{?}{=} 12$
$25 - 6 \stackrel{?}{=} 19$ $\qquad$ $10 + 2 \stackrel{?}{=} 12$
$19 = 19$ ✓ $\qquad$ $12 = 12$ ✓

**2.** $\frac{1}{3}x - \frac{1}{2}y = 1$ **(1)**

$x + 4y = -8$ **(2)**
Clear equation **(1)** of fractions. We observe that the LCD is 6.
$6\left(\frac{1}{3}x\right) - 6\left(\frac{1}{2}y\right) = 6(1)$
$2x - 3y = 6$ **(3)**

The new system is:
$2x - 3y = 6$ **(3)**
$x + 4y = -8$ **(2)**

**Step 1** Solve equation **(2)** for $x$.
$x + 4y = -8$ **(2)**
$x = -4y - 8$

**Step 2** Substitute $-4y - 8$ for $x$ in equation **(3)**.
$2x - 3y = 6$ **(3)**
$2(-4y - 8) - 3y = 6$

**Step 3** Solve this equation.
$-8y - 16 - 3y = 6$
$-11y - 16 = 6$
$-11y = 22$
$y = -2$

**Step 4** Obtain the value of the second variable.
$x + 4y = -8$ **(2)**
$x + 4(-2) = -8$
$x - 8 = -8$
$x = 0$
The solution is $(0, -2)$.

**Step 5** Check.
$\frac{1}{3}x - \frac{1}{2}y = 1$ **(1)** $\qquad$ $x + 4y = -8$ **(2)**
$\frac{1}{3}(0) - \frac{1}{2}(-2) \stackrel{?}{=} 1$ $\qquad$ $0 + 4(-2) \stackrel{?}{=} -8$
$\qquad\qquad\qquad\qquad -8 = -8$ ✓
$0 + 1 \stackrel{?}{=} 1$
$1 = 1$ ✓

## 8.3 Student Practice

**1.** $3x + y = 7$ **(1)**
$5x - 2y = 8$ **(2)**
Multiply equation **(1)** by 2.
$2(3x) + 2(y) = 2(7)$
$6x + 2y = 14$ **(3)**
$5x - 2y = 8$ **(2)**
$\overline{11x \qquad = 22}$
$x = 2$
$5x - 2y = 8$ **(2)**
$5(2) - 2y = 8$
$10 - 2y = 8$
$-2y = -2$
$y = 1$
The solution is $(2, 1)$.

**2.** $4x + 5y = 17$ **(1)**
$3x + 7y = 12$ **(2)**
Multiply equation **(1)** by $-3$ and equation **(2)** by 4.
$-3(4x) + (-3)(5y) = (-3)(17)$
$4(3x) + 4(7y) = 4(12)$
$-12x - 15y = -51$ **(3)**
$\underline{12x + 28y = 48}$ **(2)**
$13y = -3$
$y = -\frac{3}{13}$

Substitute $y = -\frac{3}{13}$ into one of the original equations.
$4x + 5\left(-\frac{3}{13}\right) = 17$ **(1)**
$4x - \frac{15}{13} = 17$
$13(4x) - 13\left(\frac{15}{13}\right) = 13(17)$
$52x - 15 = 221$
$52x = 236$
$x = \frac{59}{13}$

The solution is $\left(\frac{59}{13}, -\frac{3}{13}\right)$.

**3.** $\frac{2}{3}x - \frac{3}{4}y = 3$ **(1)**

$-2x + y = 6$ **(2)**

Multiply equation **(1)** by 12.

$12\left(\frac{2}{3}x\right) - 12\left(\frac{3}{4}y\right) = 12(3)$

$8x - 9y = 36$

We now have an equivalent system without fractions.

$8x - 9y = 36$ **(3)**

$-2x + y = 6$ **(2)**

Multiply equation **(2)** by 4 to eliminate $x$.

$4(-2x) + 4(y) = 4(6)$

$\begin{array}{r} -8x + 4y = 24 \quad \textbf{(4)} \\ 8x - 9y = 36 \quad \textbf{(3)} \\ \hline -5y = 60 \end{array}$

$y = -12$

Substitute $y = -12$ into one of the original equations.

$-2x + (-12) = 6$

$-2x = 18$

$x = -9$

The solution is $(-9, -12)$.

**4.** $0.2x + 0.3y = -0.1$ **(1)**

$0.5x - 0.1y = -1.1$ **(2)**

Multiply each term of each equation by 10.

$10(0.2x) + 10(0.3y) = 10(-0.1)$

$10(0.5x) - 10(0.1y) = 10(-1.1)$

$2x + 3y = -1$ **(3)**

$5x - y = -11$ **(4)**

Multiply equation **(4)** by 3 to eliminate $y$.

$\begin{array}{r} 2x + 3y = -1 \quad \textbf{(3)} \\ 15x - 3y = -33 \quad \textbf{(5)} \\ \hline 17x = -34 \end{array}$

$x = -2$

Substitute $x = -2$ into equation **(3)**.

$2(-2) + 3y = -1$

$-4 + 3y = -1$

$3y = 3$

$y = 1$

The solution is $(-2, 1)$.

## 8.4 Student Practice

**1. (a)** $3x + 5y = 1485$

$x + 2y = 564$

Solve for $x$ in the second equation and solve using the substitution method.

$x = -2y + 564$

$3(-2y + 564) + 5y = 1485$

$-6y + 1692 + 5y = 1485$

$-y = -207$

$y = 207$

Substitute $y = 207$ into the second equation and solve for $x$.

$x + 2(207) = 564$

$x + 414 = 564$

$x = 150$

The solution is $(150, 207)$.

**(b)** $7x - 3 = -6(y - 7)$

$-5y = -2(3x + 1)$

For each of the equations we first remove parentheses and then simplify so that all $x$ terms and all $y$ terms are on the left and all the numbers are on the right side of the equals sign.

$7x - 3 = -6(y - 7)$      $-5y = -2(3x + 1)$

$7x - 3 = -6y + 42$      $-5y = -6x - 2$

$7x + 6y = 45$         $6x - 5y = -2$

Using the addition method, multiply the first equation by $-6$ and the second equation by 7.

$-6(7x) + (-6)(6y) = (-6)(45)$

$7(6x) - 7(5y) = 7(-2)$

$\begin{array}{r} -42x - 36y = -270 \\ 42x - 35y = -14 \\ \hline -71y = -284 \end{array}$

$y = 4$

Substitute $y = 4$ into one of the original or equivalent equations and solve for $x$.

$7x + 6(4) = 45$

$7x + 24 = 45$

$7x = 21$

$x = 3$

The solution is $(3, 4)$.

**2.** $4x + 2y = 2$

$-6x - 3y = 6$

Use the addition method. Multiply the first equation by 3 and the second equation by 2.

$\begin{array}{r} 12x + 6y = 6 \\ -12x - 6y = 12 \\ \hline 0 = 18 \end{array}$

The statement $0 = 18$ is not true. There is no solution to this system of equations.

**3.** $3x - 9y = 18$

$-4x + 12y = -24$

Use the addition method. Multiply the first equation by 4 and the second equation by 3.

$\begin{array}{r} 12x - 36y = 72 \\ -12x + 36y = -72 \\ \hline 0 = 0 \end{array}$

The statement $0 = 0$ is true and identifies the equations as dependent. There is an infinite number of solutions to the system.

## 8.5 Student Practice

**1.** Let $x$ = number of gallons/hour pumped by the smaller pump and $y$ = number of gallons/hour pumped by the larger pump.

1st day          $8x + 5y = 49{,}000$

2nd day        $5x + 3y = 30{,}000$

Use the addition method.

Multiply the first equation by 5 and the second equation by $-8$.

$\begin{array}{r} 40x + 25y = 245{,}000 \\ -40x - 24y = -240{,}000 \\ \hline y = 5000 \end{array}$

Substitute $y = 5000$ into one of the original equations and solve for $x$.

$8x + 5(5000) = 49{,}000$

$8x + 25{,}000 = 49{,}000$

$8x = 24{,}000$

$x = 3000$

The smaller pump removes 3000 gallons per hour.

The larger pump removes 5000 gallons per hour.

**2.** Let $x$ = cost/gallon of unleaded premium gasoline and $y$ = cost/gallon of unleaded regular gasoline.

Last week's purchase    $7x + 8y = 47.15$

This week's purchase    $8x + 4y = 38.20$

Use the addition method. Multiply the first equation by 8 and the second equation by $-7$.

$\begin{array}{r} 56x + 64y = 377.20 \\ -56x - 28y = -267.40 \\ \hline 36y = 109.8 \end{array}$

$y = 3.05$

Substitute $y = 3.05$ into one of the original equations and solve for $x$.

$7x + 8(3.05) = 47.15$

$7x + 24.40 = 47.15$

$7x = 22.75$

$x = 3.25$

Unleaded premium gasoline costs \$3.25/gallon.

Unleaded regular gasoline costs \$3.05/gallon.

3. Let $x$ = the amount of 20% solution and $y$ = the amount of 80% solution.
Total amount of 65% solution
$x + y = 4000$
Amount of acid in 65% solution
$0.2x + 0.8y = 0.65(4000)$
Use the substitution method. Solve for $x$ in the first equation.
$x = 4000 - y$
Substitute $x = 4000 - y$ into the second equation.
$0.2(4000 - y) + 0.8y = 0.65(4000)$
$800 - 0.2y + 0.8y = 2600$
$800 + 0.6y = 2600$
$0.6y = 1800$
$y = 3000$
Now substitute $y = 3000$ into the first equation.
$x + y = 4000$
$x + 3000 = 4000$
$x = 1000$
Therefore he will need 1000 liters of 20% solution and 3000 liters of 80% solution.

4. Let $b$ = the speed of the boat in still water in miles/hour and $c$ = the speed of the river current in miles/hour.
We organize our information in a chart.

| | $D$ | = | $R$ | · | $T$ |
|---|---|---|---|---|---|
| Traveling against current | 72 | | $(b - c)$ | | 4 |
| Traveling with current | 72 | | $(b + c)$ | | 3 |

We obtain a system of equations.
$$72 = (b - c) \cdot 4$$
$$72 = (b + c) \cdot 3$$
We remove parentheses.
$$72 = 4b - 4c \quad \textbf{(1)}$$
$$72 = 3b + 3c \quad \textbf{(2)}$$
We multiply equation **(1)** by 3 and equation **(2)** by 4 and then use the addition method to solve for $b$.
$$(3)72 = (3)4b - (3)4c \rightarrow 216 = 12b - 12c$$
$$(4)72 = (4)3b + (4)3c \rightarrow \underline{288 = 12b + 12c}$$
Now we solve for $c$. $\qquad\qquad 504 = 24b$
$$21 = b$$
Substitute $b = 21$ into one of the equivalent equations and solve for $c$.
**(1)** $\quad 72 = 4b - 4c \rightarrow 72 = 4 \cdot 21 - 4c$
$$72 = 84 - 4c$$
$$-12 = -4c$$
$$3 = c$$
Thus the speed of the boat in still water was 21 miles/hour and the speed of the current was 3 miles/hour.

# Chapter 9   9.1 Student Practice

1. (a) $\sqrt{64} = 8$ because $8^2 = 64$.
   (b) $-\sqrt{121} = -11$ since $\sqrt{121} = 11$.

2. (a) $\sqrt{\dfrac{9}{25}} = \dfrac{3}{5}$ since $\left(\dfrac{3}{5}\right)^2 = \left(\dfrac{3}{5}\right)\left(\dfrac{3}{5}\right) = \dfrac{9}{25}$.
   (b) $-\sqrt{\dfrac{121}{169}} = -\dfrac{11}{13}$ since $\sqrt{\dfrac{121}{169}} = \dfrac{11}{13}$.

3. (a) $-\sqrt{0.0036} = -0.06$ since $(0.06)(0.06) = 0.0036$.
   (b) $\sqrt{2500} = 50$ since $(50)(50) = 2500$.
   (c) $\sqrt{196} = 14$ since $(14)(14) = 196$

4. (a) $\sqrt{13} \approx 3.606$
   (b) $\sqrt{35} \approx 5.916$
   (c) $\sqrt{127} \approx 11.269$

## 9.2 Student Practice

1. (a) $\sqrt{6^2} = 6$
   (b) $\sqrt{13^4} = \sqrt{(13^2)^2} = 13^2$
   (c) $\sqrt{18^{12}} = \sqrt{(18^6)^2} = 18^6$

2. (a) $\sqrt{y^{18}} = \sqrt{(y^9)^2} = y^9$
   (b) $\sqrt{x^{30}} = \sqrt{(x^{15})^2} = x^{15}$

3. (a) $\sqrt{625y^4} = \sqrt{625}\sqrt{y^4} = 25y^2$
   (b) $\sqrt{x^{16}y^{22}} = x^8 y^{11}$
   (c) $\sqrt{121x^{12}y^6} = 11x^6 y^3$

4. (a) $\sqrt{98} = \sqrt{49 \cdot 2} = \sqrt{49}\sqrt{2} = 7\sqrt{2}$
   (b) $\sqrt{12} = \sqrt{4 \cdot 3} = \sqrt{4}\sqrt{3} = 2\sqrt{3}$
   (c) $\sqrt{75} = \sqrt{25 \cdot 3} = \sqrt{25}\sqrt{3} = 5\sqrt{3}$

5. (a) $\sqrt{x^{11}} = \sqrt{x^{10}}\sqrt{x} = x^5\sqrt{x}$
   (b) $\sqrt{x^5 y^3} = \sqrt{x^4}\sqrt{x}\sqrt{y^2}\sqrt{y} = x^2\sqrt{xy}\sqrt{y} = x^2 y\sqrt{xy}$

6. (a) $\sqrt{48x^{11}} = \sqrt{16 \cdot 3 \cdot x^{10} \cdot x} = 4x^5\sqrt{3x}$
   (b) $\sqrt{121x^6 y^7 z^8} = \sqrt{121 \cdot x^6 \cdot y^6 \cdot y \cdot z^8} = 11x^3 y^3 z^4 \sqrt{y}$

## 9.3 Student Practice

1. (a) $7\sqrt{11} + 4\sqrt{11} = (7 + 4)\sqrt{11} = 11\sqrt{11}$
   (b) $4\sqrt{t} - 7\sqrt{t} + 6\sqrt{t} - 2\sqrt{t}$
   $= (4 - 7 + 6 - 2)\sqrt{t}$
   $= 1\sqrt{t}$
   $= \sqrt{t}$

2. $3\sqrt{x} - 2\sqrt{xy} - 5\sqrt{y} + 7\sqrt{xy}$
   $= 3\sqrt{x} + 5\sqrt{xy} - 5\sqrt{y}$

3. (a) $\sqrt{50} - \sqrt{18} + \sqrt{98}$
   $= \sqrt{25 \cdot 2} - \sqrt{9 \cdot 2} + \sqrt{49 \cdot 2}$
   $= 5\sqrt{2} - 3\sqrt{2} + 7\sqrt{2} = 9\sqrt{2}$
   (b) $\sqrt{12} + \sqrt{18} - \sqrt{50} + \sqrt{27}$
   $= \sqrt{4 \cdot 3} + \sqrt{9 \cdot 2} - \sqrt{25 \cdot 2} + \sqrt{9 \cdot 3}$
   $= 2\sqrt{3} + 3\sqrt{2} - 5\sqrt{2} + 3\sqrt{3} = 5\sqrt{3} - 2\sqrt{2}$

4. $\sqrt{9x} + \sqrt{8x} - \sqrt{4x} + \sqrt{50x}$
   $= \sqrt{9 \cdot x} + \sqrt{4 \cdot 2x} - \sqrt{4 \cdot x} + \sqrt{25 \cdot 2x}$
   $= 3\sqrt{x} + 2\sqrt{2x} - 2\sqrt{x} + 5\sqrt{2x}$
   $= \sqrt{x} + 7\sqrt{2x}$

5. $\sqrt{27} - 4\sqrt{3} + 2\sqrt{75}$
   $= \sqrt{9} \cdot \sqrt{3} - 4 \cdot \sqrt{3} + 2\sqrt{25}\sqrt{3}$
   $= 3\sqrt{3} - 4\sqrt{3} + 2 \cdot 5 \cdot \sqrt{3}$
   $= 3\sqrt{3} - 4\sqrt{3} + 10\sqrt{3} = 9\sqrt{3}$

6. $2a\sqrt{12a} + 3\sqrt{27a^3}$
$= 2a\sqrt{4}\sqrt{3a} + 3\sqrt{9a^2}\sqrt{3a}$
$= 2a \cdot 2 \cdot \sqrt{3a} + 3 \cdot 3a \cdot \sqrt{3a}$
$= 4a\sqrt{3a} + 9a\sqrt{3a}$
$= 13a\sqrt{3a}$

## 9.4 Student Practice

1. $\sqrt{3a}\sqrt{6a} = \sqrt{18a^2} = \sqrt{9a^2 \cdot 2} = 3a\sqrt{2}$
2. (a) $(2\sqrt{3})(5\sqrt{5}) = 10\sqrt{15}$
   (b) $(4\sqrt{3x})(2x\sqrt{6x}) = 8x\sqrt{18x^2}$
   $= 8x\sqrt{9}\sqrt{2}\sqrt{x^2} = 24x^2\sqrt{2}$
3. $(\sqrt{180})(\sqrt{150}) = (6\sqrt{5})(5\sqrt{6})$
   $= 30\sqrt{30} \text{ mm}^2$
4. $\sqrt{6}(\sqrt{3} + 2\sqrt{8})$
   $= \sqrt{6}\sqrt{3} + \sqrt{6} \cdot 2\sqrt{8} = \sqrt{18} + 2\sqrt{48}$
   $= \sqrt{9}\sqrt{2} + 2\sqrt{16}\sqrt{3}$
   $= 3\sqrt{2} + 8\sqrt{3}$
5. $2\sqrt{x}(4\sqrt{x} - x\sqrt{2})$
   $= 8\sqrt{x}\sqrt{x} - 2x\sqrt{x}\sqrt{2}$
   $= 8x - 2x\sqrt{2x}$
6. $(\sqrt{2} + \sqrt{3})(2\sqrt{2} - \sqrt{3})$
   $= 2\sqrt{4} - \sqrt{6} + 2\sqrt{6} - \sqrt{9}$
   $= 2(2) + \sqrt{6} - 3$
   $= 4 + \sqrt{6} - 3$
   $= 1 + \sqrt{6}$
7. $(\sqrt{6} + \sqrt{3})(\sqrt{6} + 2\sqrt{3})$
   $= \sqrt{36} + 2\sqrt{18} + \sqrt{18} + 2\sqrt{9}$
   $= 6 + 6\sqrt{2} + 3\sqrt{2} + 6 = 12 + 9\sqrt{2}$
8. $(3\sqrt{5} - \sqrt{10})^2$
   $= (3\sqrt{5} - \sqrt{10})(3\sqrt{5} - \sqrt{10})$
   $= 9\sqrt{25} - 3\sqrt{50} - 3\sqrt{50} + \sqrt{100}$
   $= 45 - 6\sqrt{50} + 10$
   $= 55 - 30\sqrt{2}$

## 9.5 Student Practice

1. (a) $\dfrac{\sqrt{98}}{\sqrt{2}} = \sqrt{\dfrac{98}{2}} = \sqrt{49} = 7$

   (b) $\sqrt{\dfrac{81}{100}} = \dfrac{\sqrt{81}}{\sqrt{100}} = \dfrac{9}{10}$

2. $\sqrt{\dfrac{50}{a^4}} = \dfrac{\sqrt{50}}{\sqrt{a^4}} = \dfrac{\sqrt{25}\sqrt{2}}{a^2} = \dfrac{5\sqrt{2}}{a^2}$

3. $\dfrac{9}{\sqrt{7}} = \dfrac{9}{\sqrt{7}} \cdot \dfrac{\sqrt{7}}{\sqrt{7}} = \dfrac{9\sqrt{7}}{\sqrt{49}} = \dfrac{9\sqrt{7}}{7}$

4. (a) $\dfrac{\sqrt{2}}{\sqrt{12}} = \dfrac{\sqrt{2}}{\sqrt{12}} \cdot \dfrac{\sqrt{3}}{\sqrt{3}} = \dfrac{\sqrt{6}}{\sqrt{36}} = \dfrac{\sqrt{6}}{6}$

   (b) $\dfrac{6a}{\sqrt{a^7}} = \dfrac{6a}{\sqrt{a^7}} \cdot \dfrac{\sqrt{a}}{\sqrt{a}} = \dfrac{6a\sqrt{a}}{\sqrt{a^8}} = \dfrac{6a\sqrt{a}}{a^4} = \dfrac{6\sqrt{a}}{a^3}$

5. $\dfrac{\sqrt{5}}{\sqrt{8x}} = \dfrac{\sqrt{5}}{2\sqrt{2x}} = \dfrac{\sqrt{5}}{2\sqrt{2x}} \cdot \dfrac{\sqrt{2x}}{\sqrt{2x}} = \dfrac{\sqrt{10x}}{2\sqrt{4x^2}} = \dfrac{\sqrt{10x}}{4x}$

6. (a) $\dfrac{4}{\sqrt{3} + \sqrt{5}} = \dfrac{4}{\sqrt{3} + \sqrt{5}} \cdot \dfrac{\sqrt{3} - \sqrt{5}}{\sqrt{3} - \sqrt{5}}$

   $= \dfrac{4\sqrt{3} - 4\sqrt{5}}{(\sqrt{3})^2 - \sqrt{15} + \sqrt{15} - (\sqrt{5})^2}$

   $= \dfrac{4\sqrt{3} - 4\sqrt{5}}{3 - 5} = \dfrac{4\sqrt{3} - 4\sqrt{5}}{-2}$

   $= \dfrac{4(\sqrt{3} - \sqrt{5})}{-2} = -2(\sqrt{3} - \sqrt{5})$

   (b) $\dfrac{\sqrt{a}}{\sqrt{10} - 3} = \dfrac{\sqrt{a}}{\sqrt{10} - 3} \cdot \dfrac{\sqrt{10} + 3}{\sqrt{10} + 3}$

   $= \dfrac{\sqrt{10a} + 3\sqrt{a}}{(\sqrt{10})^2 + 3\sqrt{10} - 3\sqrt{10} - 3^2}$

   $= \dfrac{\sqrt{10a} + 3\sqrt{a}}{10 - 9} = \dfrac{\sqrt{10a} + 3\sqrt{a}}{1} = \sqrt{10a} + 3\sqrt{a}$

7. $\dfrac{\sqrt{3} + \sqrt{5}}{\sqrt{3} - \sqrt{5}} = \dfrac{\sqrt{3} + \sqrt{5}}{\sqrt{3} - \sqrt{5}} \cdot \dfrac{\sqrt{3} + \sqrt{5}}{\sqrt{3} + \sqrt{5}}$

   $= \dfrac{\sqrt{9} + \sqrt{15} + \sqrt{15} + \sqrt{25}}{(\sqrt{3})^2 - (\sqrt{5})^2}$

   $= \dfrac{3 + 2\sqrt{15} + 5}{3 - 5} = \dfrac{8 + 2\sqrt{15}}{-2}$

   $= \dfrac{-2(-4 - \sqrt{15})}{-2} = -4 - \sqrt{15}$

## 9.6 Student Practice

1. $c^2 = a^2 + b^2$
   $c^2 = 9^2 + 12^2$
   $c^2 = 81 + 144$
   $c^2 = 225$
   $c = \pm\sqrt{225}$
   $c = \pm 15$
   The hypotenuse is 15 centimeters because length is not negative.

2. $c^2 = a^2 + b^2$
   $(\sqrt{17})^2 = (1)^2 + b^2$
   $17 = 1 + b^2$
   $16 = b^2$
   $\pm 4 = b$
   The leg is 4 meters long.

3. $c^2 = 3^2 + 8^2$
   $c^2 = 9 + 64$
   $c^2 = 73$
   $c = \pm\sqrt{73}$
   The support line is $\sqrt{73} \approx 8.5$ meters long.

4. $\sqrt{3x - 2} - 7 = 0$
   $\sqrt{3x - 2} = 7$
   $(\sqrt{3x - 2})^2 = (7)^2$
   $3x - 2 = 49$
   $3x = 51$
   $x = 17$
   The solution is 17.
   **Check.**
   $\sqrt{3(17) - 2} - 7 \overset{?}{=} 0$
   $\sqrt{51 - 2} - 7 \overset{?}{=} 0$
   $\sqrt{49} - 7 \overset{?}{=} 0$
   $7 - 7 \overset{?}{=} 0$
   $0 = 0$ ✓

**5.** $\sqrt{5x + 4} + 2 = 0$

$\sqrt{5x + 4} = -2$

$(\sqrt{5x + 4})^2 = (-2)^2$

$5x + 4 = 4$

$5x = 0$

$x = 0$

***Check.***

$\sqrt{5(0) + 4} + 2 \overset{?}{=} 0$

$\sqrt{4} + 2 \overset{?}{=} 0$

$2 + 2 \overset{?}{=} 0$

$4 \neq 0$

This does not check. There is no solution.

**6.** $-2 + \sqrt{6x - 1} = 3x - 2$

$\sqrt{6x - 1} = 3x$

$(\sqrt{6x - 1})^2 = (3x)^2$

$6x - 1 = 9x^2$

$0 = 9x^2 - 6x + 1$

$0 = (3x - 1)^2$

$0 = 3x - 1$

$x = \dfrac{1}{3}$

***Check.***

$-2 + \sqrt{6\left(\dfrac{1}{3}\right) - 1} \overset{?}{=} 3\left(\dfrac{1}{3}\right) - 2$

$-2 + \sqrt{2 - 1} \overset{?}{=} 1 - 2$

$-2 + \sqrt{1} \overset{?}{=} -1$

$-2 + 1 \overset{?}{=} -1$

$-1 = -1$ ✓

The solution is $\dfrac{1}{3}$.

**7.** $2 - x + \sqrt{x + 4} = 0$

$\sqrt{x + 4} = x - 2$

$(\sqrt{x + 4})^2 = (x - 2)^2$

$x + 4 = x^2 - 4x + 4$

$0 = x^2 - 5x$

$0 = x(x - 5)$

$x = 0$ or $x - 5 = 0$

$x = 5$

***Check.***

$2 - 0 + \sqrt{0 + 4} \overset{?}{=} 0$

$2 + \sqrt{4} \overset{?}{=} 0$

$2 + 2 \overset{?}{=} 0$

$4 \neq 0$

$x = 0$ does not check.

$2 - 5 + \sqrt{5 + 4} \overset{?}{=} 0$

$-3 + \sqrt{9} \overset{?}{=} 0$

$-3 + 3 \overset{?}{=} 0$

$0 = 0$ ✓

The only solution is 5.

**8.** $\sqrt{2x + 1} = \sqrt{x + 5}$

$(\sqrt{2x + 1})^2 = (\sqrt{x + 5})^2$

$2x + 1 = x + 5$

$x = 4$

***Check.***

$\sqrt{2(4) + 1} \overset{?}{=} \sqrt{4 + 5}$

$\sqrt{8 + 1} \overset{?}{=} \sqrt{9}$

$3 = 3$ ✓

The solution is 4.

## 9.7 Student Practice

**1.** Change of $100°C$ = change of $180°F$.

Let $C$ = the change on the Celsius scale

$f$ = the change on the Fahrenheit scale, and

$k$ = the constant of variation.

$C = kf$

$100 = k \cdot 180$

$\dfrac{100}{180} = \dfrac{5}{9} = k$

Thus, $C = \dfrac{5}{9}f$.

Let $f = -20$.

$C = \dfrac{5}{9}(-20) = -\dfrac{100}{9}$

$C = -11\dfrac{1}{9}$

The temperature drops $11\dfrac{1}{9}$ degrees Celsius.

**2.** $y = kx$

To find $k$ we substitute $y = 18$ and $x = 5$.

$18 = k(5)$

$\dfrac{18}{5} = k$

We now write the variation equation with $k$ replaced by $\dfrac{18}{5}$.

$y = \dfrac{18}{5}x$

Replace $x$ by $\dfrac{20}{23}$ and solve for $y$.

$y = \dfrac{18}{\overset{1}{\cancel{5}}} \cdot \dfrac{\overset{4}{\cancel{20}}}{23}$

$y = \dfrac{72}{23}$

Thus $y = \dfrac{72}{23}$ when $x = \dfrac{20}{23}$.

**3.** Let $d$ = the distance to stop the car,

$s$ = the speed of the car, and

$k$ = the constant of variation.

Since the distance varies directly as the square of the speed, we have

$d = ks^2$.

To evaluate $k$ we substitute the known distance and speed.

$60 = k(20)^2$

$60 = k(400)$

$\dfrac{60}{400} = \dfrac{3}{20} = k$

Now write the variation equation with the known value for $k$.

$d = \dfrac{3}{20}s^2$

$d = \dfrac{3}{20}(40)^2$

$d = \dfrac{3}{\overset{1}{\cancel{20}}}(\overset{80}{\cancel{1600}})$

$d = 240$

The distance to stop the car going 40 mph on an ice-covered road is 240 feet.

**4.** $y = \dfrac{k}{x}$

$8 = \dfrac{k}{15}$    Substitute known values of $x$ and $y$ to find $k$.

$120 = k$

$y = \dfrac{120}{x}$    Write the variation equation with $k$ replaced by 120.

$y = \dfrac{120}{\dfrac{3}{5}} = \dfrac{\overset{40}{\cancel{120}}}{1} \cdot \dfrac{5}{\underset{1}{\cancel{3}}} = 200$   Find $y$ when $x = \dfrac{3}{5}$.

Thus $y = 200$ when $x = \dfrac{3}{5}$.

**5.** Let $V =$ the volume of sales,
$p =$ the price of the calculator, and
$k =$ the constant of variation.
Since volume varies inversely as the price, then

$$V = \frac{k}{p}.$$

$120,000 = \dfrac{k}{30}$     Evaluate $k$ by substituting known values for $V$ and $p$.

$3,600,000 = k$

$V = \dfrac{3,600,000}{24}$    Write the variation equation with the evaluated constant and substitute $p = 24$.

$V = 150,000$

Thus the volume increased to 150,000 calculators sold per year when the price was reduced to $24 per calculator.

**6.** Let $R =$ the resistance in the circuit,
$i =$ the amount of current, and
$k =$ the constant of variation.
Since the resistance in the circuit varies inversely as the square of the amount of current,

$$R = \frac{k}{i^2}.$$

To evaluate $k$, substitute the known values for $R$ and $i$.

$$800 = \frac{k}{(0.01)^2}$$

$$0.08 = k$$

$R = \dfrac{0.08}{i^2}$    Write the variation equation with the evaluated constant of $k = 0.08$.

$R = \dfrac{0.08}{(0.02)^2}$    Substitute $i = 0.02$.

$$R = \frac{0.08}{0.0004}$$

$$R = 200$$

Thus the resistance is 200 ohms if the amount of current is increased to 0.02 ampere.

# Chapter 10    10.1 Student Practice

**1. (a)** $2x^2 + 12x - 9 = 0$
$a = 2, b = 12, c = -9$
**(b)** $7x^2 = 6x - 8$
$7x^2 - 6x + 8 = 0$
$a = 7, b = -6, c = 8$
**(c)** $-x^2 - 6x + 3 = 0$
$x^2 + 6x - 3 = 0$
$a = 1, b = 6, c = -3$
**(d)** $10x^2 - 12x = 0$
$a = 10, b = -12, c = 0$

**2.** $\qquad 2x^2 - 7x - 6 = 4x - 6$
$2x^2 - 7x - 6 - 4x + 6 = 0$
$\qquad 2x^2 - 11x = 0$
$\qquad x(2x - 11) = 0$
$x = 0 \qquad 2x - 11 = 0$
$\qquad\qquad 2x = 11$
$$x = \frac{11}{2}$$

The two roots are 0 and $\dfrac{11}{2}$.

**3.** $\qquad\qquad 3x + 10 - \dfrac{8}{x} = 0$

$$x(3x) + x(10) - x\left(\frac{8}{x}\right) = x(0)$$

$\qquad\qquad 3x^2 + 10x - 8 = 0$
$\qquad\qquad (3x - 2)(x + 4) = 0$
$\qquad\qquad 3x - 2 = 0 \qquad x + 4 = 0$
$\qquad\qquad 3x = 2$
$$x = \frac{2}{3} \qquad\qquad x = -4$$

**Check.**

$3\left(\dfrac{2}{3}\right) + 10 - \dfrac{8}{\frac{2}{3}} \stackrel{?}{=} 0 \qquad\qquad 3(-4) + 10 - \dfrac{8}{-4} \stackrel{?}{=} 0$

$2 + 10 - 12 \stackrel{?}{=} 0 \qquad\qquad\qquad -12 + 10 + 2 \stackrel{?}{=} 0$

$0 = 0 \ \checkmark \qquad\qquad\qquad\qquad 0 = 0 \ \checkmark$

Both roots check, so $x = \dfrac{2}{3}$ and $x = -4$ are the two roots that satisfy the equation $3x + 10 - \dfrac{8}{x} = 0$.

**4.** $2x^2 + 9x - 18 = 0$
$(2x - 3)(x + 6) = 0$
$2x - 3 = 0 \qquad x + 6 = 0$
$\qquad 2x = 3 \qquad\qquad x = 0 - 6$
$$x = \frac{3}{2} \qquad\qquad x = -6$$

**Check.**

$2\left(\dfrac{3}{2}\right)^2 + 9\left(\dfrac{3}{2}\right) - 18 \stackrel{?}{=} 0 \qquad 2(-6)^2 + 9(-6) - 18 \stackrel{?}{=} 0$

$2\left(\dfrac{9}{4}\right) + \dfrac{27}{2} - 18 \stackrel{?}{=} 0 \qquad\qquad 2(36) - 54 - 18 \stackrel{?}{=} 0$

$\dfrac{9}{2} + \dfrac{27}{2} - 18 \stackrel{?}{=} 0 \qquad\qquad\qquad 72 - 54 - 18 \stackrel{?}{=} 0$

$\dfrac{36}{2} - 18 \stackrel{?}{=} 0 \qquad\qquad\qquad\qquad 18 - 18 \stackrel{?}{=} 0$

$18 - 18 \stackrel{?}{=} 0 \qquad\qquad\qquad\qquad 0 = 0 \ \checkmark$

$0 = 0 \ \checkmark$

Both roots check, so $x = \dfrac{3}{2}$ and $x = -6$ are the two roots that satisfy the equation $2x^2 + 9x - 18 = 0$.

**5.** $-4x + (x - 5)(x + 1) = 5(1 - x)$
$-4x + x^2 - 4x - 5 = 5 - 5x$
$x^2 - 8x - 5 = 5 - 5x$
$x^2 - 3x - 10 = 0$
$(x + 2)(x - 5) = 0$
$x + 2 = 0 \qquad x - 5 = 0$
$x = -2 \qquad\qquad x = 5$

**6.** $\dfrac{10x + 18}{x^2 + x - 2} = \dfrac{3x}{x + 2} + \dfrac{2}{x - 1}$

$\dfrac{(x - 1)(x + 2)(10x + 18)}{(x - 1)(x + 2)} = \dfrac{(x - 1)(x + 2)3x}{x + 2}$

$\qquad\qquad + \dfrac{(x - 1)(x + 2)2}{x - 1}$   Multiply by the LCD.

$10x + 18 = 3x(x - 1) + 2(x + 2)$
$10x + 18 = 3x^2 - 3x + 2x + 4$
$10x + 18 = 3x^2 - x + 4$
$0 = 3x^2 - 11x - 14$
$0 = (3x - 14)(x + 1)$
$3x - 14 = 0 \qquad x + 1 = 0$
$$x = \frac{14}{3} \qquad\qquad x = -1$$

7. $t$ = the number of truck routes

$n$ = the number of cities they can service

$$t = \frac{n^2 - n}{2}$$

$$28 = \frac{n^2 - n}{2}$$

$$56 = n^2 - n$$

$$0 = n^2 - n - 56$$

$$0 = (n + 7)(n - 8)$$

$$n + 7 = 0 \qquad n - 8 = 0$$

$$n = -7 \qquad n = 8$$

We reject $n = -7$. There cannot be a negative number of cities. Thus our answer is 8 cities.

## 10.2 Student Practice

1. (a) $x^2 = 1$

$$x = \pm \sqrt{1}$$

$$x = \pm 1$$

(b) $x^2 - 50 = 0$

$$x^2 = 50$$

$$x = \pm \sqrt{50}$$

$$x = \pm 5\sqrt{2}$$

(c) $3x^2 = 81$

$$x^2 = 27$$

$$x = \pm \sqrt{27}$$

$$x = \pm 3\sqrt{3}$$

2. $4x^2 - 5 = 319$

$$4x^2 = 324$$

$$x^2 = 81$$

$$x = \pm 9$$

3. $(2x - 3)^2 = 12$

$$2x - 3 = \pm \sqrt{12}$$

$$2x - 3 = \pm 2\sqrt{3}$$

$$2x - 3 = +2\sqrt{3} \qquad 2x - 3 = -2\sqrt{3}$$

$$2x = 3 + 2\sqrt{3} \qquad 2x = 3 - 2\sqrt{3}$$

$$x = \frac{3 + 2\sqrt{3}}{2} \qquad x = \frac{3 - 2\sqrt{3}}{2}$$

The roots are $\dfrac{3 \pm 2\sqrt{3}}{2}$.

4. $\qquad x^2 + 10x = 3$

$$x^2 + 10x + 25 = 3 + 25$$

$$(x + 5)^2 = 28$$

$$x + 5 = \pm \sqrt{28}$$

$$x + 5 = +2\sqrt{7} \qquad x + 5 = -2\sqrt{7}$$

$$x = -5 + 2\sqrt{7} \qquad x = -5 - 2\sqrt{7}$$

The roots are $-5 \pm 2\sqrt{7}$.

5. $\qquad x^2 - 5x = 7$

$$x^2 - 5x + \left(-\frac{5}{2}\right)^2 = 7 + \left(-\frac{5}{2}\right)^2$$

$$\left(x - \frac{5}{2}\right)^2 = 7 + \frac{25}{4}$$

$$\left(x - \frac{5}{2}\right)^2 = \frac{53}{4}$$

$$x - \frac{5}{2} = \pm \sqrt{\frac{53}{4}}$$

$$x - \frac{5}{2} = \pm \frac{\sqrt{53}}{2}$$

$$x - \frac{5}{2} = +\frac{\sqrt{53}}{2} \qquad x - \frac{5}{2} = -\frac{\sqrt{53}}{2}$$

$$x = \frac{5}{2} + \frac{\sqrt{53}}{2} \qquad x = \frac{5}{2} - \frac{\sqrt{53}}{2}$$

$$x = \frac{5 + \sqrt{53}}{2} \qquad x = \frac{5 - \sqrt{53}}{2}$$

The roots are $\dfrac{5 \pm \sqrt{53}}{2}$.

6. $8x^2 + 2x - 1 = 0$

$$8x^2 + 2x = 1$$

$$x^2 + \frac{1}{4}x = \frac{1}{8}$$

$$x^2 + \frac{1}{4}x + \frac{1}{64} = \frac{1}{8} + \frac{1}{64}$$

$$\left(x + \frac{1}{8}\right)^2 = \frac{9}{64}$$

$$x + \frac{1}{8} = \pm \sqrt{\frac{9}{64}}$$

$$x + \frac{1}{8} = \pm \frac{3}{8}$$

$$x + \frac{1}{8} = \frac{3}{8} \qquad x + \frac{1}{8} = -\frac{3}{8}$$

$$x = \frac{3}{8} - \frac{1}{8} \qquad x = -\frac{3}{8} - \frac{1}{8}$$

$$x = \frac{2}{8} \qquad x = -\frac{4}{8}$$

$$x = \frac{1}{4} \qquad x = -\frac{1}{2}$$

The roots are $\dfrac{1}{4}$ and $-\dfrac{1}{2}$.

## 10.3 Student Practice

1. $x^2 - 7x + 6 = 0$

$a = 1, b = -7, c = 6$

$$x = \frac{-b \pm \sqrt{b^2 - 4ac}}{2a} = \frac{-(-7) \pm \sqrt{(-7)^2 - 4(1)(6)}}{2(1)}$$

$$= \frac{7 \pm \sqrt{49 - 24}}{2}$$

$$= \frac{7 \pm \sqrt{25}}{2}$$

$$= \frac{7 \pm 5}{2}$$

$$x = \frac{7 + 5}{2} = \frac{12}{2} = 6 \qquad x = \frac{7 - 5}{2} = \frac{2}{2} = 1$$

The solutions are 6 and 1.

2. $3x^2 - 8x + 3 = 0$

$a = 3, b = -8, c = 3$

$$x = \frac{-b \pm \sqrt{b^2 - 4ac}}{2a} = \frac{-(-8) \pm \sqrt{(-8)^2 - 4(3)(3)}}{2(3)}$$

$$= \frac{8 \pm \sqrt{64 - 36}}{6}$$

$$= \frac{8 \pm \sqrt{28}}{6}$$

$$= \frac{8 \pm 2\sqrt{7}}{6}$$

$$x = \frac{2(4 \pm \sqrt{7})}{6} = \frac{4 \pm \sqrt{7}}{3}$$

3. $\qquad x^2 = 7 - \frac{3}{5}x$

$$5x^2 = 35 - 3x$$

$$5x^2 + 3x - 35 = 0$$

$a = 5, b = 3, c = -35$

$$x = \frac{-3 \pm \sqrt{(3)^2 - 4(5)(-35)}}{2(5)}$$

$$x = \frac{-3 \pm \sqrt{9 + 700}}{10}$$

$$x = \frac{-3 \pm \sqrt{709}}{10}$$

**4.**
$$2x^2 = 13x + 5$$
$$2x^2 - 13x - 5 = 0$$
$$a = 2, b = -13, c = -5$$
$$x = \frac{-(-13) \pm \sqrt{(-13)^2 - 4(2)(-5)}}{2(2)}$$
$$= \frac{13 \pm \sqrt{169 + 40}}{4} = \frac{13 \pm \sqrt{209}}{4}$$
$$x = \frac{13 + \sqrt{209}}{4} \qquad x = \frac{13 - \sqrt{209}}{4}$$
$$x \approx 6.864 \quad \text{and} \quad x \approx -0.364$$

**5.**
$$5x^2 + 2x = -3$$
$$5x^2 + 2x + 3 = 0$$
$$a = 5, b = 2, c = 3$$
$$x = \frac{-2 \pm \sqrt{(2)^2 - 4(5)(3)}}{2(5)}$$
$$= \frac{-2 \pm \sqrt{4 - 60}}{10} = \frac{-2 \pm \sqrt{-56}}{10}$$
There is no real number that is $\sqrt{-56}$.
There is no solution to this problem.

**6. (a)**
$$5x^2 = 3x + 2$$
$$5x^2 - 3x - 2 = 0$$
$$a = 5, b = -3, c = -2$$
$$b^2 - 4ac = (-3)^2 - 4(5)(-2)$$
$$= 9 + 40 = 49$$

Because $49 > 0$ there will be two real roots that are rational numbers.

**(b)**
$$2x^2 + 5 = 4x$$
$$2x^2 - 4x + 5 = 0$$
$$a = 2, b = -4, c = 5$$
$$b^2 - 4ac = (-4)^2 - 4(2)(5)$$
$$= 16 - 40$$
$$= -24$$

Because $-24 < 0$, the equation $2x^2 + 5 = 4x$ has no real number solutions.

## 10.4 Student Practice

**1. (a)** $y = 2x^2$

| x | y |
|---|---|
| -2 | 8 |
| -1 | 2 |
| 0 | 0 |
| 1 | 2 |
| 2 | 8 |

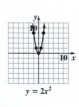

$y = 2x^2$

**(b)** $y = -2x^2$

| x | y |
|---|---|
| -2 | -8 |
| -1 | -2 |
| 0 | 0 |
| 1 | -2 |
| 2 | -8 |

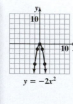

$y = -2x^2$

**2. (a)** $y = x^2 + 2x$

| x | y | |
|---|---|---|
| -4 | 8 | |
| -3 | 3 | |
| -2 | 0 | |
| -1 | -1 | Vertex |
| 0 | 0 | |
| 1 | 3 | |
| 2 | 8 | |

$y = x^2 + 2x$

**(b)** $y = -x^2 - 2x$

| x | y | |
|---|---|---|
| -4 | -8 | |
| -3 | -3 | |
| -2 | 0 | |
| -1 | 1 | Vertex |
| 0 | 0 | |
| 1 | -3 | |
| 2 | -8 | |

$y = -x^2 - 2x$

**3.** $y = x^2 - 4x - 5$
$$x = \frac{-b}{2a} = \frac{-(-4)}{2(1)} = \frac{4}{2} = 2$$
If $x = 2$, then
$$y = (2)^2 - 4(2) - 5 = -9$$
The point $(2, -9)$ is the vertex.
$x$-intercepts:
$$x^2 - 4x - 5 = 0$$
$$(x + 1)(x - 5) = 0$$
$$x + 1 = 0 \qquad x - 5 = 0$$
$$x = -1 \qquad x = 5$$
The $x$-intercepts are the points $(-1, 0)$ and $(5, 0)$.

Scale: each square equals 2 units

$y = x^2 - 4x - 5$

**4.** $h = -16t^2 + 32t + 10$

**(a)** Since the equation is a quadratic equation, the graph is a parabola. $a$ is negative, therefore the parabola opens downward and the vertex is the highest point.
$$t = \frac{-b}{2a} = \frac{-32}{2(-16)} = \frac{-32}{-32} = 1$$
If $t = 1$, then $h = -16(1)^2 + 32(1) + 10$
$$h = -16 + 32 + 10$$
$$h = 26$$
The vertex $(t, h)$ is $(1, 26)$.
Now find the $h$-intercept, the point where $t = 0$.
$$h = -16t^2 + 32t + 10$$
$$h = 10$$
The $h$-intercept is at $(0, 10)$.

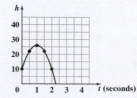

**Table Values**

| t | h |
|---|---|
| 0 | 10 |
| 0.5 | 22 |
| 1 | 26 |
| 1.5 | 22 |
| 2 | 10 |

(b) The ball is 26 feet above the ground one second after it is thrown. This is the greatest height of the ball.

(c) The ball hits the ground when $h = 0$.

$$0 = -16t^2 + 32t + 10$$

$$t = \frac{-32 \pm \sqrt{(32)^2 - 4(-16)(10)}}{2(-16)}$$

$$t = \frac{-32 \pm \sqrt{1024 + 640}}{-32}$$

$$t = \frac{-32 \pm \sqrt{1664}}{-32}$$

$$t = \frac{4 \pm \sqrt{26}}{4}$$

$$t \approx \frac{4 - 5.099}{4} \quad \text{and} \quad t \approx \frac{4 + 5.099}{4}$$

$$t \approx -0.275 \qquad t \approx 2.275$$

In this situation it is not useful to find points for negative values of $t$. Therefore $t \approx 2.3$ seconds. The ball will hit the ground approximately 2.3 seconds after it is thrown.

## 10.5 Student Practice

1. $a^2 + b^2 = c^2$ Let $a = x$, then $b = x - 6$.

$$x^2 + (x - 6)^2 = 30^2$$
$$x^2 + x^2 - 12x + 36 = 900$$
$$2x^2 - 12x - 864 = 0$$
$$x^2 - 6x - 432 = 0$$
$$(x + 18)(x - 24) = 0$$
$$x + 18 = 0 \qquad x - 24 = 0$$
$$x = -18 \qquad x = 24$$

One leg is 24 meters, the other leg is $x - 6$ or $24 - 6 = 18$ meters.

2. Let $c$ = cost for each person in the original group
$s$ = number of people
Number of people $\times$ cost per person = total cost
$$s \cdot c = 240$$
If 2 people cannot go, then the number of people drops by 2 but the cost for each increases by \$4.
$$(s - 2)(c + 4) = 240$$

If $s \cdot c = 240$, then
$$c = \frac{240}{s}$$

$$(s - 2)\left(\frac{240}{s} + 4\right) = 240$$

$$(s - 2)\left(\frac{240 + 4s}{s}\right) = 240$$

$$(s - 2)(240 + 4s) = 240s$$
$$4s^2 + 232s - 480 = 240s$$
$$4s^2 - 8s - 480 = 0$$
$$s^2 - 2s - 120 = 0$$
$$(s - 12)(s + 10) = 0$$
$$s = 12 \qquad s = -10$$

The number of people originally going on the trip was 12.

**Check.** If 12 people originally planned to go on the trip, then
$$12 \times c = 240.$$
$$c = \$20 \text{ per person}$$
If 2 people dropped out, does this mean they must increase the cost of the trip by \$4?
$$10 \times c = 240$$
$$c = \$24.00 \text{ Yes.}$$

3. Let $x$ = width
$y$ = length
$$160 = 2x + y$$
The area of a rectangle is (width)(length).
$$A = xy \qquad \text{Substitute } y = 160 - 2x$$
$$3150 = x(160 - 2x)$$
$$3150 = 160x - 2x^2$$
$$2x^2 - 160x + 3150 = 0$$
$$x^2 - 80x + 1575 = 0$$
$$(x - 45)(x - 35) = 0$$
$$x = 45 \quad \text{and} \quad x = 35$$

First solution:
If the width is 45 feet then the length is
$$y = 160 - 2(45)$$
$$= 160 - 90$$
$$= 70 \text{ feet}$$
**Check.** $\qquad 45 + 45 + 70 = 160 \quad \checkmark$
$$(45)(70) = 3150 \quad \checkmark$$

Second solution:
If the width is 35 feet then the length is
$$y = 160 - 2(35) = 160 - 70 = 90 \text{ feet}$$
**Check.** $\qquad 35 + 35 + 90 = 160 \quad \checkmark$
$$(35)(90) = 3150 \quad \checkmark$$

## Appendix B Student Practice

1. $3.4 \text{ m} \cdot 1 = 3.4 \text{ m} \cdot \dfrac{100 \text{ cm}}{1 \text{ m}} = \dfrac{(3.4)(100)}{1} \cdot \dfrac{\text{m}}{\text{m}} \cdot \text{cm} = 340 \text{ cm}$

2. $125 \text{ L} \cdot \dfrac{0.001 \text{ kL}}{1 \text{ L}} = \dfrac{(125)(0.001)}{1} \cdot \dfrac{\text{L}}{\text{L}} \cdot \text{kL} = 0.125 \text{ kL}$

3. $5.72 \text{ lb} \cdot \dfrac{1 \text{ kg}}{2.20 \text{ lb}} = 2.6 \text{ kg}$

4. $8 \text{ cm} \cdot \dfrac{1 \text{ in.}}{2.54 \text{ cm}} \approx 3.15 \text{ in.}$

5. $3 \text{ L} \cdot \dfrac{1.06 \text{ qt}}{1 \text{ L}} = 3.18 \text{ qt}$

6. $\dfrac{70 \text{ mi}}{\text{hr}} \cdot \dfrac{5280 \text{ ft}}{1 \text{ mi}} \cdot \dfrac{1 \text{ hr}}{60 \text{ min}} \cdot \dfrac{1 \text{ min}}{60 \text{ sec}} = \dfrac{(70)(5280) \text{ ft}}{(60)(60) \text{ sec}} \approx 102.7 \text{ ft/sec}$

## Appendix C Student Practice

1. (a) The lowest level of sodium, 670 mg, is in the Wendy's Grilled Chicken Sandwich.
   (b) The highest level of cholesterol, 100 g, is in the McDonald's Big Mac.
2. (a) The two activities single women spend the most time doing are spending time with friends, and reading and watching TV.
   (b) Couples spend more time per week with their family (21 hr) than any other activity.

(c) The greatest difference between single men and single women is the time spent reading and watching TV (6 hr).
3. (a) Walgreens had $16 \times 500 = 8000$ stores.
   (b) Rite Aid had $9 \times 500 = 4500$ stores. Medicine Shoppe had $3 \times 500 = 1500$ stores. Since $4500 - 1500 = 3000$, Rite Aid had 3000 more stores than Medicine Shoppe.
   (c) Medicine Shoppe had $3 \times 500 = 1500$ stores, and CVS had $14 \times 500 = 7000$ stores. There were $1500 + 7000 = 8500$ total Medicine Shoppe and CVS stores.

**4. (a)** The greatest number of accidents, 89, occurred during the 1999–2000 period.
**(b)** During the 1995–1996 period there were 61 accidents. During 1997–1998 there were 84 accidents. The increase in the number of accidents from 1995–1996 to 1997–1998 was $84 - 61 = 23$.
**(c)** During the 1999–2000 period there were 89 accidents. During 2001–2002 there were 75 accidents. The decrease in the number of accidents from 1999–2000 to 2001–2002 was $89 - 75 = 14$.

**5. (a)** There were 15 days in 2001 in which the air was considered unhealthy.
**(b)** The year with the most days in which the air was considered unhealthy was 2002.
**6. (a)** 15% of Seattle's rainfall falls between April and June.
**(b)** 40% of Seattle's rainfall occurs from October to December.
**(c)** 35% of the total rainfall occurs from January to March. This is $35\% \times 37 = 0.35 \times 37 = 12.95$ inches.

# Appendix D Student Practice

**1.**
$$y - y_1 = m(x - x_1)$$
$$y - (-2) = \frac{3}{4}(x - 5)$$
$$y + 2 = \frac{3}{4}x - \frac{15}{4}$$
$$4y + 4(2) = 4\left(\frac{3}{4}x\right) - 4\left(\frac{15}{4}\right)$$
$$4y + 8 = 3x - 15$$
$$-3x + 4y = -15 - 8$$
$$3x - 4y = 23$$

**2.** $(-4, 1)$ and $(-2, -3)$

$$m = \frac{y_2 - y_1}{x_2 - x_1} = \frac{-3 - 1}{-2 - (-4)} = \frac{-4}{-2 + 4} = \frac{-4}{2} = -2$$

Substitute $m = -2$ and $(x_1, y_1) = (-4, 1)$ into the point–slope equation.
$$y - y_1 = m(x - x_1)$$
$$y - 1 = -2[x - (-4)]$$
$$y - 1 = -2(x + 4)$$
$$y - 1 = -2x - 8$$
$$y = -2x - 7$$

**3.** First, we need to find the slope of the line $5x - 3y = 10$. We do this by writing the equation in slope–intercept form
$$5x - 3y = 10$$
$$-3y = -5x + 10$$
$$y = \frac{5}{3}x - \frac{10}{3}$$

The slope is $\frac{5}{3}$. A line parallel to this passing through $(4, -5)$ would have the equation
$$y - y_1 = m(x - x_1)$$
$$y - (-5) = \frac{5}{3}(x - 4)$$
$$y + 5 = \frac{5}{3}x - \frac{20}{3}$$
$$3y + 3(5) = 3\left(\frac{5}{3}x\right) - 3\left(\frac{20}{3}\right)$$
$$3y + 15 = 5x - 20$$
$$-5x + 3y = -35$$
$$5x - 3y = 35$$

**4.** Find the slope of the line $6x + 3y = 7$ by rewriting it in slope–intercept form.
$$6x + 3y = 7$$
$$3y = -6x + 7$$
$$y = -2x + \frac{7}{3}$$

The slope is $-2$. A line perpendicular to this passing through $(-4, 3)$ would have a slope of $\frac{1}{2}$, and would have the equation
$$y - y_1 = \frac{m(x - x_1)}{}$$
$$y - 3 = \frac{1}{2}[x - (-4)]$$
$$y - 3 = \frac{1}{2}(x + 4)$$
$$y - 3 = \frac{1}{2}x + 2$$
$$2y - 2(3) = 2\left(\frac{1}{2}x\right) + 2(2)$$
$$2y - 6 = x + 4$$
$$-x + 2y = 10$$
$$x - 2y = -10$$

# Answers to Selected Exercises

## Chapter 0

**0.1 Exercises** **1.** 12 **3.** Answers may vary. When two or more numbers are multiplied, each number that is multiplied is called a factor. In $2 \times 3$, 2 and 3 are factors. **5.** **7.** $\frac{3}{5}$ **9.** $\frac{1}{3}$ **11.** 5 **13.** $\frac{2}{3}$

**15.** $\frac{6}{17}$ **17.** $\frac{7}{9}$ **19.** $2\frac{5}{6}$ **21.** $9\frac{2}{5}$ **23.** $5\frac{3}{7}$ **25.** $20\frac{1}{2}$ **27.** $6\frac{2}{5}$ **29.** $12\frac{1}{3}$ **31.** $\frac{16}{5}$ **33.** $\frac{33}{5}$ **35.** $\frac{11}{9}$ **37.** $\frac{59}{7}$

**39.** $\frac{97}{4}$ **41.** 8 **43.** 24 **45.** 21 **47.** 12 **49.** 21 **51.** 15 **53.** 70 **55.** $23\frac{1}{2}$ **57.** $\frac{33}{160}$ **59.** $\frac{1}{4}$ **61.** $\frac{1}{2}$ **63.** $\frac{3}{5}$

**Quick Quiz 0.1** *See Examples noted with Ex.* **1.** $\frac{21}{23}$ (Ex. 1) **2.** $\frac{75}{11}$ (Ex. 6) **3.** $4\frac{19}{21}$ (Ex. 5) **4.** See Student Solutions Manual

**0.2 Exercises** **1.** Answers may vary. A sample answer is: 8 is exactly divisible by 4. **3.** 36 **5.** 20 **7.** 54 **9.** 105 **11.** 120 **13.** 120 **15.** 546 **17.** 90 **19.** $\frac{5}{8}$ **21.** $\frac{2}{7}$ **23.** $\frac{25}{24}$ or $1\frac{1}{24}$ **25.** $\frac{31}{63}$ **27.** $\frac{11}{15}$ **29.** $\frac{35}{36}$ **31.** $\frac{2}{45}$ **33.** $\frac{1}{2}$ **35.** $\frac{53}{56}$

**37.** $\frac{3}{2}$ or $1\frac{1}{2}$ **39.** $\frac{11}{30}$ **41.** $\frac{1}{4}$ **43.** $\frac{1}{2}$ **45.** $7\frac{11}{15}$ **47.** $1\frac{35}{72}$ **49.** $4\frac{11}{12}$ **51.** $6\frac{13}{28}$ **53.** $5\frac{19}{24}$ **55.** $4\frac{3}{7}$ **57.** 9

**59.** $\frac{23}{24}$ **61.** $7\frac{9}{16}$ **63.** $\frac{10}{21}$ **65.** $2\frac{7}{10}$ **67.** $19\frac{11}{21}$ **69.** $1\frac{13}{24}$ **71.** $12\frac{1}{12}$ **73.** $33\frac{3}{7}$ **75.** $24\frac{2}{3}$ mi **77.** $2\frac{7}{12}$ hr

**79.** $A = 12$ in.; $B = 15\frac{7}{8}$ in. **81.** $1\frac{5}{8}$ in. **83.** $\frac{9}{11}$ **84.** $\frac{133}{5}$

**Quick Quiz 0.2** *See Examples noted with Ex.* **1.** $\frac{5}{3}$ or $1\frac{2}{3}$ (Ex. 8) **2.** $7\frac{8}{15}$ (Ex. 12a) **3.** $2\frac{5}{18}$ (Ex. 12b) **4.** See Student Solutions Manual

**0.3 Exercises** **1.** First, change each mixed number to an improper fraction. Look for a common factor in the numerator and denominator to divide by, and, if one is found, perform the division. Multiply the numerators. Multiply the denominators. **3.** $\frac{24}{25}$ **5.** $\frac{17}{30}$ **7.** $\frac{6}{25}$

**9.** $\frac{12}{5}$ or $2\frac{2}{5}$ **11.** $\frac{1}{6}$ **13.** $\frac{6}{7}$ **15.** $\frac{18}{5}$ or $3\frac{3}{5}$ **17.** $\frac{3}{5}$ **19.** $\frac{1}{7}$ **21.** 14 **23.** $\frac{8}{7}$ or $1\frac{1}{7}$ **25.** $\frac{7}{27}$ **27.** $\frac{7}{6}$ or $1\frac{1}{6}$ **29.** $\frac{15}{14}$ or $1\frac{1}{14}$

**31.** $\frac{8}{35}$ **33.** $1\frac{1}{3}$ **35.** $8\frac{2}{3}$ **37.** $6\frac{1}{4}$ **39.** $\frac{8}{15}$ **41.** 6 **43.** $\frac{5}{4}$ or $1\frac{1}{4}$ **45.** $54\frac{3}{4}$ **47.** 40 **49.** 28 **51.** $\frac{3}{16}$ **53.** (a) $\frac{5}{63}$

**(b)** $\frac{7}{125}$ **55. (a)** $\frac{7}{6}$ **(b)** $\frac{8}{21}$ **57.** 26 shirts **59.** 136 sq mi **61.** 55 **62.** 49

**Quick Quiz 0.3** *See Examples noted with Ex.* **1.** $\frac{5}{6}$ (Ex. 2) **2.** $14\frac{5}{8}$ (Ex. 4) **3.** $1\frac{8}{25}$ (Ex. 9a) **4.** See Student Solutions Manual

**0.4 Exercises** **1.** 10, 100, 1000, 10,000, and so on **3.** 3, left **5.** 0.875 **7.** 0.2 **9.** $0.\overline{63}$ **11.** $\frac{4}{5}$ **13.** $\frac{1}{4}$ **15.** $\frac{5}{8}$ **17.** $\frac{3}{50}$

**19.** $\frac{17}{5}$ or $3\frac{2}{5}$ **21.** $\frac{11}{2}$ or $5\frac{1}{2}$ **23.** 2.09 **25.** 10.82 **27.** 261.208 **29.** 131.79 **31.** 3.9797 **33.** 122.63 **35.** 30.282 **37.** 0.0032
**39.** 0.10575 **41.** 87.3 **43.** 0.0565 **45.** 2.64 **47.** 261.5 **49.** 0.508 **51.** 3450 **53.** 0.0076 **55.** 73,600 **57.** 0.73892
**59.** 14.98 **61.** 0.01931 **63.** 8.22 **65.** 16.378 **67.** 2.12 **69.** 768.3 **71.** 52.08 **73.** 1.537 **75.** 2.6026 L **77.** 21 hr; $4
**79.** $\frac{2}{3}$ **80.** $\frac{1}{6}$ **81.** $\frac{93}{100}$ **82.** $\frac{11}{10}$ or $1\frac{1}{10}$

**Quick Quiz 0.4** *See Examples noted with Ex.* **1.** 5.7078 (Ex. 5) **2.** 3.522 (Ex. 9) **3.** 28.8 (Ex. 11) **4.** See Student Solutions Manual

**Use Math to Save Money** **1.** $100, $300, $2000, $8000, $8000, $8000, $12,000 **2.** $3 \times \$25 + \$50 + \$200 + 2 \times \$20 = \$365$
**3.** $100 loan, $300 loan, $2000 car loan **4.** $40 **5.** $2000 − $400 = $1600; $1600/$240 is 7 months if we round to the nearest whole number.
**6.** Answers will vary.

**0.5 Exercises** **1.** Answers may vary. Sample answers follow. 19% means 19 out of 100 parts. Percent means per 100. 19% is really a fraction with a denominator of 100. In this case it would be $\frac{19}{100}$. **3.** 79% **5.** 56.8% **7.** 7.6% **9.** 239% **11.** 360% **13.** 367.2% **15.** 0.03
**17.** 0.004 **19.** 2.5 **21.** 0.074 **23.** 0.0052 **25.** 1 or 1.00 **27.** 5.2 **29.** 13 **31.** 72.8 **33.** 150% **35.** 5% **37.** 6%
**39.** 40% **41.** 85% **43.** $4.92 tip; $37.72 total bill **45.** 21% **47.** 540 gifts **49. (a)** $29,640 **(b)** $35,040 **51.** 240,000
**53.** 20,000,000 **55.** 240 **57.** 20,000 **59.** 0.1 **61.** $4000 **63.** $6400 **65.** 25 mi/gal **67.** 22 mi/gal **69.** 4.0 in.

**Quick Quiz 0.5** *See Examples noted with Ex.* **1.** 96.9 (Ex. 4c) **2.** 15% (Ex. 7b) **3.** 20 (Ex. 9) **4.** See Student Solutions Manual

**0.6 Exercises**   **1.** $1287    **3. (a)** 76 ft    **(b)** He should buy the cut-to-order fencing. He will save $7.80.    **5.** Jog $2\frac{2}{3}$ mi; walk $3\frac{1}{9}$ mi; rest $4\frac{4}{9}$ min; walk $1\frac{7}{9}$ mi    **7.** Betty; Melinda increases each activity by $\frac{2}{3}$ by day 3 but Betty increases each activity by $\frac{7}{9}$ by day 3.    **9.** $4\frac{1}{2}$ mi    **11. (a)** 900,000 lb/day   **(b)** 14.29%    **13. (a)** $48,635   **(b)** $14,104.15    **15.** 18%    **17.** 69%

**Quick Quiz 0.6**   *See Examples noted with Ex.*    **1.** 350 stones (Ex. 1)    **2.** 46% (Ex. 2)    **3.** $1215 (Ex. 1)    **4.** See Student Solutions Manual

**Career Exploration Problems**    **1.** $1823.08    **2.** 43.4%    **3. (a)** $122.15   **(b)** $242.47

**You Try It**   **1. (a)** $\frac{2}{3}$   **(b)** $\frac{3}{4}$   **(c)** $\frac{2}{5}$    **2. (a)** $4\frac{1}{5}$   **(b)** $5\frac{2}{7}$    **3. (a)** $\frac{12}{5}$   **(b)** $\frac{55}{9}$    **4.** 15    **5. (a)** 60   **(b)** 100    **6. (a)** $\frac{11}{18}$   **(b)** $\frac{5}{12}$    **7. (a)** $5\frac{7}{9}$   **(b)** $2\frac{1}{2}$    **8. (a)** $\frac{10}{33}$   **(b)** $\frac{1}{2}$   **(c)** $\frac{18}{5}$ or $3\frac{3}{5}$    **9. (a)** $\frac{15}{28}$   **(b)** $\frac{5}{12}$    **10. (a)** $\frac{33}{8}$ or $4\frac{1}{8}$   **(b)** 2    **11.** 0.875    **12. (a)** $\frac{29}{100}$   **(b)** $\frac{7}{40}$    **13. (a)** 8.533   **(b)** 3.46    **14. (a)** 1.35   **(b)** 3.4304    **15.** 18.5    **16. (a)** 52%   **(b)** 0.8%   **(c)** 186%   **(d)** 7.7%   **(e)** 0.09%    **17. (a)** 0.28   **(b)** 0.0742   **(c)** 1.65   **(d)** 0.0025    **18.** 13.8    **19. (a)** 83.3%   **(b)** 125%    **20.** 300 ft$^2$    **21.** $154.06

**Chapter 0 Review Problems**   **1.** $\frac{3}{4}$    **2.** $\frac{3}{10}$    **3.** $\frac{18}{41}$    **4.** $\frac{3}{5}$    **5.** $\frac{57}{8}$    **6.** $6\frac{4}{5}$    **7.** $26\frac{2}{3}$    **8.** 15    **9.** 5    **10.** 45    **11.** 22    **12.** $\frac{17}{20}$    **13.** $\frac{29}{24}$ or $1\frac{5}{24}$    **14.** $\frac{4}{15}$    **15.** $\frac{13}{30}$    **16.** $5\frac{23}{30}$    **17.** $6\frac{9}{20}$    **18.** $2\frac{29}{36}$    **19.** $1\frac{11}{12}$    **20.** $\frac{30}{11}$ or $2\frac{8}{11}$    **21.** $10\frac{1}{2}$    **22.** 50    **23.** $\frac{20}{7}$ or $2\frac{6}{7}$    **24.** $\frac{1}{16}$    **25.** $\frac{24}{5}$ or $4\frac{4}{5}$    **26.** $\frac{3}{20}$    **27.** 6    **28.** 7.201    **29.** 7.737    **30.** 29.561    **31.** 4.436    **32.** 0.03745    **33.** 362,341    **34.** 0.07956    **35.** 125.5    **36.** 0.07132    **37.** 1.3075    **38.** 90    **39.** 1.82    **40.** 0.375    **41.** $\frac{9}{25}$    **42.** 0.014    **43.** 0.361    **44.** 0.0002    **45.** 1.253    **46.** 0.25%    **47.** 32.5%    **48.** 90%    **49.** 10%    **50.** 120    **51.** 3.96    **52.** 95%    **53.** 60%    **54.** 13,480 students    **55.** 75%    **56.** 400,000,000,000    **57.** 2500    **58.** 300,000    **59.** 9    **60.** $12,000    **61.** 25    **62.** $320    **63.** $800    **64.** 1840 mi; 1472 mi    **65.** 1500 mi; 1050 mi    **66.** $349.07    **67.** $462.80    **68.** $1585.50    **69.** $401.25

**How Am I Doing? Chapter 0 Test**   **1.** $\frac{8}{9}$ (obj. 0.1.2)    **2.** $\frac{4}{3}$ (obj. 0.1.2)    **3.** $\frac{45}{7}$ (obj. 0.1.3)    **4.** $11\frac{2}{3}$ (obj. 0.1.3)    **5.** $\frac{15}{8}$ or $1\frac{7}{8}$ (obj. 0.2.3)    **6.** $4\frac{7}{8}$ (obj. 0.2.4)    **7.** $\frac{5}{6}$ (obj. 0.2.4)    **8.** $\frac{4}{3}$ or $1\frac{1}{3}$ (obj. 0.3.1)    **9.** $\frac{7}{2}$ or $3\frac{1}{2}$ (obj. 0.3.2)    **10.** $1\frac{21}{22}$ (obj. 0.3.2)    **11.** $8\frac{1}{8}$ (obj. 0.3.1)    **12.** $\frac{7}{2}$ or $3\frac{1}{2}$ (obj. 0.3.2)    **13.** 14.64 (obj. 0.4.4)    **14.** 3.9897 (obj. 0.4.4)    **15.** 1.312 (obj. 0.4.5)    **16.** 73.85 (obj. 0.4.7)    **17.** 230 (obj. 0.4.6)    **18.** 263.259 (obj. 0.4.7)    **19.** 7.3% (obj. 0.5.1)    **20.** 1.965 (obj. 0.5.2)    **21.** 6.3 (obj. 0.5.3)    **22.** 6% (obj. 0.5.4)    **23.** 18 computer chips (obj. 0.3.2)    **24.** 100 (obj. 0.5.5)    **25.** 700 (obj. 0.5.5)    **26.** 65% (obj. 0.6.1)    **27.** 60 tiles (obj. 0.6.1)

## Chapter 1   **1.1 Exercises**

**1.** Integer, rational number, real number    **3.** Irrational number, real number    **5.** Rational number, real number    **7.** Rational number, real number    **9.** Irrational number, real number    **11.** $-20,000$    **13.** $-37\frac{1}{2}$    **15.** $+7$    **17.** $-8$    **19.** 2.73    **21.** 1.3    **23.** $\frac{5}{6}$    **25.** $-15$    **27.** $-50$    **29.** $\frac{3}{10}$    **31.** $-\frac{7}{13}$    **33.** $\frac{1}{35}$    **35.** $-19.2$    **37.** 0.4    **39.** $-14.16$    **41.** $-6$    **43.** 0    **45.** $\frac{9}{20}$    **47.** $-8$    **49.** $-3$    **51.** 59    **53.** $\frac{7}{18}$    **55.** $\frac{2}{5}$    **57.** $-2.21$    **59.** 12    **61.** $-12$    **63.** 15.94    **65.** $167 profit    **67.** $-$3800    **69.** 9-yd gain    **71.** 3500    **73.** $28,000,000    **75.** 18    **77.** $\frac{19}{16}$ or $1\frac{3}{16}$    **78.** $\frac{2}{3}$    **79.** $\frac{1}{12}$    **80.** $\frac{25}{34}$    **81.** 1.52    **82.** 0.65    **83.** 1.141    **84.** 0.26

**Quick Quiz 1.1**   *See Examples noted with Ex.*    **1.** $-34$ (Ex. 5b)    **2.** 0.5 (Ex. 11)    **3.** $-\frac{13}{24}$ (Ex. 13b)    **4.** See Student Solutions Manual

**1.2 Exercises**   **1.** First change subtracting $-3$ to adding $+3$. Then use the rules for addition of two real numbers with different signs. Thus, $-8 - (-3) = -8 + 3 = -5.$    **3.** $-22$    **5.** $-4$    **7.** $-11$    **9.** 8    **11.** 5    **13.** 0    **15.** 3    **17.** $-\frac{2}{5}$    **19.** $\frac{27}{20}$ or $1\frac{7}{20}$    **21.** $-\frac{19}{12}$ or $-1\frac{7}{12}$    **23.** $-0.9$    **25.** 4.47    **27.** $-\frac{17}{5}$ or $-3\frac{2}{5}$    **29.** $-\frac{14}{3}$ or $-4\frac{2}{3}$    **31.** $-53$    **33.** $-73$    **35.** 7.1    **37.** $\frac{35}{4}$ or $8\frac{3}{4}$    **39.** $-\frac{37}{6}$ or $-6\frac{1}{6}$    **41.** $-\frac{38}{35}$ or $-1\frac{3}{35}$    **43.** $-8.5$    **45.** $-\frac{29}{5}$ or $-5\frac{4}{5}$    **47.** 6.06    **49.** $-5.047$    **51.** 7    **53.** $-48$    **55.** $-2$    **57.** 0    **59.** 10    **61.** 1.1    **63.** 626 ft    **65.** $-21$    **66.** $-51$    **67.** $-19$    **68.** 15°F    **69.** $6\frac{2}{3}$ mi

**Quick Quiz 1.2**   *See Examples noted with Ex.*    **1.** 7 (Ex. 2)    **2.** $-1.9$ (Ex. 4)    **3.** $\frac{51}{56}$ (Ex. 3b)    **4.** See Student Solutions Manual

**1.3 Exercises**   **1.** To multiply two real numbers, multiply the absolute values. The sign of the result is positive if both numbers have the same sign, but negative if the two numbers have opposite signs.    **3.** $-40$    **5.** 0    **7.** 49    **9.** 0.264    **11.** $-4.5$    **13.** $-\frac{3}{2}$ or $-1\frac{1}{2}$    **15.** $\frac{9}{11}$

**17.** $-\frac{5}{26}$ **19.** 0 **21.** 6 **23.** 15 **25.** $-12$ **27.** $-130$ **29.** $-0.6$ **31.** $-0.9$ **33.** $-\frac{3}{10}$ **35.** $\frac{20}{3}$ or $6\frac{2}{3}$ **37.** $\frac{7}{10}$ **39.** 14

**41.** $-\frac{5}{4}$ or $-1\frac{1}{4}$ **43.** $-\frac{2}{3}$ **45.** $-24$ **47.** $-72$ **49.** 0 **51.** $-0.3$ **53.** $-\frac{8}{35}$ **55.** $\frac{2}{27}$ **57.** 9 **59.** 4 **61.** 17 **63.** $-72$

**65.** $-1$ **67.** He gave $4.40 to each person and to himself. **69.** $235.60 **71.** Approximately 20 yd gained **73.** Approximately 70 yd lost

**75.** A gain of 20 yd **76.** $-6.69$ **77.** $-\frac{11}{6}$ or $-1\frac{5}{6}$ **78.** $-15$ **79.** $-88$

**Quick Quiz 1.3** *See Examples noted with Ex.* **1.** $-\frac{15}{8}$ or $-1\frac{7}{8}$ (Ex. 1d) **2.** $-120$ (Ex. 2) **3.** 4 (Ex. 5) **4.** See Student Solutions Manual

**1.4 Exercises** **1.** The base is 3 and the exponent is 4. Thus you multiply $(3)(3)(3)(3) = 81$. **3.** The answer is negative. When you raise a negative number to an odd power the result is always negative. **5.** If you have parentheses surrounding the $-2$, then the base is $-2$ and the exponent is 4. The result is 16. If you do not have parentheses, then the base is 2. You evaluate to obtain 16 and then take the negative of 16, which is $-16$. Thus $(-2)^4 = -16$ but $-2^4 = -16$. **7.** $6^5$ **9.** $w^2$ **11.** $p^4$ **13.** $(3q)^3$ or $3^3q^3$ **15.** 27 **17.** 81 **19.** 216 **21.** $-27$ **23.** 16

**25.** $-25$ **27.** $\frac{1}{16}$ **29.** $\frac{8}{125}$ **31.** 4.41 **33.** 0.00032 **35.** 256 **37.** $-256$ **39.** 161 **41.** $-21$ **43.** $-128$ **45.** 23 **47.** 1000

**48.** $-19$ **49.** $-\frac{5}{3}$ or $-1\frac{2}{3}$ **50.** $-8$ **51.** 2.52 **52.** $1696

**Quick Quiz 1.4** *See Examples noted with Ex.* **1.** 256 (Ex. 3) **2.** 3.24 (Ex. 4b) **3.** $\frac{27}{64}$ (Ex. 4c) **4.** See Student Solutions Manual

**1.5 Exercises** **1.** $3(4) + 6(5)$ **3.** (a) 90 (b) 42 **5.** 10 **7.** 5 **9.** $-29$ **11.** 24 **13.** 21 **15.** 13 **17.** $-6$ **19.** 42

**21.** $\frac{9}{4}$ or $2\frac{1}{4}$ **23.** 0.848 **25.** $-\frac{7}{16}$ **27.** 5 **29.** $-\frac{23}{2}$ or $-11\frac{1}{2}$ **31.** 18.35 **33.** $\frac{29}{18}$ or $1\frac{11}{18}$ **35.** $1(-2) + 5(-1) + 10(0) + 2(+1)$

**37.** 2 over par **39.** 0.125 **40.** $-\frac{19}{12}$ or $-1\frac{7}{12}$ **41.** $-1$ **42.** $\frac{72}{125}$

**Quick Quiz 1.5** *See Examples noted with Ex.* **1.** $-77$ (Ex. 1) **2.** 16.96 (Ex. 1) **3.** 11 (Ex. 3) **4.** See Student Solutions Manual

**Use Math to Save Money** **1.** $(\$2500 \times 0.05) \times 12 = \$1500$ **2.** $4500 **3.** $3450 **4.** About 28 months or 2 years, 4 months
**5.** About 14 months or 1 year, 2 months **6.** $(2500 + (5800/12)) \times 0.05 = \$149.17$ *a month* **7.** $(\$2500 + (\$5800/12)) \times 0.20 = \$596.67$ *a month*
**8.** Answers will vary. **9.** Answers will vary. **10.** Answers will vary.

**1.6 Exercises** **1.** variable **3.** Here we are multiplying 4 by $x$ by $x$. Since we know from the definition of exponents that $x$ multiplied by $x$ is $x^2$, this gives us an answer of $4x^2$. **5.** Yes, $a(b - c)$ can be written as $a[b + (-c)]$. $3(10 - 2) = (3 \times 10) - (3 \times 2)$
$$3 \times 8 = 30 - 6$$
$$24 = 24$$

**7.** $10x - 25y$ **9.** $-8a + 6b$ **11.** $9x + 3y$ **13.** $-8m - 24n$ **15.** $-x + 3y$ **17.** $-81x + 45y - 72$ **19.** $-10x + 2y - 12$

**21.** $10x^2 - 20x + 15$ **23.** $\frac{x^2}{5} + 2xy - \frac{4x}{5}$ **25.** $5x^2 + 10xy + 5xz$ **27.** $13.5x - 15$ **29.** $18x^2 + 3xy - 3x$ **31.** $-3x^2y - 2xy^2 + xy$

**33.** $-5a^2b - 10ab^2 + 20ab$ **35.** $2a^2 - 4a - \frac{5}{4}$ **37.** $0.36x^3 + 0.09x^2 - 0.15x$ **39.** $-1.32q^3 - 0.28qr - 4q$

**41.** $800(5x + 14y) = 4000x + 11{,}200y$ ft$^2$ **43.** $4x(3000 - 2y) = 12{,}000x - 8xy$ ft$^2$ **44.** $-16$ **45.** 64 **46.** 14 **47.** 4 **48.** 10

**Quick Quiz 1.6** *See Examples noted with Ex.* **1.** $-15a - 35b$ (Ex. 1) **2.** $-2x^2 + 8xy - 16x$ (Ex. 4) **3.** $-12a^2b + 15ab^2 + 27ab$ (Ex. 4)
**4.** See Student Solutions Manual

**1.7 Exercises** **1.** A term is a number, a variable, a product, or a quotient of numbers and variables. **3.** The two terms $5x$ and $-8x$ are like terms because they both have the variable $x$ with the exponent of one. **5.** The only like terms are $7xy$ and $-14xy$ because the other two have different exponents even though they have the same variables. **7.** $-31x^2$ **9.** $6a^3 - 7a^2$ **11.** $-5x - 5y$ **13.** $7.1x - 3.5y$ **15.** $-2x - 8.7y$

**17.** $5p + q - 18$ **19.** $5bc - 6ac$ **21.** $x^2 - 10x + 3$ **23.** $-10y^2 - 16y + 12$ **25.** $-\frac{1}{15}x - \frac{2}{21}y$ **27.** $\frac{11}{20}a^2 - \frac{5}{6}b$ **29.** $-2rs + 2r$

**31.** $\frac{19}{4}xy + 2x^2y$ **33.** $28a - 20b$ **35.** $-27ab - 11b^2$ **37.** $-2c - 10d^2$ **39.** $11x + 20$ **41.** $7a + 9b$ **43.** $14x - 8$ ft

**45.** $-\frac{13}{12}$ or $-1\frac{1}{12}$ **46.** $-\frac{3}{8}$ **47.** $\frac{23}{50}$ **48.** $-\frac{15}{98}$

**Quick Quiz 1.7** *See Examples noted with Ex.* **1.** $\frac{13}{6}xy + \frac{5}{3}x^2y$ (Ex. 6) **2.** $0.6a^2b - 4.4ab^2$ (Ex. 4a) **3.** $20x - 2y$ (Ex. 7)
**4.** See Student Solutions Manual

**1.8 Exercises** **1.** $-7$ **3.** $-12$ **5.** $\frac{25}{2}$ or $12\frac{1}{2}$ **7.** $-26$ **9.** $-1.3$ **11.** $\frac{25}{4}$ or $6\frac{1}{4}$ **13.** 10 **15.** 5 **17.** $-24$ **19.** $-20$

**21.** 9 **23.** 39 **25.** $-2$ **27.** 44 **29.** $-2$ **31.** $-24$ **33.** 15 **35.** 32 **37.** 32 **39.** $-\frac{1}{2}$ **41.** 352 ft$^2$ **43.** 1.24 cm$^2$

**45.** 32 in.$^2$ **47.** 56,000 ft$^2$ **49.** $\approx 28.26$ ft$^2$ **51.** $-78.5°C$ **53.** $2340.00 **55.** The coldest temperature was $-396.4°F$. The warmest temperature was $253.4°F$. **57.** 16 **58.** $-x^2 + 2x - 4y$

**Quick Quiz 1.8** *See Examples noted with Ex.* **1.** 2 (Ex. 3) **2.** $\frac{9}{2}$ or $4\frac{1}{2}$ (Ex. 4) **3.** 33 (Ex. 4) **4.** See Student Solutions Manual

**1.9 Exercises** **1.** $-(3x + 2y)$ **3.** distributive **5.** $4x + 12y$ **7.** $2c - 16d$ **9.** $x - 7y$ **11.** $8x^3 - 4x^2 + 12x$

**13.** $-2x + 26y$   **15.** $2x + 11y + 10$   **17.** $15a - 60ab$   **19.** $12a^3 - 19a^2 - 22a$   **21.** $3a^2 + 16b + 12b^2$   **23.** $-13a + 9b - 1$
**25.** $6x^2 + 30x - 30$   **27.** $12a^2 - 8b$   **29.** $1947.52°F$   **30.** $453,416 \text{ ft}^2$   **31.** $54$ to $67.5$ kg   **32.** $4.05$ to $6.3$ kg

**Quick Quiz 1.9**   *See Examples noted with Ex.*   **1.** $-14x - 4y$ (Ex. 1)   **2.** $-6x + 15y - 36$ (Ex. 2)   **3.** $-8a - 24ab + 8b$ (Ex. 4)
**4.** See Student Solutions Manual

**Career Exploration Problems**   **1.** 21.6W per conductor; total power loss of 43.2W   **2.** 172.8W   **3.** 0.6A   **4.** 168 in.$^3$

**You Try It**   **1. (a)** 5   **(b)** 1   **(c)** 0.5   **(d)** $\frac{1}{4}$   **(e)** 4.57   **2.** $-14$   **3. (a)** 6   **(b)** $-6$   **4.** 3   **5.** $-1$   **6. (a)** $-54$   **(b)** $-8$   **(c)** 6   **(d)** 21
**7. (a)** 81   **(b)** 2.25   **(c)** $\frac{1}{16}$   **8. (a)** $-8$   **(b)** 256   **9.** 67   **10. (a)** $8a - 12$   **(b)** $-25x + 5$   **11.** $-3a^2 - 3a + 8ab$   **12.** 103
**13.** 2500 sq ft   **14.** $28x + 8$

**Chapter 1 Review Problems**   **1.** $-8$   **2.** $-4.2$   **3.** $-9$   **4.** 1.9   **5.** $-\frac{1}{3}$   **6.** $-\frac{7}{22}$   **7.** $\frac{1}{6}$   **8.** $-\frac{1}{2}$   **9.** 8   **10.** 13

**11.** $-33$   **12.** 9.2   **13.** $-\frac{13}{8}$ or $-1\frac{5}{8}$   **14.** $\frac{11}{24}$   **15.** $-22.7$   **16.** $-88$   **17.** $-3$   **18.** 13   **19.** 32   **20.** $\frac{5}{6}$   **21.** $-\frac{25}{7}$ or $-3\frac{4}{7}$

**22.** $-72$   **23.** 60   **24.** $-30$   **25.** $-243$   **26.** 64   **27.** 625   **28.** $-\frac{8}{27}$   **29.** $-81$   **30.** 0.36   **31.** $\frac{25}{36}$   **32.** $\frac{27}{64}$   **33.** $-44$

**34.** 20.004   **35.** 1   **36.** $-21x + 7y$   **37.** $18x - 3x^2 + 9xy$   **38.** $-7x^2 + 3x - 11$   **39.** $-6xy^3 - 3xy^2 + 3y^3$   **40.** $-5a^2b + 3bc$

**41.** $-3x - 4y$   **42.** $-5x^2 - 35x - 9$   **43.** $10x^2 - 8x - \frac{1}{2}$   **44.** $-55$   **45.** 1   **46.** $-4$   **47.** $-3$   **48.** 0   **49.** 17   **50.** 15

**51.** $810   **52.** $68°F$ to $77°F$   **53.** $75.36   **54.** $8580   **55.** $100,000 \text{ ft}^2$; $200,000   **56.** $10.45 \text{ ft}^2$; $689.70   **57.** $-2x + 42$

**58.** $-17x - 18$   **59.** $-2 + 10x$   **60.** $-12x^2 + 63x$   **61.** $5xy^3 - 6x^3y - 13x^2y^2 - 6x^2y$   **62.** $x - 10y + 35 - 15xy$   **63.** $-31a + 2b$

**64.** $15a - 15b - 10ab$   **65.** $-3x - 9xy + 18y^2$   **66.** $10x + 8xy - 32y$   **67.** $-2.3$   **68.** 8   **69.** $-\frac{22}{15}$ or $-1\frac{7}{15}$   **70.** $-\frac{1}{8}$

**71.** $-1$   **72.** $-0.5$   **73.** $\frac{81}{40}$ or $2\frac{1}{40}$   **74.** $-8$   **75.** 240   **76.** $-25.42$   **77.** $600   **78.** 0.0081   **79.** $-0.0625$   **80.** 10

**81.** $-4.9x + 4.1y$   **82.** $-\frac{1}{9}$   **83.** $-\frac{2}{3}$   **84.** No: the dog's temperature is below the normal temperature of $101.48°F$.
**85.** $3y^2 + 12y - 7x - 28$   **86.** $-12x + 6y + 12xy$

**How Am I Doing? Chapter 1 Test**   **1.** $-0.3$ (obj. 1.1.4)   **2.** 2 (obj. 1.2.1)   **3.** $-\frac{14}{3}$ or $-4\frac{2}{3}$ (obj. 1.3.1)   **4.** $-70$ (obj. 1.3.1)

**5.** 4 (obj. 1.3.3)   **6.** $-3$ (obj. 1.3.3)   **7.** $-64$ (obj. 1.4.2)   **8.** 2.56 (obj. 1.4.2)   **9.** $\frac{16}{81}$ (obj. 1.4.2)   **10.** 6.8 (obj. 1.5.1)

**11.** $-25$ (obj. 1.5.1)   **12.** $-5x^2 - 10xy + 35x$ (obj. 1.6.1)   **13.** $6a^2b^2 + 4ab^3 - 14a^2b^3$ (obj. 1.6.1)   **14.** $2a^2b + \frac{15}{2}ab$ (obj. 1.7.2)

**15.** $-1.8x^2y - 4.7xy^2$ (obj. 1.7.2)   **16.** $5a + 30$ (obj. 1.7.2)   **17.** $14x - 16y$ (obj. 1.7.2)   **18.** 122 (obj. 1.8.1)   **19.** 37 (obj. 1.8.1)

**20.** $\frac{13}{6}$ or $2\frac{1}{6}$ (obj. 1.8.1)   **21.** $\approx 96.6$ km/hr (obj. 1.8.2)   **22.** $22,800 \text{ ft}^2$ (obj. 1.8.2)   **23.** $23.12 (obj. 1.8.2)   **24.** 3 cans (obj. 1.8.2)

**25.** $3x - 6xy - 21y^2$ (obj. 1.9.1)   **26.** $-3a - 9ab + 3b^2 - 3ab^2$ (obj. 1.9.1)

## Chapter 2   **2.1 Exercises**   **1.** equals, equal   **3.** solution   **5.** Answers may vary.   **7.** $x = 7$   **9.** $x = 11$   **11.** $x = 17$
**13.** $x = -5$   **15.** $x = -13$   **17.** $x = 62$   **19.** $x = 15$   **21.** $x = 21$   **23.** $x = 0$   **25.** $x = 0$   **27.** $x = 21$

**29.** No; $x = 9$   **31.** No; $x = -14$   **33.** Yes   **35.** Yes   **37.** $x = -1.8$   **39.** $x = 0.6$   **41.** $x = 1$   **43.** $x = -\frac{1}{4}$

**45.** $x = -7$   **47.** $x = \frac{17}{6}$ or $2\frac{5}{6}$   **49.** $x = \frac{3}{7}$   **51.** $x = 5.2$   **53.** $x = 20.2$   **55.** $-2x - 4y$   **56.** $-2y^2 - 4y + 4$

**Quick Quiz 2.1**   *See Examples noted with Ex.*   **1.** $x = 14.3$ (Ex. 2)   **2.** $x = -3.5$ (Ex. 2)   **3.** $x = -8$ (Ex. 3)
**4.** See Student Solutions Manual

**2.2 Exercises**   **1.** 6   **3.** 7   **5.** $x = 48$   **7.** $x = -30$   **9.** $x = 80$   **11.** $x = -15$   **13.** $x = 4$   **15.** $x = 8$
**17.** $x = -\frac{8}{3}$   **19.** $x = 50$   **21.** $x = 15$   **23.** $x = -7$   **25.** $x = 0.2$   **27.** $x = 4$   **29.** No; $x = -7$   **31.** Yes

**33.** $y = -0.03$   **35.** $t = \frac{8}{3}$   **37.** $y = -0.7$   **39.** $x = 3$   **41.** $x = -4$   **43.** $x = -36$   **45.** $x = 1$   **47.** $m = 2$

**49.** $x = 84$   **51.** $x = -10.5$   **53.** $9xy - 8y^2$   **54.** $x - 9$   **55.** 22.5 tons   **56.** 27 earthquakes

**Quick Quiz 2.2**   *See Examples noted with Ex.*   **1.** $x = -38$ (Ex. 4)   **2.** $x = 14$ (Ex. 4)   **3.** $x = -12$ (Ex. 6)
**4.** See Student Solutions Manual

**2.3 Exercises**   **1.** $x = 9$   **3.** $x = 6$   **5.** $x = -4$   **7.** $x = 13$   **9.** $x = 3.1$   **11.** $x = 28$   **13.** $x = -27$   **15.** $x = 8$

**17.** $x = 3$   **19.** $x = \frac{11}{2}$   **21.** $x = -9$   **23.** Yes   **25.** No; $x = -11$   **27.** $x = -1$   **29.** $x = 7$   **31.** $y = -1$   **33.** $x = 16$

**35.** $y = 3$   **37.** $x = 4$   **39.** $x = \frac{1}{4}$ or 0.25   **41.** $x = \frac{5}{2}$   **43.** $x = 6.5$   **45.** $a = 0$   **47.** $x = -\frac{16}{5}$   **49.** $y = 2$   **51.** $x = -4$

**53.** $z = 5$  **55.** $a = -6.5$  **57.** $x = -\dfrac{2}{3}$  **59.** $x = -0.25$  **61.** $x = 8$  **63.** $x = -1.5$  **65.** 42  **66.** $-37$  **67.** 21

**68.** \$1184.35  **69. (a)** \$629.30  **(b)** \$647.28

**Quick Quiz 2.3** *See Examples noted with Ex.*  **1.** $x = -\dfrac{4}{11}$ (Ex. 3)  **2.** $x = 4$ (Ex. 1)  **3.** $x = -\dfrac{8}{7}$ (Ex. 6)
**4.** See Student Solutions Manual

**2.4 Exercises**  **1.** $x = -7$  **3.** $x = 1$  **5.** $x = 1$  **7.** $x = 12$  **9.** $y = 20$  **11.** $x = 3$  **13.** $x = \dfrac{7}{3}$  **15.** $x = -\dfrac{7}{2}$ or $-3.5$

**17.** Yes  **19.** No  **21.** $x = 1$  **23.** $x = 8$  **25.** $x = 2$  **27.** $x = -5$  **29.** $y = 4$  **31.** $x = -22$  **33.** $x = 2$

**35.** $x = -12$  **37.** $x = -\dfrac{5}{3}$  **39.** $x = \dfrac{10}{3}$  **41.** No solution  **43.** Infinite number of solutions  **45.** $x = 0$  **47.** No solution

**49.** $-\dfrac{52}{3}$ or $-17\dfrac{1}{3}$  **50.** $\dfrac{22}{5}$ or $4\dfrac{2}{5}$  **51.** $572 - 975$ g  **52.** 3173 seats

**Quick Quiz 2.4** *See Examples noted with Ex.*  **1.** $x = -\dfrac{7}{5}$ or $-1.4$ (Ex. 2)  **2.** $x = \dfrac{19}{31}$ (Ex. 2)  **3.** $x = \dfrac{11}{26}$ (Ex. 4)  **4.** See Student
Solutions Manual

**Use Math to Save Money**  **1.** Shell: \$4.55; ARCO: \$4.88  **2.** Shell: \$13.65; ARCO: \$13.74  **3.** Shell: \$18.20; ARCO: \$18.17
**4.** Shell: \$45.50; ARCO: \$44.75  **5.** 3.75 gallons  **6.** Shell  **7.** ARCO  **8.** Answers will vary.  **9.** Answers will vary.
**10.** Answers will vary.

**2.5 Exercises**  **1.** Multiply each term by 5. Then subtract 160 from both sides. Then divide each side by 9. We would obtain $\dfrac{5F - 160}{9} = C$.

**3. (a)** 10 m  **(b)** 16 m  **5. (a)** $y = \dfrac{18 - 2x}{9}$ or $y = -\dfrac{2}{9}x + 2$  **(b)** $y = 4$  **7.** $b = \dfrac{2A}{h}$  **9.** $P = \dfrac{I}{rt}$  **11.** $m = \dfrac{y - b}{x}$

**13.** $y = \dfrac{5}{3}x + 10$  **15.** $x = -\dfrac{5}{3}y + 10$  **17.** $y = \dfrac{c - ax}{b}$  **19.** $r^2 = \dfrac{A}{\pi}$  **21.** $r^2 = \dfrac{GM}{g}$  **23.** $t = \dfrac{A - P}{Pr}$  **25.** $h = \dfrac{S - 2\pi r^2}{2\pi r}$

**27.** $m = \dfrac{2K}{v^2}$  **29.** $L = \dfrac{V}{WH}$  **31.** $r^2 = \dfrac{3V}{\pi h}$  **33.** $W = \dfrac{P - 2L}{2}$  **35.** $a^2 = c^2 - b^2$  **37.** $C = \dfrac{5F - 160}{9}$  **39.** $R = \dfrac{E^2}{P}$

**41.** $S = \dfrac{360A}{\pi r^2}$  **43.** 0.8 mi  **45.** 30 ft  **47. (a)** $x = \dfrac{V - 55,440}{3780}$  **(b)** 2018  **49.** $A$ doubles  **51.** $A$ increases 4 times  **53.** \$16

**54.** 1  **55.** 39,000 ft$^2$  **56.** $10\dfrac{7}{12}$ hr

**Quick Quiz 2.5** *See Examples noted with Ex.*  **1.** $x = \dfrac{A - 2w}{3}$ (Ex. 3)  **2.** $a = \dfrac{3A - bh}{h}$ (Ex. 4)  **3.** $y = \dfrac{2ax + 5}{2ax}$ (Ex. 4)
**4.** See Student Solutions Manual

**2.6 Exercises**  **1.** Yes, both statements imply 5 is to the right of $-6$ on a number line.  **3.** $>$  **5.** $>$  **7.** $<$  **9. (a)** $<$  **(b)** $>$

**11. (a)** $>$  **(b)** $<$  **13.** $<$  **15.** $>$  **17.** $>$  **19.** $<$  **21.** $<$  **23.** $<$  **25.**
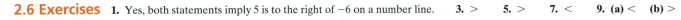

**27.**   **29.** 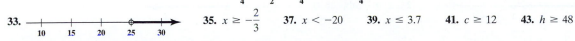  **31.**

**33.** 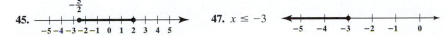  **35.** $x \geq -\dfrac{2}{3}$  **37.** $x < -20$  **39.** $x \leq 3.7$  **41.** $c \geq 12$  **43.** $h \geq 48$

**45.** 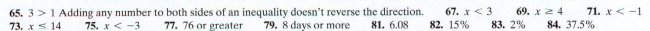  **47.** $x \leq -3$

**49.** $x \leq 5$  **51.** $x > -9$

**53.** $x \geq 8$  **55.** $x < -12$

**57.** $x < -1$  **59.** $x < \dfrac{3}{2}$

**61.** $x > -6$  **63.** $x > \dfrac{1}{3}$

**65.** $3 > 1$ Adding any number to both sides of an inequality doesn't reverse the direction.  **67.** $x < 3$  **69.** $x \geq 4$  **71.** $x < -1$
**73.** $x \leq 14$  **75.** $x < -3$  **77.** 76 or greater  **79.** 8 days or more  **81.** 6.08  **82.** 15%  **83.** 2%  **84.** 37.5%

**Quick Quiz 2.6** *See Examples noted with Ex.*  **1.** 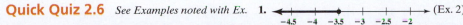 (Ex. 2)

**2.** $x \leq 6$ 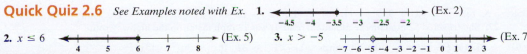 (Ex. 5)  **3.** $x > -5$ (Ex. 7)
**4.** See Student Solutions Manual

## Career Exploration Problems

1. $\dfrac{BMR - 6.25H + 5A - 5}{10} = W$; $\dfrac{BMR - 10W + 5A - 5}{6.25} = H$

2. 74 inches or 6 feet 2 inches tall    3. 125 pounds

## You Try It

1. $x = -12$    2. $y = 4$    3. $a = \dfrac{4H - b}{c}$    4. $x \le -4$

## Chapter 2 Review Problems

1. $x = -7$    2. $x = -3$    3. $x = -3$    4. $x = 40.4$    5. $x = -2$    6. $x = -7$

7. $x = 3$    8. $x = -\dfrac{7}{2}$ or $-3.5$    9. $x = 5$    10. $x = 1$    11. $x = 20$    12. $x = \dfrac{2}{3}$    13. $x = 5$    14. $x = \dfrac{35}{11}$    15. $x = 4$

16. $x = -17$    17. $x = -7$    18. $x = 3$    19. $x = 4$    20. $x = 32$    21. $x = 0$    22. $x = -17$    23. $x = -32$

24. $x = -\dfrac{17}{5}$ or $-3.4$    25. $y = 3x - 10$    26. $y = \dfrac{-5x - 7}{2}$    27. $r = \dfrac{A - P}{Pt}$    28. $h = \dfrac{A - 4\pi r^2}{2\pi r}$    29. $p = \dfrac{3H - a - 3}{2}$

30. $y = \dfrac{c - ax}{b}$    31. (a) $A = \dfrac{10x}{BC}$ (b) $A = -40$    32. (a) $w = \dfrac{P - 2l}{2}$ (b) $w = 6.5$    33. (a) $h = \dfrac{V}{lw}$ (b) $h = 6$

34. $x \le -1$    35. $x < -4$    38. $x < 5$

36. $x > -3$    37. $x \ge 6$

39. $x < 10$    40. $x > -3$    41. $h \le 32$ hr    42. $n \le 22$

43. $x = \dfrac{13}{6}$    44. $x = \dfrac{11}{8}$    45. $x = 0$    46. $x = 4$    47. $x < 2$

48. $x \le -8$    49. $x \ge \dfrac{19}{7}$

50. $x \ge -15$

## How Am I Doing? Chapter 2 Test

1. $x = 2$ (obj. 2.3.1)    2. $x = \dfrac{1}{3}$ (obj. 2.3.2)    3. $y = -\dfrac{7}{2}$ or $-3.5$ (obj. 2.3.3)

4. $y = 8.4$ or $\dfrac{42}{5}$ (obj. 2.4.1)    5. $x = 1$ (obj. 2.3.3)    6. $x = -\dfrac{6}{5}$ or $-1.2$ (obj. 2.3.3)    7. $y = 7$ (obj. 2.4.1)    8. $y = \dfrac{7}{3}$ or $2\dfrac{1}{3}$ (obj. 2.3.3)

9. $x = 13$ (obj. 2.3.3)    10. $x = 20$ (obj. 2.3.3)    11. $x = 10$ (obj. 2.3.3)    12. $x = -4$ (obj. 2.3.3)    13. $x = 12$ (obj. 2.3.3)

14. $x = -\dfrac{1}{5}$ or $-0.2$ (obj. 2.4.1)    15. $x = 3$ (obj. 2.4.1)    16. $x = 2$ (obj. 2.4.1)    17. $x = -2$ (obj. 2.4.1)    18. $w = \dfrac{A - 2P}{3}$ (obj. 2.5.1)

19. $w = \dfrac{6 - 3x}{4}$ (obj. 2.5.1)    20. $a = \dfrac{2A - hb}{h}$ (obj. 2.5.1)    21. $y = \dfrac{10ax - 5}{8ax}$ (obj. 2.5.1)    22. $B = \dfrac{3V}{h}$ (obj. 2.5.1)    23. $B = 30$ in.$^2$

(obj. 2.5.1)    24. $x \le -3$ (obj. 2.6.4)    25. $x > -\dfrac{5}{4}$ (obj. 2.6.4)

26. $x < 2$ (obj. 2.6.4)    27. $x \ge \dfrac{1}{2}$ (obj. 2.6.4)

## Chapter 3    3.1 Exercises

1. $x + 11$    3. $x - 12$    5. $\dfrac{1}{8}x$ or $\dfrac{x}{8}$    7. $2x$    9. $3 + \dfrac{1}{2}x$    11. $2x + 9$    13. $\dfrac{1}{3}(x + 7)$

15. $\dfrac{1}{3}x - 2x$    17. $5x - 11$    19. $x =$ value of a share of AT&T stock

$x + 74.50 =$ value of a share of IBM stock

21. $w =$ width

$2w + 7 =$ length

23. $x =$ number of boxes sold by Keiko

$x - 43 =$ number of boxes sold by Sarah

$x + 53 =$ number of boxes sold by Imelda

25. Measure of 1st angle $= s - 25$

Measure of 2nd angle $= s$

Measure of 3rd angle $= 3s$

27. $v =$ value of exports of Canada

$2v =$ value of exports of Japan

29. $p =$ price of the Summer on the Beach Concert tickets;

$\dfrac{1}{2}p =$ price of All Star Concert tickets

31. $x =$ number of men aged 16–24

$x + 82 =$ number of men aged 25–34

$x - 25 =$ number of men aged 35–44

$x - 110 =$ number of men aged 45 and above

33. $x = 7$

34. $x = -\dfrac{5}{2}$ or $-2\dfrac{1}{2}$

## Quick Quiz 3.1    See Examples noted with Ex.

1. $x + 10$ or $10 + x$ (Ex. 1)    2. $2x - 5$ (Ex. 2)    3. Measure of first angle: $x + 15$

4. See Student Solutions Manual

Measure of second angle: $x$

Measure of third angle: $5x$ (Ex. 5)

## 3.2 Exercises

1. 1261    3. 2368    5. 182    7. 43    9. 9    11. $-4$    13. 12    15. 40 red tablet cases    17. 7 hrs    19. 4 hrs

21. 4 items    23. 550 lb    25. 5 mi    27. She traveled 52 mph on the mountain road. It was 12 mph faster on the highway route.

29. The score on the final lab must be 92.    31. (a) $F - 40 = \dfrac{x}{4}$ (b) 200 chirps (c) 77°F    32. $10x^3 - 30x^2 - 15x$

33. $-2a^2b + 6ab - 10a^2$    34. $-5x - 6y$    35. $-4x^2y - 7xy^2 - 8xy$

**Quick Quiz 3.2** *See Examples noted with Ex.* **1.** 17 (Ex. 2) **2.** 48 (Ex. 3) **3.** 74 (Ex. 6) **4.** See Student Solutions Manual

**3.3 Exercises** **1.** The long piece is 11 feet long. **3.** James earns $39,450 per year. **5.** Dave worked 30 hours.
The short piece is 6.5 feet long. Lin earns $42,850 per year. Sarah worked 45 hours.
Kate worked 25 hours.

**7.** 11 students born in Oklahoma. **9.** Length is 13 meters. **11.** Length is 41 centimeters. **13.** Width is 32.5 centimeters.
19 students born in Texas. Width is 8 meters. Width is 7 centimeters. Length is 62.5 centimeters.
2 students born in Arizona.

**15.** The cheetah can run 70 mph. **17.** Length A = 20" **19.** Original square was 11 m × 11 m **21.** $\frac{5}{9}$ **22.** $x = 15$
The jackal can run 35 mph. Length B = 15"
The elk can run 45 mph. Length C = 10" **23.** $-19x + 2y - 2$ **24.** $9x^2y - 6xy^2 + 7xy$
Length D = 13"

**Quick Quiz 3.3** *See Examples noted with Ex.* **1.** Width is 75 yd **2.** Psychology: 39 students **3.** First side: 22 m
Length is 230 yd (Ex. 3) Art history: 13 students Second side: 33 m
Algebra: 27 students (Ex. 2) Third side: 18 m (Ex. 2)

**4.** See Student Solutions Manual

**Use Math to Save Money** **1.** The Gold plan **2.** The Silver plan **3.** The Silver plan **4.** The Bronze plan **5.** $30 **6.** $60
**7.** $90 **8.** The Silver plan **9.** $2319

**3.4 Exercises** **1.** 38 coffee products **3.** 5 hr **5.** 12 weeks **7.** $360 **9.** $22,000 **11.** $12,000 **13.** $3000 at 7%
$2000 at 5%

**15.** Conservative fund = $250,000 **17.** $12,000 **19.** 13 quarters **21.** 18 nickels **23.** 10 boxes of beige paper **25.** $10 bills = 26
Growth fund = $150,000 9 nickels 6 dimes 20 boxes of white paper $20 bills = 31
9 quarters $50 bills = 10

**27.** $925,000 worth of furniture **29.** 12 **30.** −17 **31.** 25 **32.** −26

**Quick Quiz 3.4** *See Examples noted with Ex.* **1.** 9 months (Ex. 1) **2.** $11,600 (Ex. 3) **3.** $2200 at 4%
$2800 at 5% (Ex. 5)

**4.** See Student Solutions Manual

**3.5 Exercises** **1.** distance around **3.** surface **5.** 180° **7.** 378 ft² **9.** 98 in.² **11.** 9 in. **13.** ≈ 153.86 ft² **15.** ≈ 9520.48 mi
**17.** 134,400 m² **19.** 5 cm **21.** ≈ 200.96 cm **23.** Equal angles = 72° **25.** 30° each **27.** 19 ft **29.** ≈ 2512 in.³ **31.** Yes
3rd angle = 36°

**33. (a)** $V ≈ 452.16$ cm³ **35.** ≈ 163.28 ft² **37. (a)** ≈ 73.0 ft² **39.** ≈ 6.28 ft **40.** ≈ 78.5 **41.** $x = -15$ **42. (a)** 5% **43. (a)** $390
**(b)** $S ≈ 376.8$ cm² **(b)** ≈ $182.50 **(b)** 7% **(b)** $97.50

**Quick Quiz 3.5** *See Examples noted with Ex.* **1.** 116 ft² (Ex. 3) **2.** ≈ 113 in.³ (Ex. 7) **3.** $2280 (Ex. 9) **4.** See Student Solutions Manual

**3.6 Exercises** **1.** $x > 67,000$ **3.** $x < 175$ **5.** $h ≤ 1500$ **7.** $x ≥ 93$ **9.** It must be less than or equal to 140 cm.
**11.** 22 or more deliveries **13.** Depth can be no more than 7.5 feet. **15.** They can drive no more than 81.8 miles. **17.** The range must be 68°F
or higher. **19.** Marc can produce 5000 or fewer new items. **21.** more than $130,000 in sales **23.** 455 or more discs would need to be sold.
**25.** $q = \frac{p + 3}{5}$ **26.** $t = \frac{I}{pr}$ **27.** $x ≥ 4\frac{2}{3}$ **28.** $x > -4$

**Quick Quiz 3.6** *See Examples noted with Ex.* **1.** No more than 9.9 feet (Ex. 3) **2.** Less than 18 hours per week (Ex. 2)
**3.** More than $285,000 in sales (Ex. 3) **4.** See Student Solutions Manual

**Career Exploration Problems** **1.** Length: 132 feet; width: 81 feet **2.** $4998 **3.** $15

**You Try It** **1.** Width = 16 m; length = 40 m **2.** $600 at 2%; $3000 at 4% **3.** 3768 in.³ **4.** Lexi must make sales of $30,000 or more.

**Chapter 3 Review Problems** **1.** $x + 19$ **2.** $\frac{2}{3}x$ **3.** $\frac{1}{2}x$ or $\frac{x}{2}$ **4.** $x - 18$ **5.** $3(x + 4)$ **6.** $2x - 3$ **7.** $r =$ the
number of retired people; $4r =$ the number of working people; $0.5r =$ the number of unemployed people **8.** $3w + 5 =$ the length; $w =$ the
width **9.** $b =$ the number of degrees in angle $B$; $2b =$ the number of degrees in angle $A$; $b - 17 =$ the number of degrees in angle $C$
**10.** $a =$ the number of students in algebra; $a + 29 =$ the number of students in biology; $0.5a =$ the number of students in geology **11.** 3
**12.** −7 **13.** $40 **14.** 16 years old **15.** 13.3 hr; 12.3 hr **16.** 88 **17.** 23 ft **18.** 1st angle = 32°; 2nd angle = 96°; 3rd angle = 52°
**19.** 31.25 yd; 18.75 yd **20.** George = $42,000; Heather = $23,000 **21.** 310 kilowatt-hours **22.** 280 mi **23.** $3000 **24.** $200
**25.** $7000 at 12%; $2000 at 8% **26.** $2000 at 4.5%; $3000 at 6% **27.** 18 nickels; 6 dimes; 9 quarters **28.** 24 nickels; 21 dimes; 26 quarters
**29.** ≈ 21.23 in.² **30.** 24 ft **31.** 71° **32.** 42 mi² **33.** 462 ft³ **34.** ≈ 381.51 in.³ **35.** ≈ 415.27 ft² **36.** 12 in. **37.** $3440
**38.** ≈ $23,864 **39.** 160 miles or less **40.** More than $260,000 **41.** CP distance = 630 mi **42.** Less than $16.15
UP distance = 960 mi

**43.** More than 3 years **44.** More than 60 months **45.** They made 27 field goals and 9 free throws. **46.** $13,600 at 9.75%
$15,600 at 8.5%

**47.** It can travel 640 miles in one hour. It would take approximately 169 minutes. **48.** 12 sec

## How Am I Doing? Chapter 3 Test

**1.** 35 (obj. 3.2.1)　**2.** 36 (obj. 3.2.1)　**3.** −4 (obj. 3.2.1)　**4.** first side = 20 m; second side = 30 m; third side = 16 m (obj. 3.3.1)　**5.** Width = 20 m; length = 47 m (obj. 3.3.1)　**6.** 1st pollutant = 8 ppm; 2nd pollutant = 4 ppm; 3rd pollutant = 3 ppm (obj. 3.3.1)　**7.** 15 months (obj. 3.4.1)　**8.** $32,000 (obj. 3.4.2)　**9.** $1400 at 14%; $2600 at 11% (obj. 3.4.3)　**10.** 16 nickels; 7 dimes; 8 quarters (obj. 3.4.4)　**11.** ≈ 213.52 in. (obj. 3.5.1)　**12.** 192 in.$^2$ (obj. 3.5.1)　**13.** ≈ 4187 in.$^3$ (obj. 3.5.2)　**14.** 96 cm$^2$ (obj. 3.5.1)　**15.** $450 (obj. 3.5.3)　**16.** At least an 82 (obj. 3.6.2)　**17.** More than $110,000 (obj. 3.6.2)　**18.** 600 hr (obj. 3.6.2)

## Cumulative Test for Chapters 0–3

**1.** 3.69　**2.** $\frac{31}{24}$ or $1\frac{7}{24}$　**3.** $-\frac{3}{7}$　**4.** $\frac{35}{4}$ or $8\frac{3}{4}$　**5.** $\frac{4}{7}$　**6.** 0.6　**7.** $-2x + 2y + 20$　**8.** $3x - 6xy - 21y^2$　**9.** 37　**10.** 21　**11.** $b = \frac{2H - 3a}{5}$　**12.** $x \geq -3$; 

**13.** $y = 1$　**14.** Literature students = 42; sociology students = 54　**15.** Width = 7 cm; length = 32 cm　**16.** $230,435　**17.** $3000 at 15%; $4000 at 7%　**18.** $V \approx 113.04$ in.$^3$; 169.56 lb

# Chapter 4

## 4.1 Exercises

**1.** When multiplying exponential expressions with the same base, keep the base the same and add the exponents.　**3.** $\frac{2^2}{2^3} = \frac{2 \cdot 2}{2 \cdot 2 \cdot 2} = \frac{1}{2} = \frac{1}{2^{3-2}}$　**5.** 6; $x$ and $y$; 11 and 1　**7.** $2^2 a^3 b$　**9.** $-5x^3 y^2 z^2$　**11.** $7^{10}$　**13.** $8^{21}$　**15.** $x^{12}$　**17.** $t^{16}$　**19.** $-20x^6$　**21.** $50x^3$　**23.** $18x^3 y^8$　**25.** $\frac{2}{15} x^3 y^5$　**27.** $-2.75x^3 yz$　**29.** 0　**31.** $80x^3 y^7$　**33.** $-24x^4 y^7$　**35.** 0　**37.** $-24a^4 b^3 x^2 y^5$　**39.** $-30x^3 y^4 z^6$　**41.** $y^7$　**43.** $\frac{1}{y^3}$　**45.** $\frac{1}{11^{12}}$　**47.** $2^7$　**49.** $\frac{a^8}{4}$　**51.** $\frac{x^7}{y^9}$　**53.** $2x^4$　**55.** $-\frac{x^3}{2y^2}$　**57.** $\frac{f^2}{30g^5}$　**59.** $17x^4 y^3$　**61.** $\frac{y^2}{16x^3}$　**63.** $\frac{3a}{4}$　**65.** $x^{12}$　**67.** $x^{15} y^5$　**69.** $r^6 s^{12}$　**71.** $27a^9 b^6 c^3$　**73.** $9a^8$　**75.** $\frac{x^7}{128m^{28}}$　**77.** $\frac{25x^2}{49y^4}$　**79.** $81a^8 b^{12}$　**81.** $-8x^9 z^3$　**83.** $\frac{9}{x}$　**85.** $25a^5 b^7$　**87.** $\frac{49}{a^{10}}$　**89.** $\frac{16x^4}{y^{12}}$　**91.** $\frac{7a^2 c^3}{4b}$　**93.** $\frac{1}{3xy}$　**94.** $-11$　**95.** $-46$　**96.** $\frac{2}{25}$　**97.** $-4$　**98.** ≈ 64%　**99.** 0.6%

## Quick Quiz 4.1

*See Examples noted with Ex.*　**1.** $-10x^3 y^7$ (Ex. 3)　**2.** $-\frac{4x^3}{5y^2}$ (Ex. 8)　**3.** $81x^{12} y^{20}$ (Ex. 13c)　**4.** See Student Solutions Manual

## 4.2 Exercises

**1.** $\frac{1}{x^4}$　**3.** $\frac{1}{81}$　**5.** $y^8$　**7.** $\frac{z^6}{x^4 y^5}$　**9.** $\frac{a^3}{b^2}$　**11.** $\frac{x^9}{8}$　**13.** $\frac{3}{x^2}$　**15.** $\frac{1}{9x^2 y^4}$　**17.** $\frac{3xz^3}{y^2}$　**19.** $4xy$　**21.** $\frac{b^3 d}{ac^4}$　**23.** $\frac{1}{8}$　**25.** $\frac{z^8}{9y^4}$　**27.** $\frac{1}{x^6 y}$　**29.** $1.2378 \times 10^5$　**31.** $6.3 \times 10^{-2}$　**33.** $8.8961 \times 10^{11}$　**35.** $3.42 \times 10^{-6}$　**37.** 302,000　**39.** 0.00047　**41.** 983,000　**43.** $2.37 \times 10^{-5}$ mph　**45.** 149,600,000 km　**47.** $6.3 \times 10^{15}$　**49.** $1.0 \times 10^1$　**51.** $8.1 \times 10^{-11}$　**53.** $4.5 \times 10^5$　**55.** $5.61 \times 10^4$ dollars　**57.** $6.6 \times 10^{-5}$ mi　**59.** About $3.68 \times 10^8$ mi/yr　**61.** 13.2%　**63.** $-0.8$　**64.** $-1$　**65.** $-\frac{1}{28}$

## Quick Quiz 4.2

*See Examples noted with Ex.*　**1.** $\frac{3y^2}{x^3 z^4}$ (Ex. 3c)　**2.** $\frac{a^8}{2b}$ (Ex. 4b)　**3.** $8.76 \times 10^{-3}$ (Ex. 6)　**4.** See Student Solutions Manual

## 4.3 Exercises

**1.** A polynomial in $x$ is the sum of a finite number of terms of the form $ax^n$, where $a$ is any real number and $n$ is a whole number. An example is $3x^2 - 5x - 9$.　**3.** The degree of a polynomial in $x$ is the largest exponent of $x$ in any of the terms of the polynomial.　**5.** Degree 4; monomial　**7.** Degree 5; trinomial　**9.** Degree 6; binomial　**11.** $-3x - 15$　**13.** $-2x^2 + 2x - 1$　**15.** $\frac{5}{6}x^2 + \frac{1}{2}x - 9$　**17.** $3.4x^3 + 2.2x^2 - 13.2x - 5.4$　**19.** $5x - 24$　**21.** $\frac{1}{15}x^2 - \frac{1}{14}x + 11$　**23.** $3x^3 - x^2 + 8x$　**25.** $-4.7x^4 - 0.7x^2 - 1.6x + 0.4$　**27.** $6x - 6$　**29.** $4x^2 y^2 + 4xy - 10$　**31.** $x^4 - 3x^3 - 4x^2 - 24$　**33.** 743,000　**35.** 90,400　**37.** $3x^2 + 12x$　**39.** $y = \frac{8}{3}x + \frac{2}{3}$　**40.** 

$x > 7$　**41.** $x = 7$　**42.** $x = 2$

## Quick Quiz 4.3

*See Examples noted with Ex.*　**1.** $-4x^2 - 11x + 5$ (Ex. 2)　**2.** $6x^2 - 9x - 16$ (Ex. 5)　**3.** $-3x + 8$ (Ex. 5)　**4.** See Student Solutions Manual

## Use Math to Save Money

**1.** ($450 + $425 + $460)/3 = $445　**2.** $445 − (0.06 × $445) − $30 = $388.30　**3.** $388.30 − ($388.30 × 0.2) = $310.64　**4.** $445 − $310.64 = $134.36; $134.36/445 = 30.2%　**5.** $134.36 × 12 = $1612.32

## 4.4 Exercises

**1.** $-12x^4 + 2x^2$　**3.** $24x^3 - 4x^2$　**5.** $-4x^6 + 10x^4 - 2x^3$　**7.** $x + \frac{3}{2}x^2 + \frac{5}{2}x^3$　**9.** $-2x^5 y + 4x^4 y - 5x^3 y$　**11.** $9x^4 y + 3x^3 y - 24x^2 y$　**13.** $3x^4 - 9x^3 + 15x^2 - 6x$　**15.** $-2x^3 y^3 + 12x^2 y^2 - 16xy$　**17.** $-28x^5 y + 12x^4 y + 8x^3 y - 4x^2 y$　**19.** $-6c^2 d^5 + 8c^2 d^3 - 12c^2 d$　**21.** $12x^7 - 6x^5 + 18x^4 + 54x^3$　**23.** $-16x^6 + 10x^5 - 12x^4$　**25.** $x^2 + 12x + 35$　**27.** $x^2 + 8x + 12$　**29.** $x^2 - 6x - 16$　**31.** $x^2 - 9x + 20$　**33.** $-20x^2 - 7x + 6$　**35.** $2x^2 + 6xy - 5x - 15y$　**37.** $15x^2 - 5xy + 6x - 2y$　**39.** $20y^2 - 7y - 3$　**41.** $10x^4 + 23x^2 y^3 + 12y^6$　**43.** The signs are incorrect. The result is $-3x + 6$.　**45.** $20x$　**47.** $20x^2 - 23xy + 6y^2$　**49.** $49x^2 - 28x + 4$　**51.** $16a^2 + 16ab + 4b^2$　**53.** $0.8x^2 + 11.94x - 0.9$　**55.** $\frac{1}{4}x^2 + \frac{1}{24}x - \frac{1}{12}$　**57.** $6x^4 + 16x^2 y^3 + 8y^6$　**59.** $10x^2 - 11x - 6$　**61.** $x = -10$　**62.** $w = -\frac{25}{7}$ or $-3\frac{4}{7}$　**63.** 9 twenties; 8 tens; 23 fives　**64.** $25 billion　**65.** $36.2 billion　**66.** $58.5 billion　**67.** $67.4 billion

**Quick Quiz 4.4** *See Examples noted with Ex.* **1.** $8x^3y^4 - 12x^2y^3 + 16xy^2$ (Ex. 3) **2.** $6x^2 - x - 15$ (Ex. 4) **3.** $12a^2 - 26ab + 12b^2$ (Ex. 5) **4.** See Student Solutions Manual

**4.5 Exercises** **1.** binomial **3.** The middle term is missing. The answer should be $16x^2 - 56x + 49$. **5.** $y^2 - 49$ **7.** $x^2 - 81$ **9.** $36x^2 - 25$ **11.** $4x^2 - 49$ **13.** $25x^2 - 9y^2$ **15.** $0.36x^2 - 9$ **17.** $4y^2 + 20y + 25$ **19.** $25x^2 - 40x + 16$ **21.** $49x^2 + 42x + 9$ **23.** $9x^2 - 42x + 49$ **25.** $\frac{4}{9}x^2 + \frac{1}{3}x + \frac{1}{16}$ **27.** $81x^2y^2 + 72xyz + 16z^2$ **29.** $49x^2 - 9y^2$ **31.** $9c^2 - 30cd + 25d^2$ **33.** $81a^2 - 100b^2$ **35.** $25x^2 + 90xy + 81y^2$ **37.** $x^3 - 4x^2 + 8x - 15$ **39.** $2x^4 + 7x^3 + x^2 + 7x + 4$ **41.** $a^4 + a^3 - 13a^2 + 17a - 6$ **43.** $3x^3 - 2x^2 - 25x + 24$ **45.** $2x^3 - x^2 - 16x + 15$ **47.** $2a^3 + 3a^2 - 50a - 75$ **49.** $24x^3 + 14x^2 - 11x - 6$ **51.** 19; 41 **52.** Width = 7 m; length = 10 m

**Quick Quiz 4.5** *See Examples noted with Ex.* **1.** $49x^2 - 144y^2$ (Ex. 2) **2.** $6x^3 - x^2 - 19x - 6$ (Ex. 6) **3.** $15x^4 - 16x^3 - 8x^2 + 17x - 6$ (Ex. 5) **4.** See Student Solutions Manual

**4.6 Exercises** **1.** $5x^3 - 3x + 4$ **3.** $y^3 + 7y - 3$ **5.** $9x^4 - 4x^2 - 7$ **7.** $8x^4 - 9x + 6$ **9.** $3x + 5$ **11.** $x - 3 - \frac{32}{x - 5}$ **13.** $3x^2 - 4x + 8 - \frac{10}{x + 1}$ **15.** $2x^2 - 3x - 2 - \frac{5}{2x + 5}$ **17.** $2x^2 + 3x - 1$ **19.** $2x^2 + 3x + 6 + \frac{23}{2x - 3}$ **21.** $y^2 - 4y - 1 - \frac{9}{y + 3}$ **23.** $y^3 + 2y^2 - 5y - 10 + \frac{-25}{y - 2}$ **25.** 2010: \$3.25; 2015: \$3.77 **26.** 170 and 171 **27.** (a) 7.5 (b) 7.3 (c) 6.3 (d) 2.7% decrease (e) 13.7% decrease

**Quick Quiz 4.6** *See Examples noted with Ex.* **1.** $5x^3 - 16x^2 - 2x$ (Ex. 1) **2.** $4x^2 - 5x - 2$ (Ex. 4) **3.** $x^2 + 2x + 8 + \frac{13}{x - 2}$ (Ex. 3) **4.** See Student Solutions Manual

**Career Exploration Problems** **1.** $1.4 \times 10^6$ transformants; $3.28 \times 10^4$ transformants **2.** (a) 50,100,000 (b) 0.00302 (c) $237 \times 10^6 = 237{,}000{,}000$ **3.** $0.25X^2 + 0.5Xy + 0.25y^2$

**You Try It** **1.** (a) $2^{23}$ (b) $16a^8$ (c) $-3a^5b^4$ **2.** (a) $7x^3$ (b) $-\frac{1}{3x}$ (c) $\frac{b^6}{2a^2}$ **3.** (a) 1 (b) 1 (c) 1 (d) $6a$ **4.** (a) $a^{20}$ (b) $4n^6$ (c) $\frac{27x^9}{y^3}$ (d) $25s^4t^{10}$ (e) $-a^{10}b^5$ **5.** (a) $\frac{1}{a^3}$ (b) $x$ (c) $\frac{n^6}{m^9}$ (d) $\frac{1}{9}$ **6.** (a) $3.864 \times 10^5$ (b) $5.2 \times 10^{-5}$ **7.** (a) $7.75 \times 10^{10}$ (b) $3.04 \times 10^4$ **8.** $-6x^4 - 4x^3 + x^2$ **9.** $3 - 3x^2$ **10.** (a) $-6a^3 + 10a^2 - 2a$ (b) $-4x^3y + 12x^2y^2 - 12xy^3$ **11.** (a) $4a^2 - 25b^2$ (b) $4a^2 + 20ab + 25b^2$ (c) $4a^2 - 20ab + 25b^2$ (d) $6a^2 + 7ab - 5b^2$ **12.** (a) $12x^3 - 4x^2 + x + 3$ (b) $3x^3 - 17x^2 + 11x - 5$ **13.** $3x^3 + 14x^2 - 7x - 10$ **14.** $6a^2 - 3a + 1$ **15.** $x^2 + 5x - 1$

**Chapter 4 Review Problems** **1.** $-18a^7$ **2.** $5^{23}$ **3.** $6x^4y^6$ **4.** $-14x^4y^9$ **5.** $\frac{1}{7^{12}}$ **6.** $\frac{1}{x^5}$ **7.** $y^{14}$ **8.** $\frac{1}{9^{11}}$ **9.** $-\frac{3}{5x^5y^4}$ **10.** $-\frac{2a}{3b^6}$ **11.** $x^{24}$ **12.** $\frac{2^4}{5^6b^{10}}$ **13.** $9a^6b^4$ **14.** $81x^{12}y^4$ **15.** $\frac{25a^2b^4}{c^6}$ **16.** $\frac{y^9}{64w^{15}z^6}$ **17.** $\frac{b^5}{a^3}$ **18.** $\frac{m^8}{p^5}$ **19.** $\frac{2y^3}{x^6}$ **20.** $\frac{x^{15}}{8y^3}$ **21.** $\frac{1}{36a^8b^{10}}$ **22.** $\frac{3y^2}{x^3}$ **23.** $\frac{4w^2}{x^5y^6z^8}$ **24.** $\frac{b^5c^3d^4}{27a^2}$ **25.** $1.563402 \times 10^{11}$ **26.** $1.79632 \times 10^5$ **27.** $9.2 \times 10^{-4}$ **28.** $1.74 \times 10^{-6}$ **29.** 120,000 **30.** 6,034,000 **31.** 0.25 **32.** 0.0000432 **33.** $2.0 \times 10^{13}$ **34.** $9.36 \times 10^{19}$ **35.** $7.8 \times 10^{-11}$ **36.** $7 \times 10^8$ dollars **37.** $7.94 \times 10^{14}$ cycles **38.** $6 \times 10^{12}$ operations **39.** $5.5x^2 - x - 2.3$ **40.** $7x^3 - 3x^2 - 6x + 4$ **41.** $\frac{1}{10}x^2y - \frac{13}{21}x + \frac{5}{12}$ **42.** $\frac{1}{4}x^2 - \frac{1}{4}x + \frac{1}{10}$ **43.** $-3x^2 - 15$ **44.** $15x^2 + 2x - 1$ **45.** $28x^2 - 29x + 6$ **46.** $20x^2 + 48x + 27$ **47.** $10x^3 - 30x^2 + 15x$ **48.** $-4x^2y^4 - 20x^2y^3 + 24xy^2$ **49.** $5a^2 - 8ab - 21b^2$ **50.** $8x^4 - 10x^2y - 12x^2 + 15y$ **51.** $16x^2 + 24x + 9$ **52.** $a^2 - 25b^2$ **53.** $49x^2 - 36y^2$ **54.** $25a^2 - 20ab + 4b^2$ **55.** $4x^3 + 27x^2 + 5x - 3$ **56.** $2x^3 - 7x^2 - 42x + 72$ **57.** $2y^2 + 3y + 4$ **58.** $6x^3 + 7x^2 - 18x$ **59.** $4x^2 - 6x + 8$ **60.** $3x - 2$ **61.** $3x - 7$ **62.** $2x^2 - 5x + 13 - \frac{27}{x + 2}$ **63.** $3x - 4 + \frac{3}{2x + 3}$ **64.** $x^2 + 3x + 8$ **65.** $2x^2 + 4x + 5 + \frac{11}{x - 2}$ **66.** About \$14.52 **67.** $1.847 \times 10^9$ **68.** $2.733 \times 10^{-23}$ g **69.** $4.76 \times 10^7$ lb **70.** $3xy + 2x$ **71.** $2x^2 - 4y^2$

**How Am I Doing? Chapter 4 Test** **1.** $3^{34}$ (obj. 4.1.1) **2.** $\frac{1}{25^{16}}$ (obj. 4.1.2) **3.** $8^{24}$ (obj. 4.1.3) **4.** $12x^4y^{10}$ (obj. 4.1.1) **5.** $-\frac{7x^3}{5}$ (obj. 4.1.2) **6.** $-125x^3y^{18}$ (obj. 4.1.3) **7.** $\frac{49a^{14}b^4}{9}$ (obj. 4.1.3) **8.** $\frac{3x^4}{4}$ (obj. 4.1.3) **9.** $\frac{1}{64}$ (obj. 4.2.1) **10.** $\frac{6c^5}{a^4b^3}$ (obj. 4.2.1) **11.** $3xy^7$ (obj. 4.2.1) **12.** $5.482 \times 10^{-4}$ (obj. 4.2.2) **13.** 582,000,000 (obj. 4.2.2) **14.** $2.4 \times 10^{-6}$ (obj. 4.2.2) **15.** $-2x^2 + 5x$ (obj. 4.3.2) **16.** $-11x^3 - 4x^2 + 7x - 8$ (obj. 4.3.3) **17.** $-21x^5 + 28x^4 - 42x^3 + 14x^2$ (obj. 4.4.1) **18.** $15x^4y^3 - 18x^3y^2 + 6x^2y$ (obj. 4.4.1) **19.** $10a^2 + 7ab - 12b^2$ (obj. 4.4.2) **20.** $6x^3 - 11x^2 - 19x - 6$ (obj. 4.5.3) **21.** $49x^4 + 28x^2y^2 + 4y^4$ (obj. 4.5.2) **22.** $25s^2 - 121t^2$ (obj. 4.5.1) **23.** $12x^4 - 14x^3 + 25x^2 - 29x + 10$ (obj. 4.5.3) **24.** $3x^4 + 4x^3y - 15x^2y^2$ (obj. 4.4.2) **25.** $3x^3 - x + 5$ (obj. 4.6.1) **26.** $2x^2 - 7x + 4$ (obj. 4.6.2) **27.** $2x^2 + 6x + 12$ (obj. 4.6.2) **28.** $3.77 \times 10^9$ barrels per year (obj. 4.2.2) **29.** $4.18 \times 10^6$ mi (obj. 4.2.2)

## Chapter 5   5.1 Exercises

**1.** factors   **3.** No; $6a^3 + 3a^2 - 9a$ still has a common factor of $3a$.   **5.** $8a(a + 1)$   **7.** $7ab(3 - 2b)$
**9.** $2\pi r(h + r)$   **11.** $5x(x^2 + 5x - 3)$   **13.** $4(3ab - 7bc + 5ac)$   **15.** $8x^2(2x^3 + 3x - 4)$   **17.** $7x(2xy - 5y - 9)$
**19.** $9x(6x - 5y + 2)$   **21.** $y(3xy - 2a + 5x - 2)$   **23.** $8xy(3x - 5y)$   **25.** $7x^2y^2(x + 3)$   **27.** $8x^2y(2x^2y - 3y - 1)$
**29.** $(x + 2y)(7a - b)$   **31.** $(x - 4)(3x - 2)$   **33.** $(2a - 3c)(6b - 5d)$   **35.** $(b - a^2)(7c - 5d + 2f)$   **37.** $(ab - 4)(3a - 5 - b)$
**39.** $(a - 3b)(4a^3 + 1)$   **41.** $(a + 2)(1 - x)$   **43.** $2\pi(x + y + z)$   **45.** 1,664,000 metric tons   **46.** 2,752,000 metric tons   **47.** About
41 lb/person   **48.** About 30 lb/person

**Quick Quiz 5.1**   *See Examples noted with Ex.*   **1.** $x(3 - 4x + 2y)$ (Ex. 2)   **2.** $5x(4x^2 - 5x - 1)$ (Ex. 5)   **3.** $(a + 3b)(8a - 7b)$ (Ex. 6)
**4.** See Student Solutions Manual

## 5.2 Exercises

**1.** We must remove a common factor of 5 from the last two terms. This will give us $3x(x - 2y) + 5(x - 2y)$. Then our final
answer is $(x - 2y)(3x + 5)$.   **3.** $(b - 4)(a + 6)$   **5.** $(x - 4)(x^2 + 3)$   **7.** $(a + 3b)(2x - y)$   **9.** $(3a + b)(x - 2)$
**11.** $(a + 2b)(5 + 6c)$   **13.** $(c - 2d)(6 + x)$   **15.** $(y - 2)(y - 3)$   **17.** $(9 - y)(6 + y)$   **19.** $(3x + y)(2a - 1)$
**21.** $(x + 4)(2x - 3)$   **23.** $(t - 1)(t^2 + 1)$   **25.** $(2x + 5y^2)(3x + 4w)$   **27.** We must rearrange the terms in a different order so
that the expression in parentheses is the same in each case. We use the order $6a^2 - 8ad + 9ab - 12bd$ to factor $2a(3a - 4d) + 3b(3a - 4d) =$
$(3a - 4d)(2a + 3b)$.   **29.** $-\dfrac{15}{7}$   **30.** $\dfrac{2}{15}$   **31.** $-\dfrac{a}{5b^2}$   **32.** $4x^2 - 20x + 25$   **33.** Pharmacist: \$127,000; pharmacy technician: \$35,000
**34.** About 21,323.1 thousand barrels of oil each day

**Quick Quiz 5.2**   *See Examples noted with Ex.*   **1.** $(7x + 12)(a - 2)$ (Ex. 5)   **2.** $(y^2 + 3)(2x - 5)$ (Ex. 7)   **3.** $(10y - 3)(x + 4b)$
(Ex. 6)   **4.** See Student Solutions Manual

## 5.3 Exercises

**1.** product, sum   **3.** $(x + 4)(x + 4)$   **5.** $(x + 5)(x + 7)$   **7.** $(x - 3)(x - 1)$   **9.** $(x - 7)(x - 4)$
**11.** $(x + 8)(x - 3)$   **13.** $(x - 14)(x + 1)$   **15.** $(x + 7)(x - 5)$   **17.** $(x - 6)(x + 4)$   **19.** $(x + 12)(x + 3)$
**21.** $(x - 6)(x - 4)$   **23.** $(x + 3)(x + 10)$   **25.** $(x - 5)(x - 1)$   **27.** $(a + 8)(a - 2)$   **29.** $(x - 4)(x - 8)$
**31.** $(x + 7)(x - 3)$   **33.** $(x + 7)(x + 8)$   **35.** $(y + 9)(y - 5)$   **37.** $(x + 12)(x - 3)$   **39.** $(x + 3y)(x - 5y)$
**41.** $(x - 7y)(x - 9y)$   **43.** $4(x + 5)(x + 1)$   **45.** $6(x + 1)(x + 2)$   **47.** $5(x - 1)(x - 5)$   **49.** $3(x + 4)(x - 6)$
**51.** $7(x + 5)(x - 2)$   **53.** $5(x - 1)(x - 6)$   **55.** $(12 + x)(10 - x)$   **57.** $18a^6b^9$   **58.** $25y^{12}$   **59.** $\dfrac{x^6}{y^8}$   **60.** $8x^2 + 8xy - 6y^2$
**61.** 120 mi   **62.** \$3800   **63.** 25°C   **64.** June

**Quick Quiz 5.3**   *See Examples noted with Ex.*   **1.** $(x + 7)(x + 10)$ (Ex. 2)   **2.** $(x - 6)(x - 8)$ (Ex. 3)   **3.** $2(x + 6)(x - 8)$ (Ex. 10)
**4.** See Student Solutions Manual

## 5.4 Exercises

**1.** $(4x + 1)(x + 5)$   **3.** $(5x + 2)(x + 1)$   **5.** $(4x - 3)(x + 2)$   **7.** $(2x + 1)(x - 3)$   **9.** $(3x + 1)(3x + 2)$
**11.** $(3x - 5)(5x - 3)$   **13.** $(2x - 5)(x + 4)$   **15.** $(4x - 1)(2x + 3)$   **17.** $(3x + 2)(2x - 3)$   **19.** $(5x - 1)(2x + 1)$
**21.** $(x - 2)(7x + 9)$   **23.** $(9y - 4)(y - 1)$   **25.** $(5a + 2)(a - 3)$   **27.** $(7x - 2)(2x + 3)$   **29.** $(5x - 2)(3x + 2)$
**31.** $(6x + 5)(2x + 3)$   **33.** $(6x + 1)(2x - 3)$   **35.** $(3x^2 + 1)(x^2 - 5)$   **37.** $(2x + 5y)(x + 3y)$   **39.** $(5x - 4y)(x + 4y)$
**41.** $5(2x + 1)(x - 3)$   **43.** $3x(2x - 5)(x + 4)$   **45.** $(5x - 2)(x + 1)$   **47.** $2(3x - 2)(2x - 5)$   **49.** $x(6x - 1)(2x - 3)$
**51.** $2(2x - 1)(2x + 7)$   **53.** 51.9%   **54.** $x = \dfrac{7}{2}$   **55.** (a) 17,600,000   (b) Approximately 10.8%   **56.** Approximately 27.4%
**57.** (a) 1.8 million   (b) Approximately 28.1%   **58.** (a) 2.1 million   (b) Approximately 35.6%

**Quick Quiz 5.4**   *See Examples noted with Ex.*   **1.** $(2x + 3)(6x - 1)$ (Ex. 3)   **2.** $(5x - 3)(2x - 3)$ (Ex. 4)   **3.** $3x(2x - 5)(x + 2)$
(Ex. 7)   **4.** See Student Solutions Manual

## Use Math to Save Money

**1.** \$5100 + \$3800 + \$3200 = \$12,100   **2.** \$5500 + \$4000 + \$3500 = \$13,000
**3.** \$12,100/\$13,000 ≈ 93%   **4.** \$13,000 × 0.5 = \$6500   **5.** \$12,100 − \$6500 = \$5600   **6.** \$13,000 × 0.33 = \$4290, \$6500 − \$4290 = \$2210

## 5.5 Exercises

**1.** $(10x + 1)(10x - 1)$   **3.** $(9x + 4)(9x - 4)$   **5.** $(x + 7)(x - 7)$   **7.** $(5x + 9)(5x - 9)$   **9.** $(x + 5)(x - 5)$
**11.** $(1 + 4x)(1 - 4x)$   **13.** $(4x + 7y)(4x - 7y)$   **15.** $(6x + 13y)(6x - 13y)$   **17.** $(10x + 9)(10x - 9)$   **19.** $(5a + 9b)(5a - 9b)$
**21.** $(3x + 1)^2$   **23.** $(y - 5)^2$   **25.** $(6x - 5)^2$   **27.** $(7x + 2)^2$   **29.** $(x + 7)^2$   **31.** $(5x - 4)^2$   **33.** $(9x + 2y)^2$   **35.** $(3x - 5y)^2$
**37.** $(4a + 9b)^2$   **39.** $(7x - 3y)^2$   **41.** $(8x + 5)^2$   **43.** $(12x + 1)(12x - 1)$   **45.** $(x^2 + 4)(x + 2)(x - 2)$   **47.** $(3x^2 - 4)^2$
**49.** No two binomials can be multiplied to obtain $9x^2 + 1$.   **51.** 9; one answer   **53.** $4(2x + 3)(2x - 3)$   **55.** $3(7x + y)(7x - y)$
**57.** $4(2x - 1)^2$   **59.** $2(7x + 3)^2$   **61.** $(x + 9)(x + 7)$   **63.** $(2x - 1)(x + 3)$   **65.** $3(2x + 3)(2x - 3)$   **67.** $(3x + 7)^2$
**69.** $9(2x - 1)^2$   **71.** $2(x - 9)(x - 7)$   **73.** $x^2 + 3x + 4 + \dfrac{-3}{x - 2}$   **74.** $2x^2 + x - 5$   **75.** 1.2 oz of greens; 1.05 oz of bulk vegetables;
0.75 oz of fruit   **76.** 1.44 oz of greens; 1.26 oz of bulk vegetables; 0.9 oz of fruit

**Quick Quiz 5.5**   *See Examples noted with Ex.*   **1.** $(7x + 9y)(7x - 9y)$ (Ex. 3)   **2.** $(3x - 8)^2$ (Ex. 6)   **3.** $2(9x + 10)(9x - 10)$
(Ex. 9)   **4.** See Student Solutions Manual

## 5.6 Exercises

**1.** $x(3x - 6y + 5)$   **3.** $(4x + 5y)(4x - 5y)$   **5.** Prime   **7.** $(x + 5)(x + 3)$   **9.** $(3x + 2)(5x - 1)$
**11.** $(a - 3c)(x + 3y)$   **13.** $(y + 7)^2$   **15.** $(2x - 3)^2$   **17.** $(2x - 3)(x - 4)$   **19.** $(x - 10y)(x + 7y)$   **21.** $(a + 3)(x - 5)$
**23.** $4x(2 + x)(2 - x)$   **25.** Prime   **27.** $3xy(z + 1)(z - 3)$   **29.** $3(x + 7)(x - 5)$   **31.** $5xy^3(x - 1)^2$   **33.** $-1(2x^2 + 1)(x + 2)(x - 2)$
**35.** Prime   **37.** $5x(x + 2)(x + 1)(x - 1)$   **39.** $5(x^2 + 2xy - 6y)$   **41.** $3x(2x + y)(5x - 2y)$   **43.** $2(5x - 3)(3x - 2)$   **45.** prime
**47.** \$28,000   **48.** 372 live strains   **49.** 66   **50.** $d = \dfrac{3y - 8x}{11}$   **51.** 34   **52.** $x \geq -15$

**Quick Quiz 5.6**   *See Examples noted with Ex.*   **1.** $(2x - 3)(3x - 4)$ (Ex. 1b)   **2.** $3(4x + 1)(5x - 2)$ (Ex. 1b)   **3.** Prime (Ex. 3)
**4.** See Student Solutions Manual

**5.7 Exercises**    **1.** $x = -2, x = 6$    **3.** $x = -12, x = -2$    **5.** $x = \dfrac{3}{2}, x = 2$    **7.** $x = \dfrac{2}{3}, x = \dfrac{3}{2}$    **9.** $x = 0, x = -13$

**11.** $x = 3, x = -3$    **13.** $x = 0, x = 1$    **15.** $x = \dfrac{2}{3}, x = 2$    **17.** $x = -3, x = 2$    **19.** $x = -\dfrac{1}{3}$    **21.** $x = 0, x = -5$

**23.** $x = -8, x = -2$    **25.** $x = -\dfrac{3}{2}, x = 4$    **27.** You can always factor out $x$.    **29.** $L = 14$ m; $W = 10$ m    **31.** 66 groups    **33.** 10 students

**35.** 3 sec; 12 m above ground after 2 sec    **37.** 2415 telephone calls    **39.** 136 handshakes    **41.** $-10x^5y^4$    **42.** $12a^{10}b^{13}$    **43.** $-\dfrac{3a^4}{2b^2}$    **44.** $\dfrac{1}{3x^5y^4}$

**Quick Quiz 5.7**    *See Examples noted with Ex.*    **1.** $x = \dfrac{1}{5}, x = \dfrac{1}{3}$ (Ex. 1)    **2.** $x = 3, x = -1$ (Ex. 4)    **3.** $x = 3, x = -\dfrac{3}{4}$ (Ex. 4)
**4.** See Student Solutions Manual

**Career Exploration Problems**    **1.** Rectangular piece: width is 7 ft and length is 18 ft. Triangular piece: base is 5 ft and height is 4 ft.

**2.** Base is 12 ft and height is 22 ft.    **3.** $\dfrac{44}{3}$ yd$^2$

**You Try It**    **1.** **(a)** $5a(a - 3)$   **(b)** $4x(x - 2y + 1)$   **(c)** $6x^2(x^2 - 3)$    **2.** $(3x^2 + 2)(a - 4)$    **3. (a)** $(x + 3)(x + 6)$
**(b)** $(x + 7)(x - 5)$   **(c)** $3(x + 1)(x - 4)$    **4.** $(2x + 3)(4x - 3)$    **5. (a)** $(3x + 4y)(3x - 4y)$   **(b)** $(9x^2 + 1)(3x + 1)(3x - 1)$
**(c)** $(4a + 3)^2$   **(d)** $(2x - 5y)^2$    **6. (a)** $4(x + 3)(x - 2)$   **(b)** $x(3x + 1)(x + 2)$   **(c)** $x(3x + 8)(3x - 8)$   **(d)** $3(4x - 1)^2$

**7. (a)** This is a sum of squares.   **(b)** There are no two factors of 2 that add to 1.    **8.** $x = \dfrac{3}{2}$ or $x = -1$    **9.** Length = 15 ft; width = 6 ft

**Chapter 5 Review Problems**    **1.** $4x^2(3x - 5y)$    **2.** $5x^3(2 - 7y)$    **3.** $8x^2y(3x - y - 2xy^2)$    **4.** $3a(a^2 + 2a - 3b + 4)$
**5.** $(a + 3b)(2a - 5)$    **6.** $3xy(5x^2 + 2y + 1)$    **7.** $(2x + 5)(a - 4)$    **8.** $(a - 4b)(a + 7)$    **9.** $(x^2 + 3)(y - 2)$
**10.** $3(2x - y)(5a + 7)$    **11.** $(5x - 1)(3x + 2)$    **12.** $(5w - 3)(6w + z)$    **13.** $(x + 9)(x - 3)$    **14.** $(x + 10)(x - 1)$
**15.** $(x + 6)(x + 8)$    **16.** $(x + 3y)(x + 5y)$    **17.** $(x^2 + 7)(x^2 + 6)$    **18.** $(x^2 - 7)(x^2 + 5)$    **19.** $6(x + 2)(x + 3)$
**20.** $2(x - 6)(x - 8)$    **21.** $(4x - 5)(x + 3)$    **22.** $(3x - 1)(4x + 5)$    **23.** $(2x - 3)(x + 1)$    **24.** $(3x - 4)(x + 2)$
**25.** $(10x - 1)(2x + 5)$    **26.** $(5x - 1)(4x + 5)$    **27.** $2(x - 1)(3x + 5)$    **28.** $2(x + 1)(3x - 5)$    **29.** $2(2x - 3)(x - 5)$
**30.** $4(x - 9)(x + 4)$    **31.** $(4x + 3y)(3x - 2y)$    **32.** $(3x - 5y)(2x + 5y)$    **33.** $(7x + y)(7x - y)$    **34.** $4(2x + 3y)(2x - 3y)$
**35.** $(y + 6x)(y - 6x)$    **36.** $(3y + 5x)(3y - 5x)$    **37.** $(6x + 1)^2$    **38.** $(5x - 2)^2$    **39.** $(4x - 3y)^2$    **40.** $(7x - 2y)^2$
**41.** $2(x + 4)(x - 4)$    **42.** $3(x + 3)(x - 3)$    **43.** $7(2x + 5)^2$    **44.** $8(3x - 4)^2$    **45.** $(2x + 3y)(2x - 3y)$    **46.** $(x + 15)(x - 2)$
**47.** $(3x - 4)(3x + 1)$    **48.** $10x^2y^2(5x + 2)$    **49.** $3(x - 3)^2$    **50.** $x(5x - 6)^2$    **51.** $(4x + 3)(x - 4)$    **52.** $x^3a(3a + 4x)(a - 5x)$
**53.** $2(3a + 5b)(2a - b)$    **54.** $(11a + 3b)^2$    **55.** $(a - 1)(7 - b)$    **56.** $(3x + 5y)(x + 1)(x - 1)$    **57.** $(3b - 7)(6 + c)$
**58.** $(5b + 8)(2 - 3x)$    **59.** $(b - 7)(5x + 4y)$    **60.** $(x^2 + 9y^6)(x + 3y^3)(x - 3y^3)$    **61.** $(3x^2 - 5)(2x^2 + 3)$    **62.** $yz(14 - x)(2 - x)$
**63.** $x(3x + 2)(4x + 3)$    **64.** $3(2w - 1)^2$    **65.** $2y(2y - 1)(y + 3)$    **66.** $9(x^2 + 4)(x - 2)(x + 2)$    **67.** Prime    **68.** Prime
**69.** $4y(2y^2 - 5)(y^2 + 3)$    **70.** $(4x^2y - 7)^2$    **71.** $(2x + 5)(a - 2b)$    **72.** $(2x + 1)(x + 3)(x - 3)$    **73.** $x = -5, x = 4$

**74.** $x = -6, x = \dfrac{1}{2}$    **75.** $x = 0, x = \dfrac{5}{2}$    **76.** $x = 4, x = -3$    **77.** $x = -5, x = \dfrac{1}{2}$    **78.** $x = -8, x = -3$    **79.** $x = -5, x = -9$

**80.** $x = -\dfrac{3}{5}, x = 2$    **81.** $x = -3$    **82.** $x = -3, x = \dfrac{3}{4}$    **83.** $x = \dfrac{1}{5}, x = 2$    **84.** Base = 5 in.; altitude = 10 in.    **85.** Width = 5 ft; length = 6 ft

**86.** 6 sec    **87.** 8 amperes, 12 amperes

**How Am I Doing? Chapter 5 Test**    **1.** $(x + 14)(x - 2)$ (obj. 5.3.1)    **2.** $(4x + 9)(4x - 9)$ (obj. 5.5.1)    **3.** $(5x + 1)(2x + 5)$
(obj. 5.4.2)    **4.** $(3a - 5)^2$ (obj. 5.5.2)    **5.** $x(7 - 9x + 14y)$ (obj. 5.1.1)    **6.** $(2x + 3b)(5y - 4)$ (obj. 5.2.1)    **7.** $2x(3x - 4)(x - 2)$ (obj.
5.6.1)    **8.** $c(5a - 1)(a - 2)$ (obj. 5.6.1)    **9.** $(9x + 10)(9x - 10)$ (obj. 5.5.1)    **10.** $(3x - 1)(3x - 4)$ (obj. 5.6.1)    **11.** $5(2x + 3)(2x - 3)$
(obj. 5.6.1)    **12.** Prime (obj. 5.6.2)    **13.** $x(3x + 5)(x + 2)$ (obj. 5.6.1)    **14.** $-5y(2x - 3y)^2$ (obj. 5.5.3)    **15.** $(9x + 1)(9x - 1)$ (obj. 5.5.1)
**16.** $(9y^2 + 1)(3y + 1)(3y - 1)$ (obj. 5.6.1)    **17.** $(x + 3)(2a - 5)$ (obj. 5.6.1)    **18.** $(a + 2b)(w + 2)(w - 2)$ (obj. 5.6.1)

**19.** $3(x - 6)(x + 5)$ (obj. 5.6.1)    **20.** $x(2x + 5)(x - 3)$ (obj. 5.6.1)    **21.** $x = -5, x = -9$ (obj. 5.7.1)    **22.** $x = -\dfrac{7}{3}, x = -2$ (obj. 5.7.1)

**23.** $x = -\dfrac{5}{2}, x = 2$ (obj. 5.7.1)    **24.** $x = 7, x = -4$ (obj. 5.7.1)    **25.** Width = 7 mi; length = 13 mi (obj. 5.7.2)

**Chapter 6**    **6.1 Exercises**    **1.** 4    **3.** $\dfrac{6}{x}$    **5.** $\dfrac{3x + 1}{1 - 3x}$    **7.** $\dfrac{a(a - 2b)}{2b}$    **9.** $\dfrac{x + 2}{x}$    **11.** $\dfrac{x - 5}{3x - 1}$    **13.** $\dfrac{x - 3}{x(x - 7)}$

**15.** $\dfrac{3x + 1}{x + 5}$    **17.** $\dfrac{3x - 5}{4x - 1}$    **19.** $\dfrac{x - 5}{x + 1}$    **21.** $-\dfrac{3}{5x}$    **23.** $-\dfrac{2x + 3}{5 + x}$    **25.** $\dfrac{3x + 4}{3x - 1}$    **27.** $\dfrac{3x - 2}{4 - x}$    **29.** $\dfrac{u - 2b}{3a - b}$    **31.** $9x^2 - 42x + 49$

**32.** $49x^2 - 36y^2$    **33.** $2x^2 - 5x - 12$    **34.** $2x^3 - 9x^2 - 2x + 24$    **35.** $\dfrac{23a^2}{7} + \dfrac{3b}{4}$    **36.** $-\dfrac{49}{6}$ or $-8\dfrac{1}{6}$    **37.** $1\dfrac{5}{8}$ acres    **38.** 6 hr, 25 min

**Quick Quiz 6.1**    *See Examples noted with Ex.*    **1.** $\dfrac{x}{x - 5}$ (Ex. 4)    **2.** $-\dfrac{2}{b}$ (Ex. 5)    **3.** $\dfrac{2x - 1}{4x + 5}$ (Ex. 3)    **4.** See Student Solutions Manual

**6.2 Exercises**    **1.** factor the numerators and denominators completely and divide out common factors    **3.** $\dfrac{4(x + 5)}{x + 3}$    **5.** $\dfrac{3x^2}{4(x - 3)}$    **7.** $\dfrac{x - 2}{x + 3}$

**9.** $\dfrac{(x + 6)(x + 2)}{x + 5}$    **11.** $x + 3$    **13.** $\dfrac{3(x + 2y)}{4(x + 3y)}$    **15.** $\dfrac{(x + 5)(x - 2)}{3x - 1}$    **17.** $\dfrac{-5(x + 3)}{2(x - 2)}$ or $\dfrac{5(x + 3)}{-2(x - 2)}$ or $-\dfrac{5(x + 3)}{2(x - 2)}$    **19.** 1

**21.** $\dfrac{x + 4}{x + 8}$    **23.** $x = -8$    **24.** $\dfrac{19}{28}$    **25.** $79.5(8981)\$x = \$713{,}989.5x$    **26.** Harold's was 6 ft by 6 ft. George's was 9 ft by 4 ft.

**Quick Quiz 6.2**   *See Examples noted with Ex.*   **1.** $\dfrac{2(x+4)}{x-4}$ (Ex. 1)   **2.** $\dfrac{x-25}{x+5}$ (Ex. 1)   **3.** $\dfrac{2(x-3)}{3x}$ (Ex. 3)
**4.** See Student Solutions Manual

**6.3 Exercises**   **1.** The LCD would be a product that contains each factor. However, any repeated factor in any one denominator must be repeated the greatest number of times it occurs in any one denominator. So the LCD would be $(x+5)(x+3)^2$.   **3.** $\dfrac{2(2x+1)}{2x+5}$   **5.** $\dfrac{2x-5}{x+3}$

**7.** $\dfrac{2x-7}{5x+7}$   **9.** $3a^2b^3$   **11.** $90x^3y^5$   **13.** $18(x-3)$   **15.** $(x+3)(x-3)$   **17.** $(x+5)(3x-1)^2$   **19.** $\dfrac{7+3a}{ab}$

**21.** $\dfrac{3x-13}{(x+7)(x-7)}$   **23.** $\dfrac{y(5y-3)}{(y+1)(y-1)}$   **25.** $\dfrac{43a+12}{5a(3a+2)}$   **27.** $\dfrac{4z+x}{6xyz}$   **29.** $\dfrac{9x+14}{2(x-3)}$   **31.** $\dfrac{x+10}{(x+5)(x-5)}$   **33.** $\dfrac{3a+17b}{10}$

**35.** $\dfrac{-2(2x-17)}{(2x-3)(x+2)}$   **37.** $\dfrac{-7x}{(x+3)(x-1)(x-4)}$   **39.** $\dfrac{4x+23}{(x+4)(x+5)(x+6)}$   **41.** $\dfrac{4x-13}{(x-4)(x-3)}$   **43.** $\dfrac{11x}{y-2x}$

**45.** $\dfrac{-y(17y+10)}{(4y-1)(2y+1)(y-5)}$   **47.** $\dfrac{8x+3}{(2x+1)(x-3)}$   **49.** $x=-7$   **50.** $x=4$   **51.** $x>\dfrac{5}{7}$   **52.** $81x^{12}y^{16}$

**53.** At least 17 days   **54.** 287,985 people

**Quick Quiz 6.3**   *See Examples noted with Ex.*   **1.** $\dfrac{2x+7}{(x+2)(x-4)}$ (Ex. 6)   **2.** $\dfrac{bx+by+xy}{bxy}$ (Ex. 8)   **3.** $\dfrac{5x-1}{(x+3)(x-3)(x+4)}$ (Ex. 7)
**4.** See Student Solutions Manual

**Use Math to Save Money**   **1.** $\$500 \times 0.15 = \$75$   **2.** $\$500 \times 0.25 = \$125$   **3.** $\$500/\$5000 = 10\%$   **4.** $\$9 \times 48 = \$432$

**6.4 Exercises**   **1.** $\dfrac{5x}{4x+3}$   **3.** $\dfrac{4b+a}{5}$   **5.** $\dfrac{x(x-2)}{4+5x}$   **7.** $\dfrac{2}{3x+10}$   **9.** $\dfrac{1}{xy}$   **11.** $\dfrac{2x-1}{x}$   **13.** $\dfrac{x-6}{x+5}$   **15.** $\dfrac{3(a^2+3)}{a^2+2}$

**17.** $\dfrac{3(x+3)}{2x-5}$   **19.** $\dfrac{4x-1}{7-4x}$   **21.** No expression in any denominator can be zero because division by zero is undefined. So $-3$, $5$, and $0$ are not allowed.

**23.** $\dfrac{5y(x+5y)^2}{x(x-6y)}$   **25.** $y=\dfrac{-5x+8}{6}$   **26.** $x>-1$   **27.** 6   **28.** $\$24,000$

**Quick Quiz 6.4**   *See Examples noted with Ex.*   **1.** $\dfrac{a(3a-4b)}{3(5a-16b)}$ (Ex. 2)   **2.** $ab$ (Ex. 2)   **3.** $\dfrac{2}{x-1}$ (Ex. 4)
**4.** See Student Solutions Manual

**6.5 Exercises**   **1.** $x=-12$   **3.** $x=2$   **5.** $x=15$   **7.** $x=10$   **9.** $x=-5$   **11.** $x=-\dfrac{14}{3}$ or $-4\dfrac{2}{3}$   **13.** $x=-2$

**15.** $x=8$   **17.** $x=-1$   **19.** No solution   **21.** $x=-3$   **23.** $x=-9$   **25.** No solution   **27.** $x=4$   **29.** No solution

**31.** $x=-6, x=-\dfrac{1}{3}$   **33.** $(4x+1)(2x-1)$   **34.** $x=\dfrac{5}{2}$ or $2\dfrac{1}{2}$ or $2.5$   **35.** Width $=10$ in.; length $=12$ in.   **36.** $\$946.45$

**Quick Quiz 6.5**   *See Examples noted with Ex.*   **1.** $x=\dfrac{5}{24}$ (Ex. 1)   **2.** $x=2$ (Ex. 2)   **3.** No solution (Ex. 4)
**4.** See Student Solutions Manual

**6.6 Exercises**   **1.** $x=\dfrac{88}{5}$ or $17\dfrac{3}{5}$ or $17.6$   **3.** $x=\dfrac{204}{5}$ or $40\dfrac{4}{5}$ or $40.8$   **5.** $x=6.5$   **7.** $x=\dfrac{91}{4}$ or $22\dfrac{3}{4}$ or $22.75$

**9. (a)** 650 New Zealand dollars   **(b)** 75 New Zealand dollars less   **11.** 56 mph   **13.** 29 mi   **15.** $n=\dfrac{377}{20}$ or $18\dfrac{17}{20}$ in.   **17.** $y=200$ m

**19.** $k=\dfrac{56}{5}$ or $11\dfrac{1}{5}$ ft   **21.** $k=\dfrac{768}{20}$ or $38\dfrac{2}{5}$ m   **23.** 48 in.   **25.** 35 in.   **27.** 61.5 mph   **29.** Commuter airliner $=250$ km/hr   **31. (a)** $\$0.17$
Helicopter $=210$ km/hr

**(b)** $\$0.14$   **(c)** $\$5.48$   **33.** $3\dfrac{3}{7}$ hr or 3 hr, 26 min   **35.** $8.92465 \times 10^{-4}$   **36.** 6,830,000,000   **37.** $\dfrac{w^8}{x^3y^2z^4}$   **38.** $\dfrac{27}{8}$ or $3\dfrac{3}{8}$

**Quick Quiz 6.6**   *See Examples noted with Ex.*   **1.** $x=55$ (Ex. 1)   **2.** 24 ft (Ex. 4)   **3.** 172 flights (Ex. 1)
**4.** See Student Solutions Manual

**Career Exploration Problems**   **1.** $C_{\text{ave}}=\dfrac{\$350n+\$22,000}{n}$   **2.** 8 hours   **3.** $\$98.75$ per hour for 50 treadmills; $\$71.25$ per hour for 100 treadmills

**You Try It**   **1.** $\dfrac{2(x-5)}{x-3}$   **2.** $\dfrac{x+y}{2x-y}$   **3.** $\dfrac{2x-5}{2}$   **4.** $\dfrac{x-2}{2(x-3)}$   **5.** 6   **6.** $x-1$   **7.** No solution   **8.** 15.75 ft

**Chapter 6 Review Problems**   **1.** $\dfrac{x}{x-y}$   **2.** $-\dfrac{4}{5}$   **3.** $\dfrac{x}{x+3}$   **4.** $\dfrac{2x-3}{5-x}$   **5.** $\dfrac{2(x-4y)}{2x-y}$   **6.** $\dfrac{2-y}{3y-1}$   **7.** $\dfrac{x-2}{(5x+6)(x-1)}$

**8.** $2(2x+y)$   **9.** $\dfrac{x+6}{6(x-1)}$   **10.** $\dfrac{2y-1}{5y}$   **11.** $\dfrac{4y(3y-1)}{(3y+1)(2y+5)}$   **12.** $\dfrac{3y(4x+3)}{2(x-5)}$   **13.** $\dfrac{2(x+3y)}{x-4y}$   **14.** $\dfrac{3}{16}$   **15.** $\dfrac{4(5y+1)}{3y(y+2)}$

**16.** $\dfrac{3x^2+6x+1}{x(x+1)}$   **17.** $\dfrac{2(5x-11)}{(x+2)(x-4)}$   **18.** $\dfrac{(x-1)(x-2)}{(x+3)(x-3)}$   **19.** $\dfrac{2xy+4x+5y+6}{2y(y+2)}$   **20.** $\dfrac{(2a+b)(a+4b)}{ab(a+b)}$   **21.** $\dfrac{3x-2}{3x}$

**22.** $\dfrac{2x^2+7x-2}{2x(x+2)}$   **23.** $\dfrac{3}{2(x-9)}$   **24.** $\dfrac{1-2x-x^2}{(x+5)(x+2)}$   **25.** $-\dfrac{4}{9}$   **26.** $\dfrac{22}{5x^2}$   **27.** $w-2$   **28.** 1   **29.** $-\dfrac{y^2}{2}$

**30.** $\dfrac{x+2y}{y(x+y+2)}$   **31.** $\dfrac{-1}{a(a+b)}$ or $-\dfrac{1}{a(a+b)}$   **32.** $\dfrac{-3a-b}{b}$ or $-\dfrac{3a+b}{b}$   **33.** $a=2$   **34.** $a=15$   **35.** $x=\dfrac{2}{7}$   **36.** $x=-2$

**37.** $x = \dfrac{1}{2}$   **38.** No solution   **39.** $x = -\dfrac{12}{5}$ or $-2\dfrac{2}{5}$ or $-2.4$   **40.** $y = 2$   **41.** $y = -2$   **42.** No solution   **43.** $y = \dfrac{5}{4}$ or $1\dfrac{1}{4}$ or $1.25$

**44.** $y = 0$   **45.** $x = \dfrac{14}{5}$ or $2\dfrac{4}{5}$ or $2.8$   **46.** $x = \dfrac{5}{4}$ or $1\dfrac{1}{4}$ or $1.25$   **47.** $x = \dfrac{132}{5}$ or $26\dfrac{2}{5}$ or $26.4$   **48.** $x = 6$   **49.** $x = 16$

**50.** $x = \dfrac{91}{10}$ or $9\dfrac{1}{10}$ or $9.1$   **51.** 8.3 gal   **52.** 167 cookies   **53.** 240 mi   **54.** Train = 60 mph   **55.** 182 ft   **56.** 1200 ft
Car = 40 mph

**57.** $2\dfrac{2}{5}$ hr or 2 hr, 24 min   **58.** 12 hr   **59.** $-\dfrac{a+4}{3a^2}$   **60.** $\dfrac{4a^2}{2a+3}$   **61.** $\dfrac{x-2y}{x+2y}$   **62.** $\dfrac{x+6}{x}$   **63.** $\dfrac{x+6}{x}$   **64.** $\dfrac{(b-a)(b+a)}{ab(x+y)}$

**65.** $\dfrac{1}{3}$   **66.** $-\dfrac{6}{y^2}$   **67.** $\dfrac{y(x+3y)}{2(x+2y)}$   **68.** $x = 12$   **69.** $x = -32$   **70.** $b = -2$

## How Am I Doing? Chapter 6 Test
**1.** $\dfrac{2}{3a}$ (obj. 6.1.1)   **2.** $\dfrac{2x^2(2-y)}{(y+2)}$ (obj. 6.1.1)   **3.** $\dfrac{5}{12}$ (obj. 6.2.1)   **4.** $\dfrac{1}{3y(x-y)}$ (obj. 6.2.1)

**5.** $\dfrac{2a+1}{a+2}$ (obj. 6.2.2)   **6.** $\dfrac{3a+4}{(a+1)(a-2)}$ (obj. 6.3.3)   **7.** $\dfrac{x-a}{ax}$ (obj. 6.3.3)   **8.** $-\dfrac{x+2}{x+3}$ (obj. 6.3.3)   **9.** $\dfrac{x}{4}$ (obj. 6.4.2)   **10.** $\dfrac{6x}{5}$ (obj. 6.4.1)

**11.** $\dfrac{2x-3y}{4x+y}$ (obj. 6.1.1)   **12.** $\dfrac{x}{(x+2)(x+4)}$ (obj. 6.3.3)   **13.** $x = -\dfrac{1}{5}$ (obj. 6.5.1)   **14.** $x = 4$ (obj. 6.5.1)   **15.** No solution (obj. 6.5.2)

**16.** $x = \dfrac{47}{6}$ (obj. 6.5.1)   **17.** $x = \dfrac{45}{13}$ (obj. 6.6.1)   **18.** $x = 37.2$ (obj. 6.6.1)   **19.** 151 flights (obj. 6.6.1)   **20.** \$368 (obj. 6.6.1)   **21.** 102 ft (obj. 6.6.2)

## Cumulative Test for Chapters 0–6
**1.** $-\dfrac{1}{3}$ [1.1.5]   **2.** $-\dfrac{6}{7}$ [1.3.3]   **3.** 66 [1.8.1]   **4.** \$112.50 [3.4.2]   **5.** $x = 6$ [2.3.3]

**6.** $h = \dfrac{A}{\pi r^2}$ [2.5.1]   **7.** $x > 1.25$ [2.6.4]   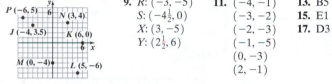   **8.** 15 [3.2.1]   **9.** Length = 13 ft, width = 8 ft [3.3.1]

**10.** \$42,000 [3.4.2]   **11.** $(4x^2 + b^2)(2x + b)(2x - b)$ [5.5.1]   **12.** $2a(4a + b)(a - 5b)$ [5.4.3]   **13.** $-\dfrac{4a^2}{b^6}$ [4.2.1]   **14.** $5.6 \times 10^{-4}$ [4.2.2]

**15.** $9x^2 - 30x + 25$ [4.5.2]   **16.** $\dfrac{(x-2)(3x+1)}{3x(x+5)}$ [6.2.1]   **17.** $\dfrac{11x-3}{2(x+2)(x-3)}$ [6.3.3]   **18.** $x = -\dfrac{9}{2}$ or $-4\dfrac{1}{2}$ [6.5.1]   **19.** $\dfrac{ab(3b+2a)}{5b^2 - 2a^2}$ [6.4.1]

## Chapter 7   7.1 Exercises
**1.** 0   **3.** The order in which you write the numbers matters. The graph of $(5, 1)$ is not the same as the graph of $(1, 5)$.   **5.** They are not the same because the $x$ and $y$ coordinates are different. To plot $(2, 7)$ we move 2 units to the right on the $x$-axis, but for the ordered pair $(7, 2)$ we move 7 units to the right on the $x$-axis. Then to plot $(2, 7)$ we move 7 units up, parallel to the $y$-axis, but for the ordered pair $(7, 2)$ we move 2 units up.   **7.**

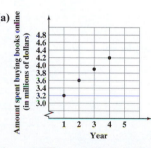

**9.** $R: (-3, -5)$
$S: (-4\frac{1}{2}, 0)$
$X: (3, -5)$
$Y: (2\frac{1}{2}, 6)$

**11.** $(-4, -1)$
$(-3, -2)$
$(-2, -3)$
$(-1, -5)$
$(0, -3)$
$(2, -1)$

**13.** B5   **15.** E1   **17.** D3

**19. (a)**

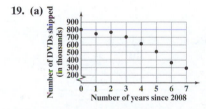

**(b)** The number of DVDs shipped decreased overall between 2008 and 2015, with a slight increase in 2010.

**21. (a)**

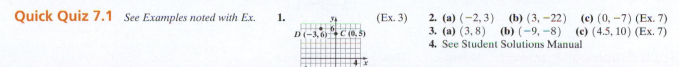

**(b)** An estimated \$4.5 billion will be spent buying books online in year 5.

**23.** No   **25.** Yes   **27.** Jon is right because for the ordered pair $(5, 3)$, $x = 5$ and $y = 3$. When we substitute these values in the equation we get $2(5) - 2(3) = 4$. For the ordered pair $(3, 5)$, when we replace $x = 3$ and $y = 5$ in the equation we get $2(3) - 2(5)$, which equals $-4$, not 4.
**29. (a)** $(0, 7)$   **(b)** $(2, 15)$   **31. (a)** $(-1, 11)$   **(b)** $(3, -13)$   **33. (a)** $(-3, -5)$   **(b)** $(5, 1)$   **35. (a)** $(-2, 0)$   **(b)** $(-4, 3)$   **37. (a)** $(7, 3)$
**(b)** $\left(-1, \dfrac{5}{7}\right)$   **39. (a)** $(2, 2)$   **(b)** $\left(\dfrac{3}{2}, 5\right)$   **41.** $\approx 1133.54$ yd$^2$   **42.** 12   **43.** $2(2x + 3)(2x - 3)$   **44.** $3(x + 6)(x - 3)$

## Quick Quiz 7.1   *See Examples noted with Ex.*
**1.**

(Ex. 3)   **2. (a)** $(-2, 3)$   **(b)** $(3, -22)$   **(c)** $(0, -7)$ (Ex. 7)
**3. (a)** $(3, 8)$   **(b)** $(-9, -8)$   **(c)** $(4.5, 10)$ (Ex. 7)
**4.** See Student Solutions Manual

## To Think About, page 411
By choosing multiples of 4 as replacements for $x$, we get integers as the corresponding $y$-values.

**7.2 Exercises** **1.** No, replacing $x$ by $-2$ and $y$ by 5 in the equation does not result in a true statement. **3.** $x$-axis

**5.** $y = x - 4$
$(0, -4)$
$(2, -2)$
$(4, 0)$

**7.** $y = -2x + 1$
$(0, 1)$
$(-2, 5)$
$(1, -1)$

**9.** $y = 3x - 1$
$(0, -1)$
$(2, 5)$
$(-1, -4)$

**11.** $y = 2x - 5$
$(0, -5)$
$(2, -1)$
$(4, 3)$

**13.** $y = -x + 3$

**15.** $3x - 2y = 0$

**17.** $y = -\frac{3}{4}x + 3$

**19.** $4x + 6 + 3y = 18$

**21.** $y = 6 - 2x$
**(a)** $(3, 0), (0, 6)$
**(b)**

**23.** $x + 3 = 6y$
**(a)** $(-3, 0), \left(0, \frac{1}{2}\right)$
**(b)**

**25.** $x = 4$

**27.** $y - 2 = 3y$

**29.** $2x + 5y - 2 = -12$

**31.** $2x + 9 = 5x$

**33.**

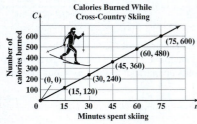

**35.**

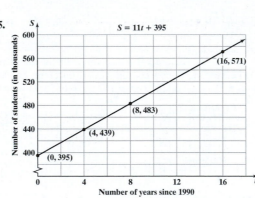

**37.** $x = -2$

**38.** $\dfrac{4y^2}{9x^4}$

**39.** $7.8 \times 10^{-5}$

**40.** $2x^2 + 6x - 4$

**Quick Quiz 7.2** *See Examples noted with Ex.* **1.** (Ex. 1)

$y = \frac{1}{4}x + 2$

**2.** (Ex. 5)

$y = 4$

**3. (a)** $(2, 0), (0, 4)$ (Ex. 3)
**(b)**

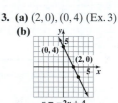

$y = -2x + 4$

**4.** See Student Solutions Manual

**To Think About, page 421** Since the slope is negative, we would expect the $y$-values decrease as you go from left to right. In other words, the line will go down to the right, as we can see in the graph.

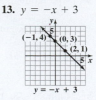

**You Try It Graphing Organizer** Method 1: Method 2: Method 3:

$m = -3; b = 2$

**7.3 Exercises**   **1.** No, division by zero is impossible, so the slope is undefined.   **3.** 2   **5.** $-5$   **7.** $\frac{3}{5}$   **9.** $-\frac{1}{4}$   **11.** $-\frac{4}{3}$   **13.** 0

**15.** $-\frac{16}{5}$   **17.** $m = 8;\ (0, 9)$   **19.** $m = -3;\ (0, 4)$   **21.** $m = -\frac{8}{7};\left(0, \frac{3}{4}\right)$   **23.** $m = -6;\ (0, 0)$   **25.** $m = 0;\ (0, -2)$

**27.** $m = \frac{7}{3};\left(0, -\frac{4}{3}\right)$   **29.** $y = \frac{3}{5}x + 3$   **31.** $y = 4x - 5$   **33.** $y = -x$   **35.** $y = -\frac{5}{4}x - \frac{3}{4}$

**37.**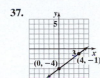
$y = \frac{3}{4}x - 4$

**39.**
$y = -\frac{5}{3}x + 2$

**41.**
$y = \frac{2}{3}x + 2$

**43.**
$y + 2x = 3$

**45.**
$y = 2x$

**47.** **(a)** $\frac{5}{6}$   **(b)** $-\frac{6}{5}$

**49.** **(a)** $-8$   **(b)** $\frac{1}{8}$   **51.** **(a)** $\frac{2}{3}$   **(b)** $-\frac{3}{2}$   **53.** Yes; $2x - 3y = 18$   **55.** **(a)** $y = 35x + 625$   **(b)** $m = 35;\ (0, 625)$

**(c)** The amount of increase (in thousands) in the number of cell phone accessories in the U.S. per year from 2010 to 2020.

**57.** $x < \frac{12}{5}$

**58.** $x \le 24$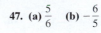

**Quick Quiz 7.3**   *See Examples noted with Ex.*   **1.** $\frac{1}{2}$ (Ex. 1)   **2.** **(a)** $m = -3;\ (0, 2)$ (Ex. 4)

**(b)** 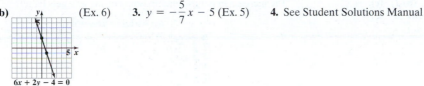 (Ex. 6)   **3.** $y = -\frac{5}{7}x - 5$ (Ex. 5)   **4.** See Student Solutions Manual

$6x + 2y - 4 = 0$

**Use Math to Save Money**   **1.** $\$42{,}000/12 = \$3500$   **2.** $\$3500 \times 0.05 = \$175$   **3.** $FV = \$175 \times [1.0067^{480} - 1]/0.0067 \approx \$618{,}044$
**4.** $\$618{,}044 \times 0.04 \approx \$24{,}722$   **5.** $\$24{,}722/12 \approx \$2060$   **6.** $\$3500 - \$2060 = \$1440$   **7.** Yes, he needs to increase the amount.

**7.4 Exercises**   **1.** $y = 3x + 6$   **3.** $y = -2x + 11$   **5.** $y = -3x + \frac{7}{2}$   **7.** $y = \frac{1}{4}x + 4$   **9.** $y = -2x - 6$   **11.** $y = -4x + 2$

**13.** $y = 5x - 10$   **15.** $y = \frac{1}{3}x + \frac{1}{2}$   **17.** $y = -3x$   **19.** $y = -3x + 3$   **21.** $y = -\frac{2}{3}x + 1$   **23.** $y = \frac{2}{3}x - 4$   **25.** $y = -\frac{2}{3}x$

**27.** $y = -2$   **29.** $y = -5$   **31.** $x = 4$   **33.** $y = \frac{1}{3}x + 5$   **35.** $y = -\frac{1}{2}x + 4$   **37.** $y = 2.4x + 227$   **39.** $t = \frac{1}{3}$   **40.** $\frac{(2x + 5)(x - 3)}{(2x - 5)^2}$

**41.** $\$61.20$   **42.** 290 min

**Quick Quiz 7.4**   *See Examples noted with Ex.*   **1.** $y = \frac{2}{3}x - 7$ (Ex. 1)   **2.** $y = 6x + 19$ (Ex. 2)   **3.** $x = 4$; the slope is undefined (Ex. 2)
**4.** See Student Solutions Manual

**7.5 Exercises**   **1.** No, all points in one region will be solutions to the inequality while all points in the other region will not be solutions. Thus testing any point will give the same result, as long as the point is not on the boundary line.

**3.**
$y \ge 4x$

**5.**
$2x - 3y < 6$

**7.**
$2x - y \ge 3$

**9.**
$y < 2x - 4$

**11.**
$y < -\frac{1}{2}x$

**13.**
$x \ge 2$

**15.**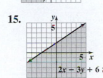
$2x - 3y + 6 \ge 0$

**17.**
$x > -2y$

**19.**
$2x > 3 - y$

**21.**
$2x \ge -3y$

**23.** 7   **24.** $-7$   **25.** $-28$   **26.** $-20$
**27.** $\$95.20$   **28.** $\$19{,}040$

**Quick Quiz 7.5**   *See Examples noted with Ex.*   **1.** Use a dashed line. If the inequality has a $<$ or a $>$ symbol, the points on the line itself are not included. This is indicated by a dashed line. (Ex. 1)   **2.** 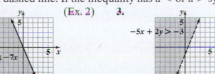 (Ex. 2)   **3.** 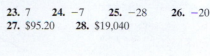 (Ex. 1)

$3y \le -7x$

$-5x + 2y > -3$

**4.** See Student Solutions Manual

**To Think About, page 443** The relation is a function since for each time, there is exactly one bus stop. Since the location of the bus depends on the time, the bus stop is the dependent variable and time is the independent variable.

**7.6 Exercises** **1.** Using a table of values, an algebraic equation, or a graph **3.** possible values; independent
**5.** If a vertical line can intersect the graph more than once, the relation is not a function. If no such line exists, then the relation is a function.

**7. (a)** Domain = $\left\{ -3, \frac{3}{7}, 3 \right\}$ **(b)** Not a function   **9. (a)** Domain = $\{0, 3, 6\}$   **(b)** Function
   Range = $\left\{ -1, \frac{3}{7}, 4 \right\}$   Range = $\{0.5, 1.5, 2.5\}$

**11. (a)** Domain = $\{1, 9, 12, 14\}$   **(b)** Function   **13. (a)** Domain = $\{3, 5, 7\}$   **(b)** Not a function
   Range = $\{1, 3, 12\}$   Range = $\{75, 85, 95, 100\}$

**15.**  $y = x^2 + 3$ **17.**  $y = 2x^2$ **19.** $x = -2y^2$ **21.** $x = y^2 - 4$ **23.** $y = \dfrac{2}{x}$ **25.** $y = \dfrac{4}{x^2}$

**27.** $x = (y+1)^2$ **29.** $y = \dfrac{4}{x-2}$

**31.** Function   **33.** Not a function   **43.** $f(0) = 31.6$, $f(4) = 32.24$, $f(10) = 34.4$
**35.** Function   **37.** Not a function   The curve slopes more steeply for larger values of $x$.
**39. (a)** 26 **(b)** 2 **(c)** $-4$   The increase in pet ownership is increasing as $x$ gets larger.
**41. (a)** 3 **(b)** 24 **(c)** 9

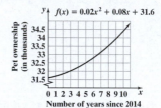

**45.** $-8x^3 + 12x^2 - 32x$   **46.** $5a^2b + 30ab - 10a^2$
**47.** $-19x + 2y - 2$   **48.** $9x^2y - 6xy^2 + 7xy$

**Quick Quiz 7.6** *See Examples noted with Ex.* **1.** No. Two different ordered pairs have the same first coordinate. (Ex. 2)

**2. (a)** 41 **(b)** 34 (Ex. 8)   **3. (a)** $-7$ **(b)** $-\dfrac{7}{8}$ (Ex. 8)   **4.** See Student Solutions Manual

**Career Exploration Problems** **1.**

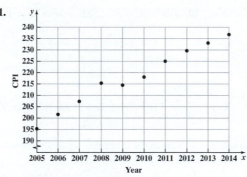

**2.** 2005–2008: \$6.67;
   2008–2009: $-$\$0.76; 2010–2014: \$4.67
**3.** $y = 4.60x - 9027.66$
**4.** 2016: $y = 4.60(2016) - 9027.66$
   $= \$245.94$; 2020: $y = 4.60(2020) - 9027.66$
   $= \$264.34$

**You Try It** **1.** $2x - y = 6$ **2. (a)** $x$-intercept: $(4, 0)$ $y$-intercept: $(0, -2)$ **(b)** $-2x + 4y = -8$ **3. (a)** $x = -2$ **(b)** $y = 5$ **4.** $-2$ **5.** Slope $= -\dfrac{5}{3}$; $y$-intercept $= (0, 3)$ **6.** $y = 2x - 3$

**7.** $y = 3x - 4$ **8. (a)** 3 **(b)** $-\dfrac{1}{3}$ **9.** $y = -\dfrac{1}{2}x + \dfrac{7}{2}$ **10.** $y = \dfrac{1}{2}x + \dfrac{1}{2}$ **11.** $y \le 2x - 4$ **12.** Yes **13.** $y = (x+2)^2$

**14.** No   **15. (a)** 20 **(b)** 6

## Chapter 7 Review Problems

**1.**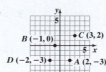

**2.**

**3.** (a) $(0, 7)$　(b) $(-1, 10)$
**4.** (a) $(1, 2)$　(b) $(-4, 4)$
**5.** (a) $(6, -1)$　(b) $(6, 3)$

**6.** 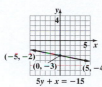　**7.** 　**8.**

**9.** $m = -\dfrac{5}{6}$　**10.** $m = -\dfrac{5}{3}$　**11.** $m = \dfrac{9}{11}; \left(0, \dfrac{15}{11}\right)$　**12.** $y = -\dfrac{1}{2}x + 3$

**13.** 　**14.** 　**15.**

**16.** $y = -6x + 14$　**17.** $y = -\dfrac{1}{3}x + \dfrac{11}{3}$　**18.** $y = x + 3$　**19.** $y = 7$

**20.** $y = \dfrac{2}{3}x - 3$　**21.** $y = -3x + 1$　**22.** $x = 5$

**23.** 　**24.** 　**25.**

**26.** Domain: $\{-6, -5, 5\}$
Range: $\{-6, 5\}$
not a function

**27.** Domain: $\{-2, 2, 5, 6\}$
Range: $\{-3, 4\}$
function

**28.** Function　**29.** Not a function　**30.** Function

**31.** 　**32.** 　**33.**

**34.** (a) 7　(b) 31　**35.** (a) $-1$　(b) $-5$　**36.** (a) 1　(b) $\dfrac{1}{5}$
**37.** (a) 0　(b) 4

**38.** 　**39.** 　**40.**

**41.** $-\dfrac{2}{5}$　**42.** $m = -\dfrac{7}{6}$; $y$-intercept $= \left(0, \dfrac{5}{3}\right)$　**43.** $y = \dfrac{2}{3}x - 7$

**44.** $y = -x + 3$　**45.** \$210　**46.** \$174　**47.** $y = 0.09x + 30$; $(0, 30)$; it tells us that if Russ and Norma use no electricity, the minimum cost will be \$30.
**48.** $m = 0.09$; the electric bill increases \$0.09 for each kilowatt-hour of use.　**49.** 1300 kilowatt-hours　**50.** 2400 kilowatt-hours
**51.** 17,020,000 people in 1994
13,254,000 people in 2008
11,640,000 people in 2014

**52.**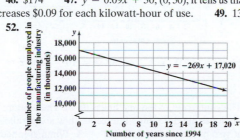

**53.** The slope is $-269$. The slope tells us that the number of people employed in manufacturing decreases each year by 269 thousand. In other words, employment in manufacturing goes down 269,000 people each year.　**54.** The $y$-intercept is $(0, 17,020)$. This tells us that in the year 1994, the number of manufacturing jobs was 17,020 thousands, which is 17,020,000.　**55.** 2016　**56.** 2019

## How Am I Doing? Chapter 7 Test

**1.** 　(obj. 7.1.1)　**2.** 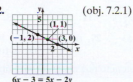　(obj. 7.2.1)　**3.** 　(obj. 7.2.3)　**4.** 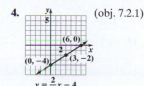　(obj. 7.2.1)

5. (a) $x$-intercept: $(-2, 0)$
   $y$-intercept: $(0, -4)$

6. $m = 1$ (obj. 7.3.1)    7. $m = -\dfrac{3}{2}; \left(0, \dfrac{5}{2}\right)$ (obj. 7.3.2)    8. $y = \dfrac{3}{4}x - 6$ (obj. 7.3.3)

(b) (obj. 7.2.2)

$4x + 2y = -8$

9. (a) $y = \dfrac{1}{2}x - 4$ (obj. 7.4.1)   (b) $-2$ (obj. 7.3.5)   10. $y = -\dfrac{3}{2}x + \dfrac{7}{2}$ (obj. 7.4.2)

11. (obj. 7.5.1)    12. (obj. 7.5.1)

$4y \le 3x$          $-3x - 2y > 10$

13. No, two different ordered pairs have the same first coordinate. (obj. 7.6.1)

14. Yes, any vertical line passes through no more than one point on the graph. (obj. 7.6.3)

15. (obj. 7.6.2)    16. (a) $-3$   (b) $-3$ (obj. 7.6.4)

$y = 2x^2 - 3$

# Chapter 8   8.1 Exercises

1. The lines are parallel. One line intersects the $y$-axis at $-5$. The other line intersects the $y$-axis at 6. There is no solution. The system is inconsistent. That is the solution to the system.    3. intersect, one    5. The graphs of the two lines will intersect at one point. That point is the $y$-intercept.

7.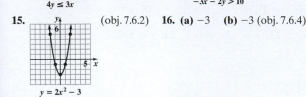
$x + y = 5$, $(4, 1)$, $x - y = 3$

9. $y = 2x + 5$   $y = -3x$, $(-1, 3)$

11. $3x - 2y = 12$, $(2, -3)$, $4x + y = 5$

13.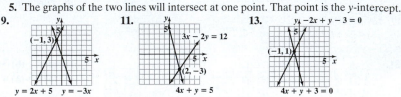
$-2x + y - 3 = 0$, $(-1, 1)$, $4x + y + 3 = 0$

15. $2x + 3y = 14$, $(-2, 6)$, $3x - 2y = -18$

17. $y = \dfrac{3}{4}x + 7$, $(-4, 4)$, $y = -\dfrac{1}{2}x + 2$

19. inconsistent system of equations, $-9x + 6y = -9$, $3x - 2y = -4$

21. $\dfrac{1}{2}y - 3 = x$, $y - 2x - 6 = 0$, dependent equations

23. $y = \dfrac{2}{3}x - 1$, $y = \dfrac{1}{2}x - 2$, $(-6, -5)$

25. (a) 10-10 Plan   $y = 7x$
    Tier-2 Plan   $y = 36 + 3x$

(b)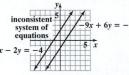
Tier - 2 Plan  $y = 36 + 3x$    10 - 10 Plan  $y = 7x$, $(9, 63)$

| $x$ | 1 | 4 | 8 | 12 |
|---|---|---|---|---|
| $y = 7x$ | 7 | 28 | 56 | 84 |

| $x$ | 1 | 4 | 8 | 12 |
|---|---|---|---|---|
| $y = 36 + 3x$ | 39 | 48 | 60 | 72 |

(c) 9 min   (d) Tier-2 Plan   27. $(54.18, -23.76)$   29. $(-23, 457)$   31. $x = 1$   32. $x = 10$   33. $x = \dfrac{2}{3}y + 2$   34. $y = -2x + 8$

## Quick Quiz 8.1   See Examples noted with Ex.

1. (Ex. 1)
$4x + 3y = 9$, $(3, -1)$, $x - 3y = 6$

2. The slope of each line is $\dfrac{3}{5}$. The $y$-intercept of the first line is $(0, 2)$. The $y$-intercept of the second line is $(0, 7)$. Since the lines have the same slopes but different $y$-intercepts, they are parallel lines. Therefore there is no solution. It is an inconsistent system of equations. (Ex. 2)   3. When graphed, the two equations yield the same line. The equations are dependent. There is an infinite number of solutions. (Ex. 3)   4. See Student Solutions Manual

## 8.2 Exercises

1. $(4, 1)$   3. $(1, 2)$   5. $(3, -5)$   7. $(0, 3)$   9. $\left(\dfrac{2}{3}, \dfrac{5}{3}\right)$   11. $(2, -3)$   13. $(-3, 6)$   15. $(3, -2)$

17. $(6, -5)$   19. $(6, 4)$   21. $(2, 3)$   23. $(-2, 5)$   25. $(3, 5)$   27. $(4, 5)$   29. (a) $y = 1500 + 900x$   (b) 10 weeks; \$10,500
$y = 500 + 1000x$   (c) 50 weeks; Boston Construction

31. 3; 7; An extra equation is needed for each additional variable to reduce the system to one equation with one unknown.   33. both

35. No solution; parallel lines have no points in common   36. (a) $m = -\dfrac{7}{4}; (0, 3)$   (b)   37.

$7x + 4y = 12$          $6x + 3y \ge 9$

38. \$7.50   39. 62%; about \$150.00

**Quick Quiz 8.2** *See Examples noted with Ex.* **1.** $\left(\frac{1}{2}, 1\right)$ (Ex. 1) **2.** $(4, -3)$ (Ex. 1) **3.** $(0, 5)$ (Ex. 1) **4.** See Student Solutions Manual

**8.3 Exercises** **1.** Eliminate $y$ by multiplying the second equation by 2, then adding the equations. **3.** Eliminate $y$ by multiplying the first equation by 4, multiplying the second equation by 3, then adding the equations. **5.** $(2, -3)$ **7.** $(2, -1)$ **9.** $\left(-\frac{1}{2}, 4\right)$ **11.** $\left(3, \frac{2}{5}\right)$

**13.** $(1, 0)$ **15.** $\left(-\frac{1}{2}, \frac{1}{3}\right)$ **17.** $(-10, 4)$ **19.** $(-2, 3)$ **21.** $\left(\frac{1}{3}, 1\right)$ **23.** Multiply equation (1) by 4. **25.** Multiply equation (1) by 6. Multiply equation (2) by 20.

**27.** $\left(2, \frac{1}{5}\right)$ **29.** $(3, -2)$ **31.** $\left(-2, \frac{4}{3}\right)$ **33.** $(-3, 3)$ **35.** Multiply equation (1) by 10.
$5x - 3y = 1$
$5x + 3y = 6$
**37.** Multiply equation (1) by 10.
Multiply equation (2) by 100.
$40x + 5y = 90$
$20x - 5y = 100$

**39.** $(-1, 2)$ **41.** $(-4, -7)$ **43.** $(8, -10)$ **45.** $\left(-\frac{2}{3}, 6\right)$ **47.** $(12, 16)$ **49.** $(0, -8)$ **51.** $\left(-\frac{4}{3}, \frac{2}{3}\right)$ **52.** between 550 and 770 airplanes

**53.** $720 **54.** $x = \frac{8}{7}$ **55.** $y = \frac{5}{3}$

**Quick Quiz 8.3** *See Examples noted with Ex.* **1.** $(3, -2)$ (Ex. 2) **2.** $(4, -1)$ (Ex. 2) **3.** $(5, 1)$ (Ex. 3) **4.** See Student Solutions Manual

**Use Math to Save Money** **1.** Private loan: $I = (10,000)(0.0465)(20) = \$9300$; Total Amount $= 10,000 + 9300 = \$19,300$; Subsidized loan: $I = (10,000)(0.06)(20 - 4.5) = \$9300$; Total Amount $= 10,000 + 9300 = \$19,300$. **2.** The total amount that Alicia must pay back is the same for each loan. **3.** Total Amount $\div$ Number of Payments $=$ Monthly Payment; Private loan: $19,300 \div (12 \text{ months} \times 20 \text{ years}) = 19,300 \div 240 = \$80.42$; Subsidized loan: $19,300 \div (12 \text{ months} \times 15.5 \text{ years}) = 19,300 \div 186 = \$103.76$ **4.** Private student loan **5.** $19,300, 240, \$80.42; \$19,300, 186, \$103.76 **6.** Answers will vary.

**8.4 Exercises** **1.** parallel lines, inconsistent **3.** intersect, independent, consistent **5.** $(4, -2)$ **7.** $(18, 13)$ **9.** No solution **11.** Infinite number of solutions **13.** $(5, 4)$ **15.** $(50, 40)$ **17.** $(8, 5)$ **19.** $\left(\frac{1}{2}, \frac{2}{3}\right)$ **21.** $\left(\frac{7}{3}, \frac{8}{3}\right)$ **23.** $(3, -3)$ **25.** $(4, 1)$

**27.** No solution. The system is inconsistent. **29.** $(-4, -3)$ **31.** $x = -\frac{2}{9}$ **32.** $x = 0.1$ **33.** $3722.50 **34.** Less: $62.50 owed

**Quick Quiz 8.4** *See Examples noted with Ex.* **1.** $(9, -8)$ (Ex. 1) **2.** No solution (Ex. 2) **3.** $(3, 2)$ (Ex. 1) **4.** See Student Solutions Manual

**8.5 Exercises** **1.** 8 coach class
7 1st class
**3.** 12 cleaners
4 supervisors
**5.** 13 hr day shift
10 hr night shift
**7.** original length $= 10$ ft
original width $= 9$ ft
new length $= 14$ ft
new width $= 14$ ft
**9.** 8 qt of 50% antifreeze originally. She added 8 qt of 80% solution.

**11.** $b =$ base pay $= \$30,000$;
$r =$ commission rate $= 20\%$
**13.** 50 km/hr $=$ wind speed
550 km/hr $=$ plane's speed
**15.** 30 lb nuts
20 lb raisins
**17.** The speed of the jet stream was 50 mph.
The speed of the plane was 450 mph.

**19.** $y =$ total weekly manager's salary at each store $= \$1450$;
$x =$ total weekly sales at each store $= \$3000$
**21.** $6 to clean one police cruiser;
$9 to clean one police SUV
**23.** $-\frac{1}{2}$ **24.** slope $= -\frac{3}{4}$;
$y$-intercept $= (0, -2)$

**25.** $y = \frac{5}{4}x + \frac{7}{2}$ **26.** $y = -2x + 10$

**Quick Quiz 8.5** *See Examples noted with Ex.* **1.** 10,000 people purchased student tickets; 8500 people purchased general admission tickets (Ex. 1) **2.** each Dell PC cost $800; each Apple iMac cost $1200 (Ex. 1) **3.** 400 mi/hr $=$ speed of plane; 100 mi/hr $=$ speed of jet stream (Ex. 4) **4.** See Student Solutions Manual

**Career Exploration Problems** **1.** Price: $106.25; quantity: 37,500 phones **2.** 37,000 units

**You Try It** **1.** (a) [graph] No solution (b) [graph] Infinitely many solutions (c) [graph] One solutions; $(1, -2)$ **2.** $(8, 1)$ **3.** $(-1, -2)$ **4.** (a) Substitution; $(0, -4)$ (b) Addition; $(-6, 9)$

**5.** (a) Infinitely many solutions
(b) No solution
**6.** 3 lb of green peppers;
5 lb of onions

**Chapter 8 Review Problems** **1.** [graph] $-x + 3y = 9$, $(-3, 2)$, $2x + 3y = 0$ **2.** [graph] $(2, 4)$, $-2x - y = -8$, $-3x + y = -2$ **3.** [graph] $2x - y = 6$, $(2, -2)$, $6x + 3y = 6$ **4.** [graph] $2x - y = 1$, $(-1, -3)$, $3x + y = -6$

**5.** $(3, 3)$ **6.** $(3, 5)$ **7.** $(15, 21)$ **8.** $(8, 12)$ **9.** $(2, 1)$ **10.** $(-1, 0)$ **11.** $\left(0, -\frac{1}{2}\right)$ **12.** $\left(\frac{1}{3}, -4\right)$ **13.** $(5, -3)$ **14.** $(12, 1)$

**15.** $\left(\frac{4}{5}, -\frac{6}{5}\right)$   **16.** $(6, -4)$   **17.** $(-1, -4)$   **18.** No solution, inconsistent system   **19.** $(9, 4)$   **20.** $\left(-3, -\frac{2}{3}\right)$

**21.** infinite number of solutions, dependent equations   **22.** $(-1, -2)$   **23.** $(4, -3)$   **24.** infinite number of solutions, dependent equations

**25.** $(8, -5)$   **26.** $\left(\frac{29}{28}, \frac{1}{7}\right)$   **27.** $(10, 8)$   **28.** $(-12, -15)$   **29.** No solution, inconsistent system   **30.** infinite number of solutions,

dependent equations   **31.** $\left(\frac{33}{14}, \frac{16}{7}\right)$   **32.** $\left(-\frac{7}{32}, \frac{33}{16}\right)$   **33.** $(-5, 4)$   **34.** $(9, -8)$   **35.** $(2, 4)$   **36.** $(3, 1)$   **37.** $(3, 2)$   **38.** 126

adults; 60 students   **39.** 21 2-lb bags; 42 5-lb bags   **40.** 275 mph in still air; 25 mph wind speed   **41.** 20 L of 20% solution; 20 L of 30%

solution   **42.** 54 balsam firs; 25 Norwegian pines   **43.** 8 tons of 15% salt; 16 tons of 30% salt   **44.** Speed of the boat, 19 km/hr; speed of the

current, 4 km/hr   **45.** The printer that broke prints 1200 labels/hour; the other printer prints 1800 labels/hour.   **46.** Lexi may be running at $2\frac{3}{4}$

miles per hour while Olivia is running at 2 miles per hour, or Lexi may be running at $4\frac{3}{4}$ miles per hour while Olivia is running at 4 miles per hour.

## How Am I Doing? Chapter 8 Test   **1.** $(-3, -4)$ (obj. 8.2.1)   **2.** $(-3, 4)$ (obj. 8.3.1)

**3.**  (obj. 8.1.1)   **4.** $(6, 2)$ (obj. 8.4.1)   **5.** $(4, 3)$ (obj. 8.4.1)   **6.** $(5, 1)$ (obj. 8.4.1)   **7.** $(3, 0)$ (obj. 8.4.1)   **8.** no solution, inconsistent system (obj. 8.4.1, 8.4.2)   **9.** infinite number of solutions, dependent equations (obj. 8.4.1, 8.4.2)   **10.** $(-2, 3)$ (obj. 8.4.1)   **11.** $(-5, 3)$ (obj. 8.4.1)   **12.** $(-3.5, -4.5)$ (obj. 8.4.1)   **13.** First number is 5, second number is $-3$ (obj. 8.5.1)   **14.** 14,500 of the \$8 tickets; 16,000 of the \$12 tickets (obj. 8.5.1)   **15.** A shirt costs \$20 and a pair of slacks costs \$24. (obj. 8.5.1)   **16.** 1000 booklets; \$450 total cost (obj. 8.5.1)   **17.** 50 km/hr; 450 km/hr (obj. 8.5.1)

## Chapter 9   9.1 Exercises   **1.** The principal square root of $N$, where $N \geq 0$, is a nonnegative number $a$ that has the property $a^2 = N$.

**3.** No; $(0.3)(0.3) = 0.09$   **5.** $\pm 3$   **7.** $\pm 7$   **9.** 4   **11.** 12   **13.** $-6$   **15.** 0.9   **17.** $\frac{6}{11}$   **19.** $\frac{7}{13}$   **21.** 25   **23.** $-100$

**25.** 13   **27.** $-\frac{1}{8}$   **29.** $\frac{3}{4}$   **31.** 0.06   **33.** 140   **35.** $-17$   **37.** 5.196   **39.** 7.874   **41.** $-13.528$   **43.** $-13.964$   **45.** 13 ft

**47.** 7.5 sec   **49.** 3   **51.** $-4$   **53.** 3   **55.** 3   **57.** No, a 4th power of a real number must be nonnegative.   **59.** $(2, 1)$   **60.** No solution,

inconsistent system   **61.** Snowboard: \$250; goggles: \$50   **62.** 470 mph

### Quick Quiz 9.1   *See Examples noted with Ex.*   **1.** 11 (Ex. 1)   **2.** $\frac{5}{8}$ (Ex. 2)   **3.** 0.3 (Ex. 3)   **4.** See Student Solutions Manual

## 9.2 Exercises   **1.** Yes, $\sqrt{6^2} = \sqrt{6 \cdot 6} = \sqrt{36} = 6$   **3.** No, $\sqrt{(-9)^2} = \sqrt{(-9)(-9)} = \sqrt{81} = 9$   **5.** 8   **7.** $18^2$   **9.** $10^4$   **11.** $33^4$

$(\sqrt{6})^2 = (\sqrt{6})(\sqrt{6}) = \sqrt{36} = 6$   but $-\sqrt{9^2} = -\sqrt{9 \cdot 9} = -\sqrt{81} = -9$

**13.** $5^{70}$   **15.** $x^6$   **17.** $t^9$   **19.** $y^{13}$   **21.** $6x^4$   **23.** $12x$   **25.** $x^3y^2$   **27.** $4xy^{10}$   **29.** $10a^5b^3$   **31.** $2\sqrt{6}$   **33.** $3\sqrt{5}$   **35.** $3\sqrt{2}$

**37.** $6\sqrt{2}$   **39.** $3\sqrt{10}$   **41.** $8\sqrt{2}$   **43.** $2x\sqrt{2x}$   **45.** $3w^2\sqrt{3w}$   **47.** $4z^4\sqrt{2z}$   **49.** $2x^2y^3\sqrt{7xy}$   **51.** $4y\sqrt{3yw}$   **53.** $5xy\sqrt{3y}$

**55.** $8x^5$   **57.** $y^3\sqrt{y}$   **59.** $x^2y^7$   **61.** $3x^2y^3\sqrt{15xy}$   **63.** $6ab^2c^2\sqrt{2c}$   **65.** $13ab^4c^2\sqrt{ac}$   **67.** $x + 5$   **69.** $4x + 1$   **71.** $y(x + 7)$

**73.** $(4, -1)$   **74.** $x = -18$   **75.** $2(x + 3y)^2$   **76.** $(2x - 7)(x + 3)$   **77.** $\frac{x + 3}{4y}$   **78.** $\frac{37x + 10}{4(x - 2)(3x + 1)}$

### Quick Quiz 9.2   *See Examples noted with Ex.*   **1.** $2\sqrt{30}$ (Ex. 4)   **2.** $9xy^2\sqrt{x}$ (Ex. 6)   **3.** $3x^4y^2\sqrt{15y}$ (Ex. 6)   **4.** See Student Solutions Manual

## 9.3 Exercises   **1.** 1. Simplify each radical term.   **3.** $6\sqrt{5}$   **5.** $5\sqrt{2} + 3\sqrt{3}$   **7.** $3\sqrt{7}$   **9.** $12\sqrt{5}$   **11.** $\sqrt{2}$   **13.** $9\sqrt{2}$

2. Combine like radicals.

**15.** $5 + 6\sqrt{2} + 6\sqrt{3}$   **17.** $6\sqrt{5} - 2\sqrt{2}$   **19.** $13\sqrt{2x}$   **21.** $0.2\sqrt{3x}$   **23.** $7\sqrt{5y}$   **25.** $-\sqrt{2}$   **27.** $6\sqrt{7x} - 15x\sqrt{7x}$

**29.** $-5x\sqrt{2x}$   **31.** $10x\sqrt{3}$   **33.** $2y\sqrt{6y} - 6y\sqrt{6}$   **35.** $38x\sqrt{5x}$   **37.** $(8\sqrt{5} + 6\sqrt{3})$ mi   **39.** $x \geq -\frac{4}{3}$ 

**40.** 8 games at \$8.99 and 4 games at \$12.49   **41.** $\frac{9y^4}{4x^2}$   **42.** $-27a^6x^{12}$

### Quick Quiz 9.3   *See Examples noted with Ex.*   **1.** $13\sqrt{2}$ (Ex. 3a)   **2.** $3\sqrt{3}$ (Ex. 5)   **3.** $10\sqrt{a} + 6\sqrt{2a}$ (Ex. 4)
**4.** See Student Solutions Manual

## 9.4 Exercises   **1.** $\sqrt{30}$   **3.** $2\sqrt{5}$   **5.** $3\sqrt{6}$   **7.** $12\sqrt{10}$   **9.** $5\sqrt{2a}$   **11.** $30x\sqrt{2}$   **13.** $16a\sqrt{6}$   **15.** $-6b\sqrt{a}$
**17.** $\sqrt{10} + \sqrt{15}$   **19.** $6\sqrt{2} + 15\sqrt{5}$   **21.** $2x - 16\sqrt{5x}$   **23.** $2\sqrt{3} - 18 + 4\sqrt{15}$   **25.** $-6 - 9\sqrt{2}$   **27.** $21 + 14\sqrt{2}$
**29.** $5 - \sqrt{21}$   **31.** $-6 + 9\sqrt{2}$   **33.** $2\sqrt{14} + 34$   **35.** $25 - 4\sqrt{6}$   **37.** $53 + 10\sqrt{6}$   **39.** $48 - 6\sqrt{15}$   **41.** $12\sqrt{10}$
**43.** $\sqrt{14} + 21\sqrt{2} - \sqrt{42}$   **45.** $9 - 4\sqrt{14}$   **47.** $-10\sqrt{xy} - 5x\sqrt{y} + 20x$   **49.** $9x^2y - 5$   **51.** 12 ft$^2$   **53.** $(8a + 5b)(8a - 5b)$

**54.** $(2x - 5)^2$   **55.** approx. 40.3 mph   **56.** $8\frac{1}{3}$ yr

### Quick Quiz 9.4   *See Examples noted with Ex.*   **1.** $24xy$ (Ex. 2b)   **2.** $2\sqrt{6} + 6 - 4\sqrt{15}$ (Ex. 4)   **3.** $-29 + 2\sqrt{6}$ (Ex. 6)
**4.** See Student Solutions Manual

## Use Math to Save Money   **1.** 8/3 or 2 2/3 cups   **2.** 56 oz   **3.** 3.5 lb   **4.** \$9.31   **5.** \$6.23   **6.** Lucy would save
approximately 67%.   **7.** \$162 × 52 × 67% = \$5644.08   **8.** \$162 − \$130 = \$32; \$32 × 52 = \$1664

**9.5 Exercises** **1.** 2 **3.** $\frac{1}{3}$ **5.** 6 **7.** $\frac{\sqrt{6}}{x^2}$ **9.** $\frac{3\sqrt{2}}{a^2}$ **11.** $\frac{3\sqrt{7}}{7}$ **13.** $\frac{8\sqrt{15}}{15}$ **15.** $\frac{x\sqrt{2x}}{2}$ **17.** $\frac{2\sqrt{2x}}{x}$ **19.** $\frac{3\sqrt{7}}{7}$ **21.** $\frac{6\sqrt{a}}{a}$

**23.** $\frac{\sqrt{2x}}{2x^2}$ **25.** $\frac{3\sqrt{x}}{x^2}$ **27.** $\frac{\sqrt{15}}{5}$ **29.** $\frac{\sqrt{210}}{21}$ **31.** $\frac{9\sqrt{2x}}{8x}$ **33.** $2(\sqrt{3}+1)$ **35.** $\sqrt{10}-\sqrt{2}$ **37.** $-\frac{2+\sqrt{14}}{5}$ or $\frac{-2-\sqrt{14}}{5}$

**39.** $2\sqrt{14}-7$ **41.** $x(2\sqrt{2}+\sqrt{5})$ **43.** $\frac{\sqrt{6x}-\sqrt{2x}}{4}$ **45.** $\frac{13-2\sqrt{30}}{7}$ **47.** $\frac{4\sqrt{35}+3\sqrt{5}+3\sqrt{2}+4\sqrt{14}}{3}$

**49.** $5\sqrt{6}+8\sqrt{2}$ **51.** $\sqrt{x}-5$ **53. (a)** $s=\frac{1}{h}\sqrt{3Vh}$ **(b)** $s=3$ in. **55.** $(3\sqrt{5}-6)$ m **57.** $x\approx-0.71$ **58.** 0

**Quick Quiz 9.5** *See Examples noted with Ex.* **1.** $\frac{3\sqrt{10}}{10}$ (Ex. 3) **2.** $\frac{15x\sqrt{2}-5x\sqrt{5}}{13}$ (Ex. 6b) **3.** $11+2\sqrt{30}$ (Ex. 7)
**4.** See Student Solutions Manual

**9.6 Exercises** **1.** 13 **3.** $2\sqrt{14}$ **5.** $7\sqrt{2}$ **7.** $5\sqrt{3}$ **9.** $2\sqrt{14}$ **11.** $\sqrt{57}$ **13.** 10.08 **15.** 9.4 ft **17.** 98.0 ft

**19.** Top cable $= 145.40$ ft; bottom cable $= 140.77$ ft **21.** $x=7$ **23.** $x=9$ **25.** $x=7$ **27.** $x=\frac{81}{2}$ **29.** $x=5$ only **31.** $y=0,3$

**33.** $x=1$ only **35.** $y=0,1$ **37.** $y=\frac{1}{2}$ only **39.** $x=\frac{2}{3}$ only **41.** We defined the $\sqrt{\ }$ symbol to mean the positive square root of the

radicand. Therefore, it cannot be equal to a negative number. **42.** $x=5$ **43.** $49x^2-28x+4$ **44.** 20 **45.** Relation

**Quick Quiz 9.6** *See Examples noted with Ex.* **1.** $2\sqrt{7}$ (Ex. 1) **2.** $x=14$ (Ex. 4) **3.** $x=8$ (Ex. 7) **4.** See Student Solutions Manual

**9.7 Exercises** **1.** $y=72$ **3.** $y=\frac{1029}{2}$ **5.** $y=1296$ **7.** 300 calories **9.** 56 min **11.** $y=\frac{45}{4}$ **13.** $y=1$ **15.** $y=\frac{40}{27}$

**17.** 3.45 min **19.** 675 lb **21. (a)** 15.0 km, 8.2 km, and 2.7 km **23. (a)** $k=1$ **(b)** $\approx 248.25$ Earth years **25.** 160
**(b)** **26.** $S=4.80h$

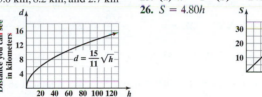

**Quick Quiz 9.7** *See Examples noted with Ex.* **1.** 6 in. (Ex. 6) **2.** 281.25 calories (Ex. 1) **3.** 40 cm$^2$ (Ex. 3)
**4.** See Student Solutions Manual

**Career Exploration Problems** **1.** 1.618 **2.** 4.05 feet **3.** Yes, the sign meets the town's requirements.
**4.** The dimensions of this sign are about 6.8 feet by 4.2 feet.

**You Try It** **1. (a)** 12 **(b)** 15 **(c)** $\approx 4.243$ **(d)** This is not a real number **(e)** $\approx 12.247$ **2. (a)** $5\sqrt{3}$ **(b)** $2abc^2\sqrt{2a}$ **(c)** $x^7\sqrt{x}$

**3.** $9\sqrt{2}-\sqrt{5}$ **4.** $19-7\sqrt{10}$ **5. (a)** 5 **(b)** $\frac{14}{3}$ **6. (a)** $\sqrt{3}+2\sqrt{6}$ **(b)** $-21$ **7. (a)** $\frac{\sqrt{6}}{2}$ **(b)** $-\sqrt{10}+\sqrt{15}$

**8. (a)** $\frac{\sqrt{6}}{\sqrt{5}}$ **(b)** $\frac{\sqrt{30}}{5}$ **9.** $b\approx 9.2$ **10.** $x=\frac{15}{2}$ **11.** $w=15$ **12.** $m=\frac{3}{4}$

**Chapter 9 Review Problems** **1.** 8 **2.** $-11$ **3.** $-12$ **4.** 17 **5.** 0.2 **6.** 0.7 **7.** $\frac{1}{10}$ **8.** $\frac{8}{9}$ **9.** 10.247 **10.** 14.071

**11.** 8.775 **12.** 9.381 **13.** $2\sqrt{7}$ **14.** $5\sqrt{5}$ **15.** $2\sqrt{10}$ **16.** $4\sqrt{5}$ **17.** $y^5$ **18.** $x^2y^3\sqrt{x}$ **19.** $4xy^2\sqrt{xy}$ **20.** $7x^2y^3\sqrt{2}$
**21.** $2x^2\sqrt{3x}$ **22.** $6x^4\sqrt{2x}$ **23.** $2ab^2c^2\sqrt{30ac}$ **24.** $11a^3b^2\sqrt{c}$ **25.** $2x^3y^4\sqrt{14xy}$ **26.** $3x^6y^3\sqrt{11xy}$ **27.** $3\sqrt{5}$ **28.** $9\sqrt{6}$
**29.** $7x\sqrt{3}$ **30.** $8a\sqrt{2}+2a\sqrt{3}$ **31.** $-7\sqrt{5}+2\sqrt{10}$ **32.** $-2\sqrt{10}+6\sqrt{7}$ **33.** $6x^2$ **34.** $-10a\sqrt{b}$ **35.** $-8x^3$ **36.** $-12$
**37.** $\sqrt{30}-10+5\sqrt{2}$ **38.** $20+3\sqrt{11}$ **39.** $27+8\sqrt{10}$ **40.** $8-4\sqrt{3}+12\sqrt{6}-18\sqrt{2}$ **41.** $15-3\sqrt{2}-10\sqrt{3}+2\sqrt{6}$

**42.** $66+36\sqrt{2}$ **43.** $74-40\sqrt{3}$ **44.** $\frac{\sqrt{3x}}{3x}$ **45.** $\frac{2y\sqrt{5}}{5}$ **46.** $\frac{\sqrt{30}}{6}$ **47.** $\frac{\sqrt{66}}{11}$ **48.** $\frac{x\sqrt{5}}{5}$ **49.** $\sqrt{5}-\sqrt{2}$ **50.** $\frac{2(\sqrt{6}+\sqrt{3})}{3}$

**51.** $-7+3\sqrt{5}$ **52.** $\frac{3-2\sqrt{3}}{3}$ **53.** $c=\sqrt{89}$ **54.** $a=\sqrt{2}$ **55.** $b=\frac{\sqrt{51}}{2}$ **56.** 30 m **57.** 186.7 mi **58.** 60 ft **59.** $x=13$

**60.** $x=-2$ **61.** $x=6$ **62.** $x=4$ **63.** $x=7$ **64.** $x=2$ **65.** $y=48$ **66.** $y=1$ **67.** $y=\frac{1}{2}$ **68.** 5 lumens

**69.** Approximately 134 ft **70.** It will be 8 times as much. **71.** $\frac{6}{11}$ **72.** 0.02 **73.** $7x^3y\sqrt{2y}$ **74.** $3\sqrt{3}$

**75.** $2\sqrt{15}+3\sqrt{6}-20-6\sqrt{10}$ **76.** $57+24\sqrt{3}$ **77.** $\frac{5\sqrt{3}}{6}$ **78.** $\frac{6-\sqrt{6}+6\sqrt{2}-2\sqrt{3}}{10}$ **79.** $x=42$ **80.** $x=-\frac{1}{2},x=2$

**How Am I Doing? Chapter 9 Test** **1.** 11 (obj. 9.1.1) **2.** $\frac{3}{10}$ (obj. 9.1.1) **3.** $4xy^3\sqrt{3y}$ (obj. 9.2.2) **4.** $10xz^2\sqrt{xy}$ (obj. 9.2.2)

**5.** $3\sqrt{3}$ (obj. 9.3.2) **6.** $8\sqrt{a}+5\sqrt{2a}$ (obj. 9.3.2) **7.** $12ab$ (obj. 9.4.1) **8.** $5\sqrt{2}+2\sqrt{15}-15$ (obj. 9.4.2) **9.** $37+20\sqrt{3}$ (obj. 9.4.3)

**10.** $19+\sqrt{10}$ (obj. 9.4.3) **11.** $\frac{\sqrt{5x}}{5}$ (obj. 9.5.2) **12.** $\frac{\sqrt{3}}{2}$ (obj. 9.5.2) **13.** $\frac{17+\sqrt{3}}{22}$ (obj. 9.5.3) **14.** $a(\sqrt{5}-\sqrt{2})$ (obj. 9.5.3)

**15.** 12.49 (obj. 9.1.2) **16.** $x=\sqrt{133}$ (obj. 9.6.1) **17.** $x=3\sqrt{3}$ (obj. 9.6.1) **18.** $x=\frac{35}{2}$ (obj. 9.6.2) **19.** $x=9$ (obj. 9.6.2)

**20.** 16 in. (obj. 9.7.2) **21.** \$85.77 (obj. 9.7.1) **22.** 21.2 cm$^2$ (obj. 9.7.1)

# Chapter 10   10.1 Exercises

**1.** $a = 1$, $b = 8$, $c = 7$   **3.** $a = 8$, $b = -11$, $c = 0$   **5.** $a = 1$, $b = 3$, $c = -15$   **7.** $x = 0$, $x = \dfrac{1}{5}$   **9.** $x = 0$, $x = -2$   **11.** $x = 0$, $x = \dfrac{3}{2}$   **13.** $x = 7$, $x = -4$   **15.** $x = \dfrac{1}{2}$, $x = -6$   **17.** $x = \dfrac{1}{3}$, $x = 5$

**19.** $x = 7$, $x = -2$   **21.** $x = -\dfrac{3}{4}$, $x = -\dfrac{2}{3}$   **23.** $a = -8$, $a = -4$   **25.** $y = \dfrac{1}{3}$, $y = -\dfrac{3}{2}$   **27.** $x = \dfrac{6}{5}$   **29.** $x = 0$, $x = 3$   **31.** $x = 0$, $x = -\dfrac{14}{3}$   **33.** $y = -2$, $y = 5$

**35.** $x = 3$, $x = 7$   **37.** $x = 1$   **39.** $x = -5$, $x = 3$   **41.** $x = -7$, $x = 2$   **43.** $x = -\dfrac{9}{2}$, $x = 3$   **45.** $x = \dfrac{5}{2}$, $x = -4$   **47.** $x = -\dfrac{10}{3}$, $x = 4$   **49.** You can always factor out $x$.

**51.** $a = \dfrac{14}{9}$, $c = -14$   **53.** The number of labels produced is 1200 and 2000.   **55.** Average = 1600, Revenue for 1600 = $360, which is more than producing 1200 or 2000 labels   **57.** Revenue for 1500 = $357.50, Revenue for 1700 = $357.50, It appears the maximum revenue is at 1600 labels.

**59.** $\dfrac{(2x - 1)(x + 2)}{2x}$   **60.** $\dfrac{2x^2 - 4x + 2}{3x^2 - 11x + 10} = \dfrac{2(x - 1)^2}{(3x - 5)(x - 2)}$

**Quick Quiz 10.1**   *See Examples noted with Ex.*   **1.** $x = \dfrac{5}{3}$, $x = -\dfrac{3}{2}$ (Ex. 2)   **2.** $x = 8$, $x = 6$ (Ex. 5)   **3.** $x = -3$, $x = \dfrac{2}{3}$ (Ex. 3)   **4.** See Student Solutions Manual

## 10.2 Exercises   **1.** $x = \pm 7$   **3.** $x = \pm 6\sqrt{2}$   **5.** $x = \pm 2\sqrt{7}$   **7.** $x = \pm 3$   **9.** $x = \pm 2\sqrt{5}$   **11.** $x = \pm 5\sqrt{5}$   **13.** $x = \pm 2\sqrt{3}$

**15.** $x = \pm\sqrt{5}$   **17.** $x = 3, x = 11$   **19.** $x = -4 \pm \sqrt{6}$   **21.** $x = \dfrac{-5 \pm \sqrt{2}}{2}$   **23.** $x = \dfrac{1 \pm \sqrt{7}}{3}$   **25.** $x = \dfrac{-1 \pm 3\sqrt{2}}{5}$

**27.** $x = \dfrac{5 \pm 3\sqrt{6}}{4}$   **29.** $x = 3 \pm 2\sqrt{5}$   **31.** $x = -3 \pm \sqrt{2}$   **33.** $x = 6 \pm \sqrt{41}$   **35.** $x = 0, x = 7$   **37.** $x = 0, x = 5$

**39.** $x = \dfrac{9}{2}, x = -1$   **41.** $a = -4, b = -4$   **42.** $x = -1, x = 4$   **43.** 200 lb/in.$^2$   **44.** A total of 16 tires were defective. 6 had both kinds of defects, 7 had defects only in workmanship, and 3 had defects only in materials.

**Quick Quiz 10.2**   *See Examples noted with Ex.*   **1.** $x = \pm\sqrt{6}$ (Ex. 2)   **2.** $x = \dfrac{2 \pm 2\sqrt{10}}{3}$ (Ex. 3)   **3.** $x = -4 \pm \sqrt{26}$ (Ex. 4)   **4.** See Student Solutions Manual

## 10.3 Exercises   **1.** $\sqrt{b^2 - 4ac} = \sqrt{100} = 10$   Therefore, there are two rational roots.   **3.** No. Standard form is $4x^2 + 5x - 6 = 0$. Then $a = 4, b = 5,$ and $c = -6$.

**5.** $x = -2 \pm \sqrt{3}$   **7.** $x = \dfrac{3 \pm \sqrt{41}}{2}$   **9.** $x = \dfrac{1}{4}, x = -2$   **11.** $x = 4, x = -\dfrac{5}{2}$   **13.** $x = \dfrac{3 \pm \sqrt{33}}{12}$   **15.** $x = \dfrac{1 \pm \sqrt{19}}{6}$

**17.** $x = 5, x = 2$   **19.** No real solution   **21.** $y = \dfrac{5 \pm \sqrt{13}}{4}$   **23.** $d = \dfrac{4}{3}, d = -3$   **25.** No real solution   **27.** $x = \dfrac{3}{2}$

**29.** $x \approx 0.372, x \approx -5.372$   **31.** $x \approx 4.108, x \approx -0.608$   **33.** $x \approx -1.894, x \approx -0.106$   **35.** $x = \dfrac{3}{2}, x = \dfrac{2}{3}$   **37.** $x = 3, x = \dfrac{1}{3}$

**39.** $t = -1 \pm \sqrt{23}$   **41.** $y = \dfrac{1 \pm \sqrt{51}}{5}$   **43.** $x = \dfrac{\pm\sqrt{39}}{3}$   **45.** $x = 1 \pm 2\sqrt{2}$   **47.** 2 ft   **49.** $x^2 + 5x + 2$   **50.** $y = -\dfrac{5}{13}x - \dfrac{14}{13}$

**51.**

**Quick Quiz 10.3**   *See Examples noted with Ex.*   **1.** $x = \dfrac{-1 \pm \sqrt{2}}{3}$ (Ex. 1)   **2.** $x = \dfrac{1 \pm 2\sqrt{5}}{2}$ (Ex. 2)   **3.** $x = \dfrac{-1 \pm \sqrt{7}}{3}$ (Ex. 2)   **4.** See Student Solutions Manual

## Use Math to Save Money   **1.** December 2009: $295.00, November 2015: $355.00   **2.** $1775.00   **3.** 4° lower = 8% savings, $1775.00 × 0.08 = $142   **4.** $355 = 20% of $1775; 20% = 10° lower, 72 − 10 = 62 degrees   **5.** Answers will vary.   **6.** Answers will vary.

## 10.4 Exercises   **1.** When you let $y = 0$ and you solve for $x$ in this equation, you obtain the square root of a negative number. Therefore, there is no real solution. So there are no $x$-intercepts.   **3.** The parabola always has a vertex at $(0, c)$ if $b = 0$. This means the vertex is always on the $y$-axis.

**5.**

$y = x^2 + 2$

**7.**

$y = -\dfrac{1}{3}x^2$

**9.**

$y = 3x^2 - 1$

**11.**

$y = (x - 2)^2$

**13.**

$y = -\dfrac{1}{2}x^2 + 4$

**15.**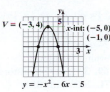
$y = \frac{1}{2}(x-3)^2$

**17.**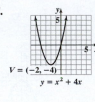
$V = (-2, -4)$
$y = x^2 + 4x$

**19.**
$V = (-2, 4)$
$y = -x^2 - 4x$

**21.**
$V = (2, 8)$
$y = -2x^2 + 8x$

**23.**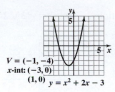
$V = (-1, -4)$
x-int: $(-3, 0)$
$(1, 0)$ $y = x^2 + 2x - 3$

**25.**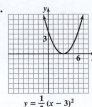
$V = (-3, 4)$
x-int: $(-5, 0)$
$(-1, 0)$
$y = -x^2 - 6x - 5$

**27.**
$V = (-3, 0)$
x-int: $(-3, 0)$
$y = x^2 + 6x + 9$

**29. (a)**
$(4, 82.4)$
$(8.1, 0)$
**(b)** 62.8 m
**(c)** 82.4 m
**(d)** About 8.1 sec

**31. (a)**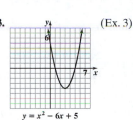
$(4.5, 20.25)$
$N = 9x - x^2$
Number of mosquitoes in millions
Number of inches of rain
**(b)** 4.5 in.
**(c)** 20.25 million

**33.** Vertex: $(-0.256, 1658.327)$
y-intercept: $(0, 1635)$
x-intercepts: $(1.905, 0)$ and $(-2.418, 0)$

**34.** $y = 1$   **35.** $y = 75$   **36.** 7 in.   **37.** 8.25 sec

**Quick Quiz 10.4**   *See Examples noted with Ex.*   **1.** $(3, -4)$ (Ex. 3)   **2.** $(1, 0)$ and $(5, 0)$ (Ex. 3)   **3.** (Ex. 3)
**4.** See Student Solutions Manual

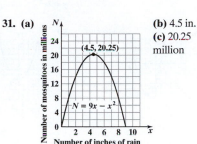
$y = x^2 - 6x + 5$

**10.5 Exercises**   **1.** Base $= 11$ cm   **3.** 10 ft by 10 ft   **5.** 21 people   **7.** 80 members were expected.   **9.** Width $= 25$ ft
altitude $= 16$ cm   100 members attended.   length $= 50$ ft

**11.** 600 mph   **13.** 50 stores   **15.** 15 stores   **17.** 2012 and 2013   **19.** About 74 mi   **21.**
**22.** $\sqrt{15} - 2\sqrt{3}$   **23.** 0.75 ft

**Quick Quiz 10.5**   *See Examples noted with Ex.*   **1.** 12 yd; 5 yd (Ex. 1)   **2.** 6 sec (Ex. 2)   **3.** 61.25 m (Ex. 2)
**4.** See Student Solutions Manual

**Career Exploration Problems**   **1.** The lowest point is 72 feet from the left tower and about 13 feet above the road.   **2.** The cable is 30 feet from the roadway at about 35 feet and at about 109 feet from the left tower.   **3.** The lot's area will be maximized when its dimensions are 55 feet by 110 feet.

**You Try It**   **1.** $x = \frac{1}{4}$ or $x = 2$   **2.** $x = \pm 6$   **3.** $x = -2$ or $x = \frac{3}{2}$   **4.** $x = \frac{-1 \pm \sqrt{21}}{5}$   **5.** One real number solution
**6. (a)** $(3, 4)$   **7.**   **8.** Length $= 11$ m; width $= 5$ m
**(b)** The parabola opens downward

$\left(\frac{3}{2}, -\frac{49}{4}\right)$
$y = x^2 - 3x - 10$

**Chapter 10 Review Problems**   **1.** $a = 14, b = 3, c = -16$   **2.** $a = 11, b = 8, c = 0$   **3.** $a = 2, b = 6, c = -3$

**4.** $a = 4, b = 5, c = -14$   **5.** $x = -25, x = -1$   **6.** $x = 5, x = -11$   **7.** $y = -7, y = 2$   **8.** $x = \frac{4}{3}$   **9.** $x = \frac{5}{3}, x = \frac{1}{2}$

**10.** $x = -\frac{3}{2}, x = \frac{4}{3}$   **11.** $x = \frac{1}{2}, x = 1$   **12.** $x = \frac{1}{4}, x = -\frac{4}{3}$   **13.** $x = 0, x = -8$   **14.** $x = 0, x = 3$   **15.** $x = 0, x = 4$

**16.** $x = -\frac{2}{5}, x = 10$   **17.** $x = \pm 7$   **18.** $x = \pm\sqrt{39}$   **19.** $x = \pm 3\sqrt{2}$   **20.** $x = \pm 2\sqrt{6}$   **21.** $x = 4 \pm \sqrt{7}$   **22.** $x = 2 \pm \sqrt{3}$

**23.** $x = \frac{-2 \pm 2\sqrt{7}}{3}$   **24.** $x = \frac{1 \pm 4\sqrt{2}}{2}$   **25.** $x = -1, x = -7$   **26.** $x = -3, x = -11$   **27.** $x = 3 \pm \sqrt{2}$   **28.** $x = -1 \pm \sqrt{3}$

**29.** $x = \frac{-5 \pm \sqrt{31}}{2}$   **30.** $x = -2 \pm \sqrt{10}$   **31.** $x = -2 \pm 2\sqrt{3}$   **32.** $x = \frac{7 \pm \sqrt{17}}{4}$   **33.** $x = \frac{-5 \pm \sqrt{73}}{4}$   **34.** $x = \frac{4 \pm 2\sqrt{7}}{3}$

**35.** No real solution   **36.** $x = \frac{1 \pm \sqrt{41}}{10}$   **37.** $x = \frac{-3 \pm \sqrt{21}}{6}$   **38.** $x = \frac{5}{2}, x = 2$   **39.** $x = \frac{3}{2}, x = -\frac{1}{2}$   **40.** $x = -\frac{1}{5}$

**41.** $x = 4, x = \frac{3}{2}$   **42.** $x = \frac{3 \pm \sqrt{3}}{3}$   **43.** $x = \frac{7 \pm \sqrt{209}}{10}$   **44.** $x = \frac{-2 \pm \sqrt{19}}{3}$   **45.** $x = \frac{-7 \pm \sqrt{29}}{10}$   **46.** $x = \frac{9 \pm \sqrt{93}}{2}$

**47.** No real solution     **48.** $x = \pm 3\sqrt{2}$     **49.** $y = -7$ only     **50.** $x = \dfrac{-4 \pm \sqrt{31}}{3}$     **51.** $y = -\dfrac{9}{8}, y = 1$     **52.** $y = -3, y = 5$

**53.** $x = 0, x = \dfrac{1}{6}$     **54.**  $y = 2x^2$     **55.**  $y = x^2 + 4$     **56.**  $y = x^2 - 3$     **57.**  $y = -\dfrac{1}{2}x^2$

**58.** 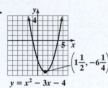 $y = x^2 - 3x - 4$     **59.**  $y = \dfrac{1}{2}x^2 - 2$     **60.**  $y = -x^2 + 4x + 5$     **61.**  $y = -x^2 - 6x - 8$

**62.** 7 ft by 4 ft or 8 ft by 3.5 ft     **63.** 16 cm; 12 cm
**64.** Width $= 18$ in.; length $= 24$ in.
**65.** $t = 30$ sec     **66.** 24 members
**67.** Going, 45 mph; returning, 60 mph
**68.** 60 pizza stores     **69.** 30 fewer pizza stores
**70.** 1990 and 2020     **71.** 2000 and 2010

## How Am I Doing? Chapter 10 Test
**1.** $x = \dfrac{-7 \pm \sqrt{129}}{10}$ (obj. 10.3.1)     **2.** $x = -5, x = \dfrac{2}{3}$ (obj. 10.1.3)     **3.** No real solution

(obj. 10.3.3)     **4.** $x = \dfrac{3}{4}, x = 5$ (obj. 10.1.3)     **5.** $x = -\dfrac{5}{4}, x = \dfrac{1}{3}$ (obj. 10.3.1)     **6.** $x = \dfrac{4}{3}$ (obj. 10.1.3)     **7.** $x = 0, x = 8$ (obj. 10.1.2)

**8.** $x = \pm 3$ (obj. 10.2.1)     **9.** $x = -\dfrac{1}{2}, x = 6$ (obj. 10.3.1)     **10.** $x = \dfrac{3}{2}, x = -\dfrac{1}{2}$ (obj. 10.3.1)

**11.**  (obj. 10.4.2) $y = 3x^2 - 6x$     **12.**  (obj. 10.4.2) $y = -x^2 + 8x - 12$     **13.** 9 m; 12 m (obj. 10.5.1)     **14.** 7 sec (obj. 10.5.1)

## Practice Final Examination
**1.** $-2x + 21y + 18xy + 6y^2$     **2.** 14     **3.** $18x^5y^5$     **4.** $2x^2y - 8xy$     **5.** $x = \dfrac{5}{2}$

**6.** $x = \dfrac{8}{5}$     **7.** $x \ge 2.6$      **8.** $2x^2 - 3x - 2$     **9.** $4x^2 + 12x + 9$     **10.** $x^2 - 9y^2$

**11.** $2x^3 - 5x^2y + xy^2 + 2y^3$     **12.** $2(2x + 1)(x - 5)$     **13.** $3x(x - 5)(x + 2)$     **14.** $\dfrac{6x - 11}{(x + 2)(x - 3)}$     **15.** $\dfrac{11(x + 2)}{2x(x + 4)}$     **16.** $-\dfrac{12}{5}$

**17.** slope $= \dfrac{5}{2}$     **18.** $3x + 4y = 14$ or $y = -\dfrac{3}{4}x + \dfrac{7}{2}$     **19.** 38.13 in.$^2$     **20.** $x = -5, y = 2$     **21.** $a = -2, b = -3$

**22.** $x\sqrt{5x}$     **23.** $6\sqrt{3} - 12 + 12\sqrt{2}$     **24.** $-\dfrac{\sqrt{15} + \sqrt{35} + \sqrt{21} + 7}{2}$     **25.** $x = \dfrac{2}{3}, x = -\dfrac{1}{4}$

**26.** $y = \dfrac{3 \pm \sqrt{7}}{2}$     **27.** $x = \pm 2$

**28.** $y = x^2 + 6x + 8$     **29.** $\sqrt{39}$     **30.** 5     **31.** Width 7 m; length 12 m     **32.** 200 reserved-seat tickets; 160 general admission tickets
**33.** Base 8 m; altitude 17 m

## Appendix B Exercises
**1.** 34,000 m     **3.** 5700 cm     **5.** 250 mm     **7.** 0.563 m     **9.** 29,400 mg     **11.** 0.0984 kg
**13.** 7000 mL     **15.** 0.000004 kL     **17.** 1.7 in.     **19.** 22.5 km     **21.** 2.8 m     **23.** 17.8 cm     **25.** 1089.6 g     **27.** 171.6 lb     **29.** 2.8 L
**31.** 193,248,000 in.     **33.** 27.27 mph     **35.** 0.11 yr

## Appendix C Exercises
**1.** 84,916 mi$^2$     **3.** 5,044,930 people     **5.** 2 representatives     **7.** Treasure State     **9.** Nevada
**11.** 2000 homes     **13.** DuPage County     **15.** 2800 homes     **17.** 2,000,000 apartment units     **19.** 8,000,000 more apartment units     **21.** about
8 million people     **23.** between 2000 and 2010     **25.** 3 million more people     **27.** 37 million people     **29.** movies and an exercise program
**31.** 5.5 million     **33.** $2.5 million greater     **35.** Between 2008 and 2009     **37.** 74 million people     **39.** 65 years and above     **41.** 19%
**43.** 81%     **45.** 845 million people     **47.** 18.9%     **49.** $2000.70

## Appendix D Exercises
**1.** $y = -\dfrac{2}{3}x + 8$     **3.** $y = 5x + 33$     **5.** $y = -\dfrac{1}{5}x + \dfrac{6}{5}$     **7.** $y = \dfrac{5}{7}x + \dfrac{13}{7}$     **9.** $y = -\dfrac{2}{3}x - \dfrac{8}{3}$
**11.** $y = -3$     **13.** $5x - y = -10$     **15.** $x - 3y = 8$     **17.** $2x - 3y = 15$     **19.** $7x - y = -27$     **21.** Neither     **23.** Parallel
**25.** Perpendicular

# Beginning Algebra Glossary

**Absolute value of a number (1.1)** The absolute value of a number $x$ is the distance between 0 and the number $x$ on the number line. It is written as $|x|$. $|x| = x$ if $x \geq 0$, but $|x| = -x$ if $x < 0$.

**Altitude of a geometric figure (1.8)** The height of the geometric figure. In the three figures shown, the altitude is labeled $a$.

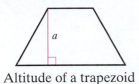

Altitude of a trapezoid

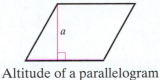

Altitude of a parallelogram

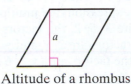

Altitude of a rhombus

**Altitude of a triangle (1.8)** The height of any given triangle. In the three triangles shown, the altitude is labeled $a$.

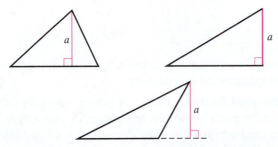

**Associative property of addition (1.1)** If $a$, $b$, and $c$ are real numbers, then

$$a + (b + c) = (a + b) + c.$$

This property states that if three numbers are added, it does not matter *which two numbers* are added first; the result will be the same.

**Associative property of multiplication (1.3)** If $a$, $b$, and $c$ are real numbers, then

$$a \times (b \times c) = (a \times b) \times c.$$

This property states that if three numbers are multiplied, it does not matter *which two numbers* are multiplied first; the result will be the same.

**Base (1.4)** The number or variable that is raised to a power. In the expression $2^6$, the number 2 is the base.

**Base of a triangle (1.8)** The side of a triangle that is perpendicular to the altitude.

**Binomial (4.3)** A polynomial of two terms. The expressions $a + 2b$, $6x^3 + 1$, and $5a^3b^2 + 6ab$ are all binomials.

**Circumference of a circle (1.8)** The distance around a circle. The circumference of a circle is given by the formula $C = \pi d$ or $C = 2\pi r$, where $d$ is the diameter of the circle and $r$ is the radius of the circle.

**Coefficient (4.1)** A coefficient is a factor or a group of factors in a product. In the term $4xy$ the coefficient of $y$ is $4x$, but the coefficient of $xy$ is 4. In the term $-5x^3y$ the coefficient of $x^3y$ is $-5$.

**Commutative property for addition (1.1)** If $a$ and $b$ are any real numbers, then $a + b = b + a$.

**Commutative property for multiplication (1.3)** If $a$ and $b$ are any real numbers, then $ab = ba$.

**Complex fraction (6.4)** A fraction that contains at least one fraction in the numerator or in the denominator or both. These three fractions are complex fractions:

$$\frac{7 + \dfrac{1}{x}}{x^2 + 2}, \qquad \frac{1 + \dfrac{1}{5}}{2 - \dfrac{1}{7}}, \qquad \text{and} \qquad \frac{\dfrac{1}{3}}{4}.$$

**Constant (2.3)** Symbol or letter that is used to represent exactly one single quantity during a particular problem or discussion.

**Coordinates of a point (7.1)** An ordered pair of numbers $(x, y)$ that specifies the location of a point in a rectangular coordinate system.

**Degree of a polynomial (4.3)** The degree of the highest-degree term of a polynomial. The degree of the polynomial $5x^3 + 2x^2 - 6x + 8$ is 3. The degree of the polynomial $5x^2y^2 + 3xy + 8$ is 4.

**Degree of a term of a polynomial (4.3)** The sum of the exponents of the variables in the term. The degree of $3x^3$ is 3. The degree of $4x^5y^2$ is 7.

**Denominator (0.1) and (6.1)** The bottom number or algebraic expression in a fraction. The denominator of

$$\frac{3x - 2}{x + 4}$$

is $x + 4$. The denominator of $\dfrac{3}{7}$ is 7. The denominator of a fraction may not be zero.

**Dependent equations (8.1)** Two equations are dependent if every value that satisfies one equation satisfies the other. A system of two dependent equations in two variables will not have a unique solution.

G-1

**Diagonal of a four-sided figure (1.8)** A line connecting two nonadjacent corners of the figure. In each of the figures shown, line $AC$ is a diagonal.

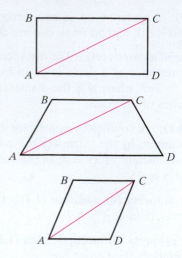

**Difference (3.1)** The result of subtracting one number or expression from another. The mathematical expression $x - 6$ can be written in words as the difference between $x$ and 6.

**Difference-of-two-squares polynomial (5.5)** A polynomial of the form $a^2 - b^2$ that may be factored by using the formula

$$a^2 - b^2 = (a + b)(a - b).$$

**Distributive property (1.6)** For all real numbers $a$, $b$, and $c$, $a(b + c) = ab + ac$.

**Dividend (0.4)** The number that is to be divided by another. In the problem $30 \div 5 = 6$, the three parts are as follows:

5 is the divisor
30 is the dividend
6 is the quotient

**Divisor (0.4)** The number you divide into another.

**Domain of a relation (7.6)** In any relation, the set of values that can be used for the independent variable is called its domain. This is the set of all the first coordinates of the ordered pairs that define the relation.

**Equilateral triangle (1.8)** A triangle with three sides equal in length and three angles that each measure $60°$. Triangle $ABC$ is an equilateral triangle.

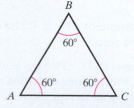

**Even integers (1.3)** Integers that are exactly divisible by 2, such as $\ldots, -4, -2, 0, 2, 4, 6, \ldots$.

**Exponent (1.4)** The number that indicates the power of a base. If the number is a positive integer, it indicates how many times the base is multiplied. In the expression $2^6$, the exponent is 6.

**Expression (4.3)** A mathematical expression is any quantity using numbers and variables. Therefore, $2x$, $7x + 3$, and $5x^2 + 6x$ are all mathematical expressions.

**Extraneous solution (6.5) and (9.6)** An obtained solution to an equation that when substituted back into the original equation, does *not* yield an identity. $x = 2$ is an extraneous solution to the equation

$$\frac{x}{x - 2} - 4 = \frac{2}{x - 2}.$$

An extraneous solution is also called an extraneous root.

**Factor (0.1) and (5.1)** When two or more numbers, variables, or algebraic expressions are multiplied, each is called a factor. If we write $3 \cdot 5 \cdot 2$, the factors are 3, 5, and 2. If we write $2xy$, the factors are 2, $x$, and $y$. In the expression $(x - 6)(x + 2)$, the factors are $(x - 6)$ and $(x + 2)$.

**Fractions**

  **Algebraic fractions (6.1)** The indicated quotient of two algebraic expressions.

$$\frac{x^2 + 3x + 2}{x - 4} \quad \text{and} \quad \frac{y - 6}{y + 8}$$

are algebraic fractions. In these fractions the value of the denominator cannot be zero.

  **Numerical fractions (0.1)** A set of numbers used to describe parts of whole quantities. A numerical fraction can be represented by the quotient of two integers for which the denominator is not zero. The numbers $\frac{1}{5}$, $-\frac{2}{3}$, $\frac{8}{2}$, $-\frac{4}{31}$, $\frac{8}{1}$, and $-\frac{12}{1}$ are all numerical fractions. The set of rational numbers can be represented by numerical fractions.

**Function (7.6)** A relation in which no two different ordered pairs have the same first coordinate.

**Hypotenuse of a right triangle (9.6)** The side opposite the right angle in any right triangle. The hypotenuse is always the longest side of a right triangle. Side $AB$ is the hypotenuse of the triangle below.

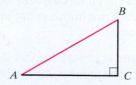

**Identity (2.1)** A statement that is always true. The equations $5 = 5$, $7 + 4 = 7 + 4$, and $x + 8 = x + 8$ are examples of identities.

**Imaginary number (9.1)** A complex number that is not a real number. An imaginary number can be created by taking the square root of a negative real number. The numbers $\sqrt{-9}$, $\sqrt{-7}$, and $\sqrt{-12}$ are all imaginary numbers.

**Improper fraction (0.1)** A numerical fraction whose numerator is larger than or equal to its denominator. $\frac{8}{3}$, $\frac{5}{2}$, and $\frac{7}{7}$ are improper fractions.

**Inconsistent system of equations (8.1)** A system of equations that does not have a solution.

**Independent equations (8.1)** Two equations that are not dependent are said to be independent.

**Inequality (2.6) and (7.5)** A mathematical relationship between quantities that are not equal. $x \le -3$, $w > 5$, and $x < 2y + 1$ are mathematical inequalities.

**Integers (1.1)** The set of numbers . . ., $-5, -4, -3, -2, -1, 0, 1, 2, 3, 4, 5, \ldots$.

**Intercepts of an equation (7.2)** The point or points where the graph of the equation crosses the $x$-axis or the $y$-axis or both. (*See* $x$-intercept, $y$-intercept.)

**Irrational number (9.1)** A real number that cannot be expressed in the form $\frac{a}{b}$, where $a$ and $b$ are integers and $b \ne 0$. $\sqrt{2}$, $\pi$, $5 + 3\sqrt{2}$, and $-4\sqrt{7}$ are irrational numbers.

**Isosceles triangle (1.8)** A triangle with two equal sides and two equal angles. Triangle $ABC$ is an isosceles triangle. Angle $BAC$ is equal to angle $ACB$. Side $AB$ is equal in length to side $BC$.

**Least common denominator of numerical fractions (0.2)** The smallest whole number that is exactly divisible by all denominators of a group of fractions. The least common denominator (LCD) of $\frac{1}{6}$, $\frac{2}{3}$, and $\frac{3}{5}$ is 30. The least common denominator is also called the lowest common denominator.

**Leg of a right triangle (9.6)** One of the two shorter sides of a right triangle. Side $AC$ and side $BC$ are legs of triangle $ACB$.

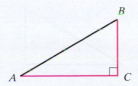

**Like terms (1.7)** Terms that have identical variables and exponents. In the expression $5x^3 + 2xy^2 + 6x^2 - 3xy^2$, the term $2xy^2$ and the term $-3xy^2$ are like terms.

**Linear equation in two variables (7.2)** An equation of the form $Ax + By = C$, where $A$, $B$, and $C$ are real numbers. The graph of a linear equation in two variables is a straight line.

**Line of symmetry of a parabola (10.4)** A line that can be drawn through a parabola such that, if the graph were folded on this line, the two halves of the curve would correspond exactly. The line of symmetry through the parabola formed by $y = ax^2 + bx + c$ is given by $x = \frac{-b}{2a}$. The line of symmetry of a parabola always passes through the vertex of the parabola.

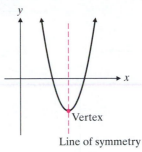

Line of symmetry

**Mixed number (0.1)** A number that consists of an integer written next to a proper fraction. $2\frac{1}{3}$, $4\frac{6}{7}$, and $3\frac{3}{8}$ are all mixed numbers. Mixed numbers are sometimes called mixed fractions or mixed numerals.

**Monomial (4.3)** A polynomial of one term. The expressions $3xy$, $5a^2b^3cd$, and $-6$ are all monomials.

**Natural numbers (0.1)** The set of numbers $1, 2, 3, 4, 5, \ldots$. This set is also called the set of counting numbers.

**Numeral (0.1)** The symbol used to describe a number.

**Numerator (0.1)** The top number or algebraic expression in a fraction. The numerator of

$$\frac{x + 3}{5x - 2}$$

is $x + 3$. The numerator of $\frac{12}{13}$ is 12.

**Numerical coefficient (4.1)** The number that is multiplied by a variable or a group of variables. The numerical coefficient in $5x^3y^2$ is 5. The numerical coefficient in $-6abc$ is $-6$. The numerical coefficient in $x^2y$ is 1. A numerical coefficient of 1 is not usually written.

**Odd integers (1.3)** Integers that are not exactly divisible by 2, such as $-3, -1, 1, 3, 5, 7, 9, \ldots$.

**Opposite of a number (1.1)** Two numbers that are the same distance from zero on the number line but lie on different sides of it are considered opposites. The opposite of $-6$ is 6. The opposite of $\frac{22}{7}$ is $-\frac{22}{7}$.

**Ordered pair (7.1)**   A pair of numbers presented in a specified order. An ordered pair is often used to specify a location on a graph. Every point in a rectangular coordinate system can be represented by an ordered pair $(x, y)$.

**Origin (7.1)**   The point $(0, 0)$ in a rectangular coordinate system.

**Parabola (10.4)**   A curve created by graphing a quadratic equation. The curves shown are all parabolas. The graph of the equation $y = ax^2 + bx + c$, where $a \neq 0$, will always be a parabola.

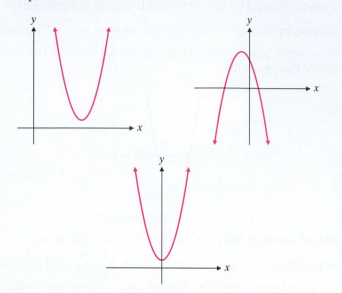

**Parallel lines (7.3) and (8.1)**   Two straight lines that never intersect. The graph of an inconsistent system of two linear equations in two variables will result in parallel lines.

**Parallelogram (1.8)**   A four-sided figure with opposite sides parallel. Figure $ABCD$ is a parallelogram.

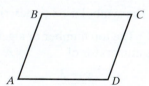

**Percent (0.5)**   Hundredths or "per one hundred"; indicated by the % symbol. Thirty-seven hundredths $\left( \dfrac{37}{100} \right) = 37\%$ (thirty-seven percent).

**Perfect square number (5.5)**   A number that is the square of an integer. The numbers 1, 4, 9, 16, 25, 36, 49, 64, 81, 100, 121, 144, … are perfect square numbers.

**Perfect-square trinomial (5.5)**   A polynomial of the form $a^2 + 2ab + b^2$ or $a^2 - 2ab + b^2$ that may be factored using one of the following formulas:

$$a^2 + 2ab + b^2 = (a + b)^2$$

or

$$a^2 - 2ab + b^2 = (a - b)^2.$$

**Perimeter (1.8)**   The distance around any plane figure. The perimeter of this triangle is 13. The perimeter of this rectangle is 20.

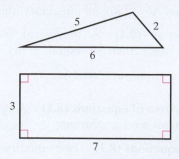

**Pi (1.8)**   An irrational number, denoted by the symbol $\pi$, that is approximately equal to 3.141592654. In most cases 3.14 can be used as a sufficiently accurate approximation for $\pi$.

**Polynomial (4.3)**   An expression that contains terms with nonnegative integer exponents. The expressions $5ab + 6$, $x^3 + 6x^2 + 3$, $-12$, and $x + 3y - 2$ are all polynomials. The expressions $x^{-2} + 2x^{-1}$, $2\sqrt{x} + 6$, and $\dfrac{5}{x} + 2x^2$ are not polynomials.

**Prime number (0.1)**   Any natural number greater than 1 whose only natural number factors are 1 and itself. The first eight prime numbers are 2, 3, 5, 7, 11, 13, 17, and 19.

**Prime polynomial (5.6)**   A prime polynomial is a polynomial that cannot be factored by the methods of elementary algebra. $x^2 + x + 1$ is a prime polynomial.

**Principal (3.4)**   In monetary problems, the principal is the original amount of money invested or borrowed.

**Principal square root (9.1)**   For any given nonnegative number $N$, the principal square root of $N$ (written $\sqrt{N}$) is the nonnegative number $a$ if and only if $a^2 = N$. The principal square root of 25 ($\sqrt{25}$) is 5.

**Proper fraction (0.1)**   A numerical fraction whose numerator is less than its denominator; $\dfrac{3}{7}$, $\dfrac{2}{5}$, and $\dfrac{8}{9}$ are proper fractions.

**Proportion (6.6)**   A proportion is an equation stating that two ratios are equal.

$$\frac{a}{b} = \frac{c}{d} \quad \text{where } b, d \neq 0$$

is a proportion.

**Pythagorean Theorem (9.6)**   In any right triangle, if $c$ is the length of the hypotenuse and $a$ and $b$ are the lengths of the two legs, then $c^2 = a^2 + b^2$.

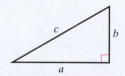

**Quadratic equation (5.7)** A quadratic equation is a polynomial equation with one variable that contains at least one term with the variable squared, but no term with the variable raised to a higher power. $5x^2 + 6x - 3 = 0$, $x^2 = 7$, and $5x^2 = 2x$ are all quadratic equations.

**Quadratic formula (10.3)** If $ax^2 + bx + c = 0$ and $a \neq 0$, then the solutions to the quadratic equation are given by

$$x = \frac{-b \pm \sqrt{b^2 - 4ac}}{2a}.$$

**Quotient (0.4)** The result of dividing one number or expression by another. In the problem $12 \div 4 = 3$, the quotient is 3.

**Radical (9.1)** An expression composed of a radical sign and a radicand. The expressions $\sqrt{5x}$, $\sqrt{\dfrac{3}{5}}$, $\sqrt{5x + b}$, and $\sqrt{10}$ are called radicals.

**Radical equation (9.6)** An equation that contains one or more radicals. $\sqrt{x + 4} = 12$, $\sqrt{x} = 3$, and $\sqrt{3x} + 4 = x + 2$ are all radical equations.

**Radical sign (9.1)** The symbol $\sqrt{\phantom{x}}$ used to indicate the square root of a number.

**Radicand (9.1)** The expression beneath the radical sign. The radicand in $\sqrt{5ab}$ is $5ab$.

**Range of a relation (7.6)** In any relation, the set of values that represents the dependent variable is called its range. This is the set of all the second coordinates of the ordered pairs that define the relation.

**Ratio (6.6)** The ratio of one number $a$ to another number $b$ is the quotient $a \div b$ or $\dfrac{a}{b}$.

**Rational numbers (1.1) and (9.1)** A number that can be expressed in the form $\dfrac{a}{b}$, where $a$ and $b$ are integers and $b \neq 0$. $\dfrac{7}{3}$, $-\dfrac{2}{5}$, $\dfrac{7}{-8}$, $\dfrac{5}{1}$, 1.62, and 2.7156 are rational numbers.

**Rationalizing the denominator (9.5)** The process of transforming a fraction that contains a radical in the denominator to an equivalent fraction that does not contain a radical in the denominator.

| | Equivalent Fraction with |
|---|---|
| **Original Fraction** | **Rationalized Denominator** |
| $\dfrac{3}{\sqrt{2}}$ | $\dfrac{3\sqrt{2}}{2}$ |

**Rationalizing the expression (9.5)** The process of transforming a radical that contains a fraction to an equivalent expression that does not contain a fraction inside a radical.

| | Equivalent Expression after |
|---|---|
| **Original Radical** | **It Has Been Rationalized** |
| $\sqrt{\dfrac{3}{5}}$ | $\dfrac{\sqrt{15}}{5}$ |

**Real number (9.1)** Any number that is rational or irrational. $2, 7, \sqrt{5}, \dfrac{3}{8}, \pi, -\dfrac{7}{5}$, and $-3\sqrt{5}$ are all real numbers.

**Rectangle (1.8)** A four-sided figure with opposite sides parallel and each interior angle measuring 90°. The opposite sides of a rectangle are equal.

**Relation (7.6)** A relation is any set of ordered pairs.

**Rhombus (1.8)** A parallelogram with four equal sides. Figure $ABCD$ is a rhombus.

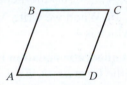

**Right angle (1.8)** An angle that measures 90°. Right angles are usually labeled in a sketch by using a small square to indicate that it is a right angle. Here, angle $ABC$ is a right angle.

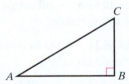

**Right triangle (1.8)** A triangle that contains a right angle.

**Root of an equation (2.1) and (5.7)** A value of the variable that makes an equation into a true statement. The root of an equation is also called the solution of an equation.

**Scientific notation (4.2)** A positive number is written in scientific notation if it is in the form $a \times 10^n$, where $1 \leq a < 10$ and $n$ is an integer.

**Slope intercept form (7.3)** The equation of a line that has slope $m$ and the $y$-intercept at $(0, b)$ is given by $y = mx + b$.

**Slope of a line (7.3)** The ratio of change in $y$ over the change in $x$ for any two different points on a nonvertical line. The slope $m$ is determined by

$$m = \frac{y_2 - y_1}{x_2 - x_1},$$

where $x_2 \neq x_1$ for any two points $(x_1, y_1)$ and $(x_2, y_2)$ on a nonvertical line.

**Solution of a linear inequality (2.6)**　The possible values that make a linear inequality true.

**Solution of an equation (2.1)**　A number that, when substituted into a given equation, yields an identity. The solution of an equation is also called the root of an equation.

**Solution of an inequality in two variables (7.5)**　The set of all possible ordered pairs that, when substituted into the inequality, will yield a true statement.

**Solution to a system of two equations in two variables (8.2)**　An ordered pair that can be substituted into each equation to obtain an identity in each case.

**Square (1.8)**　A rectangle with four equal sides.

**Square root (9.1)**　For any given nonnegative number $N$, the square root of $N$ is the number $a$ if $a^2 = N$. One square root of 16 is 4 since $(4)^2 = 16$. Another square root of 16 is $-4$ since $(-4)^2 = 16$. When we write $\sqrt{16}$, we want only the positive root, which is 4.

**Standard form of a quadratic equation (5.7)**　A quadratic equation that is in the form $ax^2 + bx + c = 0$.

**Subscript of a variable (3.2)**　A small number or letter written slightly below and to the right of a variable. In the expression $5 = 2(x - x_0)$, the subscript of $x$ is 0. In the expression $t_f = 5(t_a - b)$, the subscript of the first $t$ is $f$. The subscript of the second $t$ is $a$. A subscript is used to indicate a different value of the variable.

**System of equations (8.1)**　A set of two or more equations that must be considered together.

**Term (1.7)**　A number, a variable, or a product of numbers and variables. For example, in the expression $a^3 - 3a^2b + 4ab^2 + 6b^3 + 8$, there are five terms. They are $a^3, -3a^2b, 4ab^2, 6b^3$, and 8. The terms of a polynomial are separated by plus or minus signs.

**Trapezoid (1.8)**　A four-sided figure with two sides parallel. The parallel sides are called the bases of the trapezoid. Figure $ABCD$ is a trapezoid.

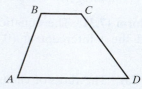

**Trinomial (4.3)**　A polynomial of three terms. The expressions $x^2 + 6x - 8$ and $a + 2b - 3c$ are trinomials.

**Variable (1.4)**　A letter that is used to represent a number or a set of numbers.

**Variation (9.7)**　An equation relating values of one variable to those of other variables. An equation of the form $y = kx$, where $k$ is a constant, indicates *direct variation*. An equation of the form $y = \dfrac{k}{x}$, where $k$ is a constant, indicates *inverse variation*. In both cases, $k$ is called the *constant of variation*.

**Vertex of a parabola (10.4)**　The lowest point on a parabola opening upward or the highest point on a parabola opening downward. The $x$-coordinate of the vertex of the parabola formed by the equation $y = ax^2 + bx + c$ is given by $= \dfrac{-b}{2a}$.

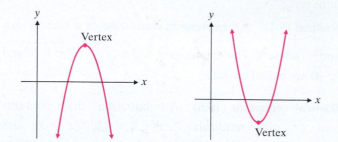

**Vertical line test (7.6)**　If a vertical line can intersect the graph of a relation more than once, the relation is not a function.

**Whole numbers (0.1)**　The set of numbers 0, 1, 2, 3, 4, 5, . . . .

**$x$-intercept (7.2)**　The ordered pair $(a, 0)$ is the $x$-intercept of a line if the line crosses the $x$-axis at $(a, 0)$. The $x$-intercept of line $l$ on the following graph is $(4, 0)$.

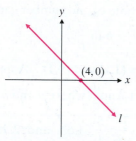

**$y$-intercept (7.2)**　The ordered pair $(0, b)$ is the $y$-intercept of a line if the line crosses the $y$-axis at $(0, b)$. The $y$-intercept of line $p$ on the following graph is $(0, 3)$.

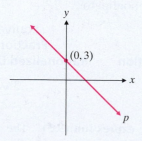

# Subject Index

# Index of Applications

# Photo Credits

## PROPERTIES OF REAL NUMBERS

| | Addition | Multiplication |
|---|---|---|
| **Commutative Properties** | $a + b = b + a$ | $ab = ba$ |
| **Associative Properties** | $(a + b) + c = a + (b + c)$ | $(ab)c = a(bc)$ |
| **Identity Properties** | $a + 0 = 0 + a = a$ | $a \cdot 1 = 1 \cdot a = a$ |
| **Inverse Properties** | $a + (-a) = -a + a = 0$ | $a \cdot \dfrac{1}{a} = \dfrac{1}{a} \cdot a = 1 (a \neq 0)$ |
| **Distributive Property** | | $a(b + c) = ab + ac$ |

## PROPERTIES OF EXPONENTS

If $x, y \neq 0$, then

$$x^a \cdot x^b = x^{a+b}$$

$$\frac{x^a}{x^b} = x^{a-b} \text{ if } a \geq b$$

$$\frac{x^a}{x^b} = \frac{1}{x^{b-a}} \text{ if } a < b$$

$$x^0 = 1$$

$$(x^a)^b = x^{ab}$$

$$(xy)^a = x^a y^a$$

$$\left(\frac{x}{y}\right)^a = \frac{x^a}{y^a}$$

## ABSOLUTE VALUE

**Definition**

$$|x| = \begin{cases} x \text{ if } x \geq 0 \\ -x \text{ if } x < 0 \end{cases}$$

## INEQUALITIES

If $a < b$, then for all values of $c$, $a + c < b + c$ and $a - c < b - c$.

If $a < b$ when $c$ is a **positive number** ($c > 0$), then $ac < bc$ and $\dfrac{a}{c} < \dfrac{b}{c}$.

If $a < b$ when $c$ is a **negative number** ($c < 0$), then $ac > bc$ and $\dfrac{a}{c} > \dfrac{b}{c}$.

## FACTORING AND MULTIPLYING FORMULAS

| **Perfect-Square Trinomials** | $a^2 + 2ab + b^2 = (a + b)^2$ $a^2 - 2ab + b^2 = (a - b)^2$ |
|---|---|
| **Difference of Two Squares** | $a^2 - b^2 = (a + b)(a - b)$ |
| **Sum of Two Squares** | $a^2 + b^2$ cannot be factored |

## PROPERTIES OF LINES AND SLOPES

The **slope** of any nonvertical line passing through $(x_1, y_1)$ and $(x_2, y_2)$ is $m = \dfrac{y_2 - y_1}{x_2 - x_1} (x_1 \neq x_2)$.

The **standard form** of the equation of a straight line is $Ax + By = C$.

The **slope–intercept form** of the equation of a straight line with slope $m$ and $y$-intercept $(0, b)$ is $y = mx + b$.

A **horizontal line** has a slope of zero. The equation of a horizontal line can be written as $y = b$.

A **vertical line** has no slope (the slope is not defined for a vertical line).

The equation of a vertical line can be written as $x = a$.